D0459876

Environment

Environment

AN INTERDISCIPLINARY ANTHOLOGY

Selected, Edited, and with Introductions by

Glenn Adelson
Lake Forest College

James Engell
Harvard University

Brent Ranalli
The Cadmus Group, Inc.

K. P. Van Anglen
Boston University

Illustrations Editor
Erin Sheley

Yale University Press
New Haven & London

The Selection Credits on pages 901–14 constitute
an extension of the copyright page.

Set in Postscript Janson OldStyle type by Integrated Publishing Solutions.
Printed in the United States of America.

Library of Congress Cataloging-in-Publication Data
Environment : an interdisciplinary anthology / selected, edited, and with
introductions by Glenn Adelson . . . [et al.]; illustrations editor, Erin Sheley.
p. cm.
Includes bibliographical references and index.
ISBN 978-0-300-12614-3 (cloth : alk. paper)—
ISBN 978-0-300-11077-7 (pbk. : alk. paper)
1. Environment—Philosophy. 2. Environmental science—Philosophy.
3. Environmental science—Methodology. 4. Environmental protection—
Philosophy. 5. Interdisciplinary research. I. Adelson, Glenn.
GE40.E548 2008
333.72—dc22 2007033169

A catalogue record for this book is available from the British Library.

The paper in this book meets the guidelines for permanence and durability
of the Committee on Production Guidelines for Book Longevity
of the Council on Library Resources.

10 9 8 7 6 5 4 3

To the memory of Rachel Carson

There is an *active* principle alive in all things;
In all things, in all natures, in the flowers
And in the trees, in every pebbly stone
That paves the brooks, the stationary rocks,
The moving waters and the invisible air.
All beings have their properties which spread
Beyond themselves, a power by which they make
Some other being conscious of their life,
Spirit that knows no insulated spot,
No chasm, no solitude; from link to link
It circulates, the soul of all the worlds.
This is the freedom of the universe;
Unfolded still the more, more visible,
The more we know; and yet is reverenced least
And least respected in the human mind,
Its most apparent home.

William Wordsworth

We are part of the Creation—the living world—in body and spirit. We belong on this planet as a biological heritage, and we have a sacred personal duty to keep it intact and healthy.

Edward O. Wilson

Brief Contents

PART ONE Concepts and Case Studies

PART TWO Foundational Disciplines and Topics

I. BIOLOGICAL INTERACTIONS

II. HUMAN DIMENSIONS

III. SOCIAL CONNECTIONS

Coda

Contents

The conservation of biological diversity (biodiversity) can focus on different levels of organization—from the biosphere and major biomes like the tropical rainforest through individual ecosystems down to populations and genes—but the level that is most tractable and understandable to the lay person is the species. Here, specific examples illustrate overexploitation, habitat destruction, and introduction of invasive species, the three major causes of species extinction.

In 1979, at Three Mile Island south of Harrisburg, Pennsylvania, on the Susquehanna River, a series of events, human errors, and equipment malfunctions triggered an accident that alarmed tens of millions. Though there were no serious health or environmental effects, TMI proved that previous expert reports and safety

from it? The concept of wilderness is as intractable and as rich as wilderness itself, as the varied, even conflicting views of these writers attest.

The environment of the city and the impact of cities on the environment span the development of urban areas in the nineteenth century to their likely future in the twenty-first. Aspects of this topic include air and water pollution, patterns of urban development and migration, population concerns, and social justice, especially with regard to women, the poor, minorities, and the Third World.

PART TWO Foundational Disciplines and Topics

Coda

The environment and environmental issues are complex, organically connected, and massively detailed. To understand them, and then to act in ways that will protect the environment, avert disasters, and maintain both global and human health, requires nothing less than a redefinition of what it means to be human. Any one program or statement is insufficient. A new environmental consciousness will alter daily habits, economic planning, politics, spiritual orientations, the dedication of educational efforts, and the direction of scientific thinking.

Interconnections

Cross-Listed Selections for Chapters in Part One

Use these listings to expand the range of each of the chapters (1–10) in Part One, as well as to establish links between selections and chapters. Chapter numbers are in parentheses at the end of each entry.

3. Nuclear Power: Three Mile Island, Chernobyl, and the Future

4. Biotechnology and Genetically Manipulated Organisms: *Bt* Corn and the Monarch Butterfly

5. The Paradox of Sustainable Development

8. Globalization Is Environmental

Why Environmental Studies?

With historically unprecedented power, we have become chief stewards of the Earth. How can we perform this task well? How can we ensure that future generations will be able to continue good stewardship?

The best answers begin with environmental literacy. Environmental education can shape every child's awareness and direct every adult's actions. But the environment is unlike all other subjects. In ecology, states Barry Commoner, "everything is connected to everything else." Like the web of interdependent relationships in an ecosystem, the web of Environmental Studies is exciting and complex. Every thread we pick up leads to unexpected places—and to other threads. We can hardly understand the causes and significance of tropical deforestation, for instance, without knowing something about biodiversity and species endangerment, climate change and atmospheric chemistry, soil science and agricultural practices, anthropology and the cultural traditions of forest peoples, and the economics and politics of road building and oil exploration. How can one person make sense of it all? How can it be learned, and how can it be taught?

In the past three centuries, with the rise of modern science—engineering, powerful technologies, advances in medicine, and specialized research—human knowledge has accelerated, in large measure by dividing into specific disciplines. Boundaries of these disciplines at times change and reform, and advances in one discipline may trigger or influence discoveries in another. (For example, biology and chemistry share a shifting boundary in biochemistry; the study of anthropology has affected both law and literature.) Discoveries at the leading edges of specific areas of study add knowledge and new disciplines arise, breaking off and growing from older ones.

Yet now, our exploration of the natural world—the study of Earth and life on Earth—as well as of the massive impacts that human beings now exert on Earth and all its life, uncovers overarching questions, problems, and concepts that transcend the boundaries of any one discipline or approach. If we are to direct intelligently how our activities affect Earth and life on Earth (including our own lives), then it becomes vital to study institutions, beliefs, values, laws, economies, and history in light of what we know about the natural world. Let's look at one specific example.

Changes in Earth's climate are so complex that simply to measure those changes and to understand why they occur requires the methods of atmospheric science, oceanography, planetary science, climate history, mathematical modeling and supercomputing, chemistry, statistics, physics, and other disciplines. But measuring any change is only the first step. Next, to grasp how climate change affects life on Earth involves biological study, including species migration, endangerment, and extinction, as well as ecology, natural history, soil science, forestry, and hydrology. Finally, because human activity contributes to climate change, we need to examine how historical, religious, and philosophical assumptions, economic policies, laws, international treaties, building codes, engineering decisions, consumer choices, renewable sources of energy, conservation, and even sequestration of carbon dioxide will, in combination, affect climate change.

As the interconnected character of these and many other issues grew increasingly evident, scientists, writers, and educators looked for ways to study them together. They also searched for a term to describe that kind of study. "Environmental" is a capacious word not tied to one discipline or way of thinking. Routinely and accurately applied to aspects of many natural sciences, "environmental" also characterizes important elements of law, politics, economics, and the other social sciences, as well as approaches and emphases in literary criticism, the arts, ethics, and spiritual values. "Environmental" can refer to the natural world, the world built by humans, or a combination of both. "Studies" in the plural embraces multiple sets of disciplines and problems, areas and issues that by definition are not self-contained. They lead to one another. In the last decade of the twentieth century, "Environmental Studies" became a flexible, common term, perhaps the most common term, for collective efforts to understand the interrelated systems and phenomena of nature, including the human presence in those systems and its effects on them.

Environmental Studies establishes knowledge of specific natural phenomena and specific human institutional beliefs and practices. It then mediates between these. Environmental Studies characterizes complex natural conditions on Earth and describes the interface of human with non-human activity. To establish such knowledge, to perform such a mediation, and to promote desirable human action, Environmental Studies relies not only on multiple disciplines but also on key terms and concepts such as "the commons," "wilderness," and "the precautionary principle." This last principle might be roughly summarized as follows: Do not undertake any (environmental) action until you know with a fair degree of certainty what its consequences, unintended as well as intended, will be. So that the action may be weighed against alternatives, including doing nothing, do not assume lack of harm unless lack of harm is demonstrated.

The word "conservation" refers more explicitly to the preservation and management of natural resources, species, or habitats. Conservation forms a key part of Environmental Studies. "Ecology" remains most often applied to biological studies, to the relationship between organisms and their environment; however, since its essence is to trace relationships and interdependencies, "ecology" has acquired a larger, metaphoric valence in areas such as "industrial ecology" or "ecoengineering." "Sustainable development," a key though problematic concept, addresses how humans might live so that we do not impair but may even enhance the ability of future generations to enjoy a standard of living and quality of life as high or higher. Yet, sustainable development does not necessarily guarantee the continued existence of certain species or the preservation of all resources. It values nature not so much for its own sake as for our ability to enjoy continued or even higher levels of use. All these terms and their associated areas of interest interpenetrate one another. For example, conservation biology involves ecology. Similarly, sustainable development invokes ecoengineering and conservation. All these exist under the larger umbrella of Environmental Studies.

A working description of Environmental Studies thus involves several levels. First, Environmental Studies pursues and draws upon individual disciplines to understand the constituent elements of larger problems and issues. Then, to grasp and define such problems and issues, Environmental Studies combines the results of an appropriate set of these disciplines and may even foster new ones. Finally, in practical terms, Environmental Studies brings together relevant disciplines and concepts to address environmental issues and to resolve environmental problems. This process may entail technological applications, institutional reforms, government regulations, individual actions, or all of these. Environmental Studies engages science, social practices, values, and beliefs. Environmental Studies is a cornerstone of education and public policy. It calls on all of us. The environment is too precious and too complex to be left to special interests or to any one group of scientists, economists, philosophers, or poets.

Embracing so much of importance, Environmental Studies offers no easy formulas or answers. It is not a way to master facts that will by themselves determine policies and actions. Yet, Environmental Studies *is* the way to see how those facts are discovered and determined, how the interpretation of those facts may be contested, and then how those judgments might lead to actions, policies, laws, and treaties. Conflicts and disagreements are inevitable. Environmental Studies engenders debates and difficult decisions. It sharpens awareness that environmental policies often entail trade-offs, compromises, and periodic revisions. While casting in doubt the effectiveness of a militant environmental purity, this process, often a political one, by no means rules out ideals or idealism. The fact that Environmental Studies now commands huge interest and is regarded as central to the education of every world citizen is grounds for great hope.

To undertake Environmental Studies is to re-envision liberal education in the arts and sciences. This book therefore poses questions of personal, civic, ethical, political, and global import, and provides perspectives from the natural sciences, the social sciences, and the humanities. A liberal education is disinterested and self-critical. It encourages the open-minded, open-ended pursuit of life's large questions wherever they may lead, including into differences of opinion, as long as those opinions are informed.

Disinterestedness and open-mindedness do not mean standing on the sideline. No course of study can claim more final practicality than what is presented in this book, or in Environmental Studies generally. It is hard to imagine a more useful course of study, or one that could lead to a greater number of careers and professions. It is impossible to glimpse something related to the fate of humanity and nature and not feel compelled to advocate certain actions over others. If liberal education is disinterested, and if it holds no direct agenda merely for personal reward or advancement, this does *not* mean that it should avoid conclusions about what is good—or harmful—for society and civic health, or for humanity and nature at large. The final goal of knowing is to act with greater foresight, to cherish what we know, and then to love what we cherish, not only in words but also in deeds.

Individuals and corporate bodies—companies, organizations, and governments—that know a lot but only about one thing, or that have an intensive interest but only in a single goal, are capable of doing a great deal of unintended harm, environmentally and otherwise. Because they know little about other things, and even less about how they all connect, they can damage the long-term prospects for humanity and nature. In every facet of society, we need leaders, citizens, and consumers who attain a broad, holistic view.

Some societies have sustained mutually beneficial relations with the environment for long periods of time. In the case of certain indigenous peoples, their science, culture, and environment are not just intertwined, they are unified. An intimate, practical awareness of the environment in a particular locality—of flora and fauna, of seasons, skies, and landscape—is the soil out of which a community's laws, trade, morals, and myths grow organically. In such a context, the notion of spirituality, justice, or commerce that does not refer to the local landscape and to the manner in which it must continue to sustain the group would appear absurd. An individual becomes attuned to fine changes in the local environment and realizes how each action affects it. Such an understanding of, and sympathy with, the environment we call holistic.

Citizens of industrial and post-industrial cultures can explore a universe much wider than the one available to many indigenous peoples—through travel, books, scientific instruments, and electronic communication. This universe is so broad and rich, the information so complex, that it is hard for anyone to take it all in at once. While we still derive our sustenance from the land, water, and air, our culture no longer automatically attunes us to that fact, and, as such, our stewardship of the environment is no longer instinctual and habitual.

The living world—the animate and inanimate bound in their larger, shared environment—is a shattered continuum. It is an evolutionary process at times broken, stressed, or partly destroyed, often by natural events but increasingly by human activity. Yet it remains a whole. If we wish to understand our relationships to the global environment as well as to the environment in our own backyards in ways that encourage us to make intelligent choices and changes—if we, too, want stewardship to become second nature—then we need to shape a new holistic vision.

We need to rediscover and to see again, as if for the first time, the connections between our physical and biological support systems, our political and economic institutions, and our habits as ethical beings. This is the task of Environmental Studies. It's not for the lazy or fainthearted. Instead, it's exciting, it represents the future, and it's the right choice.

The Design and Use of This Book

The biologist Edward O. Wilson remarks, "If I learned anything in my forty-one years of teaching, it is that the best way to transmit knowledge and stimulate thought is to teach from the top down." In other words, begin by posing large problems, questions, and concepts of the highest significance and then later, once attention and curiosity are secured, "peel off layers of causation as currently understood . . . in growing technical and philosophically disputatious detail." Wilson warns, "Do *not* teach from the bottom up, e.g., 'first we'll learn some of this, and some of that, and we'll combine the knowledge later to build a picture of something larger.'" He concludes that in teaching and learning a large subject, this means to "put it up whole as quickly as possible, show why it matters . . . and will for a lifetime, then dissect to get as close to the bottom as possible."

We believe this is the best way to approach Environmental Studies. Part One of this book, "Concepts and Case Studies" (Chapters 1–10), asks large questions and presents significant challenges regarding high-profile, complex environmental topics that cut across disciplinary boundaries. Immediately following the Contents is an alternative table of contents called "Interconnections," a cross-listing of additional selections related to the topics for each chapter in Part One. These cross-listed selections come either from other chapters in Part One or from chapters in Part Two. These listings underscore how each concept and case study brings several different kinds of knowledge more deeply into play. These cross-listings may be of as much or more use to readers and teachers than the conventional, linear Contents.

Part Two, "Foundational Disciplines and Topics," comprises three sections. These present key environmental aspects of (I) the natural sciences (Chapters 11–15); (II) humanistic fields and their cultural, artistic, and personal values (Chapters 16–20); and (III) the social sciences, such as law, economics, and political science, that explore public policy and group behavior (Chapters 21–25).

Probably the *worst* way to use this book is simply to read it through doggedly chapter by consecutive chapter. Instead, treat the book like one giant Web site with numerous internal links. (An appendix offers relevant Web sites associated with each chapter, too.) The *best* way to use this book is to start with a chapter or single selection, most likely from one of the case studies or concepts in Part One, and then group around that as many other relevant readings as seems practicable. "Interconnections," the cross-listed selections for chapters in Part One, should be useful in this regard. Many different combinations of readings will form larger units or virtual chapters not explicitly listed as such.

For example, one could make the following constellation on the dilemma of fossil fuels: the chapters on climate shock, nuclear power, air and water, and energy (two chapters from Part One and two from Part Two). One could combine readings from the chapter on species endangerment (Part One) with those on biodiversity (Part Two). Juxtaposing selections from the chapter on sustainable development with selections from chapters on urbanization, forests and deforestation, and soil and agriculture creates a set of nested concerns and problems about how we use land to grow crops, extract resources, and house large human populations. As another example, an understanding of some key environmental aspects of the American West would draw specific selections from the chapters on deforestation, wilderness, urbanization, soil and agriculture, air and water, history, and nature writing.

Countless other constellations could be constructed: for instances of the tragedy of the commons, for feminist perspectives, for countries and regions like India and sub-Saharan Africa, for the power of nature to teach life lessons. Selections can be chosen to illustrate opposing points of view: "rational" and "emotional" ways of apprehending the environment, technological optimism and pessimism, anthropocentric and biocentric perspectives, explorations of the transcendent and the bestial in human nature.

Readers and teachers will choose, mix, match, and constellate chapters and selections to suit their own needs. This book hopes to encourage varied courses of reading with variable emphases, depending which selections are drawn from the volume, and in what order. Chapter introductions distill and highlight vital information relevant to each concept, case study, or discipline. The Web sites listed at the back of the book provide added resources, as do references to further readings, widely available, given at the end of each chapter introduction.

A NOTE ON EDITORIAL PRACTICE

Texts were selected after extensive comparison. Due to space constraints, much that is worthy could not be included. This book contains some classic works of Environmental Studies but also some unfamiliar yet highly valuable additions. Selections met a test of clarity, brevity, and authority. Where selections represent excerpts, we hope that they will send readers to the full, original articles and books from which they are drawn. Spelling and punctuation have been standardized. Notes by the original authors containing citations for quotations and essential research references have been regularized. Authorial notes that are purely discursive or subsumed by generally known research in the field have usually been omitted. Editorial excisions in selections are indicated by ellipsis points (. . .) or, in the case of larger omissions, by an ornament between paragraphs. Clarifications and notes by the editors are set in square brackets.

Overture

Nature and Human Perception

Martin Buber, "I Contemplate a Tree" in *I and Thou* (1923), translated by Walter Kaufmann

I contemplate a tree.

I can accept it as a picture: a rigid pillar in a flood of light, or splashes of green traversed by the gentleness of the blue silver ground.

I can feel it as movement: the flowing veins around the sturdy, striving core, the sucking of the roots, the breathing of the leaves, the infinite commerce with earth and air—and the growing itself in its darkness.

I can assign it to a species and observe it as an instance, with an eye to its construction and its way of life.

I can overcome its uniqueness and form so rigorously that I recognize it only as an expression of the law—those laws according to which a constant opposition of forces is continually adjusted, or those laws according to which the elements mix and separate.

I can dissolve it into a number, into a pure relation between numbers, and eternalize it.

Throughout all of this the tree remains my object and has its place and its time span, its kind and condition.

But it can also happen, if will and grace are joined, that as I contemplate the tree I am drawn into a relation, and the tree ceases to be an It. The power of exclusiveness has seized me.

This does not require me to forego any of the modes of contemplation. There is nothing that I must not see in order to see, and there is no knowledge that I must forget. Rather is everything, picture and movement, species and instance, law and number included and inseparably fused.

Whatever belongs to the tree is included: its form and its mechanics, its colors and its chemistry, its conversation with the elements and its conversation with the stars—all this in its entirety.

Dream of Arcadia, c. 1838 Thomas Cole. Denver Art Museum Collection: Gift of Mrs. Lindsey Gentry, 1954.71. Photo by the Denver Art Museum.

The tree is no impression, no play of my imagination, no aspect of a mood; it confronts me bodily and has to deal with me as I must deal with it—only differently.

One should not try to dilute the meaning of the relation: relation is reciprocity.

Does the tree then have consciousness, similar to our own? I have no experience of that. But thinking that you have brought this off in your own case, must you again divide the indivisible? What I encounter is neither the soul of a tree nor a dryad, but the tree itself.

PART ONE

Concepts and Case Studies

Keystone Essay: Aldo Leopold, "The Fusion of Lines of Thought" (1935?)

This fragment that Aldo Leopold jotted in a notebook is prescient in many ways. It was written before the two great discoveries in biology and geology of the mid-twentieth century: the nature of the DNA molecule and the theory of plate tectonics and continental drift. It also prophesied the need for an interdisciplinary approach to the relationship between humans and the environment, decrying the schism between the developing fields of ecology and economics.

The two great cultural advances of the past century were the Darwinian theory and the development of geology. The one explained how, and the other where, we live. Compared with such ideas, the whole gamut of mechanical and chemical invention pales into a mere matter of current ways and means.

Just as important as the origin of plants, animals, and soil is the question of how they operate as a community. Darwin lacked time to unravel any more than the beginnings of an answer. That task has fallen to the new science of ecology, which is daily uncovering a web of interdependencies so intricate as to amaze—were he here—even Darwin himself, who, of all men, should have the least cause to tremble before the veil. . . .

One of the anomalies of modern ecology is that it is the creation of two groups, one of which seems barely aware of the existence of the other. The one studies the human community almost as if it were a separate entity, and calls its findings sociology, economics, and history. The other studies the plant and animal community, [and] comfortably relegates the hodgepodge of politics to "the liberal arts." The inevitable fusion of these two lines of thought will, perhaps, constitute the outstanding advance of the present century.

Keystone Essay: Gifford Pinchot, from "The Birth of Conservation" in *Breaking New Ground* (published posthumously, 1947)

Writing of himself in the third person, Pinchot, the pioneering head of the U.S. Forest Service at the beginning of the twentieth century, describes the moment when he realized that all natural resource needs and problems — including those of the forests — are interrelated, and so must be addressed (bureaucratically as well as programmatically) in an interrelated fashion. This interconnected, interagency approach contrasted with previous policy.

In the gathering gloom of an expiring day, in the moody month of February, some forty years ago, a solitary horseman might have been observed pursuing his silent way above a precipitous gorge in the vicinity of the capital city of America. Or so an early Victorian three-volume novelist might have expressed it.

In plain words, a man by the name of Pinchot was riding a horse by the name of Jim on the Ridge Road in Rock Creek Park near Washington. And while he rode, he thought. He was a forester, and he was taking his problems with him, on that winter's day of 1907, when he meant to leave them behind.

The forest and its relation to streams and inland navigation, to water power and flood control; to the soil and its erosion; to coal and oil and other minerals; to fish and game; and many another possible use or waste of natural resources—these questions would not let him be. What had all these to do with Forestry? And what had Forestry to do with them?

Here were not isolated and separate problems. My work had brought me into touch with all of them. But what was the basic link between them?

Suddenly the idea flashed through my head that there was a unity in this complication—that the relation of one resource to another was not the end of the story. Here were no longer a lot of different, independent, and often antagonistic questions, each on its own separate little island, as we had been in the habit of thinking. In place of them, here was one single question with many parts. Seen in this new light, all these separate questions fitted into and made up the one great central problem of the use of the earth for the good of man.

To me it was a good deal like coming out of a dark tunnel. I had been seeing one spot of light ahead. Here, all of a sudden, was a whole landscape. Or it was like lifting the curtain on a great new stage.

There was too much of it for me to take it all in at once. As always, my mind worked slowly. From the first I thought I had stumbled on something really worth while, but that day in Rock Creek Park I was far from grasping the full reach and swing of the new idea.

It took time for me to appreciate that here were the makings of a new policy, not merely nationwide but worldwide in its scope—fundamentally important because it involved not only the welfare but the very existence of men on the Earth. I did see, however, that something ought to be done about it.

But, you may say, hadn't plenty of people before that day seen the value of Forestry, of irrigation, of developing our streams, and much besides? Hadn't plenty pointed out the threat of erosion, the shame and pity of the destruction of wildlife, and the reasons against man's vandalism of many kinds? Hadn't plenty pointed out that forests, for example, affect floods, and many other cases in which one natural resource reacts upon another?

Certainly they had. But so far as I knew then or have since been able to find out, it had occurred to nobody, in this country or abroad, that here was one question instead of many, one gigantic single problem that must be solved if the generations, as they came and went, were to live civilized, happy, useful lives in the lands which the Lord their God had given them.

But, you might go on, after the new idea was born, wasn't the situation just as it had been before? Shouldn't each Government bureau go on dealing with its same old subject in its same old way?

Not by a jugful, as they say in the backwoods.

The ancient classic simile of the bundle of twigs, which the Fascists have perverted and made known throughout the world, might point the moral and adorn the tale I am trying to tell. In union there is strength. But perhaps we can find a better illustration in the birth of our own Government.

The American colonies, like the Government bureaus which have to do with the various natural resources, were founded at different times, for different reasons, and by different kinds of people. Each colony, from Georgia to New Hampshire, dealt with nature in a somewhat different form. Each had to face a problem unlike the problems of all the others, and each was itself unlike all the other colonies.

Before the Declaration of Independence they were so many weak and separate twigs. Could we have become what we are today if the thirteen colonies had remained independent, self-sufficient little nationlets, quarreling among themselves over rights, boundaries, jurisdictions, instead of merging into a single nation with a single federal purpose?

The mere fact of union produced something different and unknown before. Here were new purpose and new power, and a future infinitely greater than anything thirteen separated colonies could ever have lived to see. Union did not wipe out the thirteen separate characters of the thirteen separate states, but it did bind them together into the strength of the new nation.

E Pluribus Unum is the fundamental fact in our political affairs. *E Pluribus Unum* is and always must be the basis in dealing with the natural resources. Many problems fuse into one great policy, just as many states fuse into one great Union. When the use of all the natural resources for the general good is seen to be a common policy with a common purpose, the chance for the wise use of each of them becomes infinitely greater than it had ever been before.

The Conservation of natural resources is the key to the future. It is the key to the safety and prosperity of the American people, and all the people of the world, for all time to come. The very existence of our Nation, and of all the rest, depends on conserving the resources which are the foundations of its life. That is why Conservation is the greatest material question of all.

Moreover, Conservation is a foundation of permanent peace among the nations, and the most important foundation of all. But more of that in another place.

It is not easy for us moderns to realize our dependence on the Earth. As civilization progresses, as cities grow, as the mechanical aids to human life increase, we are more and more removed from the raw materials of human existence, and we forget more easily that natural resources must be about us from our infancy or we cannot live at all.

What do you eat, morning, noon, and night? Natural resources, transformed and processed for your use. What do you wear, day in and day out—your coat, your hat, your shoes, your watch, the penny in your pocket, the filling in your tooth? Natural resources changed and adapted to your necessity.

What do you work with, no matter what your work may be? What are the desk you sit at, the book you read, the shovel you dig with, the machine you operate, the car you drive, and the light you see by when the sunlight fails? Natural resources in one form or another.

What do you live in and work in, but in natural resources made into dwellings and shops and offices? Wood, iron, rock, clay, sand, in a thousand different shapes, but always natural resources. What are the living you earn, the medicine you take, the movie you watch, but things derived from nature?

What are railroads and good roads, ocean liners and birch canoes, cities and summer camps, but natural resources in other shapes?

What does agriculture produce? Natural resources. What does industry manufacture? What does commerce deal in? What is science concerned with? Natural resources.

What is your own body but natural resources constantly renewed—your body, which would cease to be yours to command if the natural resources which keep it in health were cut off for so short a time as one or two percent of a single year?

There are just two things on this material earth—people and natural resources.

From all of which I hope you have gathered, if you did not realize it before, that a constant and sufficient supply of natural resources is the basic human problem.

1 Climate Shock

If an asteroid hurtling toward Earth would, with strong probability, strike this planet in forty years, raise sea levels permanently between six inches to sixteen feet, force up to one-quarter of all species into extinction, inaugurate plagues and disease, inundate parts of some nations, drown populated islands whole, render coasts uninhabitable, intensify hurricanes, typhoons, and tornadoes into record-breaking storms, cause frequent floods and landslides, and kill millions of people, then every government would work furiously to discover how that asteroid might be diverted or destroyed.

There is no such asteroid (as far as we know), and there is no international cooperation, therefore, to stop it. But all the rest in this scenario is very possibly true. It is just happening more slowly (and at this very moment) than the future impact of an asteroid. The cause is not one impending catastrophic event. It is our own burning of carbon. Ironically, solutions to the problems caused by burning carbon, though difficult to achieve, may, in fact, prove easier to discover than solutions for the impact of an asteroid.

Climate is not "weather" in the usual sense, or even global weather over long periods of time. Large, multiple earth systems, physical, chemical, and biological, produce world climate. These involve the atmosphere, land, oceans, and living organisms. For the first time ever, human activity is now measurably changing global climate.

Increased levels of greenhouse gases, especially carbon dioxide (CO_2), act as the chief vehicle for human impact on Earth's climate systems. (See Chapter 13, below.) The main source of these increased levels is human combustion of fossil fuels. Burning wood, forest fires, and slash-and-burn techniques to clear land contribute, too. For a thousand years prior to 1800, the atmospheric CO_2 level measured about 270 parts per million (ppm). In the past two centuries, this has risen sharply, at an increasing rate, to 370 ppm, as high as it has been for 420,000 years. Even if we stopped using fossil fuels today, these concentrations would remain for decades. Given increasing global energy use per capita and further population growth, greenhouse gas levels will rise more. No one knows when, if, or at what level they might stabilize.

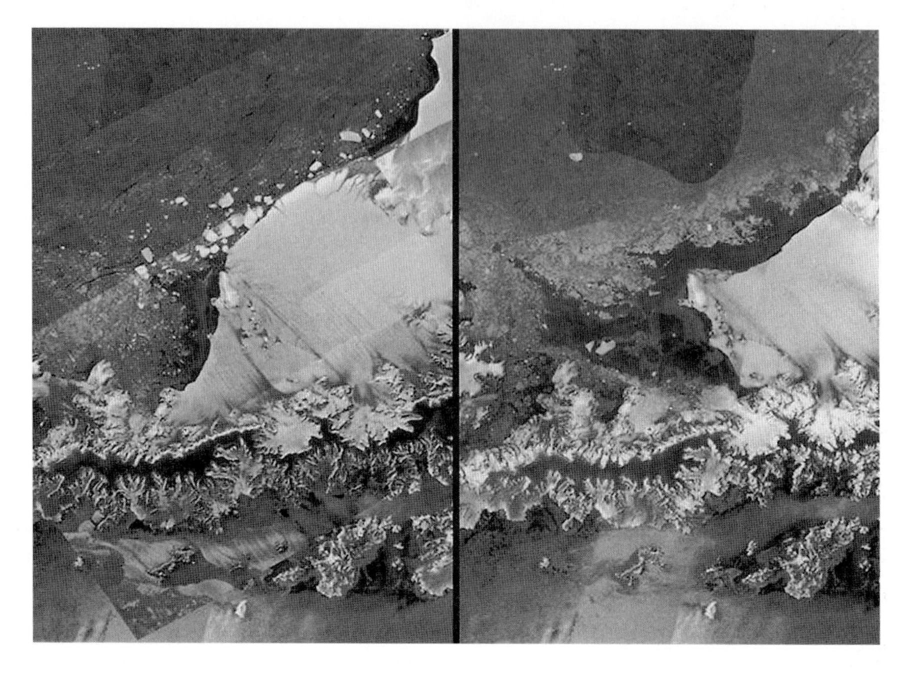

Figure 1.1 The Larsen Ice Shelf in Antarctica, before and after its collapse in 2002. Courtesy of NASA.

Greenhouse gases trap heat. As CO_2 levels rise, the atmosphere also holds more moisture and energy. CO_2 is absorbed mostly by oceans and some by forests (these are called CO_2 "sinks"), but these sinks cannot absorb atmospheric CO_2 as fast as human activity now produces it. Many countries neither regulate nor regard it as an air pollutant. The Kyoto Protocol of 1998 is designed to reduce CO_2 emissions worldwide, but the Protocol has proved divisive. In late 2004 Russia ratified it. This triggered its observance by fifty-five signatory countries, which also happened to produce 55 percent of the world's CO_2 emissions. But the United States rejected the Kyoto Protocol. China and India ratified it but with no mandated CO_2 emissions goals.

Climate change also involves the interplay of ocean temperatures, currents, chemistry, and salinity. Changes in both surface and deep ocean currents occur. Marine life including plankton and archaea are affected and set off further changes. Glacial ice everywhere is melting; a huge amount of melt occurs in Greenland. The resulting cold water runoff might disrupt ocean currents and prevent warm water from reaching northern and western Europe. From 1953 to 2003 the Arctic ice pack lost about half its thickness. Because the size of the pack has diminished, too, less heat is reflected back by white ice and more is absorbed by darker water, further reducing the ice, a positive feedback loop. Atmospheric moisture and winds are among the changing earth systems, as is the solar heat reflected by increased cloud cover and decreased ice and snow cover. Changes in forest density and location play pivotal roles. Because so many factors interact, atmospheric science, forestry, oceanography, ecology, microbiology, and meteorology are all involved.

To record changes in earth systems and in global climate is a daunting task. To predict changes into the future requires discovering what global climate was in the distant past and why it changed. Tree rings, fossils, and ice cores from glaciers and the seabed provide data. As we understand how earth systems interact and produce global climate, we model them mathematically and approximate their actual performance. However, even great computing power, like that of the National Center for Atmospheric Research in Boulder, Colorado, gives varied predictions. Why? Some data cannot be known with precision and the mathematical models themselves differ. The single most unpredictable variable in calculating and predicting climate is how much carbon humans will burn in the next decades.

Average global temperature rose by almost 1°F in the past century, but warming varied by location. Areas of Alaska and Canada increased 4° to 7°F. As CO_2 concentration accelerated, the World Meteorological Organization noted that the average global temperature rose three times faster from 1975 to 2000 than from 1900 to 1975. In Europe, the summer of 2003 was the hottest in 500 years. Authorities attributed 26,000 deaths to the heat. Globally, nine of the ten warmest years recorded between 1861 and 2000 occurred in the 1990s. Estimates of temperature increase in the present century range from 1.4 to 5.8°C (2.5 to 10.4°F).

In the past fifty years tropical storms have lasted 60 percent longer and have produced winds 50 percent stronger than known historical averages. Since 1970, category 4 and 5 hurricanes and typhoons have increased. During 1987–2004 they were 57 percent more common than during 1970–1981. These and similar statistics do not prove that global warming is the sole cause, but they create concern. Unabated, such changes will lead to climate shock and catastrophic consequences. A few models indicate the possibility of sudden reversal to a much colder climate in the Northern Hemisphere and in Europe in particular, especially if global warming shuts down ocean current (thermohaline) circulation that moves warm water from near the equator toward the poles.

Future climate change now appears to be more severe than estimated as recently as 2001. Sea levels are rising. (They rose four to eight inches in the past century.) This occurs several ways. First, seawater expands in volume as it warms (thermal expansion). Floating ice already displaces water, but ice not floating can melt, too. This is of huge concern if the vast ice sheets of Antarctica and Greenland melt. Ice not floating in the sea can also break off and, like ice cubes dropped in a drink, raise the sea level. For the next hundred years, estimates of increase in sea level elevation range from four inches to sixteen feet. Coastal wetlands will be impacted. Mountain glaciers are receding. By 2070 Glacier National Park in

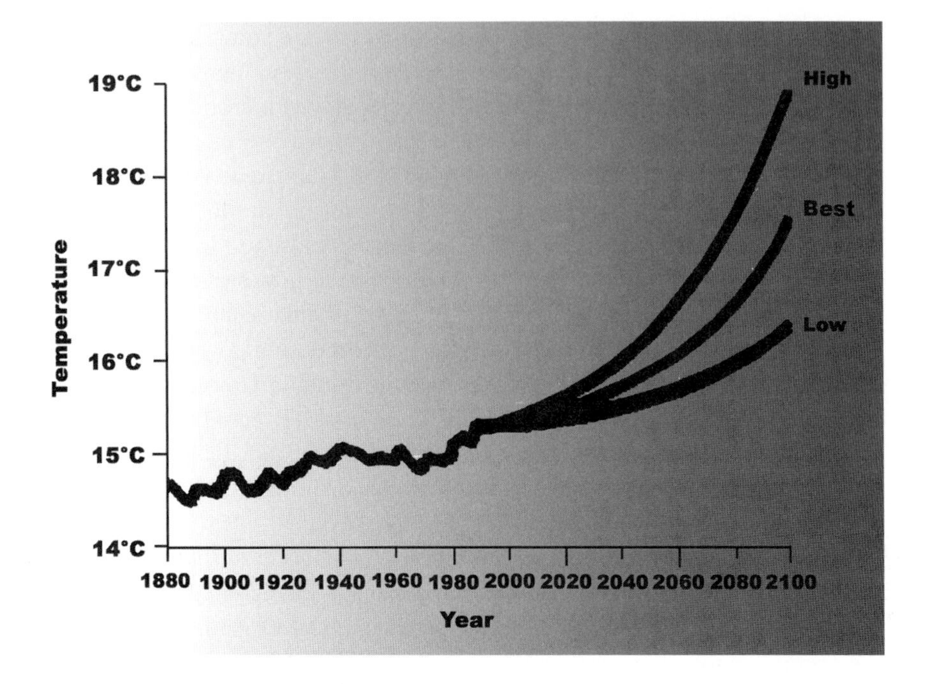

Figure 1.2 Projections of temperature increase based on continued human production of carbon dioxide (low, medium [best estimate], high). Courtesy of Philippe Rekacewicz, Grid-Arendal, Norway.

Montana may have *no* glaciers. All coastal cities will be affected, some perhaps severely. Some shorelines will disappear. Regions of poor countries such as Bangladesh, inundated with coastal floods, may cease to exist.

Weather patterns are intensifying. Storms, certainly more violent, may also be more numerous, floods and winds more destructive, droughts more severe and protracted, forest fires more common. In 2000, about 25,000 lives were lost in Venezuela's worst flooding in history. In reporting weather events, the phrase "worst on record" is becoming more common. Extremes of heat, cold, rainfall, winds, and forest fires will likely rise. Ironically, to deal with such extremes humans will probably burn more carbon.

Tropical diseases will spread more widely with greater virulence. Minute temperature changes can markedly increase disease-carrying insects and parasites. The dengue virus in South America and Rift Valley fever in Africa and the Middle East have extended their range. Malaria probably will, too. Eastern oyster virus now infects Maine coastal waters, immune when slightly colder. Coral reefs show significant stress. Whole ecologies are changing. In the 1990s, more than forty *million* Alaskan spruce trees died when warmer temperatures permitted the spruce beetle to multiply.

The science needed to chart earth systems and to predict global climate change is advanced and interdisciplinary. But just as the changes are largely the product of human behavior, so are the ways to adjust to those changes and, if desired, to abate them. Political will and cooperation on an unprecedented scale are needed to lower greenhouse gas levels. On 14 September 2004, Tony Blair, Prime Minister of Great Britain, stated that human activity "is causing global warming at a rate that began as significant, has become alarming, and is simply unsustainable in the long-term. And by long-term . . . I mean within the lifetime of my children certainly; and possibly within my own. And by unsustainable, I do not mean a phenomenon causing problems of adjustment. I mean a challenge so far-reaching in its impact and irreversible in its destructive power, that it alters radically human existence."

Many scientists and politicians promote imposing a tax on the carbon content of fuels to encourage conservation and alternate energy sources. More forest "sinks" will help, but forests globally are disappearing, not expanding (see Chapter 6, below). In an agreed international system, countries might sell carbon or carbon dioxide credits (Kyoto provides for the latter). Renewable energy sources and nuclear power will be important. Technology to remove CO_2 from the atmosphere on a massive scale might become feasible. Treating carbon before its combustion could reduce CO_2 emissions. Yet exisiting emissions patterns will be hard to curb or reverse. Developing countries require more energy if they are to escape poverty. China, with its vast population, depends on coal more than any other nation. United States citizens make up less than 5 percent of the world's population yet produce 25 percent of its CO_2. Increasingly, we are altering major earth systems. This will last for generations and may accelerate. The single most important—and unknown—factor is how human beings will act in the future. Prime Minister Blair warned, "there is no doubt that the time to act is now. It is now that timely action can avert disaster. It is now that with foresight and will such action can be taken without disturbing the essence of our way of life, by adjusting behavior not altering it entirely." If we care about the world that our children will inhabit, we can begin to change the way we live now.

FURTHER READING

Claussen, E., ed. *Climate Change: Science, Strategies and Solutions.* Boston: E. J. Brill, 2001.

Gore, Al. *An Inconvenient Truth: The Planetary Emergency of Global Warming and What We Can Do About It.* Emmaus, Pa.: Rodale, 2006.

Intergovernmental Panel on Climate Change (IPCC), reports available at www.ipcc.ch/ and published by Cambridge University Press.

Jorgenson, Dale W. *The Role of Substitution in Understanding the Costs of Climate Change Policy.* Arlington, Va.: Pew Center on Global Climate Change, 2000.

Kerr, Richard A. "Is Katrina a Harbinger of Still More Powerful Hurricanes?" *Science* 309 (16 Sept. 2005): 1807.

McElroy, Michael B., Chris P. Nielsen, and Peter Lydon, ed. *The Atmospheric Environment: Effects of Human Activity*. Princeton: Princeton University Press, 2002.
Schneider, S. H. et al., ed. *Climate Change Policy*. Washington, D.C.: Island Press, 2002.
Victor, David G. *The Collapse of the Kyoto Protocol and the Struggle to Slow Global Warming*. Princeton: Princeton University Press, 2004.
Weart, Spencer R. *The Discovery of Global Warming*. Cambridge: Harvard University Press, 2003.

John Houghton, from "The Greenhouse Effect" in *Global Warming: The Complete Briefing* (1997)

By 1995 many scientists and a significant part of the informed public believed global warming a fact. Evidence since then, including increased temperature readings of the upper as well as the surface atmosphere, confirms this conviction. John Houghton, Co-Chair of the Intergovernmental Panel on Climate Change (see Further Reading, above) explains the "greenhouse effect," the interconnected series of phenomena that cause global warming and lead to myriad environmental consequences.

The basic principle of global warming can be understood by considering the radiation energy from the sun which warms the Earth's surface and the thermal radiation from the Earth and the atmosphere which is radiated out to space. On average these two radiation streams must balance. If the balance is disturbed (for instance by an increase in atmospheric carbon dioxide) it can be restored by an increase in the Earth's surface temperature.

HOW THE EARTH KEEPS WARM

To explain the processes which warm the Earth and its atmosphere, I will begin with a very simplified Earth. Suppose we could, all of a sudden, remove from the atmosphere all the clouds, the water vapor, the carbon dioxide, and all the other minor gases and the dust leaving an atmosphere of nitrogen and oxygen only. Everything else remains the same. What, under these conditions, would happen to the atmospheric temperature?

The calculation is an easy one, involving a relatively simple radiation balance. Radiant energy from the sun falls on a surface of one square meter in area outside the atmosphere and directly facing the sun at a rate of about 1370 watts—about the power radiated by a reasonably sized domestic electric fire. However, few parts of the Earth's surface face the sun directly and in any case for half the time they are pointing away from the sun at night, so that the average energy falling on one square meter of a level surface outside the atmosphere is only one quarter of this or about 343 watts. As this radiation passes through the atmosphere a small amount, about 6 percent, is scattered back to space by atmospheric molecules. About 10 percent on average is reflected back to space from the land and ocean surface. The remaining 84 percent, or about 288 watts per square meter on average, remains actually to heat the surface—the power used by three good-sized incandescent electric light bulbs.

To balance this incoming energy, the Earth itself must radiate on average the same amount of energy back to space in the form of thermal radiation. All objects emit this kind of radiation; if they are hot enough we can see the radiation they emit. The sun at a temperature of about 6,000°C looks white; an electric fire at 800°C looks red. Cooler objects emit radiation which cannot be seen by our eyes and which lies at wavelengths beyond the red end of the spectrum—infrared radiation (sometimes called long-wave radiation to distinguish it from the short-wave radiation from the sun). On a clear, starry winter's night we are very aware of the cooling effect of this kind of radiation being emitted by the Earth's surface into space—it often leads to the formation of frost.

The amount of thermal radiation emitted by the Earth's surface depends on its temperature—the warmer it is, the more radiation is emitted. The amount of radiation also depends on how absorbing the surface is; the greater the absorption, the more the radiation. Most of the surfaces on the Earth, including ice and snow, would appear "black" if we could see them at infrared wavelengths; that means that they absorb nearly all the thermal radiation which falls on them instead of reflecting it. It can be calculated that, to balance the energy coming in, the average temperature of the Earth's surface must be –6°C to radiate the right amount. This is much colder than is actually the case. In fact, an average of temperatures measured near the surface all over the Earth—over the oceans as well as over the land—averaging, too, over the whole year, comes to about 15°C. Some factor not yet taken into account is needed to explain this discrepancy.

THE GREENHOUSE EFFECT

The gases nitrogen and oxygen, which make up the bulk of the atmosphere (Table 1.1 gives details of the atmosphere's composition), neither absorb nor emit thermal radiation. It is the water vapor, carbon dioxide, and some other minor gases present in the atmosphere in much smaller quantities which absorb some of the thermal radiation leaving the surface, acting as a partial blanket for this radiation and causing the difference of 21°C or so between the actual average surface temperature on the Earth of about 15°C and the figure of –6°C which applies when the atmosphere contains nitrogen and oxygen only. This blanketing is known as the *natural greenhouse effect* and the gases are known as greenhouse gases. It is called "natural" because all the atmospheric gases (apart from the chlorofluorocarbons—CFCs) were there long before human beings came on the scene. . . .

The basic science of the greenhouse effect has been known since early in the nineteenth century when the similarity between the radiative properties of the

TABLE 1.1
The Composition of the Atmosphere: The Main Constituents (Nitrogen and Oxygen) and the Greenhouse Gases

Gas	*Concentration in parts per million by volume (ppmv)*
Main Constituents	
Nitrogen (N_2)	780,000
Oxygen (O_2)	210,000
Greenhouse Gases (Uniform Concentration)	
Carbon dioxide (CO_2)	360
Methane (CH_4)	1.8
Nitrous oxide (N_2O)	0.3
CFCs	0.001
Greenhouse Gases (Variable Concentration)	
Water vapor (H_2O)	Between 0 and 20,000
Ozone (O_3)	Between 1 and 1,000

Note: Nitrogen and oxygen concentration values are approximate.
Based on 1995 data.

Earth's atmosphere and of the glass in a greenhouse was first pointed out—hence the name "greenhouse effect." In a greenhouse, visible radiation from the sun passes almost unimpeded through the glass and is absorbed by the plants and the soil inside. The thermal radiation which is emitted by the plants and soil is, however, absorbed by the glass which re-emits some of it back into the greenhouse. The glass thus acts as a "radiation blanket" helping to keep the greenhouse warm.

However, the transfer of radiation is only one of the ways heat is moved around in a greenhouse. A more important means of heat transfer is due to convection in which less dense warm air moves upwards and more dense cold air moves downwards. A familiar example of this process is the use of convective electric heaters in the home which heat a room by stimulating convection in it. The situation in the greenhouse is therefore more complicated than would be the case if radiation was the only process of heat transfer.

Mixing and convection are also present in the atmosphere, although on a much larger scale, and in order to achieve a proper understanding of the greenhouse effect, convective heat transfer processes in the atmosphere must be taken into account as well as radiative ones.

Within the atmosphere itself (at least in the lowest three-quarters or so of the atmosphere up to a height of about 10 km, which is called the troposphere) convection is, in fact, the dominant process for transferring heat. It acts as follows. The surface of the Earth is warmed by the sunlight it absorbs. Air close to the surface is heated and rises because of its lower density. As the air rises it expands and cools—just as the air cools as it comes out of the valve of a tire. As some air masses rise, other air masses descend, so the air is continually turning over as different movements balance each other out—a situation of convective equilibrium. Temperature in the troposphere falls with height at a rate determined by these convective processes; the drop turns out on average to be about 6°C per kilometer of height. . . .

Absorbing gases in the atmosphere absorb some of the radiation emitted by the Earth's surface and in turn emit radiation out to space. The amount of thermal radiation they emit is dependent on their temperature.

Radiation is emitted out to space by these gases from levels somewhere near the top of the atmosphere—typically from between 5 and 10 km high. Here, because of the convection processes mentioned earlier, the temperature is much colder—30 to 50°C or so colder—than at the surface. Because the gases are cold, they emit correspondingly less radiation. What these gases have done, therefore, is to absorb some of the radiation emitted by the Earth's surface but then to emit much less radiation out to space. They have, therefore, acted as a radiation blanket over the surface (note that the outer surface of a blanket is colder than inside the blanket) and helped to keep it warmer than it would otherwise be.

Thomas R. Karl and Kevin E. Trenberth, "Modern Global Climate Change" (2003)

Two American scientists present a cogent summary of recent climate change based on years of records, observations, and modeling, as well as knowledge of earlier climate history. Their work draws also on the Intergovernmental Panel on Climate Change (IPCC) and on the research of numerous scientists in many nations.

Abstract: Modern climate change is dominated by human influences, which are now large enough to exceed the bounds of natural variability. The main source of global climate change is human-induced changes in atmospheric composition. These perturbations primarily result from emissions associated with energy use, but on local and regional scales, urbanization and land use changes are also important. Although there has been progress in monitoring and understanding climate change, there remain many scientific, technical, and institutional impediments to precisely planning for, adapting to, and mitigating the effects of climate change. There is still considerable uncertainty about the rates of change that can be expected, but it is clear that these changes will be increasingly manifested in important and tangible ways, such as changes in extremes of temperature and precipitation, decreases in seasonal and perennial snow and ice extent, and sea level rise. Anthropogenic climate change is now likely to continue for many centuries. We are venturing into the unknown with climate, and its associated impacts could be quite disruptive.

The atmosphere is a global commons that responds to many types of emissions into it, as well as to changes in the surface beneath it. As human balloon flights around the world illustrate, the air over a specific location is typically halfway around the world a week later, making climate change a truly global issue.

Planet Earth is habitable because of its location relative to the sun and because of the natural greenhouse effect of its atmosphere. Various atmospheric gases contribute to the greenhouse effect, whose impact in clear skies is ~60% from water vapor, ~25% from carbon dioxide, ~8% from ozone, and the rest from trace gases including methane and nitrous oxide. Clouds also have a greenhouse effect. On average, the energy from the sun received at the top of the Earth's atmosphere amounts to 175 petawatts (PW) (or 175 quadrillion watts), of which ~31% is reflected by clouds and from the surface. The rest (120 PW) is absorbed by the atmosphere, land, or ocean and ultimately emitted back to space as infrared radiation.[1] Over the past century, infrequent volcanic eruptions of gases and debris into the atmosphere have significantly perturbed these energy flows; however, the resulting cooling has lasted for only a few years. Inferred changes in total solar irradiance appear to have increased global mean temperatures by perhaps as much as 0.2°C in the first half of the twentieth century, but measured changes in the past 25 years are small. Over the past 50 years, human influences have been the dominant detectable influence on climate change.[2] The following briefly describes the human influences on climate, the resulting temperature and precipitation changes, the time scale of responses, some important processes involved, the use of climate models for assessing the past and making projections into the future, and the need for better observational and information systems.

The main way in which humans alter global climate is by interference with the natural flows of energy through changes in atmospheric composition, not by the actual generation of heat in energy usage. On a global scale, even a 1% change in the energy flows, which is the order of the estimated change to date,[2] dominates all other direct influences humans have on climate. For example, an energy output of just one PW is equivalent to that of a million power stations of 1000-MW capacity, among the largest in the world. Total human energy use is about a factor of 9000 less than the natural flow.[3]

Global changes in atmospheric composition occur from anthropogenic emissions of greenhouse gases, such as carbon dioxide that results from the burning of fossil fuels and methane and nitrous oxide from multiple human activities. Because these gases have long (decades to centuries) atmospheric lifetimes, the result is an accumulation in the atmosphere and a build-up in concentrations that are clearly shown both by instrumental observations of air samples since 1958 and in bubbles of air trapped in ice cores before then. Moreover, these gases are well distributed in the atmosphere across the globe, simplifying a global monitoring strategy. Carbon dioxide has increased 31% since preindustrial times, from 280 parts per million by volume (ppmv) to more than 370 ppmv today, and half of the increase has been since 1965.[4] (See Figure 1.3.) The greenhouse gases trap outgoing radiation from the Earth to space, creating a warming of the planet.

Emissions into the atmosphere from fuel burning further result in gases that are oxidized to become highly reflective micron-sized aerosols, such as sulfate, and strongly absorbing aerosols, such as black carbon or soot. Aerosols are rapidly (within a week or less) removed from the atmosphere through the natural hydrological cycle and dry deposition as they travel away from their source. Nonetheless, atmospheric concentrations can substantially exceed background conditions in large areas around and downwind of the emission sources. Depending on their reflectivity and absorption properties, geometry and size distribution, and interactions with clouds and moisture, these particulates can lead to either net cooling, as for sulfate aerosols, or net heating, as for black carbon. Importantly, sulfate aerosols affect climate directly by reflecting solar radiation and indirectly by changing the reflective properties of clouds and their lifetimes. Understanding their precise impact has been hampered by our inability to measure these aerosols directly, as well as by their spatial inhomogeneity and rapid changes in time. Large-scale measurements of aerosol patterns have been inferred through emission data, special field experiments, and indirect measurements such as sun photometers.[5]

Human activities also have a large-scale impact on the land surface. Changes in land-use through urbanization and agricultural practices, although not global, are often most pronounced where people live, work, and grow food, and are part of the human impact on climate.[6,7] Large-scale deforestation and desertification in Amazonia and the Sahel, respectively, are two instances where evidence suggests there is likely to be human influence on regional climate.[8–10] In general, city climates differ from those in surrounding rural green areas, because of the "concrete jungle" and its effects on heat retention, runoff, and pollution, resulting in urban heat islands.

There is no doubt that the composition of the atmosphere is changing because of human activities, and today greenhouse gases are the largest human influence on global climate.[2] Recent greenhouse gas emission trends in the United States are upward,[11] as are global emissions trends, with increases between 0.5 and 1% per year over the past few decades.[12] Concentrations of both reflective and nonreflective aerosols are also estimated to be increasing. Because radiative forcing from greenhouse gases dominates over the net cooling forcings from aerosols,[2] the popular term for the human influence on global climate is "global warming," although it really means global heating, of which the observed global temperature increase is only one consequence.[13] (See Figure 1.3.) Already it is estimated that the Earth's climate has exceeded the bounds of natural variability,[2] and this has been the case since about 1980.

Surface moisture, if available (as it always is over the oceans), effectively acts as the "air conditioner" of the surface, as heat used for evaporation moistens the air rather than warming it. Therefore, another consequence of global heating of

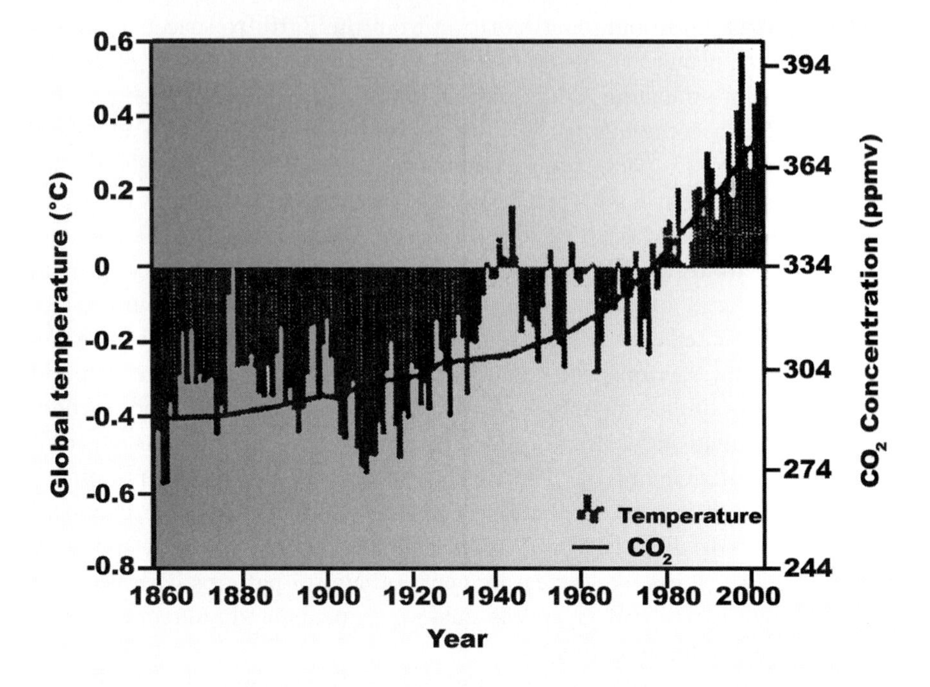

Figure 1.3 Time series of departures from the 1961 to 1990 base period for an annual mean global temperature of 14°C (bars) and for a carbon dioxide mean of 334 ppmv (solid curve) during the base period, using data from ice cores and (after 1958) from Mauna Loa. The global average surface heating approximates that of carbon dioxide increases, because of the cancellation of aerosols and other greenhouse gas effects, but this does not apply regionally. Many other factors (such as the effects of volcanic eruptions and solar irradiance changes) are also important.

the lower troposphere is accelerated land-surface drying and more atmospheric water vapor (the dominant greenhouse gas). Accelerated drying increases the incidence and severity of droughts, whereas additional atmospheric water vapor increases the risk of heavy precipitation events.[14] Basic theory,[15] climate model simulations,[2] and empirical evidence (see Figure 1.4) all confirm that warmer climates, owing to increased water vapor, lead to more intense precipitation events even when the total precipitation remains constant, and with prospects for even stronger events when precipitation amounts increase.[16–18]

There is considerable uncertainty as to exactly how anthropogenic global heating will affect the climate system, how long it will last, and how large the effects will be. Climate has varied naturally in the past, but today's circumstances are unique because of human influences on atmospheric composition. As we progress into the future, the magnitude of the present anthropogenic change will become overwhelmingly large compared to that of natural changes.

In the absence of climate mitigation policies, the 90% probability interval for warming from 1990 to 2100 is 1.7° to 4.9°C.[19] About half of this range is due to uncertainty in future emissions and about half is due to uncertainties in climate models,[2, 19] especially in their sensitivity to forcings that are complicated by feedbacks, discussed below, and in their rate of heat uptake by the oceans.[20] Even with these uncertainties, the likely outcome is more frequent heat waves, droughts, extreme precipitation events, and related impacts (such as wild fires, heat stress, vegetation changes, and sea level rise) that will be regionally dependent.

The rate of human-induced climate change is projected to be much faster than most natural processes, certainly those prevailing over the past 10,000 years.[2] Thresholds likely exist that, if crossed, could abruptly and perhaps almost irreversibly switch the climate to a different regime. Such rapid change is evident in past climates during a slow change in the Earth's orbit and tilt, such as the Younger Dryas cold event from ~11,500 to ~12,700 years ago,[2] perhaps caused by freshwater discharges from melting ice sheets into the North Atlantic Ocean and a change in the ocean thermohaline circulation.[21, 22] The great ice sheets of Greenland and Antarctica may not be stable, because the extent to which cold-season heavier snowfall partially offsets increased melting as the climate warms remains uncertain. A combination of ocean temperature increases and ice sheet melting could systematically inundate the world's coasts by raising sea level for centuries.

Given what has happened to date and is projected in the future,[2] substantial further climate change is guaranteed. The rate of change can be slowed, but it is unlikely to be stopped in the twenty-first century.[23] Because concentrations of long-lived greenhouse gases are dominated by accumulated past emissions, it takes many decades for any change in emissions to have much effect. This means the atmosphere still has unrealized warming (estimated to be at least another 0.5°C) and that sea level rise may continue for centuries after an abatement of anthropogenic greenhouse gas emissions and the stabilization of greenhouse gas concentrations in the atmosphere.

Our understanding of the climate system is complicated by feedbacks that either amplify or damp perturbations, the most important of which involve water in various phases. As temperatures increase, the water-holding capacity of the atmosphere increases along with water vapor amounts, producing water vapor feedback. As water vapor is a strong greenhouse gas, this diminishes the loss of energy through infrared radiation to space. Currently, water vapor feedback is estimated to contribute a radiative effect from one to two times the size of the direct effect of increases in anthropogenic greenhouse gases.[24, 25] Precipitation-runoff feedbacks occur because more intense rains run off at the expense of soil moisture, and warming promotes rain rather than snow. These changes in turn alter the partitioning of solar radiation into sensible versus latent heating.[14] Heat storage feedbacks include the rate at which the oceans take up heat and the currents redistribute and release it back into the atmosphere at variable later times and different locations.

Cloud feedback occurs because clouds both reflect solar radiation, causing cooling, and trap outgoing long-wave radiation, causing warming. Depending on the height, location, and the type of clouds with their related optical properties, changes in cloud amount can cause either warming or cooling. Future changes in clouds are the single biggest source of uncertainty in climate predictions. They contribute to an uncertainty in the sensitivity of models to changes in greenhouse gases, ranging from a small negative feedback, thereby slightly reducing the direct radiative effects of increases in greenhouse gases, to a doubling of the direct radiative effect of increases in greenhouse gases.[25] Clouds and precipitation processes cannot be resolved in climate models and have to be parametrically represented (parameterized) in terms of variables that are resolved. This will continue for some time into the future, even with projected increases in computational capability.[26]

Ice-albedo feedback occurs as increased warming diminishes snow and ice cover, making the planet darker and more receptive to absorbing incoming solar radiation, causing warming, which further melts snow and ice. This effect is greatest at high latitudes. Decreased snow cover extent has significantly contributed to the earlier onset of spring in the past few decades over northern-hemisphere high latitudes.[27] Ice-albedo feedback is affected by changes in clouds, thus complicating the net feedback effect.

The primary tools for predicting future climate are global climate models, which are fully coupled, mathematical, computer-based models of the physics, chemistry, and biology of the atmosphere, land surface, oceans, and cryosphere and their interactions with each other and with the sun and other influences (such as volcanic eruptions). Outstanding issues in modeling include specifying forcings of the climate system; properly dealing with complex feedback processes (see Figure 1.5) that affect carbon, energy, and water sources, sinks and transports; and improving simulations of regional weather, especially extreme events. Today's inadequate or incomplete measurements of various forcings, with the exception of well-mixed greenhouse gases, add uncertainty when trying to simulate past and present climate. Confidence in our ability to predict future climate is dependent on our ability to use climate models to attribute past and present climate change to specific forcings. Through clever use of paleoclimate data, our ability to reconstruct past forcings should improve, but it is unlikely to provide the regional detail necessary that comes from long-term direct measurements. An example of forcing uncertainty comes from recent satellite observations and data analyses of twentieth-century surface, upper air, and ocean temperatures, which indicate that estimates of the indirect effects of sulfate aerosols on clouds may be high, perhaps by as much as a factor of two.[27–29] Human behavior, technological change, and the rate of population growth also affect future emissions, and our ability to predict these must be factored into any long-term climate projection.

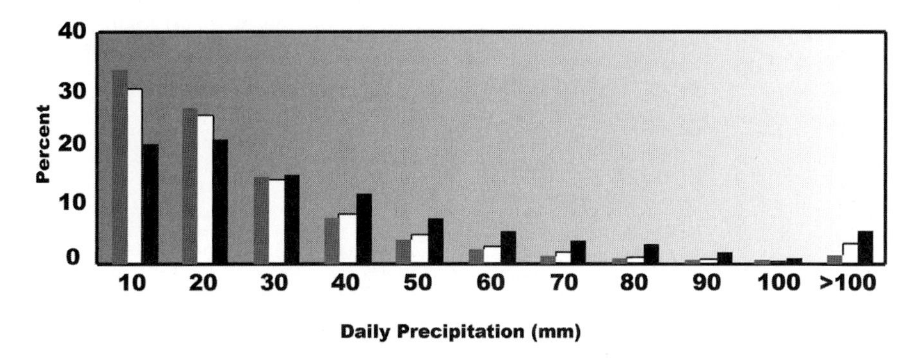

Figure 1.4 Climatology of the intensity of daily precipitation as a percentage of total amount in 10 mm/day categories for different temperature regimes, based on 51, 37, and 12 worldwide stations, respectively: gray bars, –3°C to 19°C; white bars, 19°C to 29°C; black bars, 29°C to 335°C. By selection, all stations have the same seasonal mean precipitation amount of 230 +/– 5 mm. As temperatures and the associated water-holding capacity of the atmosphere increase, more precipitation falls in heavy (more than 40 mm/day) to extreme (more than 100 mm/day) daily amounts.

Regional predictions are needed for improving assessments of vulnerability to and impacts of change. The coupled atmosphere-ocean system has a preferred mode of behavior known as El Niño, and similarly the atmosphere is known to have preferred patterns of behavior, such as the North Atlantic Oscillation (NAO). So how will El Niño and the NAO change as the climate changes? There is evidence that the NAO, which affects the severity of winter temperatures and precipitation in Europe and eastern North America, and El Niño, which has large regional effects around the world, are behaving in unusual ways that appear to be linked to global heating.[2, 31–33] Hence, it is necessary to be able to predict the statistics of the NAO and El Niño to make reliable regional climate projections.

Ensembles of model predictions have to be run to generate probabilities and address the chaotic aspects of weather and climate. This can be addressed in principle with adequate computing power, a challenge in itself. However, improving models to a point where they are more reliable and have sufficient resolution to be properly able to represent known important processes also requires the right observations, understanding, and insights (brain power). Global climate models will need to better integrate the biological, chemical, and physical components of the Earth system (see Figure 1.5). Even more challenging is the seamless flow of data and information among observing systems, Earth system models, socioeconomic models, and models that address managed and unmanaged ecosystems. Progress here is dependent on overcoming not only scientific and technical issues but also major institutional and international obstacles related to the free flow of climate-related data and information.

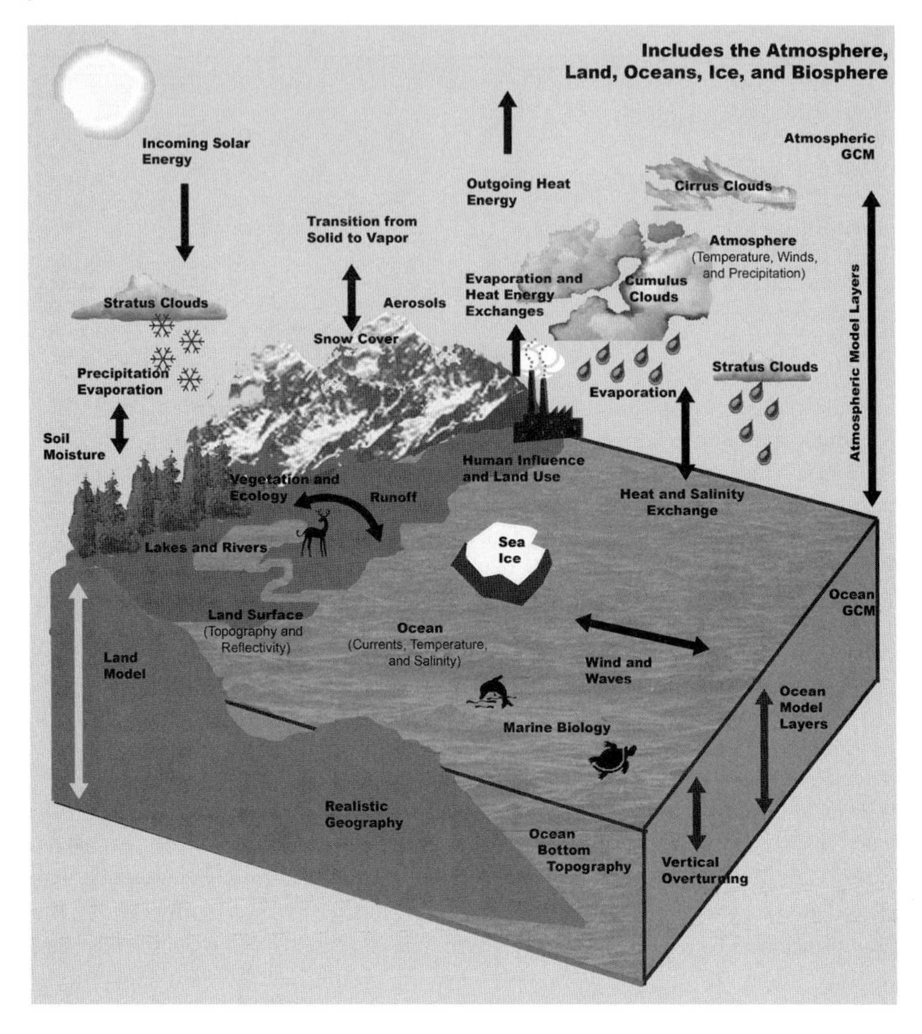

Figure 1.5 Components of the climate system and the interactions among them, including the human component. All these components have to be modeled as a coupled system that includes the oceans, atmosphere, land, cryosphere, and biosphere. GCM = General Circulation Model.

In large part, reduction in uncertainty about future climate change will be driven by studies of climate change assessment and attribution. Along with climate model simulations of past climates, this requires comprehensive and long-term climate-related data sets and observing systems that deliver data free of time-dependent biases. These observations would ensure that model simulations are evaluated on the basis of actual changes in the climate system and not on artifacts of changes in observing system technology or analysis methods.[34] The recent controversy regarding the effects that changes in observing systems have had on the rate of surface versus tropospheric warming[35, 36] highlights this issue. Global monitoring through space-based and surface-based systems is an international matter, much like global climate change. There are encouraging signs, such as the adoption in 1999 of a set of climate monitoring principles,[37] but these principles are impotent without implementation. International implementation of these principles is spotty at best.[38]

We are entering the unknown with our climate. We need a global climate observing system, but only parts of it exist. We must not only take the vital signs of the planet but also assess why they are fluctuating and changing. Consequently, the system must embrace comprehensive analysis and assessment as integral components on an ongoing basis, as well as innovative research to better interpret results and improve our diagnostic capabilities. Projections into the future are part of such activity, and all aspects of an Earth information system feed into planning for the future, whether by planned adaptation or mitigation. Climate change is truly a global issue, one that may prove to be humanity's greatest challenge. It is very unlikely to be adequately addressed without greatly improved international cooperation and action.

Notes

1. J. T. Kiehl and K. E. Trenberth, *Bulletin of the American Meteorological Society* 78 (1997): 197.
2. J. T. Houghton et al., ed., *Climate Change 2001: The Scientific Basis* (Cambridge: Cambridge University Press, 2001) (available at www.ipcc.ch/).
3. R. J. Cicerone, *Proceedings of the National Academy of Sciences U.S.A.* 100 (2000): 10304.
4. Atmospheric CO_2 concentrations from air samples and from ice cores are available at http://cdiac.esd.ornl.gov/trends/co2/sio-mlo.htm and http://cdiac.esd.ornl.gov/trends/co2/siple.htm, respectively.
5. M. Sato et al., *Proceedings of the National Academy of Sciences U.S.A.* 100 (2003): 6319.
6. A. J. Dolman, A. Verhagen, and C. A. Rovers, ed., *Global Environmental Change and Land Use* (Dordrecht: Kluwer, 2003).
7. G. B. Bonan, *Ecological Applications* 9 (1999): 1305.
8. J. G. Charney, *Quarterly Journal of the Royal Meteorological Society* 101 (1975): 193.
9. C. Nobre et al., in P. Kabot et al., ed., *Vegetation, Water, Humans, and the Climate* (Heidelberg: Springer Verlag, 2003).
10. A. N. Hahmann and R. E. Dickinson, *Journal of Climate* 10 (1997): 1944.
11. U.S. Department of State, *U.S. Climate Action Report 2002* (Washington, D.C., 2002) (available at http://yosemite.epa.gov/oar/globalwarming.nsf/content/ResourceCenterPublicationsUSClimateActionReport.html).
12. G. Marland, T. A. Boden, and R. J. Andres, at the Web site *Trends: A Compendium of Data on Global Change* (CO_2 Information Analysis Center, Oak Ridge: National Laboratory, 2002) (available at http://cdiac.esd.ornl.gov/trends/emis/em-cont.htm).
13. Global temperatures are available from www.ncdc.noaa.gov/oa/climate/research/2002/ann/ann02.html.
14. K. E. Trenberth, A. Dai, R. M. Rasmussen, and D. B. Parsons, *Bulletin of the American Meteorological Society* 84 (2003): 1205 (available at www.cgd.ucar.edu/cas/adai/papers/rainChBamsR.pdf).
15. The Clausius Clapeyron equation governs the water-holding capacity of the atmosphere, which increases by ~7% per degree Celsius increase in temperature. [See note 13 above.]

16. R. W. Katz, *Advances in Water Resources* 23 (1999): 133.
17. P. Ya. Groisman, *Climate Change* 42 (1999): 243.
18. T. R. Karl and R. W. Knight, *Bulletin of the American Meteorological Society* 78 (1998): 1107.
19. T. Wigley and S. Raper, *Science* 293 (2001): 451.
20. S. J. Levitus et al., *Science* 287 (2001): 2225.
21. W. S. Broecker, *Science* 278 (1997): 1582.
22. T. F. Stocker and O. Marchal, *Proceedings of the National Academy of Sciences U.S.A.* 97 (2000): 1362.
23. M. Hoffert et al., *Science* 298 (2002): 981.
24. U.S. National Research Council, *Climate Change Science: An Analysis of Some Key Questions* (Washington, D.C.: National Academy, 2001).
25. R. Colman, *Climate Dynamics* 20 (2003): 865.
26. T. R. Karl and K. E. Trenberth, *Scientific American* 281 (December 1999): 100.
27. P. Ya. Groisman, T. R. Karl, R. W. Knight, and G. L. Stenchikov, *Science* 263 (1994): 198.
28. C. E. Forest, P. H. Stone, A. Sokolov, M. R. Allen, and M. D. Webster, *Science* 295 (2002): 113.
29. J. Coakley, Jr., and C. D. Walsh, *Journal of Atmospheric Science* 59 (2002): 668.
30. J. Coakley, Jr., personal communication.
31. M. A. Saunders, *Geophysical Research Letters* 30 (2003): 1378.
32. M. P. Hoerling, J. W. Hurrell, and T. Xu, *Science* 292 (2001): 90.
33. K. E. Trenberth and T. J. Hoar, *Geophysical Research Letters* 24 (1997): 3057.
34. K. E. Trenberth, T. R. Karl, and T. W. Spence, *Bulletin of the American Meteorological Society* 83 (2002): 1558.
35. The Climate Change Science Program plan is available at www.climatescience.gov.
36. B. Santer et al., *Science* 300 (2003): 1280.
37. The climate principles were adopted by the Subsidiary Body on Science, Technology and Assessment of the United Nations Framework Convention on Climate Change (UNFCCC).
38. Global Climate Observing System (GCOS), *The Second Report on the Adequacy of the Global Observing Systems for Climate in Support of the UNFCCC* (GCOS-82, WMO/TD 1143, Geneva: World Meteorological Organization, 2003) (available from www.wmo.ch/web/gcos/gcoshome.html).
39. We thank A. Leetmaa, J. Hurrell, J. Mahlman, and R. Ciccrone for helpful comments, and J. Enloe for providing the calculations for Figure 1.4. This article reflects the views of the authors and does not reflect government policy. The National Climatic Data Center is part of NOAA's Satellite and Information Services. The National Center for Atmospheric Research is sponsored by the NSF [National Science Foundation].

John Gribbin, from "Earth's Temperature Trends," "CO_2 and Ice Ages," and "Oceans and Climate" in *Hothouse Earth: The Greenhouse Effect and Gaia* (1990)

In 1990 many critics and some scientists considered global warming a hypothesis or even an alarmist's fantasy. Opportunistic as some writing on the subject is, several accounts such as Gribbin's underscore the longer history and more complex record of research needed to put atmospheric and climate study in perspective. Gribbin takes a tour d'horizon *of long-term issues and emphasizes that, in conjunction with atmospheric conditions, ocean currents play a crucial role in climate regulation and change.*

TEMPERATURE TRENDS

In February 1988, the East Anglia team updated their global temperature survey, adding in new data from recording stations that had not been included in the previous survey and taking the records forward through to the end of 1987. Adding in the extra recording stations made a slight difference to the figures—the northern hemisphere average for 1981–1984, for example, was reduced by 0.04°C in the revised figures. But these minor adjustments paled into insignificance alongside the main conclusions of the study. Averaging over the whole globe, 1987 was now seen as the warmest year since the start of reliable observations in 1858, while 1981 and 1983 tied in second place, just 0.05°C cooler. The downward blip in the record of the mid-1980s in the northern hemisphere was now seen as no more than a temporary lull in the upward trend. The figures for the southern hemisphere were even more dramatic. There, seven of the eight warmest years on record occurred in the 1980s, with 1987 again the warmest. In the northern hemisphere alone, 1987 was just a little cooler than 1981 (perhaps the lingering effect of that volcanic dust [from the eruption of El Chichón in South America in 1982]). And, remember, 1988 is now breaking those records. The 1980s stand out as something quite remarkable in the historical record of temperature changes [the trend upward has continued to 2005].

These particular measurements all come from observing stations on land. But researchers at the U.K. Meteorological Office have been studying records of sea surface temperatures measured by observers on board ships since the middle of the nineteenth century. This kind of study is a saga in itself: if it is difficult comparing measurements made by different people using different instruments at different places on land, it is at least doubly difficult working out exactly what has been measured when somebody, a hundred years ago, hauled up a bucket of sea water to the heaving deck of a ship in the middle of the Atlantic or Pacific oceans and stuck a thermometer in it. Many factors affect how much the water warms up, or cools down, while it is on deck and therefore affect the measured temperature. How long was the bucket allowed to stand on the deck before the temperature was taken? Did it stand in sunlight, or in the shade? Was the bucket "pre-cooled" by giving it a quick dip in the briny before the real sample was taken? Was the bucket made of wood, canvas, or metal? And so on. The answers to some of these questions can never, in some cases, be known. Reconstructing past temperature patterns depends on obtaining many different measurements, and using statistical techniques not only to take averages in the right way (balancing, for example, a few readings from a large area of ocean against many measurements from a small land mass) but also to eliminate readings which are shown, by those statistical tests, to be unreliable. Fortunately, the bucket technique was more or less standardized by 1900, so there are relatively few problems in interpreting the all-important record of sea temperatures during the twentieth century, although the experts do disagree slightly in their estimates of sea

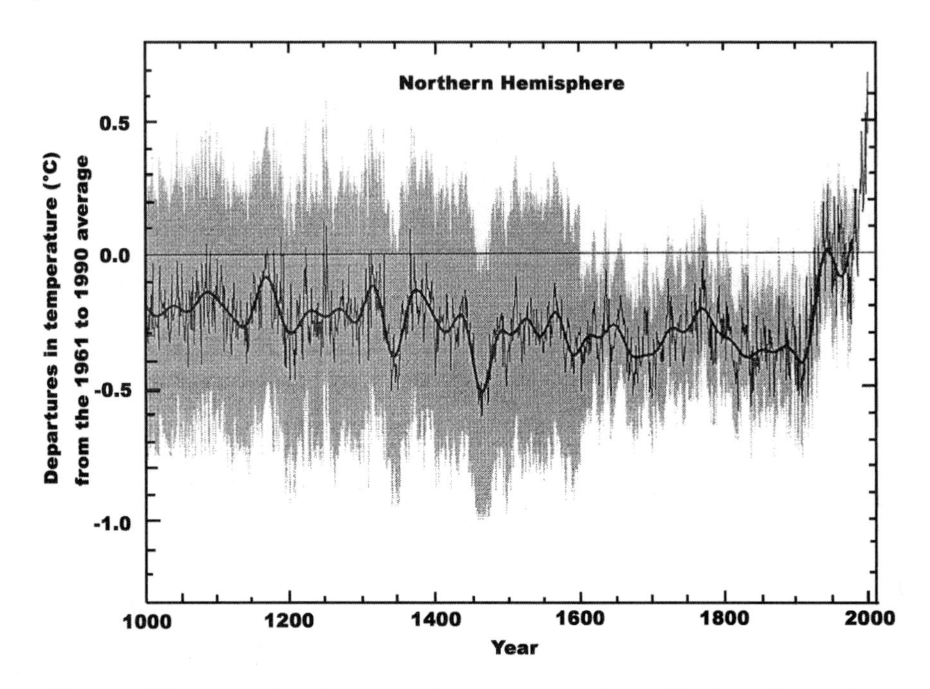

Figure 1.6 Variations of the Earth's surface temperature in the Northern Hemisphere for the past 1,000 years. Yearly data and running average from thermometers, tree rings, corals, ice cores, and historical records. Gray represents 95 percent confidence range. *Source:* Intergovernmental Panel on Climate Change.

temperatures back in the 1850s and 1860s. On all interpretations, however, the sea surface measurements also show that 1987 was the warmest year since records began in the 1850s, and they show the same two-stage warming during the twentieth century that was detected from the land-based figures. . . .

Global mean temperature increased by about 0.5°C between 1880 and 1940, actually decreased by about 0.2°C between 1940 and 1970 (almost halfway back to where it started), and then increased by about 0.3°C between 1970 and 1980, with a continuing rapid rise in the 1980s. Jones has looked in detail at the twenty-year period from 1967 to 1986. Concentrating on the land surface, where most of us live, he has found a total warming of 0.31°C in the northern hemisphere and 0.23°C in the southern hemisphere. This may come as a surprise to some readers, especially those in northwest Europe. I can imagine them thinking that their own experience doesn't seem to bear this out and that there are no signs, in their part of the globe, that the world is getting warmer. That wouldn't be surprising, because one of the most important discoveries shown up by Jones's study of temperature trends in different geographical regions is that the warming is far from evenly distributed—so far from even, in fact, that a large part of

Europe, including Britain, *cooled* by about 0.25°C between the mid-1970s and the mid-1980s.

This patchiness in the temperature record is not just a feature of Europe, or even of the northern hemisphere. In the southern hemisphere, the present warming is most pronounced over Australia, southern South Africa, the region where the tip of South America almost meets the Antarctic Peninsula, and in the region of Antarctica near Australia. In the northern hemisphere, the warming is strongest over Alaska, northwestern Canada, the Greenland Sea, most of the [former] Soviet Union (but especially Siberia), parts of southern Asia, north Africa, and southwestern Europe. There is also, though, the newly identified cooling trend over a large part of Europe, northeastern and eastern Canada, and southwest Greenland. Over Scandinavia, mean temperatures fell by about 0.6°C at a time when the northern hemisphere warmed by 0.31°C. Scandinavia has got nearly a full degree "out of step" with its surroundings.

Parts of the Antarctic coast have cooled even more dramatically, by about 1°C, while the strongest recorded warming has reached 2°C over northwestern Canada, 1.6°C over western Siberia, and just over 1.2°C at the tip of South Africa—all over the same twenty-year period. . . .

The records show that over the past thirty to forty years there has been a significant increase in precipitation at mid-latitudes (including Britain and nearby continental Europe), and corresponding decreases in precipitation at low latitudes (including the parts of northern Africa that now seem to be chronically afflicted by drought and famine). . . .

WHY ALL THE FUSS?

. . . It seems appropriate . . . to mention briefly why so many people, by the end of 1988, were concerned about the rising trend of temperatures. An increase of less than one degree Celsius in a hundred years doesn't sound very much to get excited about in everyday terms. It is less than the amount by which the temperature varies during the day, or from day to night, and far less than the seasonal variations that inhabitants of temperate latitudes are used to. So why all the fuss?

The decline in rainfall in the subtropical region of the northern hemisphere—the root cause of continuing tragic famine across northern Africa—is the most obvious example to date of why even such small shifts in climate are a cause for concern. This trend is exactly in line with calculations of how rainfall patterns should change as the world warms, and it has occurred at exactly the time the world has begun to warm rapidly. There is no proof that this is cause and effect, but the circumstantial evidence for a link is strong. In all probability, the greenhouse effect has already [by 1990] killed hundreds of thousands of people. The droughts and associated famines have happened during a rise in temperature of less than half a degree Celsius over the past twenty years, but the forecasts and

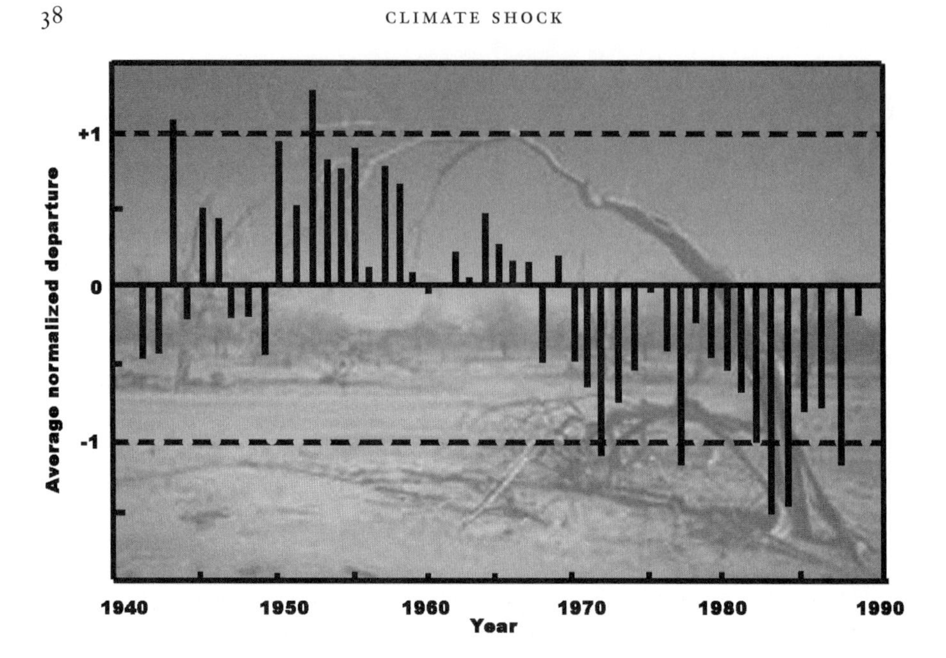

Figure 1.7 Variations in rainfall in the Sahel region of Africa, measured as deviations from the long-term average. The measurements are "normalized" in such a way that any departure within the range from −1 to +1°C might occur by chance, but any departure outside of this range is probably a sign of a change in weather patterns. On this evidence, the "normal" weather in the Sahel is drier now than in the 1950s. Adapted from R. A. Kerr, *Science* 227 (1985): 1453.

projections suggest that a rise of a full degree is on the cards over the next twenty to thirty years, with more to follow.

Putting that in perspective, although weather experts are fond of telling you that the climate is always changing, in fact the range of variation that has happened on any human timescale is small. The difference between the cold of the most recent ice age and present-day conditions corresponds only to a global warming of about 3°C. The climatic conditions on Earth today are those of an interglacial, an interval of ten to twenty thousand years between ice ages proper, which themselves last for about a hundred thousand years. The warmest period of the present interglacial occurred about six thousand years ago. It may be no coincidence that this was the time when humanity began to take steps along the road leading to civilization; but that warmth, the hottest the Earth has been since the ice age, was no more than a degree above the temperatures that prevailed in the 1880s, and half a degree warmer than the 1980s. By the first decade of the twenty-first century, the world will be warmer than it has been at any time since the latest ice age *began*, more than a hundred thousand years ago. [This, in fact, has occurred.] By any standards, this is a major human disturbance of the environment.

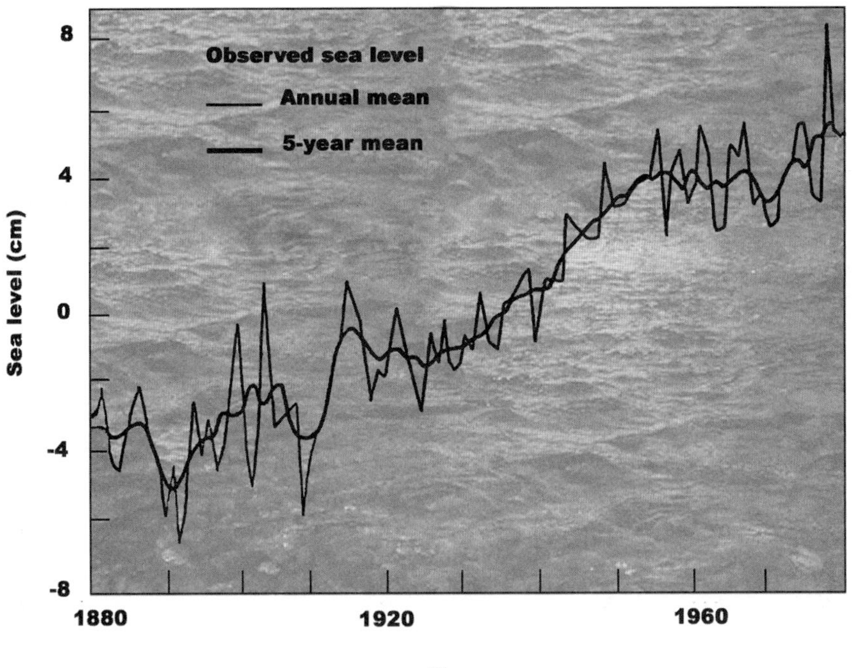

Figure 1.8 The rising trend of global sea level closely follows the rising temperature trend. *Source:* James Hansen/GISS.

. . . No matter how small the temperature differences might seem to those of us who live in centrally heated or air-conditioned houses, work in similarly insulated offices, and take skiing holidays in winter but fly off to the sunshine in summer, even the more cautious projections now predict that by the early twenty-first century the world's climate will have shifted into a state that has never existed during the entire time that modern agriculture, on which we all depend for our food, was being developed. . . .

Mention rising sea levels, and most people immediately think in terms of melting ice sheets. In fact, this is not the main cause of the increase in sea level since the nineteenth century. When the world gets warmer, the water in the sea will expand, just as a bar of iron expands when it is heated. (If you are making a mug of cocoa, don't measure out how much milk you need by filling the mug to the brim before pouring the milk into the pan to heat—the same measure of milk will be too big for the mug when it is hot.) This thermal expansion of the upper layers of the ocean is enough to account for most of the change that has occurred in sea level this century. Indeed, the rising trend of sea level closely matches up to the changing trend in average global temperatures, even including a flatten-

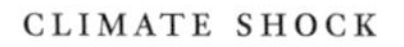

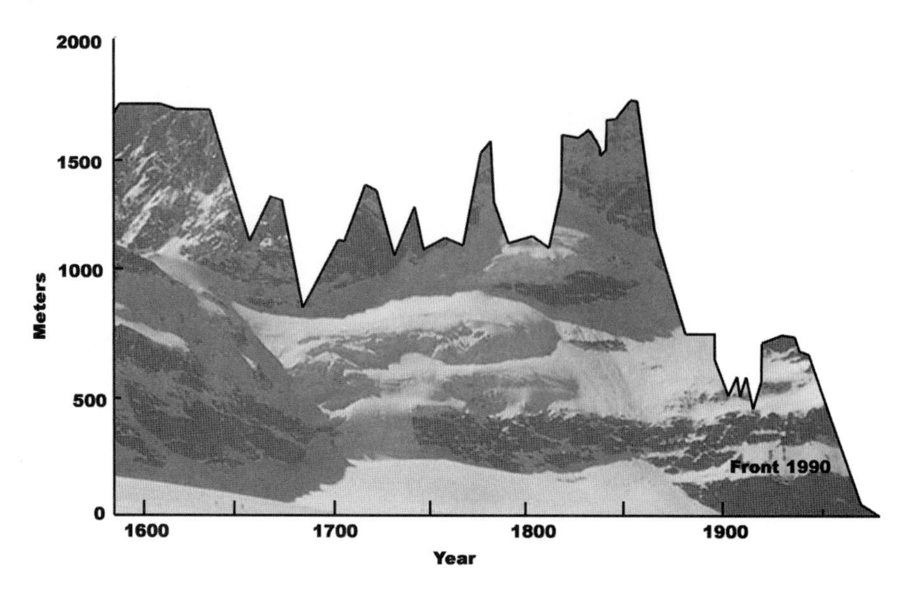

Figure 1.9 This graph plots the change in position of the front of the Grindewald glacier, in Switzerland, relative to its present position since 1500. The glacier has retreated about 1.5 kilometers since 1850, and about half of this retreat has occurred since 1940. The glacial retreat mirrors the rise in global temperature.

ing off after the 1940s and a renewed rise in the 1970s. This thermal expansion cannot account for all of the trend, and new dams and reservoirs built during the past century may have trapped enough water behind them to cause a *fall* in sea level, other things being equal, of one or two centimeters overall; so there probably has been some melting back of glaciers as well. . . . Even if we had no measurements of global temperature variations over the past hundred years, the sea level trend itself would be compelling evidence that the world had warmed by about half a degree Celsius. If the trend continues, cities around the Mediterranean will be among those affected most severely in the early part of the next [twenty-first] century. Venice is already at risk; much of Alexandria, with a population of 3.5 million, is less than a meter above sea level; the wetlands of the Camargue, in southern France, will be swamped. Whole countries, notably Bangladesh and the Netherlands, are threatened by rising sea levels, as is the state of Florida. It is because the prospects are so grim that scientists have been so cautious about asserting that the greenhouse effect is beginning to bite. But now they have proof.

Another independent piece of evidence pointing the same way emerged in 1986. As a result of the opening up of Alaska to oil exploration, many boreholes were drilled into the frozen ground of this part of the Arctic during the 1970s. In this cold permafrost, heat is transferred only by conduction—there is no liquid water to move from layer to layer, carrying heat with it and confusing the picture. If the temperature at the surface stays the same for a long time, then the pattern of temperature changes going deeper into the permafrost becomes stable, in much the same way that if a bar of iron is kept with one end in a bucket of ice and the other at a constant temperature of 20°C it will establish a stable "temperature gradient," with the temperature at each point along the bar being a constant and smooth variation from 0°C at one end to 20°C at the other.

But when researchers from the U.S. Geological Survey examined temperatures recorded at different depths down fourteen boreholes covering a region of a hundred thousand square kilometers scattered across northern Alaska, they found that this equilibrium pattern had been destroyed in the upper layers of the permafrost. The pattern of temperature variations that they found can best be explained if a pulse of warmth is slowly penetrating deeper into the permafrost, working its way downward from the surface. The pattern is the same across northern Alaska, and corresponds to a rise in average surface temperatures of about 2°C over the past hundred years. This is higher than the global average, but broadly fits the picture for high northern latitudes inferred by the NASA team (and others) on the basis of historical temperature records; it has the advantage that in the permafrost, nature is averaging out all short-term variations (including seasonal fluctuations) and presenting us with a broad picture of the main trends. Although the figures come only from one region of the globe, and are therefore not as convincing on their own as the sea-level variations, they are another completely independent confirmation of the warming.

While some researchers probe down into the ground to determine details of the warming trend, others are looking up into the air. David Karoly, of Monash University in Clayton, Australia, is one of many observers who have been taking the temperature of the atmosphere at different altitudes using instruments carried on unmanned balloons. I mention his work in particular because it covers the southern hemisphere, and helps to show that those critics who doubted the southern figures reported by Hansen's group in 1981 were wrong. Karoly's balloon measurements indicate that in the lower part of the atmosphere (the troposphere, from the surface of the Earth up to an altitude of about ten to fifteen kilometers), the southern hemisphere warmed at a rate of between 0.1 and 0.5°C per decade between 1950 and 1985. The warming became more pronounced between 1966 and 1985, when the data from thirty-one sites around the southern hemisphere indicate warming at rates in the range from 0.2 to 0.8°C per decade.

On its own, this is just icing on the cake. If you haven't already been convinced that the world is getting warmer, then these figures alone won't change your mind. But the balloons—radiosondes—also send back temperature measurements from higher altitudes, and these are particularly significant. Karoly's analysis shows that in the lower stratosphere, the region of the atmosphere immediately above the troposphere, temperatures have been *falling* since the mid-

1960s. Once again, the trend is apparent in measurements from around the southern hemisphere, and different observing stations report coolings in the range from 0.2 to 1.0°C per decade. Other researchers have found similar changes in stratospheric temperatures in other parts of the world. Karin Labitzke of the Free University of Berlin has carried out a huge study of northern hemisphere variations, based on data from three thousand observing stations (as ever, there are more observations from the north). This shows an average stratospheric cooling, between latitudes 20°N and 70°N, of 0.34°C between 1966 and 1980.

Why should this cause excitement among the climatologists? Surely if you are looking for a warming trend, and the lower atmosphere is getting hotter but the upper atmosphere is getting colder, it might seem that you only need to take an average in the right way, over the whole atmosphere, for the trend to disappear. But that isn't the case. There is much more air in the troposphere, where the atmosphere is more dense, and the ground and sea are also getting warmer. The stratospheric cooling is nowhere near enough to offset these effects. It is significant for quite another reason: this pattern of changes, with the troposphere getting warmer while the stratosphere gets cooler, is exactly the pattern that is predicted for the greenhouse effect. . . .

THE CARBON DIOXIDE CONNECTION

Solid evidence that the greenhouse effect—or the lack of it—might help to explain the pattern of ice ages came in 1982, when researchers from the University of Bern, in Switzerland, reported their analysis of bubbles of carbon dioxide trapped in the icecaps. When snow falls on the ice, it is light and fluffy and there are plenty of gaps between the snowflakes. These gaps are full of air. As more snow falls and the years pass, each layer is squeezed down in its turn and becomes ice, but some of that air stays trapped in the ice as bubbles. On a polar icecap (or, indeed, on a mountain glacier) the youngest ice is at the top, and successively older layers lie deeper below the surface. Because the snow is laid down in regular annual layers, which are still visible when it has been compressed into ice, the age of a piece of ice from a particular depth can be determined, in some cases, by drilling a core of ice from the sheet, and counting the layers down from the surface. The technique, pioneered by a Dane, Willi Dansgaard, and his colleagues in Copenhagen, is useful for climatologists because the same kind of isotope measurements that reveal temperatures of long ago from the deep sea sediments also reveal how temperature has changed over the period of time covered by the ice core. . . . But the Swiss researchers decided to investigate not the temperature record in their core sample, but the atmospheric record.

Similar efforts had been made before, but with mixed success; it was the Swiss work that opened the eyes of climatologists to the implications of the technique. Careful studies of the air trapped in the bubbles in the ice showed that about twenty thousand years ago, when the most recent ice age was at its coldest, the amount of carbon dioxide in the air was only in the range from 180 to 240 parts per million, compared with 280 ppm in the early nineteenth century, before widespread burning of fossil fuel began. About sixteen thousand years ago, at the time when, we know from other geological evidence, the ice sheets began to melt, the proportion of carbon dioxide began to increase, and by the end of the ice age it had reached roughly the same concentration as the natural level today. The evidence was persuasive, but many climatologists were worried that carbon dioxide might somehow be leaking out of the old bubbles buried in the ice, or that the tricky techniques needed to extract and measure the tiny samples of old air might be giving a false reading. These doubts were allayed when Nick Shackleton and his colleagues found exactly the same pattern of carbon dioxide variations from a completely independent technique, using deep sea sediments and another kind of isotope study, this time looking at carbon instead of oxygen.

These studies provide a beautiful example of science at work, with insights that are important to our lives in the years ahead coming from a combination of biology, physics, and chemistry applied to studies of the shells of long dead, microscopic sea creatures found in sediments hauled up from the bottom of the sea. It all starts with an understanding of how the oceans would absorb carbon dioxide if there were no life on Earth.

On a lifeless Earth, carbon dioxide would be taken out of the air and dissolved in the water of the seas, and . . . this could play a part in controlling global temperatures, although possibly not as effective a part as the processes involving life. If the world were physically the same as it is today, but lifeless, there would be about three times more carbon dioxide in the air than there actually is. Microscopic lifeforms near the surface of the ocean—plankton—act as a pump, extracting carbon dioxide from the air and depositing it on the sea floor. They take up carbon dioxide in the form of dissolved carbonate, and when they die their shells and skeletons, rich in carbonate, fall to the bottom of the sea. Some of this material is buried and becomes part of the long-term geological cycles that provide one of the feedbacks by which the temperature of Gaia is regulated. But some of the material dissolves back into carbonate in the deep water. This carbonate-enriched water, however, cannot rise back up to the surface. It is colder and more dense than the water above, and, like hot air, it is *hot* water that rises. So as well as taking some carbon dioxide out of circulation altogether for millions of years, the biological pump increases the concentration of carbonate in deep waters at the expense of surface waters and, ultimately, the atmosphere.

Now the carbon isotopes come into the story. Like oxygen, atoms of carbon come in different varieties. The important ones, in this study, are carbon-12 and carbon-13. Both are stable, and carbon-12 is the lighter of the two isotopes. Because of the way their biochemistry works, the plankton prefer to take up the lighter isotope to build their shells. So the biological carbon pump not only shifts carbonate to the bottom waters, it also preferentially shifts carbon-12 rather than carbon-13 out of the surface layers. Some organisms, however, live in the cold, deep waters, not near the surface. They build their shells out of the carbonate that is available to them, and this is already enriched with light carbon, so their shells contain even less carbon-13, proportionately, than the shells of creatures that live in the surface layers. Experts can identify the two kinds of shells, even in sediments tens of thousands of years old, and can sort them into the two categories. When the two types of shell are analyzed, they provide a measure of how the amount of carbon being shifted from the surface to the deep ocean has varied—carbon-13, rather than carbon-12, is actually used as the indicator, because there is less of it in the first place, so small variations in the balance between the two isotopes show up more strongly as a fraction of carbon-13 than of carbon-12. As the activity of the biological pump changes, so the proportions of carbon-13 being locked away in these shells changes; by measuring the carbon isotope ratio in shells from both surface creatures and bottom dwellers with exquisite precision, the researchers can determine how active these biological processes were at different times in the past, and infer how much (or how little) carbon dioxide can have been present in the air in order to match these measurements. If there were no biological activity, for example, then the ratio of isotopes would be the same everywhere, in the surface and the deep sea, and we would know that the carbon dioxide concentration was about three times the present value.

By now, it should come as no surprise to find that these studies show that there was less carbon dioxide in the air during the latest ice age than there is today. The ice core studied by the Swiss team extends back over the past forty thousand years, and over that timespan the concentrations of carbon dioxide measured in the ice core bubbles and the concentrations inferred from isotope measurements of sea bed sediments are exactly the same. But the sedimentary record goes back much further, for hundreds of thousands of years, and provides much more information. The intriguing new discovery is that the changes in carbon dioxide *precede* the changes in ice cover. Working with Nicklas Pisias, of Oregon State University, Shackleton applied the technique to analyze a sediment core covering a span of 340,000 years, more than three complete glacial/interglacial cycles. They found all three of the Milankovitch rhythms present in both the temperature record (determined from oxygen isotopes) and the carbon dioxide record (determined from carbon isotopes) in the core. [The Milankovitch rhythms are changes in the eccentricity of Earth's orbit, changes in the obliquity or "tilt" of Earth's axis, and precession of Earth's orbit, three types of orbital change credited with influencing climate over long periods of time.] But by comparing their findings with calculations of how the astronomical cycles have varied, they found that the changes in the Earth's orbit precede the changes in carbon dioxide, and that the changes in carbon dioxide in turn precede changes in climate—or, at least, changes in ice cover. Somehow, the astronomical changes cause the biological pump to increase or decrease its activity, and this produces a change in the strength of the greenhouse effect which is then a key element in switching the world into or out of an ice age—and which also helps to explain ups and downs of temperature on lesser scales, *within* an ice age or an interglacial. Carbon dioxide amplifies the changes that the Milankovitch process is trying to produce.

The effect almost certainly operates through changes in the workings of the carbon pump at high latitudes. High latitudes feel the effects of the Milankovitch rhythms more strongly than tropical regions do, and it is also at high latitudes that the deep ocean water breaks through to the surface and makes direct contact with the atmosphere. Today, there is not enough sunlight reaching high latitudes, some researchers suggest, for the plankton that live there to use all of the available nutrients; if there were more sunlight, as there is at some other times in the Milankovitch cycles, they might grow more vigorously, taking carbon out of the atmosphere as they do so. All such proposals, however, are still very tentative. The only way to get a better understanding of the links between astronomy, biology, and climate that control the pulsebeat of ice ages is by taking a more detailed look at the latest ice age. Almost as if it had been waiting in the wings for its cue, just such a more detailed picture of recent climatic changes emerged in 1987 from a painstaking analysis of ice samples (and bubbles trapped in them) from a core more than two kilometers long, drilled through the Antarctic ice cap.

THE ANTARCTIC CORNUCOPIA

The core was drilled at the Soviet Union's Vostok Station, in East Antarctica. Drilling began in 1980, and eventually reached a depth of 2.2 kilometers. The ice at the bottom of the borehole is formed from snow that fell in Antarctica more than 160,000 years ago, in the ice age before [the] last [one]. At that depth, a year's snowfall from that long gone era corresponds to a layer of ice just one centimeter thick. It is hard for anyone who has never been involved in such a project to comprehend the engineering skills required to extract a core of ice this long, especially bearing in mind that the job was being carried out in the most inhospitable continent on Earth. The best measure of the difficulty of the task is that it took from 1980 to 1985 to extract the core. . . .

The Vostok core has been analyzed by a large team of researchers in a collaboration involving scientists from France and the Soviet Union. Their main conclusions, based on analyses of the first 2,083 meters of core extracted from the hole, were reported in the journal *Nature* in October 1987; but new insights will continue to emerge in the years ahead, as the pieces of the core, carefully preserved in cold storage, are subjected to new tests.

Today, the average surface temperature at the Vostok site is −55.5°C. The Franco-Soviet team used yet another isotope technique to infer past temperatures from their analysis of the core. This time, the "thermometer" was based on

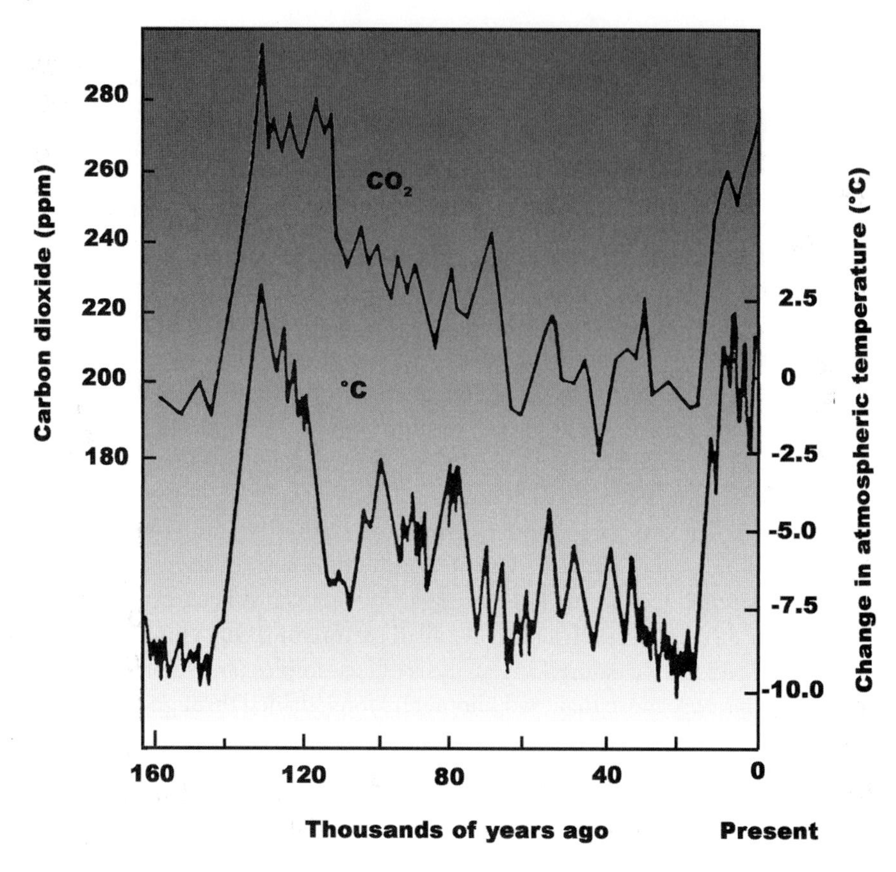

Figure 1.10 As the Earth shifts into or out of an interglacial, temperature variations and changes in the concentration of carbon dioxide in the atmosphere move in step. Data from the Vostok ice core.

the concentration of heavy hydrogen (also known as deuterium) in the ice. The principle, however, is the same as with the oxygen isotopes—heavier molecules of water (ones that contain deuterium) evaporate less easily but freeze out of water vapor in the air more easily, so the amount of heavy hydrogen in snow falling in Antarctica to be preserved as ice depends on temperature. Deuterium provides a direct measure of the temperature when and where the snow was falling, while the oxygen isotopes give a broader picture of global changes in ice cover. The deuterium record, in fact, closely matches the oxygen-18 record from ocean sediments, but the ice cores provide more detail. The analysis shows that at the coldest points of both the latest ice age and the penultimate ice age temperatures were 9°C lower than they are today, while during the warmest years of the interglacial that separated the two latest ice ages, a little over 150,000 years ago, temperatures were 2°C higher than they are today.

Samples of air from bubbles trapped in the ice show that there was a dramatic increase in the concentration of carbon dioxide in the atmosphere, from 190 ppm to 280 ppm, at the end of each of the two latest ice ages, and there was a comparable decrease in the carbon dioxide concentration at the beginning of the most recent ice age (Figure 1.10). The temperature record, in particular, also shows clear signs of the familiar Milankovitch cycles at work. But as well as confirming the existence of connections between carbon dioxide, ice ages, and the Milankovitch rhythms, ice cores from both the Antarctic and the Arctic are providing completely new insights into the climatic changes that cause the shift from ice age to interglacial and back again; those new insights point to a key contribution from the Earth's biological systems—a role for Gaia. . . .

ALL AT SEA

The oceans are the key to understanding changes in global climate—but, to mix the metaphor, they are also the joker in the pack. Most of our planet is covered by water, and interactions between the oceans and the atmosphere establish the balance of natural greenhouse gases that has maintained temperatures on Earth comfortably above those of the airless Moon, but comfortably below those of hothouse Venus. The seas also smooth out the distribution of heat across the planet, transporting warmth from the tropics, where solar heating is strongest, to higher latitudes, where the direct heat of the Sun is weaker. Unfortunately, we do not understand the workings of the oceans, and of the air/sea interface, well enough to predict exactly how these patterns will change as the world warms. But there is a disturbing possibility that any changes that do occur will be sudden, a switch in the mode of oceanic circulation from one pattern to another, rather than gradual.

Today, the atmosphere and the oceans each carry about the same amount of heat from the equator to the poles, but they do it in different ways. In the atmosphere, from the equator out to about 30° latitude in each hemisphere, the dominant feature is simple convection, with hot air rising near the equator, cooling and moving outward high in the troposphere, and sinking down to the surface again to become a cool wind blowing into the equator from higher latitudes. This circulation pattern produces the reliable trade winds that were of such prime importance to shipping in the days of sail. Further away from the equator, heat is redistributed in a more messy fashion by the weather systems—the highs and lows—which are familiar to anyone who lives in temperate latitudes. The weather systems are actually eddies, like the swirling patterns formed in flowing water when it moves past an obstruction. At these latitudes, sometimes the wind blows one way, at other times it blows from a different direction. But by and large the effect is the same—air blowing towards the poles is warm, while air blowing from the polar regions is cold.

The way oceans transport heat from the equator to the poles is different for two reasons. First, the oceans are heated from the top, while the atmosphere is heated from the bottom. Warm ocean water heated by the Sun cannot rise because it is already at the top of the ocean. So although there is convection in the oceans, it is, in a sense, upside-down convection compared with what we are used to in the atmosphere. Atmospheric convection is driven by hot air rising at the equator; but oceanic convection is driven by cold water sinking at high latitudes. Just as the warm air rising up pushes other air out of the way and sets the atmospheric convection circulating, so the sinking cold water at high latitudes pushes other water out of the way, eventually ensuring that water rises to the surface in the tropics and is warmed by the Sun as it begins to move out towards the poles.

But even then the ocean currents cannot flow as freely as the winds blow, and the second difference between oceanic and atmospheric currents comes into play. There are huge land masses which divide the oceans of the world into two main basins in the northern hemisphere (the Atlantic/Arctic and the Pacific) and three in the south (Atlantic, Indian, and Pacific). Although the southern ocean basins do join at high latitudes to form the Southern Ocean, surrounding Antarctica, the way most of the waters of the world are confined by continental masses forces the currents to flow around the basins in roughly circular patterns, called gyres. The path followed by such an oceanic current depends partly on the difference in temperature between the equator and the poles, partly on the effect of the Earth's rotation, and partly on the shape of the ocean basin itself. The Earth's rotation (from west to east) spreads out currents flowing along the eastern side of a basin, but piles up the currents into strong, relatively narrow streams in the western sides of the ocean basins—the western land mass is always moving towards the water, while the eastern continents are moving away from it (which is why sea level at the Pacific end of the Panama Canal is lower than sea level at the Caribbean end). The Gulf Stream which runs northward up the western side of the North Atlantic is the archetypal example of all these processes at work. It carries thirty million cubic meters of water every second through the Florida Straits, and that water is 8°C warmer than the water that returns south as the gyre completes its circuit of the basin. The rate at which heat is being transported northward is more than a million billion watts.

The Atlantic is, in fact, the only basin which provides heat for the Arctic Ocean in this way; the northern Pacific is almost completely blocked off at the Bering Strait, where the Soviet Union and the United States very nearly have a land frontier (the strait was, indeed, dry land during the recent series of ice ages, when sea level was lower). So in the Pacific Ocean, equatorial heat cannot reach up to the highest northern latitudes, and most of the warmth flows southward, linking up with the warm southward flow from the Indian Ocean (which is completely blocked to the north). Then it splits into two parts: a current that flows into the Southern Ocean and eventually transports heat to high southern latitudes, and a current which flows around southern Africa and into the Atlantic, where it moves northward and eventually helps to make the warm currents flowing poleward in the *North* Atlantic stronger.

In the south, meantime, the main current in the Southern Ocean flows in an almost circular path around Antarctica, as the Circumpolar Current. Although this provides a barrier to the direct flow of warm water southward, heat is carried across the Circumpolar Current in the form of eddies, rather like the swirling low pressure systems that transport atmospheric warmth northward at the latitudes of the United States and Europe. About 300,000 billion watts of heat is transported in this way—significantly less than the equivalent northward flow, because of the way the shape of the ocean basins forces the currents to flow.

Over the long span of geological time, the geographical patterns of the land masses have changed as continents have drifted about the globe. This has altered the circulation patterns of the oceans, and has caused dramatic changes in climate—we live in an epoch when there is ice at both poles, but when the dinosaurs dominated the Earth there may only have been ice on high mountains. Then ocean currents carrying warm water could reach both polar regions largely unobstructed.

Seamus Heaney, "Höfn" (2004)

Höfn is an Icelandic word for port, harbor, or haven, a place that is settled because it is safe. For decades, Seamus Heaney wrote poems about the lives of people—both living and long dead—who populate northern, often harsh climates. This and many of his verses have a tactile, sensuous appeal where sound reinforces and enriches sense. Here, for the first time, the Nobel laureate in literature entertains the possibility of climate change, of climate shock, envisioned in the melting of the glacier yet also in the ominous possibility, too, of its cold advance. There is something uneasy about the whole affair, and what is human and warm seems hard put to stop whatever melting or frigid force has been set in motion, ironically, by humans themselves.

Höfn

The three-tongued glacier has begun to melt.
What will we do, they ask, when boulder-milt
Comes wallowing across the delta flats

And the miles-deep shag ice makes its move?
I saw it, ridged and rock-set, from above,
Undead grey-gristed earth-pelt, aeon-scruff,

And feared its coldness that still seemed enough
To iceblock the plane window dimmed with breath,
Deepfreeze the seep of adamantine tilth

And every warm, mouthwatering word of mouth.

2 Species in Danger

Three Case Studies

Let us sit and tell sad stories of the death of species. Most schoolchildren know of the extinction of the dodo, a huge, flightless dove of Mauritius, whose name has become a synonym for stupidity characterized by lack of awareness. The passenger pigeon is the most unbelievable of all extinction stories, for no more than three decades prior to its demise, flocks of them existed in such numbers that they blocked out the sun. John James Audubon estimated that one flock he observed flying over the Ohio River Valley contained between two and three *billion* birds. The passenger pigeon had an evil twin, the rocky mountain locust, and the parallels between them are as astonishing as the numbers involved. In the late nineteenth century, at the same time that the passenger pigeon was blocking out the sun over the Ohio, the rocky mountain locust arose in swarms as large as the state of Colorado. Yet some time within the first five years of the twentieth century both species were extinct in the wild. Every species has a story to tell and every endangered and extinct species has a sad one.

The conservation of biological diversity can focus on different levels of biological organization—from the biosphere and major biomes such as the tropical rainforest through individual ecosystems down to populations and genes—but the level most tractable to the scientist and most understandable to the lay person is the *species*. It is also the most effective politically for many reasons, not least of which is that, in the case of most animals, species have an identifiable face and a memorable story. In the United States, if you asked citizens to name an environmental law, the first to come to mind for many individuals would be the Endangered Species Act, even though the Clean Air and Clean Water Acts have had more impact and have been more effective in accomplishing their goals. Yet the Endangered Species Act not only has changed our view of what an environmental law can do, that is, declare that a species has an incalculable value not to be outweighed by any human activity, but it has also taken hold of the imagination—we may be able to stop a unique form of life, such as the whooping crane or the grizzly bear, from disappearing from the Earth forever.

Biologists have identified three principal causes for the endangerment and extinction of species: overexploitation, habitat destruction, and the introduction of non-native species. In many cases the interaction between these factors can exacerbate the rate of endangerment. With the passenger pigeon, habitat destruction—the cutting of the Midwestern forests for farmland—and over-exploitation—hunting pigeons to be eaten as we eat chicken today—worked hand-in-hand to bring about the rapid extinction of the species. Jared Diamond's recommended reading "Overview of Recent Extinctions" collects the evidence to support the attribution of endangerment and extinction to these three causes. In order to compare these three major causes of species extinction, the remaining selections are divided into three respective case studies. First, the African elephant exemplifies *overexploitation;* second, *habitat destruction* sealed the fate of the ivory-billed woodpecker; and third, the introduction of an *invasive species* threatens the avian fauna of Guam.

The intelligence and memory of the African elephant (*Loxodonta africana*) have long been the stuff of legend. The social structure and complexity of its family unit rivals that of a human family. More biologists are studying it today than are studying any other single species, and more conservation dollars have been donated for its survival than for any other species on Earth. It differs from its sister species the Indian elephant (*Elephas maximus*) in several important ways. The African elephant is larger—it is the world's largest extant land animal—and has proportionally larger ears. The Indian elephant is trainable and has become an important beast of burden in Asia, while the African elephant remains undomesticated and dangerous to humans and their agriculture. Most important for this story, the African elephant has larger tusks, found in both the male and female, while the Indian elephant has smaller tusks, found only in the male.

The overexploitation of the African elephant has been primarily to harvest the material of its tusks—ivory. Ivory is durable, easy to carve, beautiful, and pleasing to the touch, making it the functional equivalent of a precious stone. In the 1970s, a shortage of oil caused economic collapses throughout the world, which increased the value of precious materials, such as gold and ivory. The pressure to hunt elephants for their ivory increased dramatically. Mafaniso Hara's selection *International Trade in Ivory* tells the story of the ivory ban that took place under the provisions of the Convention on International Trade in Endangered Species (see selection) and the dramatic choice of the burning of the world's ivory reserves that took place in response.

The African elephant's decline was not entirely due to overexploitation—habitat destruction played a part as well. It is difficult for humans to live and farm in elephant range, because the elephants are so destructive of habitat themselves. As a result, elephant habitat has been reduced significantly by human fences, and elephants have been shot when they come close to farms and dwellings. Despite the interactive effect of habitat destruction, most experts agree that poaching for ivory has had the greatest impact on the endangerment of the African elephant.

True ivory is produced only by the African and Indian elephants, along with their extinct relatives, the mastodons and mammoths. Other animals besides the elephants—walruses, hippopotami, sperm whales—produce substances related to and used like ivory. The ivory-billed woodpecker, however, has a bill of the same material as that of all other birds; the "ivory" here simply refers to its color. The ivory-billed woodpecker's supposed extinction represents a classic example of habitat destruction. This woodpecker's home was the great bottomland hardwood forests of the southeastern United States, almost all of which were cut down for farmland, like the forests the passenger pigeon relied on further north. There were no verified sightings of this species between 1942 and 2005.

The story of the ivory-billed woodpecker raises one of the more empirically difficult aspects of the extinction problem: how do we know that a species has actually gone extinct? Given a large enough area of possible habitat, it is virtually impossible to prove that something does not exist. Although the ivory-billed woodpecker had no verified sightings for decades, many observers, such as the biologist profiled in Jonathan Rosen's "Ghost Bird," had good reasons to believe they had been seen recently. The case of the ivory-billed woodpecker raises interesting questions about the degree of proof a governmental agency or conservation organization should require before declaring a species extinct or endangered. It also raises questions about the benefits and detriments to conservation in being overly liberal or conservative in how long one waits before listing a species as officially endangered or extinct. The good news is that in early 2005, the United States Department of the Interior verified sightings of the ivory-billed woodpecker in Arkansas and published its first set of plans for the species's recovery (see selection). The sightings remain controversial.

Historically the causes of species endangerment and extinction have moved from overexploitation to habitat destruction to non-native invasions. The story of the effect of the brown tree snake on the birds of Guam is a dramatic example of how introduction of a non-native species can damage an island ecosystem. The brown tree snake, an introduced predator, is inexorably wiping out the majority of bird species indigenous to Guam, species that evolved in the absence of aggressive predators. This is a story that has repeated itself to less dramatic effect since the age of exploration began. Especially harmful have been feral cats and goats and stowaway rats, who often have a double effect on native species, as both predators and competitors. As world travel becomes more common, the likelihood for species inadvertently to be carried to new habitats increases. Zebra mussels released from the waste-water of ships have endangered native mussels in North America. Fungal diseases transported across the oceans in shipments of wood carried the chestnut blight and Dutch elm disease. Purple loosestrife and kudzu escaped from gardens to choke out native ecosystems. Cheatgrass seed was accidentally included in seed mixes of imported prairie grasses, leading cheatgrass to overtake large areas of the American west.

The stories could go on and on, which is the point of this chapter: generalizations can and should be made about species extinctions, but to influence human action, it is more powerful to present the face and the story of individual species that have suffered endangerment and extinction. We save what we understand and love, and, as a general rule, we love the unique forms of life more than abstract conceptions of them.

FURTHER READING

Beard, Peter. *The End of the Game: The Last Word from Paradise: A Pictorial Documentation of the Origins, History, and Prospects of the Big Game in Africa.* New York: Chronicle Books, 1996.

Cokinos, Christopher. *Hope Is the Thing with Feathers: A Personal Chronicle of Vanished Birds.* New York: Tarcher/Putnam, 2000.

Diamond, Jared. "Overview of Recent Extinctions." In David Western and Mary Pearl, ed., *Conservation for the Twenty-first Century.* New York: Oxford University Press, 1989.

Hoose, Phillip. *The Race to Save the Good Lord Bird.* New York: Farrar, Straus and Giroux, 2004.

Quammen, David. *The Song of the Dodo: Island Biogeography in an Age of Extinctions.* New York: Scribner, 1996.

Weidensaul, Scott. *The Ghost with Trembling Wings: Science, Wishful Thinking, and the Search for Lost Species.* New York: North Point Press, 2002.

IVORY AND ELEPHANTS

Mafaniso Hara, from *International Trade in Ivory from the African Elephant: Issues Surrounding the CITES Ban* (1997)

The Convention on International Trade in Endangered Species (CITES) is the international analog to the United States Endangered Species Act, although its jurisdiction is limited to the movement of endangered species and their parts, such as ivory, between countries. It has no effect on actions that endanger species when those actions occur only within a single country's borders. CITES gained its greatest international attention in 1989, when the African elephant was moved to the treaty's endangered category and all international movement of the elephant's most valuable product, ivory, was outlawed. Mafaniso Hara provides a cogent analysis of the philosophical differences in elephant conservation between the southern African and eastern African countries.

International trade in live animals as well as for their parts and derivatives is an enormous and lucrative business worth an estimated U.S. $5 billion annually. According to Porter and Brown,[1] nearly one third of this trade is illegal. It is claimed that the illegal trade is an important cause of loss of species and overexploitation of natural resources. The dramatic decrease of the African elephant population from 1.3 million in 1979 to 625,000 in 1989 is attributed mainly to killing for commercial sale of its ivory tusks. In 1989, the market for ivory was estimated to have been worth U.S. $50–60 million annually.

CONVENTION ON INTERNATIONAL TRADE IN ENDANGERED SPECIES (CITES)

The international regime for dealing with illegal traffic in wildlife products is the Convention on International Trade in Endangered Species (CITES). Completed in 1973, the CITES convention is the umbrella organization for combating commercial overexploitation of wild animals and plants and it is recognized by most countries (by 1994, 122 countries were party to the convention) in the world. The main instrument it uses is imposition of trade sanctions against violators of its conventions. The CITES secretariat and the parties to the convention meet every two years to decide how to regulate trade in species according to the different degrees of danger. . . . The treaty created three categories of species (referred to as appendices) which are defined according to the threat to the species' existence. Various levels of control corresponding to the protection required to preserve a particular species are distributed among the categories.

Appendix I provides the highest protection. This category is for species that are threatened with extinction and are not to be traded except for scientific or cultural endeavors.

Appendix II provides an intermediate level of protection by regulating rather than banning trade. This category is for species that are not yet endangered but are believed to merit monitoring, thus requiring export permits from the country of origin.

Appendix III offers the least protection as it merely involves identification of potentially threatened species. The state may or may not impose trade restrictions on species in this category. Export permits are required only if the country of origin has listed the species in this appendix. . . .

THE BANNING OF TRADE IN IVORY FROM THE AFRICAN ELEPHANT

In 1977 the African elephant was identified as a threatened species and was thus listed in appendix II of CITES. During the 1980s, rising demand and soaring ivory prices resulted in an increase in poaching. As a control measure, CITES established a system of ivory export quotas for the range countries in 1985. This did little to stem the tide. Elephant populations continued to dwindle. In Kenya alone the population plunged from 167,000 in 1973 to just 16,000 by 1989. At this rate, it was estimated, the elephant could be extinct by 2010. These trends created an international public outcry. With the increasing globalization of environmental issues in the 1980s, there was a growing international feeling that no state had the right to allow destruction of a species even if its habitat was restricted to that country's own territory. To publicize the problem, president Moi of Kenya set fire to 2,500 ivory tusks (twelve tons in weight worth $3 million) in front of international television cameras on 18 July 1989. He declared all-out war on poachers; they were to be shot on sight.

The events of the 1980s raised public alarm that the world would lose the elephant completely unless something drastic was done. Pressure (from Western nations and conservation groups) was mounting to ban all trade in ivory. The World Wildlife Fund (WWF) and Conservation International (CI) commissioned a study of the African elephant problem in 1988. According to this study, the sustainable level of ivory production was fifty metric tons annually, whereas the world had been consuming 770 metric tons per year in the preceding decade. Following the findings of the study, the two organizations called for a worldwide ban on trade in ivory from the African elephant. At the time, other elephant conservation specialists countercharged that the report was based on inaccurate figures and that it had deliberately exaggerated the reduction in the elephant population. Despite several other commissioned studies, consensus on the issue could not be achieved. Countries either accepted or rejected the WWF/CI report, depending on what policy they wished to support. In June 1989, France, the U.S.A., West Germany, and the European Community imposed a moratorium on all ivory imports.

By the time of the seventh CITES conference in October 1989 two opposing coalitions had developed. The one was led by producer countries with rapidly declining elephant populations, composed mostly of the East African states Kenya, Tanzania, and Somalia together with the Gambia, Austria, Hungary, the U.S.A., and Western animal protection groups which proposed that the African elephant be listed in appendix I. Opposing the total ban was a coalition of Southern African producer countries with stable or growing elephant populations. The coalition, which at the time united opposing parties in the apartheid struggle, consisted of Botswana, Malawi, Namibia, Zimbabwe, and South Africa. Their contention was that while Africa's elephant population had indeed declined over the previous decade because of poaching and the ivory trade, Botswana, Namibia, South Africa, and Zimbabwe had succeeded in increasing their elephant herds by providing economic incentives to local communities through the quota system. At the conference, the required two-thirds majority vote for the ban officially moved the African elephant to appendix I. Even the compromise proposed by the CITES secretariat which would have distinguished between herds that needed the protection of an ivory trade ban and those that did not, was rejected as unrealistic, unworkable, and unenforceable. The total ban

Figure 2.1 African elephants. Courtesy of WWF / Edward Mendell.

model as envisaged by the appendix I listing was derived from the practical impossibility of distinguishing illegal ivory from legal ivory. Another reason for this model was that if there were a market for legal ivory, it would be easy to place illegal ivory into the stream of legal trade (according to a report published by the African Elephant Conservation Group in 1989, 80 percent of the ivory traded during the 1980s was illegal). For the proponents of the ban a total embargo was the only way to save the elephant.

Five Southern African states, Botswana, Malawi, Namibia, South Africa, and Zimbabwe, lodged reservations against the ban and announced plans to sell their ivory through a cartel, with proceeds to be used to finance conservation and rural community development. The justification was that the biggest threats to the survival of the southern African elephant were loss of habitat and conflict with legitimate human interests; not international trade in ivory. Furthermore, they argued that a ban would simply drive the price of ivory upwards and create more incentives for the poachers. Thus, in 1992 Botswana, Malawi, Namibia, and Zimbabwe established the Southern African Center for Ivory Marketing (SACIM). The name of the organization was changed to Southern African Convention on Wildlife Management (SACWM) in July 1996. The objective of SACIM was to provide a regional management framework for sustainable trade in ivory for the benefit of conservation and the development of rural areas. South Africa already had its own control and marketing system similar to SACIM. . . .

CONFLICT WITH PEOPLE

Protected areas in the form of national parks and game reserves are the main areas where elephants and other wildlife enjoy official protection. National parks and game reserves are a form of land use and thus face competition from other alternative uses of land. In the context of Southern Africa, this competition comes from the increasing need for land for subsistence farming. In rural areas of the region agriculture is the main economic activity. With an average population growth rate of 3.5 percent per annum, the pressure for land can only be expected to increase. This growing human population in rural areas entails expansion of human settlements and agricultural activities. For example, in Malawi, where parks and game reserves comprise approximately 20 percent of the land mass, problems of encroachment on these protected areas and public demand for land for farming from parts of protected areas are increasing. At the same time elephants are increasingly having to range outside protected areas into adjacent communal areas for food and water. Furthermore, human land use patterns for agricultural activities are based on cut-and-burn systems. Thus, in these unprotected communal areas elephants are losing range to essential human needs. As the elephants range into farm areas, crop damage becomes a common phenomenon. Crop destruction can be disastrous for communities as most of the crops grown are for subsistence and only one crop can be grown per season. In such areas elephants are viewed as pests and communities develop intolerance towards them.

Because it is becoming impossible to maintain elephants in protected areas or "islands of refuge," it will increasingly become important to integrate the management of elephants in protected areas and on communal lands with general land use planning and human settlement. Thus, acceptable economic incentives will have to be found for people who have to bear the consequences of having elephants entering their settlement and farming areas.

HABITAT DESTRUCTION

In some parks and protected areas the number of elephants exceeds the carrying capacity of the habitats. The rainfall variations and drought that hit southern Africa in the early 1990s also affected the recovery and regeneration of the vegetation. This food shortage in relation to growing elephant populations can have a devastating impact on the habitat. In some areas it has already necessitated culling or translocation of the elephants. As foliage and water in protected areas become scarce, foraging into communal areas becomes even more frequent, thereby increasing the conflict with human beings.

SACWM COUNTRIES' ARGUMENTS FOR TRADE IN IVORY

The view of SACWM . . . is that those Southern African states which have an abundance of elephant populations should be allowed to use the elephant and its products for the economic benefit of the people, especially those who have to bear the burden of its conservation. Conservation officials in these countries are convinced that elephants and other wildlife on communal land and, ultimately, also in protected areas are doomed, unless they are allowed to outcompete subsistence farming. . . . Cases throughout Southern Africa show that wildlife populations on communal or private land, that are in competition with other forms of land use such as agriculture, are viable over the long term only if the economic value and yield from wildlife exceed the value and yield from other land use, or at least significantly supplement the yield from other competitive forms of land use. Elephants could provide income through the sale of products and hunting licenses and also provide much needed protein for communities in the affected rural areas. In addition, elephant management is very costly and the revenue could help to offset management costs.

Secondly, many African countries have vast stockpiles of ivory recovered from elephant culls, confiscations, and natural mortality. For example, up till 30 June 1996, Zimbabwe had accrued thirty-three tons of ivory worth U.S. $4 million. About half of this belongs to rural communities managing their wildlife under the Communal Areas Management Program For Indigenous Resources (Campfire) program. It is argued that controlled sale of these tusks could fund future conservation efforts, provide incentives to communities, and also lower demand for illegal ivory. . . . In Zimbabwe loss of revenue from sale of ivory led to the scaling down of anti-poaching operations in 1990, with the result that 100 elephants were illegally killed as compared to ten in the previous year.

Thirdly, prohibition of international trade in species may even accelerate the elephant's disappearance, it is argued. Banning a species might result in the loss in economic value of the species as it may no longer be regarded as a resource. . . . Southern Africa as a whole generally attempts to maintain or increase economic value of wildlife as an incentive for its conservation. In addition . . . the Nile crocodile and South Africa's white rhino [are examples] for which international trade or the CITES exemption on the movement of trophies has contributed to the survival of the species, to the conservation of its habitat, and to the economic benefit of human beings. This could be applied to the elephant and by doing so, local human populations could support the conservation efforts because they would be benefiting from the rational utilization of their wildlife resources. . . .

[The] major grievances SACWM has with the CITES approach to the issue:

- Some parties to CITES list other countries' species with abandon as a way to protect species even when the primary source of the threat is not international trade. International trade in wild species has an insignificant effect on most fauna and flora from southern Africa compared to the threats from land use conflicts and the loss of habitat.

- Many blatant violations of CITES conventions by the same governments opposing attempts to establish well controlled trade, such as in ivory, are taking place. Instances of such violations include the open trade of ivory at airports and village markets in the very West African countries which voted for the ban. Equally, hundreds of illegal crocodile skins from South America are reported to have been entering and were being laundered in some European countries.
- SACWM and an increasing number of other developing countries believe that CITES has become the principal tool for foreign animal rights organizations agitating against any trade or any form of consumptive use of wildlife. With the boom of the animal protection industry in the 1990s these NGOs are using CITES as a mechanism for fundraising. There is a growing feeling that these and a few powerful donor governments have hijacked CITES and are manipulating it for their own ends.
- Producer countries feel indignant that the Western NGOs and governments stem from societies whose economies are based on non-sustainable economies with disproportionately high consumption rates of imported non-renewable resources, in environments marked by the worst degrees of transformation and loss of biodiversity. Conservation philosophies emerging from such conditions have to be treated with skepticism. There is also a feeling that these rich countries and their animal rights groups are more interested in making developing countries a natural history museum than in the welfare of the people, especially if the people are Africans. Bonner[2] uses the term "eco-imperialism" to refer to this attitude. The moral arguments advanced by animal rights groups against killing elephants are usually accompanied by an attitude of cultural imperialism.
- Conservationists arguing for the lifting of the ban also point to the decimation of the rhinoceros and the tiger populations as proof of the failure of approaches based on the philosophy of trade prohibition.
- And last but not least, the producer African nations, burdened with the responsibility for enforcing the ban, are being left to bear the financial cost of this conservation strategy without much help from the most ardent Western proponents of the ban.

Notes

1. G. Porter and J. W. Brown, "The Trade in Ivory from African Elephants," in *Global Environmental Politics* (Boulder, Colo.: Westview Press, 1996).
2. Raymond Bonner, "At the Hand of Man: Peril and Hope for Africa's Wildlife," *New York Times Magazine*, 7 February 1993.

Philip Muruthi, Mark Stanley Price, Pritpal Soorae, Cynthia Moss, and Annette Lanjouw, from "Conservation of Large Mammals in Africa: What Lessons and Challenges for the Future?" (2000)

This article presents four case studies of large mammal conservation in Africa — elephant, mountain gorilla, black rhinoceros, and hirola antelope — only one of which, the elephant, is excerpted here. The article presents a careful look at just one population of African elephants — those living in the Amboseli National Park in Kenya. Interestingly, the general lesson learned for the conservation of the great mammals of Africa is more a political one than a biological one: conservation efforts cannot be effective without the support of the local people who must live in the same ecosystems as the animals.

The Amboseli elephant population has been monitored continuously since 1972. The population declined in the 1970s due to poaching and drought. A reversal in the downward population trend has occurred since the beginning of 1979 with the population now growing dramatically (Figure 2.2). The Amboseli elephant population is growing at an annual rate of 3.9 percent through breeding alone.

Figure 2.2 Population of elephants at Amboseli from 1972 to 1998. The population declined in the 1970s due to poaching and drought. A reversal in the downward population trend has occurred since the beginning of 1979, with the population now growing dramatically.

Until the mid- to late 1970s, the Amboseli elephant population was little disturbed by human activity. Their movements were natural, seasonal events resulting in relatively minor effects on the habitat. Poaching in the 1970s led to the confinement of elephants inside the park. However, with increased security and tolerance by local Maasai people, Amboseli elephants are now spending much more time outside the park, ranging on Maasai Group Ranches and across the border into Tanzania.

In 1973, after many negotiations with central government, the Maasai vacated 390 km^2 of land which then became Amboseli National Park. Because most of the swamps are located inside the park, this move denied the Maasai their dry season grazing. The Maasai were promised adequate water supplies outside the park. They would also be compensated for the cost of tolerating wildlife grazing on their group ranch lands, and obtain direct economic benefits through development of wildlife viewing and tourist campsites and through services provided by the government. According to Lindsay,[1] between 1983 and 1985 three rhino and over twenty elephants were speared by Maasai in protest against the park and inadequate water supplies. The conflict situation has not abated and between January and November 1997, at least fifteen elephants, representing over seventy-five percent of the population's mortality for the period, had been killed in conflict situations with local people.

LESSONS LEARNED

As elephants continue to utilize communal lands, long-term means must be sought to alleviate antagonism between nature preservation and the needs of Maasai people for land and adequate livelihoods. The Amboseli area urgently requires a land use policy, developed with full participation of the Maasai, to cater for an expanding range of interests, and for the benefit of local people and wildlife conservation. Through such a policy, and efforts to encourage coexistence outside the park, pressure on habitat inside the park will be reduced to allow regeneration of vegetation. After all, the park was not designed to be a self-sufficient unit but as part of a multiple land-use system.

At Amboseli, conflicts occur over shared resources such as space, water, and forage, or over crop damage in cultivated fields east of the park. Wildlife managers should understand the nature of conflicts between humans and wildlife and seek innovative ways to mitigate them. One successful approach being applied by AWF [African Wildlife Federation] in the Amboseli area is a "consolation scheme" under which a landowner whose livestock is killed by an elephant is paid for the loss in order to prevent retaliatory spearing of elephants. We should also build upon the positive attitude that Maasai have towards wildlife with which they have coexisted for centuries. The increase in the elephant population since 1979 runs against the continuing collapse of elephant populations elsewhere, a measure of the protection given by the Kenya government and the Maasai around Amboseli. The task now is to maintain such a positive attitude given the transition in lifestyles that the Maasai are undergoing.

At the country level, Kenya and Tanzania should develop common transboundary initiatives aimed at conserving wildlife, and to foster cooperation and security. The population of elephants and the Maasai community straddle their common border. International forces and a strong expatriate lobby within Kenya have influenced conservation at Amboseli. International funding has enabled Cynthia Moss to conduct twenty-five years of research on Amboseli elephants. Through her work supported by AWF, the elephants are well-known all over the world with a positive benefit in increased numbers of tourists visiting Amboseli. Tourism has made a significant contribution to the conservation of Amboseli elephants. Through it Amboseli National Park and its wildlife earn both the local and national governments significant revenue. When tourism grew at twenty-two percent per annum between 1965 and 1969, it contributed more than seventy percent of the local Kajiado County Council's income. The Maasai leadership and national government see the park and its wildlife as valuable resources which they strive to protect. . . .

The conservation of a large mammal species should be considered within an overall focus on conservation of habitats and the overall threats facing the species in the area. The approach must be multidisciplinary. To de-emphasize the boundaries between protected areas and surrounding pastoral lands with the aim of ensuring that wildlife continues to have access to areas outside parks and reserves, landowners must be given incentives to tolerate or welcome the situation. For the elephant, the two fundamental conservation concerns are how habitat can be effectively secured for elephants and how to mitigate human-elephant conflicts. An effective conservation approach has to integrate ecological knowledge, working with traditional Maasai institutions (starting at the neighborhood level), training of individuals at all levels, and capacity building of local Maasai communities to improve their ability to manage and benefit from wildlife/elephant conservation. This approach is working for the outreach project in support of elephant conservation in the Amboseli area being undertaken by AWF.

All these case studies show that political support is key to the success of large mammal conservation in Africa. All stakeholders should participate fully. The success of the "consolation scheme" at Amboseli can be attributed to the participation of the local Maasai at all stages of its planning and implementation.

Note

1. W. K. Lindsay, "Integrating Parks and Pastoralists: Some Lessons from Amboseli," in D. Anderson and R. Groves, ed., *Conservation in Africa* (Cambridge: Cambridge University Press, 1987), 149–67.

IVORY-BILLED WOODPECKER

David Wagoner, "The Author of *American Ornithology* Sketches a Bird, Now Extinct" (1979)

In his poem David Wagoner captures the spirit both of the bird, the ivory-billed woodpecker, and of the early ornithologists, such as Alexander Wilson (depicted here) and John James Audubon, who attempted to catalog the still relatively unknown — at least to Eu-

Figure 2.3 Illustration of the Ivory-billed Woodpecker from *Birds of America* by John James Audubon. Collection of The New-York Historical Society, 1863.17.066.

ropean natural historians—biological diversity of the United States. The work of these naturalists is considered to be a high point in the fusion of art and science, but Wagoner shows just what was sacrificed in achieving that goal.

The Author of *American Ornithology* Sketches a Bird, Now Extinct
(Alexander Wilson, Wilmington, N.C. 1809)

When he walked through town, the wing-shot bird he'd hidden
Inside his coat began to cry like a baby,
High and plaintive and loud as the calls he'd heard
While hunting it in the woods, and goodwives stared
And scurried indoors to guard their own from harm.

And the innkeeper and the goodmen in the tavern
Asked him whether his child was sick, then laughed.
Slapped knees, and laughed as he unswaddled his prize,
His pride and burden: an ivory-billed woodpecker
As big as a crow, still wailing and squealing.

Upstairs, when he let it go in his workroom,
It fell silent at last. He told at dinner
How devoted masters of birds drawn from the life
Must gather their flocks around them with a rifle
And make them live forever inside books.

Later, he found his bedspread covered with plaster
And the bird clinging beside a hole in the wall
Clear through to already-splintered weatherboards
And the sky beyond. While he tied one of its legs
To a table leg, it started wailing again.

And went on wailing as if toward cypress groves
While the artist drew and tinted on fine vellum
Its red cockade, gray claws, and sepia eyes
From which a white edge flowed to the lame wing
Like light flying and ended there in blackness.

He drew and studied for days, eating and dreaming
Fitfully through the dancing and loud drumming
Of an ivory bill that refused pecans and beetles,
Chestnuts and sweet-sour fruit of magnolias,
Riddling his table, slashing his fingers, wailing.

He watched it die, he said, with great regret.

Jonathan Rosen, from "The Ghost Bird" (2001)

The hope to prove a hypothesis of extinction wrong is one of the great hopes left for modern conservationists. Jonathan Rosen takes us to the swamplands of Louisiana on a personal journey of one who has that hope. His plaintive anthem, "it has been only fifty years since the last authenticated sighting" of the ivory-billed woodpecker, cuts two ways: how long fifty years is in the life of a single person and how short it is in the life of a species.

The American Bird Conservancy's field guide "All the Birds of North America" lists the Ivorybill, along with the passenger pigeon, the great auk, and the Carolina parakeet, in its "Extinct Birds" section. There is no entry for the bird at all in the recently published "Sibley Guide to Birds" or in Kenn Kaufman's new "Birds of North America." My National Geographic guide refers to the bird's "probable extinction," and my Peterson guidebook dutifully describes the bird but then adds, cagily, "very close to extinction, if, indeed, it still exists."

Extinction. It is the death not merely of an individual but of all the individuals—past, present, and potential—that make up a species. Once gone, there is no retrieval. This, despite the fact that there are photographs of the Ivorybill, recordings of its voice, and even a silent movie of its nesting habits, which was made in the 1930s, when the bird was studied in one of its last redoubts—an area of old-growth forest in Louisiana that, notwithstanding a fight waged by conservationists, was ultimately felled for timber.

The possibility that any bird said to be extinct might still be around is exhilarating, but the Ivorybill, in particular, has what environmentalists refer to as "charisma," a sort of magical personality that has affected bird-watchers since they first noticed the bird. For one thing, the Ivorybill is—or was—very big. At twenty inches, it was America's largest woodpecker and the biggest woodpecker in the world after the now (presumably) extinct imperial woodpecker, which lived in Mexico, and whose chances for rediscovery are even slimmer than the Ivorybill's.

The Ivorybill also has a reputation for indomitable defiance, and that spirit may have doomed it. Its habitat was exclusively old-growth forest—trees that had lived for hundreds of years—and it simply could not withstand the encroachments of man. . . .

The Ivorybill was not only tough; it was gorgeous—boldly patterned in black and white, with an ivory-white bill that measured three inches from base to tip. The male had a brilliant, blood-red crest. One of the bird's nicknames, the Lord God bird, apparently refers to the fact that people who saw it were so impressed that they cried out, "Lord God!"

John James Audubon, who was a far more gifted painter than [Alexander] Wilson, saw the Ivorybill as somehow already existing in the realm of fine art. In his "Ornithological Biography," Audubon wrote:

> I have always imagined, that in the plumage of the beautiful Ivory-billed Woodpecker, there is something very closely allied to the style of coloring of the great Vandyke. The broad extent of its dark glossy body and tail, the large and well-defined white markings of its wings, neck, and bill, relieved by the rich carmine of the pendent crest of the male, and the brilliant yellow of its eye, have never failed to remind me of some of the boldest and noblest productions of that inimitable artist's pencil.

Audubon, who was born in 1785, died in 1851, before the frontier had been closed and when there were still American birds that had not been named. Today, it requires imagination *not* to see the Ivorybill as a painting, since artistic representations of it are virtually all we have. The bird's scientific name does more than Audubon's poetic description to put the Ivorybill back into nature—*Campephilus principalis,* or "the principal eater of grubs."

There has always been an impulse to make symbols of birds. "Hail to thee, blithe Spirit! / Bird thou never wert—" Shelley sang to his skylark. The Ivorybill has become an emblem of the now vanished American wilderness. The white bill of the bird was used in trade by Native Americans, and it has been discovered in their graves—as currency, perhaps, for the world to come. Ivory-billed woodpeckers continue to have an almost totemic force for birders today.

It is easy to see why birds, which mediate between earth and sky, are the repository of high poetic hopes. But it is precisely because birds aren't spirits that they are so compelling to look at. For me, the thrill of bird-watching is catching the glint of an alien consciousness—the uninflected, murderous eye, the aura of reptilian toughness under the beautiful, soft feathers, the knowledge that if I were the size of a sparrow, and a sparrow were as big as I am, it might rip my head off without a second's hesitation. . . .

Big trees play an important role in the life of Ivorybills. The ornithologist James T. Tanner, who, between 1937 and 1939, studied the last significant population of the birds, noted that they fed on the grubs that attacked recently dead trees. The birds scaled off the bark with their massive bills to get to the bugs beneath. To enjoy a steady supply of dying trees, they required the sort of old-growth forests in which trees are mature enough to die on a regular basis, and an area warm and moist enough to promote rapid decay. The bottomlands of the Mississippi Delta were ideal, as were parts of East Texas and the Florida Panhandle. The birds in Tanner's study lived in a region of Louisiana that is known today as the Tensas River National Wildlife Refuge. Tanner referred to the area as the Singer Tract, because it was then owned by the Singer Sewing Machine Company, which, despite protests, sold its logging rights to a Chicago lumber company that began leveling the forest in 1938.

The Ivorybill's feeding requirements are a perfect example of specialization; its restricted habitat eliminates the competition of other birds. The price a species pays for specializing, though, is vulnerability to changes in the environment.

For example, the Kirtland's warbler, which breeds exclusively in a few counties of Michigan, requires young jack pine trees with interconnected lower branches. Such finicky behavior has nearly cost the warbler its life. Fewer forest fires over the years have meant that fewer young pine trees sprout up, and the bird has been kept alive only by the efforts of environmentalists to plant new trees. But it is easier to plant new trees than to conjure up old ones. The fate of the Ivorybill may have been sealed some fifty years ago, when the last stands of virgin forest were logged out of existence in the Deep South.

According to Christopher Cokinos, in "Hope Is the Thing with Feathers," one of the last official observers of an ivory-billed woodpecker was the Audubon Society's Richard Pough, who, in 1943, went to the Singer Tract while it was being logged. After much searching, Pough found the lone female of the species that had been reported to still be in the area. She was sitting near her roosting tree and, despite the encroachment of loggers, refused to fly away. From Pough's description, one gets the feeling that the bird was simply awaiting its doom. In his report to the Audubon Society, Pough conjectured that there may have been "psychological" factors involved. Like the bird that Alexander Wilson had captured, it appeared willing to die rather than compromise.

This tragic, romantic view of the bird may be misleading. There are ornithologists, like [J. V.] Remsen at Louisiana State, who speculate that Ivorybills can live in recently dead trees that are only a hundred years old—instead of the virgin timber they have been known to favor. If that is true, there are areas where logging stopped some fifty or sixty years ago which are once again producing trees that are congenial to the Ivorybill's habits. There are also areas, like the Pearl [River], where, because of their frequent inaccessibility, enough large trees may have been left uncut to give the birds a place to roost and breed—provided, of course, that they are still around.

The question divides ornithologists, naturalists, and bird-watchers. It is notoriously difficult to prove the absence of something. The birds may have a life span of twenty years, and even the most pessimistic ornithologists concede that a few renegades could have dodged detection, like those Japanese soldiers hidden in the jungle who didn't hear the order to surrender after the Second World War.

It is also possible that the birds were more adaptable than anyone imagined, and that they are quietly reproducing in remote regions. It has been only fifty years since the last authenticated sighting, and there have been numerous unauthenticated ones, many delivered years after the bird was seen, by locals who may have been afraid that their land would be confiscated by the government if an endangered bird was discovered nesting there. Remsen told me that he has been shown credible photographs from the 1970s of nesting Ivorybills by a man who believed that he would lose his land if their presence was made known. (More than once I heard a local adage: "There's no such thing as an endangered species on private property.") As for why bird-watchers never seem to be the ones to report Ivorybill sightings, it is often observed that birders, unlike hunters—and birds—do not stray far from marked paths. . . .

Audubon, toward the end of his life, developed dementia. The birds he had so painstakingly documented began to fly out of his head. Extinction is like that desolation amplified, as if the earth itself could forget the animals that inhabit it. It is difficult to talk about the Ivorybill without resorting to the language of longing: I've heard many birders routinely refer to the Ivorybill as "the Grail."

United States Fish and Wildlife Service, "Recovery Outline for the Ivory-billed Woodpecker" (2005)

In a simplified ecological scheme, species can be divided into two categories—specialists and generalists. As with the northern spotted owl of the Pacific Northwest, the ivory-billed woodpecker was a specialist on old-growth trees, and only dead and dying ones at that. When habitats are disrupted, specialists are far more likely to become endangered or extinct than generalists, who are more able to find substitutes for their lost habitat. One by-product of human habitat modification is that we are losing specialists and finding ourselves more and more sharing the planet only with generalists, such as cockroaches, rats, and pigeons.

On 28 April 2005, the United States Department of the Interior announced that sightings of the ivory-billed woodpecker had been verified in the Cache River National Wildlife Refuge in east-central Arkansas. The Fish and Wildlife Service, the agency of the Department of the Interior responsible for endangered species protection, released a preliminary plan for the bird's recovery, the section on factors of endangerment of which is presented here.

HABITAT LOSS AND DEGRADATION (FACTOR A)

The primary reason for the decrease in Ivory-billed numbers throughout its range appears to be a reduction in suitable habitat (and indirect destruction of their food source) due to large scale conversion of forest habitats. Essential features of Ivory-billed Woodpecker habitat include: extensive, continuous forest areas, very large trees, and agents of tree mortality resulting in a continuous supply of recently dead trees or large dead branches in mature trees. According to Tanner, "In many cases their [Ivory-billed Woodpeckers] disappearance almost coincided with logging operations. In others, there was no close correlation, but there are no records of Ivory-billed inhabiting areas for any length of time after those have been cut over."[1] Noel Snyder argues that the close correlation between timber harvesting activities and the decline of the Ivory-bill may reflect an increased exposure to poaching and collecting rather than food limitation in logged over forests. . . .

Habitat loss has probably affected Ivory-billed Woodpeckers since the original cutting of virgin forest, with some losses being gradual and others occurring very rapidly. Jackson estimated that by the 1930s only isolated remnants of the original southern forest remained.[2] Forest loss continued with another period of accelerated clearing and conversion to agriculture of bottomland hardwood forests of the Lower Mississippi Valley during the 1960s and 1970s. The combined effect of those losses has resulted in reduction and fragmentation of the remaining forested lands. The conversion rate of forest to agricultural lands has reversed in the past few years. Currently, many public and private agencies are working to protect and restore forest habitat. Nevertheless, until more is learned about the Ivory-bill's habitat requirements, the extensive habitat loss and fragmentation and the lack of information on specific habitat requirements remain a threat to this species.

OVERUTILIZATION FOR COMMERCIAL, RECREATIONAL, SCIENTIFIC, OR EDUCATIONAL PURPOSES (FACTOR B)

Historical records indicate that Ivory-billed Woodpeckers (bills and the plumage) were collected and used for various purposes by native and colonial Americans. Collection of Ivory-bills for scientific purposes has been documented since the 1800s. Jackson presented data indicating that such collecting resulted in the taking of over 400 specimens, mostly between 1880 and 1910.[3] By itself, overutilization may not have caused the widespread decline of Ivory-bill numbers. However collecting in combination with the concurrent habitat loss likely hastened the decline of the species. It is possible that local populations could have been extirpated by collecting. For example, Ivory-bills are believed to have been reduced by excessive collecting, rather than as a result of the conversion of forest habitats in a small area of the Suwannee River region of Florida. In addition, Tanner indicated that many Ivory-bills were killed merely to satisfy curiosity. . . .

SMALL POPULATION SIZE AND LIMITED DISTRIBUTION (FACTOR E)

Ivory-billed Woodpecker populations appear to have been in a state of continuous fragmentation and decline since the early 1800s. Early accounts gave no accurate or definite estimates of abundance, but populations were probably never large and were limited to habitats subject to high tree mortality, e.g., areas that were regularly flooded or burned. The small population size and limited distribution of the Ivory-billed Woodpecker place this species (previously thought to be extinct) at risk from naturally occurring events and environmental factors. The Ivory-bill is currently known to occur in only one area in southeastern Arkansas. While a substantial amount of habitat is protected in the area in which the species was rediscovered, threats exist from normal environmental stochasticity. For example, sporadic natural events such as tornados or ice storms could destroy the only remaining nest or roost trees or severe weather conditions could result in nesting or fledging failures. Additionally, the exact number and genetic health of remaining birds is unknown. Ivory-bill populations are at risk from genetic and demographic stochastic events (such as normal variations in survival and mortality, genetic drift, inbreeding, etc.).

Notes

1. J. T. Tanner, *The Ivory-billed Woodpecker. National Audubon Society Research Report No. 1* (New York: Audubon Society, 1942).
2. J. A. Jackson, *In Search of the Ivory-billed Woodpecker* (Washington, D.C.: Smithsonian Books, 2004).
3. Ibid.

THE BROWN TREE SNAKE AND THE AVIAN FAUNA OF GUAM

Alan Burdick, from "It's Not the Only Alien Invader" (1994)

Two of the causes of species extinction, overhunting and habitat destruction, are easy to stop—if we have the will. The third, the introduction of non-native species, is intractable. Even if there were consensus among all humans that such introductions must stop, the rat is out of the bag, so to speak—we've already moved too many species around, both knowingly and unknowingly. Alan Burdick's essay highlights this difficulty in the case of the brown tree snake and its effect on the bird species of Guam.

Three mornings a week, Danielle Kitaoka rises at 3:00 A.M. and drives twenty minutes from her home in Kaneohe, on the island of Oahu, to Honolulu International Airport, where she works as an inspector for the Hawaii Department of Agriculture. She makes a few calls from her office, then swings by the department's dog kennel near the airport to pick up an eager beagle named Columbo. Her main task for the day begins promptly at 6:00 with the arrival from Guam of Continental Flight 934.

"I'm on the shift because the other inspector started to burn out," Kitaoka said recently as she stood on the Tarmac at the appointed hour, watching the jetliner pull up to its gate. A small, freckle-faced woman in her late twenties, Kitaoka did not sound pleased at her assignment. "I'm not a morning person," she muttered.

Kitaoka owes her job to an unsettling discovery. At 7:25 A.M. on 3 September 1991, workers at Honolulu International found a three-foot brown tree snake, dead with its head crushed, on one of the runways. At noon that same day, a second brown tree snake was found stunned but alive under a cargo plane that had just landed at Hickam Air Force Base, ten minutes south of Honolulu International. The Hawaiian Islands are one of the few tropical locales with no indigenous snakes—no indigenous reptiles of any kind, for that matter, nor any amphibians or mammals except a seal and a fruit-eating bat. Both of the brown tree snakes discovered that day three years ago had traveled from the same place: the island of Guam, 2,500 miles to the west.

The specimens of *Boiga irregularis*, as the bird-eating brown tree snake is known to biologists, had long been preceded here by reports of the snake's notoriety. Guam, wholly free of brown tree snakes a half-century ago, now hosts perhaps a million of them. Nine of Guam's eleven native species of forest birds have subsequently disappeared; the Guam rail, a flightless bird found nowhere else, reigns only from behind the safety of an electrified fence in a rearing compound near the Guam airport. Hawaii's own population of flightless birds was wiped out years ago: clubbed to death or eaten by rats—or by the mongooses introduced last century in a failed attempt to eradicate the rats. The state's remaining birds include forty percent of those on the United States' endangered species list. All are unprepared, evolutionarily speaking, for *Boiga irregularis.*

Nor are Hawaii's human residents likely to relish an encounter. Hospitals on Guam have treated more than 200 brown tree snake victims, many of them infants and toddlers bitten while they slept. (The snake's venom, relatively mild to an adult, can cause severe pain and swelling in children.) *Boiga irregularis* also displays an irrepressible urge to climb: onto power lines—causing frequent, costly, and dispiriting outages on Guam—and now, apparently, into the wheel wells of outbound airplanes. At the Honolulu airport, four brown tree snakes had been found in the eight years prior to 1991, but two in one day set alarm bells ringing. The United States Fish and Wildlife Service had already assembled a loose federation of volunteers known as the Snake Watch Alert Team—SWAT—to report any sightings of *Boiga irregularis.* In July 1992, the State Agriculture Department, with a $100,000 Federal grant, acquired three snake-sniffing dogs, trained Kitaoka and four other inspectors to handle them, and sent them to the gate to intercept the invader. . . .

Much of what is known about the natural comings and goings on Hawaii can be found on display or on file at the Bernice Pauahi Bishop Museum, a twelve-acre campus of limestone exhibit halls and research offices in the pleated foothills north of Honolulu. I stopped by the museum one afternoon to visit with Francis Howarth, an entomologist on staff there and one of several scientists on Oahu fearfully awaiting some sign of the brown tree snake.

"I've advocated placing a rim of caged chicks around the airport to monitor its arrival," Howarth told me as we sat across from each other at an oak table in the museum's library. He spoke through a dense white beard, which merged into a shock of silver hair atop his head. Burly forearms, laced with scratches earned in search of this endangered damselfly or that blind, cave-dwelling cricket, rested on the table top.

"We're fouling our nest at several levels," he continued, "and I think the most pervasive problem is alien species." Now as never before, he explained, exotic plants and creatures are traversing the world, borne on the swelling tide of human traffic to places where nature never intended them to be. "The introduction of alien species has gone up manyfold because of the ease of getting from one place to another," he said. "The situation is getting much, much worse."

His concern is widely shared among ecologists today. African killer bees have reached California; colonies of South American fire ants have settled in Texas. August brought news that *Polygonum perfoliatum*, a fast-growing vine from Asia, is quickly supplanting kudzu as the strangler of the South; it apparently arrived as seeds in the root balls of imported rhododendrons.

The most troublesome invader to have reached the United States in recent years is the zebra mussel. Native to the Caspian Sea, the striped mussel entered the St. Lawrence Seaway in the mid-1980s in ballast water released by a transoceanic cargo ship. From there it spread rapidly through the Great Lakes, clogging reservoir intake pipes, sinking buoys, and slurping up the planktonic life that local fish depend upon for nourishment. In 1990, Congress passed the Nonindigenous Aquatic Nuisance Prevention and Control Act—a bramble of task forces and requests for funds—to remedy the situation. Oblivious, larvae of the mussel last summer rode the flooded Mississippi by the billions, extending their grip as far south as New Orleans.

"Alien species are permanent," Howarth said, clasping his hands. "There's no putting the mongoose back in the cage."

The proliferation of exotic species has spawned its own field of biological inquiry—"invasion theory"—and Howarth is a leading investigator. And for better or worse, the Hawaiian archipelago is an ideal laboratory for someone with his interests. The dioramas and taxidermed contents of the Bishop Museum attest to Hawaii's status as an ecological treasure: more than ninety percent of the species found here are utterly unique on earth. Yet according to a 1992 report issued by the Nature Conservancy of Hawaii and the Natural Resources Defense Council, this ecological Eden is besieged by a "silent invasion of pest species"—not illegal immigrants or "emerging viruses" but flora and fauna. On average, five new plants and twenty species of insect, plus the occasional pathogen and predator, are introduced to Hawaii each year, a phenomenon that has led the Harvard biologist Edward O. Wilson to pronounce Hawaii "the invasion capital of the world." . . .

Most of the concern about invading plants and creatures has focused on their economic impact. In September 1993, in a hefty report titled "Harmful Non-Indigenous Species in the United States," the Congressional Office of Technology Assessment estimated that between 1906 and 1991 the seventy-nine most harmful exotic species to enter the United States—among them the gypsy moth and the Mediterranean fruit fly—cost the nation $97 billion in crop damage and pest-control efforts. The agriculture industry in Hawaii loses millions of dollars each year to alien pests like the sugar-cane aphid, which was accidentally imported a few years ago in a shipment of golf course sod.

But something less easily quantified—though ultimately more crucial—is at stake, as Howarth sees it. With plants and animals moving more readily from place to place, an increasing number of native species—residents of the planet's ecological backwaters, unable to compete with zoology's rising cosmopolitan class—are being pushed out of existence. Howarth suspects that "a lot, if not the majority, of extinctions from habitat destruction is from alien species introductions."

It is a new era, wherein the greatest threat to biological diversity may no longer be bulldozers or pesticides but, in a sense, nature itself: carried in the valises of businessmen, hidden in the orchids sent on Mother's Day, lurking in the landing gear of an airliner.

The airport shuttle driver in Honolulu laughed when I told him I was headed for Guam. "Watch out for snakes," he said.

An overnight flight across the dateline from Hawaii brings the visitor to the United States territory of Guam: nine miles wide, thirty miles long and one of the world's more unusual cultural cauldrons. The northern end of the island, a jungly limestone plateau, is owned largely by the United States military, which wrested Guam from the Japanese in 1944. Most native Guamanians inhabit the lush volcanic hills and small, palm-fringed towns of the island's southern end. The midsection of Guam is common ground to both populations: strip malls, fast-food joints, and traffic jams, although it is also home to a burgeoning resort industry, much of it Japanese-owned.

The first brown tree snake arrived on the island sometime in the late 1940s. Most biologists believe it came from one of the Admiralty Islands, near New Guinea, coiled in the dashboard of a jeep or in some other salvaged bit of wartime junk. Prior to that fateful arrival, the only snake on the island had been *Rhamphotyphlops braminus*, a blind and wriggling creature closer in size and spirit to a worm. Hence the early reports of the brown tree snake treated its arrival with benign curiosity. "Because they eat small pests and are not dangerous to man," *The Guam Daily News* reported in October 1965, alongside a photo of a seven-foot snake killed at the United Seamen's Service Center, "they may be considered beneficial to the island." Several equally large specimens were remarked upon in the local press in ensuing years, including one that a Mrs. Edith Smith found slipping across her neck at 4:00 one morning in June 1966.

Tales of the snake's exploits have since reached hysterical proportions. There is the snake that crawled in through the sink; the snake that materialized in the toilet while a dinner party hostess sat (briefly) on it; the snake that jumped from an automobile air-conditioning vent and sent the driver into a near-fatal swerve. According to a story published in *The Wall Street Journal* in 1991, the snakes "hang from trees like fat, brown strands of cooked spaghetti." But I didn't see that. A search of my hotel grounds that first morning revealed no snakes. None lurked in the grass island of the Taco Bell parking lot down the road. The closest I came was a local man who said he'd once found a snake burrowing into the nose of his pet goat, which subsequently died. "And you know, the color isn't brown," he added excitedly. "It's blue!"

I was baffled until I spoke one afternoon with Earl Campbell, Guam's resident herpetologist. . . . "Many of my friends have lived on Guam for years and never seen snakes," Campbell reassured me. "The snakes are extremely averse to light. During the day they take cover in places where people aren't likely to encounter them."

After we'd chatted for a while, he suggested that if I wanted to see a snake—or improve my odds of seeing one—I come back the following evening. . . . "Looking for a snake in the forest is sort of like being in a demented 'Where's Waldo?' book," Campbell said, scanning the gnarled branches. Although the brown tree snake has been known to grow as long as ten feet, most are no more than a yard long and thin as a pinky. Add to this the fact that it is a master climber, strong enough to stand virtually upright on its prehensile tail, and what you have is a supremely surreptitious creature. It will consume anything that smells of blood, from dog food and soiled tampons to semicooked spareribs and raw hamburger—even the foam plate that the hamburger comes on. With its flexible jaw, the snake can devour creatures more than half its own weight, a category that has come to include live poultry and at least one Labrador retriever puppy.

"It's incredible how well the brown tree snake does on Guam," Campbell marveled as we crept first down one path, then another, then another. "My average here is one an hour." When I sounded unimpressed, Campbell noted that this patch of jungle contained the highest density of brown tree snakes—of any kind of snake, for that matter—anywhere in the world. "In the Amazon," he said, "where there are maybe a dozen species of snake, people search for twelve hours to find one snake at night."

By 10:00, the heat of the day had dissolved and the humidity began to condense around us. Spider webs choked the trail and a symphony of katydids chirred from everywhere and nowhere at once. Yet I soon realized that the aura had been even more sepulchral when I'd visited on the previous afternoon, in daylight: there had been no sound but for passing wasps and the drone of an overhead plane. It struck me that, even after several days on the island, I could count on one hand the number of birds I'd seen: one turtledove—a species that immigrated from the Philippines several years ago—and a few bedraggled European sparrows that stuck close by the hotel. The silence I'd been hearing was the absence of birdsong.

The effect was unsettling: a creature apparent only by the absence of any life surrounding it. No wonder Hawaiians are concerned, I thought. It was only eleven years ago, more than three decades after the snake immigrated to Guam, that ecologists were able to rule out pesticides and habitat destruction as the cause of their island's silent spring. By the time Hawaiians know for certain whether the brown tree snake is among them, the damage will be done.

My best view of a brown tree snake came the following afternoon near the Guam International Airport. I had joined Thomas Hall, a stocky, upbeat fellow who runs the Animal Damage Control division of the United States Department of Agriculture on Guam. In his case, "damage control" means making sure the brown tree snake does not leave the island. It is a daunting task. Along with the busy commercial airport, Guam is home to a Navy air base, a commercial and a naval shipping port, and an Air Force base. All require inspection by Hall or members of his crew, which includes thirteen employees and two Jack Russell terriers.

"Detector dogs have caught several snakes," Hall told me as we circled the commercial and naval runways, which lie alongside each other in the center of Guam, in his red Isuzu Trooper. "Once, these pallets of bombs were being loaded up to go to San Diego through Hawaii and the dogs pulled a snake right out of the middle of the pallet. There's no way anyone would have seen it. The same thing with cars. We had some being loaded one morning down at Navy, going to Hawaii. The dogs pulled a snake out of the power steering unit. We caught one just the other day near an airplane. She had eight eggs in her."

We drove past hangars, cargo docks, and boxy military apartments. Hall parked near a dense thicket—beloved by brown tree snakes, he said—that would soon be cleared away.

"Right now, our goal is to make the ports snake-free," he said, wading into the brush. His current strategy is to cast a wide net of snake traps around the ports and to check them regularly. Hall lifted a branch to reveal one of the specially designed traps, which looked like two wire-mesh waste-baskets attached at their rims. The bait was a live white mouse, safely ensconced in a tiny wire cabin with a week's worth of diced potato. Federally supported researchers in Colorado and Oregon are busily experimenting with various chemical cues and sexual pheromones in the hopes of one day replacing the mouse with a synthetic, and less costly, lure. Any permanent solution—there is a federally financed plot taking shape to infect the snake with the serpentine equivalent of pneumonia, for example—is still years from fruition.

In the meantime, it is up to Hall to keep Hawaii and other places safe from brown tree snakes. "It'll be a long time before every snake is stopped from going off island," he said as we forged through the brush. But even one snake is enough. *Boiga irregularis* can survive for months on no food and little water, and females can produce fertile eggs for up to seven years after mating. Already the island of

Saipan, 150 miles to the north, reports some six snake sightings a year, indicating a firmly entrenched population. Hall thrashed onward, checking trap after trap and coming up empty. Finally, eager to demonstrate his effectiveness, Hall led me to a clearing under a large tree. A trunk-size wire cage sat on the ground. Inside, in a writhing tangle, were at least a dozen snakes. "We make sure we have some on hand for research," Hall explained.

A large one lunged at us, mouth agape, as we approached the cage, then struggled to free its fangs from the mesh. Hall lifted the lid and plucked it out with a pair of long metal tongs. He let it dangle by its neck for a moment, six feet of pure muscle coiling and uncoiling in midair. Its skin was neither brown nor blue, I noticed, but a murky green, like grass in Manhattan; its underside was yellow. Then Hall handed the snake to me. At his instruction, I placed my thumb on a bony ridge at the base of the skull and, pressing down gently, clamped the jaw safely shut against my forefinger. The rest of the snake immediately began to wreathe itself around my forearm and neck, squeezing with a gentle, almost clinical indifference, like the blood pressure gauge in a doctor's office. As the tip of the snake's tail probed my right ear, I lifted its pebbly head for a closer look—hoping to catch a glimmer of intent. But its eyes were tiny yellow beads, slit vertically like a cat's and far more impenetrable.

Late on my last night on Guam, I looked out my hotel window and beheld a more sinister manifestation of *Boiga irregularis:* utter darkness. No street lamps, no porch lights, no glaring strip malls, only a few pairs of headlights streaming down Guam's blackened main boulevard. According to a television news report the following morning, two brown tree snakes had crawled onto the power lines at the south end of Guam, cutting electricity to the entire island. It was the latest of the roughly 1,200 outages snakes have caused since 1978.

On hearing the news, I hurried over to the Guam Power Authority's headquarters, a squat building along congested Marine Drive. The press officer agreed to show me the culprits. She returned from her office with a pizza box, which she flipped open. There they were: two snakes, each roughly three feet long. The true color of their skin was now virtually impossible to discern, since it had turned mostly black from the high voltage and in some places was flayed off altogether.

Julie Savidge, from "Extinction of an Island Forest Avifauna by an Introduced Snake" (1987)

Julie Savidge treats scientifically what Alan Burdick treated anecdotally—she carefully details the decline of the birds of Guam. Underscoring the frustration of trying to reverse the process of extinction by introduced species, she notes the poignant uniqueness of the scientific opportunity: "Rarely have data been gathered during the extinction process."

In certain situations, usually following introduction of exotic species by humans, predators have driven prey to extinction. . . . Most avian extinctions within the past 200 years have been of insular species. About half of these extinctions are presumed to have been caused by introduced predators, particularly rats. Predation as the cause for extinction has generally been surmised only after the fact, and is thus based primarily on correlations derived from attempts to reconstruct past events. Rarely have data been gathered during the extinction process. Consequently, the dynamics of the process usually have not been documented.

STUDY AREA AND HISTORICAL BACKGROUND

The island of Guam offered a rare opportunity to study the decline and impending extinction of an entire forest avifauna. Guam is one of the Mariana Islands, a north-south chain of fifteen islands halfway between Japan and New Guinea. It is about 45 km long and 6–13 km wide. Second-growth forest dominates the northern limestone plateau, but primary forest remains in cliff areas; the vegetation in southern Guam is largely savanna with ravine forest in the river valleys. Cocos Island is a small (1800 × 250 m) island 2.5 km south of Guam. The western half of the island is dominated by mixed forest and strand vegetation, whereas the eastern half is a tourist development. Before the recent declines and extinctions, the resident avifauna of Guam consisted of eighteen native birds: twelve land birds, three breeding seabirds, two wetland species, and a reef-heron, and seven introduced species (Table 2.1). Seven of these species (Pacific Reef-heron, *Egretta sacra;* Yellow Bittern, *Ixobrychus sinensis;* Brown Noddy, *Anous stolidus;* White Tern, *Gygis alba;* Philippine Turtledove, *Streptopelia bitorquata;* Eurasian Tree-sparrow, *Passer montanus;* and Micronesian Starling, *Aplonis opaca*) also occur on Cocos Island.

Historically, most forest birds were common to abundant throughout Guam. However, populations of Guam's forest species plummeted in recent decades, while bird populations in savanna habitat on Guam, on Cocos Island, and on Rota, Tinian, and Saipan, islands 71–127 km north of Guam, remained relatively stable. Bridled White-eye (*Zosterops conspicillatus rotensis*) populations on Rota have declined also but the pattern of decline on Rota (disappearance from low elevations) is different from that on Guam. Except for the Vanikoro Swiftlet (*Aerodramus vanikorensis*), a cave nester, all the other ten forest species of birds on Guam have followed a similar pattern of decline (Table 2.1). These birds disappeared from southern ravine forests in the 1960s and gradually their ranges contracted and populations declined progressively to the north. . . . Ralph and Sakai[1] described southern Guam as the most massive avian desert they had seen. By early 1983, all ten forest species occurred together only in 160 ha of mature forest beneath the cliffline at the northern tip of Guam, with a few species still occupying parts of the northern plateau. Bridled White-eyes (*Z. c. conspicillatus*) have not been seen on Guam since summer 1983, and are presumed to be locally extinct. Guam Flycatchers (*Myiagra freycineti*), an endemic species, and Rufous Fantails (*Rhipidura rufifrons*) may also be extinct on Guam. Since July 1985, only three Guam Rails (*Rallus owstoni*), an endemic species found in forest and second-growth vegetation on Guam, have been located in the wild. The remaining forest avifauna is extremely rare (see Table 2.1).

Other native and introduced species have declined in a pattern similar to that of the forest birds (see Table 2.1). The endemic White Tern was described as common over the whole island in 1961. By the early 1980s, it was restricted mainly to the northern coastline of Guam. This species has remained abundant on Cocos Island. Introduced Black Drongos (*Dicrurus macrocercus*) are rare in southern and central Guam. Philippine Turtledoves, a species introduced by the Spanish, have undergone a decline of over eighty percent since the 1960s in southern and central Guam. . . .

THE MECHANISMS OF OVEREXPLOITATION OF PREY BY *BOIGA*

Murdoch and Oaten[2] summarize a variety of mechanisms that can lead to stability in natural predator-prey systems, including refuges for prey, invulnerable prey classes, barriers to predator dispersal, and the existence of alternative prey combined with predator switching. How then has *Boiga* been able to overexploit the avian and mammalian fauna on Guam and not in its native range?

1) *Boiga* has few competitors and no significant predators to limit its populations on Guam. Surveys and field trapping indicate the snake occurs in high densities, with a conservative estimate of about sixteen snakes/ha in northern forests.[3]

2) Theoretically, if refuges of adequate size were present, some number of prey would be safe. However, on Guam, refuges are present only in urban areas, on human-made structures and in savanna, habitats unsuitable for resident forest species. Additionally, the forest canopy on Guam is relatively low, usually less than fifteen meters. This simplified vertical structure may facilitate *Boiga*'s access to prey species throughout the canopy. In the native range of *Boiga* (e.g., Solomon Islands and New Guinea), forest canopy may extend above forty meters. Though no data are available on snake densities at increasing canopy heights, the greater complexity of taller forests may provide additional prey refuges.

3) Bird densities appear naturally lower on Guam than in the native range of *Boiga*, possibly due in part to the simplified vertical structure of the forest. J. Engbring and F. L. Ramsey,[4] using the circular count method, estimated 24.9 birds/ha in Ritidian before snakes invaded and 17.2 birds/ha in comparable habitat on Rota. In contrast, using a variety of census techniques, Bell[5] estimated 69 birds/ha in lowland rainforest in New Guinea. Even if predator densities were

comparable between Guam and New Guinea (no data are available on *Boiga* densities in New Guinea), with a lower density of birds on Guam, the impact of predation on bird populations would be greater.

4) *Boiga*'s generalized food habits and the availability of alternative prey have contributed to the maintenance of high populations of the snake. Small reptiles, particularly skinks and geckos, are major constituents of the diet of the brown tree snake. Small lizards are exceedingly common on Guam and apparently their reproductive potential is high enough to withstand the predation pressure. Young snakes probably grow and maintain themselves on small lizards until they are large enough to take birds and small mammals. Even then, lizards make up a major portion of the snake's diet in areas where birds and small mammals have declined. For the larger snakes, the birds (eggs to adults) and small mammals are probably preferred prey items because they provide a greater reward. By utilizing this abundant lizard prey resource, *Boiga* can maintain high densities while decimating its more vulnerable prey (birds and small mammals). As indicated by the high predation intensity (ninety percent within the exposure period), the high population levels of *Boiga* in Naval Magazine, an area supporting snakes since the 1950s and without birds for nearly two decades, attest to the importance of alternative prey in maintaining snake densities.

Two additional characteristics make *Boiga* an effective predator. First, *Boiga* may have the ability to locate bird prey other than by random encounter. As indicated by Guamanians responding to the questionnaires, snakes predictably enter chicken and pigeon coops shortly after eggs are laid. Secondly, a characteristic that makes *Boiga*, and snakes in general, effective predators even if prey populations decline is that snakes can go for long periods without feeding. If taken alone, this latter characteristic would interject an increased level of instability into the system. However, since *Boiga* can utilize alternative prey, the lag time before snake populations decline is even further accentuated.

THE FUTURE OF GUAM'S AVIFAUNA

It is questionable whether any of the forest species will maintain populations on Guam. Even the introduced Philippine Turtledove, which utilizes forest edge, is experiencing high nest mortality. P. Conry monitored twelve turtledove nests in forest habitat and concluded that the loss of reproduction for turtledoves caused by predation appears to be so severe (the daily mortality rate of nests was calculated at 13.5 percent and nest success at 0.7 percent) that annual recruitment probably cannot replace natural attrition. Only three native forest species, the Vanikoro Swiftlet, Mariana Crow, and Micronesian Starling, are not near extinction. The population of swiftlets, a cave nester, is estimated at approximately 400 birds. Swiftlets have declined on Guam and throughout the Mariana Islands. Since the pattern of decline of swiftlets on Guam was different from the patterns of the rest of the forest birds, it is unclear what role, if any, *Boiga* played. How-

TABLE 2.1

Resident Birds of Guam

Scientific name	*Common name*	*Status on Guam*[1]	*Nesting habitat*
Phaethon lepturus[2]	White-tailed Tropicbird	N; R	Coastal cliffs
Egretta sacra	Pacific Reef-heron	N; U	Coastal areas
Ixobrychus sinensis	Yellow Bittern	N; C	Wetlands, grasslands, scrub
Francolinus francolinus	Black Francolin	I; C	Savanna
Coturnix chinensis[2]	Blue-breasted Quail	I; C	Grasslands/agricultural areas
Rallus owstoni[3]	Guam Rail	N; <50	Forest, second-growth, scrub
Gallinula chloropus guamii[4]	Common Moorhen	N; R	Wetlands
Anous stolidus	Brown Noddy	N; U	Coastal islets
Gygis alba[5]	White Tern	N; R	Coastal areas (nests in trees)
Columba livia	Rock Dove	I; U	Urban
Streptopelia bitorquata[5]	Philippine Turtledove	I; C(n)	Second-growth, forest
Gallicolumba x xanthonura	White-throated Ground-dove	N; Last seen in Jan. 1986	Forest, second-growth
Ptilinopus roseicapilla	Mariana Fruit-dove	N; Last seen in 1985	Forest, second-growth
Aerodramus vanikorensis bartschi[2]	Vanikoro Swiftlet	N; R	Caves
Halcyon c. cinnamomina	Micronesian Kingfisher	N; <50	Forest, second-growth
Acrocephalus l. luscinia[6]	Nightingale Reed-warbler	N; Extinct by 1970	Wetlands
Rhipidura rufifrons uraniae	Rufous Fantail	N; Last seen in 1984	Forest[7]
Myiagra freycineti[3]	Guam Flycatcher	N; Last seen in 1984	Forest[7]
Zosterops c. conspiciliatus	Bridled White-eye	N; Last seen in 1983	Forest[7]
Myzomela cardinalis saffordi	Cardinal Honeyeater	N; Last seen in Jan. 1986	Forest, second-growth[7]
Lonchura malacca[2]	Chestnut Mannikin	I; U	Grasslands
Passer montanus	Eurasian Tree-sparrow	I; C	Urban

Aplonis opaca guami	Micronesian Starling	N; <100	Forest, second-growth[7]
Dicrurus macrocercus[5]	Black Drongo	I; C(n)	Second-growth, forest edge, urban
Corvus kubaryi	Mariana Crow	N; <100	Forest, second-growth

Native forest species that have followed a similar pattern of decline on Guam are in bold print.

Notes

1. Information taken from unpublished data provided by the Guam Division of Aquatic and Wildlife Resources and J. M. Jenkins, *The Native Forest Birds of Guam*, Ornithological Monographs no. 31 (Washington, D.C.: American Ornithologists' Union, 1983). I = Introduced; N = Native; C = Common; C(n) = Common in northernmost Guam, uncommon to rare in central and southern Guam; R = Rare; U = Uncommon.
2. Species appears to have declined but causal factors have not been investigated.
3. Species endemic to Guam.
4. Species appears to have declined because of loss of wetland habitat.
5. Species has followed a pattern of decline similar to that of the forest birds.
6. Reasons for extinction have not been identified. Baker [R. H. Baker, *The Avifauna of Micronesia: Its Origin, Evolution, and Distribution* (University of Kansas Publications of the Museum of Natural History) 3 (1951): 1–359] comments that the species was never very abundant on Guam and that the absence of natural enemies, especially snakes, may have been one of the principal reasons why the Reed-warbler had been able to survive.
7. Historically reported in most habitats except savanna.

ever, the present nesting cave appears relatively immune to snake predation because it has a smooth ceiling. Crows are estimated at less than 100, and no successful reproduction has been recorded at nests monitored since 1985. One might predict that crow populations will eventually decline to extinction from lack of recruitment. Starlings have recently been observed nesting on artificial structures, and populations might be augmented using nest-boxes placed on concrete telephone poles. Both the Micronesian Kingfisher and Guam Rail appear to be breeding successfully in captivity. However, snake populations will need to be reduced on Guam before these species can be reintroduced. Since snakes have never been controlled at this scale, it will probably be many years before effective control measures are developed and enacted.

In summary, the data clearly suggest that predation by the introduced snake *Boiga irregularis* is responsible for the decline of ten species of native forest birds as well as several other types of birds on Guam. This is the first time a reptile has been implicated in the decimation of an avifauna, and the example shows how rapidly extinction can ensue under the appropriate ecological circumstances.

Notes

1. C. J. Ralph and H. F. Sakai, "Forest Bird and Fruit Bat Populations and Their Conservation in Micronesia: Notes on a Survey," *Elepaio* 40 (1979): 20–25.
2. W. W. Murdoch and A. Oaten, "Predation and Population Stability," *Advances in Ecological Research* 9 (1975): 1–131.
3. J. A. Savidge, "The Role of Disease and Predation in the Decline of Guam's Avifauna," Ph.D. Dissertation, University of Illinois, Urbana-Champaign, 1986.
4. J. Engbring and F. L. Ramsey, "Distribution and Abundance of the Forest Birds of Guam: Results of a 1981 Survey." United States Fish and Wildlife Service FWS/OBS-84/20. 1984.
5. H. L. Bell, "A Bird Community of Lowland Rainforest in New Guinea. I. Composition and Density of the Avifauna," *Emu* 82 (1982): 24–41.

3 Nuclear Power

Three Mile Island, Chernobyl, and the Future

Advocates of nuclear power argue that it produces no greenhouse gases except negligible water vapor. Because generating electricity currently puts into the atmosphere twice the CO_2 produced by motor vehicles, this is a key argument. Furthermore, these supporters claim that while nuclear power may leave radioactive waste and may heat some bodies of water minimally, it does not pollute air or water in the usual sense. It has a worker safety record far superior to the oil and coal industries and emits negligible levels of radiation safe for the general public. These supporters also claim that, especially as new technologies emerge, nuclear waste can be handled with no more, perhaps even less, danger than the toxins left by burning coal and oil. Nuclear power in this view is cleaner and safer than any means of generating electricity except renewable sources—and those cannot possibly meet full electric demand for decades, if ever. Compared with fossil fuels, even natural gas, nuclear power exerts a much lower impact on air, water, human health, and, above all, climate. So, why are many countries not building any new reactors? Why have they not done so for twenty years? Against this, China plans to build thirty nuclear power plants between 2006 and 2025, commissioning, on average, one plant every eight months.

Some history, including the story of two accidents, is in order. After World War II, which ended when the United States dropped two atomic bombs on Japan, and while stockpiling more than 60,000 atomic weapons, the major nuclear powers—the United States, the United Kingdom, Russia, France, and eventually China—developed peaceful atom programs, especially for generating electricity. Japan, Germany, and some other nations built reactors but no bombs. India, Pakistan, Israel, and (apparently) North Korea later built both. Iran may develop the capacity to produce nuclear weapons. In the minds of many, a link between nuclear weapons and nuclear power persists.

In the 1950s, governments and private companies promised unlimited, cheap atomic energy. In the U.S., subsidized and regulated by the federal government, private enterprise developed reactors to harness fission reactions and heat water, creating steam to drive turbine generators. Countries developed different reactor designs, for example, graphite, pressurized water, and heavy water.

From 1950 to 1978 accidents at several reactors released radiation. In one instance, three workers died. Some "incidents" (this word officially describes serious accidents) were costly. In a few cases, authorities covered up the worst news. Despite one engineer's statement about the incident at Lagoona Beach, Michigan, "We almost lost Detroit," the public generally regarded nuclear power as safe and non-polluting. Dangers posed by certain radiation levels were not fully understood, even by scientists. Film crews marched into Nevada deserts after A-bomb tests. Through the 1950s, infants received heavy radiation doses to shrink "enlarged" thyroid glands; some nuclear plant workers and medical x-ray technicians received levels of radiation later deemed illegal.

Total energy use worldwide grew dramatically. From 1960 to 1970, promoted by utility companies, electric consumption in the U.S. doubled. Cheap oil and economical coal mining methods drove prices down. (By 1985 power companies, often hitting the limit of generation capacities, were actively aiding energy conservation.) Backed by the Nuclear Regulatory Acts of 1946 and 1954, American companies constructed about eighty commercial reactors. Opponents, fearing radiation, accidents, and a link with nuclear arms, accused companies of siting reactors in rural locales where resistance would be minimal. No comprehensive procedures dealt with spent uranium fuel or other radioactive waste.

Realization of the dangers of radiation grew; it was admitted that catastrophic accidents could occur; thermal pollution became an issue. The problem of radioactive waste was temporarily addressed but ultimately unresolved by storing much of it at reactor sites. Critics said, "no level of radiation is safe." Advocates countered that radiation occurs naturally and that nuclear power was far less polluting and dangerous than fossil fuels. Revelations about radiation releases at Windscale (1957), a graphite reactor, and the subsequent contamination of milk dampened British enthusiasm. The Soviet government silenced opposition. In the U.S., groups won the legal right to environmental reviews that delayed or prevented plant construction. Challenges added costs; so did problems of waste and the decommissioning of aging reactors. (Today, many reactor owners, citing safety and equipment upgrades, are applying to extend their operating licenses far beyond original dates.) Revelations of safety violations long after the fact eroded public confidence.

In 1974, a blue ribbon federal panel issued a report that reckoned the statistical chances of serious nuclear accident, injury, and death as extremely, almost infinitesimally, low. The Union of Concerned Scientists and the EPA criticized the report. They charged that the possibility of a reactor meltdown was minimized. By the late 1970s, the debate was popular, polarized, and political. Bumper stickers read "No Nukes" or "Go Nukes"; in Europe, "*Atomkraft, nein danke*" (nuclear power, no thanks).

At this volatile and politically crucial juncture, an incident occurred in March 1979 at Three Mile Island, a twin reactor site on an island in the Susquehanna River south of Harrisburg, Pennsylvania. Through a combination of mechanical failure and operator misjudgment, the reactor core partially melted. How close to total meltdown remains a matter of debate. No serious cancer or health effects were detected, even in studies of more than 30,000 people from 1979 to 1998. However, the accident could easily have been far worse. The partial meltdown alarmed tens of millions and belied expert predictions about the rarity of such an event. The reactor remains inoperative. Just prior to the incident, a popular movie, *The China Syndrome*, depicted a near meltdown. An operator fighting industry corruption (played by Jack Lemmon) stated that such an event would "render uninhabitable an area the size of Pennsylvania." Three Mile Island raised fears to new highs and diminished nuclear industry credibility to new lows. Plans for future reactors soon were scrapped.

The disaster in Ukraine at Chernobyl in 1986 was much worse. An unnecessary, simulated emergency test backfired and caused a steam explosion that damaged the graphite reactor and vented massive radiation. Within days, thirty-one staff, rescue, and containment workers died. Elevated birth defects and cancer rates, including at least twelve fatal cases of thyroid cancer, followed and persist. All Europe was alerted. In the late 1990s Germany foreclosed further nuclear power development. It would not be until 2002 that a European nation, Finland, would approve new nuclear plant construction.

Nuclear power fills 19 percent of U.S. electricity needs, 80 percent in France. It has polluted and damaged far less than would the use of fossil fuels to generate an equal amount of power. It produces no greenhouse gases. Moreover, new plant designs, whether relying on fuel pellets rather than rods, or lithium rather than water, are more reliable and safer than older ones. Thirty-five or more American miners die yearly in mine accidents, and black lung disease kills an estimated 1,500. No available alternative could replace the power generated by nuclear plants without significant increases in air pollution and global warming. Yet, such pollution, health hazards, and climate change are incremental, slow, and insidious rather than potentially catastrophic and sudden. Psychologically and politically, it appears easier, at least so far, to live with the former rather than with the latter. That does not mean it is better environmentally. A decreased appetite for power consumption overall and better alternatives, such as wind, solar, and geothermal, need to be developed.

While nuclear power has regained promise, problems remain. The amount of fissionable material, like oil, is finite. Building breeder reactors to create more fuel has been successfully opposed everywhere except Japan. Nuclear facilities entice terrorists. Spent fuel—less than was once thought—can be used to create nuclear devices. The problem of waste is significant. The U.S. has 52,000 tons of high-level radioactive waste increasing at 2,000 tons per year from commercial reactors alone. The repository inside Yucca Mountain, Nevada, ninety miles from Las Vegas, is designed to hold more than that for 10,000 years in specially designed canisters. Yet no one can guarantee how long the canisters will withstand corrosion, how much heat will be generated, or that the site will remain dry so that ground water will be unpolluted. A half dozen dormant, not extinct, volcanoes are nearby. No state wants such a facility. Nevada vetoed it, but Congress overrode that veto and in 2002 the Nuclear Regulatory Commission began studying final approval of the site on which, by 2004, $8 billion had been spent. Some waste will remain hazardous for hundreds of human generations. Meanwhile, promising new technologies to melt waste at 2,000°C and vitrify it permanently are being tested.

Global warming, demands for power, and air pollution are all on the rise. Science will contribute to what decisions are made, but politics will ultimately dictate a continuing slow demise—or rapid renaissance—of nuclear power.

FURTHER READING

Garwin, Richard, and Georges Charpak. *Megawatts and Megatons.* New York: Alfred A. Knopf, 2001.

Monroe, James G., and Edward J. Woodhouse. *The Demise of Nuclear Energy: Lessons for Democratic Control of Technology.* New Haven: Yale University Press, 1989.

Mounfield, Peter R. *World Nuclear Power.* London: Routledge, 1991.

Osif, Bonnie A., Anthony J. Baratta, and Thomas W. Conkling. *TMI Twenty-five Years Later.* University Park: Penn State Press, 2004.

Rees, Joseph V. *Hostages of Each Other: The Transformation of Nuclear Safety Since Three Mile Island.* Chicago: University of Chicago Press, 1994.

Weart, Spencer R. *Nuclear Fear: A History of Images.* Cambridge: Harvard University Press, 1988.

Wolf, Susan M. "Abortion, Chernobyl, and Unanswered Genetic Questions." In *Embodying Bioethics: Recent Feminist Advances,* ed. Anne Donchin and Laura M. Purdy. Lanham, Md.: Rowman and Littlefield, 1999.

John Jagger, from *The Nuclear Lion* (1991)

While quietly supporting a pro-nuke attitude, Jagger gives a detached summary of major accidents. His account can be supplemented by, among others, the one found in TMI Twenty-five Years Later *(see Further Reading, above).*

THREE MILE ISLAND[1]

In the early morning of 28 March 1979, a maintenance crew accidentally shut off the flow in the secondary water system at Unit 2 of the Three Mile Island nuclear power plant, near Harrisburg, Pennsylvania (*human error #1*). Normally, this action would not have created a problem. But in this case, it escalated into a near-disaster.

Emergency secondary-system pumps started up fourteen seconds later (*proper safety function #1*), but during a routine test two days earlier, the valves in this secondary system had been shut but not opened up again (*human error #2*). The reactor operator on duty did not notice the red panel lights indicating this blockage (*human error #3*).

With no water flowing through the secondary system, the primary system—which cools the reactor core—heated up. This caused a pressure relief valve above the reactor core to blow open to relieve the pressure (*proper safety function #2*), accompanied by automatic dropping of the control rods into the reactor, which stopped the nuclear reaction (*proper safety function #3*). However, at this point, the residual heat from the fission products in the fuel rods was building up, which made it necessary to continue circulating water through the primary core-cooling system. The emergency core-cooling system (ECCS) pumps turned on within two minutes of the accident, thus cooling the primary system (*proper safety function #4*). Up to this point, there was still no real problem.

But, unknown to the operators, the pressure relief valve that had opened above the reactor core was stuck open (*equipment failure #1*), releasing steam into the containment building and leading the operators to think that there was too much water flow in the primary system. Therefore, they turned off the ECCS pumps shortly after they had started (*human error #4*). They also stopped the *regular* primary-cooling-system pumps (*human error #5*) and opened a drain line to remove even more water from the reactor (*human error #6*). Now the core became very hot, and much of the remaining water flashed into steam, preventing effective cooling of the fuel rods. Some of the fuel rods began to melt, releasing radioactivity into the cooling water. The pressure rose rapidly, keeping the pressure-relief valve open, and shooting more water and steam, now radioactive, into the containment building.

Many hours later, the operators managed to restore the cooling-water supply and get things under control, but by then, large amounts of water had been released into the containment building and had been pumped into storage tanks. The volume of water was so great that these storage tanks eventually overflowed, releasing a small amount of radioactive gas into the atmosphere. The top of the reactor core had been uncovered by water for several hours, and 40 percent of the core melted and broke apart, debris falling to the bottom of the reactor vessel. In total, 70 percent of the core was damaged.[2] Water in the core reacted with the hot zirconium-alloy fuel rods, producing hydrogen; this burned, but there was insufficient oxygen to cause an explosion. The pressure vessel and the containment building remained intact.

It is important to recognize that, although one major equipment failure did occur, this accident involved many *human errors*. . . . Since Three Mile Island, great efforts have been made to train operators more thoroughly and to acquaint them with previously experienced problems. Great progress has also been made in providing operators with relevant data and with a clear analysis of these data, so that problems can be quickly diagnosed. Consequently, it is far less likely that such an accident will occur again. Indeed, to date, we have had no other major accident in U.S. civilian reactors. It must also be remembered that, in spite of the human errors at Three Mile Island, all but one of the failsafe systems *did* work, preventing a complete meltdown.

No civilian was hurt as a result of the Three Mile Island accident. Yet this is hardly the impression that one got from newspaper and TV reports. There was a large release of the radioactive noble gas xenon-133 (half-life 5 days), amounting to 10 million curies (a unit of radioactive decay rate, or radioactivity . . .). The xenon exposure was of little biological significance, as noble gases are not taken up by human tissues—although the beta rays could have produced some slight exposure of the skin. The accident also released 15 curies of iodine-131. Such a quantity of this isotope would be dangerous if it were confined to a room, but injected into the atmosphere, it was quickly diluted to very low levels. The average dose from this release to individuals living within a 10-mile radius of the plant was only 0.008 rem[3]—equivalent to the dose of extra cosmic radiation that a passenger receives on a roundtrip jet flight from Dallas to London. Some of the reactor personnel who had to shut things down, and later to clean things up, got fairly high exposures, but not beyond those considered acceptable for radiation workers. Two million people live within a 50-mile radius of Three Mile Island, and 325,000 of them are expected to develop cancer from natural causes in the next 30 years. It is estimated that the radiation released at Three Mile Island will add one person to this number.[4]

In terms of biological impact, Three Mile Island was certainly a minor accident. But the reactor core was left in a shambles, and many structures in the reactor building were contaminated with radioactivity. It took a decade to clean it all up, and the expense of this cleanup was horrendous. The result, then, was that Three Mile Island was primarily an *economic* disaster.

Perhaps even more important, it was also a *psychological* disaster. The press went wild. The *Philadelphia Evening Bulletin* ran a three-part series, with headlines proclaiming "It's Spilling All Over the U.S.," "Nuclear Grave Is Haunting Kentucky," and "There's No Hiding Place." . . .

WINDSCALE

The worst nuclear accident in the Western world happened in 1957, early in the history of nuclear reactors, at Windscale (now Sellafield) in Cumbria, England. It involved a gas-cooled graphite-moderated reactor designed to produce plutonium for the military. A faulty maneuver by an operator caused a fuel cartridge to split, releasing its contents, which then oxidized in the air, igniting the graphite moderator. The graphite burned furiously for almost two days. The fire was put out by flooding the reactor with water, and the whole antiquated system has now been sealed off.

The accident released 20,000 curies of iodine-131 into the atmosphere. This should have caused some 260 thyroid cancers in the exposed population—an area about 200 square miles around the plant—of which about 13 would be fatal. Because this is only a one percent increase over the natural level of thyroid cancer [*sic*—this figure posits an abnormally high "normal" cancer rate], it has not been statistically detectable.[5] The accident also released 240 curies of the uranium daughter isotope polonium-210. . . . A recent report estimates that this exposure may have caused up to thirty-three people to develop other cancers, but as with the thyroid cancers, the percentage increase over normal levels of cancer are far too small to be actually detected.[6]

The Windscale accident was environmentally significant, but harm to the population was small. Three Mile Island was not environmentally significant and there was no harm to the population. I have dwelt on the details of Three Mile Island because of the great publicity, and the false perceptions, that surround this accident. The response to Windscale has been more rational, although military secrecy prevented the public from learning details until recently. Neither of these accidents was of great consequence. Only one nuclear accident in the world has been a true disaster: Chernobyl.

CHERNOBYL . . .

Chernobyl was vastly different from other reactor accidents. In terms of fallout from the radioisotopes iodine-131 and cesium-137, Chernobyl was 2,000 times worse than Windscale and 500,000 times worse than Three Mile Island. At Three Mile Island, only 15 curies of radioiodine were released. At Chernobyl, about 100 million curies of dozens of isotopes were released. At Three Mile Island, no one was hurt, whereas 31 people died at Chernobyl, 200 more were badly irradiated, and perhaps 17,000 will die prematurely of cancer. Three Mile Island was an economic and psychological disaster but cannot even be considered an accident from a biological standpoint. Chernobyl, on the other hand, was a serious accident, economically *and* biologically. We do not want to have any more Chernobyls. But it is most unlikely that we shall, in view of the phasing out of Chernobyl-type reactors, the superior design of Western reactors, and the development of new "inherently safe" reactors.

It is of critical importance that the public have an accurate perspective on nuclear power. Many people have argued for the demise of nuclear power because of Chernobyl, which killed thirty-one people. We should remember that *Chernobyl is the only civilian nuclear reactor accident that has killed anyone, anywhere in the world.* The Japan Air Lines crash of a Boeing 747 nine months earlier that killed 520 people was not followed by a cry to stop commercial air travel. The acci-

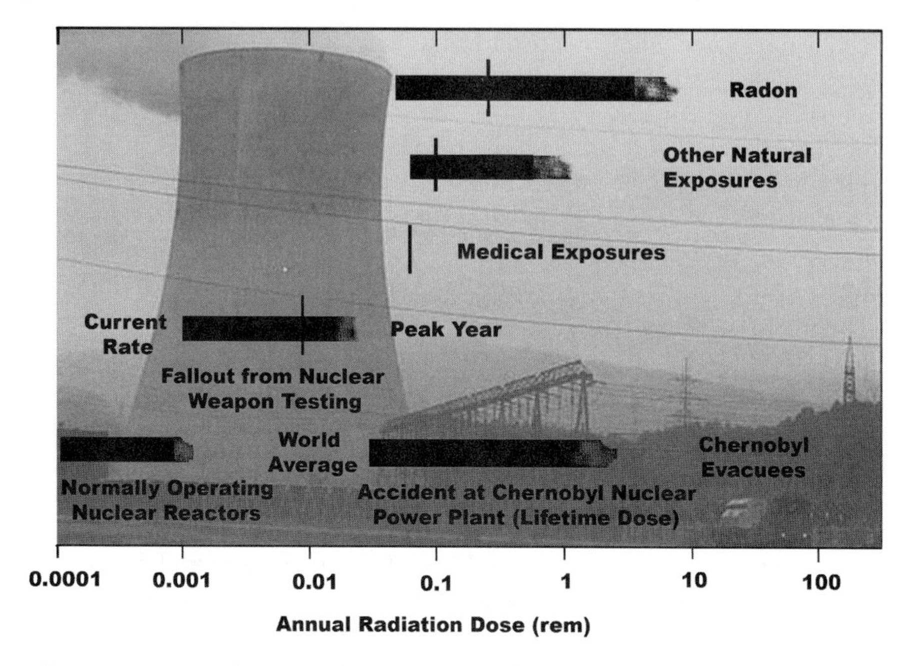

Figure 3.1 Range of *annual* radiation exposures from various sources, and the *lifetime* exposures of evacuees from the Chernobyl power plant accident. Short vertical lines indicate average indoor levels in the U.S., except that the line for nuclear weapons testing is a world average. The horizontal scale is logarithmic. From Anthony V. Nero, Jr., "Controlling Indoor Air Pollution." Copyright 1988 by Scientific American, Inc. Courtesy of George Retseck. Background shows the nuclear reactor in Goesgen, Switzerland. Photo courtesy of Greenpeace.

dental release of the deadly chemical methyl isocyanate in 1984 at a Union Carbide plant in Bhopal, India, killed at least 3,700 people outright and injured 30,000 more. [See Chapter 22.] Yet there is no move to stop the building of chemical plants. Our perceptions of danger seem to be badly skewed.

Notes

1. Daniel F. Ford, *Three Mile Island: Thirty Minutes to Meltdown* (New York: Penguin Books, 1982). This book has an excellent summary and discussion of the events that occurred at Three Mile Island, but the conclusions are unduly pessimistic.
2. W. Booth, "Postmortem on Three Mile Island," *Science* 238 (1987): 1342–45.
3. NCRP Report 93, September 1987. . . . The *maximum* individual dose outside the plant at Three Mile Island was less than 0.10 rem, equivalent to 4 months of background radiation.
4. Merril Eisenbud, *Environmental Radioactivity: From Natural, Industrial, and Military Sources* (Orlando: Academic Press, 1987), 373.
5. "Resurrecting a Nuclear Accident," *Nature* 302 (1983): 207.
6. David Dickson, "Doctored Report Revives Debate on 1957 Mishap," *Science* 239 (1988): 556–57.

L. Ray Silver, from *Fallout from Chernobyl* (1987)

One year after the catastrophic event at the old Ukrainian graphite reactor, Silver constructed a gripping narrative of what went wrong. It is quite incredible to learn that the cause of the accident was poor handling of a badly designed emergency drill ironically conceived to maintain energy output.

They should have been attentive but relaxed. In fact the mood in the Chernobyl-4 control room was tense; the crew were distracted and restive. A higher authority had decided to run an electrical test while the turbine generator wound down. The test was being conducted by electrical engineers from a supplier firm. Dom Tech Energo people were outsiders unfamiliar with nuclear operations and unknown to the Chernobyl personnel. The objective was to determine how long the momentum of the spinning turbine would generate emergency power if the reactor was suddenly shut off. A cyclist seeing his dynamo-powered bicycle light dim out as he lost speed on an uphill climb would understand. . . .

Valery Khodemchuk and Vladimir Shashenok had heard the test procedure outlined when they began the previous night's shift. As they understood it, there were four stages to coincide with the reactor shutdown. First, reactor power would be reduced to less than a third of its capacity. Then the emergency cooling system—the circuit to douse a runaway reactor with cold water—was to be blocked off to prevent its inadvertent use. The cooling pumps normally powered by the station's electrical supply would then be connected to one of the two turbine generators to see if it could sustain them. Finally, the reactor's steam would be cut off from that turbine to let it spin on its own momentum. ("Flywheel energy," Model T Ford drivers would have called that momentum early in the century.) In an emergency some thirty to forty seconds would be needed to switch in the standby diesel generators. Would a free-spinning turbine generate sufficient pump power during that interval to prevent the reactor from overheating? At 1:00 A.M. Friday they took steps to find out. They began the gradual shutdown of Chernobyl-4.

It took twelve hours—well into another shift—to cut the reactor power in half, to switch off one of the two turbine generators and connect pumps in the test arrangement. An hour after that they isolated the emergency core cooling system. There was no valid reason to disconnect it except that the test manager wanted things done that way. When the crunch came "the blocked-off emergency system might have been useful," a fourteen-member review team told the International Atomic Energy Agency [IAEA]. Valery Legasov, deputy director of the Soviet's Kurchatov Institute of Atomic Energy and the man primarily responsible for finding out what went wrong at Chernobyl, was more emphatic. They had "violated the most sacred rule" by disconnecting a protective device, he said. "If at least one violation of the six committed had not been done the accident would not have happened."

As events transpired, regional authorities needed all the power they could get that Friday. So station management was told to keep Chernobyl-4 operational with one generator supplying 500 megawatts of electricity for the next nine hours. Khodemchuk and Shashenok, returning with the night crew, would have no reason to know that they were operating the reactor without emergency-coolant protection. It had been shut off in the early hours of Friday morning. "The fact the emergency system was not reset reflects the attitude of the operating staff in respect to violating normal procedures," the IAEA review team observed. "Their attitude seems to have been conditioned by overconfidence stemming from successful, trouble-free operation and an urge to conduct the test. The presumption that this was an electro-technical test with no effect on reactor safety seems to have minimized the attention given in safety terms."

Shortly after 11:00 P.M. Chernobyl management was told it could take unit 4 out of service and test preparations were resumed. The reactor at this point was at half power with 20 million liters of water pouring into its 1661 boiler tubes every hour to make 1600 megawatts worth of steam or a third that much electrical energy. This was equal to the pull of 2.1 million horses in harness. To control that activity the chief operator had two regulatory systems. With one eye on temperature, pressure, and steam gauges, he kept the mix of steam and water in each of the reactor tubes within precise limits. His other eye was on the heat-making, atom-splitting process which he controlled by moving 211 neutron absorber rods in and out of the reactor. The deeper these rods penetrate into the reactor core the more neutrons they intercept to prevent their atom-splitting, heat-making action. Two dozen of these rods were more or less permanently positioned for stability; another two dozen were automatically controlled in response to ongoing nuclear activity within the reactor vessel. There were 115 manual control rods to be regulated by control personnel such as Vladimir Shashenok while twenty-four "scram" rods were poised to plunge the depth of the reactor for emergency shutdown. The control room chief had a half-dozen senior operators, veterans like Valery Khodemchuk, to share duties. They were assisted, guided, and checked by an array of semi-automated controls, digital computer data, programmed systems, protective devices, and shutdown mechanisms.

Like the operators of big engines everywhere since the dawn of the Industrial Revolution, the Chernobyl crew was also subject to operational rules. One of these ordained that their reactor was not to be operated for any sustained period below 700 megawatts or twenty-two percent of its heat-energy capacity. There are good reasons for this. For one thing the RBMK [graphite] reactor tends to "poison" out like a car choking on a too-lean mixture at stalling speed. . . .

More pertinent, a nuclear reactor responds like a giant airliner. At low speed it becomes unstable and may suddenly fall out of control. The RBMK design gives this instability another dimension. Nuclear engineers call it a "positive void coefficient" which means that if too much cooling water turns to steam it leaves a perilous vapor-filled void in the reactor's pressure tubes, those fuel channels where the water flows over the finger-thick rods of uranium fuel. The water's prime purpose is to transfer the heat from the nuclear reaction in the fuel channels to the steam system. But the hot water also captures a large percentage of neutrons before they can trigger a chain reaction. Once the water vaporizes to steam it no longer has this moderating effect. The RBMK reactor is designed to heat the water in the pressurized fuel channels from 270 to 285 degrees Celsius. Within that temperature range about fifteen percent of the water vaporizes to steam; the remaining water is adequate to keep the atom-splitting, heat-making process under control. In effect the steam-making heat is proportional to neutron multiplication. More steam means less water, leaving a void through which more neutrons pass unrestrained to split more atoms and make more steam ad infinitum. With trillions of atoms splitting each second, the machine can accelerate at incredible speed. It did.

The turbine test was to be conducted while their reactor was between 1000 and 700 megawatts of output. As power reduction resumed the control room staff had a common thought. Watching the digital numbers spin lower they anticipated the end of a year-long record run. It seemed that nothing would mar Chernobyl-4's record performance now. In fact, the first misstep to disaster occurred a half-hour into Saturday morning. In throttling back this giant steam engine, the chief operator let the pressure drop drastically. The throttle slipped its holding notches, so to speak, and the power level plummeted. In a quarter-hour the reactor's output fell from nearly 1000 megawatts to thirty. It was as if the crew chief had more than a million horses in harness when he dropped the reins and just 40,000 when he picked them up again fifteen minutes later. In trying to recover power he put the Chernobyl reactor badly out of balance. When he picked up horsepower again it was like Ben Hur trying to drive his chariot on one wheel.

By 1:00 A.M. Saturday Chernobyl-4 had been brought back to a 200-megawatt level. This was less than a third of the minimum permissible for sustained operation. Yet to get even to this level meant withdrawing an excessive number of manual control rods. The chief operator was now in the position of the balloonist who throws too many ballast bags overboard, buying altitude in mid-flight at the cost of later control. As the IAEA review team observed, the reactor was being operated at a dangerous level with little room to maneuver and an inherent tendency to zoom out of control. He was at the point "where small power changes lead to large steam volume and hence void changes that make power and feedwater control very difficult," the review team said. "The combination of too many control rods withdrawn from the core and operation at this low power level violated a number of procedures. It also created the conditions which both accelerated the reactor's response to plant perturbations and reduced the effectiveness of the protection system."

Still the chief operator persisted. All four coolant pumps were now put into operation as the test demanded. Yet at this power, the reactor could convert only a fraction of the normal water supply to steam. Feedwater flooded the system without making more steam; it stressed pumps and vibrated pipes. Water rose and steam pressure fell in the steam drums, both beyond tolerable levels. In turn, this imbalance caused more control rods to be withdrawn, further reducing the response margin. His radius for reaction was diminishing.

Challenged, pressured, impelled by the sheer force of desperation, the crew chief made a fourth move beyond authority, beyond reason. A protective switch would shut off the reactor if the water level and steam pressure dropped beyond limits. Aware that this safeguard would be tripped at any moment and abort the turbine test, he disconnected the protective device. It was 1:19 A.M. A half-minute later the feedwater was pouring into the separator three times as fast as the reactor could make steam. Attempting to compensate, the automatic control rods withdrew as far as they could go. Now almost all the remaining manual rods were withdrawn as well. The balloonist had virtually exhausted his ballast control. There were no reins left on the horsepower. Still the steam pressure fell. Two seconds before 1:20 A.M., control room records show that the steam supply to the turbine bypass was shut off. Still steam pressure dropped. Abruptly the chief operator cut the feedwater flow. Two minutes later he had the water-steam mix in relative balance. The automatic control rods had returned to the reactor interior. But the manual rods had not.

At 1:22:30 A.M., confident that he could proceed with the turbine test, the chief operator called for a computer printout. Infallibly it mapped the position of all 211 neutron-absorbing rods and measured the neutron flow from each of the 1661 reactor channels. Under normal conditions not less than thirty of the manually operated control rods must be left in the reactor core to sop up excessive neutron production. In exceptional circumstances, with express permission of station management, reactor operation could be continued with a minimum of fifteen rods kept in reserve. But as Academician Legasov told the Vienna post-mortem conference, "Not even General Secretary Gorbachev could authorize the removal of more rods than that." The 1:22:30 A.M. power map showed six to eight control rods left for reaction. The chief operator—or the test manager—decided to proceed anyway.

The final misstep defied all logic. The test was to determine whether the momentum of a free-spinning turbine generator would power emergency pumps *after* the nuclear plant shut off. Indeed, the reactor was equipped with a safety trip, comparable to a safety valve on a conventional boiler, that automatically triggered a shutdown the moment steam was blocked from getting to the turbine. The test program called for the reactor to trip off the moment the turbine trial began. But at 1:23:04 A.M. that Saturday—just as steam power was cut off from the turbine

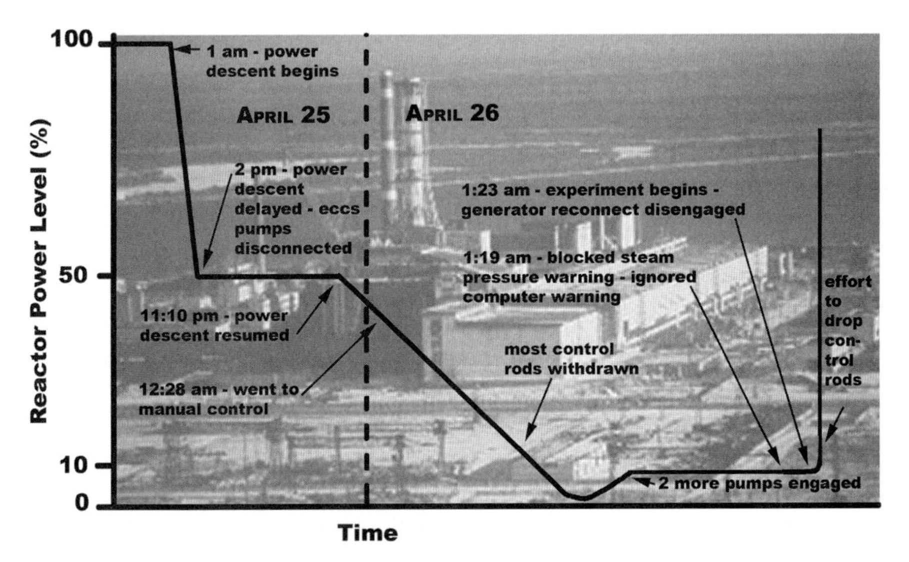

Figure 3.2 Major events leading to the Chernobyl accident. The time axis is not to scale. Background shows the Chernobyl reactor. Photo courtesy of Greenpeace.

generator to start the test—they closed off the emergency stop valve so the reactor would keep running in case the turbine test had to be repeated.

As the stop valve was closed, steam no longer flowed to the turbine generator. So pressure built in the reactor. As the generator slowed so did the four main cooling pumps that were now powered by it. "The effect on feedwater flow, steam pressure, and main coolant flow perturbed the system and introduced rapid void formation in a large part of the core," the IAEA review group concluded. "This led to a rapid increase in reactor power with which emergency shutdown arrangements could not cope."

Enrico Fermi's first experimental nuclear assembly at Chicago had a single absorber rod to soak up excess neutrons if things went wrong. It was suspended by a rope and Fermi reputedly handed a colleague an axe before they began. "If it gets away on us cut the rope and scram," he advised. The boffins [British slang: scientist, technocratic, or "nerd"] have referred to the emergency shutdown control as the "scram" button ever since. At Chernobyl it would plunge twenty-four scram rods into the reactor depths within ten seconds. If thirty of the regular control rods were already providing minimal stability then the emergency rods would prove effective. But no more than eight control rods were now in position.

At 1:23:30 A.M., a quarter-minute since they had bottled up the reactor's steam supply, a power surge began. In three seconds steam production went from 268,000 to 710,000 horsepower. Six seconds after the test began a group of automatic control rods were driven out of the reactor. Eleven seconds after that two groups of automatic control rods began driving in to curb the fast-rising power. Thirty-six seconds after he had begun the test the shift foreman pressed the scram button. It was 1:23:40 A.M. and it was too late. In the following four seconds the steam generated in the runaway reactor reached a hundred times its full power rating. It was as if 428 million horses were bursting from it.

The kettle in which such awesome forces brewed was not large. Less than twelve meters in diameter, seven meters high, it was about a sixth the size of one of those million-gallon water tanks that loom on suburban horizons. Like any steam plant the Chernobyl reactor was essentially a furnace full of boiler tubes. The furnace was built of graphite blocks, 1700 tons of pure carbon to slow neutrons to atom-splitting speed. Some 175 tons of uranium fuel, stacked in finger-thick, meter-long cans, was vertically suspended. Here water trapped the primal heat from disintegrating uranium atoms to become steam to drive the generator.

Like all boilers since Hero of Alexandria's day, it was built to deliver steam under pressure. But when the RBMK reactor was designed nuclear hazards, not steam explosions, were the engineers' prime concern. Their first thought was to safely confine the lethal radiation. So they wrapped the reactor core in a graphite blanket nearly a meter thick to contain the neutron flux. They enclosed it with a steel shroud and surrounded it with a meter-wide wall of water, another meter of sand, and two meters of high-density concrete-mineral mix. Bursting boilers never crossed their minds. The reactor was topped by steel blocks a third of a meter thick and roofed by a 520-ton, three-meter slab of minerals and concrete in which the vertical fuel channels were imbedded. The floor above the reactor [was] comprised of removable blocks to admit the snout of the fuelling machine. These blocks, nearly a meter thick, were made of a dense compound of iron, barium, serpentine, concrete, and stone. Exploding steam fired them upward like shrapnel.

. . . While the heat-making capacity of fossil fuels has been well within the range of man's control since he first fired a boiler, the nuclear age brought a new dimension to steam plant perils. "In a nuclear plant there is no limit to how far up the power can go as long as the fuel hangs together," Ontario Hydro's chief engineer William R. Morison pointed out. "The only thing that stopped it from going to a thousand times full power at Chernobyl was the fact that the fuel disintegrated."

Valery Khodemchuk and Vladimir Shashenok were among a dozen night-shift members in the control room when the turbine test began at 1:23:04 A.M. The control room was thirty meters away from the reactor and on the same level as its base. Six metal and concrete walls isolated them from the reactor and its multi-layered containment structure. The disconcerting behavior of the automatic control rods in the next quarter-minute may have sparked Khodemchuk and Shashenok to investigate. Voluntarily—or on instruction—they went towards the reactor hall in the half-minute before Chernobyl-4 blew up.

Skala, the computer control system assessing data from several hundred sensors throughout the reactor second by second, tracked the first evidence of disaster at 1:23:41 A.M. Like lightning promising thunder to follow, neutron activity more than doubled in the next second and a half. At that moment the pen recorder went close to vertical and vanished from the graph paper. Two seconds later the fuel temperature was paralleling the neutron path like the second track of a railway. In an instant it went from 280 to 1,700 degrees Celsius, and in another half second it was past 3,000 degrees Celsius. Steam production was commensurate with the radioactivity and the heat rise in the fuel. Steam burst the pressure tubes, stripped the fuel of its last coolant, cascaded into the hot graphite, reacted with zirconium tubing and near-molten uranium. In the control room crew members were stupefied by the spin of digital displays, the violent swing of instrument needles, the frenzy of monitor lights, printer-head chatter, klaxon-horn warnings, alarm bells.

At 1:23:48 A.M. a steam explosion of unimaginable energy shattered the reactor core. Its force lifted the slab top of the vault rupturing every tube of the reactor. Imperceptibly diminished, it fired the concrete and metal blocks of the fuelling hall floor like shell fragments through the building roof thirty meters above. In the control room there were audible reverberations then an awesome moment of silence. Racing toward the reactor hall in the last and worst moment of his life, Valery Khodemchuk heard the ripping, rending, wrenching, screeching, scraping, tearing sounds of a vast machine breaking apart.

Now air poured through the shattered roof and the opened top of the vault. Within the core, steam reacted with zirconium to produce that first explosive in nature's arsenal, hydrogen. Near-molten fuel fragments shattered nearly incandescent graphite, torching chunks of it, exploding the hydrogen. The second blast followed the first by three seconds. For Vladimir Shashenok it was not as terrifying as the first explosion. The first had showered them with fiery debris, buried his crew mate almost at his feet, shattered steam pipes, burned and scalded him.

Fast losing consciousness, the control technician could discern rescuers running towards him. "There, Valery's there," he told them.

David R. Marples, from "Introduction" to *No Breathing Room: The Aftermath of Chernobyl* by Grigori Medvedev (1993)

Still facing Soviet censorship and denial of the worst effects of the 1986 Chernobyl disaster, Grigori Medvedev three years later published in Russian an exposé of the event. It was translated into English in 1991 as The Truth About Chernobyl. *Medvedev then moved on to Chernobyl's longer-term effects, particularly those on human health and well-being. David Marples, who has written extensively on the former Soviet Union, Russia, nuclear power, security, and economics, here introduces and summarizes Medvedev's impassioned story.*

The Soviet authorities subsequently employed their traditional methods to blanket Chernobyl in an official interpretation that was at best a gross distortion, at worst a dangerous myth that cost the lives of more citizens than were killed outright by the explosion. All information on health statistics became classified. The victim count reached thirty-one and was deliberately halted. It has never risen, despite overwhelming and documented evidence that more than 2,000 people died as a result of Chernobyl, either from the direct effects of radiation or from diseases directly related to high-level exposure, in the first few months after the disaster. A government commission was appointed to deal with the accident's consequences. A variety of government ministries sent teams of workers into the area. And a ten-kilometer (eventually thirty-kilometer) zone around the damaged reactor was designated an evacuation area after two Politburo officials, Egor Ligachev and Nikolai Ryzhkov, visited Chernobyl on 2 May.

Although Chernobyl was no longer a secret, there was very little frankness either about the magnitude of what had occurred or about the continuing cleanup and decontamination operations. As Medvedev points out, the government commission issued orders that the public should be "persuaded" that Chernobyl was just a minor accident. Approximately 135,000 people had been evacuated. In August a Soviet delegation led by Valerii Legasov presented a Soviet account of events to the IAEA in Vienna, a report that received wide praise in the West for its openness. Some contrasted so-called Soviet frankness about Chernobyl with U.S. secrecy about the Three Mile Island accident in Pennsylvania seven years earlier. The report in Vienna laid the blame on a series of "incredible blunders" by Chernobyl operators, who had dismantled seven different safety systems before conducting a wretched experiment, evidently to see whether enough power was still being generated during shutdown to keep safety systems in operation. By December 1986 a concrete shell had been erected over the damaged reactor, a tomb—hence the name "sarcophagus." By November 1986 the first two Chernobyl reactors were back in operation, and in December 1987 the third unit was also returned to the grid system. At this juncture the outsider gained the impression that a great battle had been won. Soviet accounts compared this battle with the victory over Germany in the Great Patriotic War (World War II). . . .

The Ministry of Atomic Energy was founded after Chernobyl, in July 1986. As Medvedev notes, this gave rise to some hope that its officials would prove more enlightened than some of their Moscow counterparts. It became clear from the first, however, that the ministry wished to continue and even to expand the nuclear energy program and that all other considerations were secondary to that aim. The ministry called in the IAEA to inspect plants that were suspected of having safety problems, and they were all declared to be satisfactory. But by 1988 the lack of official reaction to Chernobyl had given rise to a groundswell of discontent that introduced a new element into Soviet life: national alienation. In addition to a movement against nuclear power—a genuine grassroots movement, in contrast to the officially sponsored movement to remove nuclear weapons from the earth by the year 2000—there was a feeling in republics such as the Ukraine, Lithuania, and Belarus, that Chernobyl was a disaster that had resulted from a central planning system operated from Moscow (the "center") and that the continuing adverse ramifications of the accident were a consequence of that same center "covering up" the real story. Legasov, the fifty-year-old spokesperson for the Soviet side in Vienna, was reportedly a bitterly disillusioned man; he committed suicide on 27 April 1988. In Kiev, Yurii Shcherbak prepared a manuscript based on interviews with eyewitnesses to Chernobyl, just as Grigori Medvedev was planning to unveil a quite different version of events. The reality was so disturbing that it was scarcely credible.

CHERNOBYL: THE REAL STORY

The Chernobyl disaster occurred at night on a weekend. The experiment was not one that would have entered the logbooks of any genuine scientists. The plant director, Viktor Bryukhanov, was not present. Nor was the chief engineer, Mykola Fomin. The operator in charge was an electrical engineer without previous experience at a nuclear power plant. Yet the actual cause of the disaster appears to have been the faulty design of the control rods, which set off a chemical reaction when they were inserted into the reactor core at the time of the power surge that was to blow the roof off the reactor. Fortunately, there were only a handful of personnel at the reactor site. Unfortunately, there was no protective clothing for the fire fighters who arrived from Pripyat. The roof of the reactor was made of combustible material, and the fire was therefore more severe than it might otherwise have been. When Bryukhanov arrived, he could not believe that the chunks of graphite scattered around the turbine room were from the reactor core. But he was hardly alone in his ignorance. The failure to follow basic safety procedures caused many deaths that weekend.

In Pripyat, that Saturday morning, the population was oblivious to the nature of the disaster. Some people, having been awakened in the night by the shudder from the power surge, knew that a mishap had occurred. But there were no official reports being issued, no warnings on the radio. Pripyat went about its normal business: wedding parties, soccer games, fishing trips (including in the cooling pond of the reactor, evidently one of the best places in the area for catching fish). Children wandered off to their Saturday schools in neighborhoods where radiation levels were already dangerously high. This surrealistic life continued for nearly forty hours. Many party officials, who were aware of what had happened, simply fled the city. Months later some of them were declared to be "still on the run," a sad but somehow appropriate reflection of their declining role as a moral force in Soviet life.

. . . Pripyat was eventually evacuated, but ostensibly for only three days, and very little else was done to deal with the situation at Chernobyl before the arrival of Ligachev and Ryzhkov.

A wider evacuation was then ordered, but it took place to the west and northwest, so that the evacuees formed a bizarre procession, following almost exactly the path of the radiation cloud. They were herded from one danger spot to another. The reactor continued to throw out radioactive elements at least until 10 May. Although more than 450 different types of radionuclides entered the atmosphere and began to fall wherever the wind took them, the most dangerous initially was iodine-131 (responsible for the growth in thyroid tumors); its half-life, however, is only eight days. A more protracted problem was caused by other elements, especially cesium-137 and strontium-90, with half-lives spanning some three decades. Hotspots of plutonium posed and continue to pose problems that will last, for all intents and purposes, an eternity. By 1 May 1986, as the traditional May Day parade took place in Kiev, the wind had changed direction and an ominous cloud appeared over the city. Kievans were exposed unawares to levels of radiation nearly 100 times the natural background levels.

Even more serious was the impact on Belarus, a small republic of ten million people just north of Chernobyl (the border is about thirteen kilometers from the nuclear station), which had seen little serious reform. Nikolai Slyunkov, the local party leader, played a role in the cleanup work after Chernobyl, but the public at large remained woefully ignorant of its predicament. Some seventy percent of the radioactive fallout landed on Belarusian territory, especially the southern part of the Gomel region. Residents in this region were soon receiving radiation doses that were probably higher than in the evacuated zone, although the exact amounts will never be known. Southern Gomel was a quiet farming community where potatoes were the main crop. If the farmers could not see the radiation and were not informed about it, there was virtually no chance that they would see and heed the newspaper accounts published in Moscow or Minsk. Today there is very little attention paid to these unfortunate early victims of Chernobyl; some two million people live in this most severely irradiated zone.

The decontamination and reconstruction process can be compared to a military operation in terms of its willful sacrifice of human life. The process began with helicopter crews flying kamikaze missions to drop debris into the gaping hole of the reactor core. Some on board took photographs, although none survived to recall the event. "Volunteers" were sent to the reactor roof to throw spadefuls of graphite into the reactor core. Coal miners dug a tunnel beneath the damaged reactor to construct a concrete base in order to prevent the shell from falling into the earth's core (the dreaded "China syndrome"). Military reservists were flown in after the first month, and some were stationed in the zone for a full six months, at emergency levels of radiation exposure—perhaps 75 rems, perhaps more; the geiger counters often did not register such high levels. The Ministry of Defense claimed on one occasion that it kept no records of the reservists who served in the zone in the summer of 1986. And there was little attention paid to basic safety procedures. Mario Dederichs, former Moscow Bureau chief of the German magazine *Stern*, took a photograph shortly after the accident of some of the reservists who, wearing no protective clothing, were nonchalantly eating lunch below the open reactor. In May 1989, on the way to the Chernobyl station, I myself saw such men sitting in ditches, smoking, just below signs that warned, "Danger! Radiation!" Why, I wondered, do they do this? Don't they care about survival?

What happened to these reservists? By 1991, as a result of civilian protests (the mothers of the young men being the most vociferous), they were withdrawn from the thirty-kilometer zone. According to official reports from the headquarters of the cleanup operation in Chernobyl, many have since died of heart attacks. According to their own union, more than 7,000 of their number had died by the spring of 1991. Like much else about this tragedy, the figures are almost impossible to corroborate. Doctors point out that there is no known correlation between increased radiation exposure and heart disease, and there is no official data bank on the number of victims. A list of highly contaminated persons that was maintained in Belarus was mysteriously "stolen." A press release from the Ukrainian Ministry of Health issued in the spring of 1990 noted that malignant and evidently incurable diseases of the skin had developed among the cleanup crews. . . .

Similarly, the correspondence between low-level radiation and increased cancers or leukemias remains a moot issue. In Belarus, the scientific investigation into the consequences of the disaster has indicated that there has been a significant rise in anemias, as well as a notable (though not dramatic) increase in leukemias in the irradiated regions, particularly among boys five to nine years old. The issue is a sensitive one because the relationship between exposure to increased radiation and the illnesses endemic in the Chernobyl zone is far from clear. The situation has hardly been helped by the former Center for Radiation Medicine, whose director, Anatolii Romanenko, made the glib statement in 1997 that after Chernobyl the health of the population in the affected areas was "even better than before." The Belarusian public is so disillusioned with the government's efforts to deal with the tragedy that, according to a poll of April 1992, only ten percent of the population had faith in the effectiveness of such aid. Conversely, more than sixty percent had confidence in a private charitable trust called Children of Chernobyl.

In short, the Chernobyl tragedy continues and expands. Radionuclides are entering the food chain. Radioactive particles in some cases transform themselves into more toxic substances. The economic crisis is responsible for a lack of nutritious food; the lifestyle of the population is in general far from healthy. In the Ukraine, in 1991, the mortality rate exceeded the birth rate—an alarming statistic. . . .

Charles Perrow, from *Normal Accidents* (1984)

With Three Mile Island in mind but two years before Chernobyl, Perrow argues how inherently unpredictable and unique any *major accident is. Each escaped prior analyses and prevention. In that respect all accidents are "normal" or, one might say, "normally abnormal." Contributing causes to nuclear accidents have included a repair worker using a candle instead of a flashlight, gross operator error, faulty instruments, stuck valves, poor welds, failure of redundancy systems, desire for commercial gain, and bizarrely contrived emergency tests more dangerous than the accidents they were designed to combat. Perrow shows that any catastrophic accident—nuclear ones are no exception, they prove the rule—is* by definition *unpredictable, its causes not single but multiple, occurring in unforeseen and unforeseeable combinations. There is no way to guard against all possible scenarios in their individual, nefarious combinations. While the probability of each is impossibly small, their number is remarkably large.*

NUCLEAR POWER AS A HIGH-RISK SYSTEM: WHY WE HAVE NOT HAD MORE TMIs—BUT WILL SOON

Operating Experience

We have not given the nuclear power generation system enough time to express itself; and we are only just beginning to uncover the potential dangers that make any prediction of risk very uncertain. We are about twenty years into the era of commercial plant operation [in 1984], but our experience is not all with one type or size of plant. Indeed, the oldest plant in operation in 1982 was a 430 megawatt (Mw) reactor operating more or less continuously since 1967. We do not build this size any more, so its sixteen years of operating experience is of somewhat limited value.

The small plants of around 400 Mws are different in many respects from the larger ones of around 1,000 Mws; changes in scale produce surprising results. For example, the larger plants appear to be less reliable; there is more downtime after the first two or three years. In addition to size, there are two different types of reactors, the pressurized water reactors (PWR) and the boiling water reactors (BWR). Experience accumulated in one does not necessarily enable us to judge the reliability of the other; some aspects are similar, some different. In addition to size and type, there are four different manufacturers. General Electric builds only BWRs, while Westinghouse, Babcock and Wilcox, and Combustion Engineering all build PWRs. The designs differ, of course, limiting the accumulation of experience to some degree.

Thus, to say, as proponents of nuclear power often do, that we have 500 "reactor years" of experience with commercial plants (summing up the number of plants times the number of years each has been operating) is quite misleading. There is no consensus on what would be adequate experience for such a complex and novel transformation process as controlled nuclear fission creating steam that drives turbines; there are thousands of years operating experience with large turbines, but very little with nuclear fission. The condensate polisher problem on the turbine side of the plant at TMI would have been trivial in a coal-fired plant but was not in a nuclear plant. We have been building large pressure vessels since the late nineteenth century, but are only beginning to learn the problems with welded stainless steel vessels forty feet high that are bombarded with neutrons. Every few months new problems appear in nuclear plants, including the failure of supposedly failure-proof emergency scram systems. At the time of TMI we had only thirty-five years experience with reactors the size of Unit 2; that is infancy for a system of this size and complexity. . . .

For example, steam generators are a problem with all power plants; the pipes rust. Special care and materials are used in nuclear plants, but in 1981 it appeared that seventeen reactors, some only five or six years old, had serious rusting problems. The repairs on two plants owned by the Virginia Electric and Power Company cost a total of $112 million. Rusting is a special problem in nuclear plants since the thin tubes in the generators are immersed in water continuously, and leaks will allow radioactive water to get in the secondary (nonradioactive) cooling system. Various steps were taken to reduce the rust, but apparently without success in some plants.[1] The point is, in a nuclear plant leaks in the generator are failures that can interact with other failures, and thus be a source of system accidents; repairs to such a system can be enormously expensive (in contrast to a conventionally fueled power plant); and there was no way to anticipate these problems in a new technology with such large design and construction lead times.

More serious is the problem of core embrittlement. The bombardment of the containment vessel by the nuclear reaction going on within it has had a greater impact than anticipated. The forty-foot stainless steel vessel is designed to last forty years, but there are already potential brittleness problems in forty-seven plants, the Nuclear Regulatory Commission (NRC) announced in 1981, and of these, thirteen have serious problems. One of these is only three years old; three others, four years old. The problem is that the core is very hot—about 550°F—and if you have an emergency and must force thousands of gallons of cold water into the core, the inside of the eight-inch-thick vessel will shrink faster than the outside, creating cracks. In an accident, the pressure must be kept high, further straining the core. These problems apply to PWR systems only, but PWRs account for two-thirds of the operating reactors.[2]

These are technical reasons why we have not had sufficient time to have a truly serious nuclear accident—the system is quite new and has not been given a chance to reveal its full potential for danger. Unknown potential cannot be corrected, except by running the plants and taking the risks; without experience, we cannot be sure of the potential for damage inherent in the system's characteristics. . . .

Defense in Depth

. . . We can be glad that we have containment buildings. These are concrete shells that cover the reactor vessel and other key pieces of equipment, and are maintained at negative pressures—that is, at a lower air pressure than the atmosphere outside of them—so that if a leak occurs, clean air will flow in rather than radioactive air flowing out. The Soviet Union, which did not begin a large nuclear generating program until about 1970, is far less concerned about the chance of large accidents, so they did not build containment structures for their early reactors, nor do they yet require emergency core cooling systems. Had the accident at Three Mile Island taken place in one of the plants near Moscow, it would have exposed the operators to potentially lethal doses, and irradiated a large population. [This happened at Chernobyl, two years after Perrow's book appeared.]

At TMI, the hydrogen explosion (or "burn") that took place in the containment building generated a pressure surge equal to one-half that which the building was designed to handle. The building was built this strong only because the state of Pennsylvania insisted that it meet the criterion of being able to withstand a direct hit from a jet airliner (it is close to the Harrisburg airport). The initial plans did not call for this. Even if the building were not reinforced, it is unlikely, I am told, that the hydrogen burn would have breached containment and allowed the radioactive particles to escape. However, such a disaster might occur in a plant with all those flaws in the concrete we heard of; the explosion might have taken place thirty minutes later when there would have been much more hydrogen to burn; and it could have happened in a part of the building where more missiles would have been created, which could have ruptured the many penetrations required in the building for controls and pipes. While containment is absolutely necessary, it may not be sufficient. It *can* be ruptured. . . .

Fortunately, we tend to build our plants in sparsely populated areas, though they are generally near big cities. The ideal spot for a nuclear plant cannot exist. It should be far from any population concentration in case of an accident, but close to one because of transmission economies; it has to be near a large supply of water, but that is also where people like to live; it should be far from any earthquake faults, but these tend to be near coastlines or rivers or other desirable features; it should be far from agricultural activities, but that also puts it far from the places that need its power. The result has been that most of our plants are near population concentrations, but in farming or resort areas just outside of them. The Indian Point nuclear stations, for example, are on the Hudson River, but just thirty-five miles upwind [assuming a northerly wind] of Manhattan. The owner of one of the plants there, Consolidated Edison, once proposed building a nuclear plant in the middle of Queens—truly one of the most densely populated areas in the United States. Some plants are built on earthquake-prone coastlines, others on rivers that supply fresh water for large cities and for irrigation. . . .

Despite all these problems, semi-remote siting has no doubt increased the safety of nuclear plants. Many have had small emissions of radioactive materials as a result of accidents. Were they located in Queens, the long-term dangers would be higher. . . .

Finally, there is the Emergency Core Cooling System (ECCS). Should there be a danger of a core melt, this system will flood the core with water, cooling it. It is in the nature of the beast that we cannot use full-scale testing to see how effectively ECCS will work. In a series of tests with a nine-inch model reactor core, all tests failed.[3] Some critics, such as the Union of Concerned Scientists, believe that as presently constituted, ECCS is an inadequate safeguard. At the Browns Ferry nuclear station, the fire that shut down two reactors and burned out of control for several hours rendered the ECCS system inoperative. Fortunately, other means were used to prevent massive fuel melting. The assessment of the ECCS made in the most ambitious safety study commissioned and carried out by the Atomic Energy Committee (forerunner of the NRC), the *Reactor Safety Study* (RSS or WASH 1400, or Rasmussen Report as it is variously referred to), failed to consider that anything else might be wrong in a plant when there was an emergency that required ECCS. That is, the study ignored the possibility that there could be a variety of failures that in themselves would defeat this safety device. For example, steam generators are a continuous problem with nuclear plants; should many of the tubes in them fail in an accident, so would ECCS. There are also problems with other major subsystems. The integrity of the reactor vessel itself has been questioned, drawing upon industrial experience with vessels in nonnuclear systems.[4] Finally, in 1981 the Browns Ferry control rods failed to drop on command, and in 1983 the automatic shutdown system at the Salem plant in South Jersey failed twice. Both events were assumed to have extremely low probability; both could easily defeat the ECCS.

It is true we can be glad that containment, siting, and major emergency systems exist to reduce the dangers. No doubt there would have been more severe accidents without them. But they are unlikely to prevent all future disasters. Siting is not remote enough; containment is vulnerable to hydrogen explosions, missiles, and faulty construction; and the "defense in depth" major emergency systems such as ECCS are defenses with perhaps not that much depth.

Trivial Events in Nontrivial Systems

. . . Let's start with a trivial event like the ones that plague us all. In 1980 a worker in the North Anna Number 1 plant of the Virginia Electric and Power Company (VEPCO) was cleaning the floor in an auxiliary building. His shirt caught on a three-inch handle of a circuit breaker protruding from a wall. He pulled it free, and apparently was unaware that in doing so he activated the breaker. This shut off the current to the control rod mechanism, and the reactor scrammed (shut off) automatically. This trivial event caused a four-day shutdown, which cost consumers several hundred thousand dollars. Fortunately, the weather was mild, so demand was low. The executive vice president of VEPCO termed the accident embarrassing, but suggested there was a fortunate lesson for us all: The incident "clearly demonstrates the sensitivity of nuclear station systems to the slightest deviation from normal and the ability of these systems to perform safely as designed in immediately stopping the unit."[5] Shutting off current to a major safety system is hardly a slight deviation from normal, and that it can be done so casually suggests an undue degree of sensitivity.

Piping is always a problem in any plant. In a nuclear plant this problem is a bit more severe. During the TMI accident, operators sent radioactive water to the wrong places because the plumbing was so complex and pressures could cause reverse flows. At one plant a small error sent radioactive waste water into the drinking water system that went to the fountains!

Clams are another problem. The filters used on cooling water intake systems from rivers and bays do not keep out the clam larvae, which then lodge in the cooling pipes in the plants and begin reproducing. Eventually, the pipes become clogged with thousands of clams. A report on one plant in Arkansas suggested a week-long shutdown to remove them. Clams foul non-nuclear plants too, but stopping and starting them is not as dangerous.

Even changing light bulbs has its dangers in these highly engineered, complex systems. In 1978 a worker changing a light bulb in a control panel at the Rancho Secco 1 reactor in Clay Station, California, dropped the bulb. It created a short circuit in some sensors and controls. Fortunately, the reactor scram controls were not among those affected, and the reactor automatically scrammed. But the loss of some sensors meant the operators could not determine the condition of the plant, and there was a rapid cooling of the core. As we have already noted, normally the inside temperature of the reactor vessel is at 550°F. Within an hour it had dropped to 280°. The colder, internal walls tried to shrink but the hotter, external ones would not allow shrinkage. This put strong internal stresses on the core. Meanwhile, to prevent a meltdown of the fuel rods, the internal pressure must remain high—2,200 pounds per square inch—while the temperature must drop. At the lower temperature of 280°, the strength of the vessel is reduced, but the pressure remains high. This rapid cooling, which can occur with high pressure injection, or with a loss of instrumentation and control, did not in this case damage the core. But this is probably only because the plant had been operating at full power for less than three years. A spokesman for the NRC said: "If it had been ten to fifteen full power years, instead of two to three, which it was, that vessel might have cracked."[6] A cracked vessel would result in a loss of coolant and a meltdown; no emergency system would be available to cool the core. . . .

CONCLUSION

We have not had more serious accidents of the scope of Three Mile Island simply because we have not given them enough time to appear. But the ingredients for such accidents are there, and unless we are very lucky, one or more will appear in the next decade and breach containment. [Two decades later, in 2004, no such accident had yet occurred, except at Chernobyl (see above).] Large nuclear plants of 1,000 or so megawatts have not been operating very long—only about thirty-five to forty years of operating experience exists, and that constitutes "industrial infancy" for complicated, poorly understood transformation systems. There is ample evidence that problems abound in these large systems, and that they are different from the problems of the smaller units where we have a bit more experience. For all nuclear power plants, the steam generator and the core embrittlement problems are awesome. Small failures can interact and render inoperative the safety systems designed to prevent a steam generator failure from being catastrophic. Trivial events can place stress on the embrittled core in ways unimagined by designers. The sources of other errors and failures appear all too numerous, judging from the events covered in this chapter.

The catastrophic potential of nuclear plant accidents is acknowledged by all, but defense in depth is held by experts to reduce accident probabilities to nearly zero. Yet core containment, emergency cooling systems, and isolated siting all appear to be inadequate; all have been threatened. Nor can we have any confidence whatsoever that quality control in construction and maintenance is near the heroic levels necessary to make these dangerous systems safe. A long list of construction failures, cover-ups, threats, and sheer ineptitude plagues the industry. I have argued that construction problems are probably no worse than in most other industries, but that is no comfort; it has to be much better. Nor has the actual operation of nuclear plants appeared to be as far above normal industrial standards as would be required of such a dangerous undertaking. If anything, it is somewhat below industrial standards. These statements regarding construction, maintenance, and organizational management are based upon the reviews and statements of the Nuclear Regulatory Commission itself, including its chairman. . . .

Notes

1. Matthew L. Wald, *New York Times*, 21 September 1981.
2. Ibid.
3. Olson McKinley, *Unacceptable Risk: The Nuclear Power Controversy* (New York: Bantam Books, 1976), 22.
4. Union of Concerned Scientists, *The Risks of Nuclear Power Reactors* (Cambridge, Mass.: CS, 1977), 44–51.
5. *Washington Post*, 29 February 1980.
6. *New York Times*, 26 September 1981.

Hans Blix, from "Nuclear Power and the Environment" (1989)

Now more famous for his role as chief U.N. weapons inspector prior to the U.S. 2003 invasion of Iraq, Blix previously headed the International Atomic Energy Agency (IAEA). His statement on nuclear power and the environment, especially regarding the green-

house effect, remains one of the best informed, sane, responsible, and politically neutral engagements with that pressing issue.

I am not going to suggest to you that nuclear power is the solution to the environmental damages and threats which our excessive and careless use of fossil fuels have led to: acid rains, dying lakes and forests, and global warming. It is a fact, however, that nuclear power reactors emit no sulphur dioxide (SO_2), no nitrogen oxides (NO_x) and no carbon dioxide (CO_2). The wastes to which they do give rise are minuscule in volume compared to the wastes of the fossil fuels and they can be isolated almost in their entirety from the biosphere. My aim is to show that a continued and expanded use of nuclear power must be *one* among several measures, all of which must be relied on to restrain our use of fossil fuels and thereby to limit the emissions which their burning gives rise to, including those of carbon dioxide.

Energy is the lifeblood of our societies. An enormous increase in the use of hydropower, coal, oil, and gas has helped to raise the standards of living in many countries to unprecedented levels. The difference in the levels of energy consumption between the states with the highest living standards and those with the lowest is staggering. The electricity consumption of Norway is about 23,000 kWh [kilowatt-hour] per capita and year, while that of Bangladesh is 50 kWh per capita and year. Still, when we are now examining the environmental consequences of the accelerated use of energy by the rich countries with increasing alarm, we must note, sadly, that low per capita energy use has by no means protected developing countries from severe environmental damage. Rapid population growth has led to an ever growing total use of firewood, and to deforestation and desertification. More and more people have to seek their fuel from an ever less plentiful supply. Thus, when we focus on the environmental threats caused by the rich countries' consumption of energy, we must remember that there is another environmental crisis caused by energy use in many parts of the underdeveloped world. . . .

Attention is naturally concentrated on the emissions of carbon dioxide (CO_2). About fifty percent of the greenhouse effect is attributed to the increasing carbon dioxide levels in the atmosphere. A smaller part of these emissions is the result of deforestation. Forceful programs for reforestation and a halt to deforestation are desirable measures. We should remember, however, that such programs may not be easy to implement in poverty-stricken countries with rapid population growth. There remains the major emissions for which the industrialized countries carry the greatest responsibility, namely the emission of CO_2 through the burning of fossil fuels, e.g., for heating, transportation, and electricity generation. These emissions cannot be prevented by technical means, as can be done with emissions of SO_2 and NO_x, which have been linked to acid rain and dying forests. CO_2 emissions can only be reduced by limiting the burning of coal, oil, and gas. . . .

For more than a decade, most environmentalists have been advocating two main responses to the environmental threats from energy use, whether acid rains, dying forests, or the arrival of the greenhouse, and these are energy conservation and an expanded use of renewable energy sources, in particular solar power, wind power, and biomass.

At the time when I participated in the Swedish public debate before the referendum on nuclear power in 1980, the opponents proposed an immediate halt to the construction of further nuclear plants and a closure of operating plants by 1990. They denied that any increased fossil fuel use would be required and argued that energy conservation, solar and wind power, and biomass could be used instead. Today Sweden uses thirty-five percent more electricity than in 1980. About forty-five percent of the total electricity generation comes from nuclear power, and about fifty percent from hydropower. Solar and wind power provide 0.004 percent. Biomass in the form of wood chippings and other waste products is used by industry, but mainly for heating purposes. A victory for the anti-nuclear option in 1980 would have been an unmitigated environmental and economic disaster for Sweden.

When we are now faced with the contentions that conservation, solar power, wind power, biomass use, and various other energy sources are adequate to enable us both to reduce the burning of fossil fuels and to do away with nuclear power, let us have a full-scale public discussion of these options to determine their real value—and to discard their illusory value. We cannot plan substantial reductions in the use of coal, oil, or gas on the basis of dreams. We need to escape from the greenhouse, yes, but we need also to escape from the dreamhouse! Let us respect those who withdraw to the countryside to cultivate biodynamic carrots and get their electricity from a windmill. But don't let us believe that they have the recipes for the rest of us.

I shall start by discussing conservation, which is a significant, realistic, and necessary element in any effort to reduce the burning of fossil fuels. Conservation embraces both a more efficient and a more discriminate use of energy. In both respects much can be achieved, but it must also be recognized that some means of achieving conservation may be unacceptably costly and others—like tax penalties, for example, on CO_2 production—may be unpalatable. How much can conservation realistically contribute to a reduction of CO_2 emissions? It is easier to gain support for squeezing more miles out of a gallon of petrol than for restricting driving.

Since the automotive sector is significant and often focused on, let me add some comment on it. It may well be that we should stimulate the use of the electricity-powered means of communication, like trams, trolley buses, and trains and thereby, hopefully, somewhat restrain the use of oil-based means of transportation. If the electricity were produced by hydropower or nuclear power, such policies would help restrain CO_2 emissions. More efficient engines in cars would have the same effect. However, the total emissions of CO_2 from fossil fuel burning would be reduced—it has been calculated—by only some 4.5 percent, or 900 million tons per year, if all cars, trucks, and buses in the world could be changed overnight to the best engine performance standards which now exist. Even more could be achieved, of course, if the fuel economy were to improve even further. However, none of these changes occur overnight and the chances are that even with promotion of electrically driven transport and more energy efficient engines, the number of cars in the world will increase to such an extent that total emissions do not fall. This does not, of course, speak against the measures I mentioned. Without them there would be further increases in the emissions.

If predictions about future energy savings are necessarily hypothetical, we do have some past experience to go by. The first oil price shock in 1973 caused a fundamental change in the energy demand in the industrialized Western countries. Primary energy demand—mainly for oil—had up till then followed the increase in the gross domestic product very closely, but since then, it has remained almost constant in spite of an increase in GDP of more than thirty percent between 1974 and 1986. This means that it was possible, through higher oil prices and government encouragement, to save more than thirty percent of the primary energy during twelve years in industrialized countries. Worldwide, the picture was different. Globally, primary energy use continued to increase by two percent per year.

During the same period, another significant change occurred in industrialized countries: electricity use began to grow with GDP in an almost one to one ratio. Electricity became the preferred form for energy use in many industrial processes and heating applications. Through its precision in achieving effects exactly where wanted, and in the quantity wanted, it also became a significant factor in achieving the savings in primary energy uses. This was true for those industrialized countries which could most easily use and adopt advanced technologies.

When we now look at actual national plans for the future, we find that they do not generally foresee a decreasing primary energy use—particularly not in the developing world. Indeed, there is a pathetic gap between the frequent claims as to what conservation can achieve and actual energy plans. China, for instance, already the world's biggest burner of coal, plans to double its coal use from the mid-1980s to 2000 [see Chapter 14], and India plans to triple its coal use. These two nations, with more than one third of the world's population, will then use more coal than all of the OECD countries together, including the U.S.A., the U.K., and the F.R.G. [the then West Germany]. They are typical of many countries in the Third World, most of which are not able to use advanced and demanding technologies. I am not saying that there is no room for greater energy efficiency in developing countries. I am only saying that the plans of these countries are for a sharp increase in the use of fossil fuels. This means that if we are to succeed in diminishing the greenhouse effect, the industrialized countries must make the major effort.

My conclusion is that although there may be a significant potential for saving both primary energy and electricity, especially in high-consumption countries, we must prepare to face increasing global demands. Knowledgeable organizations—even taking conservation measures into account—forecast 30–45 percent higher primary energy demand in 2000 than in the mid-1980s, 60–70 percent higher electricity demand, and about 40 percent increase in the use of coal alone over the same time period. These forecasts may be wrong, as earlier forecasts have often been. A prospect of catastrophe might affect them. Nevertheless, we must tackle the question particularly whether and how we can contain, indeed, reduce the share of CO_2-producing fossil fuels in the world's energy balance.

Let me turn first to the scope for an expanded use of renewable energy sources. If you except hydropower, these sources now contribute less than 0.5 percent of the global energy supply—and a major part of that is geothermal energy which is important for instance in Iceland and some parts of Italy and the U.S.A. Solar and wind energy have so far shown little promise of economic competitiveness.

As to solar power, there have been promising developments in photovoltaic cells, but their price is still too high to make them economically acceptable, except for use in satellites, isolated locations, watches, calculators, etc. It might also be noted that in order to produce the same amount of electric energy as a 1000 MW(e) power station operating at eighty percent load factor, some ninety square kilometers would need to be covered by solar cells, assuming that they were to be placed in Central Europe and that the best existing types of cells were used.

There exists a long experience with wind generators. In California alone, there are some 15,000 wind generators now in operation. They seem to have an optimum size in the range of 100 to 200 kW. Larger units in the range of 1 to 3 MW have so far proved unsatisfactory. Experience shows that even in very favorable locations, a load factor of thirty percent from a wind generator is a very good result. That means that it would take at least some 13,000 wind generators of 200 kW capacity each to provide the same electric energy as a 1000 MW(e) thermal power station (with eighty percent load factor). The surface needed would again be of the order of 100 square kilometers.

It seems clear that much work is still needed to reduce the costs of solar and wind generators considerably if they are ever to become generally economically competitive. I am not surprised that Mrs. Helga Steeg, Director General of the International Energy Agency in Paris, recently stated that "the contribution of renewable energies, besides hydropower, in most of the IEA countries (i.e., most of the Western industrialized countries) for the year 2010 can be estimated to be a maximum of five percent." This does not negate the fact that these energy forms can be very important in specific regions or situations. Nor do present modest results suggest that we should withhold research and development funds for these energy sources. But they do suggest that it would be irresponsible to count on major global contributions from energy systems that have so far not turned out to be viable on a large scale. . . .

My conclusion is that conservation and renewable energy forms certainly should be supported, but they alone will not be sufficient to counter the greenhouse effect. Before focusing on the relevance of nuclear power, I must briefly discuss hydrogen, which is sometimes mentioned as an alternative to electricity for use in heating and transport. There is a large body of experience with hydrogen. City gas, which has now been replaced by natural gas in many cities, was a mixture of hydrogen and carbon monoxide. In the Soviet Union, an aircraft has been propelled with hydrogen as fuel, and reports reach us that automobile makers experiment with cars fueled by hydrogen. This is all very understandable and interesting. Petrol will have to be replaced one day. Cars may well have to run on electricity or hydrogen. However, hydrogen has to be produced industrially and for that, either electricity or very high temperatures are needed. Both could be achieved through nuclear reactors. Ideas of producing hydrogen by the use of electricity from solar cells covering large areas in the Sahara seem fanciful and, at any rate, decades away. And even if we were to succeed in producing large quantities of hydrogen, there would still remain the major safety problems of transporting, storing, and using it.

Now let me turn to nuclear power. It is recognized that nuclear power plants are capable of producing very large amounts of electric energy without adding CO_2 or SO_2 and NO_x to the atmosphere. Yet, a number of arguments are now mobilized to try to convince the world that nuclear power is totally irrelevant to our needs to generate energy without generating CO_2. Let me discuss some of these arguments.

We hear, for instance, that electricity now only accounts for 29 percent of the total energy consumption and that fossil-fired power plants today contribute only some 12–15 percent to the greenhouse effect. It is implied that it is not worthwhile doing anything in this sector, except saving electric energy.

I submit that this is an uncaring attitude. I shall not claim that a continued and expanded use of nuclear power is a panacea for the problem of CO_2 emissions, but I shall try to show that it can make a contribution that, along with others, is most helpful to contain those emissions. Let me point out in the first place that if the electric energy that was generated from nuclear power last year had instead been produced by coal-fired power plants, this would have given rise to additional emissions of about 1600 million tons of CO_2. This figure is not small when you compare it with the 4000 million tons which the Toronto Conference [in 1988] recommended as a target for reductions by 2005. Let me point out, in the second place, that electricity is a constantly growing part of energy consumption and that it is helpful in efficiently substituting for the direct use of fossil fuels. Even extreme low energy scenarios assume that we will use about twice as much electricity in the world around 2020 as we use now. That would require an *addition* of a 1000 MW(e) plant every 4.4 days on the average

[2500 GW(e)]. There would further be a need for replacements for old and obsolete plants—both nuclear and fossil-fueled. It is not without interest from the viewpoint of CO_2 emissions whether these new plants and replacement plants will be fueled by coal, gas, uranium, or driven by hydro where available. If all the additional electricity that is assumed in the low energy scenario to be needed by 2020 were to be generated by coal, there would be an annual added emission of some 16,500 million tons of CO_2. If, instead, all of this new electricity were to be [generated] by nuclear power or hydropower, there would be no further CO_2 emissions. Such extreme scenarios are not realistic, of course. Both among additional plants and replacement plants there will be a mix of coal, nuclear, gas, and hydro. The figures help to clarify, however, that it is of considerable importance how this mix is made. The more fossil, the more CO_2, the more nuclear and hydro, the less CO_2.

A current standard objection to nuclear power's relevance for CO_2 abatement is that it would require a new nuclear power plant to be put into operation every 2.5 days over thirty-five years to replace coal-fired plants. Such a perspective is evidently meant to deter everybody from devoting a further thought to the possible relevance of nuclear power. As I said a moment ago, however, there will be a mix and it matters how it is constituted. Let me also remind you that in each of the years 1984 and 1985, thirty-three new nuclear power plants actually came into operation, or one new plant every eleven days on the average. There exists a capacity to build many more nuclear plants than we do today.

A further objection to future increased reliance on nuclear power is that the world's uranium resources should be too limited to permit nuclear power to be anything but a short parenthesis. The reality is that already known low-cost uranium resources are sufficient to sustain a much larger nuclear power sector than we now have. If we include the resources which geologists consider likely to exist, we might well fuel ten times more reactors than we now have over their whole lifetimes, even with the present type of reactors and without plutonium recycling.

A perennial argument is that nuclear power plants are unacceptably expensive. There are indeed cases of plants which have become very expensive. The nuclear industry is not the only one in which this happens. Moreover, the statement conveniently overlooks the highly successful and economic nuclear programs in, for example, Canada, Belgium, France, Japan, Sweden, and Switzerland. There have also been studies of the economics of nuclear power by the Nuclear Energy Agency of the OECD in Paris, in cooperation with the Union of Producers and Distributors of Electricity (UNIPEDE) and the IAEA. The results show that nuclear power remains competitive in most OECD countries. Nuclear plants generally cannot compete economically with coal-fired plants built near coal mines.

My conclusion is that the arguments which have been advanced against the use of nuclear power in the face of the CO_2 problem and which I have cited are all thin and contrived. The reality which is becoming increasingly recognized by all who do not choose to close their eyes is that a continued and expanded use of nuclear power will be an indispensable contribution to the efforts to restrain CO_2 emissions.

The real objections to nuclear power—present or expanded—are not new. They relate to safety, waste disposal, and the risk of proliferation of nuclear weapons. Let me deal with them in the reverse order.

The proliferation risk exists whether we have a nuclear power industry that is of today's size or larger, or none. Even a worldwide closing of nuclear power stations would not eliminate the risk and probably not significantly reduce it. Let me remind you that all nuclear-weapon States had their weapons first and power reactors second. Further disarmament, on the other hand, may well promote a fuller commitment to nonproliferation, verified by the IAEA. It should also be remembered that the current nuclear power use, apart from relieving the world of huge quantities of CO_2, reduces the pressures on oil resources. It would take the peak oil production of Saudi Arabia of 1974 to generate the electricity we obtain from nuclear power today (about 400 Mtoe [megatons oil-equivalent]). This effect is not insignificant in a world where competition for oil resources is a major security issue—like the risk of proliferation.

Let me now turn to the question of wastes. All fuel cycles give rise to wastes, for instance, tailings from uranium mining, slag heaps from coal mining, leakage of methane gas, oil spills in the ocean, ashes, etc.

In the nuclear fuel cycle, the amounts of highly radioactive wastes are small. To be a bit provocative, one might say that the wastes are one of the great assets of nuclear power compared to other energy sources. The limited quantities make it possible for us to safely manage and dispose of practically all the waste that arises, and certainly all that is hazardous, in a controlled manner. The operation of all nuclear power plants in the world last year [1988] gave rise to some 7000 tons of spent fuel. If the electricity generated by that fuel had been generated by the combustion of coal, it would have resulted—as I have already mentioned—in the emission of 1600 million tons of CO_2 and tens of millions of tons of SO_2 and NO_x, even with the best flue gas cleaning equipment available. In addition, there would have been some 100,000 tons of poisonous heavy metals, including arsenic, cadmium, chromium, copper, lead, and vanadium. These, of course, remain poisonous forever, and are not isolated from the biosphere.

It is often argued by opponents of nuclear power that we do not know how to isolate the nuclear waste safely. That is not correct. Detailed designs for packaging the wastes, whether in the vitrified form after reprocessing, or as compacted spent fuel assemblies, have been worked out and have been approved by safety authorities. When most countries have decided to store such spent fuel for some thirty to fifty years before either disposing of it as waste or reprocessing it, it is for the reason that allowing the radioactivity to decay simplifies the design of both the waste container and the storage and makes it easier to achieve the safe isolation which is required for a very long time period. . . .

Let me end my comments on nuclear waste by saying that if other industries had as good methods for waste management and disposal as the nuclear power industry does, the world would have far fewer environmental problems.

The concern mostly voiced about nuclear power relates to safety. In the Chernobyl accident large amounts of radioactive substances were released. Some 200–300 persons—operational and firefighting staff—received high doses and suffered radiation sickness. Twenty-nine of them died of their radiation injuries [but see Marples, above], but most of the others are back in productive work after hospital treatment. Over 100,000 persons have been evacuated from their homes and have been settled in other areas. In the restricted zone that was established within a radius of thirty km around the power plant, comprehensive decontamination has been carried out. Some people have moved back, but no general return has been authorized.

Chernobyl was a grave and extremely costly industrial accident by any standard but it was certainly not unique in the number of deaths or injuries. Other ways of generating energy also take their tolls. The oil platform that exploded last year in the North Sea took 165 lives. An explosion in a coal mine in the Federal Republic of Germany, likewise in 1988, killed 57 miners. A gas explosion in Mexico City in 1984 left some 450 dead and thousands of people injured. A dam that burst in India in 1979 killed 15,000 people. And last month, an explosion caused by a leak from a gas pipeline destroyed two trains in the Soviet Union and caused more than 600 deaths. . . .

Although the Chernobyl reactor was very different from most power reactors operating in the world, and this kind of release is implausible on reactors which, like the Three Mile Island reactor, have containments designed to stop releases into the environment, no one contends that there are *no* risks in using nuclear power. However, these risks should be compared to the risks and damage connected with the main alternative way of generating the electricity, namely through the use of fossil fuels.

We should, of course, compare the whole fuel cycles. The health and environmental consequences of the mining of uranium must be compared with those of the mining of coal and the extraction of gas and oil; one should also compare the transportation of each, the fission of the uranium with the burning of fossil fuels, and the emissions from the burning of these fuels with the disposal of nuclear waste. One should compare the number of casualties and health injuries and the amount of environmental damage per quantity of electricity generated. Such examinations are, in fact, under way and from what we have seen so far, they point to a positive picture for nuclear power.

Whether such rational comparisons will help to overcome anxiety about nuclear power is less certain. We have not accepted and tucked away the risks of nuclear power in our minds as we have done with, say, the risks of coal mine accidents. Perhaps the ambition need not be to eliminate all anxiety. After all, lots of people rationally accept flying despite feelings of anxiety. We certainly need to demystify nuclear power—but this may take time. Meanwhile perhaps we should be content merely to get broad acceptance of the use of nuclear power. The anxiety which continues to exist should continue to spur all those who are connected with nuclear power to reduce even further the risks of significant accidents. There must be an increased awareness, however, that if broad acceptance is *not* achieved for nuclear power, we shall have to face the risks that are connected with the alternatives—including their environmental consequences.

Let me conclude: Nuclear power can help significantly to meet growing needs of electricity without contributing to global warming, acid rains, or dying forests.

A responsible management and disposal of nuclear wastes is entirely feasible.

The safety of nuclear power, like the safety of any other industrial activity, must be continuously strengthened through technological improvements and methods of operation.

There is a dire need for more factual information to narrow the gap between the real and the perceived risks of nuclear power and to demonstrate the risks of the alternatives.

4 Biotechnology and Genetically Manipulated Organisms

Bt Corn and the Monarch Butterfly

The late nineteenth century brought about a revolution in biology: Charles Darwin unveiled the idea of evolution by natural selection, and Gregor Mendel provided insight into the nature of inheritance, giving birth to the science of genetics. While these discoveries were profound and startling, human culture had developed the practical application of the mechanisms of selection and genetics for ten millennia. Human manipulation of the genetics of non-human organisms began with the introduction of agriculture 10,000 years ago. The first field cultivators noticed that particular crosses (hybridization) between plants, especially cereal grains like wheat, provided food that was more reliable and more easily harvested than wild-growing plants.

Human cultures have changed the genetics and morphology of plants and animals through the process of artificial selection. Well-known examples include dog and cattle breeds and the many varieties of apples. Natural selection is the process through which different individuals survive longer and reproduce more often because they have different genetically controlled characteristics that cause them to thrive in certain environments. Artificial selection can be thought of as the human-driven subset of natural selection: humans determine which individuals will survive by deciding which seeds to plant or which animals to breed. These decisions are based on whether the organisms possess characteristics favorable to humans, such as sweeter flavor, greater biomass, or gentler disposition. In fact, in his research leading to *The Origin of Species*, Darwin spent years observing commercial experiments in artificial selection, especially pigeon breeding.

The development of strains of wheat and corn, varieties of fruit, and breeds of dogs, pigeons, and cattle were all accomplished through human genetic manipulation before anyone knew what a gene was. The readings in this chapter concentrate on the history of corn (*Zea mays*, known outside of North America as maize), which is the most universally grown and marketed of the food plants to originate in the New World. The indigenous people of Mexico and Central America originally used artificial selection on a wild grass called teosinte to produce plants with larger and more numerous seeds. As the selection from Paul Mangelsdorf relates, the history of the "improvement" of corn is complex, with both indigenous peoples and modern agriculturalists exerting strong influence.

The nature of genetic manipulation has taken an important turn in the past three decades with the introduction of the technology of genetic recombination. Moving beyond cross-breeding between closely related plants or animals, scientists now take genetic material from one species—in the readings below the species is the bacterium *Bacillus thuringiensis* (*Bt*)—and, using a virus as a carrier, insert that genetic material into a species unrelated to it—in this case corn. With *Bt* genes inserted into its genome, the corn plant becomes toxic to caterpillars, several of which are its most common agricultural pests. This allows a greater yield with less negative environmental impact from pesticides.

Rapidly advancing technological progress can bring with it consequences unintended by its developers. Sometimes these unintended consequences can be beneficial, as with the Internet, which was first developed to enable sharing of data and programs on a network. More often, they are detrimental, as with the pesticide DDT, which not only failed to eliminate the insect pests it targeted, but contaminated the natural food chain and caused tremendous reductions in the numbers of top predators such as bald eagles, osprey, and peregrine falcons. Our understanding of the possibility of detrimental unintended consequences has led to the "precautionary principle," which warns against the public use of potentially dangerous technology without adequate testing.

The precautionary principle applies to *Bt* corn because, unfortunately, *Bt* genes are toxic to every sort of caterpillar, not just agricultural pests. Corn, a wind-pollinated plant, sheds large amounts of pollen into the air, some of which lands on the flowers of other corn plants and pollinates them. However, much of the pollen lands on the ground, on the water, or on the leaves of other plants in the vicinity, including milkweeds. In its caterpillar stage, the monarch butterfly, much loved in North America, depends on milkweed leaves for nutrition. The monarch's migration, a story uncovered by Lincoln Brower, is one of the great biological feats. Millions of monarchs, flying from as far away as Canada, descend on the Oyamel fir forests in central Mexico to overwinter. This migration is so unusual, so beautiful, and so awe-inspiring, that the World Conservation Union, which publishes the definitive books on endangered species (known as the "Red Books"), established a unique category for it—the monarch's migration is referred to as an "endangered biological phenomenon." Deforestation of its Mexican habitat raised the first concern for the monarch's endangerment. However, there is a new danger: wind-borne *Bt* corn pollen lands on the milkweed leaves that the monarch larvae eat. Losey, Raynor, and Carter (see below) conducted the initial experiments establishing that *Bt* pollen on milkweed damages the growth and survival of monarch larvae.

The selections provide information on genetic engineering followed by discussions of *Bt* corn and the monarch. This case study is emblematic of the larger issue of genetic manipulation of the world's crop plants and domesticated animals. Is genetic modification of these organisms a panacea for the world's malnourished billions or an ecological disaster in the making?

FURTHER READING

Lewontin, Richard. "Genes in the Food." *New York Review of Books*, 21 June 2001, 84–90.

McHughen, Alan. *Pandora's Picnic Basket: The Potential and Hazards of Genetically Modified Foods*. Oxford: Oxford University Press, 2000.

Mendel, Gregor. *Experiments in Plant Hybridization*. 1865; reprint, Cambridge, Mass.: Harvard University Press, 1963.

Pollan, Michael. *The Botany of Desire: A Plant's Eye View of the World*. New York: Random House, 2001.

Shiva, Vandana. *Stolen Harvest: The Hijacking of the Global Food Supply*. Boston: South End Press, 2000.

Paul C. Mangelsdorf, from "Modern Breeding Techniques" in *Corn: Its Origin, Evolution, and Improvement* (1974)

Improving seeds through experimentation has been one of the dominant trends in the history of agriculture. The change in the seed of the corn plant from its wild relative, teosinte, to its current state is one of the most dramatic examples of this trend. Paul Mangelsdorf, a preeminent corn geneticist, provides a history of this improvement, from pre-Columbian Native Americans to the industrial technologists of the early twentieth century.

The "improvement" of corn kernels has continued unabated. The "Green Revolution" of the 1960s and 1970s has both increased yield per acre and decreased susceptibility to a wide array of corn diseases. Other manipulations of the plant, such as the genetic induction of male sterility, have decreased the labor-intensiveness of crop harvesting, because the male flower tassels no longer need to be removed. As technology advances, however, the ownership of that technology by large corporations has concentrated the profits from corn-growing in the hands of the few, usually corporations and governments, and taken it away from small landholders.

Ironically, corporate growers are now so successful at maximizing corn yields that supply has outstripped demand, prices have fallen, and producers rely on government assistance to stay solvent. Cheap, subsidized corn is used to fatten animals on industrial feedlots, and as Manning describes in Chapter 12, a diet based on corn syrup spells obesity for humans, too.

The mother of maize domestication, like that of many other inventions, was necessity. The plant was probably never abundant in the wild—never in any region the principal component of the native vegetation. And unlike the sweet potato and cassava, which when harvested renewed the colony from fragments of roots overlooked by the diggers, a colony of maize might be completely extinguished by a single thorough picking. As the Indians found their resources of wild maize diminishing on the one hand and on the other observed how much more productive than wild plants were the "volunteer" maize plants which grew up in the well-fertilized soil around their dwellings from accidentally dropped seeds, it was but a step to the conscious sowing of seed in selected sites. Removal of competing vegetation was an obvious second step or indeed might have actually have been the first, with sowing coming later. . . .

At some stage in the development of Indian agriculture . . . the practice of artificial selection began and has persisted to this day. The Hopi, it is said, never save seed from an ear which shows evidence of mixture. The care and maintenance of particular varieties was, in earlier times, kept largely within families, the responsibility for a variety being handed down within the family from generation to generation. . . .

Among pure-blooded non-Spanish-speaking Indians of Guatemala, rigid selection for type of seed is often practiced. . . . The practice of preserving distinct types accounts in part for the fact that in Guatemala there is a high correlation between the diversity of maize and the percentage of Indians in the population. In the nine departments of Guatemala in which the proportion of Indians, "Indigenas," as they are called in the Guatemalan census, is two-thirds or more, are found all except one of the Guatemalan races of maize.

The Guatemalan practice of rigorous selection for type, like the Guatemalan maize itself, may be a cultural trait introduced from South America. There, especially in Peru, selection has become an art. I have seen at intermediate altitudes in Peru the harvest of fields of Kculli, the black corn used for dying and for coloring nonalcoholic *chichas* and the puddings, *mazomorras*, made from corn and tapioca flours, laid out in the sun to dry with not a single color deviant in sight. These have already been removed from the drying area and set off in a pile to one side. The pile of off-type ears is usually remarkably small and testifies to the effectiveness of previous generations of rigid selection. In the race Cuzco, seven types differing in endosperm texture or pericarp [seed coat] color are maintained in states of relative purity. Several of these involve mosaic or variegated pericarp which is controlled by an unstable locus, and there is little possibility of ever establishing a completely pure race with such a pericarp pattern. Yet varieties are maintained in which the great majority of the ears are of this type.

The Peruvian Indian must be given credit also for developing other special-purpose types of corn such as the sweet corn race Chullpi, which is especially prized for making the native beer *chicha*, its high sugar content contributing to developing a high alcoholic content in the final product. The Peruvian Chullpi is the progenitor of the Colombian and Mexican Maíz Dulce, and . . . the latter may have played a role in the origin of modern sweet corn varieties. The most spectacular product of selection in Peru, however, is the race Cuzco Gigante, the large-seeded flour corn of the Urabamba Valley of the Cuzco region. Repeated hybridization has undoubtedly played a part in the evolution of this race, but human selection for two characteristics, floury endosperm and size of seed, have also been important. The Peruvians have given particular attention to these two characteristics because most of the maize they use for food is consumed as individual, unprocessed kernels to which heat is applied. In contrast, Mexican Indians, whose staff of life is an unleavened bread, the *tortilla*, make it by steeping the maize in limewater, rubbing off the pericarp, grinding the hulled kernels, and fashioning the dough into round, flat sheets which are baked on hot metal surfaces. The Peruvian method, when used on small corneous [hard] kernels of wild maize or maize in the early stages of domestication, causes the kernels to explode or "pop," thus instantly converting the hard, bony seed, too difficult to chew, into a tender, tasty morsel, nutritionally the counterpart of a whole-grain bread.

As kernel size increased, largely because of hybridization between races, the kernels lost their ability to pop and when exposed to heat merely became parched. Parched flour corn is more easily chewed than parched flint corn, and when parching became the predominant method of processing, selection for floury endosperm became a rewarding practice, and, since the individual kernel was still the unit of consumption, selection for large seeds was undoubtedly also practiced. . . . Following the creation of new genetic diversity resulting from the hybridization of a large-seeded Peruvian type with a fairly large-seeded race, Tabloncillo from Mexico, several more centuries of selection have culminated in the production of Cuzco Gigante, the corn with the world's largest kernels.

By this time boiling had largely replaced parching as a means of preparing maize for food. I have seen groups of Peruvian Indian farm laborers making their entire noonday meal from the boiled grains, eaten individually, of Cuzco maize. Should the modern corn breeder seek to improve kernel size in maize, he will find the genes needed for that purpose assembled for him in the large-seeded varieties of Peru, the product of centuries of selection on the part of the Indians. . . .

Although it may have been unconscious on his part, the Indian was nevertheless a maize hybridizer on a grand scale. Through the centuries he brought distinct races together repeatedly, with the result that some have very complex pedigrees. The modern corn breeder will wish to examine these to identify, and in some instances to employ for breeding material, those races which have appeared repeatedly in the ancestry of highly successful modern races. The breeder may also profitably give some attention to those races which, highly successful in some parts of the world, seem not yet to have been generally employed in modern breeding operations. . . .

THE HISTORY OF HYBRID CORN

. . . Two distinct lines of descent have converged to make hybrid corn an accomplished fact, and a third line has had a marked influence. . . .

One line of descent in the genealogy of hybrid corn begins with Charles Darwin, who made extensive investigations on the effects of self- and cross-pollination in plants. Included among his experimental subjects was the corn plant. His were the first controlled experiments in which crossed and self-pollinated individuals were compared under identical environmental conditions. He was the first to see that it was the crossing between unrelated varieties of a plant, not the mere act of crossing itself, that produced hybrid vigor, for he found that when separate flowers on the same plant and different plants of the same strain were crossed their progeny did not possess such vigor. He concluded quite correctly that the phenomenon occurred only when diverse heredities were united.

Darwin's experiments [became known to the American William Beal]. At Michigan State College Beal undertook the first controlled experiments aimed at the improvement of corn for the utilization of hybrid vigor, or as he preferred to call it, "controlled parentage." He selected some of the varieties of flint and dent corn, then commonly grown, and planted them together in a field isolated from other corn. He removed the tassels—the pollen-bearing male flower clusters—from one variety before pollen was shed. The female flowers of these emasculated plants had then to receive their pollen from the tassels of another variety growing in the same field. The seed borne on the detasselled plant, being crossed seed, produced only hybrid plants the following season.

The technique Beal invented for crossing corn—planting two kinds in the same field and removing the tassels of one—proved highly successful. . . . As a device for increasing yields of corn, this method of crossing two open-pollinated varieties—each genetically heterogeneous—was not completely effective. Many of the crosses were more productive than their parents but seldom enough so to justify the time and care spent in producing the crossed seed. The missing requirement—the basic principle that made hybrid corn practical—was discovered by George H. Shull at the Carnegie Institution. . . .

[Danish botanist Wilhelm Johannsen had already shown] that in self-fertilized plants, such as the bean, the progeny of a single plant represents a "pure line" in which all individuals are genetically identical and in which any residual variation is environmental in origin. . . .

Johannsen's pure-line theory has been widely applied to the improvement of cereals and other self-fertilized plants. Many of the varieties of wheat, oats, barley, rice, sorghum, and flax grown today are the result of sorting out the pure lines in mixed agricultural races and identifying and multiplying the superior ones. We also owe to Johannsen the concept—a working tool of every plant and animal breeder—of the "genotype," which represents the individual's hereditary endowment, and the "phenotype," which represents its physical characteristics.

Shull's contribution was to apply the pure-line theory to corn with spectacular though unpremeditated results. His experiments started with the objective of analyzing the inheritance of quantitative or "blending" characteristics, and he chose as an inherited quantitative character suitable for study the number of kernel rows on ears of corn. Since corn is naturally a cross-pollinated plant, he practiced artificial self-pollination in order to produce lines breeding true for various numbers of kernel rows. These lines, as a consequence of the inbreeding resulting from self-pollination, declined in vigor and productiveness and at the same time each became quite uniform. Shull concluded correctly that he had isolated pure lines of corn . . . Then as a first step in studying the inheritance of kernel row number he crossed these pure lines. The results were surprising and highly significant. The hybrids between two pure lines were quite uniform like their inbred parents, but, unlike their parents, they were vigorous and productive. Some hybrids were definitely superior to the original open-pollinated varieties from which they had been derived. Inbreeding had isolated from a single heterogeneous species the diverse germinal entities whose union Darwin had earlier postulated as the cause of hybrid vigor.

Shull recognized at once that inbreeding followed by crossing offered an entirely new method of improving the yield of corn. . . . He proposed the isolation of inbred strains through self-pollination as the first step and the crossing of two selected inbred strains as the second. Only the seed of the first generation cross was to be used for crop production because hybrid vigor is always at its maximum in the first generation.

Shull's idea of maintaining otherwise useless inbred strains of corn solely for the purpose of utilizing the heterosis [hybrid vigor] resulting from their hybridization was revolutionary as a method of corn-breeding. As a genetic system it is comparable to types of adaptive polymorphism [benefiting by having many forms] . . . in which natural selection (in the case of hybrid corn, selection acting in a manmade environment) preserves certain chromosomes, not primarily because of their intrinsic worth but because they interact effectively with other chromosomes similarly preserved to produce a highly successful Mendelian population. . . .

One further major development was needed to make hybrid corn practical and the great boon to agriculture which it has become. . . .

[Donald] Jones's solution was simply to use seed from a double cross instead of a single cross. The double cross, which combines four inbred strains, is a hybrid of two single crosses; for example, two inbred strains, A and B, are combined to produce the single cross A × B. Two additional inbred strains, C and D, are combined to produce a second single cross, C × D. All four strains are now brought together in the double cross (A × B) × (C × D). At first glance it may seem paradoxical to solve the problem of hybrid seed production by making three crosses instead of one. But the double cross is actually an ingenious device for making a small amount of scarce single-crossed seed go a long way. Whereas single-crossed seed is produced on undersized ears borne on stunted inbred plants, double-crossed seed is produced on normal-sized ears borne on vigorous single-cross plants. A few bushels of single-crossed seed can be converted in one generation to several thousand bushels of double-crossed seed. The difference in cost of the two kinds of seed is reflected in the units in which they were formerly sold: double-crossed seed was priced by the bushel, single-crossed seed by the thousand seeds. Double-cross hybrids are never as uniform as single crosses, but they may be just as productive or more so.

John E. Losey, Linda S. Raynor, and Maureen E. Carter, "Transgenic Pollen Harms Monarch Larvae" (1999)

For plant reproduction to take place, the pollen from a male flower must find its way to the pistil of a female flower. If all is successful, the ovary of that pistil matures into the fruit. Pollen transport tends to take one of two forms—wind dispersal or animal-borne (usually bee, moth, fly, or bird) dispersal. Plants that are wind pollinated tend to have separate male and female flowers and almost always produce far more pollen than is necessary for the number of fertilizations that actually take place. In other words, a tremendous abundance of pollen is scattered to the winds, landing on lakes, roadside plants, car windows, and the flowers of plants of dozens of different species. Corn is a wind-pollinated plant, which has a cluster of male flowers (the tassel) at the top, and clusters of female flowers (on the ears) below.

Milkweed, on the other hand, is bee pollinated, and produces pollen in small packages for the bees to carry between milkweeds of the same species. It is nearly impossible to find milkweed pollen anywhere but on a milkweed plant or a bee. The monarch caterpillar, though an occasional pollinator of the milkweed plant, is not an effective pollinator; on the contrary, it is a predator, spending all of its immature life chomping away on milkweed leaves. These leaves are highly concentrated with cardiac glycosides, a chemical toxic to most animals. This specialized relationship between the monarch and milkweed is the reason that monarch butterflies are avoided by birds, who are poisoned when they try to eat one.

Although plants transformed with genetic material from the bacterium *Bacillus thuringiensis* (*Bt*) are generally thought to have negligible impact on non-target organisms, *Bt* corn plants might represent a risk because most hybrids express the *Bt* toxin in pollen, and corn pollen is dispersed over at least sixty meters by wind. Corn pollen is deposited on other plants near corn fields and can be ingested by the non-target organisms that consume these plants. In a laboratory assay we found that larvae of the monarch butterfly, *Danaus plexippus*, reared on milkweed leaves dusted with pollen from *Bt* corn, ate less, grew more slowly, and suffered higher mortality than larvae reared on leaves dusted with untransformed corn pollen or on leaves without pollen.

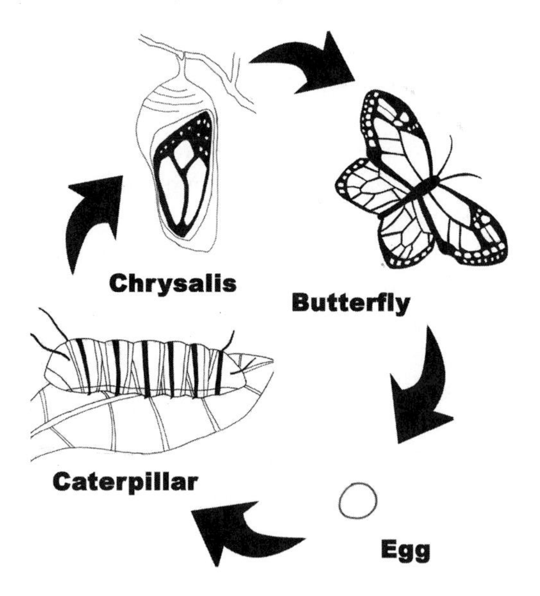

Figure 4.1 The monarch butterfly lifecycle. Drawing by Erin Sheley.

Larval survival (fifty-six percent) after four days of feeding on leaves dusted with *Bt* pollen was significantly lower than survival either on leaves dusted with untransformed pollen or on control leaves with no pollen. Because there was no mortality on leaves dusted with untransformed pollen, all of the mortality on leaves dusted with *Bt* pollen seems to be due to the effects of the *Bt* toxin.

There was a significant effect of corn pollen on monarch feeding behavior. The mean cumulative proportion of leaves consumed per larva was significantly lower on leaves dusted with *Bt* pollen and on leaves dusted with untransformed pollen compared with consumption on control leaves without pollen. The reduced rates of larval feeding on pollen-dusted leaves might represent a gustatory response of this highly specific herbivore to the presence of a "non-host" stimulus. However, such a putative feeding deterrence alone could not explain the nearly twofold decrease in consumption rate on leaves with *Bt* pollen compared with leaves with untransformed pollen.

The low consumption rates of larvae fed on leaves with *Bt* pollen led to slower growth rates: the average weight of larvae that survived to the end of the experiment on *Bt*-pollen leaves was less than half the average final weight of larvae that fed on leaves with no pollen.

These results have potentially profound implications for the conservation of monarch butterflies. Monarch larvae feed exclusively on milkweed leaves; the common milkweed, *A. syriaca*, is the primary host plant of monarch butterflies in the northern United States and southern Canada. Milkweed frequently occurs in and around the edges of corn fields, where it is fed on by monarch larvae. Corn fields shed pollen for eight to ten days between late June and mid-August, which is during the time when monarch larvae are feeding. Although the northern range of monarchs is vast, fifty percent of the summer monarch population is concentrated within the midwestern United States, a region referred to as the "corn belt" because of the intensity of field corn production. The large land area covered by corn in this region suggests that a substantial portion of available milkweeds may be within range of corn pollen deposition.

With the amount of *Bt* corn planted in the United States projected to increase markedly over the next few years, it is imperative that we gather the data necessary to evaluate the risks associated with this new agrotechnology and to compare these risks with those posed by pesticides and other pest-control tactics.

Lincoln Brower, from "Canary in the Cornfield: The Monarch and the *Bt* Corn Controversy" (2001)

The protection of the environment in the United States is the responsibility of a wide range of institutions, including Congress, the Presidency, the Supreme Court, the Environmental Protection Agency (EPA), the United States Department of Agriculture (USDA), the United States Fish and Wildlife Service, parallel agencies and departments from the fifty states and their municipal subdivisions, corporations, non-governmental organizations, citizens groups, universities, and the media. In the story that follows, Brower claims that a consortium of biotech corporations, the EPA, the USDA, academia, and the media worked together to block the research of Losey, Raynor, and Carter (see selection above) about the adverse effects of Bt *pollen on monarch caterpillars. Brower's title alludes to the fact that miners used canaries to test for toxic air; the canary has come to symbolize early warning.*

In their article, "Transgenic Pollen Harms Monarch Larvae" [see selection above], the Cornell authors asked: Could windblown corn pollen accumulate on plants that grow extensively in and adjacent to cornfields and, like conventional insecticides, inadvertently kill native insects that are not pests? To test this question, they chose the monarch as their nontarget species. Female monarchs lay eggs on wild milkweed plants, the only plants that their caterpillars can eat. In their experiment, conducted in the laboratory, the authors dusted pollen gathered from one of the *Bt*-corn strains onto the leaves of the common milkweed. They established that caterpillars that fed on the dusted leaves ate less, grew more slowly, and suffered higher mortality than caterpillars reared on milkweed leaves dusted with pollen from a non-*Bt*-corn strain. The scientists were circumspect about their results and stated clearly that more research was needed to determine the impact of the toxic pollen on monarchs in the natural environment.

According to a private communication, an ashen-faced president of a major biotech company marched into a board meeting shortly after the article appeared and stated, "I have only one thing to say about the Cornell publication: Bambi." Had the scientists chosen a different insect, it is likely that few people would have responded to the *Nature* article [Losey, Raynor, and Carter's paper]. They used the monarch, however, loved by schoolchildren, gardeners, and millions of other people throughout the world. The monarch instantly became a bête noire for the field of biotechnology. The world press latched onto the study even before the article was in print, and soon protesters wearing corn and butterfly costumes were marching in the streets.

The agricultural industry's reaction to the news was immediate and vigorous. Criticisms belittling the Cornell study appeared widely in the U.S. and on British television. Agricultural companies launched web pages (for example, on monsanto.com, novartis.com, and farmsource.com) downplaying and, in some instances, ridiculing the study. The principal argument they put forward was that the benefits of using *Bt* corn far outweigh the environmental costs of the pesticides it replaces. Their most common assertion—that *Bt* corn reduces the need for other insecticides in cornfields by two orders of magnitude [about 100-fold], a gross exaggeration—was repeated in press releases and uncritically accepted by numerous scientists. This same justification was used in articles favoring the new technology that appeared in respected journals, such as the *Proceedings of the National Academy of Sciences, U.S.A.*

The Cornell study mobilized the environmental community at a critical time because the earlier approval of *Bt* corn was about to expire, and the EPA was required to undertake a reassessment process before renewing the registration. The Union of Concerned Scientists and the Environmental Defense Fund petitioned the EPA to restrict the planting of *Bt* corn and to reassess the environmental risks of genetically engineered crops. The environmentalists' initiative made it clear that further scientific study of the relationship between the monarch and *Bt* corn was needed before a ruling could be made. From this point on, however, scientific efforts to define that relationship would be overshadowed by the agricultural industry's efforts to control the information on which the EPA decision would be based.

The industry's early responses to the Cornell paper were designed to cast doubt on whether the scientists' laboratory findings were applicable to monarch caterpillars in the field. Many statements were misleading, fanciful, and betrayed an ignorance of the monarch's natural history. Incorrect or speculative pronouncements fed to the media included that the major geographic area of monarch reproduction lies outside the corn belt; that monarchs breed before pollen is released from the corn tassels; and that pollen release occurs over too short a time to have a major impact on the caterpillars. All these industry-released statements ignored the extensively documented literature on the monarch's lifecycle, including information known since the nineteenth century that multiple overlapping generations of the monarch occur throughout the summer breeding range, virtually assuring that the monarch caterpillars would be widely exposed to the shedding corn pollen. Other press reports asserted that few pollen grains land on milkweed leaves, that monarchs lay most of their eggs on the undersurfaces of the leaves, that milkweed leaves have slick surfaces to which corn pollen grains will not stick, that the toxicity of the pollen grains is below the threshold that kills monarch larvae, and that one hundred times more monarchs are killed by cars and trucks than by *Bt* corn. The most flagrant lack of scholarship exhibited by the *Bt*-corn proponents was their failure to cite the current scientific literature documenting that extensive monarch breeding occurs throughout the North American corn belt.

The agricultural industry's manipulation of the press was soon made even clearer. Several corporations, including the Monsanto Company, Novartis AG of Switzerland, and the Pioneer Hi-Bred of DuPont Company formed a soothingly named consortium, the Agricultural Biotechnology Stewardship Working Group (ABSWG). The ABSWG contacted university scientists and provided funding for studies that would address issues raised by the Cornell findings. U.S. and Canadian scientists conducted a research program during the summer of 1999, the results of which were to be presented at a scientific symposium in Chicago on November 2, hosted by the ABSWG, and also attended by representatives of the EPA and the USDA. The avowed purpose of this symposium was for the scientists to present and discuss their findings, review their methodologies, and determine through consensus what information was inconclusive or missing. . . .

Because of the hurried nature of their summer research, all of the meeting participants prefaced their scientific presentations with the caveat that their data and conclusions were preliminary. Some results indicated possible major impact, others suggested minor impact, and most agreed that the current research base could not resolve the problem. . . .

At the meeting, Carol Yoon, a *New York Times* science journalist, made a stunning announcement: she had just received a fax from her *Times* editor indicating that a media advisory had been released earlier in the day. The headline describing the still-in-progress meeting stated: "Scientific Symposium to Show No Harm to Monarch Butterfly." Several of the participating scientists whose studies were supported by ABSWG had apparently agreed on the contents of the misleading press statement prior to the symposium. There was now no doubt that the symposium had been co-opted by the ABSWG, and that the press was being manipulated. Yoon's report exposing this fiasco, "No Consensus on the Effects of Engineering on Corn Crops," was published in the *Times* on November 4.

A little more than a month later, on December 8, 1999, the EPA held a Scientific Advisory Panel meeting, a requirement of the EPA regulatory process leading to renewal or denial of re-registering *Bt* corn for commercial use. . . . I related that the results of the Chicago meeting had been inconclusive and obfuscated by the Agricultural Biotechnology Stewardship Working Group. Another testimony, by a scientist representing one of the agricultural companies, was a vituperative commentary on both the Cornell results and another recent *Nature* paper documenting that *Bt* toxin can leach from the corn plants into the soil. A clear pattern was emerging: corporate spokespeople will attack scientists who discover any potentially adverse environmental effects of GMO crops. . . .

Aware that new data and more sophisticated analyses would be forthcoming, the Union of Concerned Scientists and eleven other public-interest organizations made a request to the EPA: to postpone the next Scientific Advisory Panel (SAP) meeting until more data were collected and made available to the public, including the scientists' findings gathered over the summer of 2000. The EPA, however, held the SAP meeting on October 19, a month before the scientific symposium was scheduled to take place. Prior to the SAP meeting, the EPA had allowed several corporations to review the agency's preliminary assessment and suggest modifications. In addition, the EPA allowed the companies to withhold important data as confidential business information. One of the principal documents contained approximately forty deletions of so-called "proprietary" data. It was therefore impossible for the EPA panel or independent scientists to evaluate the data. Both industry and the EPA documents also ignored relevant data readily available in the scientific literature. Thus, without considering the new information that would be presented the following month, and drawing passages almost verbatim from documents prepared by industry, the EPA's interim assessment of the risks and benefits presented to the SAP stated that "the published preliminary monarch toxicity information is not sufficient to cause undue concern of harmful widespread effects to monarch butterflies at this time."

The Summer 2000 research results were presented in November at a second Chicago symposium, attended by many of the same industry, academic, and governmental groups that had been present at the 1999 symposium. Investigations examined the toxicity of the various strains of *Bt* pollen, when and where monarchs feed and breed, and where they encounter the pollen. Some of the findings seemed reassuring. Toxicity studies appeared to indicate that the pollen of some strains of *Bt* corn was less lethal than that of others and that most of the strains currently in use may be in the less-toxic group. Several studies indicated that corn pollen does not drift very far from the cropfield, and a risk analysis using the new data predicted little effect on larvae feeding on milkweeds beyond a few meters from the edge of a field. Other studies warned of new threats. One determined *Bt* pollen to be toxic to later-stage monarch larvae—significant because older caterpillars had been assumed to be less sensitive than the young ones. Clarifying a contentious point of the 1999 symposium, new data fed into revised computer models now led to predictions that pollen shedding and monarch breeding happen simultaneously over wide geographic areas. This finding was made all the more important by new data showing that extensive monarch breeding occurs on milkweed growing inside cornfields. This, in turn, underscored the devastating effects that the long-term use of herbicides, and ge-

Figure 4.2 Monarch butterflies killed by a frost in Guanajuato, Mexico. Courtesy of Naomi Pierce.

netically manipulated organisms such as Round-up Ready crops, will have as their use totally eliminates milkweeds from the fields.

The papers presented at this symposium reflected the complexities of the *Bt*-corn issue. Working with different methodologies even in areas where their investigations overlapped, the scientists' findings were not easily compared. The studies, for example, used different techniques for collecting and testing pollen samples and for controlling contamination by other vegetable matter. In addition, none of the studies addressed. . . . recommendations that toxicology tests were needed to determine whether sublethal doses of pollen ingested by larvae affect reproduction or migratory capacities of adult butterflies. In summary, despite the EPA's interim assessment, the overall database that had been assembled through November 2000 was not adequate to resolve whether *Bt* pollen is a significant detriment to the monarch butterfly.

A major issue that emerges from the *Bt*-corn debate is the way in which scientific information is obtained and used in the federal regulatory process—a question with consequences far greater than the decision to register or ban *Bt* corn. As the handling of the monarch saga has shown, the EPA's October 2000 decision was based on scientific information that was largely controlled by the industry and failed to measure up to even minimum standards adhered to by the international scientific community. These standards require peer review of manuscripts by independent scientists chosen by the editorial boards of scholarly scientific journals. Peer review assures that experiments are reproducible, that the data are statistically valid, that the conclusions are logically derived from the data, and that they state clearly what is and what is not resolved. This independent evaluation of scientific evidence is a *sine qua non* [essential condition] for the integrity of science. By ignoring the standard of peer-reviewed science and by relying on information supplied by the same corporations that it means to regulate, the current U.S. federal regulatory system is severely flawed.

The *Bt*-corn issue has raised public concerns about the system by which the federal government evaluates the safety of genetically engineered products. The process that will finally determine the commercial fate of *Bt* corn is the same one that is applied to every one of the thousands of toxic chemical products and genetically modified organisms that fall under the jurisdiction of our nation's regulatory system. This is the system warned of in *Silent Spring*. It is the system that Wendell Berry described more than thirty years ago. Will North American society ever face up to the environmental and cultural erosion caused by the cozy economic relationships of agriculture, business, government, and large segments of academia? . . .

It seems certain that the profit-driven mindset of our political and corporate leaders will continue to promote biotechnology, and to fuel unsustainable human population growth with its consequent usurpation of natural habitats and their rich arrays of natural creatures, large and small.

Molly Lesher, "Seeds of Change" (2004)

While U.S. consumers have largely accepted genetically modified ingredients in their diet, consumers abroad and environmental groups have reacted differently. To some, the decision to eat genetically modified foods may seem like a straightforward choice. But as time passes, creating truly segmented markets for conventional and genetically modified foods becomes increasingly difficult and costly.

Biotechnology was born almost 150 years ago in the monastery garden of Gregor Mendel, who bred and crossbred pea plants to create new combinations of height, color, and shape. The insights he derived about genetic inheritance eventually allowed plant breeders in the twentieth century to create higher-yielding "hybrid" seeds. When combined with chemical inputs, hybrid seeds increased many crop yields dramatically.

In the 1970s, scientists pushed past Mendel's crossbreeding techniques when they discovered how to remove genes from one organism and insert them into another completely unrelated one, creating life forms that could not otherwise occur. This new, more precise approach opened the door to a wide range of possibilities for new and improved agricultural products by gene swapping among plants, animals, and organisms such as bacteria. Modern biotechnology now allows lab technicians to implant an Arctic flounder gene that resists cold temperatures into a strawberry plant to defend against frost.

Yet the technology also raised concerns. In 1994, a small biotech company introduced the first genetically engineered food into U.S. supermarkets—the FlavrSavr tomato. The novelty of this tomato was that it would continue to ripen after being picked, without softening and while maintaining a deep red color and sweet taste. But, as with many genetically modified organisms, scientists spliced a gene marker into the tomato to indicate whether the target trait implanted correctly. In the case of the FlavrSavr tomato, the gene marker consisted of the target trait (delayed ripening) and the marker trait (antibiotic resistance). Researchers then grew the tomato plant in a mixture of water and antibiotics; if the plant lived, they knew that the delayed ripening trait had inserted correctly.

But gene markers contain proteins that become part of the plant, and unless processing destroys them, we consume the new proteins in our food. This led some consumer advocates to worry that people eating the FlavrSavr tomato might develop resistance to medication. So while the tomato tasted better than the average grocery-store variety, safety concerns dominated, and the tomato disappeared. Concerns about gene markers largely abated over time, but most research into producing higher-quality foods shifted to the back burner, focusing instead on making production cheaper, easier, and less polluting.

Seed developers fared better with two other types of genetic modifications—herbicide tolerance and insect resistance. Herbicide-tolerant crops contain an extra enzyme that renders the plant resistant to a particular herbicide. This allows farmers to remove weeds by spraying herbicide over an entire field, rather than taking care to distinguish the weeds from the soybean plants. Insect-resistant crops are genetically modified to contain the soil bacterium *Bacillus thuringiensis* (*Bt*), which kills corn borers and cotton worms. Because applying the insecticide can harm farmers, many prefer to plant genetically modified seeds to reduce the danger from spraying and inhaling the chemicals.

Today, herbicide-tolerant soybeans and canola and insect-resistant corn and cotton dominate genetically modified agriculture. These crops are grown in Argentina, China, South Africa, Canada, and the United States. In 2003, U.S. producers planted two-thirds of the global harvest of genetically modified crops, with 81 percent of U.S. soybean and 40 percent of U.S. corn acreage devoted to some form of genetically modified production. But how much do genetically modified seeds actually reduce farmers' costs? . . .

Soybean farmers decreased their exposure to the most toxic chemicals and reduced the number of herbicide applications, both cutting labor costs and markedly diminishing harm to themselves and consumers. However, genetically modified soybean yields are currently about three percent less than some conventional hybrids because it takes time to integrate the trait into higher-yielding varieties. Moreover, agricultural policy analysts note an increase in the amount of herbicides used in soybean production between 1995 and 1998—the most current data available—controlling for the growth in soybean acreage. At least part of this increase could be attributed to the fact that herbicide-tolerant soybean farmers spray herbicide less selectively and thus use more of the chemical. But while lower yields and higher overall herbicide use decreased farmers' anticipated cost savings, farmers still cut costs, as genetically modified soybean acreage continues to increase.

Farmers planting insect-resistant corn, on the other hand, anticipated cost reductions from decreased pesticide use and increased yields. But the evidence on actual cost savings is less clear. These farmers face a complicated cost calculation because corn borer infestations, unlike weed levels, fluctuate widely. From year to year, farmers do not know how much damage corn borers will cause, making it hard to know whether buying the higher-priced seeds will be profitable. Moreover, unpredictable insect levels led U.S. farmers to spray corn borer insecticides on only five percent of their fields prior to the introduction of insect-resistant seeds. This may explain why insect-resistant corn seeds have not significantly reduced average pesticide use—and thus average costs. . . .

Scientists are also looking closely at the effects of genetically modified agricultural products. The most serious environmental threats involve a loss of valued species and the movement of genetically modified genes to non-target plants, insects, and animals. Concerns about the disappearance of certain species first emerged when Cornell University entomologist John Losey published a 1999 study in *Nature* suggesting that the pollen from insect-resistant corn harms the monarch butterfly, which like the corn borer and the cotton worm, evolves through a caterpillar stage of development. This finding galvanized both environmental and consumer advocacy groups. They worried that other insects and animals could also be at risk, which could lead to an uncertain alteration to the delicate balance of ecosystems. [See selection above.]

Gene flow—the transfer of genetically modified genes to non-target organisms by natural processes, such as drifting pollen—also has raised concerns. Some worry that insects could develop a resistance to the *Bt* insecticide, or that herbicide-tolerant genes could spread to wild weeds to produce new breeds that would be increasingly difficult to eradicate. Although resistance is a natural part of evolution, and organisms instinctively become immune to chemicals that would have killed their ancestors, opponents of genetic modification raise the specter of "super" insects and weeds impervious to traditional chemicals. Gene flow also makes conventional and organic farmers uneasy, as pollen from genetically modified plants could drift into their fields rendering them unable to sell their products as "non-genetically modified" or "organic." And farmers planting genetically modified seeds worry about being sued if pollen drift from their fields is responsible for this intermingling.

Gene flow of this sort also reduces the possibility that marketers can untangle the mix of genetically modified and conventional products in the U.S. grain distribution system. And handlers compound the problem because they are not always equipped to accurately segregate the grains once they arrive at storage and transport facilities. The inadvertent mixing of genetically modified and conventional stocks has already caused trouble. In 1998, the Food and Drug Administration (FDA) approved genetically modified Starlink corn for use in animal feed, but it withheld approval for human consumption because of concerns that humans might be allergic to a new protein that it contained. Yet Starlink corn somehow found its way into taco shells (October 2000) and into bread rolls (March 2001). The ensuing controversies forced the maker of Starlink to discontinue its production at a loss of millions of dollars and to recall almost 300 food products from around the globe.

The Starlink episode raises another key issue—food safety. Gene markers, the downfall of the FlavrSavr tomato, were the first food safety concern associated with genetically modified foods. Gene markers can code for just about any trait, but antibiotic-resistant markers are inexpensive and easy to use, making them standard in agricultural biotechnology research. Today, most experts believe that gene markers pose few risks to humans, including resistance to antibiotics.

Yet scientists don't always know how new genes will function within a plant, in other organisms up the food chain, and ultimately in the human body. Because interaction effects are not always predictable, some worry that newly formed proteins will cause unforeseen and possibly dangerous human allergic reactions. And if products are not labeled, as in the United States, it is difficult to guard against allergens. This led researchers at the National Academy of Sciences (NAS) to study whether genetically modified foods are more dangerous than foods altered by other means. They conclude that all foods containing new genetic combinations should be examined for safety, regardless of whether the changes occurred by conventional breeding, genetic engineering, or another such method. But NAS researchers also find that the chances of unanticipated genetic changes—like new allergens—increase as the relationship between the target gene and the host grows more distant. . . .

In the United States, consumer sentiment appears favorable. The International Food Information Council (IFIC), an organization that communicates scientific information about food safety and nutrition to consumers, has surveyed how the U.S. public feels about genetically modified foods since their introduction. The IFIC asks participants whether they think "[agricultural] biotechnology will provide benefits for you and your family within the next five years." In 2003, a majority of those surveyed—62 percent—believed that the technology would provide benefits. However, this is down from 78 percent in 1997. U.S. consumers also think that firms should inform them if they are eating genetically modified products. Researchers at California Polytechnic State and the National University of Ireland found that 81 percent of respondents feel that mandatory labeling for genetically modified foods is "somewhat" to "very" important.

Other parts of the world seem more skeptical—like residents of the European Union, a major U.S. trading partner. The latest Eurobarometer survey on biotechnology sampled 16,000 European consumers and found that acceptance of genetically modified foods has continued to decline. Yet support also varies a great deal by country—81 percent of Greeks oppose genetically modified foods, as compared with only 30 percent of Spaniards. This consumer opposition led E.U. officials to introduce a stringent regulatory regime for genetically modified imports this year, although other motives aimed at punishing the United States may have also played a role.

Different cultural values at least partially explain the disparate consumer attitudes in the United States and Europe. Longstanding cultural mores about food, for instance, affect consumer sentiment. In Europe, native dishes and cooking styles are traditions that residents hold dear. Europeans spend more time than their U.S. cousins do on food preparation and a larger percentage of their budget on food (controlling for higher food prices), and three-hour meals are not uncommon. In contrast, U.S. consumers value quick service and convenience.

Cultural beliefs about how society should balance technological innovation with the environment also influence consumers' purchasing decisions. Researchers at Ghent University found that Belgian consumers rejected genetically modified foods primarily because of negative attitudes toward biotechnology generally, rather than from a consideration of the pros and cons of a particular genetically modified food item. Further, Europeans appear to value environmental preservation more than those in the United States do, as demonstrated by Europe's early support for recycling and the Kyoto Protocol. Meanwhile, U.S. consumers tend to adopt new technologies more quickly than do Europeans (apart from the cell phone), as shown by high U.S. adoption rates of home computers and personal digital assistants.

Recent experiences may be another key factor. Many U.K. consumers are dubious about their government's ability to regulate food, in part because their memories of its mishandling of the mad cow epidemic still resonate. In fact, a 2001 survey by researchers at the University of Illinois asked U.K. and U.S. consumers whether they believed that "the government ensures the safety of the overall food supply." The survey found that only 25 percent of U.K. consumers trust their government to guarantee food safety, compared to 76 percent in the United States.

Differences like these will make selling genetically modified foods in Europe—and perhaps in other parts of the world—difficult. Consumers in Japan, Australia, New Zealand, South Korea, and Indonesia have also been wary of biotech foods, and as a result their governments have imposed import restrictions.

Moreover, these attitudes may be slow to change. Across cultures, food choices are bound up in many parts of life—religion (the sacred cow for Hindus), cultural identity (apple pie for U.S. residents), and social cohesion (cappuccino for Italians). As a result, the foods we eat often evoke an intimate and deeply emotional response. For instance, in the United States guinea pigs are pets; in parts of South America they are gastronomic delicacies. And advertising may not be able to easily change such closely held views. In fact, U.S. and European researchers find that increased media coverage about genetically modified products—even if it is positive—heightens concern.

So what does this mean for U.S. farmers and firms in the food business? In surveys on both sides of the Atlantic, consumers say that they are willing to pay significantly more for non–genetically modified alternatives—16 to 38 percent more in the United States and up to 50 percent more in parts of Europe. These premiums are much higher than the cost reductions associated with genetically modified seeds. However, perhaps talk is cheap, as people could be overestimating their willingness to pay more for non–genetically modified products. . . .

The United States and the European Union, along with other countries such as Japan and Australia, have each developed procedures for testing and introducing foods produced by genetic engineering into the marketplace. Most developing countries do not have such rules, and no overarching framework of international law exists.

In the United States, the creation of genetically modified organisms did not prompt new rules or substantial changes to existing laws governing food and environmental safety. U.S. regulators look at the chemical properties of genetically modified foods with a mass spectrometer, and if they line up with the conventional variety—as all genetically modified crops have to date—regulators determine that the plants are "substantially similar." This means that the regulation of genetically modified foods falls under three different jurisdictions—the Environmental Protection Agency (EPA), the United States Department of Agriculture (USDA), and the Food and Drug Administration (FDA).

The EPA regulates pesticide, herbicide, and fertilizer use and sets limits on the amount of harmful chemicals allowed in agriculture production. So, if a genetically modified food incorporates a chemical into the plant, as with insect-resistant corn, then the EPA plays a role in the approval process. The USDA issues permits for field trials and reviews petitions by seed developers to commercially release genetically modified seeds. The FDA has the broadest mandate as the supervisor and coordinator of licensing and testing of genetically modified foods (excluding meat, poultry, and dairy, which the USDA monitors). Seed developers must also consult other U.S. laws, including various state seed certification rules. U.S. law currently does not require labels for genetically modified food products.

In contrast to U.S. regulators, E.U. authorities consider genetically modified crops "novel foods," in part because they contain proteins that do not exist in conventional varieties. In 1998, the European Union temporarily banned most new genetically modified foods while officials created a new regulatory regime. In the meantime, they continued to import a small number of crops approved prior to the ban, but with detailed labeling requirements. The European Union recently unveiled their new regime, which includes strict admission, labeling, and tracking requirements, as well as a 0.9 percent threshold for "accidental contamination" of both food and feed grains as they move through the supply chain. Individual E.U. countries may also have additional rules, such as crop registration procedures. In May 2004, a variety of insect-resistant corn became the first new genetically modified food allowed into the European Union in six years. But biotechnology supporters should be cautious, as the approval process was slow, difficult, and costly.

Two sets of international agreements also apply to genetically modified foods. The first is the Cartagena Protocol on Biodiversity, ratified in 2003 by 82 countries including European nations and Japan, but not the United States. The Protocol seeks to protect countries from risks associated with imports of genetically engineered organisms. The Protocol requires exporting countries to provide information about the way scientists modified the food item, label all genetically modified products, and adhere to the importing country's national biosafety laws and risk assessment procedures.

The other body of international law concerning genetically modified foods resides in the World Trade Organization (WTO). To prevent discrimination based on nationality, the WTO requires a country refusing imports to base its decision on scientific evidence of food or environmental safety. It was on this basis that in the summer of 2003, the United States, Canada, and Argentina began the WTO process of challenging the legality of the European Union's moratorium on new genetically modified organisms. Such disputes, a direct result of the patchwork approach to regulating genetically modified foods, will endure so long as nations cannot agree to a single set of standards.

Segmented markets for genetically altered and conventional foods may seem like an easy alternative, but it could cause problems, including price increases for consumers, as crops must be stored and shipped separately. And as different ideologies and attitudes continue to clash, the debate about biotechnology in the food supply will persist.

5 The Paradox of Sustainable Development

In 1987, the World Commission on Environment and Development (also known as the Brundtland Commission) issued *Our Common Future*, and gave sustainable development its canonical definition: "development that meets the needs of today without inhibiting the ability of future generations to meet their own needs."

The concept was not new in 1987. The Iroquois have a long tradition of considering the effects of communal decisions on "seven generations." Thomas Jefferson wrote in 1789 that since the earth "belongs to each . . . generation during its course, fully and in its own right, no generation can contract debts greater than may be paid during the course of its own existence." The word "sustainability" had actually been used for nearly a decade in this sense in environmental circles before *Our Common Future* appeared.

The Brundtland Commission, however, took the concept to a whole new level. Sustainable development was more than resource conservation, it was a vision of humankind's social and economic future toward which government, business, and civil society could all contribute. The Commission argued that economic growth, social development, and environmental protection should properly be seen as interrelated. None of these goals could be achieved and sustained unless all three were achieved together.

For instance, preservation of a resource like a rainforest is neither a realistic nor a humane goal unless some alternative means of livelihood is devised for those who subsist by cutting down the trees. By the same token, a fishing community can not afford to ignore the effects of toxins and overfishing on the resource that provides its livelihood. And even the most powerful corporation cannot hope to sustain profits in the long term unless it supports the health and prosperity of the communities in which it operates, and secures the long-term viability of the natural resource base upon which it depends.

The Brundtland Commission's call to action was heeded by the international community. As the ideological divisions of the Cold War receded, sustainable development provided a useful new platform for international collaboration. Local and national governments, multinational corporations and industry groups, and non-governmental organizations of all stripes signed on to the concept. The 1992 Earth Summit in Rio de Janeiro established "Agenda 21" as an action plan for achieving sustainable development. Since then, the U.N. and other organizations have continued to monitor progress at the local, national, and global levels. One must be impressed by the considerable successes the sustainable development movement has achieved, in light of the deep distrust that so often characterizes relations between the environmental community and the business community, or human rights advocates and multinational corporations, or international conservationists and third-world governments. Under the banner of sustainable development, real progress has been made in improved water supply and quality, sanitation, irrigation, reforestation, and the development of viable tourist industries based on wildlife protection, tangibly improving the lives of millions of the world's most desperately poor.

Yet the apparent consensus masks a wide divergence of opinions on what sustainable development actually entails. As Sharachchandra Lélé points out (see below), the conceptual ambiguity of sustainable development has allowed the movement to succeed in bringing parties with widely divergent perspectives to the table, but it also threatens to turn the movement into something vapid and ineffectual.

Conceptual confusion begins with the underlying terms. "Development" might mean industrialization after the fashion of the old colonial powers, or it might mean economic growth generally, or raised living standards (measured, for example, in infant mortality or literacy), or simply socioeconomic change (see Daly selection in Chapter 23, below). It might be understood strictly in economic terms, or it might include a component of political and institutional reform. And depending on one's perspective, it might apply to the "developed" as well as to the "developing" nations.

The word "sustainable" too can be understood in multiple ways. Lélé distinguishes between "ecological" and "social" sustainability. Others count "economic" sustainability as a third type. There is a lively debate between proponents of "strong" sustainability, which requires preserving a minimum level of natural capital, and "weak" sustainability, which permits substitution among different types of capital (e.g., natural, manufactured, intellectual, social), as long as the total package enables future generations to maintain a minimum standard of living. Weak sustainability would permit the destruction of a forest, for example, if the economic infrastructure built with the revenue will provide equivalent value to society. Despite the apparently stark contrast between the two positions, there is room for compromise. Nobel prize–winning economist Robert Solow, a prominent proponent of weak sustainability, freely acknowledges that certain features of the landscape are unique and irreplaceable and should be preserved; while defenders of strong sustainability like Herman Daly must grapple with the problem of non-renewable resources—a sustainable society must substitute some other kind of capital for the non-renewable resources it consumes.

(See the book by Bryan Norton listed in Further Reading for a thoughtful "hybrid" position.)

Several conceptual models and tools have been offered to make sustainable development approachable. One simplistic model is the three-legged stool. Economic growth, environmental protection, and equity are sometimes said to be the three legs that hold up the stool of sustainable development. Just as a stool requires all three legs to be of equal length in order to function properly, development is said to require equal measures of economics, environment, and equity in order to be sustainable. This model has been criticized for reifying the three components of sustainable development as separate and equal, when in fact they are interdependent and might vary in importance from case to case. (Indeed, it has been argued, the environment might more properly be seen as the floor on which the sustainable development stool sits.)

The IPAT model is a tool that allows us to conceptualize the relationship among population, consumption, technology, and environmental degradation. If "I" is environmental impact, "P" is population, "A" is affluence (e.g., rate of material consumption), and "T" is technology, then $I = P * A * T$, or perhaps more intuitively, $I = (P * A) / T$. That is, the environmental impact of a society increases with population growth, and increases with greater affluence, but can be mitigated by improvements in technology (e.g., wireless communication, cleaner fuels, improved recycling capabilities). This model reminds us that both the affluent, population-stable nations and the poorer, growing nations bear responsibility for global environmental impact, and it also suggests the importance of technology as a safety valve. Yet this model too has its limitations. Technological innovations can be destructive as well as constructive, and the concept of "affluence" invites even more divergent interpretations, as the selection by Kumarappa illustrates.

A third conceptual tool, the Ecological Footprint, is described in the selection by Wackernagel and Rees.

The role of economic growth in sustainable development can be problematic. Part of the problem, it should not surprise us, is conceptual ambiguity. Conventionally, economic growth is measured in aggregate, for instance as change in GDP. Since the benefits of aggregate growth are not always distributed equitably, and they often bypass the poor communities most in need of development aid, Lélé considers economic growth irrelevant to sustainable development at best, and a distraction at worst. When economist Benjamin Friedman defends growth in *The Moral Consequences of Economic Growth*, on the other hand, he defends a definition more nuanced than change in GDP: "a rising standard of living for the clear majority of citizens."

Thoughtful observers on all sides can agree, at least, that what we are really after is improved living standards, and thus distribution matters as well as growth. This conceptual clarity can help us evaluate policies and think critically about the standard growth-oriented model of economic development. Are there viable alternative models? What indicators of progress are available besides GDP?

The standard model of development also requires large-scale capital investment and extensive government borrowing. Much international lending and borrowing in past decades was ill-conceived and suffered from poor oversight, leaving third-world governments today with large debts to multilateral institutions and Northern banks. Third-world indebtedness is a major obstacle to domestic investments and the elimination of poverty, and often indebted governments are forced to resort to liquidating their natural capital—forests, soils, and mineral wealth—at discount rates to pay their debts. For the sake of the environment and the alleviation of poverty—and for reasons of fairness, considering the questionable circumstances under which much of the debt was incurred—many see forgiveness of third-world debt as a sensible proposition.

One important role of economic growth in sustainable development is that the surplus it produces can continually be invested into further research and development, producing a virtuous cycle of environmental and economic benefits. The Brundtland Commission urges all nations to promote the development and diffusion of environmentally sound technologies. There is clearly much room for improvement in this area. But, as noted above, it would be a mistake to see technology as a magic bullet. There is every reason to believe that technological progress, like so many other areas of human endeavor, is ultimately subject to the law of diminishing returns: once we exhaust the most readily available options, each successive step in the direction of progress becomes more difficult and costly. Tainter (see below) extends this analysis beyond technological progress to resource depletion, economic growth, and socioeconomic complexity, and reaches conclusions that challenge our very understanding of sustainability.

FURTHER READING

Hall, Charles A. S., ed. *Quantifying Sustainable Development: The Future of Tropical Economies.* San Diego: Academic Press, 2000.

Merkel, Jim. *Radical Simplicity: Small Footprints on a Finite Earth.* Gabriola, B.C.: New Society Publishers, 2003.

Norton, Bryan G. *Sustainability: A Philosophy of Adaptive Ecosystem Management.* Chicago: Chicago University Press, 2005.

Peterson, Tarla Rai. *Sharing the Earth: The Rhetoric of Sustainable Development.* Columbia: University of South Carolina Press, 1997.

Princen, Thomas. *The Logic of Sufficiency.* Cambridge, Mass.: MIT Press, 2005.

Redclift, Michael. *Sustainable Development: Exploring the Contradictions.* London: Methuen, 1987.

Weiss, Edith Brown. *In Fairness to Future Generations: International Law, Common Patrimony, and Intergenerational Equality.* Tokyo: United Nations University, 1988.

World Commission on Environment and Development, from *Our Common Future* (1987)

The Brundtland Commission's 1987 report to the U.N. General Assembly put sustainable development at the forefront of the global agenda.

Environment and development are not separate challenges; they are inexorably linked. Development cannot subsist upon a deteriorating environmental resource base; the environment cannot be protected when growth leaves out of account the costs of environmental destruction. These problems cannot be treated separately by fragmented institutions and policies. They are linked in a complex system of cause and effect.

First, environmental stresses are linked one to another. For example, deforestation, by increasing run-off, accelerates soil erosion and siltation of rivers and lakes. Air pollution and acidification play their part in killing forests and lakes. Such links mean that several different problems must be tackled simultaneously. And success in one area, such as forest protection, can improve chances of success in another area, such as soil conservation.

Second, environmental stresses and patterns of economic development are linked one to another. Thus agricultural policies may lie at the root of land, water, and forest degradation. Energy policies are associated with the global greenhouse effect, with acidification, and with deforestation for fuelwood in many developing nations. These stresses all threaten economic development. Thus economics and ecology must be completely integrated in decision-making and lawmaking processes not just to protect the environment, but also to protect and promote development. Economy is not just about the production of wealth, and ecology is not just about the protection of nature; they are both equally relevant for improving the lot of humankind.

Third, environmental and economic problems are linked to many social and political factors. For example, the rapid population growth that has so profound an impact on the environment and on development in many regions is driven partly by such factors as the status of women in society and other cultural values. Also, environmental stress and uneven development can increase social tensions. It could be argued that the distribution of power and influence within society lies at the heart of most environment and development challenges. Hence new approaches must involve programs of social development, particularly to improve the position of women in society, to protect vulnerable groups, and to promote local participation in decision making. . . .

The concept of sustainable development provides a framework for the integration of environment policies and development strategies—the term "development" being used here in its broadest sense. The word is often taken to refer to the processes of economic and social change in the Third World. But the integration of environment and development is required in all countries, rich and poor. The pursuit of sustainable development requires change in the domestic and international policies of every nation.

Sustainable development seeks to meet the needs and aspirations of the present without compromising the ability to meet those of the future. Far from requiring the cessation of economic growth, it recognizes that the problems of poverty and underdevelopment cannot be solved unless we have a new era of growth in which developing countries play a large role and reap large benefits. . . .

Growth has no set limits in terms of population or resource use beyond which lies ecological disaster. Different limits hold for the use of energy, materials, water, and land. Many of these will manifest themselves in the form of rising costs and diminishing returns, rather than in the form of any sudden loss of a resource base. The accumulation of knowledge and the development of technology can enhance the carrying capacity of the resource base. But ultimate limits there are, and sustainability requires that long before these are reached, the world must ensure equitable access to the constrained resource and reorient technological efforts to relieve the pressure.

Economic growth and development obviously involve changes in the physical ecosystem. Every ecosystem everywhere cannot be preserved intact. A forest may be depleted in one part of a watershed and extended elsewhere, which is not a bad thing if the exploitation has been planned and the effects on soil erosion rates, water regimes, and genetic losses have been taken into account. In general, renewable resources like forests and fish stocks need not be depleted provided the rate of use is within the limits of regeneration and natural growth. But most renewable resources are part of a complex and interlinked ecosystem, and maximum sustainable yield must be defined after taking into account system-wide effects of exploitation.

As for non-renewable resources, like fossil fuels and minerals, their use reduces the stock available for future generations. But this does not mean that such resources should not be used. In general the rate of depletion should take into account the criticality of that resource, the availability of technologies for minimizing depletion, and the likelihood of substitutes being available. Thus land should not be degraded beyond reasonable recovery. With minerals and fossil fuels, the rate of depletion and the emphasis on recycling and economy of use should be calibrated to ensure that the resource does not run out before acceptable substitutes are available. Sustainable development requires that the rate of depletion of non-renewable resources should foreclose as few future options as possible. . . .

Critical objectives for environment and development policies that follow from the concept of sustainable development include:

- reviving growth;
- changing the quality of growth;
- meeting essential needs for jobs, food, energy, water, and sanitation;
- ensuring a sustainable level of population;
- conserving and enhancing the resource base;
- reorienting technology and managing risk; and
- merging environment and economics in decision making.

Sharachchandra Lélé, from "Sustainable Development: A Critical Review" (1991)

Lélé analyzes the competing ways the concept of "sustainable development" is used, and challenges the sustainable development movement to focus on the fundamentals of ecological sustainability.

There are those who believe that one should not try to define SD [Sustainable Development] too rigorously. To some extent, the value of the phrase does lie in its broad vagueness. It allows people with hitherto irreconcilable positions in the environment-development debate to search for common ground without appearing to compromise their positions.

. . . The absence of a clear theoretical and analytical framework, however, makes it difficult to determine whether the new policies will indeed foster an environmentally sound and socially meaningful form of development. Further, the absence of semantic and conceptual clarity is hampering a fruitful debate over what this form should actually be.

. . . Sustainable development is sometimes interpreted as "sustained growth," "sustained change," or simply "successful" development. . . . Depending on what characterization of the process is implicit this interpretation is either impossible or trivial. . . . This usage is therefore not very useful; moreover, it is confusing, because sustainability has already acquired other specific connotations.

. . . The concept of sustainability originated in the context of renewable resources such as forests or fisheries, and has subsequently been adopted as a broad slogan by the environmental movement. Most proponents of sustainability therefore take it to mean "the existence of the ecological conditions necessary to support human life at a specified level of well-being through future generations," what I call *ecological sustainability* (Figure 5.1).

Since ecological sustainability emphasizes the constraints and opportunities that nature presents to human activities, ecologists and physical scientists frequently dominate its discussion. But what they actually focus on are the ecological conditions for ecological sustainability—the biophysical "laws" or patterns that determine environmental responses to human activities and humans' ability to use the environment. The major contribution of the environment-development debate is, I believe, the realization that in addition to or in conjunction

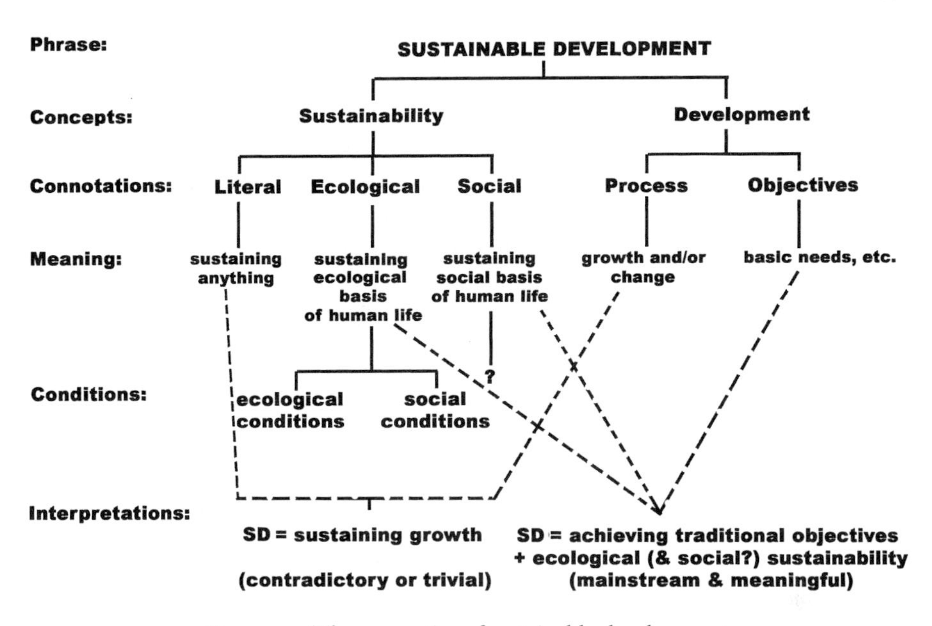

Figure 5.1 The semantics of sustainable development.

with these ecological conditions, there are social conditions that influence the ecological sustainability or unsustainability of the people-nature interaction. To give a stylized example, one could say that soil erosion undermining the agricultural basis for human society is a case of ecological (un)sustainability. It could be caused by farming on marginal lands without adequate soil conservation measures—the ecological cause. But the phenomenon of marginalization of peasants may have social roots, which would then be the social causes of ecological unsustainability.

Sometimes, however, sustainability is used with fundamentally social connotations. For instance, [E. B.] Barbier defines social sustainability as "the ability to maintain desired social values, traditions, institutions, cultures, or other social characteristics."[1] This usage is not very common, and it needs to be carefully distinguished from the more common context in which social scientists talk about sustainability, viz., the social aspects of ecological sustainability. . . .

In the mainstream interpretation of SD, ecological sustainability is a desired attribute of any pattern of human activities that is the goal of the developmental process. In other words, SD is understood as "a form of societal change that, in addition to traditional developmental objectives, has the objective or constraint of ecological sustainability." Given an ever-changing world, the specific forms of and priorities among objectives, and the requirements for achieving sustainability, would evolve continuously. But sustainability—as it is understood at each stage—would remain a fundamental concern. Ecological sustainability is, of course, not independent of the other (traditional) objectives of development.

Tradeoffs may sometimes have to be made between the extent to and rate at which ecological sustainability is achieved vis-à-vis other objectives. In other cases, however, ecological sustainability and traditional developmental objectives (such as satisfaction of basic needs) could be mutually reinforcing. This interpretation of SD dominates the SD debate; I shall therefore focus on it. . . .

The major impact of the SD movement is the rejection of the notion that environmental conservation necessarily constrains development or that development necessarily means environmental pollution—certainly not an insignificant gain. Where the SD movement has faltered is in its inability to develop a set of concepts, criteria, and policies that are coherent or consistent—both externally (with physical and social reality) and internally (with each other). . . .

POVERTY AND ENVIRONMENTAL DEGRADATION: AN INCOMPLETE CHARACTERIZATION

The fundamental premise of mainstream SD thinking is the two-way link between poverty and environmental degradation, shown schematically in Figure 5.2.

In fact, however, even a cursory examination of the vast amount of research that has been done on the links between social and environmental phenomena suggests that both poverty and environmental degradation have deep and complex causes. While substantive disagreements still exist regarding the primacy of these causes and the feasibility and efficacy of different remedies, the diagram in Figure 5.3 is probably a reasonable approximation of the general consensus on the nature of the causes and their links.

To say that mainstream SD thinking has completely ignored these factors would be unfair. But it would be fair to say that it has focused on an eclectically chosen few. In particular, inadequate technical know-how and managerial capabilities, common property resource management, and pricing and subsidy policies have been the major themes addressed, and the solutions suggested have been essentially techno-economic ones. . . . Deeper socio-political changes (such as land reform) or changes in cultural values (such as overconsumption in the North) are either ignored or paid lip-service.

This is not to say that problems of the global commons or of the lack of techno-managerial expertise are unimportant. But the intellectual discourse needs to begin with an acknowledgment that the big picture in Figure 5.3 (or something similar) essentially holds in all cases, and then proceed to developing analytical methods to help estimate the relative importance of each causal factor in specific cases and identify means of and scope for change.

CONCEPTUAL WEAKNESSES

Removal of poverty (the traditional developmental objective), sustainability, and participation are really the three fundamental objectives of the SD para-

Figure 5.2 The mainstream perception of the link between poverty and environmental degradation. Photos courtesy of Greenpeace.

digm. Unfortunately, the manner in which these objectives are conceptualized and operationalized leaves much to be desired. On the one hand, economic growth is being adopted as a major operational objective that is consistent with both removal of poverty and sustainability. On the other hand, the concepts of sustainability and participation are poorly articulated, making it difficult to determine whether a particular development project actually promotes a particular form of sustainability, or what kind of participation will lead to what kind of social (and consequently, environmental) outcome.

The Role of Economic Growth

By the mid-1970s, it had seemed that the economic growth and trickle-down theory of development had been firmly rejected, and the "basic needs approach"[2] had taken root in development circles. Yet economic growth continues to feature in today's debate on SD. In fact, "reviving [economic] growth" heads WCED's

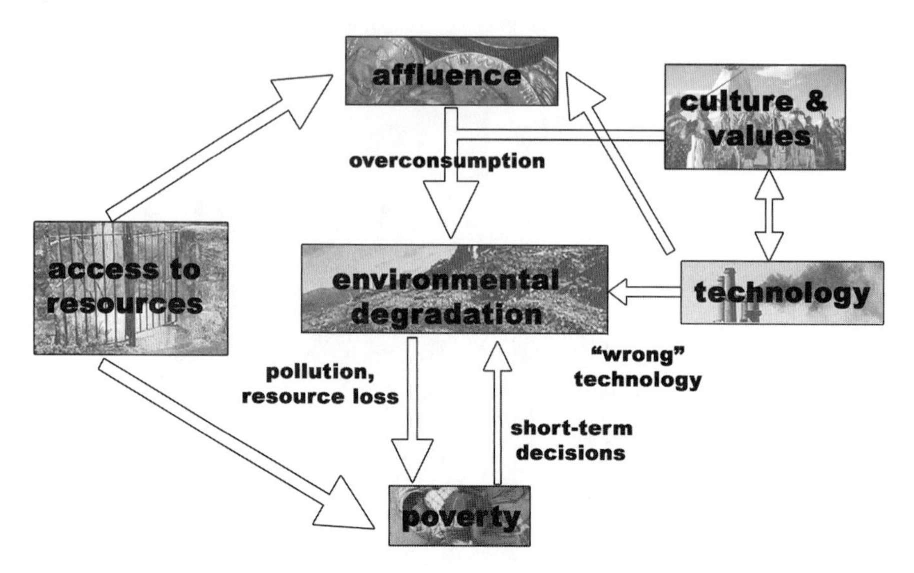

Figure 5.3 A more realistic representation of the link between poverty and environmental degradation. Photos courtesy of Greenpeace.

list of operational objectives [see Brundtland Commission, above]. Two arguments are implicit in this adoption of economic growth as an operational objective. The first, a somewhat defensive one, is that there is no fundamental contradiction between economic growth and sustainability, because growth in economic activity may occur simultaneously with either an improvement or a deterioration in environmental quality. . . . But one could turn this argument around and suggest that, if economic growth is not correlated with environmental sustainability, there is no reason to have economic growth as an operational objective of SD.

The second argument in favor of economic growth is more positive. The basic premise of SD is that poverty is largely responsible for environmental degradation. Therefore, removal of poverty (i.e., development) is necessary for environmental sustainability. This, it is argued, implies that economic growth is absolutely necessary for SD. The only thing that needs to be done is to "change the quality of [this] growth" [see Brundtland Commission, above] to ensure that it does not lead to environmental destruction. In drawing such an inference, however, there is the implicit belief that economic growth is necessary (if not sufficient) for the removal of poverty. But was it not the fact that economic growth per se could not ensure the removal of poverty that led to the adoption of the basic needs approach in the 1970s?

Thus, if economic growth by itself leads to neither environmental sustainability nor removal of poverty, it is clearly a "nonobjective" for SD. The converse is a possibility worth exploring, viz., whether successful implementation of policies for poverty removal, long-term employment generation, environmental restoration, and rural development will lead to growth in GNP, and, more important, to increases in investment, employment, and income generation. This seems more than likely in developing countries, but not so certain in developed ones. In any case, economic growth may be the fallout of SD, but not its prime mover.

Sustainability

The World Conservation Strategy was probably the first attempt to carry the concept of sustainability beyond simple renewable resource systems. It suggested three ecological principles for ecological sustainability (see the nomenclature developed above), viz., "maintenance of essential ecological processes and life-support systems, the preservation of genetic diversity, and the sustainable utilization of species and resources."[3] This definition, though a useful starting point, is clearly recursive as it invokes "sustainability" in resource use without defining it. Many subsequent attempts to discuss the notion are disturbingly muddled. There is a very real danger of the term becoming a meaningless cliché, unless a concerted effort is made to add precision and content to the discussion. . . .

Any discussion of sustainability must first answer the questions "What is to be sustained? For whom? How long?" The value of the concept (like that of SD), however, lies in its ability to generate an operational consensus between groups with fundamentally different answers to these questions, i.e., those concerned either about the survival of future human generations, or about the survival of wildlife, or human health, or the satisfaction of immediate subsistence needs (food, fuel, fodder) with a low degree of risk. It is therefore vital to identify those aspects of sustainability that do actually cater to such diverse interests, and those that involve tradeoffs.

Differentiating between ecological and social sustainability could be a first step toward clarifying some of the discussion. Further, in the case of ecological sustainability, a distinction needs to be made between renewable resources, nonrenewable resources, and environmental processes that are crucial to human life, as well as to life at large. The few researchers who have begun to explore the idea of ecological sustainability emphasize its multidimensional and complex nature.

In the context of sustainable use of renewable resources, it is necessary to go beyond the conventional simplistic notion of "harvesting the annual increment," and take into consideration the dynamic behavior of the resource, stochastic properties of and uncertainties about environmental conditions (e.g., climatic variations), the interactions between resources and activities (e.g., among forests, soils, and agriculture), and between different uses or features of the "same" resource (e.g., tree foliage and stemwood).

In the rush to derive ecological principles of (ecological) sustainability, we cannot afford to lose sight of the social conditions that determine which of these principles are socially acceptable, and to what extent. Sociologists, eco-Marxists, and political ecologists are pointing out the crucial role of socioeconomic structures and institutions in the pattern and extent of environmental degradation globally. Neoclassical economists, whose theories have perhaps had the greatest influence in development policy making in the past and who therefore bear the responsibility for its social and environmental failures, however, have been very slow in modifying their theories and prescriptions. The SD movement will have to formulate a clear agenda for research in what is being called "ecological economics" [see Chapter 23] and press for its adoption by the mainstream of economics in order to ensure the possibility of real changes in policy making.

Social sustainability is a more nebulous concept than ecological sustainability. . . . Detailed analyses of the concept . . . seem to be nonexistent. Perhaps achieving desired social situations is itself so difficult that discussing their maintainability is not very useful; perhaps goals are even more dynamic in a social context than in an ecological one, so that maintainability is not such an important attribute of social institutions/structures. There is, however, no contradiction between the social and ecological sustainability; rather, they can complement and inform each other.

Participation

A notable feature of "ecodevelopment"—SD's predecessor—as well as some of the earlier SD literature was the emphasis placed on equity and social justice. . . . Subsequently, however, the mainstream appears to have quietly dropped these terms (suggesting at least a deemphasizing of these objectives), and has instead focused on "local participation."

There are, however, three problems with this shift. First, by using the terms equity, participation, and decentralization interchangeably, it is being suggested that participation and decentralization are equivalent, and that they can somehow substitute for equity and social justice. This suggestion is at best a naive one. While all of these concepts are quite complex, it seems clear that some form of participation is necessary but not sufficient for achieving equity and social justice.

Second, the manner in which participation is being operationalized shows up the narrow-minded, quick-fix, and deceptive approach adopted by the mainstream promoters of SD. . . . Mainstream SD literature blithely assumes and insists that "involvement of local NGOs" in project implementation will ensure project success.

Third, there is an assumption that participation or at least equity and social justice will necessarily reinforce ecological sustainability. Attempts to test such assumptions rigorously have been rare. But preliminary results seem to suggest that equity in resource access may not lead to sustainable resource use unless new institutions for resource management are carefully built and nurtured. For instance, [N. S.] Jodha[4] describes how land reform in Rajasthan (India) led to the neglect of village pastures that were well-maintained under the earlier feudal structure. Similarly, communal irrigation tanks in Tamil Nadu (India) fell into disrepair with the reduction in the feudal powers of the village landlords.[5] This should not be misconstrued as an argument against the need for equity, but rather as a word of caution against the tendency to believe that social equity automatically ensures environmental sustainability (or vice-versa). . . .

CONCLUDING REMARKS: DILEMMAS AND AGENDAS

The proponents of SD are faced with a dilemma that affects any program of political action and social change: the dilemma between the urge to take strong stands on fundamental concerns and the need to gain wide political acceptance and support. Learning from the experience of ecodevelopment, which tended toward the former, SD is being packaged as the inevitable outcome of objective scientific analysis, virtually a historical necessity, that does not contradict the deep-rooted normative notion of development as economic growth. In other words, SD is an attempt to have one's cake and eat it too.

It may be argued that this is indeed possible, that the things that are wrong and need to be changed are quite obvious, and there are many ways of fixing them without significantly conflicting with either age-old power structures or the modern drive for a higher material standard of living. Therefore, it is high time that environmentalists and development activists put aside their differences and joined hands under the banner of sustainable development to tackle the myriad of problems facing us today. If, by using the politically correct jargon of economic growth and development and by packaging SD in the manner mentioned above, it were possible to achieve even fifty percent success in implementing this bundle of "conceptually imprecise" policies, the net reduction achieved in environmental degradation and poverty would be unprecedented.

I believe, however, that (analogous to the arguments in SD) in the long run there is no contradiction between better articulation of the terms, concepts, analytical methods, and policymaking principles, and gaining political strength and broad social acceptance—especially at the grassroots. In fact, such clarification and articulation is necessary if SD is to avoid either being dismissed as another development fad or being co-opted by forces opposed to changes in the status quo. . . .

In a sense, if SD is to be really "sustained" as a development paradigm, two apparently divergent efforts are called for: making SD more precise in its conceptual underpinnings, while allowing more flexibility and diversity of approaches in developing strategies that might lead to a society living in harmony with the environment and with itself.

Notes

1. E. B. Barbier, "The Concept of Sustainable Economic Development," *Environmental Conservation* 14, no. 2 (1987): 101–10.

2. P. Streeten, "Basic Needs: Premises and Promises," *Journal of Policy Modeling* 1 (1979): 136–46.
3. IUCN, *World Conservation Strategy: Living Resource Conservation for Sustainable Development* (Gland, Switzerland: International Union for Conservation of Nature and Natural Resources, United Nations Environment Program and World Wildlife Fund, 1980).
4. N. S. Jodha, "A Case Study of the Decline of Common Property Resources in India," in P. Blaikie and H. Brookfield, ed., *Land Degradation and Society* (New York: Methuen, 1987), 196–207.
5. M. von Oppen and K. V. Subba Rao, *Tank Irrigation in Semi-Arid India* (Patancheru, Andhra Pradesh, India: International Crop Research Institute for the Semi-Arid Tropics, 1980).

Mathis Wackernagel and William E. Rees, from *Our Ecological Footprint* (1996)

Committed to strict ecological sustainability, Wackernagel and Rees view the affluent, consumption-oriented lifestyle typical of the developed countries not as a goal to be emulated by the world's poor, but as a major obstacle to sustainability. They reach this conclusion using the "ecological footprint," a popular conceptual tool that demonstrates in an intuitive way how much productive land is required to sustain a person, a product, a city, or a country.

Ecological footprint analysis is an accounting tool that enables us to estimate the resource consumption and waste assimilation requirements of a defined human population or economy in terms of a corresponding productive land area.

... [Footprint size] depends on such factors as income, personal values and behavior, consumption patterns, and the technologies used to produce consumer goods. There is, therefore, wide variation in Footprint size both among countries and individuals around the world. We can illustrate these points by summarizing our detailed calculations for the average Canadian's Ecological Footprint (Table 5.1) and contrasting the result with those for several other countries (Table 5.2). Note that while U.S. consumption patterns are roughly similar to Canada's per capita totals, their average Ecological Footprints are larger.

... Estimating the area of ecologically productive land needed to produce the natural resources and services used by an average Canadian involves several major steps: first we compile annualized statistics on five major categories of consumption and waste production and divide the totals for items in these categories by total population to determine average levels. (Consumption includes: direct household consumption; indirect consumption such as the energy "embodied" in consumer goods; and consumption by businesses and government, which ultimately benefits the households. Services refers to schooling, policing, governance, or health care.) Second, we convert these data on average consumption ("ecological load") into their corresponding land areas based on the ecological productivity of relevant ecosystem types. The average Canadian's Ecological Footprint is then obtained by summing the land requirements for the various consumption/waste categories. Since this area represents that portion of plane-

TABLE 5.1

The Consumption / Land-use Matrix for the Average Canadian

Cell entries = ecologically productive land in ha/capita	*A* *ENERGY*	*B* *DEGR.*	*C* *GARDEN*	*D* *CROP*	*E* *PASTURE*	*F* *FOREST*	*TOTAL*
1 FOOD	0.33		0.02	0.60	0.33	0.02	1.30
11 fruit, vegetable, grain	*0.14*		*0.02*	*0.18*		*0.01?*	
12 animal products	*0.19*			*0.42*	*0.33*	*0.01?*	
2 HOUSING	0.41	0.08	0.002?			0.40	0.89
21 construction/maintenance	*0.06*					*0.35*	
22 operation	*0.35*					*0.05*	
3 TRANSPORTATION	0.79	0.10					0.89
31 motorized private	*0.60*						
32 motorized public	*0.07*						
33 transport of goods	*0.12*						
4 CONSUMER GOODS	0.52	0.01		0.06	0.13	0.17	0.89
41 packaging	*0.10*					*0.04*	
42 clothing	*0.11*			*0.02*	*0.13*		
43 furniture & appliances	*0.06*					*0.03?*	
44 books/magazines	*0.06*					*0.10*	
45 tobacco & alcohol	*0.06*			*0.04*			
46 personal care	*0.03*						
47 recreation equipment	*0.10*						
48 other goods	*0.00*						

(*continued*)

TABLE 5.1 (*continued*)
The Consumption / Land-use Matrix for the Average Canadian

Cell entries = ecologically productive land in ha/capita	*A* *ENERGY*	*B* *DEGR.*	*C* *GARDEN*	*D* *CROP*	*E* *PASTURE*	*F* *FOREST*	*TOTAL*
5 SERVICES	0.29	0.01					0.30
51 government (+military)	*0.06*						
52 education	*0.08*						
53 health care	*0.08*						
54 social services	*0.00*						
55 tourism	*0.01*						
56 entertainment	*0.01*						
57 bank/insurance	*0.00*						
58 other services	*0.05*						
TOTAL	2.34	0.20	0.02	0.66	0.46	0.59	4.27

Based on 1991 data (0.0 = less than 0.005 ha or 50 m^2; blank = probably insignificant; ? = lacking data).
ABBREVIATIONS
a) ENERGY = **fossil energy consumed** expressed in the land area necessary to sequester the corresponding CO_2.
b) DEGR. = **degraded land** or built-up environment.
c) GARDEN = **gardens** for vegetable and fruit production.
d) CROP = **crop land.**
e) PASTURE = **pastures** for dairy, meat and wool production.
f) FOREST = prime **forest** area. An average roundwood harvest of 163 m^3/ha every 70 years is assumed.

TABLE 5.2
Comparing People's Average Consumption in the U.S., Canada, India, and the World

Consumption per person in 1991	*Canada*	*USA*	*India*	*World*
CO_2 emission [in tons per yr]	15.2	19.5	0.81	4.2
Purchasing power [in $U.S.]	19,320	22,130	1,150	3,800
Vehicles per 100 persons	46	57	0.2	10
Paper consumption [in kg/yr]	247	317	2	44
Fossil energy use [in Gigajoules/yr]	250	287	5	56
Fresh water withdrawal [in m^3/yr]	1,688	1,868	612	644
Ecological footprint [ha/person]	4.3	5.1	0.4	1.8

Source: Food and Agriculture Organization of the United Nations (FAO), *FAO Yearbook: Production*, vol. 43 (Rome: FAO, 1990). World Resources Institute, *World Resources: Data Base Diskette* (Washington, D.C.: World Resources Institute, 1992). Statistics Canada data.

tary productivity needed to support a single individual we sometimes refer to it as the average "personal planetoid." The results of these calculations are summarized in the consumption/land-use matrix shown in Table 5.1.

It seems that Canadians are formidable consumers! For example, on average, each Canadian eats about 3,450 kilocalories worth of food each day, 1,125 in the form of animal products. Most of this food is produced by energy-intensive agriculture and is highly processed before it reaches the dinner table. According to the World Resources Institute, Canadian settlements cover about 55,000 square kilometers—0.2 ha [hectare] per capita—and have been built mainly on agricultural land. On average, Canadians drive a car 18,000 kilometers per year, use approximately 200 kilograms of packaging, spend about $2,700 on consumer goods and another $2,000 on services. Energy and material consumption in Canada is typically four to five times the world average and, in most categories, the average American's consumption is even higher (see Table 5.2). . . .

A GLOBAL COMPARISON OF FOOTPRINT SIZES—COULD EVERYBODY ON EARTH TODAY ENJOY NORTH AMERICANS' CURRENT ECOLOGICAL STANDARD OF LIVING?

As long as there is adequate ecologically productive land on Earth, local consumption that exceeds local production in any given region can be sustained by "importing" the surplus output of other regions. Of course this raises the question of just how much surplus ecoproductivity there is on the planet.

To answer this, we need to know first how much land there is. Planet Earth has a surface area of 51 billion hectares of which 13.1 billion hectares are land not covered by ice or fresh water. Of this just under 8.9 billion hectares are ecologically productive: cropland, permanent pastures, forests, and woodland. Of the remaining 4.2 billion hectares, 1.5 billion hectares are occupied by large deserts (excluding Antarctica) and 1.2 billion hectares by mostly semi-arid areas. The remaining 1.5 billion hectares include grasslands not used for pasture, wastelands, and 0.2 billion hectares of built-up areas and roads (0.03 ha/capita).

It seems, therefore, that about 8.9 billion hectares of land are potentially available for human exploitation. However, approximately 1.5 billion hectares of this is "wilderness" that arguably should remain in its near-pristine state. In fact, this mostly forested area is already fully engaged as it stands providing a variety of important life-support services and should not be otherwise exploited. It serves, among other things, as a biodiversity reserve, climate regulator, and carbon storehouse. (Harvesting this forest would lead to a net release of CO_2.) This would mean that only 7.4 of the 8.9 billion hectares of ecologically productive land are actually available for other more active forms of human use.

Since the beginning of this century, the "available" per capita ecological space on Earth has decreased from between 5 and 6 hectares to only 1.5 hectares. Meanwhile, as material welfare has increased, the Ecological Footprints of people in some industrialized countries have *expanded* to more than four hectares. These opposing trends illustrate the fundamental conflict confronting humanity and the real challenge of sustainability today: the Ecological Footprints of average citizens in rich countries exceed their "fair earthshares" by a factor of two to three; thus, if everybody on Earth enjoyed the same ecological standards as North Americans, we would require three Earths to satisfy aggregate material demand using prevailing technology. While this may seem like an astonishing result, our underlying assumptions and empirical evidence suggest that our calculated Footprint areas are actually considerable underestimates. In short, there are real biophysical constraints on material growth after all. Not even the present world population of 5.8 billion people—let alone the 10 billion expected by 2040—can hope to achieve North America's material standard of living without destroying the ecosphere and precipitating their own collapse.

These findings beg another question: what is the present aggregate demand by people on the ecosphere? A rough assessment based on four major human requirements shows that current appropriations of natural resources and services already exceed Earth's long-term carrying capacity. Agriculture occupies 1.5 billion hectares of cropland and 3.3 billion hectares of pasture. Sustainable production of the current roundwood harvest (including firewood) would require a productive forest area of 1.7 billion hectares. To sequester the excess CO_2 released by fossil fuel combustion, an additional 3.0 billion hectares of carbon sink lands would have to be set aside. This adds up to a requirement of 9.5 billion hectares compared to the 7.4 billion hectares of ecologically productive land actually available for such purposes. In other words, these four functions alone exceed available carrying capacity by close to thirty percent. (Even if all 8.9 billion hectares of ecologically productive land were included, present "overshoot" exceeds ten percent.) These simple footprint extrapolations alone are enough to suggest that present levels of material throughput in the global economy cannot be sustained. The global ecological deficit, unlike those of individual regions, obviously cannot be financed by trade; it depends instead on the liquidation of natural capital stocks. (Many people have concluded as much intuitively from reading the newspapers in recent years.)

What can we make of present international development objectives in light of these findings? The major agreed goal is to raise the developing world to present First World material standards. The Brundtland Commission, for example, argued for "more rapid economic growth in both industrial and developing countries" and suggested that "a five- to ten-fold increase in world industrial output can be anticipated by the time world population stabilizes some time in the next century."

Let's examine this prospect using Ecological Footprint analysis. If the present world population requires at least 9.6 billion hectares to sustain its activities, a five- to ten-fold increase would correspond to a total productive land requirement of 48 to 96 billion hectares (assuming the use of present technology). Thus, to accommodate sustainably the anticipated increase in population and economic output of the next four decades we would need six to twelve additional planets. The only alternative, if we continue to insist on economic growth as our major instrument of social policy, is to develop technologies that can provide the same levels of service with six to twelve times less energy and material. This is indeed a daunting task considering that the energy consumption of average households in industrialized countries is still increasing. One thing is certain, however: we cannot sustain development on phantom planets. . . .

We should emphasize at this point that small Ecological Footprints do not necessarily imply a low quality of life. In fact, Kerala, a southern state in India, has a *per capita* income of about one dollar a day (less than a sixtieth of North American incomes). However, life expectancy, infant mortality, and literacy rates are similar to those of industrialized countries. The people in Kerala enjoy good health care and educational systems, a vibrant democracy, and a stable population size. It seems that Kerala's exceptional standard of living is based more on accumulated social capital than on manufactured capital. The world has much to learn from the people of Kerala.

J. C. Kumarappa, from "Standards of Living" in *The Economy of Permanence* (1946)

Environmental limitations and a sense of fairness challenge those of us in wealthy nations to moderate our material consumption, but the drive to increase consumption is deeply engrained in our economic system. Indeed, as long as poorer nations pin their development hopes on export-driven GDP growth, reduced consumption in the North will not improve their lot, but worsen it. Economist J. C. Kumarappa, working in India during that nation's struggle for independence and self-determination in the mid-twentieth century, was in a unique position to ask whether economic development could be made to serve other ends than the standard "Western" goals — both capitalist and communist — of maximizing production and consumption. Kumarappa envisioned an economic system that would maximize human potential instead. Working with Mahatma Gandhi, Kumarappa promoted a type of "village-based" development, designed to enhance the traditional subsistence economy organically by meeting local needs with local skills, as an alternative to large, socially disruptive, export-driven, capital-intensive industrial development. Kumarappa's legacy lives on in the "appropriate technologies" movement founded by his British protégé E. F. Schumacher, author of Small Is Beautiful. *The highly successful micro-loans movement of recent years reflects similar ideals. In the passages below, Kumarappa invites us to reconsider our assumptions about "high" and "low" standards of living.*

The standard of life in England is generally spoken of as being high. There a gardener may live in a two storied-cottage with three or four bedrooms upstairs, with a flush lavatory and a bathroom. Downstairs there may be a living room and a dining room along with a kitchen store and washroom attached. All windows will have glass shutters sheltered by curtains and blinds. The doors will have heavy curtains to keep out the draft. The floors will be carpeted and the walls well-papered. Every room will have its appropriate and adequate furniture though simple and inexpensive. For instance, the dining room will have a dining table with proper armless chairs, sideboard, perhaps fitted with a mirror, with a requisite supply of table-linen, crockery, etc. The table service itself, though not very elaborate, will furnish appropriate dishes, plates, forks, spoons, etc., for the various courses such as soup, fish, meat, sweet, and dessert, for it is not the proper thing to eat one course with the equipment for another. Knife and fork for fish is of one kind, knife and fork for meat is of another, while the service sets are still different. When one person sits down to a meal there will be at least about fifty pieces to wash up. Such is generally accepted as a "high" standard of living.

In India, a really cultured man, perhaps a Dewan or a Prime Minister of a State presiding over the destinies of millions of people, may have hardly any furniture in his house though it may be of palatial dimensions. His reception rooms may have floors of marble, mosaic, or polished tiles and will be washable and clean. There may be hardly any carpets to accumulate dust and dirt. The Dewan himself will go about barefooted at home as the best of persons do in the South. Our Dewan may squat on an *asan* on the floor and eat, perhaps off a plantain leaf. He may not have been initiated into the art of wielding knives and forks, for it is an art not easily acquired, following sacred rules not meant for the common folk. He may use his nature-bestowed fingers, and when he has finished his repast, the leaf will not have to be washed but may be thrown away and may be readily disposed of by a goat which will turn it into milk for its owner! There will be only his fingers to be washed. By contrast this will be termed a "low" standard of living.

Is this an appropriate use of terms "high" and "low"? If the standard or norm must contain a multiplicity of material wants artificially created, then only these terms will have any significance. But if we choose to be perverse and regard as desirable that which calls into play the highest faculties in man, then the Dewan's life follows a higher standard than the British gardener's whose standard now becomes "low." For a standard based on material considerations the more suitable terms will be "complex" and "simple" rather than "high" and "low." We may then say that the Dewan's standard of life is "high" but "simple" and the British gardener's is "low" but "complex." It would appear as though the present terms have been specially devised to convey a psychological preference for the "complex" standard which is the foundation of a good market for the manufacturers. Who will rationally fall for a standard which is dubbed "complex"?

The complex standard converts its devotee into a drudge. From dawn till nightfall the British gardener's wife, if she means to be reasonably clean, has to toil away at sweeping the carpets with a vacuum cleaner, polishing the window panes, washing the curtains, bed and table linen, the dishes, plates, etc., and cooking utensils, apart from attending to her daily round of duties such as shopping and kitchen work. To clean even one fork properly between the prongs will take more time and labor than washing one's hands. . . .

All this for what? Her time is filled up with work that brings little of real life. . . .

For our country no one standard can be fixed. Any form chosen will have to be selected after fully taking into consideration the local demands of nutrition, climate, facilities for human progress, opportunities for expressing personality, etc.

In South India, rice as staple food may be adequate but it must be unpolished [to retain, for example, vitamin A—see selection by Vandana Shiva in Chapter 8] and balanced with other articles like milk, dhal, vegetables, fat, etc. The climate here may not call for much clothing or any footwear and a mat may be sufficient bedding. In the North, wheat may do duty as a staple with other articles to balance the diet. The severe cold of winter may call for more clothing and footwear, charpoys, or cots, etc. So what is a necessity in one place may be a superfluity in another. Hence the need to judge the mode of life in close relationship with local circumstances and environment. . . .

The British gardener's beds may be equipped with spring mattresses. These are manufactured in factories with the labor of those who formerly were helpers in the gardener's household cleaning carpets and washing dishes—but were displaced by labor-saving devices and drawn away by the factory owner by the lure of a complex standard of life. Such mattresses are made of steel springs which are themselves factory products. If any part of the mattress requires attention the factory's "service squad" will have to be called. There is no organic unity between the life of the people and the production of such a mattress.

Our Dewan, leading a high but simple life, may sleep on a mat, not necessarily a coarse one. It may well be a "Patumadai" creation [Pattamadai, a town in Tamil Nadu state, is famous for its hand-woven mats] with silk warp and made of reeds split into thirty-twos or even finer. Those mats are cooler than quilts or mattresses and they are local products. The making of these provides scope for the mat-weavers to develop their sense of art and skill in workmanship and affords an outlet for their creative faculty; thus it helps in building and expressing their personality. These mats have various artistic designs worked into them and are so supple that they can be folded like silk. They are clean being washable. Of course, the high quality ones are expensive. Mats may range from eight annas a pair to 200 rupees each according to the material used and workmanship involved. What the Dewan may spend on these will go directly to support and maintain the artisans and their families and so forms a complete cycle with the locally available reeds which constitute the raw materials. Such an economy does not require the Army, the Navy, and the Air Force to secure their raw materials, find or make the markets, and to keep the long ocean lanes open and safe. Hence, they have no need of violence, as would be the case if the Dewan patronized spring mattresses made in Britain and included them in his "Standard of Living." . . .

The British gardener's standard of living was strictly individualistic in that it was not correlated to the life of the people around him. It was confined to the four walls of his house. It is said "An Englishman's house is his castle." Yes, it effectively shuts out the world however much of material creature comforts it may provide for those inside. Such isolation from the life currents around them is caused in our country also by those who follow western modes of life.

The norm we seek for is not for a single family or even a class or group but for the local population as a whole. This means the norm will interlink the life of everyone. In a way, our ancient village organization attempted something on these lines when it tried to assure every inhabitant his subsistence by allocating an annual share to each from out of the total produce of the village, in the form of "Baluta," "Padi," etc., to its members who serve as barbers, chamars, mochis. This system recognized that they all formed one corporate whole. But what we want is not merely provision for bodily existence, but a provision also for opportunities of development of the higher creative faculties of man.

To refer again to our Dewan, when he wants a leather case for his papers, he would call in the mochi, specify the quality of leather he requires, and the shape, size, and accommodation needed. The mochi, in his turn may get the chamar to tan the required quality leather. All this will present several problems which will have to be solved. This provides scope for ingenuity and resourcefulness. Thus the Dewan's demand opens up opportunity for the exercise of the creative faculty of those around him. If instead, the Dewan walked into a British store and bought a ready made article, such a brief case may not be exactly what he wants as he had only to choose out of the ready stock. Besides, he may not even have exercised his mind as to what he wants. The thinking would have been done in advance for him by the manufacturers, not for him in particular, but as a general proposition. When he orders a thing locally he himself thinks of the various details and decides the kind of article he wants and then directs those around him to produce it. In this way, the life and thought of the consumer is closely entwined with the life and creative faculty of the producer, each attempting to solve the problems formulated by the other. Our lives are not independent entities but are closely [connected] one with another. A proper standard of life will then be the silken strand which strings together the goodly pearls of life—individual members of society. Such is the Dewan's standard of living in that it connected up his life not only with those of the spinners, weavers, mat-makers, chamars, mochis, etc., but also with his dumb fellow creatures such as the goat that fed on his dining leaf. No man liveth unto himself.

Joseph Tainter, from *The Collapse of Complex Societies* (1988)

Sustainability, yes—but for how long? Entropy poses an ultimate limit: someday in the unimaginably distant future, all processes requiring energy will have ceased. But even on a scale of decades and centuries, human enterprises and institutions face obstacles to sustainability. Diminishing returns are a familiar phenomenon. In mining, for example, the easiest and most accessible veins of ore are exhausted first, and as producers move on to less accessible veins it becomes progressively more difficult and expensive to produce each additional ton of ore. In agriculture, a doubling of production per acre requires an investment of labor and energy far greater than double. Tainter, an archaeologist, finds that diminishing returns on aggregate investments in the many facets of social complexity—from agricultural production and mineral extraction to public administration and scientific research—may help explain why civilizations collapse, and may pose the gravest threat to the sustainability of cherished institutions.

More complex societies are more costly to maintain than simpler ones, requiring greater support levels per capita. As societies increase in complexity, more networks are created among individuals, more hierarchical controls are created to regulate these networks, more information is processed, there is more centralization of information flow, there is increasing need to support specialists not directly involved in resource production, and the like. All of this complexity is dependent upon energy flow at a scale vastly greater than that characterizing small groups of self-sufficient foragers or agriculturalists. The result is that as a society evolves toward greater complexity, the support costs levied on each individual will also rise, so that the population as a whole must allocate increasing portions of its energy budget to maintaining organizational institutions. This is

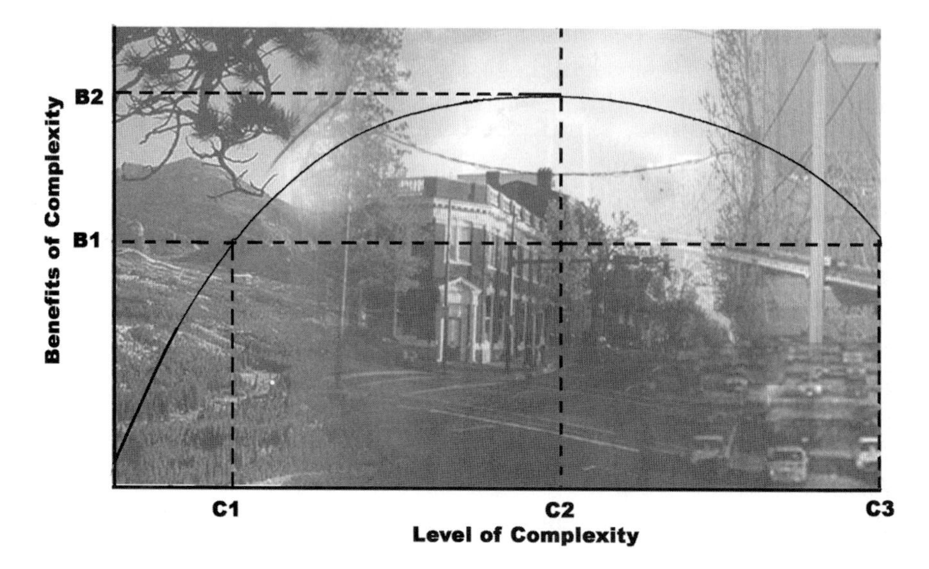

Figure 5.4 The marginal product of increasing complexity. Photos courtesy of USDA.

an immutable fact of societal evolution, and is not mitigated by type of energy source.

. . . The cost-benefit curve for these investments increases at first favorably, for the easiest, most general, most accessible, and least expensive solutions are attempted first. As these solutions are exhausted, however, continued stresses require further investments in complexity. The least costly solutions having been used, evolution now proceeds in a more expensive direction. The hierarchy expands in size, complexity, and specialization; resource production focuses increasingly on sources of supply that are more difficult to acquire and process; agricultural labor intensifies; information processing and training requirements become less generalized; and most likely, an increased military apparatus is seen as the solution to these problems. . . .

Thus a growing sociocultural system ultimately reaches a point such as B1, C1 on the curve in Figure 5.4, whereafter investment in further complexity yields increased returns, but at a declining marginal rate. When this point is reached, a complex society enters the phase where it becomes increasingly vulnerable to collapse. . . .

Under such conditions, the option to decompose (that is, to sever the ties that link localized groups to a regional entity) becomes attractive to certain components of a complex society. As marginal returns deteriorate, tax rates rise with less and less return to the local level. Irrigation systems go untended, bridges and roads are not kept up, and the frontier is not adequately defended. The population, meanwhile, must contribute ever more of a shrinking productive base to support whatever projects the hierarchy is still able to accomplish. Many of the social units that comprise a complex society perceive increased advantage to a strategy of independence, and begin to pursue their own immediate goals rather than the long-term goals of the hierarchy. Behavioral interdependence gives way to behavioral independence, requiring the hierarchy to allocate still more of a shrinking resource base to legitimization and/or control.

Thus, when the marginal cost of participating in a complex society becomes too high, productive units across the economic spectrum increase resistance (passive or active) to the demands of the hierarchy, or overtly attempt to break away. Both the lower ranking strata (the peasant producers of agricultural commodities) and upper ranking strata of wealthy merchants and nobility (who are often called upon to subsidize the costs of complexity) are vulnerable to such temptations. . . .

And so, societies faced with declining marginal returns for investment in complexity face a downward spiral from problems that seem insurmountable. Declining resources and rising marginal costs sap economic strength, so that services to the population cannot be sustained. As unrest grows among producers, increased resources from a dwindling supply must be allocated to legitimization and/or control. The economic sustaining base becomes weakened, and its members either actively or passively reduce their support for the polity. Reserve resources to meet unexpected stress surges are consumed for operating expenses. Ultimately, the society either disintegrates as localized entities break away, or is so weakened that it is toppled militarily, often with very little resistance. In either case, sociopolitical organization is reduced to the level that can be sustained by local resources. . . .

The region on the marginal product curve (Figure 5.4) between B1, C1 and B2, C2 depicts a realm in which a complex society experiences increased adversity and dissatisfaction. Stress begins to be increasingly perceived, and if modern history is any guide, ideological strife (for example, between growth and no-growth factions) may become noticeable. The system as a whole engages in "scanning" behavior, seeking alternatives that might provide a preferable adaptation. This scanning may result in the adoption by segments of the society of a variety of new ideologies and life-styles, many of them of foreign derivation (such as the proliferation of new religions in Imperial Rome). Some of these may be perceived by the hierarchy as hostile and subversive, others become briefly fashionable. At the same time, in an industrial society facing declining marginal returns, there may be increased investment in research and development (to the extent that declining resources permit), as solutions to declining productivity are sought, and in education, as individuals position themselves to reap a maximum share of a perceptibly faltering economy. Taxes rise, and inflation becomes noticeable. Prior to point B2, C2 investment and intensification can still produce positive benefits, but collapse becomes increasingly likely.

The region between B2, C2 and B1, C3 is critical. In this part of the curve increasing complexity may actually bring decreased overall benefits, as the economic system and the sustaining base are taxed to the point where productivity declines. All segments of the society compete for a shrinking economic product. This is a realm of extreme vulnerability, as a major perturbation or stress surge will impinge on a society that has inadequate reserves. Surplus production for investment in research and development has declined. The scanning behavior of the previous stage may be terminated, as the hierarchy imposes rigid behavioral controls (as in later Imperial Rome) in an attempt to increase efficiency. . . .

Much of the foregoing may read like the doom and gloom that issues from the Club of Rome.[1] Economists and others will rightly ask whether all this is really inevitable, or whether some salvation such as technical innovation can stave off collapse and permit continued growth. Tied up in all this is the question of the future of contemporary complex societies. . . .

In industrial societies, technical innovation responds to market factors, particularly physical needs and economic distress.[2] It is not, though, always the panacea that is imagined. . . . Technological innovation [too] is subject to the law of diminishing returns, and this tends to reduce (but not eliminate) its long-term potential for resolving economic weakness. Using the data cited by Wolfle,[3] Scherer observes that if R&D expenditures must grow at 4–5 percent per year to boost productivity 2 percent, such a trend cannot be continued indefinitely or the day will come when we must all be scientists. He is accordingly pessimistic about the prospects for long-term productivity growth.[4] Colin Renfrew correctly points out (in the context of discussing the development of civilization in the Aegean) that economic growth is itself susceptible to declining marginal productivity.[5]

For human societies, the best key to continued socioeconomic growth, and to avoiding or circumventing (or at least financing) declines in marginal productivity, is to obtain a new energy subsidy when it becomes apparent that marginal productivity is beginning to drop. Among modern societies this has been accomplished by tapping fossil fuel reserves and the atom. Among societies without the technical springboard necessary for such development, the usual temptation is to acquire an energy subsidy through territorial expansion. The occurrence of this temptation runs the gamut from simple agriculturalists[6] to great empires. Whenever the marginal cost of financing a social system's needs out of local yearly productivity becomes perceptibly too high, this solution must seem attractive.

The force of this attraction need hardly be argued, for the rise and expansion of empires provides one of the unequivocal touchstones of history. Such expansion, where successful, has at least the short-term advantage of providing the subsidy sought, as the accumulated reserves of the subject population, and a portion of their yearly productivity, are allocated to the dominant polity.

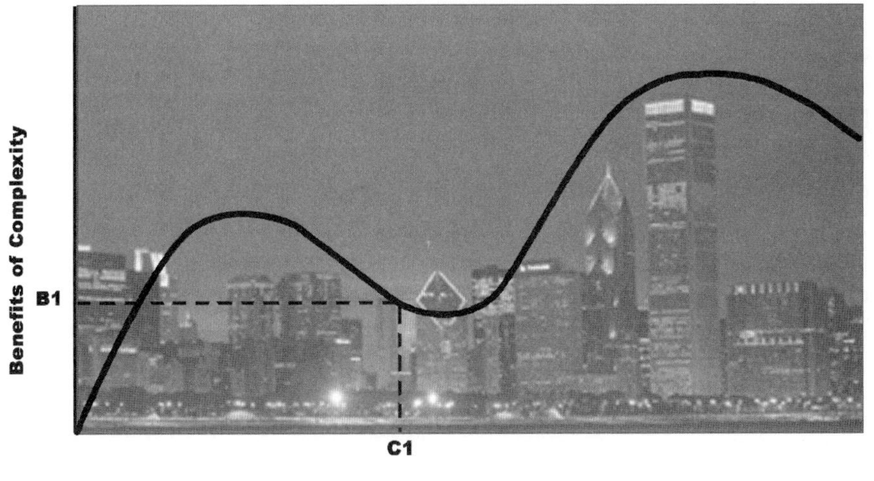

Figure 5.5 The marginal product of increasing complexity, with technological innovation or acquisition of an energy subsidy. Photo courtesy of the City of Chicago.

When some new input to an economic system is brought on line, whether a technical innovation or an energy subsidy, it will often have the potential at least temporarily to raise marginal productivity. In the long run, however, marginal returns will ultimately begin to decline again, for the reasons discussed throughout this chapter. This process is illustrated in Figure 5.5. In the curve produced here, B1, C1 represents the point where, under the pressure of diminishing returns, a productive technical innovation or a new energy subsidy is adopted. Marginal productivity starts to rise for those aspects of complexity related to acquiring and initially developing the subsidy. . . . Ultimately, though, another point of declining marginal returns is reached, presaging further innovation and expansion, or collapse. The curve shown in Figure 5.5 presents a more realistic expression of the economic history of some societies than does Figure 5.4, but this only emphasizes the recurring problem of marginal decline.

A complex society pursuing the expansion option, if it is successful, ultimately reaches a point where further expansion requires too high a marginal cost. Linear miles of border to be defended, size of area to be administered, size of the required administration, internal pacification costs, travel distance between the capital and the frontier, and the presence of competitors combine to exert a depressing effect on further growth. Thus, as Taagepera has demonstrated, empire growth tends to follow a logistic curve.[7] Growth begins slowly, accelerates as the energy subsidy is partially invested in further expansion, and falls off when the marginal cost of further growth becomes too high.

Once conquered, subject lands and their populations must be controlled, administered, and defended. Given enough time, subject populations often achieve, at least partially, the status of citizens, which entitles them to certain benefits in return for their contributions to the hierarchy, and makes them less suitable for exploitation. The energy subsidy obtained from a conquest is highest initially, due to plunder of accumulated surpluses, and then begins to decline. It declines as administrative and occupation costs rise, and as the subject population gains political rights and benefits. Ultimately the marginal returns for the conquest start to fall, whereupon the society is back to its previous predicament. Now, however, the marginal cost of further expansion has risen even higher.

Notes

1. E.g., D. H. Meadows et al., *The Limits to Growth* (New York: Universe Books, 1972).
2. R. G. Wilkinson, *Poverty and Progress: An Ecological Model of Economic Development* (London: Methuen, 1973); G. Mensch, *Stalemate in Technology: Innovations Overcome the Depression* (Cambridge: Ballinger, 1979).
3. D. Wolfle, "How Much Research for a Dollar?" *Science* 132 (1960): 517.
4. F. M. Scherer, *Innovation and Growth: Schumpeterian Perspectives* (Cambridge: MIT Press, 1984), 239, 268–69.
5. C. Renfrew, *The Emergence of Civilization: The Cyclades and the Aegean in the Third Millennium B.C.* (London: Methuen, 1972), 36–37.
6. A. P. Vayda, "Expansion and Warfare Among Swidden Agriculturalists," *American Anthropologist* 63 (1961): 346–58.
7. R. Taagepera, "Growth Curves of Empires," *General Systems* 13 (1968): 171–75.

John Clare, from "The Lament of Swordy Well" (1832–37)

Sustainability may be challenging to define, but we recognize unsustainability when we see it. John Clare, a modestly educated Englishman to whom poetry came almost as naturally as common conversation (and which he treated with almost as little revision), bore witness to the ravages caused by enclosure on landscapes beloved from his childhood in Northhamptonshire. As it had already done elsewhere, Parliament here subdivided and privatized communal agricultural, pastoral, and "waste" lands in 1809. Enclosure succeeded in making farming more profitable (indeed, there were fortunes to be made by the new owners during the grain shortages of the Napoleonic wars) and in generating economic growth. But it also undercut the economic security of many who had previously relied on common land for grazing, fuelwood, etc. No less important to the mind of the poet, the spirit of enclosure, with its rash of fences and "no trespassing" signs, undercut individual liberty and the vibrancy of traditional rural life. Still worse, in the case of Swordy Well, was the fate of the land itself. As the site of an ancient Roman quarry, Swordy Well had been a picturesque location and a haven for wildlife, valuable as pastureland but probably not well suited to agriculture. Following enclosure, it fell into the unscrupulous hands of parish (town) officials. Speaking in the voice of Swordy Well, Clare forcefully ac-

counts the losses borne by the poor, by the flora and fauna, and by the exploited landscape itself, in the name of greed, profit, and "gain."

The Lament of Swordy Well

. . . . .

I'm swordy well, a piece of land
That's fell upon the town,
Who worked me till I couldn't stand
And crush me now I'm down.

In parish bonds I well may wail,
Reduced to every shift;[1]
Pity may grieve at trouble's tale
But cunning shares the gift.
Harvests with plenty on his brow
Leaves losses' taunt with me,
Yet gain comes yearly with the plough
And will not let me be.

. . . . .

Though I'm no man, yet any wrong
Some sort of right may seek,
And I am glad if e'en a song
Gives me the room to speak.
I've got among such grubbling[2] gear
And such a hungry pack,
If I brought harvest twice a year,
They'd bring me nothing back.

When war their tyrant prices got
I trembled with alarms;
They fell and saved my little spot
Or towns had turned to farms.
Let profit keep an humble place
That gentry may be known;
Let pedigrees their honors trace
And toil enjoy its own.

The silver springs grown naked dykes
Scarce own a bunch of rushes;

When grain got high the tasteless tykes
Grubbed up trees, bank, and bushes.
And me, they turned me inside out
For sand and grit and stones,
And turned my old green hills about
And picked my very bones.

These things that claim my own as theirs
Were born but yesterday,
But ere I fell to town affairs
I were as proud as they.
I kept my horses, cows, and sheep
And built the town below
Ere they had cat or dog to keep.
And then to use me so!

. . . . .

The bees fly round in feeble rings
And find no blossom by,
Then thrum their almost weary wings
Upon the moss and die.
Rabbits that find my hills turned o'er
Forsake my poor abode;
They dread a workhouse like the poor
And nibble on the road.

If with a clover bottle[3] now
Spring dares to lift her head,
The next day brings the hasty plough
And makes me misery's bed.
The butterflies may whirr and come,
I cannot keep 'em now;
Nor can they bear my parish home
That withers on my brow.

. . . . .

I couldn't keep a dust of grit
Nor scarce a grain of sand,
But bags and carts claimed every bit
And now they've got the land.
I used to bring the summer's life

To many a butterfly,
But in oppression's iron strife
Dead tussocks bow and sigh.

I've scarce a nook to call my own
For things that creep or fly;
The beetle hiding 'neath a stone
Does well to hurry by.
Stock[4] eats my struggles every day
As bare as any road;
He's sure to be in something's way
If e'er he stirs abroad.

I am no man to whine and beg
But fond of freedom still,
I hang no lies on pity's peg
To bring a grist to mill.
On pity's back I needn't jump,
My looks speak loud alone;
My only tree they've left a stump
And nought remains my own.

My mossy hills gain's greedy hand
And more then greedy mind
Levels into a russet land
Nor leaves a bend behind.
In summers gone I bloomed in pride,
Folks came for miles to prize
My flowers that bloomed nowhere beside
And scarce believed their eyes.

Yet worried with a greedy pack
They rend and delve and tear
The very grass from off my back:
I've scarce a rag to wear.
Gain takes my freedom all away
Since its dull suit I wore,
And yet scorn vows I never pay
And hurts me more and more.

. . . . .

And should the price of grain get high—

Lord help and keep it low—
I shan't possess a single fly
Or get a weed to grow;
I shan't possess a yard of ground
To bid a mouse to thrive,
For gain has put me in a pound
I scarce can keep alive.

. . . . .

I own I'm poor like many more
But then the poor must live
And many came for miles before
For what I had to give.
But since I fell upon the town
They pass me with a sigh;
I've scarce the room to say sit down
And so they wander by.

The town that brought me in disgrace
Have got their tales to say
I ha'n't a friend in all the place
Save one and he's away.[5]
A grubbling man with much to keep
And nought to keep 'em on
Found me a bargain offered cheap
And so my peace was gone.

But when a poor man is allowed
So to enslave another
Well may the world's tongue prate aloud
How brother uses brother.
I couldn't keep a bush to stand
For years but what was gone;
And now I ha'n't a foot of land
To keep a rabbit on.

. . . . .

Lord bless ye I was kind to all
And poverty in me
Could always find a humble stall,
A rest and lodging free.

Poor bodies with a hungry ass
I welcomed many a day
And gave him tether-room and grass
And never said him nay.

There was a time my bit of ground
Made freemen of the slave;
The ass no pindar'd[6] dare to pound
When I his supper gave.
The gypsy's camp was not afraid;
I made his dwelling free
Till vile enclosure came and made
A parish slave of me.

. . . . .

And if I could but find a friend
With no deceit to sham
Who'd send me some few sheep to tend
And leave me as I am,
To keep my hills from cart and plough
And strife of mongrel men
And as spring found me find 'em now,
I should look up again.

And save his Lordship's woods that past
The day of danger dwell,
Of all the fields I am the last
That my own face can tell.
Yet what with stone pits, delving holes,
And strife to buy and sell,
My name will quickly be the whole
That's left of swordy well.

Notes

1. [extremity, effort in difficult circumstances]
2. [digging, uprooting]
3. [bundle]
4. [livestock]
5. [arguably, Clare himself]
6. [one who impounds stray animals]

6 Deforestation

The world's forests and native grasslands are vital to human life and the planet's ecological well-being. In addition to providing material for housing, paper, and fuel, forests serve as CO_2 sinks, recreation areas, and habitat for countless species of birds, other animals, and plants. Our economic prosperity and our psychological health, as well as the biodiversity of the planet, rest on the preservation, good management, and sustainability of the Earth's woodlands. Michael Williams and George M. Woodwell illustrate the current state of the world's forests, in a North American and global context respectively: a prospect that is encouraging in some respects but discouraging in most others.

Whether one defines deforestation as the wholesale destruction of extensive woodland areas or as the pursuit of selective but ecologically unsound logging practices, its devastating effects are clear. William Dietrich's "The Cutter" demonstrates that the local impact of even one act of clear-cutting in a forest is tremendous, in its effects on plant and animal species, but also with regard to soil, water retention, microclimate, and the like. These results in turn directly and indirectly influence the human community within or near the forest. On a different scale, Thomas Rudel and Bruce Horowitz's *Tropical Deforestation* gives evidence of the broader problems of regional or continental deforestation, particularly with the logging techniques used in much of the industrializing world. Here too, often disastrous ecological consequences arise from and further exacerbate the greed, financial need, social inequality, and communal tension that characterize the societies in which forests are located.

Yet observers resist recognizing the obvious, principally because historical, ideological, economic, ethnocentric, and even gender assumptions shape how forests and deforestation are perceived. Elizabeth Bishop's poem about the European discovery of Brazil re-creates the theologically and gender-driven vision of the New World that European explorers tried to impose upon the landscape before them: images of the forest as Edenic, fecund, and female or as fallen, wasted, and demonic predict later polarized responses to issues of conservation and forest preservation. Similarly, Alexis de Tocqueville summarizes the triumphalist ideology of America's continental expansion; acknowledges the capi-

Figure 6.1 *Le Douanier* (*The Dream*), 1910, by Henri Rousseau (1844–1910). Oil on canvas, 6' 8.5" × 9' 9.5". Gift of Nelson A. Rockefeller (252.1954). The Museum of Modern Art, New York, NY. Digital Image © The Museum of Modern Art/Licensed by SCALA/Art Resource, NY.

talist, democratic spirit behind it; and yet refuses to admit its permanent effects on the landscape. Even defenders of forests (like John Muir, who invokes romantic and Christian values) have historically viewed them and their loss through the distorting prism of their own perspectives.

Today, as well, those convinced of deforestation's danger to the future of the planet often reveal problematical assumptions. As the headnote to Robert Pogue Harrison's *Forests* reminds us (see below), his argument rhetorically engages the translation of empire theme—the view that nations rise and fall from primitivism to great civilization and power in a cyclical pattern—in order to claim that the lack of virtue that caused the Roman Empire to fall was ecological, leading to the destruction of the forests around the Mediterranean. Given the translation of empire's long-standing cultural resonance, this is effective, but it also ties ecological understanding to a moralistic, deterministic, and cyclical view of history now almost universally rejected by historians. And while the articles by Charles Peters and Robert K. Anderberg reveal faulty capitalist assumptions underlying deforestation, Dietrich reminds environmentalists that they ignore the effects of capitalism on workers in the forest industry at their political and moral peril. Indeed, as Rudel and Horowitz demonstrate, much of the misunderstanding by environmentalists of the nature, pattern, effects, and significance of Latin American deforestation stems from unexamined Marxist assumptions about economic and social structures in that region.

This case study is, therefore, an opportunity both to discuss the phenomenon of deforestation and its environmental impacts, and to examine our assumptions about what forests signify, and how that signification influences our understanding. The hope is that by being culturally self-reflective, even as the crisis of the world's disappearing forests intensifies, human reason can discern a way forward that will appeal alike to conscience and to the common interest.

FURTHER READING

Brown, Katrina, and David W. Pearce, ed. *The Causes of Tropical Deforestation: The Economic and Statistical Analysis of Factors Giving Rise to the Loss of the Tropical Forests.* Vancouver: University of British Columbia Press, 1994.

Earley, Lawrence S. *Looking for Longleaf: The Fall and Rise of an American Forest.* Chapel Hill: University of North Carolina Press, 2004.

Leopold, Aldo. "The Maintenance of Forests." In Susan L. Flader and J. Baird Caldecott, ed. *"The River of the Mother of God" and Other Essays by Aldo Leopold.* Madison: University of Wisconsin Press, 1991. 37–39.

Perlin, John. *A Forest Journey: The Role of Wood in the Development of Civilization.* New York: W. W. Norton, 1989.

Sedjo, Roger A. "Forests: Conflicting Signals." In Ronald Bailey, ed. *The True State of the Planet.* New York: Free Press, 1994.

Whitman, Walt. "Song of the Redwood-Tree" (1874).

Michael Williams, from *Americans and Their Forests* (1989)

Williams subtitles his book A Historical Geography. *He means this in the broadest sense, to cover not only the biological history of the American forests and their past and present geographical dispersion, but also their economic and material impact on American society, as well as their cultural history. The selection here illustrates the synthetic power of his account.*

> No other economic and geographical factor has so profoundly affected the development of the country as the forest. It forms the background of our early history . . . it enters into the everyday life of every American citizen.
>
> —Raphael Zon, "The Vanishing Heritage"

The immensity of the subject of the Americans and their forests owes much to the central role that the forest plays in American geography, economy, and culture. It is as large and complex a subject as the nation itself. The area originally occupied by the forest and the changes wrought by its removal play a large part in the evolution of the visual landscape and of the geographical organization of the country. For the development of the economy through the provision of building material and fuel and in the creation of agricultural land, the forest had no rival until the end of the twentieth century. The forest must also be accorded a high place in American cultural history as the words and phrases "woodsman," "frontier," "outdoor recreation," "environmental awareness," and "conservation" all bear out.

. . . It is the evolution of the space content, the creation of the visual scene, that is the main focus of what follows. It is well to remember, however, that in no place on Earth can the geography be separated from the economy and ethos of the society that produced it. The hands and the minds of people have made the geography of any place as surely as have climate and relief and soils and vegetation. Therefore, the role of the forest in American economy and American culture cannot be overlooked.

Possibly the greatest single activity in the evolution of the rural landscape of the United States has been the clearing of the forest. In a subject that is bedeviled by a multitude of statistics that commonly run into millions and billions, and therefore are difficult to appreciate and compare, a few figures are basic. The land area of the coterminous United States is 1,903 million acres. Of that land area it is calculated that between 822 and 850 million acres, or about forty-five percent, were covered originally by well-formed "commercial" forests, which made the forest the most important, and certainly the most visually dominant, vegetation on the continent. Approximately one-fifth of that forest cover was located west of the Great Plains in the Pacific coastal states of Washington, Oregon, and California and those portions of the Rocky Mountain states not too arid to support trees, such as upland Colorado, Wyoming, and Idaho. Over four-fifths, the greatest bulk of the forest, lay east of the Great Plains. It stretched from eastern Texas through the South, to a lesser extent from the extreme eastern portions of Kansas and Nebraska through the center of the continent, and in great profusion and density from mid-Minnesota through the Lake States to the Atlantic Coast and New England. In fact, the land east of the Mississippi to the Atlantic Ocean was an almost unbroken expanse of forest, although, as will be pointed out later, local studies reveal a greater variation in the forest than has been appreciated up to now.

The existence of that original forest is striking enough, but its gradual and then rapidly accelerating denudation is equally impressive. Clearing had far-reaching results. One vegetation was replaced by another—usually cropland—but often by weeds and different trees. Sometimes clearing produced permanent changes in the soils, runoff, hydrology, wildlife, and a multitude of other ecological characteristics of the landscape. New patterns of settlement and transport arose in the cleared areas. New geographies were created, starting with the ge-

Figure 6.2 A forest waterfall in Strafford, Vermont. Photo by Glenn Adelson.

ography of Indian occupation and alteration of the forest and moving through the various geographies of European occupation, all of which created the landscape of much of the United States today.

As a result of the agricultural clearing, lumbering, and other industrial and domestic impacts on the forests, it is estimated that by 1920 the original cover had been reduced to approximately 470 million acres. Of that remainder, only 138 million acres were original forest; 250 million acres were radically disturbed through grazing, cutting, and burning and did not sustain second growth; and 81 million acres were "wasted," that is to say, were nonrenewable and nonrestoring. By 1977, however, the story of destruction had changed to one of regrowth and birth. Trees have a remarkable capacity to grow if left alone. The commercial forest had grown to 483 million acres in extent and was still growing slightly in size and vigorously in volume. In addition, 250 million acres of noncommercial forest, that is, forest where the timber stand per acre is too low or its potential for growth is too small, also existed. In all, forests of some kind still occupy one-third of the land in the country.

Precision in these sorts of figures is difficult. Initially, the forest was so common that no one bothered to write about it, let alone collect statistics, and the extent of the forest and the amount of clearing went unrecorded. The very use

Figure 6.3 A controlled burn in the Apalachicola National Forest in Florida consumes palmetto and other woody underbrush, creating more natural habitat for trees and wildlife, and reducing fuel for unintended fires. Courtesy of USDA.

of the word "lumber" in North America for rough-cut wood and felled trees rather than "timber" was indicative of attitudes, for these materials were something that "lumbered the landscape"; they were useless and cumbrous. Standing timber was regarded as a waste material, of which there was an overabundance; in many places it had a value of less than zero. Bare land was worth more than land and trees. Even when statistics become increasingly refined, from about 1870 onward, the problem of determining regrowth and accompanying loss of cleared land that offered the opportunity for a second cutting (to say nothing of deciding what density of trees constituted a forest on the grassland edges of the prairies and elsewhere) adds to the complication of precise computation. The drawing up of a budget of the forests with debits and credits is fraught with all sorts of accounting difficulties.

In view of the importance of the effort to use and subdue the forest in the evolution of the landscape of the United States, it is surprising that it has not commanded more attention. But while visual change is going on, it is difficult to comprehend and appreciate: "Man gets accustomed to everything," wrote Tocqueville in 1831.

> He gets used to every sight. . . . [He] fells the forests and drains the marshes. . . . The wilds become villages, and the villages towns. The Amer-

ican, the daily witness of such wonders, does not see anything astonishing in all this. This incredible destruction, this even more surprising growth, seem to him the usual progress of things in this world. He gets accustomed to it as to the unalterable order of nature.[1]

The same sort of insensitivity to the imperceptible and gradual changes affecting the forest has typified the years since Tocqueville wrote. Barely anyone has attempted either to quantify or to synthesize the various impacts that have been made on the forest.

Note

1. Alexis de Tocqueville, "A Fortnight in the Wilds," *Journey to America*, trans. George Lawrence, ed. J. P. Mayer (London: Faber and Faber, 1960), 329.

George M. Woodwell, from "Forests at the End of the Second Millennium" in *Forests in a Full World* (2001)

This selection summarizes the current global situation of forests, focusing on their role in helping the Earth cope with carbon emissions and global warming. Woodwell's general view of the prospects of the forests is realistic in admitting the difficulties of prediction, but dire on the basis of the evidence we have. Despite the reforestation, renewal, and protection that have occurred in certain parts of the world, other areas suffered extensive forest devastation, with profound consequences.

. . . The most successful civilizations have decayed as forests were destroyed and replaced with shrublands. The shrublands were further impoverished by intensive grazing and erosion until underlying rock emerged and became the landscape. A recent chronicle by Perlin follows the course of Western civilization and shows persuasively the correlation between the decline in forests and the impoverishment of the landscape and the people.[1] The story is similar in ancient Syria, Persia, throughout the Levant, and the Mediterranean Basin. The decline of the Mycenaean state and, later, ancient Greece paralleled the loss of forests and the impoverishment of the landscape; so, too, with Carthage in North Africa, once a rival of ancient Rome, and, still later, Rome itself. The process continues. We have the contemporary examples of Haiti and Madagascar, now overpopulated and deforested and degraded to the point where their landscapes are dysfunctional, incapable of supporting organized society. Rivers no longer flow in established channels; water from storms in the mountains appears as floods in the lowlands and carries silt to fill harbors and destroy the coastal fisheries. The stability of the landscape is a sine qua non of the restoration of water supplies, the potential of irrigated agriculture, and the capacity of the islands to feed themselves. The restoration of a functional landscape will begin, if it happens at all, with massive aid to reforest and stabilize the uplands to regain control of water flows and water quality and to reestablish an essential and potentially infinite source of timber and fuel. Although economic and political reform may be necessary, economic and political reform that ignores the basic requirement of a stable and habitable landscape, and the necessity for public support for establishing and maintaining it, will fail. A government and a modern economy require a functional landscape to start with. . . .

Forests are thought to have covered at one time 6×10^9 ha [hectares] of the approximately 13.5×10^9 ha of land, more than 44 percent of the land area. They affect the Earth's great cycles of carbon, nitrogen, and sulfur; they affect the color of the Earth and therefore its reflectivity and its temperature; they affect local and regional water supplies; they are the major reservoir of plants and animals; and they are the home of millions and the livelihood of millions more. Their plants and soils contain enough carbon so that any change in the area or stature of forests affects the composition of the atmosphere. The expansion of agriculture into forested regions and the impoverishment of forests from intensive harvests, fires, grazing, and toxification have reduced the forested area of the Earth substantially. Current estimates of this transition are uncertain. One, based on FAO [the Food and Agriculture Organization of the United Nations] data,[2] suggests that "woodlands," a term that includes varying stages of the impoverishment of forest,[3] currently cover 5.3×10^9 ha globally or about 39 percent of the land area. The numbers do not reveal the extent of human disturbance and impoverishment. A more recent appraisal suggests that forests and woodlands currently cover 28 percent of the land surface and are declining at about 11 million hectares per year.[4] Still more recent aerial surveys of forests in the Amazon Basin have revealed extensive "cryptic deforestation," not easily detected on satellite imagery. The phrase describes the removal of large trees and the development of a network of roads that leads to further deforestation. The conclusion is that earlier estimates of deforestation were low, perhaps by as much as a factor of two. "Woodlands" include extensive areas of abandoned agricultural land, impoverished pasture, and cutover land in varying stages of succession. There is, moreover, virtually no forested area that is free of recent human influence, often severe.

Despite these changes in the area and reductions in the vigor of forests . . . geophysical data seem to be consistent in suggesting that there is a net withdrawal of carbon from the atmosphere as carbon dioxide and that the withdrawal is large in the middle latitudes of the northern hemisphere. The assumption, supported by limited data, is that the forests of these latitudes are increasing their storage of carbon in plants and soils. The analyses are becoming generally accepted as correctly defining the direction, but not the magnitude, of the net exchange of carbon between the atmosphere and land in that region at present. The cause of the net accumulation is the successional status of the forests in that region following the heavy cutting of the nineteenth and early twentieth centuries. There is also an expansion of forests into regions abandoned from agriculture during the last century. Houghton et al. have shown that the uptake during the 1980s in North America offset 10–30 percent of U.S. fossil fuel emissions.[5] . . .

Whatever the present circumstance with respect to carbon storage in forests, the probability is high that continued warming globally, if it proceeds at rates experienced in recent decades and anticipated for ensuing decades, will reverse the process by destroying the present forests and speeding the decay of organic matter in soils of forests and tundra. The result is expected to be a large amount of carbon released into the atmosphere over a period of decades, depending on the rate of warming. Forests and their soils plus the soils of the tundra probably contain twice as much carbon as the atmosphere currently, possibly as much as 1,500 billion tons. If the warming proceeds as rapidly as 0.1 to 0.2 or more degree [Celsius] per decade for the Earth as a whole, as occurred during the 1980s and 1990s and is projected for the next years, the warming will constitute an extreme chronic disturbance and the dominant cause among several of progressive biotic impoverishment. The problem has several additional aspects, including the potential reduction in the capacity of the oceans for absorbing carbon dioxide as the thermo-haline circulation moves toward stability under global warming. This transition has been anticipated as a further positive feedback that will intensify and speed the climatic disruption, the process of biotic impoverishment of forests, and the release of carbon from them into the atmosphere.

The current interest in forests is based in part on just this concern about global warming. The role of forests as a cause and, even more importantly, a potential cure of the warming has brought new attention to retaining sufficient forests globally to protect public interests in a habitable Earth.

But at the same time forests are in demand for more conventional purposes. The interests of the rural poor, the forest dwellers, and subsistence farmers globally lie in a stark and narrow dependence on the land for food and fuel and fiber. The fuel, even today, for untold millions, perhaps for more than one-third of the people in the world, is wood. The extent of this dependence is difficult to measure, even to recognize, until one discovers women in many African nations and in sections of Mexico and elsewhere trudging miles daily across a desolate landscape that was once forested to gather a bundle of sticks for fuel. Haiti's forests have been reduced to 3 percent or less of the original forest area. The land has been turned to pasture or small-plot agriculture to feed the population of an insular nation that is currently 30 percent deficient in food. Mangroves, essential in support of the coastal fisheries, are now being cut for fuelwood or for charcoal to be sold in the cities. In other parts of the world where fuelwood is lacking, as for many millions in India, dung is dried to be used for fuel. Again, the process is self-defeating: the land is deprived of the organic matter and nutrients required for success in agriculture and in reforestation.

Figure 6.4 View from across a lake in Brunswick, Vermont. Photo by Glenn Adelson.

The FAO study of 1980 dealt with the tropics only and, while it covered the needs of more than two billion people who depend on wood for fuel, it found more than half of these were afflicted by shortages of wood.[6] The greatest problems were in Africa but difficulties were common in the most densely populated segments of the world, including the coldest zones of the Himalayas, the Indus and Ganges plains, and the lowlands and islands of Southeast Asia. In the period since then the human population has grown by 30–50 percent and the problems of 1980 have become acutely severe.

Local demands on forests for timber and fiber range from these urgent daily demands for domestic fuel, through local needs for lumber, and the persistent pressure to join the cash economy by selling timber, wood, or charcoal. These demands are grossly different regionally, substantially unmeasured and uncontrolled. They are highly dependent on population density, affected by history, climate, soils, economic circumstance, and governmental policy as well as other factors. They are large and may be overwhelming regionally to the point of making difficult or impossible any larger plan for economic development designed to restore a dysfunctional landscape. They will increase as long as population increases and the downward cycle of poverty and biotic and environmental impoverishment continues. If there is any action in amelioration possible, it is short-term as long as the population is not controlled.

The net effect is an expanding demand for wood for fuel that is acute in the most densely populated tropics and subtropics globally. The demand may have little influence on the price of wood, however, because most of the fuel is outside the cash economy.

Inhabitants of the industrialized world tend to think of energy from biomass as a novelty, overlooking the heavy dependence on wood of the earlier years of the century and all previous centuries and the continuing dependence of the poor globally on wood for cooking and heating. In the wealthy industrialized world forests are worked for profit, whether the profit lies in land or timber or fiber. That purpose invades the less developed world wherever sufficient control over land can be accumulated to allow commerce to displace the interests of forest dwellers or a diffuse public: the forests are harvested for profits in timber and land. Corruption is common enough to be recognized as a major factor influencing these trends.

The issues are strongly technical, but their resolution will have profound political and economic consequences in a world that is progressively packed with people. The scientific community, preoccupied with its own interests, not entirely public, has neglected such topics and has little to say at the moment. The widely accepted general assumption that forests are good and necessary has never been adequate to preserve them. Nor have the quality and continuity of the human habitat been equated previously so clearly with the continuity of forests globally. The emergent "public interests" as opposed to corporate interests or other "stakeholder" interests require recognition, definition, and defense. They are increasingly to be defined not only qualitatively as in the past, but now, quantitatively, by science. . . .

It is the emergence and general recognition of such issues of the public interest globally that mark this moment as different from any previous moment. The Earth is clearly undergoing changes due to human activities that are accruing as systematic reductions in its capacity for sustaining the present civilization. A new focus on the details of the preservation of the human habitat is appropriate. The technical and scientific challenge is straightforward and simple by comparison with the political, but the scientific community is barely in the earliest stages of addressing it. Both will require careful cultivation outside the normal channels of formal international meetings to find new ways of defining and protecting the emergent public interests. But the first steps lie with science.

We recognize that, while the new challenges may be global, the management of forests is local. Local activities must sum to a world that works and will continue to work indefinitely. Three immediate charges emerge: (1) a definition and an appraisal of the emergent global issues that involve forests and their management now in contrast to the immediate past; (2) an appraisal of potential demands on forests and forested land for fiber (including energy) and for alternative land uses such as agriculture over the next decades; and (3) a suggestion of what will work in an increasingly crowded world.

Notes

1. John Perlin, *A Forest Journey: The Role of Wood in the Development of Civilization* (New York: W. W. Norton, 1989).
2. N. P. Sharma, R. Rowe, K. Openshaw, and M. Jacobson, "World Forests in Perspective," in *Managing the World's Forests*, ed. N. P. Sharma (Dubuque, Iowa: World Bank, Kendall/Hunt, 1992), 17–31.
3. G. M. Woodwell and R. A. Houghton, "The Experimental Impoverishment of Natural Communities: Effects of Ionizing Radiations on Plant Communities, 1961–1976," in *The Earth in Transition: Patterns and Processes of Biotic Impoverishment*, ed. G. M. Woodwell (New York: Cambridge University Press, 1990), 9–24.
4. Food and Agriculture Organization (FAO), *State of the World's Forests 1997* (Rome: FAO, 1997).
5. R. A. Houghton, J. L. Hackler, and K. T. Lawrence, "The U.S. Carbon Budget: Contributions from Land-use Change," *Science* 285 (1999): 574–78.
6. FAO, Map of the Fuelwood Situation in the Developing Countries (Rome: FAO, 1981).

Robert Pogue Harrison, from *Forests: The Shadow of Civilization* (1992)

Harrison invokes the ancient concept of the translation of empire: the belief—popularized again in the eighteenth century—that nations cyclically rise to predominance and then fall due to a gain or loss in virtue. He argues that it was a lack of environmental virtue, resulting in the human deforestation of the Mediterranean littoral, that caused the original translatio imperii: *the fall of Greece and then Rome.*

. . . In their drives to promote their civilizations both the Greeks and the Romans also promoted a mindless deforestation of the Mediterranean. Already by the fourth century B.C. Plato recalls with nostalgia a time when forests still covered much of Attica. Speaking of the hills around Athens, Plato writes in the *Critias:*

> In comparison of what then was, there are remaining only the bones of the wasted body . . . all the richer and softer parts of the soil having fallen away, and the mere skeleton of the land being left. But in the primitive state of the country, its mountains were high hills covered with soil . . . and there was abundance of wood in the mountains. Of this last the traces still remain, for although some of the mountains now only afford sustenance to bees, not so very long ago there were still to be seen roofs of timber cut from trees growing there, which were of a size sufficient to cover the largest houses; and there were many other high trees, cultivated by man and bearing abundance of food for cattle.[1]

The deforestation Plato alludes to in this passage came about largely as a result of the Athenian navy's need for wood. Forests became fleets, sinking to the bottom of the wine-dark sea. Trees became masts, drifting among the waves of Poseidon. The temple to Poseidon at Cape Sounion, overlooking the waterway that leads into and out of the bay of Piraeus, is an inspiring monument still today, but the barren mountain on which it stands, as well as the entire surrounding landscape, now drenched with that brilliant Hellenic light, shows no traces of the forests that once covered them.

As for the agrarian Romans, the insatiable mouth of empire devoured the land, clearing it for agriculture and leading to irreversible erosion in regions that were once the most fertile in the world. It is hard to imagine that a civilization as brilliant as that of the Greeks, or an empire engineered and administered so efficiently as that of the Romans, could remain so blind in their practices as to bring about the ruin of the ground on which their survivals were based. In the following passage David Attenborough describes the ecological legacy of our "antiquity":

> To them [the Romans], it seemed that nature could be ravished and plundered as men wished. They saw no reason why men should not take what they wanted as often as they wanted. The state gave legal title to undeveloped land to anyone who cleared it of forest. As the human population around the Mediterranean grew, so more and more of the forests that had once girdled it with green were destroyed. . . . When states went to war, entire forests were devastated to provide the armies with vehicles and the navies with ship[s]. So, as the classical empires spread from east to west along the Mediterranean and north into Europe, the forests were demolished.
>
> The consequences were most severely felt on the southern and eastern shores, where the rainfall was low. Here the forests had been a key factor in maintaining the health of the land. They absorbed the rain when it fell in winter, and retained it in the soil around their roots. In summer they released it slowly, so that the shaded land never dried out entirely, and springs flowed throughout the year. Their removal was catastrophic. The provinces of North Africa were, originally, among the richest in all the Empire. Six hundred cities flourished along the African shore between Egypt and Morocco. . . . By the end of the first century A.D., North Africa was producing half a million tons of grain every year and supplying the huge city of Rome, which had outstripped its own agricultural resources, with two-thirds of its wheat.
>
> The end was not long in coming. There is still argument as to how much a change in climate contributed to the final collapse. The balance of opinion seems to be that, though rainfall did diminish, the crucial blow was the stripping away of trees and the relentless plowing and replowing to extract maximum tonnage of crops. Year after year the soil of the fields was lost. In summer it was baked by the sun and blown away by the hot winds. In the

> winter, rain storms swilled it away and rivers carried it down to the coast and deposited it in their deltas. . . .
>
> All along the African coast, the land dried out. Wheat could no longer be grown; olives, which had once been prohibited by law lest they should displace the more highly valued wheat, were the only crops that would grow. Then even they began to fail. The human population dwindled. Sand blew through the stony fields and the grandiose buildings tumbled into ruins. Today, the harbor at Leptis, where once great ships came to fill their holds with grain, is buried beneath sand dunes.[2]

The syndrome described by Attenborough is best summarized by the fate of Artemis, goddess of the forests and superabundant fertility. Her temple at Ephesus was one of the Seven Wonders of the World, but it lies in ruin now, as does her city, which two thousand years ago was one of the most prosperous of the ancient world. The ruin came about not as a result of wars or some violent calamity but by the steady degradation of its surrounding environment. Samples of the pollen grains in the sedimentary strata around Ephesus indicate that four thousand years ago, around the time of the first settlements, the hills were covered with forests of oak. A few centuries later the oak gave way to plantain weed, which typically colonizes land that has been cleared for animal grazing. By 100 B.C. it is wheat pollen that predominates in the samples, indicating that pasture had given way to intensive agriculture. Transformed from forests to pasture to cultivated fields, the land around Ephesus became more productive, to be sure, but the loss of the outlying forests eventually led to disaster. As the hills could no longer retain water, the runoff rushed down into the valley. With the plowing of the land, soil erosion was exacerbated and led to a severe buildup of silt in the great harbor of Ephesus, so severe in fact that the city was eventually forced to relocate itself farther along the coast. At least four times the city's harbor silted up in this fashion, and by the ninth century A.D. it was too shallow to receive the Byzantine fleet. The city of Artemis declined into oblivion. Today it lies some three miles from the sea, prostrate under the rays of Apollo's glory.

And here we may finally return to [Giambattista] Vico. . . . Vico believed that nature and history followed two fundamentally different laws. Civilizations rise according to the "ideal eternal history" of institutional evolution. They eventually fall by virtue of a law of entropy which brings about disorder in the system as a whole. Once the cities fall, the forests return and reclaim the ground on which they were founded. For Vico nature was a closed and stable system of self-regeneration. He never suspected that civilization's law of entropy could contaminate or compromise the domain of nature as a whole, nor was he in a position, historically speaking, to suspect such a thing.

Some two-and-a-half centuries later, we now know that what Vico says about the reforestation of the civic clearings is not only inaccurate but also ironic. While forests did indeed reclaim part of Rome's civic space during the early

Middle Ages, the same is by no means true for most of the illustrious ancient cities that had their origins in the once densely forested environment of the Mediterranean. It suffices to travel around Asia Minor today and visit such cities—Ephesus, Miletus, Aphrodisias, Priene, Pergamum, Side, Kaunos, Halikarnasos, etc.—to see how nakedly they lie under the open sky. There is little in the vicinity to hide the celestial auspices now. The *lucus* [sacred grove] long ago lost its limits, and from its wide-open eye one can see today not only the ruins of a great ancient city but also those of an even more ancient forest. One face, one race. So many deserts.

Notes

1. Plato, *Critias*, in *The Complete Dialogues*, trans. and ed. B. Jowett (New York: Random House, 1937), 3.75.
2. David Attenborough, *The First Eden: The Mediterranean World and Man* (Boston: Little, Brown, 1987), 117–18.

Alexis de Tocqueville, from *Democracy in America* (1835–1840), translated by Henry Reeve

Tocqueville expounds the ideology of nineteenth-century America's expansion — that the continent was terra nullius *("a land belonging to no one"), inhabited by savages, destined for a democratic future unlike that of other nations — and then describes both the economic motives underlying Manifest Destiny, and the permanency of the forests versus the transitory nature of American civilization. This last somber note is as romantically evocative of the translation of empires as it is historically and ecologically inaccurate.*

The chief circumstance which has favored the establishment and the maintenance of a democratic republic in the United States is the nature of the territory that the Americans inhabit. Their ancestors gave them the love of equality and of freedom; but God himself gave them the means of remaining equal and free, by placing them upon a boundless continent. General prosperity is favorable to the stability of all governments, but more particularly of a democratic one, which depends upon the will of the majority, and especially upon the will of that portion of the community which is most exposed to want. When the people rule, they must be rendered happy or they will overturn the state; and misery stimulates them to those excesses to which ambition rouses kings. The physical causes, independent of the laws, which promote general prosperity are more numerous in America than they ever have been in any other country in the world, at any other period of history. In the United States not only is legislation democratic, but Nature herself favors the cause of the people.

In what part of human history can be found anything similar to what is passing before our eyes in North America? The celebrated communities of antiquity were all founded in the midst of hostile nations, which they were obliged to subjugate before they could flourish in their place. Even the moderns have found, in some parts of South America, vast regions inhabited by a people of inferior civilization, who nevertheless had already occupied and cultivated the soil. To found their new states it was necessary to extirpate or subdue a numerous population, and they made civilization blush for its own success. But North America was inhabited only by wandering tribes, who had no thought of profiting by the natural riches of the soil; that vast country was still, properly speaking, an empty continent, a desert land awaiting its inhabitants.

Everything is extraordinary in America, the social condition of the inhabitants as well as the laws; but the soil upon which these institutions are founded is more extraordinary than all the rest. When the earth was given to men by the Creator, the earth was inexhaustible; but men were weak and ignorant, and when they had learned to take advantage of the treasures which it contained, they already covered its surface and were soon obliged to earn by the sword an asylum for repose and freedom. Just then North America was discovered, as if it had been kept in reserve by the Deity and had just risen from beneath the waters of the Deluge.

That continent still presents, as it did in the primeval time, rivers that rise from never failing sources, green and moist solitudes, and limitless fields which the plowshare of the husbandman has never turned. In this state it is offered to man, not barbarous, ignorant, and isolated, as he was in the early ages, but already in possession of the most important secrets of nature, united to his fellow men, and instructed by the experience of fifty centuries. At this very time thirteen millions of civilized Europeans are peaceably spreading over those fertile plains, with whose resources and extent they are not yet themselves accurately acquainted. Three or four thousand soldiers drive before them the wandering races of the aborigines; these are followed by the pioneers, who pierce the woods, scare off the beasts of prey, explore the courses of the inland streams, and make ready the triumphal march of civilization across the desert. . . .

It would be difficult to describe the avidity with which the American rushes forward to secure this immense booty that fortune offers. In the pursuit he fearlessly braves the arrow of the Indian and the diseases of the forest; he is unimpressed by the silence of the woods; the approach of beasts of prey does not disturb him, for he is goaded onwards by a passion stronger than the love of life. Before him lies a boundless continent, and he urges onward as if time pressed and he was afraid of finding no room for his exertions. . . .

Sometimes the progress of man is so rapid that the desert reappears behind him. The woods stoop to give him a passage, and spring up again when he is past. It is not uncommon, in crossing the new states of the West, to meet with deserted dwellings in the midst of the wilds; the traveler frequently discovers the vestiges of a log house in the most solitary retreat, which bear witness to the power, and no less to the inconstancy, of man. In these abandoned fields and over these ruins of a day the primeval forest soon scatters a fresh vegetation; the beasts resume the haunts which were once their own; and Nature comes smiling to cover the traces of man with green branches and flowers, which obliterate his ephemeral track. . . .

John Muir, from "Save the Redwoods" (published posthumously, 1920)

Muir was affected by romantic views of nature. He was also deeply religious. His plea to save the redwoods bespeaks both influences. Rejecting the arguments of utility and progress that had influenced logging exponents, Muir calls upon Americans to fulfill their divinely ordained duty to preserve these trees. He likewise finds a spiritual presence in nature here, deifying the redwoods as Christ-figures being crucified by loggers.

We are often told that the world is going from bad to worse, sacrificing everything to mammon. But this righteous uprising in defense of God's trees in the midst of exciting politics and wars is telling a different story, and every Sequoia, I fancy, has heard the good news and is waving its branches for joy. The wrongs done to trees, wrongs of every sort, are done in the darkness of ignorance and unbelief, for when light comes the heart of the people is always right. Forty-seven years ago one of these Calaveras King Sequoias was laboriously cut down, that the stump might be had for a dancing-floor. Another, one of the finest in the grove, more than three hundred feet high, was skinned alive to a height of one hundred and sixteen feet from the ground and the bark sent to London to show how fine and big that Calaveras tree was—as sensible a scheme as skinning our great men would be to prove their greatness. This grand tree is of course dead, a ghastly disfigured ruin, but it still stands erect and holds forth its majestic arms as if alive and saying, "Forgive them; they know not what they do." Now some millmen want to cut all the Calaveras trees into lumber and money. But we have found a better use for them. No doubt these trees would make good lumber after passing through a sawmill, as George Washington after passing through the hands of a French cook would have made good food. But both for Washington and the tree that bears his name higher uses have been found.

Could one of these Sequoia kings come to town in all its god-like majesty so as to be strikingly seen and allowed to plead its own cause, there would never again be any lack of defenders. And the same may be said of all the other Sequoia groves and forests of the Sierra with their companions and the noble *Sequoia sempervirens*, or redwood, of the coast mountains. . . .

Any fool can destroy trees. They cannot defend themselves or run away. And few destroyers of trees ever plant any; nor can planting avail much toward

Figure 6.5 Redwood trees in Northern California. Photo by Glenn Adelson.

restoring our grand aboriginal giants. It took more than three thousand years to make some of the oldest of the Sequoias, trees that are still standing in perfect strength and beauty, waving and singing in the mighty forests of the Sierra. Through all the eventful centuries since Christ's time, and long before that, God has cared for these trees, saved them from drought, disease, avalanches, and a thousand storms; but he cannot save them from sawmills and fools; this is left to the American people. The news from Washington is encouraging. On March third [1905?] the House passed a bill providing for the Government acquisition of the Calaveras giants. The danger these Sequoias have been in will do good far beyond the boundaries of the Calaveras Grove, in saving other groves and forests, and quickening interest in forest affairs in general. While the iron of public sentiment is hot let us strike hard. In particular, a reservation or national park of the only other species of Sequoia, the *sempervirens*, or redwood, hardly less wonderful than the *gigantea*, should be quickly secured. It will have to be acquired by gift or purchase, for the Gov-

Figure 6.6 Loggers with fallen tree. Photo by Dan Todd.
Courtesy of U.S. Forest Service.

ernment has sold every section of the entire redwood belt from the Oregon boundary to below Santa Cruz.

William Dietrich, from "The Cutter" in *The Final Forest: The Battle for the Last Great Trees of the Pacific Northwest* (1992)

In these passages from "The Cutter," Dietrich describes the hard and dangerous work of logging. However, he never forgets that logging is an industry affected by an economic system. The impact of that uncaring system on the lives of loggers concerns him as much as clear cutting's negative ecological effects—hence, his challenging questions for the environmental movement.

[Russ] Poppe and [Joe] Helvey are doing the cutting on a blown-down twenty-acre quadrangle of timber in the foothills of the Olympic Mountains. It is a snarled, dangerous mess. The blowdown occurred after the U.S. Forest Service clearcut a fan-shaped bite of trees on the slope. Clearcutting—taking down every tree on a swathe of land—has been the practiced method of logging in this region for two generations. It is cheap, efficient, and leaves a clearing so barren that it gives the most desired tree to replant, Douglas fir, a chance to outgrow its shade-tolerant competitors. However, clearcutting also opens gaps in the forest as potentially calamitous to the neighboring ranks of shallow-rooted conifers as cracking the shield wall of a Roman legion. As the Northwest's forests have become increasingly fragmented by clearcutting, more and more miles of "edge" are exposed to the wind. In this case, the wind took a straight shot at the hillside from the restless North Pacific out on the horizon, striking the next draw full of trees and promptly knocking most of them over. Some had been growing two or three centuries. Now the government wants the downed trees salvaged before they decay, along with most of the survivors still standing. Just to complicate things, however, some of the biologists have spray-painted big blue W's on the bark of a few trees they want left for wildlife. Of course, what will happen to those wildlife trees or the next rank of unprotected trees, when the next big winter storm hits, no one professes to know.

Poppe had looked at the site in early summer and turned it down. The hillside is steep. The wind created a chaotic tangle of trunks and branches. The downed logs must be bucked, or cut into correct length, to fit on logging trucks and produce dimension lumber. Bucking can be even more hazardous than falling standing trees. When sawed into length, the downed logs have a tendency to lose their grip on the slope and careen downhill. The surviving trees must also come down. Some are half-rotten, and others hung up and leaning from the weight of fallen logs around them. The roots that exploded out of the ground in the storm created craters of dirt and broken sandstone, walled on one side by the upturned root ball. Other sections of the site are thick with sword fern, blackberry bramble, thistle, and fireweed.

Twice Poppe told the man who won the Forest Service bid for this mess, Rick Hurn, that he didn't want the job. Hurn persisted. It looked worse from the logging road on top than it really was, Hurn assured. He called several times, an indication he was having difficulty finding a good cutter. Finally Poppe, afraid his refusal could cost him future jobs in what promised to be a lean winter, reluctantly agreed.

"For $300 a day, a man will do just about anything," Steve [a fellow logger] jokes to Joe. The figure is a deliberate exaggeration—good cutters only earn half or a third that—but the meaning is clear. The woods are being locked up by the preservationists. A man can't afford to be too choosy. Take work where it comes.

"It's going to be a tough winter," says Steve. . . .

Poppe, breathing hard, takes a break to talk about what he does. His grandfather was a logger. His father was a manager for Crown Zellerbach timber in Cathlamet, a small pulp and paper town on the Columbia River about 150 miles to the south. Poppe's father had hoped his son might go to forestry school and get on the paperwork side of logging. But Poppe had a wife and child to support upon graduation from high school. The best money was in the woods. "It was more out of necessity than desire" that he became a logger, he said.

Now it is threatened. This kind of timber job, cutting trees centuries old, is an increasing rarity. Most old-growth trees have been put off limits while scientists try to figure out how much "ancient forest" need survive to ensure the survival of its ecosystem. The allowable cut on Olympic National Forest has plunged 90 percent in a decade, much of the decline bunched into the last two years. Logging on adjacent state of Washington land has declined 50 to 60 percent. The private lands are also short of trees: the biggest local landowner, ITT-Rayonier, has seen its annual cut plunge from 200 million board feet in the 1970s to 80 million now and perhaps down to 40 million in a few years. The number of logging contracts around Forks is down by two-thirds.

Maybe things will balance out. Certainly some logging will go on. But Poppe has a small farm in the Bogachiel River Valley and has begun buying pack horses. He is thinking maybe he could be a back-country guide to tourists if logging fades out.

"I like to feel I'm as environmentally sensitive as the next guy," said Poppe. "On the other hand, I don't like to see my way of life eliminated. Environmentalists have some idealistic notion of the way the world ought to be, and as long as it don't affect their pocketbook, it's okay. Well, it would be real romantic to think the world can be some kind of paradise, but paper and lumber is going to be needed. The way I feel, if it ain't natural, it ain't good. It ain't natural to make things out of synthetics. Wood is natural. This is the most natural alternative we have, plus it's renewable."

Poppe understands some of the alarm of urbanites who come to these woods. He said he hates to see the new clearcuts along Highway 101, sawn in panic before some new round of regulation forbids their harvest. He says the industry has overcut these forests the past five years. He approves of leaving strips of trees along streams to preserve water quality. He thinks there is a lot of work that could be done in these woods to correct old mistakes, to rehabilitate forest streams.

"That kind of work would be a good alternative for a logger," he muses. "None of us wants a handout."

What Poppe can't understand is the hysteric sweep of proposals to curb logging, the drastic cutbacks, the urban callousness toward thousands of timber families. A few years ago, the timber industry was helping lead the Northwest out of a severe recession. Now it is the bad guy, reviled on national television. Loggers have flailed out almost blindly at this shift, holding rallies with hundreds of logging trucks. They called for a boycott of Burger King because the fast-food chain used salad dressing sold by Paul Newman, the actor who had narrated a critical Audubon Society special on public television. Peninsula loggers petitioned the U.S. Fish and Wildlife Service to be declared an endangered species.

Dead spotted owls have been nailed to signs on Olympic National Park. Where do people think their houses come from? Their newspapers? Their toilet paper? The fibers in their clothes? Why do they scorn the rugged, artful grace of fallers like Poppe and Helvey?

Poppe considers the environmentalists a bit too self-righteous, as if they had invented love for the outdoors with their weekend hikes. Well, these loggers live in the woods and work in the woods and play in the woods. They are out in the chill rain and the hot sun. They see deer while driving to the logging job in the morning and eagles on the way home. They fish, they hunt, they hike. These forests are to them a mosaic of memory a city dweller can't imagine, a hundred places cut and regrown. To them, a clearcut isn't an end. It is a beginning.

Robert K. Anderberg, "Wall Street Sleaze: How the Hostile Takeover of Pacific Lumber Led to the Clear-Cutting of Coastal Redwoods" (1988)

Anderberg further illustrates North American deforestation's implication in capitalist economics. He describes how the "family-run Pacific Lumber Company . . . widely respected as the most environmentally sensitive timber company in the industry," was taken over in 1985 using junk bonds. The corporate raider reversed the firm's policies in order to maximize profits, causing significant human and environmental damage. Legislative and administrative responses—as well as the discipline of economic markets—have left matters in the industry unresolved.

Debate has been vigorous over the effects of so-called junk-bond financing on American corporations and the United States economy in general, and whether it forces a healthy reassessment by managers and investors of a company's business or rather simply benefits the deal makers who undertake highly debt-leveraged corporate acquisitions.

But, many people on both sides of the debate agree, one junk bond–financed acquisition in northern California may result in a horror story in which mighty coastal redwoods are clear-cut in order to amortize junk debt incurred in the hostile takeover of a once venerated timber company.

Prior to 1985, the family-run Pacific Lumber Company was widely respected as the most environmentally sensitive timber company in the industry. While other lumber companies often overharvested their inventory of redwoods and firs and depleted their long-term yield (scarring the landscape and wreaking havoc on the local economy), Pacific Lumber selectively cut old-growth redwoods under a sustained-yield policy that allowed new lumber to grow faster than old lumber was cut. For over forty years the company refused to clear-cut any of its forest holdings. The company was even known to have worked closely with the Save the Redwoods League, agreeing to preserve critical acreage until it could be incorporated into the national park system.

The family-run company also exhibited an air of paternalism towards its employees, many of whom lived in Scotia, the last continuously run company town in California. Examples of the company's paternalism, such as free college tuition for employees' children, are common in Scotia. Even after the company went public in 1975 (with its stock trading on the New York Stock Exchange), the company was still considered by environmentalists and employees alike as a "bulwark of responsibility."

Pacific Lumber was also unique in one other respect. While its 189,000-acre holdings do not even make it the largest timber company in its area, Pacific Lumber owns by far the largest majority of the virgin old-growth redwoods (*Sequoia sempervirens*, or Coast redwoods) that are not already incorporated into federal, state, or regional park systems. A tall relic of a vast forest system that prior to the Ice Age covered millions of acres, virgin old-growth redwoods today [1988] occupy some 110,000 acres. The trees, some more than twenty centuries old, grow in magnificent and serene groves. About seventy percent of this acreage is protected as parkland, and the lion-sized share that is not (some 16,000 to 17,000 acres) is owned by Pacific Lumber and was thought to be relatively safe.

A certain Charles E. Hurwitz changed all this one early September morning in 1985. While denominating himself as a Texas farm boy, Hurwitz is, in fact, a corporate raider, who through his investment vehicle MAXXAM Group, Inc. of New York (the former Simplicity Pattern Company) has built a large fortune by making heavily debt-financed acquisitions of companies whose assets (as measured by the price of their stock on an established stock exchange) are undervalued and thus can be purchased at a relative bargain.

The very factors that made Pacific Lumber so venerated among environmentalists and others may have made it a sitting duck for Hurwitz. Pacific Lumber in effect had stockpiled its inventory of redwoods, allowing trees to grow for decades before selectively cutting them in a manner designed to maintain consistent harvests. A corporate raider could obtain control of such vast timber reserves, radically increase harvests to increase cash flow and short-term profits, and subsequently sell the company. Augmenting the situation was Pacific Lumber's conservative management ("undermanaged," according to Hurwitz) which resulted in the company's having a fairly low stock price (perhaps even significantly lower than the net asset value of its timber), flat earnings, and no significant debt load.

Beginning on that fateful September morning in 1985, Hurwitz moved to obtain control of Pacific Lumber, initially through a hostile tender offer (an offer to purchase on the open market approximately 22 million shares of Pacific Lumber that were traded on the New York Stock Exchange). By January 1986, Hurwitz's MAXXAM Group had completed its acquisition of the outstanding stock of

Pacific Lumber, but at a cost: some $868 million, approximately $680.5 million of which was debt-financed by three issues of high-interest, low-grade investment (risky) "junk bonds," debt instruments that are the forte of Wall Street investment banking house Drexel Burnham Lambert. They are backed solely by the assets of Pacific Lumber, not by MAXXAM. Perhaps more telling, the purchasers of the junk bonds (primarily large institutional investors) can look for repayment of interest and principal on such bonds *only* from the sale or other use of Pacific Lumber's assets.

After obtaining control of Pacific Lumber, and as is common in a highly debt-leveraged acquisition, Hurwitz first sold the company's non-timber assets (the corporate headquarters for $30 million and an unrelated cutting and welding operation for about $250 million). Hurwitz then turned his attention to the remaining large asset of Pacific Lumber: its inventory of coastal redwoods. In 1986, Hurwitz stepped up harvests from 137 million to 248 million board feet per year (including clear-cutting of selected tracts of up to 500 acres and the harvesting of virgin redwoods), conceding in testimony before the California legislature that the increase in harvest was primarily to pay off the massive debt load incurred in acquiring Pacific Lumber. Many have noted that the interest payments on the junk debt used to acquire Pacific Lumber will leap in 1989 from $39 million a year to $83 million a year, more than Pacific Lumber's preacquisition cash flow.

Many environmentalists, residents, and even company loggers worry that not only is Pacific Lumber destroying a national treasure, but that the trees are being harvested at a rate that will ultimately jeopardize the sustained yield of the forests and the economy of the local region. Said one Pacific Lumber worker: "They're just leveling everything. . . . They're destroying the future, leaving nothing for the next generation." And Woody Murphy, great-grandson of the Murphy that built Pacific Lumber into a venerated logging company, said, "And when they're through, it'll be a moonscape."

Pacific Lumber executives claim that if they reduce timber cutting after twenty years of intense harvesting to preacquisition harvest levels, Pacific Lumber's lands will still have a substantial inventory of redwoods and Douglas fir left for the future. . . .

In the very least, Pacific Lumber is a vignette of what Representative John D. Dingell (D.-Michigan) calls "the takeover and dismemberment of a good corporate citizen"; a hostile corporate acquisition that turned the most respected timber company in America into the least respected while radically increasing the debt-load and jeopardizing the economic health of a soundly run company. "This case has enormous implications," warns Representative Ron Wyden (D.-Oregon), a member of the House Banking, Finance and Urban Affairs Subcommittee that is investigating the Pacific Lumber acquisition. "I'm not convinced it's responsible for management to cut trees to pay debts to people who live thousands of miles away."

Elizabeth Bishop, "Brazil, January 1, 1502" (1965)

Bishop recalls Amerigo Vespucci's landing at Guanabara Bay, which he named—with the date in mind—Rio de Janeiro. She does so in order to contrast the tapestried reality of the biologically diverse Brazilian forest with the Eurocentric interpretive constructs the explorers tried to impose on it: the forest as Edenic but sinful, virginal yet seductive, and so on. She implies that this moralistic and patriarchal interpretive pattern reveals the psychic roots of the European urge to destroy the continent.

Brazil, January 1, 1502

. . . embroidered nature . . . tapestried landscape.
—*Landscape into Art*, by Sir Kenneth Clark

Januaries, Nature greets our eyes
exactly as she must have greeted theirs:
every square inch filling in with foliage—
big leaves, little leaves, and giant leaves,
blue, blue-green, and olive,
with occasional lighter veins and edges,
or a satin underleaf turned over;
monster ferns
in silver-gray relief,
and flowers, too, like giant water lilies
up in the air—up, rather, in the leaves—
purple, yellow, two yellows, pink,
rust red and greenish white;
solid but airy; fresh as if just finished
and taken off the frame.

A blue-white sky, a simple web,
backing for feathery detail:
brief arcs, a pale-green broken wheel,
a few palms, swarthy, squat, but delicate;
and perching there in profile, beaks agape,
the big symbolic birds keep quiet,
each showing only half his puffed and padded,
pure-colored or spotted breast.
Still in the foreground there is Sin:
five sooty dragons near some massy rocks.
The rocks are worked with lichens, gray moonbursts
splattered and overlapping,
threatened from underneath by moss

Figure 6.7 Elizabeth Bishop. Courtesy of the Library of Congress.

in lovely hell-green flames,
attacked above
by scaling-ladder vines, oblique and neat,
"one leaf yes and one leaf no" (in Portuguese).
The lizards scarcely breathe; all eyes
are on the smaller, female one, back-to,
her wicked tail straight up and over,
red as a red-hot wire.

Just so the Christians, hard as nails,
tiny as nails, and glinting,
in creaking armor, came and found it all,
not unfamiliar:
no lovers' walks, no bowers,
no cherries to be picked, no lute music,
but corresponding, nevertheless,
to an old dream of wealth and luxury
already out of style when they left home—
wealth, plus a brand-new pleasure.
Directly after Mass, humming perhaps
L'Homme armé or some such tune,

they ripped away into the hanging fabric,
each out to catch an Indian for himself—
those maddening little women who kept calling,
calling to each other (or had the birds waked up?)
and retreating, always retreating, behind it.

Ranee K. L. Panjabi, from *The Earth Summit at Rio: Politics, Economics, and the Environment* (1997)

Panjabi affirms the importance of tropical forests, and assesses their status and future in the face of increasing deforestation. Elsewhere, he characterizes the political and economic perceptions that often lead nations with tropical forests to reject demands that they be preserved as hypocritical or self-interested. His conclusion is that the developed North must address problems of world poverty and trade as well as clean up its own environmental act before the South will follow suit.

THE FOREST AS AN ISSUE IN THE DEVELOPMENT/ENVIRONMENT DICHOTOMY

The dilemma of choosing between environmental and developmental priorities is most noticeable in the debate about the Earth's forests. It is this issue which most clearly highlights the Southern position and its suspicion of the motives of the North. Having denuded much of its forest cover in the process of industrialization and expansion of its urban population, the developed world has discovered that it has now become dependent on the developing nations, which contain approximately 90 percent of the world's tropical forests.[1] Although by one estimate developed countries have a 34 percent forest cover as against a 29 percent forest cover for developing states,[2] in between 50 and 90 percent of the treasure trove of biodiversity is to be found in the tropical forests of the South.[3] "Half of Latin America is covered by forests, as is 33 percent of Asia and 27 percent of Africa."[4] Some Southern nations are very fortunate in the extent of this vital resource. As the Malaysian Prime Minister, Dr. Mahathir bin Mohamad, pointed out at the Earth Summit, the land of his nation "is almost 60 percent covered with self-regenerating tropical rainforest with an additional 15 percent covered by tree plantations."[5] By way of contrast at the time of the Rio Summit, Britain had only 10 percent of its land under forest cover.[6] It is important to note that because of reforestation, forest acreage in the developed world continues to expand, but in developing nations the opposite is the case, and the rate of deforestation has accelerated.[7] Paul Harrison explains in *The Third Revolution* that "in some countries the losses are dramatic. In Madagascar only 34 percent of the original forest cover is left. In the Côte d'Ivoire and the Philippines four-fifths of the original forest has been cleared. Ethiopia's forest cover dwindled from 40 percent of her land area in 1940 to only 4 percent fifty years later."[8] Almost 40 percent of the world's forest cover had disappeared by 1980.[9] The Earth's remaining acres of rainforest are found mainly in the Amazon region, in central Africa, and in or near Southeast Asia (specifically in Malaysia, Indonesia, and New Guinea).[10] Of these, the Amazonian forest is the largest, covering "nearly three million square miles, an area nearly as large as the entire United States. It spans nine South American countries from Brazil in the east to Peru in the west, and from Venezuela in the north to Bolivia in the south."[11]

The Earth's forests are important for more than just their timber. Besides providing shelter and a way of life for thousands of the world's indigenous people, forests are a treasure trove of biological diversity. "Those portions of the earth that are covered with forests play a critical role in maintaining its ability to absorb carbon dioxide (CO_2) from the atmosphere and are thus essential in stabilizing the global climate balance."[12] Ben Jackson, an environmentalist with the World Development Movement, explains that "the forest plays a key part in regulating climate by its effects on the reflectivity of the land surface and by the way in which it recycles water from the earth back into the atmosphere."[13] The Earth's forests are one important element in the present war against global pollution. With the loss of one and a half acres of forest every second to felling and land clearance,[14] the entire planet stands to lose a vital resource. "The destruction of a forest can affect the hydrological cycle (the natural water distribution system) in a given area just as surely as the disappearance of a large inland sea."[15] Forests contribute to the production of rain clouds[16] and are therefore vital in bringing moisture for agricultural crops. "More water is stored in the forests of the earth—especially the tropical rainforests—than in its lakes."[17] Other benefits of forest cover include cleansing of the air and water and balancing of the climate.[18] The Earth's forests "also stabilize and conserve the soil, recycle nutrients through the shedding of their leaves and seeds (and eventually their trunks when they die), and provide the most prolific habitats for living species of any part of the earth's land surface. As a result, when we scrape the forests away, we destroy these critical habitats along with the living species that depend on them."[19]

The nation with the largest concentration of tropical forest, Brazil,[20] is already suffering the consequences of earlier deforestation. "Logging and agricultural and urban expansion have destroyed more than 95 percent of the once-vast Atlantic coastal rainforests and the coniferous Araucaria forests of southern Brazil."[21] The poverty of the majority of Brazilians and the population's pressure on the rainforest can be attributed in part to serious economic inequity within that nation. "Just 2 percent of Brazil's landowners control 60 percent of the nation's arable land. At least half of this land lies idle."[22] Although there has been some reduction in the logging of the Amazon rainforest since 1990, before then the destruction proceeded at an alarming pace. The previous military regime in Brazil (1964–85) allowed and even encouraged (with tax incentives) wholesale destruction via slash-and-burn clearing of large sections of the Amazon forest. "In the past decade the forest has declined by 12 percent,"[23] a victim of the situation created by the world's inequitable economic order. Since it provides a readily marketable commodity, wood, the logging continues, albeit not at the frantic pace of the past. Environmentalists around the world fear that economic "pressures have turned the Amazon into a battleground and may eventually turn it into a desert."[24]

There are varied statistical estimates concerning the extent of deforestation and even about the amount of forest cover left on Earth. One estimate suggests that since 1972 the world has lost approximately five hundred million acres of forests, "an area roughly one-third the size of the continental U.S."[25] The United Nations has calculated that by the year 2000 developing nations will lose approximately 40 percent of their remaining forests if present trends continue.[26] Given an annual loss of fifteen million hectares, much of it in Africa, Asia, and Latin America, the outlook for the next century is bleak indeed.[27] The Food and Agriculture Organization has concluded that tropical deforestation accelerated during the 1980s to destruction of approximately 1 percent of the forest cover every year.[28] "This amounts to two-thirds of the area of the United Kingdom. Every ten weeks an area the size of the Netherlands. A Barbados every day."[29] Vice-President Al Gore has commented that "wherever rainforests are found, they are under siege."[30] He explained the situation affecting the world's forests: "They are being burned to clear land for pasture; they are being clear-cut with chain saws for lumber, they are being flooded by hydroelectric dams to generate power. They are disappearing from the face of the earth at the rate of one and a half acres a second, night and day, every day, all year round."[31]

The disappearance of forests is also attributable to the great increase in the demand for land for cattle grazing. In India, the need to feed 196 million head of cattle has resulted in an unprecedented assault on forested acreage.[32] Forests are being felled daily and the land converted to pasture. The environmentalist Jeremy Rifkin has calculated that an average cow consumes about 410 kilos of vegetation per month and is "literally a hoofed locust."[33] As global meat production has nearly quadrupled since 1950,[34] the amount of land devoted to growing animal fodder has also increased, by as much as fivefold in Mexico between 1960 and 1980, to give only one example.[35] "Without question, ranching is a factor in tropical deforestation."[36] It has to be remembered that the reason for much of the emphasis on cattle rearing is directly related to the global demand for beef. Although beef consumption has fallen in some nutrition-conscious societies, this product is still a vital commodity in world agricultural trade.

It has been argued that the incentive to destroy tropical forests arises from consumer demand in the industrialized nations of the North.[37] "Markets in the North demand a wide range of tropical products, sometimes wood and wood derivatives, sometimes animal or plant products grown in the clearings. The hard currency earned by the export of these items is a powerful incentive to entrepreneurs and governments in the South alike. . . . The insatiable desire of consumers in the North for wood paneling, for hamburgers, and for cocaine is imperiling the future of forests in several tropical countries."[38] Whether the demand springs from the North or from the South itself, the destruction of forests is still the consequence.

The case of El Salvador is indicative of the possible future of a number of countries which are now richly forested. This Central American nation was once covered by forests over 90 percent of its land.[39] After the depredations of population encroachment and agricultural expansion, these vast tracts have been reduced to "a single 20 square kilometer plot of cloud forest."[40] Even though political and economic instability have forced approximately one million Salvadorans to flee their homes,[41] the country can barely support its still growing population. Poverty, malnutrition, and consequent diseases are part and parcel of the life of the majority of those who still live in that country.

It is estimated that Malaysia, presently rich in forests, will deplete this resource by the year 2000.[42] India, Sri Lanka, Nicaragua, Honduras, and Guinea are only some of the developing nations which have almost lost or will soon lose their forest resources.[43] For Bangladesh, the loss of forest cover has resulted in extensive flooding.[44] Thailand lost almost half of its forests between 1961 and 1985.[45]

The area of land suffering an annual loss of forest cover is approximately equal to the size of the State of Washington.[46] As with all environmental problems, it would be naive to assume that the consequences are confinable within national borders. "The presence or absence of trees anywhere is a major factor in the world's environmental health everywhere."[47] It is now apparent that "deforestation of a watershed in one country can lead to flooding in a neighboring country. Destruction of a forest habitat for migrating wildlife can lead to the loss of a species in another country. Loss of tropical forests, which act as major absorbants of greenhouse gases, can speed the warming of the world climate."[48] At the present time, forests cover only 7 percent of the land surface of the Earth.[49] However, this small area is home to between 50 and 90 percent of the ten million or more species which inhabit the planet.[50] Although it is internationally recognized that "forest conservation is the key to preserving the Earth's heritage of biological diversity and harbors the secrets of new lifesaving drugs and other products,"[51] acting on this awareness is another matter altogether.

Notes

1. Bruce Babbitt, *World Monitor*, 30 June 1992, 31:2.
2. Paul Harrison, *The Third Revolution* (London: I. B. Tauris, 1992), 90.
3. *Christian Science Monitor*, 2 June 1992, 10:2.
4. *The Times* [London], 10 June 1992, 12:8.
5. Statement by Dr. Mahathir bin Mohamad, Prime Minister, Malaysia, United Nations Conference on Environment and Development [UNCED], Rio de Janeiro, 13 June 1992.

6. *The Times* [London], 2 June 1992, 10:4.
7. Harrison, 91.
8. Ibid.
9. George J. Mitchell, *World on Fire* (New York: Scribner's, 1991), 38.
10. Al Gore, *Earth in the Balance* (Boston: Houghton Mifflin, 1992), 117.
11. Mitchell, 37.
12. Gore, 115–16.
13. Ben Jackson, *Poverty and the Planet* (London: Penguin, 1990), 29.
14. *New York Times*, 2 June 1992, A10:6.
15. Gore, 106.
16. Ibid.
17. Ibid.
18. Michael Renner, "Creating Sustainable Jobs in Industrial Countries," *State of the World, 1992* (New York: W. W. Norton, 1992), 150.
19. Gore, 116.
20. John C. Ryan, "Conserving Biological Diversity," *State of the World, 1992*, 10.
21. Ibid., 10–11.
22. Jackson, 26.
23. *Globe and Mail* [Toronto], 2 June 1992, B26:4.
24. Augusta Dwyer, *Into the Amazon: The Struggle for the Rain Forest* (Toronto: McClelland-Bantam, 1990), xi.
25. *Time*, 1 June 1992, 22.
26. *Global Outlook 2000* (New York: United Nations, 1990), 85.
27. Ibid.
28. United Nations Food and Agricultural Organization [FAO] Forest Resources Assessment Project, *Second Interim Report on the State of Tropical Forests*, 10th World Forestry Congress, Paris, September 1991; cited Harrison, 92.
29. Harrison, 92.
30. Gore, 117.
31. Ibid., 117–18.
32. Alan Durning and Holly Brough, "Reforming the Livestock Economy," *State of the World, 1992*, 73.
33. Jeremy Rifkin, *Time*, 20 April 1992, 58:1.
34. *Globe and Mail* [Toronto], 4 June 1992, A19:2.
35. Ibid., A19:3.
36. *Time*, 20 April 1992, 58:3.
37. Ivan L. Head, *On a Hinge of History* (Toronto: University of Toronto Press, 1991), 102.
38. Ibid.
39. Ibid., 125.
40. H. J. Leonard, "Managing Central America's Renewable Resources," *International Environmental Affairs* 1 (1990): 38–56; cited Head, 125.
41. Head, 126.
42. *Newsweek*, 1 June 1992, 23.
43. Ibid.
44. Ibid.
45. Ibid.
46. *New York Times*, 2 June 1992, A10:6.
47. Head, 101.
48. *Globe and Mail* [Toronto], 6 June 1992, A4:6.
49. *Christian Science Monitor*, 2 June 1992, 10:1.

50. Ibid.
51. *The Times* [London], 2 June 1992, 1:6.

Thomas K. Rudel with Bruce Horowitz, from *Tropical Deforestation: Small Farmers and Land Clearing in the Ecuadorian Amazon* (1993)

Rudel and Horowitz challenge the conventional political, economic, demographic, and cultural explanations for deforestation, both generally and in a Latin American context, where socialist critiques of society have greatly influenced analyses. Here they use two case studies from the Amazon region of Ecuador to illustrate what they believe are the three most typical patterns of deforestation.

. . . Tropical deforestation typically takes the three forms outlined in Table 6.1. Two types occur in places with large blocks of rainforest. One of them features a lead institution that opens a region up for development, usually by building a penetration road into the forest. Free riding groups and individuals take advantage of the increased access, stake out claims to land, and begin to clear it. In the other pattern peasants, investors, and government officials pool their resources in coalitions that develop rainforest regions for agricultural production. The internal structure of the coalitions varies considerably; some coalitions, es-

TABLE 6.1
Varieties of Tropical Deforestation

Size of forest	*Process*	*Precipitating factors*[1]
Large: blocks of forest	Lead institutions; free riding peasants	Natural resource extraction; demographic pressures; rising agricultural prices and proletarianization
Large: blocks of forest	Growth coalitions	Funds for infrastructure; demographic pressures; rising agricultural prices and proletarianization
Small: islands of forest	Encroachment	Demographic pressures; rising agricultural prices; proletarianization

1. The precipitating factors are listed in their order of importance in initiating each of the three processes of deforestation. The importance of these factors will vary from place to place. These judgments are based on reports of deforestation in Africa, Asia, and Latin America.

pecially those based on kinship, involve a union of equals; other groups have a leader around whom its members coalesce. In most coalitions the members divide up the labor in uneven, but understandable ways. In many coalitions the leader provides the infrastructure while the peasants, with help from investors, occupy and clear the land. A series of these coalitions changed the landscape in Morona Santiago between 1920 and 1990.

In the third pattern, which characterizes places with small, remnant patches of rainforest, cultivators nibble at the edges of the forest. Working alone or with family members, farmers expand their fields at the expense of the forests. The present chapter summarizes the three patterns, addresses questions about their prevalence in rainforest regions throughout the world, and spells out their implications for the design of policies to reduce rates of deforestation.

THE CASE STUDY

The history of development and deforestation in Morona Santiago provides multiple examples of growth coalitions that developed and deforested places. When priests and colonists joined forces to build agricultural communities around missions during the first half of the twentieth century, they cleared large areas in the valleys at the base of the Andes. When a regional development agency began to build roads in Morona Santiago during the 1960s, knowledgeable individuals formed coalitions to occupy and exploit the newly accessible lands. Public officials, colonists, and investors from the Sierra joined forces to open up the Upano-Palora region for cattle ranching during the late 1960s. Twenty years later some of the same people cooperated in schemes to develop the more remote parts of the Upano-Palora plain. The Salesians and Shuar leaders formed counter-coalitions that secured global titles to land and channeled funds from European foundations to those Shuar intent on converting forests into pastures for cattle.

The coalitions' contributions to land clearing varied with their composition and the context within which they worked. Shuar ambivalence about deforestation made their coalitions less effective than colonist coalitions in clearing land. Kin-based coalitions of Cuyes valley colonists cleared land slowly, but they persisted in clearing land even after a development agency withdrew its support. Coalitions with an agency's support, like the Upano-Palora groups, cleared land more rapidly. Changes in the macroeconomic context affected the coalitions' ability to raise capital and clear land. With the abundance of capital that accompanied the oil boom in the 1970s, coalitions could secure funds for road building, so they succeeded in opening up areas for settlement, and deforestation increased. A dearth of capital, as occurred in the late 1980s, slowed the road building and deforestation. Even with rising prices for agricultural commodities, colonists cleared little land.

The rapid land clearing of the 1970s fragmented the forest on the Upano-Palora plain and changed the dynamics of deforestation. Individual factors became more visible as determinants of deforestation. Some smallholders secured credit from the banks more easily, and they cleared larger proportions of their land. Some families had more children than other families, and the larger families cleared a greater proportion of their land.

These differences in household size also had an impact on deforestation in distant places. Large households joined coalitions searching for new lands more frequently than did small households. Although the pressures of providing for large families persuaded some fathers to look for new lands, agricultural expansion did not involve a desperate attempt to maintain a certain level of subsistence. The households which searched for new lands, in addition to being larger in size, had higher incomes than other households. The heads of these large, relatively affluent households talked about the acquisition of new lands as an investment in the younger generation rather than as an attempt to maintain their standard of living. The entrepreneurial inclinations of these peasants implies high rates of geographical mobility. Having moved once with some success, they will move again. The sons and daughters of the original colonists in Sinai are just as likely to move to an urban area as they are to move farther into the rainforest.

While the history of colonization in the Upano-Palora zone provides a clear example of the two-stage sequence in land clearing, the increased salience of the urban alternative for the second generation suggests the historical limits of this generalization. By changing the array of choices available to the younger generation, urbanization has affected decisions about clearing additional land. A comparison between Cuyes River colonization during the 1960s and Upano-Palora colonization during the 1980s illustrates this effect. In both places road construction proceeded slowly, but the Upano-Palora colonists seemed more sensitive in the 1980s to changes in the pace of construction than the Cuyes valley colonists were during the 1960s. The Upano-Palora colonists scrapped their plans to clear additional land as soon as road construction and trail maintenance ceased in the early 1980s. During the 1960s the construction of the Gima-Cuyes penetration road started and stopped, but the colonization of the Cuyes valley continued.

The different reactions of the two groups of colonists to changes in road building plans may reflect the increased importance of urban labor markets during the later period. During the 1960s the much smaller urban economies did not provide peasants with an alternative set of opportunities, so they committed themselves to colonization and deforestation. The presence of a larger urban economy and the Upano-Palora peasants' greater familiarity with it may have made them more sensitive to marginal changes in the economic opportunities afforded by colonization in the 1980s. When the growing inaccessibility of the remaining unclaimed lands diminished the attractions of colonization, the

Upano-Palora colonists immediately began to explore urban economic opportunities. If unclaimed lands become even more inaccessible and urban expansion continues, fewer examples of the first-stage group ventures would occur, and the two-stage sequence would become a less accurate description of how deforestation occurs.

HOW WIDESPREAD IS THIS PATTERN?

The case study provides numerous examples of how growth coalitions stimulate tropical deforestation, but questions remain about the number of rainforest regions characterized by these processes. The development and deforestation in Morona Santiago resembles processes that have occurred in other rainforest regions around the world. Development in Morona Santiago progressed through several recognizable stages like frontier development elsewhere.[1] The successive periods of in-migration from outside the province and migration from west to east within the province correspond to the first two stages in models of frontier development. The farm abandonment on the fringes of the Upano-Palora region during the 1980s may mark the onset of consolidation, the third stage of frontier development.

Morona Santiago's deforestation, like its development, follows a general pattern observed elsewhere in the world. Growth coalitions appear to play an important role in processes of deforestation in other places with large forests, just as they do in Morona Santiago. Studies of deforestation in Southeast Asia, West Africa, and Latin America describe the activities of growth coalitions, but the reports are more suggestive than conclusive because they do not focus on the coalitions' activities. The chief alternative form of land clearing in large forests involves lead institutions that open up regions for free riding peasants who clear land along newly constructed roads. Just how different are processes initiated by lead institutions and growth coalitions? While lead institutions and growth coalitions develop regions in quite different ways, regions opened up by lead institutions may, as they undergo settlement, give rise to growth coalitions that stimulate further deforestation.

If this historical sequence occurs frequently, then initial differences in land clearing between the two types of region will not persist, and the growth coalition model of deforestation will apply to a wider range of regions than our original discussion suggested. A brief case study of regional development initiated by a lead institution, in this instance an oil company in Ecuador's northern Oriente, addresses this question.

Missionaries established the first settlements in the northern Oriente much as they did in the southern Oriente. Carmelite Brothers in the Aguarico basin along the northern border with Colombia established missions that became the foci for settlement and land clearing by colonists from the Sierra. The priests did not have the economic resources necessary to build access roads into the missions, so few colonists followed the priests into the rainforests.[2]

The pattern of regional development changed completely with the discovery of oil in the late 1960s. Between 1969 and 1972 a consortium of oil companies airlifted approximately 40,000 workers into the region to drill wells and build access roads; 90 percent of the workers entered the region with the intention of claiming a parcel of land along one of the many new roads.[3] They spent their days off inspecting unclaimed lands.

To file individual claims to land, workers had to join a state-mandated colonization cooperative. Composed of part-time colonists and absentee owners, most of the cooperatives proved to be "fantasmas"—only in exceptional cases did the members work together to clear land. One of the more honest part-time colonists admitted when he filed his claim that he did not have any tools to work the land![4] Most part-time colonists, like this person, sought land for speculative purposes—they had no interest in clearing it. In the first two years of settlement about one quarter of all of the claimed land in the region changed hands in speculative transactions. The second owners began to clear the land.[5]

A classic corridor-shaped pattern of deforestation emerged during the next ten years. Late-arriving colonists acquired land in the second, third, and fourth rows back from the roads. With long, narrow lots, 250 × 2000 meters, farmers in the last row had to walk at least six kilometers to reach the road. As in Morona Santiago the extent of cleared land declined sharply with distance from the road.[6] The colonists with claims to these interior lots faced the same set of disadvantages as the colonists with farms far from the road in Morona Santiago. Conflicts with colonists close to the road about rights of passage complicated the marketing of products from the interior farms, and governments refused to provide funds for the construction of feeder roads.[7] Under these circumstances colonists cleared little land in the interior lots, and significant numbers of them began to abandon their farms during the mid-1970s.[8]

These changes in the fortunes of settlers affected rates of deforestation. Rapid deforestation occurred between 1969 and 1973 when the oil companies constructed a network of roads, and colonists cleared large amounts of land along new roads. It slowed down after 1973 as arriving colonists claimed interior lots, which they developed slowly, if at all.[9] The extent of cleared land continued to increase during the late 1970s because the oil companies extended their network of access roads into new areas, and colonists continued to settle along new roads. Several large enterprises—a cattle ranch, two oil palm plantations, and a logging company—also began to clear land during this period.[10]

For colonists with interior lots the efforts of local leaders offered some hope for the eventual construction of farm to market roads through their lands. Almost half of the colonists in the Lago Agrio region came from drought-afflicted regions in the highland province of Loja. A number of them formed groups and elected leaders prior to their arrival in the Oriente.[11] When the leaders arrived in the Oriente, they acquired land, set themselves up as merchants, and took an intense interest in local politics.[12]

Without a municipal government or a representative in the provincial government, the leaders could not find politicians who would support their pleas for local initiatives. When the national government created new municipalities and eventually a new province in the late 1980s, it created political posts for colonist leaders, who could then provide funds for feeder roads in return for political support. By creating local governments, the central government created sources that coalitions could tap for funds. Modest increases in the construction of feeder roads and in rates of deforestation followed. The lead institution's initial efforts created conditions which, a decade later, gave rise to growth coalitions.

A similar sequence of events occurred among private sector producers along the Napo River. A lead institution, in this instance, an African palm plantation, began to clear land along the Napo River in the mid-1970s. A few years later the company stopped clearing land and instituted a policy of indirect expansion. If small farmers with lands adjoining the plantations would promise to grow African palm and sell their harvests to the plantations, their owners would build feeder roads for the farmers and provide them with technical assistance in cultivating African palms.[13] Through this arrangement the small landowners got assistance in intensifying production and the plantation owners increased the size of their enterprises without becoming embroiled in land conflicts with neighboring peasants. The resulting expansion in African palm cultivation came at the expense of the surrounding rainforest. In this case, as in the oil fields, a lead institution initiated the land clearing, but growth coalitions made vital contributions to subsequent land clearing in the region.

Comparisons of deforestation in the northern and southern Oriente suggest that, while the processes differed dramatically at their inception, they converged over time. The speed with which the oil companies built roads in the northern Oriente compared with the slower moving efforts by the Salesians and CREA [Centro de Reconversion Economica del Azuay, Cañar, y Morona Santiago] in Morona Santiago explains both the greater prevalence of land speculation and the faster pace of deforestation in the north.

These differences in the pattern of deforestation began to disappear after 1975 in large part because similar patterns of land distribution developed in the two regions. In both places smallholders acquired most of the land, and they sought alliances with powerful actors who could help them obtain improvements in infrastructure. This similarity suggests that one of the main differences in patterns of deforestation, between places opened up by lead institutions and places developed by growth coalitions, may be temporary. After an initial surge of deforestation started by either a lead institution or a growth coalition, further land clearing depends largely on the ability of local elites and smallholders to form new growth coalitions.

The degree of similarity between land use patterns in the northern and southern Oriente lends credence to the convergence thesis. Large landowners leave a higher proportion of their lands in forest in both regions.[14] The differences between colonists' and Amerindian land use exist in both regions. Lowland Quichua in the north, like the Shuar in the south, clear a lower proportion of their lands and devote a higher percentage of their cleared land to the cultivation of crops than do the colonists. The colonists devote most of their cleared land (90 percent) to pasture for cattle.[15] Like the Shuar, the Quichua cleared land in some locales at rapid rates during the early 1970s in an attempt to insure continued control over the land.[16] The more acculturated Quichua, like the acculturated Shuar, want to expand agricultural production for the market, and to that end they lobby for the construction of feeder roads into their villages.[17] Like the Shuar, the lowland Quichua have in recent years migrated in large numbers to areas being opened up for settlement and deforestation. Large numbers of lowland Quichua have moved from the Tena-Puyo region to the oil rich Lago Agrio region in order to work for the oil companies and acquire land.[18] In sum, the ethnic differentials in rates of deforestation in the southern Oriente also characterize the northern Oriente.

The comparison of places opened up by lead institutions and growth coalitions suggests divergence in land clearing patterns during the initial period of development and convergence in subsequent periods. Although the land clearing patterns may converge over time, the initial differences in the organization of regional development may continue to shape land clearing because lead institutions and growth coalitions respond so differently to political pressures to preserve rainforests. The success of policies to prevent further deforestation will vary from region to region, and it depends in part on the type of organization that drives deforestation in a place.

Notes

1. E. Bylund, "Theoretical Considerations Regarding the Distribution of Settlement in Inner North Sweden," *Geografiska Annaler* B 42 (1960): 225–31; J. C. Hudson, "A Location Theory for Rural Settlement," *Annals of the Association of American Geographers* 59 (1969): 365–81; and Lawrence A. Brown, R. Sierra, and D. Southgate, "Complementary Perspectives as a Means of Understanding Regional Change: Frontier Settlement in the Ecuadorian Amazon," *Environment and Planning* A 24 (1992): 955.
2. Lorenzo Garcia, *Historia de las Misiones en la Amazonia Ecuatoriana* (Quito: Ediciones Abya-Yala, 1985).
3. Anonymous interview, IERAC administrator, Lago Agrio, 1971.
4. Field Notes, Lago Agrio office of IERAC, 1971.
5. Anonymous interview, IERAC topographer, Lago Agrio, 1971.
6. These estimates of deforestation come from aerial photographs of the region, taken in 1978, and analyzed by Mario Hiroaka and S. Yamamoto, "Agricultural Development in the Upper Amazon of Ecuador," *Geographical Review* 70, no. 4 (1980): 431.
7. Hiroaka and Yamamoto, 432; and S. R. Gonzalez and R. Morocho, "Biografia de una Colonizacion: Km. 7–80 Lago Agrio—Coca" (Quito: Centro de Investigaciones de la Amazonia Ecuatoriana [CICAME], 1976), 21.

8. H. Barral, R. Oldeman, and M. Sourdet, *Reflexiones acerca del Estado Actual y del Porvenir de la Colonizacion del Nororiente* (Quito: Ministerio de Agricultura y Ganaderia-Orstrom, 1976), 7.
9. Ibid., 7.
10. F. Sandoval, "Petroleo y medio oriente en la Amazonia Ecuatoriana," in F. Larrea, ed., *Amazonia: Presente y . . . ?* (Quito: Ediciones Abya-Yala, 1987), 168–72.
11. H. Barral and C. Orrego, *Informe sobre de la Colonizacion en la Provincia del Napo y las Transformaciones en las Sociedades Indigenas* (Quito: Ministerio de Agricultura y Ganaderia-Orstrom, 1978), 6.
12. Anonymous interviews, colonists, Cooperative Nueva Loja, Lago Agrio, 1971.
13. F. Guerrero, "Problemas Ecologicos y Sociales Relacionados con el Cultivo de Palma Africana: El Caso de Palmoriente," in F. Larrea, ed., *Amazonia: Presente y . . . ?* (Quito: Ediciones Abya-Yala, 1987), 246–49.
14. "Las Zonas Socio-economicas Actualmente Homogeneas de la Region Amazonica Ecuatoriana," Diognostico Socio-economico del Medio Rural Ecuatoriano (Quito: Ministerio de Agricultura y Ganaderia-Pronareg, 1980), 23–28.
15. "Las Zonas Socio-economicas," 73; Barral and Orrego, 29.
16. Theodore MacDonald, *De Cazadores a Ganaderos* (Quito: Ediciones Abya-Yala, 1984).
17. F. Villaroel G., "La Fiesta Nativa," *El Comercio* [Quito], 7 March 1988.
18. Barral and Orrego, 27.

Charles M. Peters, Alwyn H. Gentry, and Robert O. Mendelsohn, "Valuation of an Amazonian Rainforest" (1989)

Peters and his fellow researchers document a great irony: that tropical forests are more economically valuable left standing than they are when felled for timber production. By their analysis, the value of renewable resources like latex and fruit—not to mention ecotourism and other benefits—outweigh the short-term profit from cutting down trees.

Tropical forest resources have traditionally been divided into two main groups: timber resources, which include sawlogs and pulpwood; and non-wood or "minor" forest products, which include edible fruits, oils, latex, fiber, and medicines. Most financial appraisals of tropical forests have focused exclusively on timber resources and have ignored the market benefits of non-wood products. The results from these appraisals have usually demonstrated that the net revenue obtainable from a particular tract of forest is relatively small, and that alternative uses of the land are more desirable from a purely financial standpoint. Thus there has been a strong market incentive for destructive logging and widespread forest clearing.

We contend that a detailed accounting of non-wood resources is required before concluding *a priori* that tropical deforestation makes financial sense. To illustrate our point, we present data concerning inventory, production, and current market value for all the commercial tree species occurring in one hectare of species-rich Amazonian forest. These data indicate that tropical forests are worth considerably more than has been previously assumed, and that the actual market benefits of timber are very small relative to those of non-wood resources. Moreover, the total net revenues generated by the sustainable exploitation of

Figure 6.8 Logging in the Brazilian Amazon. Courtesy of Greenpeace/Porto de Moz.

"minor" forest products are two to three times higher than those resulting from forest conversion.

Our findings are based on an appraisal of an area along the Rio Nanay near to the small village of Mishana (3°47′S, 73°30′W), thirty km south-west of the city of Iquitos, Peru. . . .

Based on the assumption of sustainable timber harvests and annual fruit and latex collection for perpetuity, we estimate that the tree resources growing in one hectare of forest at Mishana possess a combined financial worth of $6,820. Fruits and latex represent more than 90 percent of the total market value of the forest, and the relative importance of non-wood products would increase even further if it were possible to include the revenues generated by the sale of medicinal plants, lianas, and small palms. In view of the disproportionately low NPV [net present value] of wood resources and the impact of logging on fruit and latex trees, timber management is a marginal financial option in this forest. . . .

Our results indicate that the financial benefits generated by sustainable forest use tend to exceed those that result from forest conversion. Using identical investment criteria, that is, discounting a perpetual series of net revenues at a 5 percent interest rate, the NPV of the timber and pulpwood obtained from a 1.0-ha plantation of *Gmelina arborea* in Brazilian Amazonia is estimated[1] at $3,184 or less than half that of the forest. Similarly, gross revenues from fully stocked cattle pastures in Brazil are reported[2] to be $148 per hectare per year. The present value of a perpetual series of such pastures discounted back to the present is only $2,960, and deducting the costs of weeding, fencing, and animal care would lower this figure significantly. Both these estimates are based on the optimistic assumption that plantation forestry and grazing lands are sustainable land-use practices in the tropics.

Notes

1. R. A. Sedjo, *The Comparative Economics of Plantation Forestry* (Washington, D.C.: Resources for the Future, 1983).
2. R. J. Buschbacher, *Biotropica* 19 (1987): 200–207.

7 War and Peace

Security at Stake

From ancient combat through two world wars, from Vietnam through the Cold War, from the Gulf War to the current battle against terrorism, the environment has played an important role as both cause and effect. Historically, most wars have multiple causes, with the ecological pressures of overpopulation and resource scarcity often underscoring more proximate ideological, ethnic, personal, and political tensions. The French Revolution and Napoleonic Wars on the one hand, and the Russian Revolution on the other, correlated with demographic booms in the French and Russian countryside. War fulfills some of the same functions as trade; it is a way of obtaining resources, both natural and human-produced. The Treaty of Paris that ended the American Revolution not only brought independence, it specified the right of the new nation to control resources west to the Mississippi River and to make use of the fisheries off Newfoundland. France and Germany dueled in the Franco-Prussian War and then in two world wars for reasons of national pride, but also for control of the rich mining region that lay between them. Today, access to petroleum reserves in Central Asia, the Middle East, and Africa plays a crucial role in all forms of geopolitical maneuvering: military, economic, and diplomatic. In these regions, and others, control over scarce fresh water is an additional source of tension, for rivers flow across the landscape either without regard to political boundaries or as the boundaries themselves.

Like trade, but more dramatically, wars can cause great environmental destruction. One strategy in the series of English wars on the Irish was to starve them into submission by burning their crops. This, along with inherent population pressures, led the Irish to plant only a single crop, one that could not be burned—the potato. Ironically, in the nineteenth century, the unsound ecological effects of this monocultural practice came back to haunt the Irish: their potato crop became a prime target for the attack of an enemy more deadly than the English, the potato blight (*Phytophora infestans*). Across the landscape of Southeast Asia in the 1960s and 1970s, the United States military widely spread Agent Orange, a defoliant used to strip away the natural cover of guerilla fighters. This left swaths of countryside biologically devastated and killed or seriously injured many people, Vietnamese and Americans alike. These two cases illustrate environmental destruction as a deliberate military strategy. Environmental damage can also be an unintended consequence of military action—for example, depleted uranium in Bosnia and Iraq, landmines in Mozambique, and commonly and tragically, the effects of displaced populations. Preparation for war can be devastating as well: U.S. nuclear testing once rendered the Marshall Islands virtually uninhabitable, and U.S. Navy bombing exercises in Vieques, Puerto Rico, have sparked public health and environmental protests.

Ironically, primitive warfare often helped preserve ecological balance by creating "no-man's lands" that acted as buffer zones between neighboring tribes. This phenomenon remains today in less technically advanced areas such as New Guinea, as well as in more modern incarnations like the Demilitarized Zone separating North from South Korea, an area that is now a safe haven for many species of wildlife pushed to the margins in the heavily industrialized South.

The meaning of "national security" has changed radically in the past two decades. During the Cold War, the specter that most haunted the world community was a nuclear holocaust that would not only kill hundreds of millions of people but also completely disrupt natural processes, from the radioactive contamination of crops and water to the creation of a nuclear winter, in which fallout and radioactive dust would permeate the Earth's atmosphere, greatly diminishing the intensity of solar radiation reaching the ground. Since the dissolution of the Soviet Union in 1991 and the attacks of September 11, 2001, the likelihood of all-out nuclear war has faded and the focus of security has shifted, for many nations, to defense against terrorism. Engaging in "asymmetrical warfare," loosely organized and ideologically motivated terrorist groups avail themselves of tactics and weapons that nations would not or could not use in ordinary warfare. These tactics include treating the environment as both a weapon and a target, by poisoning community water supplies, for example, or aiming suicide missions against large dams, oil refineries, and nuclear power plants. While fear of retaliation has often prevented nations from using the nuclear, chemical, and biological weapons in their arsenals, terrorists have little or no compunction about using such "weapons of mass destruction," any of which could have serious environmental consequences.

That chemical weapons can have devastating effects on plant and animal life is no accident. The chemical warfare and agricultural pesticide industries grew up together and cross-fertilized generously during the twentieth century. Biowarfare has an ancient lineage. Besieging armies commonly sought to spread disease among the enemy by means of infected carcasses. The ease with which very small numbers of Europeans conquered the New World can be credited in large part to decimation of indigenous peoples by Old World germs. Often the spread of disease was accidental but not always: in the French and Indian Wars, the British sent hostile Native Americans blankets infected with smallpox. The army

Figure 7.1 A sign warns of danger from land mines in Mozambique.
Photo by Glenn Adelson.

that wins the war is often the one that fares best against the common microbiological enemy. The American Revolution would likely not have succeeded had Washington not introduced mandatory inoculations in the colonial army. The campaign against malarial mosquitoes that brought DDT into widespread use gave Allied forces an edge in the Pacific theater in World War II. Because germs evolve and hosts develop new defenses and immunities, the war against the microbe is never over. A biotechnology revolution that began in the 1970s has accelerated the pace of that war, creating horrible new possibilities of "supergerms" far in advance of the body's natural defenses, and "designer pathogens" that can target specific genetic groups or even individuals.

A new understanding of the link between security and the environment prompts action on several levels. Thwarting terrorism requires enhanced protection of vulnerable targets, including environmental ones. But security also requires rethinking structural issues. Shifting from centralized to distributed generation, for example, makes energy infrastructure less susceptible to attack, and reducing reliance on fossil fuels by promoting alternative sources and modernizing the transportation sector can diminish entanglements with the often volatile politics of oil-exporting countries. Promoting best ecological practices in agriculture preserves self-sufficiency by reversing soil erosion and depletion, and reduces the risk of devastating blight and famine by favoring genetic diversity.

The spread of AIDS and SARS proves that public health and sanitation in Africa and China are vital to all nations; overpopulation and ecological collapse in any one country or region burden the entire world community, and the biosphere itself. At the highest level, an enlightened understanding of security is an ecological one. It reminds each of us of our interconnectedness and mutual dependence, both with other nations and with other species: in the long run, everyone's prosperity depends on the health of the whole.

FURTHER READING

Homer-Dixon, Thomas, and Jessica Blitt. *Ecoviolence: Links Among Environment, Population, and Security.* Lanham, Md.: Rowman and Littlefield, 1998.

Miller, Judith. *Biological Weapons and America's Secret War.* New York: Simon and Schuster, 2001.

Myers, Norman. *Ultimate Security: The Environmental Basis of Political Stability.* New York: W. W. Norton, 1993.

Russell, Edmund. *War and Nature: Fighting Humans and Insects with Chemicals from World War I to Silent Spring.* Cambridge: Cambridge University Press, 2001.

Robert Kaplan, from "The Coming Anarchy" (1994)

One important way people divide up the environment is through political boundaries. Robert Kaplan questions the stability of these boundaries and paints a grim picture of a future in which poverty, environmental degradation, and tribal hatreds fuel a vicious downward spiral toward anarchy in many parts of the world. More than a decade later, to what extent have his fears been confirmed, and to what extent have events taken a different course? Is it too soon to judge?

A PREMONITION OF THE FUTURE

West Africa is becoming *the* symbol of worldwide demographic, environmental, and societal stress, in which criminal anarchy emerges as the real "strategic" danger. Disease, overpopulation, unprovoked crime, scarcity of resources, refugee migrations, the increasing erosion of nation-states and international borders, and the empowerment of private armies, security firms, and international drug cartels are now most tellingly demonstrated through a West African prism. West Africa provides an appropriate introduction to the issues, often extremely unpleasant to discuss, that will soon confront our civilization. To remap the political earth the way it will be a few decades hence—as I intend to do in this article—I find I must begin with West Africa.

There is no other place on the planet where political maps are so deceptive—where, in fact, they tell such lies—as in West Africa. Start with Sierra Leone. According to the map, it is a nation-state of defined borders, with a government in control of its territory. In truth the Sierra Leonian government, run by a twenty-seven-year-old army captain, Valentine Strasser, controls Freetown by day and by day also controls part of the rural interior. In the government's territory the national army is an unruly rabble threatening drivers and passengers at most checkpoints. In the other part of the country units of two separate armies from the war in Liberia have taken up residence, as has an army of Sierra Leonian rebels. The government force fighting the rebels is full of renegade commanders who have aligned themselves with disaffected village chiefs. A premodern formlessness governs the battlefield, evoking the wars in medieval Europe prior to the 1648 Peace of Westphalia, which ushered in the era of organized nation-states.

As a consequence, roughly 400,000 Sierra Leonians are internally displaced, 280,000 more have fled to neighboring Guinea, and another 100,000 have fled to Liberia, even as 400,000 Liberians have fled to Sierra Leone. The third largest city in Sierra Leone, Gondama, is a displaced-persons camp. With an additional 600,000 Liberians in Guinea and 250,000 in the Ivory Coast, the borders dividing these four countries have become largely meaningless. Even in quiet zones none of the governments except the Ivory Coast's maintains the schools, bridges, roads, and police forces in a manner necessary for functional sovereignty. The Koranko ethnic group in northeastern Sierra Leone does all its trading in Guinea. Sierra Leonian diamonds are more likely to be sold in Liberia than in Freetown. In the eastern provinces of Sierra Leone you can buy Liberian beer but not the local brand.

In Sierra Leone, as in Guinea, as in the Ivory Coast, as in Ghana, most of the primary rain forest and the secondary bush is being destroyed at an alarming rate. I saw convoys of trucks bearing majestic hardwood trunks to coastal ports. When Sierra Leone achieved its independence, in 1961, as much as sixty percent of the country was primary rain forest. Now six percent is. In the Ivory Coast the proportion has fallen from thirty-eight percent to eight percent. The deforestation has led to soil erosion, which has led to more flooding and more mosquitoes. Virtually everyone in the West African interior has some form of malaria.

Sierra Leone is a microcosm of what is occurring, albeit in a more tempered and gradual manner, throughout West Africa and much of the underdeveloped world: the withering away of central governments, the rise of tribal and regional domains, the unchecked spread of disease, and the growing pervasiveness of war. West Africa is reverting to the Africa of the Victorian atlas. It consists now of a series of coastal trading posts, such as Freetown and Conakry, and an interior that, owing to violence, volatility, and disease, is again becoming, as Graham Greene once observed, "blank" and "unexplored." However, whereas Greene's vision implies a certain romance, as in the somnolent and charmingly seedy

Freetown of his celebrated novel *The Heart of the Matter*, it is Thomas Malthus, the philosopher of demographic doomsday [see Chapter 24, below], who is now the prophet of West Africa's future. And West Africa's future, eventually, will also be that of most of the rest of the world. . . .

Africa may be as relevant to the future character of world politics as the Balkans were a hundred years ago, prior to the two Balkan wars and the First World War. Then the threat was the collapse of empires and the birth of nations based solely on tribe. Now the threat is more elemental: *nature unchecked.* Africa's immediate future could be very bad. The coming upheaval, in which foreign embassies are shut down, states collapse, and contact with the outside world takes place through dangerous, disease-ridden coastal trading posts, will loom large in the century we are entering. (Nine of twenty-one U.S. foreign-aid missions to be closed over the next three years are in Africa—a prologue to a consolidation of U.S. embassies themselves.) Precisely because much of Africa is set to go over the edge at a time when the Cold War has ended, when environmental and demographic stress in other parts of the globe is becoming critical, and when the post–First World War system of nation-states—not just in the Balkans but perhaps also in the Middle East—is about to be toppled, Africa suggests what war, borders, and ethnic politics will be like a few decades hence. . . .

THE ENVIRONMENT AS A HOSTILE POWER

For a while the media will continue to ascribe riots and other violent upheavals abroad mainly to ethnic and religious conflict. But as these conflicts multiply, it will become apparent that something else is afoot, making more and more places like Nigeria, India, and Brazil ungovernable.

Mention "the Environment" or "diminishing natural resources" in foreign-policy circles and you meet a brick wall of skepticism or boredom. To conservatives especially, the very terms seem flaky. Public-policy foundations have contributed to the lack of interest, by funding narrowly focused environmental studies replete with technical jargon which foreign-affairs experts just let pile up on their desks.

It is time to understand "the Environment" for what it is: *the* national-security issue of the early twenty-first century. The political and strategic impact of surging populations, spreading disease, deforestation and soil erosion, water depletion, air pollution, and, possibly, rising sea levels in critical, overcrowded regions like the Nile Delta and Bangladesh—developments that will prompt mass migrations and, in turn, incite group conflicts—will be the core foreign-policy challenge from which most others will ultimately emanate, arousing the public and uniting assorted interests left over from the Cold War. In the twenty-first century water will be in dangerously short supply in such diverse locales as Saudi Arabia, Central Asia, and the southwestern United States. A war could erupt between Egypt and Ethiopia over Nile River water. Even in Europe tensions have arisen between Hungary and Slovakia over the damming of the Danube, a classic case of how environmental disputes fuse with ethnic and historical ones. The political scientist and erstwhile Clinton adviser Michael Mandelbaum has said, "We have a foreign policy today in the shape of a doughnut—lots of peripheral interests but nothing at the center." The environment, I will argue, is part of a terrifying array of problems that will define a new threat to our security, filling the hole in Mandelbaum's doughnut and allowing a post–Cold War foreign policy to emerge inexorably by need rather than by design.

Our Cold War foreign policy truly began with George F. Kennan's famous article, signed "X," published in *Foreign Affairs* in July of 1947, in which Kennan argued for a "firm and vigilant containment" of a Soviet Union that was imperially, rather than ideologically, motivated. It may be that our post–Cold War foreign policy will one day be seen to have had its beginnings in an even bolder and more detailed piece of written analysis: one that appeared in the journal *International Security*. The article, published in the fall of 1991 by Thomas Fraser Homer-Dixon, who is the head of the Peace and Conflict Studies Program at the University of Toronto, was titled "On the Threshold: Environmental Changes as Causes of Acute Conflict." Homer-Dixon has, more successfully than other analysts, integrated two hitherto separate fields—military-conflict studies and the study of the physical environment.

In Homer-Dixon's view, future wars and civil violence will often arise from scarcities of resources such as water, cropland, forests, and fish. Just as there will be environmentally driven wars and refugee flows, there will be environmentally induced praetorian regimes—or, as he puts it, "hard regimes." Countries with the highest probability of acquiring hard regimes, according to Homer-Dixon, are those that are threatened by a declining resource base yet also have "a history of state [read 'military'] strength." Candidates include Indonesia, Brazil, and, of course, Nigeria. Though each of these nations has exhibited democratizing tendencies of late, Homer-Dixon argues that such tendencies are likely to be superficial "epiphenomena" having nothing to do with long-term processes that include soaring populations and shrinking raw materials. Democracy is problematic; scarcity is more certain.

Indeed, the Saddam Husseins of the future will have more, not fewer, opportunities. In addition to engendering tribal strife, scarcer resources will place a great strain on many peoples who never had much of a democratic or institutional tradition to begin with. Over the next fifty years the earth's population will soar from 5.5 billion to more than nine billion. Though optimists have hopes for new resource technologies and free-market development in the global village, they fail to note that, as the National Academy of Sciences has pointed out, ninety-five percent of the population increase will be in the poorest regions of the world, where governments now—just look at Africa—show little ability to function, let alone to implement even marginal improvements. Homer-Dixon writes, ominously, "Neo-Malthusians may underestimate human adaptability in *today's* environmental-social system, but as time passes their analysis may become ever more compelling."

While a minority of the human population will be, as Francis Fukuyama would put it, sufficiently sheltered so as to enter a "post-historical" realm, living in cities and suburbs in which the environment has been mastered and ethnic animosities have been quelled by bourgeois prosperity, an increasingly large number of people will be stuck in history, living in shantytowns where attempts to rise above poverty, cultural dysfunction, and ethnic strife will be doomed by a lack of water to drink, soil to till, and space to survive in. In the developing world environmental stress will present people with a choice that is increasingly among totalitarianism (as in Iraq), fascist-tending mini-states (as in Serb-held Bosnia), and road-warrior cultures (as in Somalia). [Significantly, the United States intervened in all three countries.] Homer-Dixon concludes that "as environmental degradation proceeds, the size of the potential social disruption will increase."

Tad Homer-Dixon is an unlikely Jeremiah. Today a boyish thirty-seven, he grew up amid the sylvan majesty of Vancouver Island, attending private day schools. His speech is calm, perfectly even, and crisply enunciated. There is nothing in his background or manner that would indicate a bent toward pessimism. A Canadian Anglican who spends his summers canoeing on the lakes of northern Ontario, and who talks about the benign mountains, black bears, and Douglas firs of his youth, he is the opposite of the intellectually severe neoconservative, the kind at home with conflict scenarios. Nor is he an environmentalist who opposes development. "My father was a logger who thought about ecologically safe forestry before others," he says. "He logged, planted, logged, and planted. He got out of the business just as the issue was being polarized by environmentalists. They hate changed ecosystems. But human beings, just by carrying seeds around, change the natural world." As an only child whose playground was a virtually untouched wilderness and seacoast, Homer-Dixon has a familiarity with the natural world that permits him to see a reality that most policy analysts—children of suburbia and city streets—are blind to.

"We need to bring nature back in," he argues. "We have to stop separating politics from the physical world—the climate, public health, and the environment." Quoting Daniel Deudney, another pioneering expert on the security aspects of the environment, Homer-Dixon says that "for too long we've been prisoners of 'social-social' theory, which assumes there are only social causes for social and political changes, rather than natural causes, too. This social-social mentality emerged with the Industrial Revolution, which separated us from nature. But nature is coming back with a vengeance, tied to population growth. It will have incredible security implications.

"Think of a stretch limo in the potholed streets of New York City, where homeless beggars live. Inside the limo are the air-conditioned post-industrial regions of North America, Europe, the emerging Pacific Rim, and a few other iso-

Figure 7.2 An American tank passes a burning oil field in Iraq. © Newscom.

lated places, with their trade summitry and computer-information highways. Outside is the rest of mankind, going in a completely different direction."

We are entering a bifurcated world. Part of the globe is inhabited by Hegel's and Fukuyama's Last Man, healthy, well fed, and pampered by technology. The other, larger, part is inhabited by Hobbes's First Man, condemned to a life that is "poor, nasty, brutish, and short." Although both parts will be threatened by environmental stress, the Last Man will be able to master it; the First Man will not.

The Last Man will adjust to the loss of underground water tables in the western United States. He will build dikes to save Cape Hatteras and the Chesapeake beaches from rising sea levels, even as the Maldive Islands, off the coast of India, sink into oblivion, and the shorelines of Egypt, Bangladesh, and Southeast Asia recede, driving tens of millions of people inland where there is no room for them, and thus sharpening ethnic divisions.

Homer-Dixon points to a world map of soil degradation in his Toronto office. "The darker the map color, the worse the degradation," he explains. The West African coast, the Middle East, the Indian subcontinent, China, and Central America have the darkest shades, signifying all manner of degradation, related to winds, chemicals, and water problems. "The worst degradation is generally where the population is highest. The population is generally highest where the soil is the best. So we're degrading earth's best soil." [See Chapter 12, below.]

China, in Homer-Dixon's view, is the quintessential example of environmental degradation. Its current economic "success" masks deeper problems. "China's fourteen percent growth rate does not mean it's going to be a world power. It means that coastal China, where the economic growth is taking place, is joining the rest of the Pacific Rim. The disparity with inland China is intensifying." Referring to the environmental research of his colleague, the Czech-born ecologist Vaclav Smil, Homer-Dixon explains how the per capita availability of arable land in interior China has rapidly declined at the same time that the quality of that land has been destroyed by deforestation, loss of topsoil, and salinization. He mentions the loss and contamination of water supplies, the exhaustion of wells, the plugging of irrigation systems and reservoirs with eroded silt, and a population of 1.54 billion by the year 2025: it is a misconception that China has gotten its population under control. Large-scale population movements are under way, from inland China to coastal China and from villages to cities, leading to a crime surge like the one in Africa and to growing regional disparities and conflicts in a land with a strong tradition of warlordism and a weak tradition of central government—again as in Africa. "We will probably see the center challenged and fractured, and China will not remain the same on the map," Homer-Dixon says. . . .

A NEW KIND OF WAR

To appreciate fully the political and cartographic implications of postmodernism—an epoch of themeless juxtapositions, in which the classificatory grid of nation-states is going to be replaced by a jagged-glass pattern of city-states, shanty-states, nebulous and anarchic regionalisms—it is necessary to consider, finally, the whole question of war.

"Oh, what a relief to fight, to fight enemies who defend themselves, enemies who are awake!" André Malraux wrote in *Man's Fate.* I cannot think of a more suitable battle cry for many combatants in the early decades of the twenty-first century. The intense savagery of the fighting in such diverse cultural settings as Liberia, Bosnia, the Caucasus, and Sri Lanka—to say nothing of what obtains in American inner cities—indicates something very troubling that those of us inside the stretch limo, concerned with issues like middle-class entitlements and the future of interactive cable television, lack the stomach to contemplate. It is this: a large number of people on this planet, to whom the comfort and stability of a middle-class life is utterly unknown, find war and a barracks existence a step up rather than a step down.

"Just as it makes no sense to ask 'why people eat' or 'what they sleep for,'" writes Martin Van Creveld, a military historian at the Hebrew University in Jerusalem, in *The Transformation of War,* "so fighting in many ways is not a means but an end. Throughout history, for every person who has expressed his horror of war there is another who found in it the most marvelous of all the experiences that are vouchsafed to man, even to the point that he later spent a lifetime boring his descendants by recounting his exploits." When I asked Pentagon officials about the nature of war in the twenty-first century, the answer I frequently got was "Read Van Creveld." The top brass are enamored of this historian not because his writings justify their existence but, rather, the opposite: Van Creveld warns them that huge state military machines like the Pentagon's are dinosaurs about to go extinct, and that something far more terrible awaits us.

The degree to which Van Creveld's *Transformation of War* complements Homer-Dixon's work on the environment, [Samuel] Huntington's thoughts on cultural clash, my own realizations in traveling by foot, bus, and bush taxi in more than sixty countries, and America's sobering comeuppances in intractable-culture zones like Haiti and Somalia is startling. The book begins by demolishing the notion that men don't like to fight. "By compelling the senses to focus themselves on the here and now," Van Creveld writes, war "can cause a man to take his leave of them." As anybody who has had experience with Chetniks in Serbia, "technicals" in Somalia, Tontons Macoutes in Haiti, or soldiers in Sierra Leone can tell you, in places where the Western Enlightenment has not penetrated and where there has always been mass poverty, people find liberation in violence. In Afghanistan and elsewhere, I vicariously experienced this phenomenon: worrying about mines and ambushes frees you from worrying about mundane details of daily existence. If my own experience is too subjective, there is a wealth of data showing the sheer frequency of war, especially in the developing world since the Second World War. Physical aggression is a part of being human. Only when people attain a certain economic, educational, and cultural standard is this trait tranquilized. In light of the fact that ninety-five percent of the earth's population growth will be in the poorest areas of the globe, the question is not whether there will be war (there will be a lot of it) but what kind of war. And who will fight whom? . . .

Future wars will be those of communal survival, aggravated or, in many cases, caused by environmental scarcity. These wars will be subnational, meaning that it will be hard for states and local governments to protect their own citizens physically. This is how many states will ultimately die. As state power fades—and with it the state's ability to help weaker groups within society, not to mention other states—peoples and cultures around the world will be thrown back upon their own strengths and weaknesses, with fewer equalizing mechanisms to protect them. Whereas the distant future will probably see the emergence of a racially hybrid, globalized man, the coming decades will see us more aware of our differences than of our similarities. . . .

The Indian subcontinent offers examples of what is happening. For different reasons, both India and Pakistan are increasingly dysfunctional. The argument over democracy in these places is less and less relevant to the larger issue of governability. In India's case the question arises, Is one unwieldy bureaucracy in

New Delhi the best available mechanism for promoting the lives of 866 million people of diverse languages, religions, and ethnic groups? In 1950, when the Indian population was much less than half as large and nation-building idealism was still strong, the argument for democracy was more impressive than it is now. Given that in 2025 India's population could be close to 1.5 billion, that much of its economy rests on a shrinking natural-resource base, including dramatically declining water levels, and that communal violence and urbanization are spiraling upward, it is difficult to imagine that the Indian state will survive the next century. India's oft-trumpeted Green Revolution has been achieved by overworking its croplands and depleting its watershed. Norman Myers, a British development consultant, worries that Indians have "been feeding themselves today by borrowing against their children's food sources."

Pakistan's problem is more basic still: like much of Africa, the country makes no geographic or demographic sense. It was founded as a homeland for the Muslims of the subcontinent, yet there are more subcontinental Muslims outside Pakistan than within it. Like Yugoslavia, Pakistan is a patchwork of ethnic groups, increasingly in violent conflict with one another. While the Western media gushes over the fact that the country has a woman Prime Minister, Benazir Bhutto, Karachi is becoming a subcontinental version of Lagos. In eight visits to Pakistan, I have never gotten a sense of a cohesive national identity. With as much as sixty-five percent of its land dependent on intensive irrigation, with wide-scale deforestation, and with a yearly population growth of 2.7 percent (which ensures that the amount of cultivated land per rural inhabitant will plummet), Pakistan is becoming a more and more desperate place. As irrigation in the Indus River basin intensifies to serve two growing populations, Muslim-Hindu strife over falling water tables may be unavoidable.

"India and Pakistan will probably fall apart," Homer-Dixon predicts. "Their secular governments have less and less legitimacy as well as less management ability over people and resources." Rather than one bold line dividing the subcontinent into two parts, the future will likely see a lot of thinner lines and smaller parts, with the ethnic entities of Pakhtunistan and Punjab gradually replacing Pakistan in the space between the Central Asian plateau and the heart of the subcontinent.

None of this even takes into account climatic change, which, if it occurs in the next century, will further erode the capacity of existing states to cope. India, for instance, receives seventy percent of its precipitation from the monsoon cycle, which planetary warming could disrupt. [See Chapter 1, above.]

Not only will the [demographic and political] aspects of the Last Map be in constant motion, but its two-dimensional base may change too. The National Academy of Sciences reports that "as many as one billion people, or twenty percent of the world's population, live on lands likely to be inundated or dramatically changed by rising waters. . . . Low-lying countries in the developing world such as Egypt and Bangladesh, where rivers are large and the deltas extensive and densely populated, will be hardest hit. . . . Where the rivers are dammed, as in the case of the Nile, the effects . . . will be especially severe."

Egypt could be where climatic upheaval—to say nothing of the more immediate threat of increasing population—will incite religious upheaval in truly biblical fashion. Natural catastrophes, such as the October, 1992, Cairo earthquake, in which the government failed to deliver relief aid and slum residents were in many instances helped by their local mosques, can only strengthen the position of Islamic factions. In a statement about greenhouse warming which could refer to any of a variety of natural catastrophes, the environmental expert Jessica Tuchman Matthews warns that many of us underestimate the extent to which political systems, in affluent societies as well as in places like Egypt, "depend on the underpinning of natural systems." She adds, "The fact that one can move with ease from Vermont to Miami has nothing to say about the consequences of Vermont acquiring Miami's climate."

Organization for Economic Co-operation and Development (OECD), "Water and Security in the Middle East" (1999)

Continuing degradation of the world's fresh water supply, in both quantity and quality (see Chapter 13, below), has led many to suppose the next great war will be fought over access to this precious resource. In contrast, observers note that though conflicts over water have been common throughout history, such conflicts have generally been local, and have rarely escalated to violence. The fact that water resources are so easy to share—unlike a piece of disputed land—may make compromise solutions likely. One case where water plays an important though underreported role, and where compromise solutions remain elusive, is the Israeli-Palestinian conflict.

A recent workshop sponsored by the DCI [Director of Central Intelligence] Environmental Center in the U.S. identified seventeen critical environmental flashpoints that could lead to regional instability in the world over the next two decades. Six of these flashpoints focused on water supply, and three of these were in the Middle East. The Jordan River basin has often been presented as one of the key examples where environment and security issues overlap. Central to the tensions that exist between Israel and the Palestinians is the availability of adequate fresh water supplies. In addition to the obvious water scarcity/conflicts problem, the existence of refugees—Palestinian, Ethiopian, Russian, and others—is stressing political, social, and environmental systems. There are also significant constraints on the level of economic achievement of certain sectors of national or regional economies due to a lack of resources and increased mining and deterioration of the groundwater supply. The situation has become so extreme that King Hussein of Jordan identified water as the only issue that would lead him to go to war with Israel. Despite the recent advances made in the peace discussions, the water issue remains a major stumbling block to a lasting peace in the region.

Virtually all of Israel's fresh water comes from two sources: surface water supplied by the Jordan River, or ground water fed by recharge from the West Bank to one of three major aquifers. There is a long legacy of controversy over fresh water in the region, dating back thousands of years. In recent times, there was a proposed comprehensive plan for cooperative use of the Jordan River (the Johnston Plan) as early as the 1950s, but this was derailed by mistrust among the four riparian states (Israel, Jordan, Lebanon, and Syria). Each nation has tended to follow its own water policies since the failure of that agreement, often to the detriment of other nations.

Water has long been considered a security issue in the region, and on numerous occasions Israel and its neighboring Arab states have feuded over access to Jordan River waters. A number of authors have argued that a major contributing factor in the tensions leading to the 1967 War was the water issue. At the time, Israel was consuming almost 100 percent of its available fresh water supplies. Occupation of the three territories (the West Bank, the Golan Heights, and the Gaza Strip) after the war changed this situation in two ways. First, it increased the fresh water available to Israel by almost fifty percent. Second, it gave the country almost total control over the headwaters of the Jordan River and its tributaries, as well as control over the major recharge region for its underground aquifers. Control of water resources in the West Bank and the Golan Heights is now integrated into Israel's economy and, accordingly, essential to its future.

Presently, Israel draws over forty percent of its fresh water supplies from the West Bank alone, and the country would face immediate water shortages and a significant curtailment of its agricultural and industrial development if it lost control of these supplies. Former Israeli agricultural minister Rafael Eitan stated in November of 1990 that Israel must never relinquish the West Bank because a loss of its water supplies would "threaten the Jewish state." The growing number of settlements in the region poses an additional problem. The water in the West Bank is now used in a ratio of 4.5 percent by Palestinians and 95.5 percent by Israelis (while the population is over ninety percent Palestinian). The U.N. Committee on Palestinian Rights concluded in 1980 that Israel had given priority to its own water needs at the expense of the Palestinian people.

To ensure security of water supply from the West Bank aquifers, Israel has put in place quite restrictive policies regarding Palestinian use of water. Israel's application of restrictions on Palestinian development and use of water not only improves its access to West Bank water, but also extends its control throughout the territory. It is this inequitable situation with respect to water allocations which increases resentment and adds to tensions in the region.

Jeffrey A. McNeely, from "Biodiversity, War, and Tropical Forests" (2003)

Modern warfare can be destructive to the environment but, like primitive warfare, sometimes it can have beneficial effects too, especially by creating "no-man's lands." McNeely describes the positive and negative effects of modern warfare on tropical forests.

NEGATIVE IMPACTS OF WAR ON BIODIVERSITY

The negative impacts of war on biodiversity in tropical forests result from the collective actions of large numbers of people for whom war is a dispensation to ignore normal restraints on activities that cause environmental damage. War, and preparations for it, have negative impacts on all levels of biodiversity, from genes to ecosystems. These impacts can be direct—such as hunting and habitat destruction by armies—or indirect, for example, through the activities of refugees.

Sometimes these impacts can be deliberate, and a new word has been added to the military vocabulary: "ecocide," the destruction of the environment for military purposes; it clearly builds on the "scorched earth" approach of earlier times. Westing[1] divides deliberate environmental manipulations during wartime into two broad categories: those involving massive and extended applications of disruptive techniques to deny to the enemy any habitats that produce food, refuge, cover, training grounds, and staging areas for attacks; and those involving relatively small disruptive actions that in turn release large amounts of "dangerous forces" or become self-generating. An example of the latter is the release of exotic microorganisms or spreading of landmines (of which over 100 million now litter active and former war zones around the world).[2]

This discussion could be long and dreary, but only a few illustrative cases will be mentioned. Perhaps the most outstanding example is Vietnam, where U.S. forces cleared 325,000 ha [hectares] of land and sprayed 72,400 m^3 [cubic meters] of herbicides in the name of security.[3] The impact on biodiversity was severe; spreading herbicides on ten percent of the country (including fifty percent of the mangroves) led to extensive low-diversity grasslands replacing high-diversity forests, mudflats instead of highly productive mangroves, major declines in both freshwater and coastal fisheries, and so forth.[4]

Other problems are more systemic. The State Law and Order Restoration Council (SLORC), the military government in Myanmar (formerly Burma), has been involved in violent confrontations with many of the tribal groups who inhabit the densely forested mountain regions along the country's borders with

Bangladesh, India, China, Laos, and Thailand. Some of these tribal groups, such as the Karen, have turned to intensive logging to fund their war effort, even though such over-exploitation will eventually destroy the forest cover and make them more open to attack.[5] The general lawlessness along the Thai border has greatly increased the flow of logs, both with and without government permission, leading to the virtual clear felling of many of the country's most productive forests.

Africa provides several recent war-related disasters for biodiversity in tropical forests. Like the upper Amazon, the Virunga Volcanoes region (including parts of the Central African countries of Rwanda, Democratic Republic of the Congo, and Uganda) is exceptionally rich in species diversity, including the rare and endangered mountain gorilla (*Gorilla gorilla beringei*) whose total population is approximately 600. The civil war against the government of Rwanda was launched in 1990 from within the Virunga Volcanoes region, spreading deeper into Rwanda until 1994 and sending large numbers of refugees fleeing to North Kivu District in what was then Zaire, which then began a civil war of its own. The headquarters of several tropical forest World Heritage sites in Zaire were taken over by the military, including Virunga National Park, Kahuzi-Biega National Park, and the Okapi Wildlife Reserve. In 1994, some 850,000 refugees were living around Virunga National Park, partly or completely deforesting some 300 km^2 of the park in a desperate search for food and firewood. Up to 40,000 people entered the park every day, taking out between 410 and 770 tons of forest products. The bamboo forests have been especially seriously damaged, and the populations of elephants, buffalo, and hippos have been much reduced. Organizations such as the Red Cross, Médecins Sans Frontières, and CARE have supported well-meaning relief operations on the park boundaries and have even established a dump for medical wastes inside the park, with obvious disease transmission risks associated with such practices.[6] At least eighty of Virunga's park staff have been killed in battle with insurgents since 1996. . . .

POSITIVE IMPACTS OF WAR ON BIODIVERSITY

However, war, or the threat of war, can also be good for biodiversity, at least under certain conditions. As Myers put it, "In some respects, indeed, wildlife benefits from warfare: combatant armies effectively designate war zones as 'off limits' to casual wanderers, thus quarantining large areas of Africa from hunters and poachers."[7] Of course, any benefits of war to biodiversity are incidental, inadvertent, and accidental rather than a planned side effect of conflict. But even so, it is useful to review some cases where war, or preparations for war, has benefited biodiversity, perhaps supporting the views of some anthropologists that war helps societies adapt to their environmental constraints.

For example, the border between Thailand and Peninsular Malaysia was a hotbed of insurgency during the mid-1960s to mid-1970s. On the Malaysian side of the border, the military closed off all public access and potential logging ac-

Figure 7.3 An indigenous Filipino performs a war dance in front of Camp Aguinaldo in Manila during a 2003 anti-government protest. © Newscom.

tivity in the Belum Forest Reserve. As a result, this extensive area of some 160,000 ha has remained untouched by modern logging pressures and therefore is rich in wildlife resources. Malaysia is now converting this into a national park that will form a trans-boundary protected area with matching protected areas in southern Thailand.

While the second [i.e., American] Vietnam War was an ecological disaster, it also led to some important biological research, such as the extensive, long-term, review of migratory birds in eastern Asia carried out by the Migratory Animals Pathological Survey.[8] The excuse for this research was its relevance to the war effort, but it has yielded data that are useful for numerous civilian conservation applications. The watersheds through which ran the Ho Chi Minh Trail, some of the most heavily bombed parts of Indochina during the second Vietnam War, have more recently been remarkably productive in new discoveries. The new large mammal discoveries include two species of muntjak or barking deer (*Megamuntiacus vuquangensis* and *Muntiacus truongsonensis*), a unique variety of forest antelope (*Pseudoryx nghetinhensis*), and a bovid ultimately related to wild cattle (*Pseudonovibos spiralis*),[9] as well as the rediscovery of a species of pig that formerly was known only by a few fragmentary specimens. That such species could survive in such a heavily bombed area is testimony to the recuperative power of nature, and the ability of wildlife to withstand even the most extreme kinds of human pressure during warfare. Interestingly, these species now are even more

TABLE 7.1
Impacts of War on Biodiversity

Negative impacts	*Positive impacts*
• Deforestation	• Creates "no-go" zones
• Erosion	• Slows or stops developments that lead to loss of biodiversity
• Wildlife poaching	• Focuses state resolve
• Habitat destruction	• Reduces pressure on some habitats
• Pollution of land and water	• Allows vegetation to recover in some areas
• Reduces funds for conservation	• Disarms rural populations, thereby reducing hunting
• Stops conservation projects	• Can increase biodiversity-related research
• Forces people onto marginal lands	
• Creates refugees who destroy biodiversity	

severely threatened by the peacetime activities of development than they were by the Indochina wars.

Some other species are likely to have benefited from the war in Vietnam. Orians and Pfeiffer say that tigers "have learned to associate the sounds of gunfire with the presence of dead and wounded human beings in the vicinity. As a result, tigers rapidly move toward gunfire and apparently consume large numbers of battle casualties. Although there are no accurate statistics on the tiger populations past or present, it is likely that the tiger population has increased much as the wolf population in Poland increased during World War II."[10]

Fairhead and Leach report that parts of the Ziama region of Guinea, which includes an extensive biosphere reserve, became forested following a series of wars that affected the area from 1870 to 1910.[11] The resident Toma people first fought with Mandinka groups from the north and subsequently with the French colonial armies, causing major depopulation and economic devastation that in turn allowed the forest to reclaim agricultural land. The human disaster of war enabled nature to recover.

The impact of war on biodiversity is often decidedly mixed, with a complex combination of damages and benefits (Table 7.1). Nicaragua provides an outstanding example. Engaged in civil war for over twenty years, the country suffered 100,000 casualties, and nearly half of its population was relocated in one way or another. The human tragedy was immense, but biodiversity was able to recover from a long history of exploitation, as trade in timber, fish, minerals, and wildlife was sharply reduced. The domestic cattle population, which was roughly equivalent to the human population when the war started, was reduced by two-thirds, freeing pastures for re-colonization by forests. This enabled the recovery of animal populations such as white-tailed deer (*Odocoileus virginianus*), peccaries (*Tayassu angulatus*), four species of monkeys (family Cebidae), caiman (*Caiman crocodilus*), iguanas (*Iguana iguana*), large birds, and various mammalian predators. Fishing boats were destroyed and fishermen fled, leading to drastic declines in the catches of fish, shrimp, and lobsters, which in turn revitalized these fisheries. On the other hand, some hunting by soldiers had at least local negative impacts on wildlife, and new military bases and roads were established in formerly remote areas, opening them up to exploitation. Further, the country's once outstanding system of protected areas fell into neglect, and new areas planned were not established; the collapsing economy forced villagers into environmentally destructive activities, including clearing forest for firewood and harvesting wildlife for food. Nietschmann concludes that a significant portion of this conflict was over resources and territory, not ideology.[12] Biodiversity rejuvenated by the war came under renewed threat by people whom the war had impoverished; the postwar period saw a great acceleration of such impacts and now that peace has broken out, biodiversity is under renewed pressure.

On the other side of the world, the war in Indochina was disastrous to Cambodia, in both human and ecosystem terms. Years of fighting have created a climate of lawlessness in which those who control the guns also control the country's most valuable natural resources, namely forests and fisheries. Overturning any feeble efforts at control, both are being depleted at dangerous rates, according to studies being carried out by the World Bank and the Asian Development Bank (ADB). Uncontrolled logging, much of it illegal, could virtually deforest the country within five years, according to ADB, with current harvesting over three times the sustainable yield. The fish, especially from Cambodia's Tonle Sap (Great Lake), are being over-harvested, primarily for export to surrounding—and wealthier—countries. The ecological productivity of the lake was based largely on the 10,000 km^2 of flooded forest that ensured a healthy flow of nutrients into the lake. However, less than forty percent of the flood forest remains under natural vegetation. Since 1993, military commanders have come to regard the forest resources as their own resources, treating them as a supplemental source of finance irrespective of the long-term impact on the country's security. Continuing loss of forests will further affect the climate, cause erosion that fills irrigation channels and fishing grounds with silt, and leave Cambodian farmland more vulnerable to both drought and flooding. This complex of problems is very similar to that which faced Cambodia some 400 years ago, when the great civilization centered on Angkor Wat collapsed under environmental pressure.[13]

So, while war is bad for biodiversity, peace can be even worse. In the 1960s, when Indonesia and Malaysia were fighting over border claims on the island of Borneo, they did relatively little damage to its vast wilderness, but in the 1990s, they peacefully competed to cut down and sell its forests. In Indonesia, the 1997–98 forest fires that caused US$4.4 billion in damage were set primarily by businesses and military to clear forests in order to plant various cash crops. Ironically, the prices of these commodities that were to be grown have fallen considerably in recent years, making them even less profitable. Vietnam's forests are under greater pressure now that peace has arrived than they ever were during the country's wars; and Nicaragua's forests are now under renewed development pressures. Laos is paying at least part of its war debts to China and Vietnam with timber concessions. I was told in Laos that the Chinese and Vietnamese timber merchants and logging companies are able to operate with impunity there, irrespective of logging regulations, protected area boundaries, or any other considerations. This is perhaps not surprising given the dependence of the Pathet Lao on the support of Vietnam and China during the Indochina wars. The motivations may be more noble in times of peace, but the impacts of inappropriate development on biodiversity often are even worse than the impacts of war. Market forces may be more destructive than military forces.

Notes

1. A. H. Westing, *Ecological Consequences of the Second Indo-China War* (Stockholm: Almqvist and Wiksell, 1976).
2. G. Strada, "The Horror of Land Mines," *Scientific American* 274, no. 5 (1996): 26–31.
3. A. H. Westing, "The Environmental Aftermath of Warfare in Vietnam." *World Armaments and Disarmament: SIPRI Year Book 1982* (London: Taylor and Francis, 1982), 363–89.
4. B. Nietschmann, "Battlefields of Ashes and Mud," *Natural History* 11 (1990): 35–37.
5. R. Harbinson, "Burma's Forests Fall Victim to War," *The Ecologist* 22, no. 2 (1992): 72–73.
6. F. Pearce, "Soldiers Lay Waste to Africa's Oldest Park," *New Scientist* (3 December 1994): 4.
7. N. Myers, "Wildlife and the Dogs of War," *Daily Telegraph* (London), 8 December 1979.
8. H. E. McClure, *Migration and Survival of the Birds of Asia* (Bangkok: U.S. Army Component, Seato Medical Research Laboratory, 1974).
9. T. C. Dillon and E. D. Wikramanayake, "Parks, Peace and Progress: A Forum for Transboundary Conservation in Indo-China," *PARKS* 7, no. 3 (1997): 36–51.
10. G. H. Orians and E. W. Pfeiffer, "Ecological Effects of the War in Vietnam," *Science* 168 (1970): 553.
11. J. Fairhead and M. Leach, "False Forest History, Complicit Social Analysis: Rethinking Some West African Environmental Narratives," *World Development* 23, no. 6 (1995): 1023–35.
12. B. Nietschmann, "Conservation by Conflict in Nicaragua," *Natural History* 11 (1990): 42–49.
13. J. A. McNeely and P. S. Wachtel, *Soul of the Tiger: Searching for Nature's Answers in Southeast Asia* (Singapore: Oxford University Press, 1988).

Elizabeth L. Chalecki, from "A New Vigilance: Identifying and Reducing the Risks of Environmental Terrorism" (2002)

Terrorists don't only attack buildings and crowded buses: reservoirs, oilfields, and other environmental resources are also potential targets that need to be protected. Chalecki distinguishes environmental terrorism from "eco-terrorism," and provides examples of both.

Terrorism as a concept first appeared in [English] . . . in 1795 as, "a government policy intended to strike with terror those against whom it is adopted."

Over the course of the next two centuries, terrorism went from a government-sponsored policy to being regarded as anti-government policy. The word "terrorism" began to have an exclusively negative connotation by the middle of the twentieth century, so that terrorist groups wishing to avoid bad publicity began calling themselves "freedom fighters," and governments employing terrorist tactics against opponents began calling them "police actions." Today, the word "terrorism" brings to mind hijacked airliners, the World Trade Center, and the violent death of unsuspecting people.

The FBI definition of terrorism states that terrorism is "the unlawful use of force or violence against persons or property to intimidate or coerce a government, the civilian population, or any segment thereof, in furtherance of political or social objectives." Title 22, Section 2656 of the U.S. Code states that terrorism "means premeditated, politically motivated violence perpetrated against non-combatant targets by subnational groups or clandestine agents, usually intended to influence an audience." Both of these definitions concern themselves primarily with the motive behind terrorist actions and not with the selection of target, other than defining it as "non-combatant persons or property."

. . . For the purposes of this paper, environmental terrorism can be defined as the unlawful use of force against *in situ* environmental resources so as to deprive populations of their benefit(s) and/or destroy other property.

ENVIRONMENTAL TERRORISM VS. ECO-TERRORISM

Most readers, when they hear the term "environmental terrorism," are actually thinking of eco-terrorism. Not to be confused with environmental terrorism, eco-terrorism is the violent destruction of property perpetrated by the radical fringes of environmental groups in the name of saving the environment from further human encroachment and destruction. Based in deep ecology theory, the professed aim of eco-terrorists is to slow or halt exploitation of natural resources and to bring public attention to environmental issues such as unsustainable logging or wildlife habitat loss through development.

Earth First! is the organization that first brought eco-terrorism to the public debate. Founded in 1980, Earth First! is known for tree-spiking in the Pacific Northwest (although they have since repudiated this tactic), and protests against old-growth logging, road building in wilderness areas, and dam construction. Its furor wound down under pressure from law enforcement groups. In addition, when the environmentally friendly Clinton Administration took office, the group believed that its agenda would receive positive attention. The modern inheritor of the eco-terrorist mantle is the Environment Liberation Front (ELF), an Earth First! splinter group formed in 1993 in England. In an action purportedly aimed at saving lynx habitat, the American wing of ELF burned down a ski lodge in Vail, Colorado, in October 1998, resulting in $12 million in property damage, an act ironically repudiated by Earth First! itself. ELF made headlines in January 2001, when members set fire to newly built homes on Long Island to protest what they view as humans' unceasing encroachment on nature, and again in March 2001, when they set fire to a warehouse containing transgenic cotton seed and a biogenetic research facility at the University of Washington. . . .

At first glance, the distinction between environmental terrorism and eco-terrorism might seem academic. However, operationally there is a significant difference. Environmental terrorism involves targeting natural resources. Eco-terrorism involves targeting built environment such as roads, buildings, and trucks, ostensibly in defense of natural resources. Earth First!, ELF, and other eco-terrorists do not practice environmental terrorism per se if they do not choose environmental resources as their targets. ELF has targeted a ski resort, houses on Long Island, logging trucks, and office buildings, but has damaged no resources and killed no one. (In fact, ELF claims to go out of its way to avoid human casualties.) It could be argued that ELF did commit environmental terrorism by burning the transgenic cotton seed. However, the members would likely argue that genetically engineered plants are not "natural" and hence are not an environmental resource. Their stated aim remains to inflict economic damage to built facilities in defense of the environment, an important distinction to be made when considering the intersection of environmental issues and terrorism. . . .

WATER RESOURCE SITES

Sites involving water resources are vulnerable to environmental terrorist attacks in the form of explosives or the introduction of poison or disease-causing agents. The damage is done by rendering the water unusable and/or destroying the purification and supply infrastructure. Water resource sites have many vulnerable physical attributes that make them attractive to terrorists. Most water infrastructure, such as dams, reservoirs, and pipelines, are easily accessible to the public at various points. Many dams such as Hoover or Glen Canyon are tourist attractions and offer tours to the public, while many reservoirs such as the Triadelphia outside Washington, D.C., are open to the public for recreational boating and swimming. A terrorist carrying a small but powerful explosive device would not necessarily be conspicuous among tourists, sport fishermen, or hikers.

Water resource sites are also attractive to environmental terrorists because water is a vitally necessary resource for which there is no substitute. Whether its lack is due to a physical supply interruption or not, a community of any size that lacks fresh water will suffer greatly. Furthermore, a community does not have to *lack* water to suffer. Too much water at the wrong time in the form of a flood can cause greater damage, and flooding towns and settlements is a time-tested tactic in warfare.

Finally, water itself is an attractive terrorist weapon. Not only can it cause great damage in large quantities, as mentioned above, but the ability of water to spread agents downstream makes it a perfect method of transporting poison or disease-causing agents such as *Cryptosporidium* bacteria. As an example of the economic and personal chaos this type of attack can cause, the April 1993 *Cryptosporidium* outbreak in Milwaukee killed more than a hundred people, affected the health of more than 400,000 more, and cost $37 million in lost wages and productivity. The outbreak was thought to be due to a combination of an improperly functioning water treatment plant and illegal pollution discharges upstream from the water intake point, and not from a terrorist attack, but a similar outbreak in a large city such as New York or Chicago might cost billions and kill thousands of people.

... In July 2000, workers at the Cellatex chemical plant in northern France dumped 790 gallons of sulfuric acid into the Meuse River when they were denied workers' benefits. Whether they were trying to kill wildlife, people, or both is unclear, but a French analyst pointed out that this was the first time "the environment and public health were made hostage in order to exert pressure, an unheard-of situation until now."[1] Leaving aside the question of whether or not the workers would be considered terrorists, they certainly appear to have committed a terrorist act, since there was no way to isolate the effects of the acid from the general population, nor did they attempt to do so.

... In July of 1999, engineers discovered a homemade bomb in a water reservoir near Pretoria, South Africa. The dam personnel felt that the fifteen kg bomb, which had malfunctioned, would have been powerful enough to damage the twelve million liter reservoir, thereby depriving farmers, a nearby military base, and a hydrological research facility of water. Police recognized this action as deliberate sabotage and began searching the country's other reservoirs for similar devices. Even a simulated terrorist attack on a water resources site, the destruction of the Lake Nacimento Dam, caused some panic in central California until the media was notified that the situation was merely a disaster preparedness drill. And most recently, Palestinians attacked and vandalized water pipes leading to the Israeli settlement of Yitzhar to force the Israelis out of the settlement....

MINERAL AND PETROLEUM SITES

Sites involving non-renewable resources with high economic value make attractive targets for violence of every kind. Oil refineries serve as attractive military targets during wartime, depriving the enemy of energy. Vulnerable areas of mineral and petroleum sites include the extensive and necessary infrastructure for the processing and transportation of the resource. Oil derricks, wellheads, pipelines, loading terminals, and tankers are all vulnerable to fire or conven-

Figure 7.4 A U.S. soldier guards an Iraqi oil pipeline. © Newscom.

tional terrorist explosives. Attacks on this infrastructure can create extensive environmental damage before being contained. In addition, oil spills can interfere with the normal workings of power stations and desalination plants by fouling intake water.

Mineral and petroleum sites are also particularly attractive targets because of the necessity of the resource to the health and growth of the national economy. Fossil fuels in particular have no readily available substitutes, especially on short notice and in necessary quantities. An attack on a loading terminal at a main oil field can halt commerce for weeks, causing shortages in fuel and natural gas, and costing millions of dollars in higher prices. An attack on a site where a strategic metal is mined (e.g., vanadium or tungsten) can cause economic damage, and possible strategic concerns, by forcing mine customers to buy necessary supplies on the less-reliable and often more expensive spot market.

Oil pipelines in Colombia have regularly been the targets of attack. In 1997, over forty-five separate attacks on the Cano Limon–Covenas pipeline, reputedly by leftist guerrillas from the National Liberation Army (ELN), caused Colombia's national oil company Ecopetrol to declare *force majeur* on all exports from the Cano Limon field. In 1998, the ELN bombed the OCENSA pipeline, spilling over 30,000 barrels of oil and triggering a blaze that killed more than seventy people when the fire spread through nearby villages. The Trans-Alaska Pipeline has had its share of attention: an episode of the TV drama "Seven Days" included a terrorist attack on the pipeline, and in August of 1999, the Royal Canadian Mounted Police arrested a man in British Columbia who had planned to bomb the pipeline out of political and financial motives.

Categorizing the attacks as resource-as-tool or resource-as-target attacks is sometimes difficult for these types of sites: a single attack can have both types of consequences. For example, a hypothetical mineral and petroleum site for an environmental terrorist attack would be the Alaskan oil fields. An exploded pipeline or wellhead would not only waste the oil (resource-as-target terrorism), but would befoul the delicate and fragile Arctic ecosystem (resource-as-tool terrorism). The most obvious example of the crippling disruption resulting from such an attack is the destruction of the Kuwaiti oil fields at the end of the [First] Gulf War. While it remains debatable as to whether this was true terrorism or a legitimate act of war, the damage done from this attack was almost incalculable: approximately six million barrels of oil were burned *per day* until all the well fires were capped, and the people of Kuwait suffered health and ecological problems for years afterward.

Note

1. *Christian Science Monitor*, 21 July 2000, 8.

Amory B. Lovins and L. Hunter Lovins, "What Is Real Security?" (2002)

Nations concerned about security need to "harden" potential terrorist targets, including environmental targets, but they also need to make structural changes. Amory and Hunter Lovins, who prepared a report on the subject for the Pentagon in the 1980s, recount here that shorter supply chains, decentralized distribution, and supply redundancy can help reduce the vulnerability of the U.S. energy infrastructure. Similar lessons apply to food, water, and other infrastructures.

America's security faces many serious threats. Strategic planners, however, have tended to focus almost exclusively on the *military* threat. They have largely ignored equally grave vulnerabilities in vital life-support systems such as our energy, water, food, data processing, and telecommunications networks. And they have likewise neglected to safeguard the national assets that form the foundation of our security.

In our 1982 Pentagon study *Brittle Power: Energy Strategy for National Security*, we found that a handful of people could shut down three-quarters of the oil and gas supplies to the eastern states, cut the power to any major city, or kill millions by damaging a nuclear power plant. Such hazards remain real today. Between April 25 and May 11, 2001, for example, infiltrators accessed the computer system of the California Independent System Operator, the agency that operates California's power distribution network, potentially gaining the capability to black out whole cities, and cause physical damage to equipment.

Reliance on fossil fuels and their extended pipelines contributes to our insecurity. Even where fuel is extracted from politically stable regions, it must be safely transported via accident-prone ships, trucks, rail, or pipeline. On October 4, 2001, a drunk shot a bullet through the Trans-Alaska Pipeline, shutting it down for sixty hours and spilling 285,000 gallons of oil. Previously, the pipeline has been shot at on over fifty occasions. A disgruntled engineer's plot to blow up critical points then profit from oil futures trading was thwarted by luck two years ago.

How, then, can America become less vulnerable to attack and more resilient to mishaps that do occur? How can we prepare for a future that may hold increasing uncertainty, unrest, and even violence? The answer may be found by basing engineering on nature. Natural systems are efficient, diverse, dispersed, and renewable, hence, inherently resilient.

The most resilience per dollar invested comes from using energy very efficiently. Minimizing energy waste both eliminates dependence on the most vulnerable sources (such as oil from the Persian Gulf) and makes energy failures milder, slower, and easier to fix. Efficiency is also the cheapest way to meet our energy needs. [See Chapter 14, below.] During 1979–85, energy savings enabled GNP to rise by 16 percent while oil use fell 15 percent and Persian Gulf imports fell 87 percent. This was primarily achieved by making cars more efficient. Just making cars about three miles per gallon more efficient could eliminate all Persian Gulf oil imports. Did we put our young people in 0.6 mile per gallon army tanks because we did not put them in 32 mile per gallon cars?

Another key to resilience is to replace centralized energy sources gradually with many richly interconnected dispersed ones. This is the strategy of a tree that has many leaves, each with many veins, so that the random nibbling of insects won't disrupt the vital flow of nutrients. The value of dispersion was proved in the Northeast Blackout of 1965, when a power engineer in Holyoke, Massachusetts, was able to unhook the city from the collapsing grid and connect instead to a local gas turbine. The money saved by not having to black out Holyoke paid off the cost of building that power plant in four *hours.* More recently, in Sacramento, citizens suffered none of the power shortages or price spikes that other Californians faced. About ten years ago the city voted to shut down the troubled nuclear plant that provided nearly half its power. Instead, Sacramento invested in efficiency and a diverse supply mix emphasizing renewables and distributed generation. These investments boosted county economic output by $185 million and added 2,946 employee-years of net jobs. Efficiency plus a diverse, often decentralized, supply portfolio kept electricity supplies reliable and constant-price during California's power emergencies.

As the Sacramento example shows, dispersed energy systems don't cost more; indeed, they're already winning in the marketplace. Major homebuilders nationwide expect to enjoy a marketing edge by providing hundreds of grid-connected rooftop-solar systems on new housing developments; indeed, five Sacramento projects already offer solar power as standard equipment.

Central power stations, no matter how well engineered, can't supply really cheap electricity and simply cannot be made secure. The power lines that deliver the electricity cost more than the generators and cause almost all power failures. On-site and neighborhood micro-power is cheaper and eliminates grid losses and glitches. Rooftop photovoltaic systems, fuel cells, or biomass-fed microturbine or engine generators can be built on site to provide power for individual buildings or neighborhoods. When such systems fail, the effect is small and localized. If several small systems are interconnected, one failure may hardly be noticed. Widespread disruption of such a network would be difficult because it would require too many agents and too much coordination.

Dispersed systems are even more reliable when they use renewable energy sources. Thus, Department of Energy officials in 1980 had just cut the ribbon on a West Chicago solar-powered gas station when a thunderstorm blacked out the city. That was the only station pumping gas that afternoon. Manhattan's Condé Nast office tower recruited tenants at premium rents by offering the two most reliable known power supplies, fuel and solar cells, incorporated into the building.

The importance of energy resilience to national security may hold wider lessons. First, focusing exclusively on centralized military planning to counter overt military threats may create costly frontal fortification while the back door stands ajar. Indeed, there are many back doors. The average molecule of food is shipped some 1,300 miles before an American eats it. Damage a few Mississippi River bridges and easterners will soon starve. A malicious PC user could probably crash the whole financial system. There are doubtless other key vulnerabilities not yet discovered, and security experts are only now starting to think about how to reduce them.

Whatever military might has accomplished, then, it has not yet made us truly secure. Perhaps it never will. The roots of real security go deeper than armies and missiles alone. The parable of energy security reminds us that real security in its widest sense begins at home and is strengthened by self-sufficient, decentralized, sustainable communities.

Even more basic in our quest for real security, we should understand the role of our nation's strategic assets. These include a geography that shields us against physical invasion from overseas; a freedom of expression that shields us from ideological invasion by exposing concepts to the critical scrutiny of an informed public; an ecosystem much of whose once unique fertility can still be rescued from degradation; a diverse, ingenious, and independent people; and a richly inspiring body of political and spiritual values. To mature within these outward strengths—strengths more fundamental and lasting than any inventory of weaponry—will require us to remain inwardly strong, confident in our lives and liberties no matter what surprises may occur. This in turn will demand a continuing American revolution that expresses in works a sincere faith in individual and community effort. It was this faith that inspired our Republic, long before strategists became preoccupied with the narrower and more evanescent kinds of security that only a faraway government could provide. It is that faith today, the very marrow of our political system, which alone can give us *real* security.

Jeremy Rifkin, from *The Hydrogen Economy* (2002)

Rifkin takes a broad view of security, and asks how our cultural understanding of that concept has changed over time. He suggests that the most recent understanding is one that views security as a web of interdependence. One might call this an "ecological" understanding of security.

The great transformations in energy regimes throughout history have always forced a rethinking of the most basic categories of human existence. Hunter-gatherer cultures, although each unique, share a commonality of spirit, as do agricultural and industrial societies. The way human beings gather, transform, and use the dominant forms of energy at their disposal—whether it be wild animals and plants; domesticated food crops; human slaves; or coal, oil, and natural gas harnessed to machines—becomes imprinted in the various ways they define their ideas about personal and collective security.

For example, consider the notion of security held by Christians in the late medieval era with that of the bourgeoisie of twentieth-century industrial capitalism. The two very different ideas about security tell us a great deal about the very different natures of the energy regimes that people have depended on for their existence.

. . . Medieval Europe was loosely organized under the aegis of the Catholic Church, in consort with local warlords—kings, princes, and manor lords. Society was conceived of as an intimate microcosm of God's grand creation, which descended from heaven in the form of a great ladder or chain of being with God on top; and below him his emissaries on Earth, the pope and priesthood; and then lesser personages, including kings, lords, knights, farmers, tenants, and serfs; and finally our fellow creatures all the way down to the lowest regions and to every "creeping thing that creepeth upon the Earth." Every rung of the divine ladder, explained St. Thomas Aquinas, was occupied by one of God's creatures, and all the rungs were full, leaving no room for innovation, surprises, or changes in God's master plan. The Church's world was a tightly knit structure with a careful gradation of ranks and a detailed catechism of instructions governing mutual obligations. Security depended on human beings executing the tasks assigned to them by Providence and accepting their roles and responsibilities in the natural hierarchy. By faithfully fulfilling their obligations and duties in this world, medieval men and women could be assured a modicum of security on Earth as well as everlasting security in the next world.

The land, in the medieval scheme of things, was particularly important in defining security, as it was the place where one served as a steward of God's creation. In an era when agriculture was the dominant energy regime, security naturally flowed from the land. The sense of place was all the more poignant because in the feudal order people, for the most part, belonged to the land, rather than the land belonging to the people. Security, then, was a vertical affair that began in this world with a sense of attachment to the ancestral home, to which one was born, and ascended up to the next world, where one was rewarded with everlasting life.

The coal economy fundamentally changed the security equation. Now steam-driven machines allowed society to augment human slaves and animals with mechanical substitutes. The new machine-slaves made "man" feel less dependent on God and the forces of nature for his well-being. Autonomy gradually replaced divine deliverance, at least for those who controlled and benefited from what historian Arnold Toynbee later called "The Industrial Revolution."

The coal economy also greatly quickened the pace, speed, flow, diversity, and intensity of human life. Whereas, during the Middle Ages, few people wandered much beyond a day's walk from their homes, by the late nineteenth century, millions of people were regularly crossing oceans and whole continents by steamships and trains in just a few days. The railroad and telegraph shortened distances, compressed time, and added a new dimension to human life—"mobility." Millions of farmers, uprooted from their ancestral lands and feudal bondage, migrated to sprawling, bustling cities where they became "free labor," working in the factories alongside steam-driven machinery. The new urban age was increasingly characterized by a sense of restlessness and perpetual change. Innovation became the hallmark of the industrial economy. To be modern was to be avant-garde, to be continually experimenting with new ideas, fashions, and lifestyles. A younger generation castigated its elders for being too "old-fashioned."

The growing sense of self-reliance and the quickened pace of life that so marked the new era evolved into a wholly new idea about security. The medieval view of security that was bound up with belonging to the land in this world and being saved in the next world was gradually replaced with a new sense of security based increasingly on the notion of individual autonomy and mobility. These two values were to become the dominant virtues of an emerging bourgeois class in every industrial country.

Autonomy became synonymous with freedom. To be autonomous was to be in control of one's destiny and not dependent on others. The key to gaining one's autonomy was to be propertied. The new role of government, in turn, was to defend and expand markets and thus assure everyone's right to amass property so that they could gain their autonomy and be truly free.

Mobility, in the new era, meant far more than safe passage. To be mobile was always to have new choices and options available. Just as autonomy became synonymous with freedom, mobility became synonymous with opportunity. The fossil-fuel era freed human beings from the slower seasonal rhythms of an agricultural period and thus also from dependency on nature's constraints and divine intervention. Each new convert to modernity was endowed with a Promethean spirit. Armed with autonomy and mobility, every human being could become a minor god, securing a small, personal piece of paradise in this material world.

By the twentieth century, the middle class, whose sense of security was so bound up in both autonomy and mobility, found the ultimate object of their affection in the invention of the "auto-mobile." Here was a machine, fueled by gasoline, that in one stroke provided a complete sense of autonomy and mobility for its driver and passengers. It is no wonder that it so quickly became the prime metaphor for the age and the centerpiece of the industrial economy. Henry Ford once mused that his car should be regarded as a "living room on wheels." What youngster has not experienced the rush of freedom and opportunity that comes the first time he or she takes the driver's seat and puts hands on the wheel and a foot to the accelerator? In the car, on the road, one feels not only both autonomous and mobile but also as secure as it is possible to be in the modern era, and all this was made possible by the harnessing of fossil fuels.

The modern idea about security provided a framework for countries as well as for individuals. The new "science" of geopolitics, first outlined in 1882 by Friedrich Ratzel, a German professor of geography, became a mirror image of the new bourgeois sensibility.[1] Borrowing heavily from Darwin's theory of the origin and development of species, with its emphasis on biological competition for scarce space and resources, Ratzel argued that, in seeking to enlarge its commercial opportunities and military rule, the nation-state was merely fulfilling its biological destiny. Nations seek to secure their autonomy by continually expanding their reach, and the key to their success is the degree of their mobility. Ratzel believed that mobility could best be realized by maintaining a commanding presence on the world's oceans. "Germany must be strong on the seas," wrote Ratzel, "to fulfill her mission in the world."[2]

Sir Halford Mackinder, a British geographer, agreed with Ratzel that control over the world's oceans, which make up nine-twelfths of the surface of the Earth, would give any nation the mobility it needed to fend off aggressors while allowing it to advance its own expansionist plans. Later, after World War II, geopolitical theorists extended their thinking to include mobility in the air as well.[3]

Architects of the new geopolitics were quick to emphasize the strategic importance of coal, oil, iron, and other minerals, without which modern warfare could not be fought and an industrial way of life could not be advanced. Nicholas Spykman, one of the leading geopolitical figures of the early twentieth century, wrote that control over natural resources was a critical factor in formulating foreign policy.[4] . . .

Modern notions of security and geopolitics made sense in a world in which time was still linear and space was still expansive. The temporal and spatial playing field was still big enough to warrant putting a premium on autonomy and mobility.

But, in a world in which technology allows people to interact with one another and conduct business and social life at the speed of light, 24/7, duration narrows to near simultaneity, and distances nearly disappear. The Earth becomes less of a field of encounter and more of an indivisible entity. The first realization of the new reality came in 1945 when the U.S. military dropped atomic bombs on the Japanese cities of Hiroshima and Nagasaki. Suddenly, humanity came to understand that it wielded sufficient power to destroy itself. We began to talk, for the first time, about living in a single world in which everyone's security was either mutually secured or lost. The very idea of somehow being able to secure individual or national autonomy in a nuclear world seemed quaint and utterly naive.

The first photos of the Earth taken from outer space in the 1960s also helped change our ideas about security. We saw ourselves, in a visceral way, as collectively inhabiting a tiny sphere revolving around a small star somewhere in the giant universe. Whereas autonomy assumes the one against the many and mobility assumes vast space in which to maneuver, the new view of Earth from a distance suggested that there was only a single Earth community confined to a very small space in the cosmic theater.

Perhaps the most compelling new reality that has forced us to rethink our conventional notions of security is the rise in the Earth's temperatures caused by the burning of fossil fuels during the Industrial Age. [See Chapter 1, above.] We spent the better part of the fossil-fuel era attempting to enclose the Earth's ecosystems into commercial arenas of engagement so that we could advance our individual and collective sense of autonomy and mobility. Now the Earth is enclosing us in our own spent energy, undermining both the autonomy and the mobility that we so desperately sought during the whole of the modern age.

We now find ourselves caught up in a new scientific, technological, and commercial revolution that can, in theory, connect the whole of humanity and our fellow creatures into a single, indivisible web of existence. Yet our ideas about security are still wedded to an age based on the narrow Darwinian assumptions that each organism is an island seeking only to optimize its mobility and secure its autonomy. . . .

The hierarchical commercial networks, with their centralized command-and-control systems that went hand in hand with a fossil-fuel energy regime and an industrial way of life, were designed with the modern notion of security in mind. The corporate enterprises that rule the global economy in the last stages of the hydrocarbon age are a paean to the idea that institutional security is best advanced by pushing for ever greater autonomy and mobility.

Now, however, the worldwide communications web and the prospective hydrogen energy web make possible the reorganization of commercial life. In the new era of decentralized command-and-control mechanisms and potentially democratic communications and energy utilization, in which everyone is increasingly linked to everyone and everything else in multiple networks of engagement, at the speed of light, we can begin to rethink our ideas about security.

Notes

1. Ratzel cited in Stephen Kern, *The Culture of Time and Space, 1880–1918* (Cambridge: Harvard University Press, 1983), 224.
2. Friedrich Ratzel, *Das Meer als Quelle der Völkergrösse* (Munich, 1900), 1, 5; trans. Kern, 226.
3. Sir Halford Mackinder, *Democratic Ideas and Reality* (New York: W. W. Norton, 1962).
4. Nicholas J. Spykman, *The Geography of Peace* (New York: Harcourt, Brace, 1944), 5.

Gordon West and John Wilson (U.S. Agency for International Development), from "The United States and the Iraqi Marshlands: An Environmental Response," Testimony Before U.S. Congress (2004)

Marshes are unique ecosystems, and they sustain unique economies and cultures. All too often, such ecosystems, economies, and cultures have been threatened or destroyed in recent centuries by distant central governments that view the marshes as mere "swamps" waiting to be drained and "developed." Carolyn Merchant's The Death of Nature *describes the pressures that led to the eviction of Britain's fen dwellers in the 1600s, for example, despite fierce resistance, to make way for commercial agriculture. Today, the unique value of marshlands and the rights of local cultures are more widely recognized. But the recent fate of Iraq's "Marsh Arabs" is a continuing reminder of the power of state terrorism, ethnic hatred, and war mobilization on the one hand, and the importance and fragility of unique local ecosystems and cultures on the other.*

In little more than a decade, Saddam Hussein's regime systematically destroyed one of the world's largest wetlands ecosystems. This environmental disaster, perpetrated in the roughly 20,000 square kilometer marshlands of southern Iraq, an area more than twice the size of the Florida Everglades, has been compared in scale to the drying up of the Aral Sea in Central Asia and the deforestation of the Amazon. The area was once famous for its cultural richness and biodiversity. The marshes were the permanent habitat for millions of birds and a flyway for billions more migrating between Siberia and Africa. Sixty-six bird species may now be at risk. Other populations are thought to be in serious decline. Coastal fisheries in the Persian Gulf used the marshlands for spawning migrations, and they served as nursery grounds for shrimp and fish. Now fish catches have been significantly decreased. The marshlands also once served as a natural filter for waste and other pollutants in the Tigris and Euphrates rivers, protecting the Gulf which has now become noticeably degraded along the coast of Kuwait.

The indigenous marsh dwellers already have a special place in the anthropological and travel literature for their alluring way of life, living in harmony with the environment on manmade reed islands and along the periphery of the marshes in relative isolation. They may have numbered a half a million in the 1950s and a quarter of a million in the early 1990s. In 1991, a populist Shi'a uprising at the end of the Gulf War brought down the full and brutal weight of the Baghdad regime. The military raided settlements, killed tens of thousands of Marsh Arabs—although the actual number may be higher—burned settlements, and killed livestock, destroying the core of the local economy.

The period from 1991 to 1997 was marked by engineering programs which drained the marshes through the construction of manmade rivers and canals of massive proportions and overblown names. They diverted water from the marshes to irrigate vast areas for uneconomical and unsustainable wheat production, fill huge depressions or ponds to evaporate, or drain into the Shatt Al Arab. A disproportionate share of the country's limited resources was channeled into these works. By 1999, the draining of the marshes was largely over. The only remaining marsh of any note was the northern portion of Hawizeh which straddles the Iran-Iraq border. The other two marshes, Hammar and Central to the west, were totally desiccated.

At the beginning of 2003, only seven percent of the original marshlands remained. However, there has been some recent reflooding throughout the marshes. This water appears to be from a combination of heavier than usual snows in the north, the deliberate destruction of structures by people in the area after the war, the opening of gates by local government officers, and the release of water by Iran to the east.

MARSH DWELLERS: IDENTITY, SETTLEMENT, AND RIGHTS

As recently as the 1970s, there was a limited government presence in the marshlands. The Iran-Iraq War of the 1980s brought the Saddam Hussein regime in full military strength, and the displacement of tribes closest to the border began. Marsh dwellers were moved from Hawizeh in 1984 so that a dike for gun emplacements and a large army base could be built. The dike effectively drained large areas of the marsh and started the people on a series of forced moves over the succeeding fifteen years. It was also the most sustained contact with the government to date. The marsh dwellers were said to have strongly supported the Iraq government during that war. The situation became more unstable with the Gulf War, and there are reports that both marsh dwellers and outsiders escaped into the marshes for refuge, this time earning the wrath of the Iraqi army.

Following the Gulf War in February 1991, the Shi'a population rebelled against the regime after active outside encouragement, apparently taking control of most of the South. In March, the government brought in tanks and helicopters and regained control through the brutal killing of 100,000 or more people and the wholesale destruction of cities and towns. The drainage of the marshes was then put into high gear, becoming one of the highest government priorities, despite the huge investment required. The period also marked an expanded effort to force the marsh dwellers into internal displacement or foreign exile. Roughly 100,000 southern Iraqis are in border refugee camps in Iran; an uncertain number are in Saudi Arabia. No one is certain how many marsh dwellers live within the former or still existing marshlands, but estimates suggest 100,000 to 150,000 other marsh dwellers have moved farther outside the area, usually to the cities in the South. . . .

PUBLIC HEALTH CONDITIONS AND CONCERNS

People in the marshlands suffer from an absence of primary health care, malnutrition, and contaminated drinking water. There are no government services. Schistosomiasis, worms, and cholera are prevalent. Clean drinking water is a problem throughout the area. Some people purchase tanker water, having some access to treated water, but those within the marshes drink directly from the untreated source. Drinking water quality was consistently mentioned by people as their first priority. Mosquitoes, which plagued people throughout the marshes, are the second priority heard from the marsh dwellers. The mosquito problem may have been worsened by the reflooding of the marshes which do not have adequate fish in number to eat the larvae.

ECONOMIC ASSISTANCE: LOCAL ECONOMY AND OPPORTUNITIES

The marshlands witnessed massive government investment over the past two decades to first drain the area and second develop an irrigation infrastructure for the cultivation of wheat. But there has been no investment to improve the lives of the local population for decades. All government policies and actions were directed toward making the people more subservient to and dependent upon the ruling clique. As a result, there is little economic opportunity within the marshes and on their periphery. The large numbers of displaced people were almost wholly dependent on monthly food allocations. Even those people who were not displaced and depended on wheat growing sold their harvests at a crippling loss to the government. The traditional economic pursuits, including commercial fishing and birding, were brought to a standstill. Mat-making continued in the larger towns of Al Chibayish and Hammar City, but was severely undermined by the drying of the marshes.

Today, economic activity in the marshlands revolves around subsistence and limited market wheat-growing. Agricultural activity in the region is largely a monoculture, although there are a few pockets of date palm orchards and vegetables grown on the river banks. . . .

Some portions of Hawizeh and Hammar marsh still retain native vegetation and good water quality. These regions may be a seed source and faunal population base for restoring the drained marshes. By contrast, Central marsh has suffered massive drainage, and little wetland remains. . . .

USAID PROGRAM OBJECTIVES

In response to the human and ecological conditions described above, USAID, in concert with international and local Iraqi stakeholders, has developed the following objectives for its twelve-month program.

- To construct an accurate environmental, social, and economic baseline of the remaining and former marshlands to plan interventions and measure their success;
- To assist with the repatriation and resettlement of marshland dwellers in the region, who will require viable economic opportunities and social institutions that are fair and equitable and give them a voice;
- To improve the management of existing marshlands and explore options to restore adjacent drained marshes; and
- To develop and reach a broad consensus on a long-term comprehensive wetland restoration strategy integrated with a regional social and economic development program.

8 Globalization Is Environmental

The famous picture taken from the moon proclaims Earth to be a single globe, one world. The activities of globalization interconnect regions, countries, cities, villages, and farms, but also jobs, products, regulatory practices, services, banking, markets, and industries. Globalization means different things to different people because it affects them in markedly different ways. It is a contested concept with supporters, skeptics, and opponents. Generally, it is the free movement of products, resources, plants, animals, and, in some cases, people around the world. It can include the free movement of ideas and knowledge. Freedom is a cherished value but creates conflicts and dilemmas: one person's freedom to travel or migrate may be another person's threat of communicable disease or job insecurity. Economic globalization benefits many, but its freedom is not distributed equally. Increasingly, important decisions about which raw materials, products, jobs, and even ideas will cross borders are made by powerful, sometimes unaccountable multinational corporations, many of which command more resources than do most sovereign states.

This reality is fairly new, for a global economy requires secure transport, communication, and trade over long distances. After European explorations in the late Renaissance, some optimists foresaw a prosperous, peaceful world linked by commerce and even the abolition of slavery. Yet global trade is subject not only to economic but also to political and military power. Historically, it is inseparable from the checkered legacies of imperialism and colonialism. The British Empire achieved a form of globalization with mixed results: significant benefits to many regions of Asia, Africa, Europe, and North America, but also exploitation, slavery for a protracted period, and colonial rule. Globalization has never guaranteed equality. Whether current patterns of trade benefit all parties, or whether wealthier nations use them to exploit less fortunate ones, is fiercely debated. "Free trade" may suggest universal progress and prosperity, but when Russia, Argentina, and many other middle-income countries opened their markets to increased foreign competition in the 1990s, they saw economic and social conditions stagnate. When it was surpassing Great Britain on the world stage a century ago, the United States rejected free trade, choosing instead to cultivate its domestic industries and guard its economic borders with tariffs.

Figure 8.1 Earthrise from the moon. Courtesy of NASA.

However, nations aloof or isolated from globalization and international trade now neither prosper nor achieve improved environmental protections. Globalization and world trade must receive some credit for dramatic increases in life expectancy, incomes, and literacy in China and India, together 40 percent of world population. Globalization and greater access to technology can promote education. When poor women become better educated, poverty and birth rates both decline.

Today, despite—or, arguably, sometimes because—of trade agreements, unified currencies, and strictures imposed by the World Bank and IMF (International Monetary Fund), massive debt is hobbling many nations in sub-Saharan Africa and elsewhere. Perhaps the rules of the game should be changed to permit each country a more individual path that could lead to improvements? Perhaps world labor markets could be opened to permit the poor to earn incomes through temporary migration, something at present highly restricted if not impossible? Perhaps some or all of the debt of these struggling nations, debt often assumed by irresponsible or corrupt governments at the urging or insistence of economists and bankers from wealthy countries, should be forgiven? (This offers limited benefits if corruption remains.) Despite strong growth in some developing nations, the gap in per capita income between the richest and poorest countries continues to grow, as does the gap between the rich and poor in poorer nations. The WHO (World Health Organization) has even coined a term for the problem of 300 million affluent people being gravely overweight while several times that number suffer malnutrition. There is a growing epidemic of "globesity." Globalization is about health, prosperity, poverty, debt, labor, trade, and capital markets.

But globalization is also about the environment and, in the long run, it is about the environment as much as it is about *anything* else, for a world economy must run on resources that can be sustained. Economic globalization cannot be separated from environmental globalization. Endemic illnesses and poverty cannot be alleviated if the environment is severely degraded. Wherever water, air, soil, forests, mineral resources, and fish stocks are taken unwisely or depleted faster than they can recover, the recipe may for a time boost a country's GDP (gross domestic product). Yet this boost may be temporary and is not guaranteed, as the toxic ravages and flat economies of the Eastern Bloc and Soviet Union once revealed. Such exploitation financially benefits those who control extraction of resources, but for the local environment and its inhabitants the products can be waste and sorrow.

Environmental practices in developing nations are often poor because regulations are few or bypassed. Yet environmental policies and practices of wealthier nations often fail, too, though in different ways, including the exploitation of environmental vulnerabilities and resources in poorer nations. An impoverished African nation emitting little CO_2 might experience deforestation, soil erosion, and blatant pollution from mining or oil exploration. The United States emits large amounts of CO_2, has wanted to claim its forests as sinks for that gas, yet at home prosecutes the kind of pollution found in African or South American oil fields. However, it is companies from the United States, or multinationals that trade with the United States, that often pollute the African environment in ways they would not attempt elsewhere. The worst fear is that globalization encourages environmental exploitation, that it does not nurture sustainable development, and that as the rich get richer, the poor get poorer.

Environmental abuses and tragedies can certainly occur in isolation. But globalization can offer wider opportunities for degradation. The Aral Sea is drying up due to massive planting of cotton as a cash export crop. The wreckage of ships and shipping containers litters ocean floors. Crude oil spills from supertankers have fouled the sea and ruined delicate coastal habitats. Travel and trade force previously isolated ecosystems into contact, and often out of balance, with invasive species; biological diversity suffers. Culturally, increasing economic interconnectedness tends to disrupt traditional ways of life that have proven sustainable. In addition, as distant bureaucracies, multinational corporations often lack local expertise on best environmental practices, and frequently lack strong incentives for safeguarding a community's long-term sustainability. To the extent that huge multinationals can influence poor, unwary, or corrupt governments, they can weaken or ignore environmental regulations and safeguards. Governments of more developed nations can be complicit, for instance, by banning the use of harmful pesticides at home but allowing production of those substances for export. Finally, transporting goods around the world instead of producing them locally releases huge quantities of greenhouse gases.

Yet, with vigilant international environmental organizations, active government agencies, and free, energetic press and media, globalization might prevent environmental tragedies that often characterize despotic, secretive regimes.

This is because globalization has elevated a global environmental consciousness connected to worldwide economic activities. Globalization can mesh with enforceable international environmental law and treaties. The Rio Earth Summit, the Kyoto Treaty, international agreements on whaling, and the banning of certain commercial chlorofluorocarbons—these efforts, some more successful than others, stem from global awareness. More efforts are needed. If it is objected that global environmental regulations lag behind global realities—and they do—it should be admitted that regulations almost always follow the actions that they are designed to control. It is necessary to globalize environmental education and to identify good local practices that can be adapted elsewhere. Globalization has encouraged the growth of international environmental movements not merely through opposition but also through education, technology, and awareness.

Considerable evidence suggests that as nations climb the ladder of prosperity, the first few rungs of industrialization will worsen environmental conditions. This occurred in the United States and Western Europe, and it is now occurring in many developing nations. However, as average annual per capita income edges over $8,000, environmental improvements and legal protections seem to kick in. This "environmental Kuznets curve," named after the economist Simon Kuznets, supports the counterintuitive, proverbial wisdom that sometimes things have to get worse before they get better.

More local control and autonomy may be the answer to certain issues of human rights, labor equity, agricultural policy, and environmental sustainability. However, in order to regulate multinational actors and companies, the world community may also require an international agency as influential and as well funded in environmental affairs as the World Bank and IMF are in financial matters, and as powerful as the WTO (World Trade Organization) is in trade agreements.

While markets can benefit environmental concerns, for example, driving up the price of scarce resources, markets can permit or even promote environmental damage, too. Though a self-professed optimist, the economist Benjamin M. Friedman nevertheless concludes in *The Moral Consequences of Economic Growth* that for the globalization of the environment, "the net outcome will probably hinge to a large extent on the role played by policy. Even when economic growth fostered by globalization causes living standards to rise, the environment will not simply take care of itself" (390).

Globalization is a hotbed of debate because it affects everyone in some way. It exerts profound impacts on the environment. Few degrees of separation exist between the state of Amazonian rainforests, the thickness of the Arctic icecap, the price of fruit in Prague, the cost of AIDS medicine in Ghana, the regulation of coal fields in China, and the fuel efficiency of cars rolling off assembly lines on five continents. (This recognition informs the title of Thomas Friedman's book *The Lexus and the Olive Tree.*) The actions of any one person or corporate body soon affect, indirectly but powerfully, all other people. Everyone holds a stake in the common environment.

Communities of hunter-gatherers and cultivators can achieve connection with the environment on a small scale. Local gods, festivals, and customs honor the fragility and necessity of maintaining close relationships to the environment, including the duty of intergenerational stewardship. As difficult as it is to establish such reverence and habit—and they are not always effective—it is harder and more complicated to do so for a global environment. On 10 June 1963, in a speech that helped prompt an international ban on atmospheric nuclear testing but also envisioned the challenges of globalization, President John Kennedy remarked, "In the final analysis our most basic common link is that we all inhabit this small planet. We all breathe the same air. We all cherish our children's future, and we are all mortal." To thrive now and in future generations means to accept the responsibilities of a globalized environment.

FURTHER READING

Carraro, Carlo, ed. *Governing the Global Environment.* Cheltenham, U.K.: Edward Elgar, 2003.

Friedman, Benjamin M. "Growth and the Environment." In *The Moral Consequences of Economic Growth.* New York: Knopf, 2005.

French, H. *Vanishing Borders: Protecting the Planet in the Age of Globalization.* New York: W. W. Norton, 2000.

Le Carré, John. *The Constant Gardener.* New York: Scribner, 2001.

Mander, Jerry, and Edward Goldsmith, ed. *The Case Against the Global Economy and a Turn Toward the Local.* San Francisco: Sierra Club, 1996.

Mol, Arthur P. J. *Globalization and Environmental Reform: The Ecological Modernization of the Global Economy.* Cambridge, Mass.: MIT Press, 2001.

Speth, James G., ed. *Worlds Apart: Globalization and the Environment.* Washington, D.C.: Island Press, 2003.

Stiglitz, Joseph E. "The Way Ahead." In *Globalization and Its Discontents.* New York: W. W. Norton, 2002.

Stille, Alexander. "Saving Species in Madagascar." In *The Future of the Past.* New York: Farrar, Straus and Giroux, 2002.

Tisdell, Clem, and Raj Kumar Sen, ed. "Environmental Issues and Impacts." In *Economic Globalization: Social Conflicts, Labor and Environmental Issues.* Cheltenham, U.K.: Edward Elgar, 2004.

FILMS

Life and Debt (2001), how globalization affects Jamaica.

The Gods Must Be Crazy (1986), humorous and full of insight, depicts the effects of globalized culture on indigenous populations in Africa.

Trinkets and Beads (1996), the story of Huaorani natives in Ecuador and how oil exploration ravaged their land. The battle to drill lasted three years. The fields will produce enough to satisfy full U.S. oil consumption for thirteen days.

Thomas L. Friedman, "Politics for the Age of Globalization" from *The Lexus and the Olive Tree* (1999)

A popular columnist for the New York Times, *Friedman brought globalization front and center in this bestseller. He argues that free trade and global communication demand new structures and efforts to ensure democratic results. He also appears sanguine about the ultimate benefits of a globalization that is apparently inevitable, and our ability to control and minimize its negative environmental impacts. The energy of his prose and the quickness of his intellect have won admirers and converts.*

What to do? Can we develop a method of environmentally sustainable globalization? One hope is clearly that technology will evolve in ways that will help us preserve green areas faster than the Electronic Herd can trample them. As Robert Shapiro of Monsanto likes to say: "Human population multiplied by human aspirations for a middle-class existence divided by the current technological tool kit is putting unsustainable strains on the biological systems that support life on our planet. When three guys living on a lake dumped garbage in it, it was not a big deal. When 30,000 do it, you'd better find a way not to make so much garbage, or treat that garbage, or to have fewer people who make garbage—otherwise the lake is not going to be there."

That is going to require some real breakthroughs in information technology, biotechnology, and nanotechnology (miniaturization down to the molecular and atomic levels that enables tiny power sources to run huge systems) so that we can create value on a smaller and smaller scale, while using less and less stuff. For instance, it is an encouraging sign that, thanks to biotechnology, we can now go into a plant and change the base pairs in its DNA so that it naturally repels insects without having to use fertilizers or pesticides. [See Chapter 4, above.] It is an encouraging sign that, thanks to information technology, things such as recording tape and film are now being turned into digits—1's and 0's—which are not endangered, produce no garbage, and are infinitely reusable.

But technological breakthroughs alone will not be enough to neutralize the environmental impact of the herd, because the innovations simply are not happening fast enough—compared to how fast the herd moves, grows, and devours. You can see it just in the way environmental destruction statistics are now expressed. *Time* magazine reported in 1998 that fifty percent of the world's 233 known primate species are now threatened with extinction and that fifty-two acres of the world's forests are lost *every minute.*

Because of this, conservationists must also learn to move faster. They need to quickly develop the regulatory software and conservation-enforcing procedures to ensure sustainable development and to preserve the most pristine areas. They need to intensify work with local farmers and native peoples whose livelihoods depend on healthy forests and other natural systems. They need to quickly cultivate local elites ready to build and maintain parks and nature preserves that the new bourgeoisie and urban underclasses don't have the time, resources, or inclination to bother with. And of course, they need to promote effective birth control, immediately, because unbridled population growth will explode any environmental protection filters. Howard Youth, writing in *World Watch* magazine about how the Caribbean nation of Honduras had developed a green consciousness over the years, noted that this whole painstaking effort was being undermined by a shortage of condoms. "Flying over the Honduran countryside," he wrote, "you can almost see the country grow: spreading brushfires, new towns, new roads, new swatches of forest cut from the slopes create a patchwork of human activity. . . . The biggest population growth is occurring in the countryside—in villages scattered widely over rugged terrain—and in many of these places contraceptives are not readily available. . . ."

But while it would be nice if conservationists were able to move faster in all these areas, it is unrealistic to believe they will. So where does that leave us? It leaves us with this fact: For now, the only way to run as fast as the herd is by riding the herd itself and trying to redirect it. We need to demonstrate to the herd that being green, being global, and being greedy can go hand in hand. If you want to save the Amazon, go to business school and learn how to do a deal. . . .

* * *

The other method for greening globalization is to demonstrate to corporations and their shareholders that their profits and share prices will increase if they adopt environmentally sound production methods.

Jim Levine, an environmental engineer who sits on the San Francisco Bay Conservation and Development Commission and teaches companies to be both green and greedy, explained to me how it works: "What you have to do is get companies, shareholders, and Wall Street analysts to realize that poor environmental performance equals wasted profits. Up until ten years ago, environmental performance in manufacturing was not a design objective. But now, with the government hitting companies over the head with both new regulations and new tax incentives to be green, and with the SEC [Securities and Exchange Commission] telling companies they have to start accurately portraying their environmental liabilities to shareholders—such as where they are being sued for dumping and what the cleanup could cost—there has been a paradigm shift. Companies are starting to realize that if they go into Bangkok and build a plant that pollutes the environment and then the Thai government finally gets around to passing the laws and regulatory software to clean it all up, it will be a lot more expensive to deal with later, rather than building in green procedures from the start."

One of the leading companies in this new paradigm is the Chicago-headquartered health products company Baxter International, Inc. In 1997 Baxter had sales of $6.1 billion, on output from its sixty manufacturing plants worldwide. As part of its annual reporting to shareholders, Baxter includes an Environmental Financial Statement for all its operations. Its 1997 Environmental Financial Statement reported that green production practices implemented that year had saved the company $14 million, which more than covered the costs of the programs. In addition, it said that cost avoidance from green production practices since 1990 had already saved another $86 million. "This means," the report said, "that Baxter would have spent $100 million more in 1997 for raw materials production processes, disposal costs, and packaging if no environmentally beneficial actions had been implemented by the company since 1990."

Most countries right now don't have effective "polluter pays" laws, but one day many will. Which is why Baxter says in its 1997 annual report that "it is better to have all our international waste today go to reputable waste sites. Thus we can be in better shape to avoid big potential liabilities in the future." Executives who don't think this way are not taking care of their shareholders and are robbing themselves of higher bonuses. . . .

* * *

POLITICS FOR THE AGE OF GLOBALIZATION

Let's start with a politics of sustainable globalization. It has to consist of two things: one is a picture of the world, so people understand where they are; and the other is a set of Integrationist Social-Safety-Net policies for dealing with it.

You need a picture of the world because no policy is sustainable without a public that broadly understands why it's necessary and sees the world the way you do. I've always believed that Bill Clinton defeated both George Bush and Bob Dole because a majority of American voters intuitively sensed that they were in a new era and that Clinton grasped this as well and had some credible ideas of how to deal with it—while Dole and Bush didn't get it at all. Unfortunately, once in office, Clinton never fully developed that intuitive sense and made it explicit, with a real picture of the world that he needed to repeat over and over again. It started from his first week on the job when he defined America's central problem as affordable health care—not sustainable globalization.

What should Mr. Clinton have said at his first inaugural? Something like this: "My fellow Americans, my tenure as your President is coinciding with the end of the Cold War system and the rise of globalization. Globalization is to the 1990s and the next millennium what the Cold War was to the 1950s through the 1980s: If the Cold War system was built around the threat and challenge of the Soviet Union, which was dividing the world, the globalization system is built around the threat and challenge of rapid technological change and economic integration that are uniting the world.

"But as it unites the world, globalization is also transforming everyone's workplace, job, market, and community—rapidly destroying old jobs and harvesting new ones, rapidly eliminating old lifestyles and producing new ones, rapidly eliminating old markets and creating new ones, rapidly destroying old industries and inventing new ones. Foreign trade, which represented just thirteen percent of gross domestic product in 1970, is now up to nearly thirty percent of American GDP—and rising. Technological change is now so fast that American computer companies manufacture three different models of each computer every year. Not only is this a new world; it is, for the most part, a better one. Even if they have to struggle at times with this system, China, Indonesia, Korea, Thailand, Malaysia, Brazil, Argentina have seen their standards of living rise faster, for more of their citizens, than at any time in their history—thanks to the increasing effectiveness of financial markets in facilitating trade and investment by people in one country into the factories of another. Indeed, as one of my top economic advisers, Larry Summers, points out, thanks in large part to increasing globalization more than one quarter of humanity is now enjoying growth at rates in which their living standards will quadruple within a generation. Quadruple! That is unprecedented in economic history. And far from coming at the expense of the United States, this worldwide growth has led to the lowest unemployment rate in America in nearly fifty years.

"Given these challenges and opportunities, the United States needs a strategy to make globalization sustainable and to ensure that we will always be able to compete effectively in this world. Therefore, think of the world as a wheel with spokes. At the hub of this wheel is what I would label 'globalization and rapid economic and technological change.' That is, in plain language, the ONE BIG THING that is going on out there. Because it is at the center, we need a different approach to health care, welfare, education, job training, the environment, market regulation, social security, campaign finance, and expansion of free trade. Each one of these areas has to be adjusted, adapted, or reformed to enable us as a society to get the most out of this globalization system and to cushion its worst aspects. For instance, in a world where people now work at a dozen different jobs for a dozen different companies over their lifetime, they need to have portable pensions, portable health care, and more opportunities for lifelong learning. Globalization demands that our society move faster, work smarter, and take more risks than at any time in our history. As your President I make you two promises. One is that I will make it my business to equip each of you, and our society at large, to meet this challenge, with the right combination of integrationist policies and social safety nets. The other is that I will be a tireless defender of our trade laws to ensure that globalization, while it challenges the American worker, does not allow others to take advantage of our openness by dumping their products here while limiting our access to their markets.

"I'm not here to tell you this is all going to be easy. In fact, I'm here to tell you that it is going to be really hard. But if we can strike the right balance—and I think we can—we can be the vanguard for the world on how to manage integration in the age of globalization, just as we were the vanguard for the world on how to manage containment in the age of the Cold War. God Bless America."

This is what Clinton believed, but it is not what he always said. And one reason his health-care proposals got chewed apart by his opponents—not the only reason, but one reason—was that they were not effectively located in a clear and constantly repeated picture of the world, with globalization at the center and the requisite policies flowing from it. As a result, notes Harvard University economist Dani Rodrik, "the connections and complementarities in all these areas got lost in the public debate," and it made it much easier for ideologues and extremists, as well as economic populists, nationalists, know-nothings, nativist xenophobes, and opportunists, to often skew the debate on any one issue—like trade or health-care reform—and drive it into a dead end.

Politicians have to be aware that, for a lot of reasons, it is very easy to distort and demonize globalization and end up, as Clinton did, where even if you are right on the economics, you lose control of the politics, so it works against you instead of for you. People who are the biggest losers from globalization, workers who have lost their jobs to robots or foreign factories, know exactly who they are. This makes them very easy to mobilize against more integration, technology, or free trade. People who are beneficiaries of globalization, of more open trade and of foreign investment, often don't know it. They often don't make the connections between globalization and their rising standards of living, and therefore they are difficult to mobilize. Have you ever heard a worker in a microchip factory say, "Boy, am I lucky. Thanks to globalization, soaring demand for American high-tech exports, the labor shortage for skilled workers in this country, and rising expectations in the developing world, my boss had to give me a raise."

Another reason globalization is easy to distort is that people don't understand that it is largely a technology-driven phenomenon, not a trade-driven one. We had a receptionist at the Washington bureau of the *New York Times*, but the company eliminated her job. She didn't lose her job to a Mexican, she lost it to a microchip—the microchip that operates the voicemail device in all our office phones. The fact is, that microchip would have taken away her job if we had had *no* trade with Mexico. That microchip would have taken away her job had we had a thirty-foot-high wall stretching from one end of the Mexican-American border to the other. But politicians don't want to acknowledge this. None of them is going to stand up and say: "I want you to get up now, unplug your phone, go to your window, throw your phone out and shout: '*I'm not going to take it anymore! Save American jobs! Ban voice mail! Potato chips, yes! Microchips, no!*'" That's not a winning political message. It's much easier instead to rail against Mexicans and foreign factories. And, of course, foreign factories do, in some instances, take away jobs (but not nearly as many as technology destroys and creates), so there is just enough truth there to make for some very emotive, dangerous politics. And because foreign workers and factories are easy to see, and microchips are not, they loom much larger in our consciousness as the problem.

If we don't educate the public about the real nature of the world today and demystify globalization—the Separatists will always exploit this confusion for their own ends. In 1998, President Clinton could not get NAFTA [North American

Free Trade Agreement] expanded to Chile because a union-led minority, who believed that they were not benefiting from more free trade, were very active in opposing the expansion of NAFTA, while the majority that was benefiting from expanded free trade never understood who they were and therefore never got mobilized to defend their interests.

A politics of sustainable globalization, though, needs more than just the correct picture of what is happening in the world. It also needs the right balance of policies. This to me is what Integrationist Social-Safety-Nettism is all about. We Integrationist Social-Safety-Netters believe that there are a lot of things we can do in this era of globalization that are not all that expensive, do not involve radical income redistribution—or lavish compensatory welfare spending programs that would violate the economic rules of the Golden Straitjacket—but are worth doing to promote social stability and to prevent our own society from drifting into one of high walls and tinted windows more than it already has.

My Integrationist Social-Safety-Nettism would focus on democratizing globalization educationally, financially, and politically for as many people as possible, but in ways that are still broadly consistent with integration and free markets. Here's what I mean:

Democratizing globalization educationally: The Integrationist Social-Safety-Netters favor a combination of trampolines and safety nets to deal with those who get left behind, either permanently or temporarily, by globalization. As a society our Golden Straitjacket is producing enough gold—a $70 billion surplus in 1998—to afford both. The social safety nets for those who cannot compete are familiar—social security, Medicare, Medicaid, food stamps, and welfare and they need to be sustained to catch those who may never be up to the demands of the Fast World. But with the right trampolines, it should be possible constantly to shrink that pool of left-behinds.

To that end, I believe that each White House should offer in this era of globalization an annual piece of legislation that I would call "The Rapid Change Opportunity Act." It would go alongside whatever integrationist policy the Administration was pursuing that year—whether it be NAFTA expansion, renewal of Most-Favored-Nation status for China, or any other free-trade arrangements. The Rapid Change Opportunity Act would vary each year, but its goal would be to create both the reality and the perception that the government understands that globalization is inevitable but that it spreads its blessings most unevenly, and therefore the government is constantly going to be adjusting its trampolines to get as many people as possible up to speed for the Fast World.

For example, in 1997–98 my Rapid Change Opportunity Act would have included the following: pilot projects for public employment for temporarily displaced workers; tax breaks for severance pay for displaced workers; free government-provided résumé consultation for anyone who loses a job, and a further extension of the Kassebaum-Kennedy Act, so that laid-off workers could keep their health-insurance policies longer; and a national advertising campaign for one of the best, but most underreported, bipartisan achievements of the Clinton era, the Workforce Investment Act. Signed in August 1998, it consolidated the government's 150 different job-training programs into three broad grants: Individual Training Accounts that workers can use for any training they believe will most advance their job opportunities; one-stop career centers for every job-training program; and an increase in youth-training programs by $1.2 billion over five years. In addition, I would have included in my Rapid Change Opportunity Act some increased U.S. lending to the Asian, African, and Latin American development banks to promote training of women, microlending to women and small businesses, and environmental cleanup in every developing country with which America has significant trade. I would include an increase in funding for the International Labor Organization's new initiative for building alternatives to child labor in countries where children are most abused. I would include an increase in the already existing Trade Adjustment Assistance program, which provides some small income support and training for anyone who can show that his or her job was eliminated because of trade. I would expand the already existing Dislocated Worker Training program (which served 660,000 people in 1997) to assist anyone who lost a job because of new technology. Finally, I would launch a national advertising campaign to better inform people about the already existing Lifelong-Learning Tax Credit, which allows citizens to write off from their income taxes up to $1,000 of the cost of any education or training program they sign up for to upgrade their education or technical skills.

Democratizing globalization financially: There is no better way to make globalization sustainable than to give more people a financial stake in the Fast World. Whenever I think about this issue, I always recall a story that Russian journalist Aleksei Pushkov told me back in April 1995 about one of his neighbors in Moscow. "He was this poor driver who lived in the apartment off the entryway. Every Friday night he would get drunk and sing along—over and over in a very loud voice—to two English songs: 'Happy Nation' and 'All She Wants Is Another Baby.' He had no idea what the words meant. When he got really drunk, he'd start beating his wife and she would start screaming. He was driving us crazy. I wanted to throw a grenade at him. Anyway, about eight months ago, I don't know how, he got a share in a small car-repair shop. Since then, no more 'Happy Nation,' no more singing all night, no more beating his wife. He leaves every morning at 8:30 for work and he is satisfied. He knows he has some prospects in life now. My wife said to me the other day, 'Hey, look at Happy Nation'—that's what we call him— 'he's an owner now.'"

A strategy for making more people owners also has to be part of any country's Rapid Change Opportunity Act. In America, that means initiatives that will improve access to investment capital in the most distressed, low-income communities, so we are not just training people for jobs that are not available. American inner cities are emerging markets every bit as much as Bangladesh, and they sometimes need some of the same sort of market-oriented assistance programs. As Deputy Treasury Secretary Larry Summers has pointed out: "The world over, private financial markets fail when it comes to the very poor. Mainstream banks do not seek out poor communities—because that's not where the money is. Other barriers tend artificially to restrict the flow of capital to certain neighborhoods or minority groups, creating clear market failures. Yet if you deprive the people of these districts of the chance to lend or save, they are a good deal more likely to stay that way [i.e., very poor]."

One way to begin to democratize access to capital in America is to revitalize the Community Reinvestment Act, which uses government pressure to encourage commercial banks to make affordable credit to distressed neighborhoods. But there are some loans commercial banks will never make. That is why my Rapid Change Opportunity Act would also include funding for a new government-supported venture capital fund for low- and moderate-income neighborhoods. Known as the Community Development Financial Institutions Fund, it gives start-up financing to entrepreneurs ready to make risky investments in underdeveloped districts, where they see the market possibilities—for anything from private day-care centers to low-income housing to a beauty salon to entertainment facilities—but where venture capital financing wouldn't normally be made available.

These sorts of initiatives say to citizens: "While the government is asking you to leap from trapeze to trapeze, higher and higher, faster and faster, farther and farther, it is going to build a net under you at the same time. It's not a net anyone can live on for long, but it can bounce a lot of people back into the game." A hand-up really is better than a handout. Even if we waste some money on these hand-up programs, the costs are so small compared to the benefits and efficiencies that come from keeping our markets as free and open to the world as possible. My Rapid Change Opportunity Act is a tiny price to pay for maintaining the social cohesion and political consensus for integration and free trade. Hence my mottoes: "Protection, not protectionism. Cushions, not walls. Floors, not ceilings. Dealing with the reality of the Fast World, not denying it."

Democratizing globalization politically: While democratizing access to globalization is critical, particularly for developing countries, simultaneously democratizing their political systems is just as important. This is one of the real lessons of globalization's first decade: Getting your society up to speed for globalization is an enormously wrenching process and because of that it requires more democracy in the long run rather than less. In the Cold War, the leaders of developing countries had superpower patrons who would sustain them, no matter how they ran their countries. But those patrons are gone now and the masses are not going to sustain failed governments for long. (See dictionary entry for Indonesia.) If you fail now, you will fall—and unless your people catch you and support you, you will fall hard. (See dictionary entry for Suharto.)

As democracy scholar Larry Diamond points out: "We have now seen a number of examples where countries in Latin America, Eastern Europe, and East Asia have voted out of office governments they associated with the pain of globalization reforms. The new governments that came in made some adjustments but kept more or less the same globalizing, marketizing policies. How did they get away with that? Because the democratic process gave the public in these countries a sense of ownership over the painful process of economic policy reform. It was no longer something completely alien that was being done to them. They were being consulted on it and given a choice about at least the speed of the process, if not the direction. Moreover, as a result of the opportunity to participate in the process and throw out people who they felt had moved too harshly and abruptly, or too corruptly or insensitively, the whole process had much greater political legitimacy, and thus more sustainability."

Moreover, where parties and leaders have alternated in power—and political oppositions have come in and pursued pretty much the same policies of economic liberalization and globalization as their predecessors—the message sinks in to the public that there really is no alternative to the Golden Straitjacket. How many Latin American, Eastern European, and now East Asian opposition leaders have come into office in the last decade and said, "Gee, it turns out we really are bankrupt. We really do have to open up. In fact, things are even worse than I thought, and we are going to have to accelerate these reforms because there's no way out. But we'll put a human face on them." Democratization helps make that coming to terms with reality possible. And that's why the countries that are adjusting best to globalization today are often not the naturally richest ones—Saudi Arabia, Nigeria, or Iran—but rather the most democratic ones—Poland, Taiwan, Thailand, Korea. Russia is such a mess today precisely because its democratic development is stalled. It lacks not only the software and operating systems to attract the herd—but also a credible enough democracy to persuade its own people that there will be fairness, equity, and accountability in managing the pains and gains of adjusting to globalization.

Democratizing globalization—it's not only the most effective way to make it sustainable, it's the most self-interested and moral policy that any government can pursue.

Paul Hawken, from "The WTO: Inside, Outside, All Around the World" (2000)

Under the current world economic system, the power of multinational corporations animates anti-globalization activists far more than the other phenomena of globalization. Paul Hawken of the Natural Capital Institute writes his essay in the aftermath of the WTO 1999 meeting in Seattle. Protesters disrupted the proceedings. Police reacted harshly and at times created rather than prevented violence. Seattle's police chief was soon on his way out. In this excerpt, Hawken points out flaws in a world trade structure that he believes favors the few, the rich, and a few rich corporations, two hundred of which combined "have twice *the assets of 80 percent of the world's people." He views this corporatized version of globalization as undemocratic and environmentally blind. Its speedy power bypasses national, let alone regional, governments and often ignores or pulverizes local cultures.*

More than 700 organizations and between 40,000 and 60,000 people took part in the protests against the WTO's Third Ministerial on November 30 [1999]. These groups and citizens sense a cascading loss of human, labor, and environmental rights in the world. Seattle was not the beginning but simply the most striking expression of citizens struggling against a worldwide corporate-financed oligarchy—in effect, a plutocracy. Oligarchy and plutocracy often are used to describe "other" countries where a small group of wealthy people rule, but not the "First World"—the United States, Japan, Germany, or Canada.

The World Trade Organization, however, is trying to cement into place that corporate plutocracy. Already, the world's top 200 companies have twice the assets of 80 percent of the world's people. Global corporations represent a new empire whether they admit it or not. With massive amounts of capital at their disposal, any of which can be used to influence politicians and the public as and when deemed necessary, all democratic institutions are diminished and at risk. Corporate free market policies, as promulgated by the WTO, subvert culture, democracy, and community, a true tyranny. The American Revolution occurred because of crown-chartered corporate abuse, a "remote tyranny," in Thomas Jefferson's words. To see Seattle as an isolated event, as did most of the media, is to look at the battles of Concord and Lexington as meaningless skirmishes. . . .

Thomas Friedman, the *New York Times* columnist and author of an encomium to globalization entitled *The Lexus and the Olive Tree* [see above], angrily wrote that the demonstrators were "a Noah's ark of flat-earth advocates, protectionist trade unions, and yuppies looking for their 1960s fix." Not so. They were organized, educated, and determined. They were human rights activists, labor activists, indigenous people, people of faith, steel workers, and farmers. They were forest activists, environmentalists, social justice workers, students, and teachers. And they wanted the World Trade Organization to listen. They were speaking on behalf of a world that has not been made better by globalization. Income disparity is growing rapidly. The difference between the top and bottom quintiles has doubled in the past thirty years. Eighty-six percent of the world's goods go to the top fifth, the bottom fifth get one percent. The apologists for globalization cannot support their contention that open borders, reduced tariffs, and forced trade benefit the poorest three billion people in the world. . . .

Despite Friedman's invective about "the circus in Seattle," the demonstrators and activists who showed up there were not against trade. They do demand proof

Figure 8.2 March in Seattle protests the 1999 World Trade Organization meeting. © Newscom / Getty Images.

that shows when and how trade—as the WTO constructs it—benefits workers and the environment in developing nations, as well as workers at home. Since that proof has yet to be offered, the protesters came to Seattle to hold the WTO accountable.

. . . When the United States attempted to block imports of shrimp caught in the same nets that capture and drown 150,000 sea turtles each year, the WTO called the block "arbitrary and unjustified." [But see Wilson, below.] Thus far in every environmental dispute that has come before the WTO, its three-judge panels, which deliberate in secret, have ruled for business, against the environment. The panel members are selected from lawyers and officials who are not educated in biology, the environment, social issues, or anthropology.

. . . WTO rules run roughshod over local laws and regulations. The corporations operating through the WTO relentlessly pursue the elimination of any restriction on the free flow of trade including those based on local, national, or international laws that distinguish between products based on how they are made, by whom, or what happens during production. By doing so, the WTO is eliminating the ability of countries and regions to set standards, to express values, or to determine what they do or don't support. Child labor, prison labor, forced labor, substandard wages and working conditions cannot be used as a basis to discriminate against goods. Nor can a country's human rights record, environmental destruction, habitat loss, toxic waste production, or the presence of transgenic materials or synthetic hormones. Under WTO rules, the Sullivan Principles against apartheid and the boycott of South Africa would not have existed.

If the world could vote on the WTO rules, would they pass? Not one country of the 135 member-states of the WTO has held a plebiscite to see if its people support the WTO mandate. The people trying to meet in the Green Rooms at the Seattle Convention Center were not elected. Even [WTO Director] Michael Moore was not elected. . . .

This is what I remember about the violence. There was almost none until police attacked demonstrators that Tuesday in Seattle. Michael Meacher, environment minister of the United Kingdom, said afterward, "What we hadn't reckoned with was the Seattle police department, who single-handedly managed to turn a peaceful protest into a riot." There was no police restraint, despite what Mayor Paul Schell kept proudly assuring television viewers all day. Instead, there were rubber bullets, which Schell kept denying all day. In the end, more copy and video was given to broken windows than broken teeth. . . .

It's not inapt to compare the pointed lawlessness of the anarchists with the carefully considered ability of the WTO to flout laws of sovereign nations. When "The Final Act Embodying the Results of the Uruguay Round of Multilateral Trade Negotiations" was enacted April 15, 1994, in Marrakech, it was recorded as a 550-page agreement that was then sent to Congress for ratification. Ralph Nader offered to donate $10,000 to any charity of a congressman's choice if any of them signed an affidavit saying they had read it and could answer several questions about it. Only one—Senator Hank Brown, a Colorado Republican—took him up on it. After reading the document, Brown changed his opinion and voted against the Agreement. There were no public hearings, dialogues, or education. What was approved was an Agreement that gives the WTO the ability to overrule or undermine international conventions, acts, treaties, and agreements. The WTO directly violates "The Universal Declaration of Human Rights" adopted by member nations of the United Nations, not to mention Agenda 21 of the 1992 Earth Summit. (The proposed draft agenda presented in Seattle went further in that it would require Multilateral Agreements on the Environment such as the Montreal Protocol, the Convention on Biological Diversity, and the Kyoto Protocol to be in alignment with and subordinate to WTO trade polices.) The final Marrakech Agreement contained provisions that most of the delegates, even the heads-of-country delegations, were not aware of, statutes that were drafted by sub-groups of bureaucrats and lawyers, some of whom represented transnational corporations.

The police mandate to clear downtown was achieved by 9 P.M. Tuesday night. But police, some of whom were fresh recruits from outlying towns, didn't want to stop there. They chased demonstrators into neighborhoods where the distinctions between protesters and citizens vanished. The police began attacking bystanders, residents, and commuters. They had lost control. When President Clinton sped from Boeing Airfield to the Westin Hotel at 1:30 A.M. Wednesday, his limousines entered a police-ringed city of broken glass, helicopters, and boarded windows. He was too late. The mandate for the WTO had vanished sometime that afternoon. . . .

Patricia King, one of two *Newsweek* reporters in Seattle, called me from her hotel room at the Four Seasons and wanted to know if this was the '60s redux.

No, I told her. The '60s were primarily an American event; the protests against the WTO are international. Who are the leaders? she wanted to know. There are no leaders in the traditional sense. But there are thought leaders, I said. Who are they? she asked. I began to name some: Martin Khor and Vandana Shiva of the Third World Network in Asia [see Shiva, below], Walden Bello of Focus on the Global South, Maude Barlow of the Council of Canadians, Tony Clarke of Polaris Institute, Jerry Mander of the International Forum on Globalization, Susan George of the Transnational Institute, David Korten of the People-Centered Development Forum, John Cavanagh of the Institute for Policy Studies, Lori Wallach of Public Citizen, Mark Ritchie of the Institute for Agriculture and Trade Policy, Anuradha Mittal of the Institute for Food & Development Policy, Helena Norberg-Hodge of the International Society for Ecology and Culture, Owens Wiwa of the Movement for the Survival of the Ogoni People, Chakravarthi Raghavan of the Third World Network in Geneva, Debra Harry of the Indigenous Peoples Coalition Against Biopiracy, José Bové of the Confédération Paysanne Européenne, Tetteh Hormoku of the Third World Network in Africa, Randy Hayes of Rainforest Action Network. Stop, stop, she said. I can't use these names in my article. Why not? Because Americans have never heard of them.

Instead, *Newsweek* editors put the picture of the Unabomber, Theodore Kaczynski, in the article because he had at one time purchased some of John Zerzan's writings. . . .

Those who marched and protested opposed the tyrannies of globalization, uniformity, and corporatization, but they did not necessarily oppose internationalization of trade. Economist Herman Daly has long made the distinction between the two. Internationalization means trade between nations. Globalization refers to a system of uniform rules for the entire world, a world in which capital and goods move at will without the rule of individual nations. Nations, for all their faults, set trade standards. Those who are willing to meet those standards can do business with them. Do nations abuse this power? Always and constantly, the U.S. being the worst offender. But nations do provide, where democracies prevail, a means for people to set their own policy, to influence decisions, and determine their future. Globalization supplants the nation, the state, the region, and the village. While eliminating nationalism is indeed a good idea, the elimination of sovereignty is not.

One recent example of the power of the WTO is Chiquita Brands International, a two billion dollar corporation that recently made a large donation to the Democratic Party. Coincidentally, the United States filed a complaint with the WTO against the European Union because European import policies favored bananas coming from small Caribbean growers instead of the banana conglomerates. The Europeans freely admitted their bias and policy: they restricted imports from large multinational companies in Central America (plantations whose lands were secured by U.S. military force during the past century) and favored small family farmers from former colonies who used fewer chemicals. It seemed like a decent thing to do, and everyone thought the bananas tasted better. For the banana giants, this was untenable. The United States prevailed in this WTO-arbitrated case. So who won and who lost? Did the Central American employees at Chiquita Brands win? Ask the hundreds of workers in Honduras who were made infertile by the use of dibromochloropropane on the banana plantations. Ask the mothers whose children have birth defects from pesticide poisoning. Did the shareholders of Chiquita win? At the end of 1999, Chiquita Brands was losing money because it was selling bananas at below cost to muscle its way into the European market. Its stock was at a thirteen-year low, the shareholders were angry, the company was up for sale, but the prices of bananas in Europe are really cheap. Who lost? Caribbean farmers who could formerly make a living and send their kids to school can no longer do so because of low prices and demand.

Globalization leads to the concentration of wealth inside such large multinational corporations as Time-Warner, Microsoft, GE, Exxon, and Wal-Mart. These giants can obliterate social capital and local equity, and create cultural homogeneity in their wake. Countries as different as Mongolia, Bhutan, and Uganda will have no choice but to allow Blockbuster, Burger King, and Pizza Hut to operate within their borders. Under WTO rules, even decisions made by local communities to refuse McDonald's entry (as did Martha's Vineyard) could be overruled. The as-yet unapproved draft agenda calls for WTO member governments to open up their procurement process to multinational corporations. No longer could local governments buy preferentially from local vendors. Proposed rules could force governments to privatize medical care by allowing foreign companies to bid on delivering national health programs. The draft agenda could privatize and commodify education, and ban cultural restrictions on entertainment, advertising, or commercialism as trade barriers. Globalization kills self-reliance, since smaller local businesses can rarely compete with highly capitalized firms that seek market share instead of profits. Thus, developing regions may become more subservient to distant companies, with more of their income exported rather than re-spent locally. . . .

The WTO was a clash of chronologies or time frames, at least three, probably more. The dominant time frame was commercial. Businesses are quick, welcome innovation in general, and have a bias for change. They need to grow more quickly now than ever before. They are punished, pummeled, and bankrupted if they do not. With worldwide capital mobility, companies and investments are rewarded or penalized instantly by a network of technocrats and money managers who move two trillion dollars a day seeking the highest return on capital. The Internet, greed, global communications, and high-speed transportation are all making businesses move faster than before.

The second time frame is cultural. It moves more slowly. Cultural revolutions are resisted by deeper, historical beliefs. The first institution to blossom under *perestroika* was the Russian Orthodox Church. In 1989, I walked into a church near Boris Pasternak's dacha and heard priests and *babushkas* reciting the litany with perfect recall as if seventy-two years of repression had never happened. Culture provides the slow template of change within which family, community, and religion prosper. Culture provides identity and, in a fast-changing world of displacement and rootlessness, becomes ever more important. In between culture and business is governance, faster than culture, slower than commerce.

At the heart, the third and slowest chronology is Earth, nature, the web of life. As ephemeral as it may seem, it is the slowest clock ticking, always there, responding to long, ancient evolutionary cycles that are beyond civilization.

These three chronologies often conflict. As Stewart Brand points out, business unchecked becomes crime. Look at Russia. Look at Microsoft. Look at history. What makes life worthy and allows civilizations to endure are all the things that have "bad" payback under commercial rules: infrastructure, universities, temples, poetry, choirs, literature, language, museums, terraced fields, long marriages, line dancing, and art. Most everything we hold valuable is slow to develop, slow to learn, and slow to change. Commerce requires the governance of politics, art, culture, and nature, to slow it down, to make it heedful, to make it pay attention to people and place. It has never done this on its own. The extirpation of languages, cultures, forests, and fisheries is occurring worldwide in the name of speeding up business. The rate of change is unnerving to all, even to those who are supposedly benefiting. To those who are not, it is devastating. . . .

In the Inuit tradition, there is a story of a fisherman who trolls an inlet. When a heavy pull on the fisherman's line drags his kayak to sea, he thinks he has caught the "big one," a fish so large he can eat for weeks, a fish so fat that he will prosper ever after. As he daydreams about his coming ease, what he reels up is Skeleton Woman, a woman flung from a cliff long ago, her fish-eaten carcass left to rot at the bottom of the sea. Skeleton Woman is so snarled in his fishing line that she is dragged behind the fisherman wherever he goes. She is pulled across the water, over the beach, and into his house, where he collapses in terror.

In the retelling of this story by Clarissa Pinkola Estes, the fisherman has brought up a woman who represents life and death, a specter who reminds us that with every beginning there is an ending, for all that is taken, something must be given in return, that the earth is cyclical and requires respect. The fisherman, feeling pity for her, slowly disentangles her, straightens her bony carcass, and finally falls asleep. During the night, Skeleton Woman scratches and crawls her way across the floor, drinks the tears of the dreaming fisherman, and grows anew her flesh and heart and body.

This myth applies to business as much as it does to a fisherman. The apologists for the WTO want sleeker planes, engineered food, computers everywhere, golf courses that are preternaturally green. They see no limits; they know of no downside. But Life always comes with Death, with a tab, a reckoning. They are each other's consorts, inseparable and fast. The expansive dreams of the world's future wealth were met with perfect symmetry by Bill Gates III, the world's rich-

Figure 8.3 The Golden Arches rise in Tokyo. © Newscom / Photographers Showcase.

est man and co-chair of the Seattle Host Committee. But Skeleton Woman also showed up in Seattle, the uninvited guest, and the illusion of wealth, the imaginings of unfettered growth and expansion, became small and barren. Dancing, drumming, ululating, marching in black with a symbolic coffin for the world, she wove through the sulfurous rainy streets of the night. She couldn't be killed or destroyed, no matter how much gas or pepper spray or how many rubber bullets were used. She kept coming back and sitting in front of the police and raising her hands in the peace sign, and was kicked and trod upon, and it didn't make any difference. Skeleton Woman told corporate delegates and rich nations that they could not have the world. It is not for sale. The illusions of world domination have to die, as do all illusions. Skeleton Woman was there to say that if business is going to trade with the world, it has to recognize and honor the world, her life, and her people. Skeleton Woman was telling the WTO that it has to grow up and be brave enough to listen, strong enough to yield, courageous enough to give. Skeleton Woman has been brought up from the depths. She has regained her eyes, voice, and spirit. She is about in the world and her dreams are different. She believes that the right to self-sufficiency is a human right; she imagines a world where the means to kill people is not a business but a crime, where families do not starve, where fathers can work, where children are never sold, where women cannot be impoverished because they choose to be mothers and not whores. She cannot see in any dream a time when a man holds a patent to a living seed, or animals are factories, or people are enslaved by money, or water belongs to a stockholder. Hers are deep dreams from slow time. She is patient. She will not be quiet or flung to sea any time soon.

Arlene Wilson, from "The World Trade Organization: The Debate in the United States" (2000)

This dispassionate report from the Congressional Research Service (CRS) underscores how the U.S. government and American commerce participate in globalization. However, Congress and the Executive Branch are wary of complications regarding tariffs, trade, labor, and the environment. Both NAFTA (North American Free Trade Agreement) and the WTO (World Trade Organization) have created skeptics and opponents, even in Congress. Any country cedes some power when it agrees to global rules whose violations are judged outside its own borders yet whose strictures it must obey. Having said that, it is also fair to state that in the formulated trade policies of the United States, environmental concerns infrequently take priority over more narrowly defined economic and political interests.

In recent years, and especially after the WTO's Seattle Ministerial Conference in late 1999, criticisms of the WTO by labor, environment, consumer, and human rights groups in the United States have increased. . . .

The World Trade Organization (WTO) went into effect in 1995, replacing the General Agreement on Tariffs and Trade (GATT) which had been in existence since 1948. Under the WTO, the governments of the 136 member countries agree on a set of rules and principles for trade, negotiate periodically to reduce trade barriers, and participate in the dispute settlement procedure. Economists believe that, over the past fifty years, the more predictable environment for trade as well as the reduction in trade barriers has contributed to unprecedented economic prosperity for the majority of countries. On the other hand, trade liberalization under the WTO has resulted in economic costs to those whose jobs have been adversely affected, although they are relatively few compared to total employment in the United States. . . .

In environmental and food safety issues, critics charge that the WTO's dispute settlement procedure can (and has) adversely affected environmental and health laws. Supporters counter that the dispute settlement process only requires that domestic measures be implemented in a non-trade-restrictive way and do not discriminate among trading partners. . . .

BRIEF HISTORY OF THE GATT/WTO

. . . One important motivation for the GATT in the immediate postwar period was to prevent a return to the era of the 1930s where, many believed, restrictions on trade contributed to the depth and duration of the worldwide economic depression as well as the political instabilities that led to World War II. In an attempt to protect its own economy, each country imposed trade restrictions, which were often followed by trade restrictions in other countries. The Smoot Hawley tariff in the United States, which raised U.S. tariffs to an all time high in 1930, has often been cited as an important part of that process. Ultimately, economic activity declined in all countries.

The GATT, and now the WTO, has three broad functions. First, the governments of the member countries agree on a set of multilateral rules and principles for trade, which provide a stable and predictable basis for trade. The benefits of agreed-on rules and principles are difficult to quantify precisely, but are likely substantial. This is suggested by the fact that membership in the GATT/WTO has grown from 23 countries in 1948 to 136 countries at present [2000] (with about 30 more countries waiting to join). Also, as discussed later, world trade over the past half century, by growing more rapidly than world production, has served as an engine of economic growth for many countries.

The second function of the WTO is to provide a mechanism to enforce the rules. An important part of this mechanism is the dispute settlement procedure, which provides a way in which disagreements among countries over the interpretation of the rules can be resolved. Broadly, a country with a complaint requests a consultation and, if the dispute is not resolved during the consultation, the complaining country may request establishment of a panel. After the panel (or the Appellate Body if the panel's decision is appealed) issues its ruling, the Dispute Settlement Body (representatives of all the WTO members) adopts the report unless the Dispute Settlement Body decides by consensus to reject it.

Third, the WTO provides a forum for negotiations to reduce trade barriers. In the early years, negotiating rounds focused on reducing tariffs, which, on average, are now very low. As tariffs were reduced, countries sometimes turned to nontariff barriers (for example, subsidies, government procurement regulations, antidumping procedures) to restrict imports. More recently, if health, safety, or environmental standards are used primarily to restrict trade (and not for legitimate health, safety, or environmental purposes), they are called "disguised trade restrictions" or "disguised protectionism." Some WTO rules have been agreed on to prevent countries from using such standards as disguised protectionism, although it is important to emphasize that each country retains the right to set its own standards. . . .

ECONOMIC BENEFITS AND COSTS

The WTO's rules and principles for trade, trade liberalization, and dispute settlement procedures are all designed to encourage trade. Increased trade, over time, improves productivity and raises the standard of living for all countries through increased specialization, economies of scale, and improved competitiveness. Consumers, in particular, benefit from lower prices, as well as from higher-quality products and a broader variety of goods to choose from.[1]

Economists believe that, over the past fifty years, the more predictable environment for trade as well as the reduction in trade barriers has contributed to unprecedented economic prosperity for the majority of countries. For example, world exports in 1998 were eighteen times those in 1950, while world GDP in 1998 was six times that of world GDP in 1950.[2] Looked at another way, the average annual rate of growth in the volume of world exports was higher than growth in GDP in the three time periods: 1965–80, 1980–90, 1990–98. The rapid growth in world exports likely contributed substantially to the growth in world GDP.

Trade has been particularly important in raising standards of living for many developing countries, such as Korea, Taiwan, and many in Latin America. Ultimately the developed countries' economies are expected to benefit; as the developing countries' incomes rise, so do their imports from developed countries. . . .

ENVIRONMENTAL ISSUES[3]

Environmentalists first became concerned about trade issues during the Uruguay Round negotiations, which lasted from 1986 to 1994. As a result, the preamble in the agreement that established the WTO includes goals such as sustainable development [see Chapter 5, above] and seeking to protect and preserve the environment. A few environmental concerns were also addressed in the Uruguay Round agreements on agriculture, sanitary and phytosanitary measures, subsidies and countervailing measures, and technical barriers to trade. Nevertheless, the main areas of concern to environmentalists were left for future negotiations.

In April 1994, the WTO ministers established a Committee on Trade and the Environment (CTE). The purposes of the CTE are to study the relationship between trade and environmental measures and to make recommendations as to whether modifications of the WTO are required. The CTE has been meeting several times each year and has issued several reports. While considerable information and analysis have been exchanged, no recommendations regarding changes in the WTO have been made. Environmentalists have expressed disappointment that the CTE has not made more progress. Others believe that the CTE has been a useful forum for information sharing, analyzing issues, and improving understanding.

Perhaps the main issue of concern to environmentalists in the past few years is the WTO dispute settlement process in which, some claim, environmental laws have been challenged in the WTO. For example, according to the World Wide Fund for Nature International, "over the past four and a half years, the panel process has become the arena for decision-making on some of the most substantive issues in the international debate on trade and sustainable development."[4] Supporters of the WTO counter that such statements are unsubstantiated. They say that in the relatively few cases where environmental measures have been at issue, the WTO dispute panels have ruled on narrow aspects of these laws or regulations that were implemented in a discriminatory manner with regard to other WTO members and that environmental protection in the United States has not been compromised.

The well-publicized dispute regarding the U.S. import ban on shrimp harvested in a manner harmful to endangered species of sea turtles illustrates how a law or part of a law can be challenged under the WTO and what the implications of such a challenge are. Article 609 of P.L. 101-162 (a law making appropriations to several government agencies) contained the import ban, and also waived the ban if the harvesting methods of foreign countries could be shown to be not harmful to sea turtles. India, Malaysia, Pakistan, and Thailand argued that the U.S. ban violated the WTO rules requiring nondiscrimination among countries and WTO rules prohibiting quantitative restrictions on imports.

After a WTO dispute panel issued its report in April 1998, which was unfavorable to the United States, the United States appealed the ruling to the Appellate Body. The Appellate Body determined that the U.S. procedures for deciding whether countries met the requirements of the law did not provide adequate due process. It also found that the United States had unfairly discriminated between the complaining countries and Western Hemisphere nations.

The Appellate Body ruled that the import ban was administered in a way that violated WTO rules on nondiscrimination. The Appellate Body found that the environmental aspect of the statute was valid under Article XX(g) (which provides an exception from WTO rules for measures to conserve exhaustible resources), but that the United States applied the ban in a discriminatory way among countries.

The United States complied with the Appellate Body's recommendations, and issued revised import guidelines to conform to WTO obligations. According to the Administration, U.S. import restrictions on shrimp harvested in a manner harmful to sea turtles have remained fully in effect and the U.S. commitment to protect endangered sea turtles was not compromised by the Appellate Body's ruling.[5] [But see Hawken, above.]

WTO critics also claim that some multilateral environmental treaties which include trade sanctions might be challenged by a WTO dispute panel. The Convention on International Trade in Endangered Species of Wild Fauna and Flora (CITES), the Montreal Protocol on Substances that Deplete the Ozone Layer, and the Basel Convention on the Control of Transboundary Movements of Hazardous Wastes and Their Disposal all have provisions to restrict trade to achieve their environmental objectives. Given that there is some legal uncertainty, environmentalists argue for a statement(s) that would clarify WTO rules in this area. Supporters of the WTO argue that no multilateral environmental agreement has ever been challenged in the WTO, and it is unlikely that there would be such a challenge.

Another issue of concern to environmentalists is eco-labelling, the process of attaching a label to a product providing consumers with environmental information related to how it was produced. If well designed, eco-labelling can provide information to consumers who can then make an informed choice about which products to purchase and can further environmental goals. Trade analysts are concerned that eco-labelling could be used, in some circumstances, primarily to discriminate between domestic and foreign products, or between products made by different trading partners, and not mainly for environmental purposes. To some extent, transparency in the design, adoption, and application of eco-labels can alleviate some of these concerns.

Lower environmental standards (or lax enforcement) in foreign countries raise some of the same issues as lower labor standards do. WTO critics are concerned that lower environmental standards abroad might contribute to a loss of jobs in the United States and/or lower standards in the United States. Empirical evidence to support this claim is not available. Many environmental problems, however, are international in scope. Issues such as ozone depletion, climate change, and species protection are not limited to one country. In some cases, actions in one country result in higher pollution levels in one or more other countries; acid rain is an example. In other cases, actions of countries spill over into the global commons, such as the atmosphere and the oceans.

Although a strong case may be made for international negotiations on environmental issues, there is no consensus as to where such negotiations should occur. In the case of issues where there is no multilateral environmental treaty, a number of WTO critics want the WTO to address environmental concerns. Others suggest that efforts should be made to negotiate a multilateral environmental agreement, since the WTO lacks the expertise to negotiate such issues. One study concluded that although the WTO is not the appropriate institution to negotiate environmental standards, a separate institutional structure, such as a World Environmental Organization, is worthy of consideration.[6]

Notes

1. See also U.S. Council of Economic Advisors, *America's Interest in the World Trade Organization: An Economic Assessment*, 16 November 1999, 11–14.
2. World Trade Organization, *Annual Report 1999; International Trade Statistics*, 11.
3. See also Susan R. Fletcher, *Environment and the World Trade Organization (WTO) at Seattle: Issues and Concerns*, CRS Report RS20417.
4. "Environmental Group Urges Reform of WTO Dispute Settlement Process," *International Trade Reporter*, 18 August 1999, 1362.
5. Office of the U.S. Trade Representative, *2000 Trade Policy Agenda and 1999 Annual Report of the President of the United States on the Trade Agreements Program*, 45.
6. Keith E. Maskus, *Regulatory Standards in the WTO*, Working Paper 00–1, Institute for International Economics, January 2000, 9.

Vandana Shiva, from "Economic Globalization Has Become a War Against Nature and the Poor" (2000)

Author of Stolen Harvest *and* Water Wars: Privatization, Pollution and Profit, *Shiva is a prominent spokesperson not so much against global trade as against its prevailing rules and governing bodies. If Thomas Friedman's argument (above) suggests that greens should get MBAs in order to understand and influence the global economy, then Shiva and others counter that greens must lobby to change the regime of the global economy because it runs down local communities, particularly in the Third World. Shiva's roots are in the soil and water of the older—and often successful—environmental practices of India.*

Recently, I was visiting Bhatinda in Punjab because of an epidemic of farmers' suicides. Punjab used to be the most prosperous agricultural region in India. Today every farmer is in debt and despair. Vast stretches of land have become waterlogged desert. And, as an old farmer pointed out, even the trees have stopped bearing fruit because heavy use of pesticides has killed the pollinators—the bees and butterflies.

And Punjab is not alone in experiencing this ecological and social disaster. Last year I was in Warangal, Andhra Pradesh, where farmers have also been committing suicide. Farmers who traditionally grew pulses and millets and paddy have been lured by seed companies to buy hybrid cotton seeds referred to as "white gold," which were supposed to make them millionaires. Instead they became paupers.

Their native seeds have been displaced with new hybrids which cannot be saved and need to be purchased every year at a high cost. Hybrids are also very vulnerable to pest attacks. Spending on pesticides in Warangal has increased 2,000 percent from £2.5 million in the 1980s to £50 million in 1997. Now farmers are consuming the same pesticides as a way of killing themselves so that they can escape permanently from unpayable debt. . . .

On March 27, twenty-five-year-old Betavati Ratan took his life because he could not pay back debts for drilling a deep tube well on his two-acre farm. The wells are now dry, as are the wells in Gujarat and Rajasthan where more than fifty million people face a water famine.

The drought is not a "natural disaster." It is "man-made." It is the result of mining of scarce ground water in arid regions to grow thirsty cash crops for export instead of water-prudent food crops for local needs.

It is experiences such as these which tell me that we are so wrong to be smug about the new global economy. It is time to stop and think about the impact of globalization on the lives of ordinary people. This is vital if we want to achieve sustainability.

Seattle and the World Trade Organization protests last year have forced everyone to think again. For me it is now time to re-evaluate radically what we are doing. For what we are doing in the name of globalization to the poor is brutal and unforgivable. This is especially evident in India as we witness the unfolding disasters of globalization, especially in food and agriculture.

Who feeds the world? My answer is very different from that given by most people.

It is women and small farmers working with biodiversity who are the primary food providers in the Third World and, contrary to the dominant assumption, their biodiversity-based small farm systems are more productive than industrial monocultures. . . .

Yield usually refers to production per unit area of a single crop. Output refers to the total production of diverse crops and products. Planting only one crop in the entire field as a monoculture will, of course, increase its individual yield. Planting multiple crops in a mixture will have low yields of individual crops, but will have high total output of food. Yields have been defined in such a way as to make the food production on small farms, by small farmers, disappear.

This hides the production by millions of women farmers in the Third World—farmers like those in my native Himalaya who fought against logging in the Chipko movement, who in their terraced fields grow Jhangora (barnyard millet), Marsha (amaranth), Tur (pigeon pea), Urad (black gram), Gahat (horse gram), soy bean (glycine max), Bhat (glycine soya), Rayans (rice bean), Swanta (cow pea), Koda (finger millet). From the biodiversity perspective, biodiversity-based productivity is higher than monoculture productivity. I call this blindness to the high productivity of diversity a "Monoculture of the Mind," which creates monocultures in our fields.

The Mayan peasants in the Chiapas are characterized as unproductive because they produce only two tons of corn per acre. However, the overall food output is twenty tons per acre when the diversity of their beans and squashes, their vegetables and fruit trees is taken into account.

In Java, small farmers cultivate 607 species in their home gardens. In sub-saharan Africa, women cultivate as many as 120 different plants in the spaces left alongside the cash crops, and this is the main source of household food security.

A single home garden in Thailand has more than 230 species, and African home gardens have more than sixty species of tree. Rural families in the Congo eat leaves from more than fifty different species of tree.

A study in eastern Nigeria found that home gardens occupying only two percent of a household's farmland accounted for half the farm's total output. Similarly, home gardens in Indonesia are estimated to provide more than twenty percent of household income and forty percent of domestic food supplies.

Research done by FAO [Food and Agricultural Organization] has shown that small biodiverse farms can produce thousands of times more food than large, industrial monocultures.

And diversity is the best strategy for preventing drought and desertification.

What the world needs to feed a growing population sustainably is biodiversity intensification, not chemical intensification or genetic engineering. While women and small peasants feed the world through biodiversity, we are repeatedly told that without genetic engineering and globalization of agriculture the world will starve. In spite of all empirical evidence showing that genetic engineering does not produce more food and in fact often leads to a yield decline, it is constantly promoted as the only alternative available for feeding the hungry.

That is why I ask, who feeds the world?

This deliberate blindness to diversity, the blindness to nature's production, production by women, production by Third World farmers, allows destruction and appropriation to be projected as creation.

Take the case of the much-flaunted "golden rice" or genetically engineered vitamin A rice as a cure for blindness. It is assumed that without genetic engineering we cannot remove vitamin A deficiency. However, nature gives us abundant and diverse sources of vitamin A. If rice were not polished, rice itself would provide vitamin A. If herbicides were not sprayed on our wheat fields, we would have bathua, amaranth, mustard leaves as delicious and nutritious greens.

Women in Bengal use more than 150 plants as greens. But the myth of creation presents biotechnologists as the creators of vitamin A, negating nature's diverse gifts and women's knowledge of how to use this diversity to feed their children and families.

The most efficient means of rendering the destruction of nature, local economies, and small autonomous producers is by rendering their production invisible.

Women who produce for their families and communities are treated as "nonproductive" and "economically inactive." The devaluation of women's work, and of work done in sustainable economies, is the natural outcome of a system constructed by capitalist patriarchy. This is how globalization destroys local economies and the destruction itself is counted as growth.

And women themselves are devalued, because for many women in the rural and indigenous communities their work co-operates with nature's processes, and is often contradictory to the dominant market-driven "development" and trade policies, and because work that satisfies needs and ensures sustenance is devalued in general. There is less nurturing of life and life support systems. The devaluation and invisibility of sustainable, regenerative production is most glaring in the area of food. While patriarchal division of labor has assigned women the role of feeding their families and communities, patriarchal economics and patriarchal views of science and technology magically make women's work in providing food disappear. "Feeding the World" becomes disassociated from the women who actually do it and is projected as dependent on global agribusiness and biotechnology corporations. . . .

According to the McKinsey corporation, "American food giants recognize that Indian agro-business has lots of room to grow, especially in food processing. India processes a minuscule one percent of the food it grows compared with seventy percent for the U.S., Brazil and Philippines." It is not that we Indians eat our food raw. Global consultants fail to see the ninety-nine percent [of] food processing done by women at household level, or by small cottage industry, because it is not controlled by global agribusiness. Ninety-nine percent of India's agroprocessing has been intentionally kept at the household level. Now, under the pressure of globalization, things are changing. Pseudo hygiene laws that shut down the food economy based on small-scale local processing under community control are part of the arsenal of global agribusiness for establishing market monopolies through force and coercion, not competition.

In August 1998, small-scale local processing of edible oil was banned in India through a "packaging order" which made sale of open oil illegal and required all oil to be packed in plastic or aluminum. This shut down tiny "ghanis" or cold-pressed mills. It destroyed the market for our diverse oilseeds—mustard, linseed, sesame, groundnut, and coconut.

The take-over of the edible oil industry has affected ten million livelihoods. The take-over of "atta" or flour by packaged branded flour will cost 100 million livelihoods. These millions are being pushed into new poverty. The forced use of packaging will increase the environmental burden of millions of tons of plastic and aluminum. The globalization of the food system is destroying the diversity of local food cultures and local food economies. A global monoculture is being forced on people by defining everything that is fresh, local, and handmade as a health hazard. Human hands are being defined as the worst contaminants, and work for human hands is being outlawed, to be replaced by machines and chemicals bought from global corporations. These are not recipes for feeding the world, but for stealing livelihoods from the poor to create markets for the powerful. People are being perceived as parasites, to be exterminated for the "health" of the global economy. In the process new health and ecological hazards are being forced on Third World people through dumping genetically engineered foods and other hazardous products. . . .

Patents and intellectual property rights are supposed to be granted for novel inventions. But patents are being claimed for rice varieties such as the basmati for which the Doon Valley—where I was born—is famous, or pesticides derived from the neem which our mothers and grandmothers have been using. Rice Tec, a U.S.-based company, has been granted Patent No. 5,663,484 for basmati rice lines and grains. Basmati, neem, pepper, bitter gourd, turmeric . . . every aspect of the innovation embodied in our indigenous food and medicinal systems is now being pirated and patented. The knowledge of the poor is being converted into the property of global corporations, creating a situation where the poor will have to pay for the seeds and medicines they have evolved and have used to meet their needs for nutrition and health care.

Such false claims to creation are now the global norm, with the Trade Related Intellectual Property Rights Agreement of the WTO forcing countries to introduce regimes that allow patenting of life forms and indigenous knowledge.

Instead of recognizing that commercial interests build on nature and on the contribution of other cultures, global law has enshrined the patriarchal myth of creation to create new property rights to life forms just as colonialism used the myth of discovery as the basis of the take-over of the land of others as colonies. . . .

As humans travel further down the road to non-sustainability, they become intolerant of other species and blind to their vital role in our survival. In 1992, when Indian farmers destroyed Cargill's seed plant in Bellary, Karnataka, as a protest against seed failure, the Cargill Chief Executive stated: "We bring Indian farmers smart technologies which prevent bees from usurping the pollen." When I was participating in the United Nations Biosafety Negotiations, Monsanto circulated literature to defend its Roundup herbicide–resistant crops on grounds that they prevent "weeds from stealing the sunshine." But what Monsanto calls weeds are the green fields that provide vitamin A rice and prevent blindness in children and anemia in women.

A world-view that defines pollination as "theft by bees" and claims that biodiversity "steals" sunshine is a world-view which itself aims at stealing nature's harvest by replacing open, pollinated varieties with hybrids and sterile seeds, and at destroying biodiverse flora with herbicides such as Monsanto's Roundup. The threat posed to the Monarch butterfly by genetically engineered bt. crops [see Chapter 4, above] is just one example of the ecological poverty created by the new biotechnologies. As butterflies and bees disappear, production is undermined. As biodiversity disappears, with it go sources of nutrition and food.

When giant corporations view small peasants and bees as thieves, and through trade rules and new technologies seek the right to exterminate them, humanity has reached a dangerous threshold. The imperative to stamp out the smallest insect, the smallest plant, the smallest peasant comes from a deep fear—the fear of everything that is alive and free. And this deep insecurity and fear is unleashing violence against all people and all species. The global free-trade economy has become a threat to sustainability. The very survival of the poor and other species is at stake not just as a side effect or as an exception but in a systemic way through a restructuring of our world-view at the most fundamental level. Sustainability, sharing, and survival are being economically outlawed in the name of market competitiveness and market efficiency.

We need urgently to bring the planet and people back into the picture. The world can be fed only by feeding all beings that make the world. In giving food to other beings and species we maintain conditions for our own food security. In feeding the earthworms we feed ourselves. In feeding cows, we feed the soil, and in providing food for the soil, we provide food for humans. This world-view of abundance is based on sharing and on a deep awareness of humans as members of the earth family. This awareness that in impoverishing other beings, we impoverish ourselves and in nourishing other beings, we nourish ourselves is the basis of sustainability.

The sustainability challenge for the new millennium is whether global economic man can move out of the world-view based on fear and scarcity, monocultures and monopolies, appropriation and dispossession and shift to a view based on abundance and sharing, diversity and decentralization, and respect and dignity for all beings.

9 What Is Wilderness and Do We Need It?

At first glance, wilderness does not seem a problematic concept: "Wilderness" is derived from "wild"—the opposite of "tame." Thus, "wilderness" can be defined as any place that humans have not tamed. But there is a difference between being tame and having been tamed; some areas appear tame in their natural state. As William Cronon points out in "The Trouble with Wilderness," people tend to view a natural grassland as something other than wilderness. A rugged mountainside is wilderness, a dark and tangled forest is wilderness, a barren desert may be wilderness, but an untamed grassland is not. Why not? Because it does not engender the same sense of fear? Of reverence? Of strangeness? Because it offers, in its native grains, the possibility of nourishment that the other areas do not?

This example highlights the fact that understanding the concept of wilderness is not as easy as it initially seems. Robert Marshall begins "The Problem of the Wilderness" with three definitions:

- wilderness is a tract of solitude and savageness;
- it is a tract of land, whether forest or wild barren plain, uncultivated and uninhabited by human beings; or,
- in Marshall's own definition, it is a region that contains no permanent human inhabitants, possesses no possibility of conveyance by any mechanical means, and is sufficiently spacious that a person in crossing it must have the experience of sleeping out.

The United States Congress offered the following definition in Section 2(C) of the Wilderness Act (1964): "an area where the earth and its community of life are untrammeled by man, where man himself is a visitor and does not remain." By contrast, Henry David Thoreau, in "Walking," speaking for wildness, regards man as "an inhabitant, or a part and parcel of Nature."

As other selections make clear, the concept of wilderness has changed dramatically over time. William Bradford, in his journal of the Pilgrims' experience during their first winter in America, asks, "what could they see but a hideous and desolate wilderness, full of wild beasts and wild men?" He expresses what up until that point had been a universal reaction to mountains and forests among Europeans: fear and loathing.

Thoreau's excursion "Walking," written 240 years after Bradford's journal, is one of the first to break with this long tradition of viewing wilderness as something to be feared, hated, and conquered. Thoreau gives us the phrase that has been taken up as canonical by the modern wilderness movement: "In Wildness is the preservation of the world." (Note the discussion of the distinction between *wildness* and *wilderness* in the headnote to this selection, below.) However, Thoreau sets himself off from the majority of people. His justification for preservation of wildness is an elitist one, that "all men are not equally fit for civilization; and because the majority, like dogs and sheep, are tame by inherited disposition, this is no reason why the others should have their natures broken."

What, then, is the value of wilderness? To Thoreau and Marshall, it seems, wilderness is a place in which real men (white men?), unbroken by civilization, can find fulfillment. Marshall invokes William James and finds in wilderness James's "moral equivalent of war," a peaceful stimulation of the same impulses of bravery and competence that men find in war. Is wilderness the province of men, and white American men, in particular?

A more detached observer, Roderick Nash carefully lays out eight strong reasons why wilderness is valuable. Two are distinctly American: "Wilderness as a formative influence on American National Character" and "Wilderness as a Nourisher of American Arts and Letters." Nash is the first to bring a modern scientific perspective into the wilderness debate, even restating Marshall's virile justifications in these terms: "The gut-level fears associated with survival drove the wheels of evolution." In addition, Nash provides the scientific rationales of ecological processes, biological diversity, human diversity, and education.

The contemporary discussion of wilderness, which has picked up on modern academic themes of deconstruction and diversity, is exemplified in the debate between William Cronon and Donald Waller. Waller invokes a scientific concept of wilderness to counter Cronon's claim that wilderness is nothing more than a social construct. At issue is whether wilderness has become overvalued to the point that the current emphasis on its preservation is actually a hindrance to sound conservation strategy.

Gary Snyder, in his deceptively simple poem "Trail Crew Camp at Bear Valley," brings us back from a rarified academic discussion to the realization that the wilderness experience is an unmediated, physical response to a natural world that is always of a unique place, time, and condition. Snyder's poem prompts the intuition that perhaps the wilderness debate is itself inimical to wilderness, that if we think about wilderness too much, it goes away.

FURTHER READING

Krakauer, Jon. *Into the Wild.* New York: Villard Books, 1996.

Figure 9.1 *Pocahontas Saving the Life of Captain John Smith* and *Landing of Pilgrims at Plymouth Plantation* by Constantino Brumidi. The Rotunda of the United States Capitol Building. By permission of the Architect of the Capitol.

Nash, Roderick. *Wilderness and the American Mind.* 1967. 4th ed. New Haven: Yale University Press, 2001.

Oelschlager, Max. *The Idea of Wilderness.* New Haven: Yale University Press, 1991.

Sax, Joseph. *Mountains Without Handrails.* Ann Arbor: University of Michigan Press, 1980.

Soulé, Michael, and Gary Lease. *Reinventing Nature? Responses to Postmodern Deconstruction.* Washington, D.C.: Island Press, 1995.

Spence, Mark David. *Dispossessing the Wilderness: Indian Removal and the Making of the National Parks.* Oxford: Oxford University Press, 1999.

Stevens, Wallace. "The Anecdote of the Jar." *Collected Poems of Wallace Stevens.* New York: Alfred A. Knopf, 1972. 76.

William Bradford, "A Hideous and Desolate Wilderness" from *Journal* (1620–35)

William Bradford was a leader of the Pilgrims who sailed on the Mayflower *in 1620. In 1621, his wife died by drowning in Cape Cod Bay. Bradford was elected governor of the Plimouth Bay Colony thirty times over a thirty-four-year period. In this journal entry, he takes on a biblical, third-person plural voice in describing the experiences of the Pilgrims during their first winter.*

It is recorded in Scripture as a mercy to the Apostle and his shipwrecked company, that "the barbarians showed them no small kindness" in refreshing them. But these savage barbarians, when they met with them (as after will appear), were readier to fill their sides full of arrows, than otherwise. And for the season, it was winter; and they that know the winters of that country, know them to be sharp and violent, and subject to violent storms, dangerous to travel to known places, much more to search out unknown coasts. Besides, what could they see but a hideous and desolate wilderness, full of wild beasts and wild men? And what multitudes there might be of them they knew not. Neither could they, as it were, go up to the top of Pisgah, to view from this wilderness a more goodly country to feed their hopes. For which way soever they turned their eyes (save upward to the heavens) they could have little solace or content in respect of any outward objects. For summer being done, all things stand for them to look upon with a weather-beaten face; and the whole country being full of woods and thickets, represented a wild and savage hue. If they looked behind them, there was the mighty ocean which they had passed, and was now as a main bar and gulf to separate them from all the civil parts of the world. If it be said they had a ship to succor them, it is true; but what heard they daily from the master and company but that with speed they should look out a place with their shallop, where they would be at some near distance; for the season was such as he would not stir from thence until a safe harbor was discovered by them, where they would be and he might go without danger; and that victuals consumed apace, but he must and would keep sufficient for himself and company for their return. Yea, it was muttered by some, that if they got not a place in time, they would turn them and their goods on shore, and leave them. Let it be also considered what weak hopes of supply and succor they left behind them, that might bear up their minds in this sad condition and trials they were under, and they could not but be very small. It is true, indeed, the affections and love of their brethren at Leyden [in the Netherlands] were cordial and entire; but they had little power to help them, or themselves; and how the case stood between them and the merchants at their coming away hath already been declared. What could now sustain them but the spirit of God and his grace?

May not and ought not the children of these fathers rightly say, "Our fathers were Englishmen, which came over this great ocean, and were ready to perish in this wilderness. But they cried unto the Lord, and he heard their voice, and looked on their adversity." And let them therefore praise the Lord because he is good, and his mercies endure forever. Yea, let them which have been thus redeemed of the Lord show how he hath delivered them from the hand of the oppressor. When they wandered in the desert wilderness, out of the way, and found no city to dwell in, both hungry and thirsty, their soul was overwhelmed in them. Let them confess before the Lord his loving kindness and his wonderful works before the children of men.

Henry David Thoreau, from "Walking" (1862)

The phrase so commonly repeated from this essay is "In Wildness *is the preservation of the world," not "In* Wilderness *is the preservation of the world." Does Thoreau's vision of wildness correspond to the modern idea of wilderness? Is wildness a condition of humans, of nature, or of both? Compare this selection with Thoreau's "Ktaadn" in Nature Writing (see Chapter 20, below). Does Thoreau, who, in most of his writings, describes nature only a mile away from Concord Village, have a problem with* real *wilderness?*

I wish to speak a word for Nature, for absolute freedom and wildness, as contrasted with a freedom and culture merely civil—to regard man as an inhabitant, or a part and parcel of Nature, rather than a member of society. I wish to make an extreme statement, if so I may make an emphatic one, for there are enough champions of civilization: the minister, and the school-committee, and every one of you will take care of that.

... What I have been preparing to say is, that in Wildness is the preservation of the world. Every tree sends its fibers forth in search of the Wild. The cities import it at any price. Men plow and sail for it. From the forest and wilderness come the tonics and barks which brace mankind. Our ancestors were savages. The story of Romulus and Remus being suckled by a wolf is not a meaningless fable. The founders of every State which has risen to eminence have drawn their nourishment and vigor from a similar wild source. It was because the children of the Empire were not suckled by the wolf that they were conquered and displaced by the children of the Northern forests who were.

I believe in the forest, and in the meadow, and in the night in which the corn grows. We require an infusion of hemlock-spruce or arbor-vitae in our tea. There is a difference between eating and drinking for strength and from mere gluttony. The Hottentots eagerly devour the marrow of the kudu and other antelopes raw, as a matter of course. Some of our Northern Indians eat raw the marrow of the Arctic reindeer, as well as various other parts, including the summits of the antlers, as long as they are soft. And herein, perchance, they have stolen a march on the cooks of Paris. They get what usually goes to feed the fire. This is probably better than stall-fed beef and slaughterhouse pork to make a man of. Give me a wildness whose glance no civilization can endure—as if we lived on the marrow of kudus devoured raw. ...

The African hunter Cummings tells us that the skin of the eland, as well as that of most other antelopes just killed, emits the most delicious perfume of trees and grass. I would have every man so much like a wild antelope, so much a part and parcel of Nature, that his very person should thus sweetly advertise our senses of his presence, and remind us of those parts of Nature which he most

Figure 9.2 Henry David Thoreau. Courtesy of Concord Free Public Library.

haunts. I feel no disposition to be satirical when the trapper's coat emits the odor of musquash [muskrat] even; it is a sweeter scent to me than that which commonly exhales from the merchant's or the scholar's garments. When I go into their wardrobes and handle their vestments, I am reminded of no grassy plains and flowery meads which they have frequented, but of dusty merchants' exchanges and libraries rather. . . .

Ben Jonson exclaims,

"How near to good is what is fair!"

So I would say,

"How near to good is what is *wild!*"

Life consists with wildness. The most alive is the wildest. Not yet subdued to man, its presence refreshes him. One who pressed forward incessantly and never rested from his labors, who grew fast and made infinite demands on life, would always find himself in a new country or wilderness, and surrounded by the raw material of life. He would be climbing over the prostrate stems of primitive forest-trees.

Hope and the future for me are not in lawns and cultivated fields, not in towns and cities, but in the impervious and quaking swamps. When, formerly, I have analyzed my partiality for some farm which I had contemplated purchasing, I have frequently found that I was attracted solely by a few square rods of impermeable and unfathomable bog—a natural sink in one corner of it. That was the jewel which dazzled me. I derive more of my subsistence from the swamps which surround my native town than from the cultivated gardens in the village. There are no richer parterres to my eyes than the dense beds of dwarf andromeda (*Cassandra calyculata*) which cover these tender places on the earth's surface. Botany cannot go farther than tell me the names of the shrubs which grow there—the high-blueberry, panicled andromeda, lamb-kill, azalea, and rhodora—all standing in the quaking sphagnum. I often think that I should like to have my house front on this mass of dull red bushes, omitting other flower plots and borders, transplanted spruce and trim box, even graveled walks, to have this fertile spot under my windows, not a few imported barrow-fulls of soil only to cover the sand which was thrown out in digging the cellar. Why not put my house, my parlor, behind this plot, instead of behind that meager assemblage of curiosities, that poor apology for a Nature and Art, which I call my front-yard? It is an effort to clear up and make a decent appearance when the carpenter and mason have departed, though done as much for the passerby as the dweller within. The most tasteful front-yard fence was never an agreeable object of study to me; the most elaborate ornaments, acorn-tops, or what not, soon wearied and disgusted me. Bring your sills up to the very edge of the swamp, then (though it may not be the best place for a dry cellar), so that there be no access on that side to citizens. Front-yards are not made to walk in, but, at most, through, and you could go in the back way.

Yes, though you may think me perverse, if it were proposed to me to dwell in the neighborhood of the most beautiful garden that ever human art contrived, or else of a dismal swamp, I should certainly decide for the swamp. How vain, then, have been all your labors, citizens, for me!

My spirits infallibly rise in proportion to the outward dreariness. Give me the ocean, the desert, or the wilderness! In the desert, pure air and solitude compensate for want of moisture and fertility. The traveler Burton says of it, "Your *morale* improves; you become frank and cordial, hospitable and single-minded. . . . In the desert, spirituous liquors excite only disgust. There is a keen enjoyment in a mere animal existence." They who have been traveling long on the steppes of Tartary say, "On re-entering cultivated lands, the agitation, perplexity, and turmoil of civilization oppressed and suffocated us; the air seemed to fail us, and we felt every moment as if about to die of asphyxia." When I would recreate myself, I seek the darkest wood, the thickest and most interminable, and, to the citizen, most dismal swamp. I enter a swamp as a sacred place, a *sanctum sanctorum*. There is the strength, the marrow of Nature. The wildwood covers the virgin mold, and the same soil is good for men and for trees. A man's health requires as many acres of meadow to his prospect as his farm does loads of muck. There are the strong meats on which he feeds. A town is saved not more by the righteous men in it than by the woods and swamps that surround it. A township where one primitive forest waves above, while another primitive forest rots below, such a town is fitted to raise not only corn and potatoes, but poets and philosophers for the coming ages. In such a soil grew Homer and Confucius and the rest, and out of such a wilderness comes the Reformer eating locusts and wild honey.

To preserve wild animals implies generally the creation of a forest for them to dwell in or resort to. So is it with man. A hundred years ago they sold bark in our streets peeled from our own woods. In the very aspect of those primitive and rugged trees, there was, methinks, a tanning principle which hardened and consolidated the fibers of men's thoughts. Ah! already I shudder for these comparatively degenerate days of my native village, when you cannot collect a load of bark of good thickness—and we no longer produce tar and turpentine.

The civilized nations—Greece, Rome, England—have been sustained by the primitive forests which anciently rotted where they stand. They survive as long as the soil is not exhausted. Alas for human culture! little is to be expected of a nation when the vegetable mold is exhausted and it is compelled to make manure of the bones of its fathers. There the poet sustains himself merely by his own superfluous fat, and the philosopher comes down on his marrow-bones. . . .

In Literature it is only the wild that attracts us. Dullness is but another name for tameness. It is the uncivilized free and wild thinking in "Hamlet" and the "Iliad," in all the Scriptures and Mythologies, not learned in the schools, that delights us. As the wild duck is more swift and beautiful than the tame, so is the wild—the mallard—thought, which 'mid falling dews wings its way above the fens. A truly good book is something as natural, and as unexpectedly and unaccountably fair and perfect, as a wild flower discovered on the prairies of the West or in the jungles of the East. Genius is a light which makes the darkness visible, like the lightning's flash, which perchance shatters the temple of knowledge itself—and not a taper lighted at the hearth-stone of the race, which pales before the light of common day. . . .

In short, all good things are wild and free. There is something in a strain of music, whether produced by an instrument or by the human voice—take the sound of a bugle in a summer night, for instance—which by its wildness, to speak without satire, reminds me of the cries emitted by wild beasts in their native forests. It is so much of their wildness as I can understand. Give me for my friends and neighbors wild men, not tame ones. The wildness of the savage is but a faint symbol of the awful ferity with which good men and lovers meet.

I love even to see the domestic animals reassert their native rights—any evidence that they have not wholly lost their original wild habits and vigor; as when my neighbor's cow breaks out of her pasture early in the spring and boldly swims the river, a cold, gray tide, twenty-five or thirty rods wide, swollen by the melted snow. It is the buffalo crossing the Mississippi. This exploit confers some dignity on the herd in my eyes, already dignified. The seeds of instinct are preserved under the thick hides of cattle and horses, like seeds in the bowels of the earth, an indefinite period. . . .

I rejoice that horses and steers have to be broken before they can be made the slaves of men, and that men themselves have some wild oats still left to sow before they become submissive members of society. Undoubtedly, all men are not equally fit subjects for civilization; and because the majority, like dogs and sheep, are tame by inherited disposition, this is no reason why the others should have their natures broken that they may be reduced to the same level. Men are in the main alike, but they were made several in order that they might be various. If a low use is to be served, one man will do nearly or quite as well as another; if a high one, individual excellence is to be regarded. Any man can stop a hole to keep the wind away, but no other man could serve so rare a use as the author of this illustration [Shakespeare] did. Confucius says, "The skins of the tiger and the leopard, when they are tanned, are as the skins of the dog and the sheep tanned." But it is not the part of a true culture to tame tigers, any more than it is to make sheep ferocious; and tanning their skins for shoes is not the best use to which they can be put.

Robert Marshall, "The Problem of the Wilderness" (1930)

Robert Marshall is one of the great unknown heroes of the American Wilderness movement. Along with Aldo Leopold and several other conservation leaders, Marshall founded the Wilderness Society in 1935. This essay is the first piece to raise the issues that make up the ongoing debate about wilderness found in the Cronon and Waller selections, below.

For the ensuing discussion I shall use the word *wilderness* to denote a region which contains no permanent inhabitants, possesses no possibility of conveyance by any mechanical means, and is sufficiently spacious that a person in crossing it must have the experience of sleeping out. The dominant attributes of such an area are: first, that it requires any one who exists in it to depend exclusively on his own effort for survival; and second, that it preserves as nearly as possible the primitive environment. This means that all roads, power transportation, and settlements are barred. But trails and temporary shelters, which were common long before the advent of the white race, are entirely permissible.

When Columbus effected his immortal debarkation, he touched upon a wilderness which embraced virtually a hemisphere. The philosophy that progress is proportional to the amount of alteration imposed upon nature never seemed to have occurred to the Indians. Even such tribes as the Incas, Aztecs, and Pueblos made few changes in the environment in which they were born. . . . Consequently, over billions of acres the aboriginal wanderers still spun out their peripatetic careers, the wild animals still browsed in unmolested meadows, and the forests still grew and moldered and grew again precisely as they had done for undeterminable centuries.

It was not until the settlement of Jamestown in 1607 that there appeared the germ for that unabated disruption of natural conditions which has characterized all subsequent American history. At first expansion was very slow. The most intrepid seldom advanced further from their neighbors than the next drainage. At the time of the Revolution the zone of civilization was still practically confined to a narrow belt lying between the Atlantic Ocean and the Appalachian valleys. But a quarter of a century later, when the Louisiana Purchase was consummated, the outposts of civilization had reached the Mississippi, and there were foci of colonization in half a dozen localities west of the Appalachians, though the unbroken line of the frontier was east of the mountains.

It was yet possible as recently as 1804 and 1805 for the Lewis and Clark Expedition to cross two thirds of a continent without seeing any culture more advanced than that of the Middle Stone Age. The only routes of travel were the uncharted rivers and the almost impassable Indian trails. And continually the expedition was breaking upon some "truly magnificent and sublimely grand object, which has from the commencement of time been concealed from the view of civilized man."[1]

This exploration inaugurated a century of constantly accelerating emigration such as the world had never known. Throughout this frenzied period the only serious thought ever devoted to the wilderness was how it might be demolished. To the pioneers pushing westward it was an enemy of diabolical cruelty and danger, standing as the great obstacle to industry and development. Since these seemed to constitute the essentials for felicity, the obvious step was to excoriate the devil which interfered. And so the path of empire proceeded to substitute for the undisturbed seclusion of nature the conquering accomplishments of man. Highways wound up valleys which had known only the footsteps of the wild animals; neatly planted gardens and orchards replaced the tangled confusion of the primeval forest; factories belched up great clouds of smoke where for centuries trees had transpired toward the sky, and the groundcover of fresh sorrel and twinflower was transformed to asphalt spotted with chewing-gum, coal dust, and gasoline. . . .

The benefits that accrue from the wilderness may be separated into three broad divisions: the physical, the mental, and the aesthetic.

Most obvious in the first category is the contribution which the wilderness makes to health. This involves something more than pure air and quiet, which are also attainable in almost any rural situation. But toting a fifty-pound pack over an abominable trail, snowshoeing across a blizzard-swept plateau, or scaling some jagged pinnacle which juts far above timber all develop a body distinguished by a soundness, stamina, and élan unknown amid normal surroundings.

More than mere heartiness is the character of physical independence which can be nurtured only away from the coddling of civilization. In a true wilderness if a person is not qualified to satisfy all the requirements of existence, then he is bound to perish. As long as we prize individuality and competence it is imperative to provide the opportunity for complete self-sufficiency. This is inconceivable under the effete superstructure of urbanity; it demands the harsh environment of untrammeled expanses.

Closely allied is the longing for physical exploration which bursts through all the chains with which society fetters it. . . . Adventure, whether physical or mental, implies breaking into unpenetrated ground, venturing beyond the boundary of normal aptitude, extending oneself to the limit of capacity, courageously facing peril. Life without the chance for such exertions would be for many persons a dreary game, scarcely bearable in its horrible banality. . . .

One of the greatest advantages of the wilderness is its incentive to independent cogitation. This is partly a reflection of physical stimulation, but more inherently due to the fact that original ideas require an objectivity and perspective seldom possible in the distracting propinquity of one's fellow men. It is necessary to "have gone behind the world of humanity, seen its institutions like toadstools by the wayside."[2] This theorizing is justified empirically by the number of America's most virile minds, including Thomas Jefferson, Henry Thoreau, Louis Agassiz, Herman Melville, Mark Twain, John Muir, and William James, who have felt the compulsion of periodical retirements into the solitudes. Withdrawn from the contaminating notions of their neighbors, these thinkers have been able to meditate, unprejudiced by the immuring civilization.

Another mental value of an opposite sort is concerned not with incitement but with repose. In a civilization which requires most lives to be passed amid inordinate dissonance, pressure, and intrusion, the chance of retiring now and then to the quietude and privacy of sylvan haunts becomes for some people a psychic necessity. It is only the possibility of convalescing in the wilderness which saves them from being destroyed by the terrible neural tension of modern existence.

There is also a psychological bearing of the wilderness which affects, in contrast to the minority who find it indispensable for relaxation, the whole of human kind. One of the most profound discoveries of psychology has been the demonstration of the terrific harm caused by suppressed desires. To most of mankind a very powerful desire is the appetite for adventure. But in an age of machinery only the extremely fortunate have any occasion to satiate this hankering, except vicariously. As a result people become so choked by the monotony of their lives that they are readily amenable to the suggestion of any lurid diversion. Especially in battle, they imagine, will be found the glorious romance of futile dreams. And so they endorse war with enthusiasm and march away to stirring music, only to find their adventure a chimera, and the whole world miserable. It is all tragically ridiculous, and yet there is a passion there which can not be dismissed with a contemptuous reference to childish quixotism. William James has said that "militarism is the great preserver of ideals of hardihood, and human life with no use for hardihood would be contemptible."[3] The problem, as he points out, is to find a "moral equivalent of war," a peaceful stimulation for the hardihood and competence instigated in bloodshed. This equivalent may be realized if we make available to every one the harmless excitement of the wilderness. Bertrand Russell has skillfully amplified this idea in his essay on "Machines and the Emotions." He expresses the significant conclusion that "many men would cease to desire war if they had opportunities to risk their lives in Alpine climbing."[4] . . .

Of the myriad manifestations of beauty, only natural phenomena like the wilderness are detached from all temporal relationship. All the beauties in the creation or alteration of which man has played even the slightest role are firmly anchored in the historic stream. They are temples of Egypt, oratory of Rome, painting of the Renaissance, or music of the Classicists. But in the wild places nothing is moored more closely than to geologic ages. The silent wanderer crawling up the rocky shore of the turbulent river could be a savage from some prehistoric epoch or a fugitive from twentieth-century mechanization.

The sheer stupendousness of the wilderness gives it a quality of intangibility which is unknown in ordinary manifestations of ocular beauty. These are always very definite two or three dimensional objects which can be physically grasped and circumscribed in a few moments. But "the beauty that shimmers in the yellow afternoons of October, who ever could clutch it."[5] Any one who has looked across a ghostly valley at midnight, when moonlight makes a formless silver unity out of the drifting fog, knows how impossible it often is in nature to distinguish mass from hallucination. Any one who has stood upon a lofty summit and gazed over an inchoate tangle of deep canyons and cragged mountains, of sunlit lakelets and black expanses of forest, has become aware of a certain giddy sensation that there are no distances, no measures, simply unrelated matter rising and falling without any analogy to the banal geometry of breadth, thickness, and height. A fourth dimension of immensity is added which makes the location of some dim elevation outlined against the sunset as incommensurable to the figures of the topographer as life itself is to the quantitative table of elements which the analytic chemist proclaims to constitute vitality.

Because of its size the wilderness also has a physical ambiency about it which most forms of beauty lack. One looks from outside at works of art and architecture, listens from outside to music or poetry. But when one looks at and listens to the wilderness he is encompassed by his experience of beauty, lives in the midst of his aesthetic universe.

A fourth peculiarity about the wilderness is that it exhibits a dynamic beauty. A Beethoven symphony or a Shakespearean drama, a landscape by Corot, or a Gothic cathedral, once they are finished become virtually static. But the wilderness is in constant flux. A seed germinates, and a stunted seedling battles for decades against the dense shade of the virgin forest. Then some ancient tree blows down and the long-suppressed plant suddenly enters into the full vigor of delayed youth, grows rapidly from sapling to maturity, declines into the conky senility of many centuries, dropping millions of seeds to start a new forest upon the rotting debris of its own ancestors, and eventually topples over to admit the sunlight which ripens another woodland generation.

Another singular aspect of the wilderness is that it gratifies every one of the senses. There is unanimity in venerating the sights and sounds of the forest. But what are generally esteemed to be the minor senses should not be slighted. No one who has ever strolled in springtime through seas of blooming violets, or lain at night on boughs of fresh balsam, or walked across dank holms in early morning can omit odor from the joys of the primordial environment. No one who has felt the stiff wind of mountaintops or the softness of untrodden sphagnum will forget the exhilaration experienced through touch. "Nothing ever tastes as good as when it's cooked in the woods" is a trite tribute to another sense. Even equilibrium causes a blithe exultation during many a river crossing on tenuous foot log and many a perilous conquest of precipice.

Finally, it is well to reflect that the wilderness furnishes perhaps the best opportunity for pure aesthetic enjoyment. This requires that beauty be observed as a unity, and that for the brief duration of any pure aesthetic experience the cognition of the observed object must completely fill the spectator's cosmos. There can be no extraneous thoughts—no question about the creator of the phenomenon, its structure, what it resembles, or what vanity in the beholder it gratifies. "The purely aesthetic observer has for the moment forgotten his own soul";[6] he has only one sensation left and that is exquisiteness. In the wilderness, with its entire freedom from the manifestations of human will, that perfect objectivity which is essential for pure aesthetic rapture can probably be achieved more readily than among any other forms of beauty.

Notes

1. Reuben G. Thwaites, *Original Journals of the Lewis and Clark Expedition, 1804–1806*, 13 June 1805.
2. Henry David Thoreau, *Journals*, 2 April 1852.
3. William James, "The Moral Equivalent of War."
4. Bertrand Russell, *Essays in Scepticism.*
5. Ralph Waldo Emerson, *Nature.*
6. Irwin Edman, "The World, the Arts and the Artist."

Roderick Nash, from "The Value of Wilderness" (1977)

Roderick Nash can properly be called the first historian of Wilderness. His 1967 book, Wilderness and the American Mind, *is considered a masterpiece. The following selection is taken from a speech Nash gave to Italian conservationists who were advocating national parks in their homeland. An interesting point, not included here, that Nash makes*

Figure 9.3 Winter covers Glacier National Park in Montana. By permission of the California Academy of Sciences / Reuel R. Sutton.

at the beginning of this speech is that very few languages have a word that means what "wilderness" means in English. In fact, Italian translators use the telling phrase scene di disordine o confusione *(a place of disorder or confusion).*

. . . Several arguments have emerged to become the staples in the contemporary defense of nature protection in the United States. While they are presented here in terms of wilderness, such as exists in the larger national parks like Yellowstone and Yosemite, they may be applied in slightly altered form to any open space or nature reserve. They might also be applied, with appropriate alteration, to Italy or any other nation. The summary that follows is in outline form.

ARGUMENT 1: WILDERNESS AS A RESERVOIR OF NORMAL ECOLOGICAL PROCESSES

Aldo Leopold, wildlife manager and philosopher whose efforts led in 1924 to creation of the first reserved wilderness on National Forest land in the United States, once said that wilderness reveals "what the land was, what it is, and what it ought to be." He added that nature reserves conceivably had more importance for science than they did for recreation. What Leopold meant was that wilderness is a model of healthy, ecologically balanced land. At a time when so much of the environment is disturbed by technological man, wilderness has vital importance as a criterion against which to measure the impact of civilization. Without it we have no way of knowing how the land mechanism functions under normal conditions. The science of ecology needs nature reserves as medical science needs healthy people.

ARGUMENT 2: WILDERNESS AS A SUSTAINER OF BIOLOGICAL DIVERSITY

It is axiomatic in the biological sciences that there is strength in diversity. The whole evolutionary miracle is based on the presence over time of an almost infinite diversity of life forms. Maintenance of the full evolutionary capacity that produced life as we know it and, we may suppose, will continue to shape life on earth, means that the size of the gene pool should be maximized. But with his agriculture and urban growth, modern man has made extensive inroads on biological diversity. Some of the changes, to be sure, have been desirable. But many are carried too far. More species have been exterminated in the last three hundred years than in the previous three million. Many other species, including some of the most awesome life forms on earth, are threatened. The whales fall into this category. The problem is that man in his shortsighted pursuit of what he believes to be his self-interest has branded some forms of life as "useless" and therefore expendable. The creative processes that produced these life forms in the first place did not regard them as such. Modern man frequently appears to be a clumsy mechanic, pounding on a delicate and complex machine with a sledgehammer.

Wildernesses and nature reserves constitute refuges where biological diversity is maintained. In such areas life forms are preserved, banked, so to speak, against the time when they may be needed, perhaps desperately. As David Brower, the American president of Friends of the Earth and a leading contemporary defender of wilderness put it, wild places hold the answers to questions man does not yet know how to ask. Putting aside for the moment the "right" of all life to exist as it was created, there is the very practical matter of the importance of biological diversity to medical science, to agriculture, and to the perpetuation of the life-sustaining forces we are only just beginning to understand. Man pounds clumsily against a delicate machine which is nothing less than the spaceship earth—the only home he has. Nature reserves represent a step away from this potentially suicidal shortsightedness.

ARGUMENT 3: WILDERNESS AS FORMATIVE INFLUENCE ON AMERICAN NATIONAL CHARACTER

It was not until the census report of 1890 pronounced the frontier era ended that many Americans began to ponder the significance of wilderness in shaping them as individuals and as a society. The link between American character or identity, and wilderness, was forged, as historian Frederick Jackson Turner argued so persuasively in 1893, during three centuries of pioneering. Independence and individualism were two heritages; a democratic social and political theory and the concept of equal opportunity were other frontier traits. So was the penchant for practical achievement that marks the American character so distinctly.

If wilderness shaped our national values and institutions, it follows that one of the most important roles of nature reserves is keeping those values and institutions alive. Theodore Roosevelt, President of the United States from 1901 to 1909 and the leader of the first period of great achievement in conservation, was keenly aware of this relationship. "Under the hard conditions of life in the wilderness," Roosevelt wrote, those who migrated to the New World "lost all remembrance of Europe" and became new men "in dress, in customs, and in mode of life." But the United States by 1900 was becoming increasingly like the more civilized and longer settled parts of the world. Consequently Roosevelt declared that "as our civilization grows older and more complex, we need a greater and not a less development of the fundamental frontier virtues." The Boy Scouts of America was just one of the responses of Roosevelt's contemporaries to the problem he described. Without wilderness areas in which successive generations can relearn the values of their pioneer ancestors, the American culture will surely change. Perhaps it should, but many remain concerned about cutting off the roots of their national character. And merely from the standpoint of safeguarding an historical document, a part of the national past, we should save wilderness. Once all America was wild; without remnants to refresh our memories we run the risk of cultural amnesia.

ARGUMENT 4: WILDERNESS AS NOURISHER OF AMERICAN ARTS AND LETTERS

Time and again in the course of history the native land has been the inspiration for great works of music, painting, and literature. What the American painter, Alan Gussow, calls "a sense of place" is as vital to the artistic endeavor as it is to patriotism and national pride. And "place," it should be clear, has to do with the natural setting. Subdivisions, factories, and used car lots rarely inspire artistic excellence. Nature commonly does. Parks and reserves, as reservoirs of scenic beauty that touches the soul of man, have a crucial role in the quality of a nation's culture.

Certainly the United States would have a poorer artistic heritage without the existence of wild places of inspiring beauty. James Fenimore Cooper in literature, Thomas Cole and Albert Bierstadt in painting, and, to take a recent example, John Denver in music, have based their art on wilderness. In the case of the United States, wilderness had a special relationship to culture. It was the one attribute the young nation had in abundance, the characteristic that set it apart from Old World countries. Ralph Waldo Emerson and Henry David Thoreau were among the many who, by the mid-nineteenth century, called on America to attain cultural self-reliance by basing its art on the native landscape. Nature, for these philosophers, was intellectual fertilizer. Blended in the proper proportion with civilization, it produced cultural greatness. . . .

ARGUMENT 5: WILDERNESS AS A CHURCH

With the aid of churches and religions, people attempt to find solutions to, or at least live with, the weightiest mental and emotional problems of human existence. One value of wilderness for some people is its significance as a setting for what is, essentially, religious activity. In nature, as in a church, they attempt to bring meaning and tranquility to their lives. They seek a sense of oneness, of harmony, with all things. Wilderness appeals as a place to knot together the unity that civilization tends to fragment. Contact with the natural world shows man his place in systems that transcend civilization and inculcates reverence for those systems. The result is peace.

The Transcendental philosophers, Ralph Waldo Emerson and Henry David Thoreau, were among the first Americans to emphasize the religious importance of nature. Moral and aesthetic truths seemed to them to be more easily observed in wild places than in regions where civilization interposed a layer of artificiality between man and nature. John Muir, a leading force in the preservation of Yosemite National Park and first president of the Sierra Club, also believed that to be closer to nature was to be closer to God. The wild Sierra that he explored and lived in was simply a "window opening into heaven, a mirror reflecting the Creator." Leaves, rocks, and lakes were "sparks of the Divine Soul." Muir spent little time in a building called a church, but his enjoyment of wilderness was religious in every sense of the word.

ARGUMENT 6: WILDERNESS AS A GUARDIAN OF MENTAL HEALTH

Sigurd Olson, veteran guide and interpreter of the canoe country extending northward from Lake Superior, noted in 1946 that "civilization has not changed emotional needs that were ours long before it arose." Sigmund Freud had the same idea when he said that civilization bred "discontents" in the form of repressions and frustrations. One of the most distressing for modern man is the bewildering complexity of events and ideas with which civilization obliges him to deal. The price of failing to cope with the new "wilderness" of people and paper is psychological problems. The value of wilderness and outdoor recreation is the opportunity it extends to civilized man to slip back, occasionally,

Figure 9.4 *Yosemite Valley, Glacier Point Trail* by Albert Bierstadt, 1873. Yale University Art Gallery. Gift of Mrs. Vincenzo Ardenghi, 1931.

into what Olson calls "the grooves of ancestral experience." The leading advocate of wilderness protection in the 1930s, Robert Marshall [see above], spoke of the "psychological necessity" for occasional escape to "the freedom of the wilderness."

Olson and Marshall were referring to the fact that wild country offers people an alternative to civilization. The wilderness is different. For one thing, it simplifies. It reduces the life of those who enter it to finding basic human needs and satisfactions, such as unmechanized transportation, water, food, and shelter. Civilization does not commonly permit us this kind of self-sufficiency and its dividend, self-confidence. A hike of ten miles has more meaning in this respect than a flight of ten thousand. Wilderness also reacquaints civilized people with pain and fear. Surprising to some, these are ancient energizing forces—springboards to achievement long before monetary success and status were even conceived. The gut-level fears associated with survival drove the wheels of evolution. At times, of course, they hurt and even killed, but we pay a price in achievement for entering the promised land of safety and comfort. For many it is horribly dull. They turn to crime or drugs or war to fill their needs for risk and challenge. Others find beds in mental institutions the only recourse. Wilderness recreation is a better alternative.

ARGUMENT 7: WILDERNESS AS A SUSTAINER OF HUMAN DIVERSITY

Just as it promotes biological diversity (see Argument 2), the preservation of wilderness helps to preserve human dignity and social diversity. Civilization means control, organization, homogenization. Wilderness offers relief from these dehumanizing tendencies; it encourages individuality. Wild country is an arena where man can experiment, deviate, discover, and improve. Was not this the whole meaning of the New World wilderness for those settlers who migrated to it from Europe? Wilderness meant freedom. Aldo Leopold put it this way: "of what avail are forty freedoms without a blank spot on the map?" For novelist Wallace Stegner wild country was "a place of perpetual beginnings" and, consequently, "a part of the geography of hope." Somehow the preservation of wild places seemed to Americans inextricably linked to the preservation of free people. If there was wilderness, there could not be a technologically powered police state observing one's every move and thought. Total control of nature and human nature were equally suspect in American eyes. The naturalist Joseph Wood Krutch may have said it best of all when he observed that "wilderness and the idea of wilderness is one of the permanent homes of the human spirit."

There is another sense in which wilderness preservation joins hands with the perpetuation of human diversity. The very existence of wilderness is evidence of respect for minority rights. Only a fraction, although a rapidly growing one, of the American people seek scenic beauty and wilderness recreation. Only a fraction care about horse racing or opera or libraries. The fact that these things can exist is a tribute to nations that cherish and defend minority interests as part of their political ideology. Robert Marshall of the United States Forest Service made it plain in the 1930s that protection of minority rights is one of the hallmarks of a successful democracy. The majority may rule, said Marshall, but that does not mean it can impose its values universally. Otherwise art galleries (a minority interest) would be converted into hamburger stands and amusement parks. The need was for a fair division—of land, for instance—to accommodate a variety of tastes and values.

ARGUMENT 8: WILDERNESS AS AN EDUCATIONAL ASSET IN DEVELOPING ENVIRONMENTAL RESPONSIBILITY

To experience wilderness is to discover natural processes and man's dependency upon them. It is to discover man's vulnerability and, through this realization, to attain humility. Life in civilization tends to promote antipodal qualities: arrogance and a sense of mastery. Not only children believe that milk comes from bottles and heat from radiators. "Civilization," Aldo Leopold wrote, "has so cluttered [the] elemental man-earth relation with gadgets and middlemen that awareness of it is growing dim. We fancy that industry supports us, forgetting what supports industry." Contact with wilderness is a corrective that mod-

Figure 9.5 Tourists ride a people-mover through Yosemite National Park. © Newscom / Getty Images.

ern man desperately needs if he is to achieve long-term harmony between himself and his environment.

Wilderness can also instruct man that he is a member, not the master, of a community that extends to the limits of life and the earth itself. Because wild country is beyond man's control, because it exists apart from human needs and interests, it suggests that man's welfare is not the primary reason for or purpose of the existence of the earth. This seemingly simple truth is not easily understood in a technological civilization whose basis is control and exploitation. In wilderness we appreciate other powers and interests because we find our own limited.

A final contribution of wilderness to the cause of environmental responsibility is a heightened appreciation of the meaning and importance of restraint. When we establish a wilderness reserve or national park we say, in effect, thus far, and no farther to development. We establish a limit. For Americans self-limitation does not come easily. Growth has been our national religion. But to maintain an area as wilderness is to put other considerations before material growth. It is to respect the rights of nonhuman life to habitat. It is to challenge the wisdom and moral legitimacy of man's conquest and transformation of the entire earth. This acceptance of restraint is fundamental if people are to live within the limits of the earth.

William Cronon, from "The Trouble with Wilderness; or, Getting Back to the Wrong Nature" (1995)

William Cronon makes the counter-intuitive argument that there is nothing natural about the concept of wilderness, it is "entirely a creation of the culture that holds it dear, a product of the very history it seeks to deny." He claims that the more extreme cult of wilderness is counterproductive for effective conservation, causing the overvaluation of the vast and wild along with the undervaluation of nature in its less sublime forms.

Go back 250 years in American and European history, and you do not find nearly so many people wandering around remote corners of the planet looking for what today we would call "the wilderness experience." As late as the eighteenth century, the most common usage of the word "wilderness" in the English language referred to landscapes that generally carried adjectives far different from the ones they attract today. To be a wilderness then was to be "deserted," "savage," "desolate," "barren"—in short, a "waste," the word's nearest synonym. Its connotations were anything but positive, and the emotion one was most likely to feel in its presence was "bewilderment"—or terror.

. . . The movement to set aside national parks and wilderness areas followed hard on the heels of the final Indian wars, in which the prior human inhabitants of these areas were rounded up and moved onto reservations. The myth of the wilderness as "virgin," uninhabited land had always been especially cruel when seen from the perspective of the Indians who had once called that land home. Now they were forced to move elsewhere, with the result that tourists could safely enjoy the illusion that they were seeing their nation in its pristine, original state, in the new morning of God's own creation. Among the things that most marked the new national parks as reflecting a post-frontier consciousness was the relative absence of human violence within their boundaries. The actual frontier had often been a place of conflict, in which invaders and invaded fought for control of land and resources. Once set aside within the fixed and carefully policed boundaries of the modern bureaucratic state, the wilderness lost its savage image and became safe: a place more of reverie than of revulsion or fear. Meanwhile, its original inhabitants were kept out by dint of force, their earlier uses of the land redefined as inappropriate or even illegal. To this day, for instance, the Blackfeet continue to be accused of "poaching" on the lands of Glacier National Park that originally belonged to them and that were ceded by treaty only with the proviso that they be permitted to hunt there.

The removal of Indians to create an "uninhabited wilderness"—uninhabited as never before in the human history of the place—reminds us just how invented, just how constructed, the American wilderness really is. To return to my opening argument: there is nothing natural about the concept of wilderness. It is entirely a creation of the culture that holds it dear, a product of the very history it seeks to deny. Indeed, one of the most striking proofs of the cultural invention of wilderness is its thoroughgoing erasure of the history from which it sprang. In virtually all of its manifestations, wilderness represents a flight from history. Seen as the original garden, it is a place outside of time, from which human beings had to be ejected before the fallen world of history could properly begin. Seen as the frontier, it is a savage world at the dawn of civilization, whose transformation represents the very beginning of the national historical epic. Seen as the bold landscape of frontier heroism, it is the place of youth and childhood, into which men escape by abandoning their pasts and entering a world of freedom where the constraints of civilization fade into memory. Seen as the sacred sublime, it is the home of a God who transcends history by standing as the One who remains untouched and unchanged by time's arrow. No matter what the angle from which we regard it, wilderness offers us the illusion that we can escape the cares and troubles of the world in which our past has ensnared us.

This escape from history is one reason why the language we use to talk about wilderness is often permeated with spiritual and religious values that reflect human ideals far more than the material world of physical nature. Wilderness fulfills the old romantic project of secularizing Judeo-Christian values so as to make a new cathedral not in some petty human building but in God's own creation, Nature itself. Many environmentalists who reject traditional notions of the Godhead and who regard themselves as agnostics or even atheists nonetheless express feelings tantamount to religious awe when in the presence of wilderness—a fact that testifies to the success of the romantic project. Those who have no difficulty seeing God as the expression of our human dreams and desires nonetheless have trouble recognizing that in a secular age Nature can offer precisely the same sort of mirror.

Thus it is that wilderness serves as the unexamined foundation on which so many of the quasi-religious values of modern environmentalism rest. The critique of modernity that is one of environmentalism's most important contributions to the moral and political discourse of our time more often than not appeals, explicitly or implicitly, to wilderness as the standard against which to measure the failings of our human world. Wilderness is the natural, unfallen antithesis of an unnatural civilization that has lost its soul. It is a place of freedom in which we can recover the true selves we have lost to the corrupting influences of our artificial lives. Most of all, it is the ultimate landscape of authenticity. Combining the sacred grandeur of the sublime with the primitive simplicity of the frontier, it is the place where we can see the world as it really is, and so know ourselves as we really are—or ought to be.

But the trouble with wilderness is that it quietly expresses and reproduces the very values its devotees seek to reject. The flight from history that is very nearly the core of wilderness represents the false hope of an escape from responsibility, the illusion that we can somehow wipe clean the slate of our past and return to the *tabula rasa* that supposedly existed before we began to leave our marks on the world. The dream of an unworked natural landscape is very much the fantasy of people who have never themselves had to work the land to make a living—urban folk for whom food comes from a supermarket or a restaurant instead of a field, and for whom the wooden houses in which they live and work apparently have no meaningful connection to the forests in which trees grow and die. Only people whose relation to the land was already alienated could hold up wilderness as a model for human life in nature, for the romantic ideology of wilderness leaves precisely nowhere for human beings actually to make their living from the land.

This, then, is the central paradox: wilderness embodies a dualistic vision in which the human is entirely outside the natural. If we allow ourselves to believe that nature, to be true, must also be wild, then our very presence in nature represents its fall. The place where we are is the place where nature is not. If this is so—if by definition wilderness leaves no place for human beings, save perhaps as contemplative sojourners enjoying their leisurely reverie in God's natural cathedral—then also by definition it can offer no solution to the environmental and other problems that confront us. To the extent that we celebrate wilderness as the measure with which we judge civilization, we reproduce the dualism that sets humanity and nature at opposite poles. We thereby leave ourselves little hope of discovering what an ethical, sustainable, *honorable* human place in nature might actually look like.

Worse: to the extent that we live in an urban-industrial civilization but at the same time pretend to ourselves that our *real* home is in the wilderness, to just that extent we give ourselves permission to evade responsibility for the lives we actually lead. We inhabit civilization while holding some part of ourselves—what we imagine to be the most precious part—aloof from its entanglements. We work our nine-to-five jobs in its institutions, we eat its food, we drive its cars (not least to reach the wilderness), we benefit from the intricate and all too invisible networks with which it shelters us, all the while pretending that these things are not an essential part of who we are. By imagining that our true home is in the wilderness, we forgive ourselves the homes we actually inhabit. In its flight from history, in its siren song of escape, in its reproduction of the dangerous dualism that sets human beings outside of nature—in all of these ways, wilderness poses a serious threat to responsible environmentalism at the end of the twentieth century.

By now I hope it is clear that my criticism in this essay is not directed at wild nature per se, or even at efforts to set aside large tracts of wild land, but rather at the specific habits of thinking that flow from this complex cultural construction called wilderness. It is not the things we label as wilderness that are the problem—for nonhuman nature and large tracts of the natural world *do* deserve protection—but rather what we ourselves mean when we use that label. Lest one doubt how pervasive these habits of thought actually are in contemporary environmentalism, let me list some of the places where wilderness serves as the ideological underpinning for environmental concerns that might otherwise seem quite remote from it. Defenders of biological diversity, for instance, although sometimes appealing to more utilitarian concerns, often point to "untouched" ecosystems as the best and richest repositories of the undiscovered species we must certainly try to protect. Although at first blush an apparently more "scientific" concept than wilderness, biological diversity in fact invokes many of the same sacred values, which is why organizations like the Nature Conservancy have been so quick to employ it as an alternative to the seemingly fuzzier and more problematic concept of wilderness. There is a paradox here, of course. To the extent that biological diversity (indeed, even wilderness itself) is likely to survive in the future only by the most vigilant and self-conscious management of the ecosystems that sustain it, the ideology of wilderness is potentially in direct conflict with the very thing it encourages us to protect. . . .

Indeed, my principal objection to wilderness is that it may teach us to be dismissive or even contemptuous of . . . humble places and experiences. Without our quite realizing it, wilderness tends to privilege some parts of nature at the expense of others. Most of us, I suspect, still follow the conventions of the romantic sublime in finding the mountaintop more glorious than the plains, the ancient forest nobler than the grasslands, the mighty canyon more inspiring than the humble marsh. Even John Muir, in arguing against those who sought to dam his beloved Hetch Hetchy valley in the Sierra Nevada, argued for alternative dam sites in the gentler valleys of the foothills—a preference that had nothing to do with nature and everything with the cultural traditions of the sublime. Just as problematically, our frontier traditions have encouraged Americans to define "true" wilderness as requiring very large tracts of roadless land—what Dave Foreman calls "The Big Outside." Leaving aside the legitimate empirical question in conservation biology of how large a tract of land must be before a given species can reproduce on it, the emphasis on big wilderness reflects a romantic frontier belief that one hasn't really gotten away from civilization unless one can go for days at a time without encountering another human being. . . .

If the core problem of wilderness is that it distances us too much from the very things it teaches us to value, then the question we must ask is what it can tell us about *home*, the place where we actually live. How can we take the positive values we associate with wilderness and bring them closer to home? I think the answer to this question will come by broadening our sense of the otherness that wilderness seeks to define and protect. In reminding us of the world we did not make, wilderness can teach profound feelings of humility and respect as we confront our fellow beings and the earth itself. Feelings like these argue for the importance of self-awareness and self-criticism as we exercise our own ability to transform the world around us, helping us set responsible limits to human mastery—which without such limits too easily becomes human hubris. Wilderness is the place where, symbolically at least, we try to withhold our power to dominate. . . .

The task of making a home in nature is what Wendell Berry has called "the forever unfinished lifework of our species." "The only thing we have to preserve nature with," he writes, "is culture; the only thing we have to preserve wildness with is domesticity."[1] Calling a place home inevitably means that we will *use* the nature we find in it, for there can be no escape from manipulating and working and even killing some parts of nature to make our home. But if we acknowledge the autonomy and otherness of the things and creatures around us—an auton-

Figure 9.6 *The American Wilderness* by Asher Durand, 1864. Oil on canvas. Cincinnati Art Museum, The Edwin and Virginia Irwin Memorial, 1968.260. Photo by Tony Walsh, 2003.

omy our culture has taught us to label with the word "wild"—then we will at least think carefully about the uses to which we put them, and even ask if we should use them at all. Just so can we still join Thoreau in declaring that "in Wildness is the preservation of the World," for *wild*ness (as opposed to wilderness) can be found anywhere: in the seemingly tame fields and woodlots of Massachusetts, in the cracks of a Manhattan sidewalk, even in the cells of our own bodies. As Gary Snyder has wisely said, "A person with a clear heart and open mind can experience the wilderness anywhere on earth. It is a quality of one's own consciousness. The planet is a wild place and always will be."[2] To think ourselves capable of causing "the end of nature" is an act of great hubris, for it means forgetting the wildness that dwells everywhere within and around us.

Learning to honor the wild—learning to remember and acknowledge the autonomy of the other—means striving for critical self-consciousness in all of our actions. It means that deep reflection and respect must accompany each act of use, and means too that we must always consider the possibility of non-use. It means looking at the part of nature we intend to turn toward our own ends and asking whether we can use it again and again and again—sustainably—without its being diminished in the process. It means never imagining that we can flee into a mythical wilderness to escape history and the obligation to take responsibility for our own actions that history inescapably entails. Most of all, it means practicing remembrance and gratitude, for thanksgiving is the simplest and most basic of ways for us to recollect the nature, the culture, and the history that have come together to make the world as we know it. If wildness can stop being (just) out there and start being (also) in here, if it can start being as humane as it is natural, then perhaps we can get on with the unending task of struggling to live rightly in the world—not just in the garden, not just in the wilderness, but in the home that encompasses them both.

Notes

1. Wendell Berry, *Home Economics* (San Francisco: North Point, 1987), 138, 143.
2. Gary Snyder, quoted in *New York Times*, "Week in Review," 18 September 1994, 6.

Donald Waller, from "Getting Back to the Right Nature: A Reply to Cronon's 'The Trouble with Wilderness'" (1998)

In direct response to Cronon, above, Donald Waller attempts to place the concept of wilderness on a firm scientific footing. Waller picks up on the "wildness/wilderness" distinction that grew out of Thoreau's "Walking," above, and claims that an organism's wildness can be traced directly back to its evolutionary context. If this view is accepted, Cronon's social construction argument becomes irrelevant.

I begin by questioning an initial premise of Cronon's: that by idolizing wilderness and working for its protection we tend to diminish our concern for, and protection of, nearer and more mundane environments such as our cities and farms. Does concern for distant big wilderness areas necessarily decrease our concern for local environmental quality or environmental justice? An implicit assumption here appears to be that our overall efforts to protect the environment represent a zero-sum game so that additional concern for one area diminishes resources available to protect or restore other areas. If these premises are correct, we might be justified in refocusing our attention to more local issues.

What evidence exists that people who care more for remote wilderness care less for nearby or more mundane examples of nature? A qualified social scientist should address this issue in earnest (as well as the zero-sum assumption). My own admittedly subjective experience suggests just the opposite: individuals who care strongly about their nearby oaks and wetlands seem more likely to work passionately to preserve remote wild places such as the Arctic Wildlife Refuge. While ecologically aware and concerned citizens are inevitably torn among a wide set of worthy environmental causes and may decide to allocate their limited resources to remote and wild areas, this should never be taken to imply a lack of concern with local conditions (or for environmental social justice). Those few who do loudly proclaim their "misanthropic" preference for the protection of big wild areas over other human values are often doing so simply to provoke others to consider their point of view. Even Dave Foreman, co-founder of the radical group Earth First! and current Sierra Club Board member, notes that: "Wilderness advocates are not anti-people. Most of us support campaigns for human health and for social and economic justice."[1]

We might also consider the words and actions of early wilderness advocates such as John Muir and Aldo Leopold. Did not Muir plead passionately with his family to protect the small wetland on their farm? Aldo Leopold, in his essay "Illinois Bus Ride," laments the "success" of modern agriculture: "There are not hedges, brush patches, fencerows, or other signs of shiftless husbandry. The cornfield has fat steers, but probably no quail. The fences stand on narrow ribbons of sod; whoever plowed that close to barbed wires must have been saying, 'Waste not, want not.'"[2] Far from devaluing local conditions, Leopold pleads for us to extend our ecological sensibility even to these seemingly marginal scraps of habitat. . . .

DUALISTIC FALLACIES AND THE DIFFERENCE BETWEEN WILDNESS AND WILDERNESS

So what makes something wild? We tend to think in terms of things and places as being either wild or tamed, artificial or natural. Yet rarely are things and places so easily categorized. More often, degrees of wildness exist. Part of Cronon's critique centers on our historic tendency to draw dichotomies between wild vs. tame, or natural vs. unnatural. In this epoch of global climate change and the long-distance transport of heavy metals, persistent pesticides, and other pollutants, we must accept the fact that no area on Earth remains pristine or fully free of human influence. . . . Humans have become a biologic and even geologic force across the globe. Similarly, because humans are one of a related set of primate species, we must also accept the fact that humans are natural and forever a part of, and dependent upon, natural ecosystems.

While Cronon makes these points clearly and effectively, he leaves further questions unresolved. If degrees of wildness exist, and if humans must be accepted as an integral part of the systems we seek to protect and maintain, how are we to establish criteria for evaluating human behavior? What boundaries shall we place upon our own tendency to expand and subvert other biotic systems to our own ends? If no boundaries exist between wild and tame, natural and unnatural, why shouldn't we establish parks to protect rock quarries, dammed rivers, and hog farms? If all areas are considered as natural and wild, or denatured and tamed, as any other, why should we concern ourselves with conserving nature at all? This is the dilemma of environmental relativism raised, yet not resolved, in Cronon's essay.

. . . Although Cronon's essay begins with Thoreau's famous assertion that "In Wildness is the preservation of the World," it goes on to blur this distinction by focusing on the historical and cultural roots of our ideas about wilderness. If wilderness is, admittedly, a very human construct laden with cultural meaning,

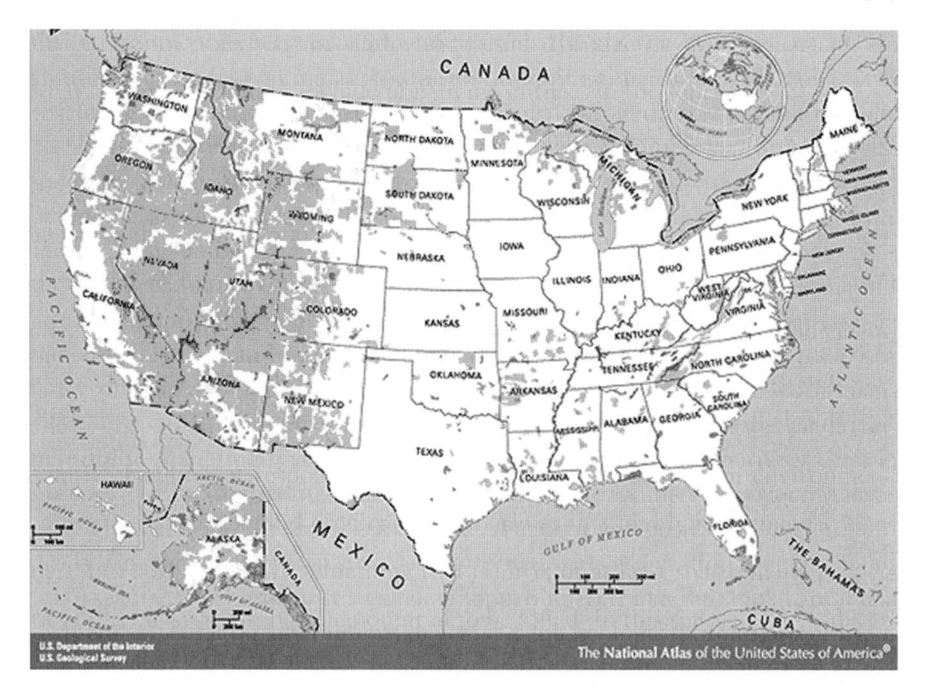

Figure 9.7 United States federal lands. Shaded areas are controlled by the Bureau of Indian Affairs, Bureau of Land Management, Bureau of Reclamation, Department of Defense, Forest Service, Fish and Wildlife Service, National Park Service, or the Tennessee Valley Authority. National Atlas of the United States, April 1, 2005, http://nationalatlas.gov.

wildness is just the opposite: that which is not, and cannot be, a human construct. Wildness existed before human cultures expanded and will exist long after human cultures have vanished. Wildness clearly also persists in many corners of our acculturated cities and farms (for example, the hedgerows mentioned by Leopold in "Illinois Bus Ride"). Because ecological systems change constantly, no single static state can be considered wild. Nevertheless, the rate and extent of human-induced changes now greatly exceed the ability of many organisms to adapt evolutionarily to these changes, meaning that we are robbing them of their wild context. However romanticized and idealized our notions of "wild" and "wilderness" are, there is and always will be a gap separating the artificial from the wild—the "otherness" that Cronon refers to. . . .

A surprising number of scientific justifications for conserving large and undeveloped lands have emerged from conservation biology in recent years. Most conspicuously, we continue to discover how many of the ecological interactions crucial for sustaining plant and animal species depend critically on how large, connected, and intact areas of habitat are. Smaller natural areas, or those subject to human disturbance, are highly prone to losing a substantial fraction of their species and these losses occur via several distinct but related mechanisms. Large wide-ranging carnivores and ungulates are directly sensitive to human activity; their requirement for extensive home ranges causes them to disappear quickly from smaller fragmented habitats. Many smaller species of amphibians, mammals, and plants are also quite sensitive to human disturbance, or to human-altered disturbance regimes. Many species are also incapable of dispersing across open or inhospitable habitats such as clear-cuts or roads, which dissect their populations into smaller subunits that are increasingly vulnerable to genetic and demographic hazards. Similarly, many neotropical migrant songbird species are suffering serious declines across eastern North America in apparent response to progressive habitat losses and fragmentation. With the increase in edge habitats, nest predators and parasites favored by such edges have also increased, drastically reducing nest success. Smaller areas also fail to sustain or survive historically dominant patterns of natural disturbance such as fire and windthrow, which causes further losses of species dependent on these disturbances. In their stead, farming, roads, channelization, and other forms of development enhance opportunities for the weedy and often exotic species that increasingly dominate our landscapes, further displacing many native species. While many of these changes are delayed or occur too slowly to catch our attention, they have already caused dramatic losses. Their eventual effects will surely be catastrophic.

Because habitat loss, fragmentation, and other forms of degradation threaten such a large fraction of our biota, we must pursue strategies that can succeed not only in conserving populations now but also in perpetuating the ecological conditions that will sustain their evolutionary future. Interestingly, all the mechanisms reviewed above strongly support approaches to conservation that emphasize large and/or connected wild areas, relatively free of human disturbance. Only large areas support larger, more viable, and interconnected populations of rare and threatened species and perpetuate the ecological processes that sustain other elements of biodiversity. While some elements of diversity can be sustained in small areas, and certain species clearly need localized and particular habitats, conservation biologists agree on the fundamental importance of allocating large blocks of suitable habitat as a first defense against further species losses. These findings have recast the agenda of most conservation organizations, including The Nature Conservancy, which now embraces the importance of protecting and restoring areas much larger than the small (50–100 acre) remnants it originally sought to protect.

While conserving wildlife has always existed as an accessory justification for wilderness, few of our wilderness areas and only one of our national parks (Everglades National Park) have been expressly designated to preserve biotic values. In addition, scientific and lay concerns for these biotic values have themselves expanded from a few vertebrate species (the "charismatic megafauna") to embrace the much broader concerns for other species now termed biodiversity.

Notes

1. Dave Foreman, "Wilderness Areas Are Vital: A Reply to Callicott," *Wild Earth* 4 (1994): 65.
2. Aldo Leopold, *A Sand County Almanac and Sketches Here and There* (New York: Oxford University Press, 1949), 119.

Gary Snyder, "Trail Crew Camp at Bear Valley, 9000 Feet. Northern Sierra—White Bone and Threads of Snowmelt Water" (1968)

Gary Snyder is one of the foremost American nature poets (see Chapter 10, below). Does this poem, in its attempt to re-create wild nature in an unmediated sense, undermine the arguments of Thoreau, Marshall, Nash, Cronon, and Waller, above; or does wilderness, in the end, resist all human attempts to express its nature? It simply is, *and must be experienced directly, through the senses, not in an intellectual debate.*

Trail Crew Camp at Bear Valley, 9000 Feet. Northern Sierra—White Bone and Threads of Snowmelt Water

Cut branches back for a day—
trail a thin line through willow
   up buckbrush meadows,
   creekbed for twenty yards
winding in boulders
   zigzags the hill
into timber, white pine.

gooseberry bush on the turns.
hooves clang on the riprap
   dust, brush, branches.
   a stone
cairn at the pass—
strippt mountains hundreds of miles.

sundown went back
   the clean switchbacks to camp.
bell on the gelding,
stew in the cook tent,
black coffee in a big tin can.

Figure 9.8 Scene from the Bükk Mountains, Hungary. Photo by Brent Ranalli.

10 The Urban Environment

Calcutta and Los Angeles

In 1900, some 150 years after the start of the industrial revolution (described by Wordsworth below, in 1805), there was only one nation (Great Britain) in which the majority of the population lived in cities. Today, dwelling in urban areas or in suburbs is fast becoming the global norm. Many consequences of urbanization have been positive. The prospect of a higher standard of living, more cultural and educational opportunities, and escape from the limitations of traditional societies are but a few of the benefits enjoyed by the millions who have moved to cities. Environmentalist Bill McKibben reminds us that there are even urban success stories, like Curitiba, Brazil. Yet the hopes of millions of others for a better life have been dashed, as they have found only urban poverty; economic exploitation; and a level of social and psychological displacement that leads, as Anup Shah argues, to crime, stress, and anomie. Then there are the environmental costs of urbanization, described here in the selections by the World Commission on Environment and Development, and by Michael Carley and Philippe Spapens. Loss of agricultural and forest land, and the depletion of natural resources in order to build, maintain, and expand cities; air, noise, and water pollution; and greater incidence of disease, among others, are all consequences of the rise of modern cities.

The histories of Calcutta (now officially Kolkata) and Los Angeles in the twentieth century illustrate these environmental impacts. At first glance, the two cases could not seem more different. Calcutta is, after all, a by-word for Third World urban poverty, the daily struggle of its inhabitants to survive and carve out a life for themselves made famous in books like Dominique Lapierre's *The City of Joy* or through the story of Mother Teresa. Despite an extensive older industrial base and a vibrant tradition of high culture, the bright Indian high-tech future represented by cities like Hyderabad—let alone the riches of Silicon Valley—seems far away from its teeming neighborhoods. The differing accounts of Lapierre and Manimanjari Mitra present a picture of environmental and human disaster caused by war, famine, storms, and an unjust economic system that has physically, socially, and ecologically warped the city's development. By contrast, even with much poverty (by American standards) and significant social problems, Los Angeles is one of the premier cities of the wealthiest nation on Earth. Commonly regarded as the popular culture capital of the world, it would appear to be all that Calcutta is not.

Yet there are broad resemblances. Like Calcutta, Los Angeles has developed through the triumph of individual political and economic self-interest over communal and environmental concerns—as well as through a more general failure to follow Richard Forman's advice to "Think globally, plan regionally, and then act locally." As the selection from Mike Davis makes clear, attempts to regulate growth, preserve green space and farmland, limit the city's vulnerability to natural disasters, prevent pollution, and respect the region's fragile ecology have historically been pushed aside in the rush to get rich out of the tide of people who continue to move into Southern California. Environmental victories have been the exception. Even when it comes to the single most precious resource, water, upon which the whole future of the city and region depends, expediency and short-term personal profit have, as Marc Reisner relates in *Cadillac Desert*, generally won out. If Calcutta is a cautionary tale, then so too is Los Angeles.

The futures of the two metropolises, admittedly, look different. North American wealth and technology seem better placed to deal with the environmental challenges facing Southern California than do the resources of population-laden West Bengal. But one wonders. In the face of its looming water crisis and the global depletion of energy resources, how long can Los Angeles continue expanding, reversing Saint Augustine's definition of God by becoming a city whose circumference is everywhere and whose center is nowhere? And if Calcutta (in contrast to Curitiba) is a pattern for the urban catastrophe threatening many cities in developing countries, may not the environmental reckoning soon to be presented to Los Angeles be one on its way to many other American cities as well, particularly in the Sunbelt? In the spirit of Gary Snyder's powerful "Night Song of the Los Angeles Basin," this case study asks readers to explore such questions.

FURTHER READING

Keil, Roger. *Los Angeles: Globalization, Urbanization, and Social Struggles.* Chichester, England: J. Wiley, 1998.

McKibben, Bill. *Hope, Human, and Wild.* Boston: Little, Brown, 1995.

Scott, Allen John, and Edward W. Soja, ed. *The City: Los Angeles and Urban Theory at the End of the Twentieth Century.* Berkeley: University of California Press, 1996.

Soja, Edward W. *Postmetropolis: Critical Studies of Cities and Regions.* Oxford, England: Blackwell Publishers, 2000.

William Wordsworth, "London" from *The Prelude*, Book VII (1805)

Wordsworth (see Chapters 18 and 20, below) bookends the history of the modern city with his personal—admittedly, pessimistic—perspective on the dehumanizing effects of the economic, social, and material processes transforming London during the industrial revolution. He invokes the natural beauty of his native Lake District in the face of this experience.

London

. . . O Friend, one feeling was there which belonged
To this great city by exclusive right:
How often in the overflowing streets
Have I gone forwards with the crowd, and said
Unto myself, "The face of everyone
That passes by me is a mystery!"
Thus have I looked, nor ceased to look, oppressed
By thoughts of what and whither, when and how,
Until the shapes before my eyes became
A second-sight procession such as glides
Over still mountains, or appears in dreams,
And all the ballast of familiar life—
The present and the past, hope, fear, all stays,
All laws, of acting, thinking, speaking man—
Went from me, neither knowing me, nor known.
And once, far travelled in such mood, beyond
The reach of common indications, lost
Amid the moving pageant, 'twas my chance
Abruptly to be smitten with the view
Of a blind beggar who, with upright face,
Stood propped against a wall, upon his chest
Wearing a written paper to explain
The story of the man and who he was.
My mind did at this spectacle turn round
As with the might of waters, and it seemed
To me that in this label was a type
Or emblem of the utmost that we know
Both of ourselves and of the universe;
And, on the shape of this unmoving man,
His fixèd face and sightless eyes, I looked
As if admonished from another world. . . .
Oh blank confusion! and a type not false

Of what the mighty city is itself
To all except a straggler here and there—
To the whole swarm of its inhabitants—
An undistinguishable world to men,
The slaves unrespited of low pursuits
Living amid the same perpetual flow
Of trivial objects, melted and reduced
To one identity by differences
That have no law, no meaning, and no end—
Oppression under which even highest minds
Must labor, whence the strongest are not free!
But though the picture weary out the eye,
By nature an unmanageable sight,
It is not wholly so to him who looks
In steadiness, who hath among least things
An under-sense of greatest, sees the parts
As parts, but with a feeling of the whole.
This (of all acquisitions first) awaits
On sundry and most widely different modes
Of education; nor with least delight
On that through which I passed. Attention comes,
And comprehensiveness and memory,
From early converse with the works of God
Among all regions, chiefly where appear
Most obviously simplicity and power.
By influence habitual to the mind
The mountain's outline and its steady form
Gives a pure grandeur, and its presence shapes
The measure and the prospect of the soul
To majesty. Such virtue have the forms
Perennial of the ancient hills; nor less
The changeful language of their countenances
Gives movement to the thoughts, and multitude,
With order and relation. This (if still
As hitherto with freedom I may speak,
And the same perfect openness of mind—
Not violating any just restraint,
As I would hope, of real modesty),
This did I feel in that vast receptacle.[1]
The spirit of nature was upon me here;
The soul of beauty and enduring life
Was present as a habit, and diffused—

Through meager lines and colors, and the press
Of self-destroying, transitory things—
Composure and ennobling harmony.

Note

1. [London]

World Commission on Environment and Development, from "The Urban Challenge" in *Our Common Future* (1987)

Though two decades old, the report of the World Commission on Environment and Development still accurately assesses the challenges facing humanity in the twenty-first century due to continuing urbanization, especially in developing countries, where urban growth is rapidly transforming societies. These challenges are either directly environmental (e.g., air and water pollution) or have important environmental dimensions (e.g., a rise in energy consumption levels, public health problems).

By the turn of the century [2000], almost half the world will live in urban areas—from small towns to huge megacities.[1] The world's economic system is increasingly an urban one, with overlapping networks of communications, production, and trade. This system, with its flows of information, energy, capital, commerce, and people, provides the backbone for national development. A city's prospects—or a town's—depend critically on its place within the urban system, national and international. So does the fate of the hinterland, with its agriculture, forestry, and mining, on which the urban system depends.

In many nations, certain kinds of industries and service enterprises are now being developed in rural areas. But they receive high-quality infrastructure and services, with advanced telecommunications systems ensuring that their activities are part of the national (and global) urban-industrial system. In effect, the countryside is being "urbanized."

THE GROWTH OF CITIES

This is the century of the "urban revolution." In the thirty-five years since 1950, the number of people living in cities almost tripled, increasing by 1.25 billion. In the more developed regions, the urban population nearly doubled, from 447 million to 838 million. In the less developed world, it quadrupled, growing from 286 million to 1.14 billion (Table 10.1 with more recent estimates reflected).

Over only sixty years, the developing world's urban population increased tenfold, from around 100 million in 1920 to close to 1 billion in 1980. At the same time, its rural population more than doubled.

TABLE 10.1
Population Living in Urban Areas, 1950–2015

Region	*1950*	*1985*	*2000*	*2015 (projected)*
			(percent)	
World total	29.1	41.1	47.1	53.6
More developed regions	52.5	70.5	73.9	77.3
Less developed regions	17.9	32.3	40.5	48.7
Africa	14.9	29.6	37.1	45.1
Latin America and the Caribbean	41.9	68.1	75.5	80.9
Asia	16.6	29.0	37.1	45.5
China	12.5	23.0	35.8	49.5
India	17.3	24.3	27.7	32.2
			(million)	
World total	732.7	1,984.5	2,856.9	3,855.9
More developed regions	426.7	786.1	882.5	951.7
Less developed regions	306.1	1,198.4	1,974.5	2,904.2
Africa	32.9	160.5	295.3	489.1
Latin America and the Caribbean	69.9	273.3	393.0	508.3
Asia	232.2	836.4	1,367.0	1,990.1

[*Note:* This table has been revised to reflect more recent UN estimates than those available when *Our Common Future* was originally published. These estimates are from the "World Population Prospects: 2002 Revision" Population Database.

- In 1940, only one person in eight lived in an urban center, while about one in 100 lived in a city with a million or more inhabitants (a "million city").
- By 1960, more than one in five persons lived in an urban center, and one in sixteen in a "million city."
- By 1980, nearly one in three persons was an urban dweller and one in ten a "million city" resident.[2]

The population of many of sub-Saharan Africa's larger cities increased more than sevenfold between 1950 and 1980—Nairobi, Dar es Salaam, Nouakchott, Lusaka, Lagos, and Kinshasa among them (Table 10.2). During these same thirty years, populations in many Asian and Latin American cities (such as Seoul, Baghdad, Dhaka, Amman, Bombay, Jakarta, Mexico City, Manila, São Paulo, Bogota, and Managua) tripled or quadrupled. In such cities, net immigration has usually been a greater contributor than natural increase to the population growth of recent decades.

In many developing countries, cities have thus grown far beyond anything

TABLE 10.2
Examples of Rapid Population Growth in Third World Cities (in Millions)

City	*1950*	*1985*	*2000*	*2015 (projected)*
Mexico City, Mexico	2.883	14.109	18.066	20.647
São Paulo, Brazil	2.313	13.395	17.099	19.963
Mumbai (Bombay), India	2.981	10.341	16.086	22.645
Jakarta, Indonesia	1.452	6.788	11.018	17.498
Cairo, Egypt	2.463	8.326	10.398	13.123
Delhi, India	1.390	6.769	12.441	20.946
Manila, Philippines	1.544	6.888	9.950	12.637
Lagos, Nigeria	0.288	3.500	8.665	17.036
Bogota, Colombia	0.676	4.373	6.771	8.900
Nairobi, Kenya	0.087	1.090	2.233	4.016
Dar es Salaam, Tanzania	0.078	1.041	2.116	4.123
Khartoum, Sudan	0.183	1.611	3.949	5.638
Amman, Jordan	0.090	0.736	1.147	1.550
Manaus, Brazil	0.089	0.761	1.392	2.134
Santa Cruz, Bolivia	0.042	0.447	1.061	1.932

[*Note:* This table has been revised to reflect more recent UN estimates than those available when *Our Common Future* was originally published. These estimates are from the report "World Population Prospects: 2003 Revision," available on the Internet at: http://www.un.org/esa/population/publications/wup2003/WUP2003Report.pdf.

imagined only a few decades ago—and at speeds without historic precedent. But some experts doubt that developing nations will urbanize as rapidly in the future as in the last thirty to forty years, or that megacities will grow as large as U.N. projections suggest. Their argument is that many of the most powerful stimuli to rapid urbanization in the past have less influence today, and that changing government policies could reduce the comparative attractiveness of cities, especially the largest cities, and slow rates of urbanization.

The urban population growth rate in developing countries as a whole has been slowing down—from 5.2 percent per annum in the late 1950s to 3.4 percent in the 1980s.[3] It is expected to decline even further in the coming decades. Nevertheless, if current trends hold, Third World cities could add a further three-quarters of a billion people by the year 2000. Over the same time, the cities of the industrial world would grow by a further 111 million.[4]

These projections put the urban challenge firmly in the developing countries. In the space of just fifteen years (or about 5,500 days), the developing world will have to increase by sixty-five percent its capacity to produce and manage its urban infrastructure, services, and shelter—merely to maintain present conditions. And in many countries, this must be accomplished under conditions of great economic hardship and uncertainty, with resources diminishing relative to needs and rising expectations.

THE CRISIS IN THIRD WORLD CITIES

Few city governments in the developing world have the power, resources, and trained staff to provide their rapidly growing populations with the land, services, and facilities needed for an adequate human life: clean water, sanitation, schools, and transport. The result is mushrooming illegal settlements with primitive facilities, increased overcrowding, and rampant disease linked to an unhealthy environment.

In most Third World cities, the enormous pressure for shelter and services has frayed the urban fabric. Much of the housing used by the poor is decrepit. Civic buildings are frequently in a state of disrepair and advanced decay. So too is the essential infrastructure of the city; public transport is overcrowded and overused, as are roads, buses and trains, transport stations, public latrines, and washing points. Water supply systems leak, and the resulting low water pressure allows sewage to seep into drinking water. A large proportion of the city's population often has no piped water, storm drainage, or roads.[5]

A growing number of the urban poor suffer from a high incidence of diseases; most are environmentally based and could be prevented or dramatically reduced through relatively small investments (Table 10.3). Acute respiratory diseases, tuberculosis, intestinal parasites, and diseases linked to poor sanitation and contaminated drinking water (diarrhea, dysentery, hepatitis, and typhoid) are usually endemic; they are one of the major causes of illness and death, especially among children. In parts of many cities, poor people can expect to see one in four of their children die of serious malnutrition before the age of five, or one adult in two suffering intestinal worms or serious respiratory infections.[6]

Air and water pollution might be assumed to be less pressing in Third World cities because of lower levels of industrial development. But in fact hundreds of such cities have high concentrations of industry. Air, water, noise, and solid waste pollution problems have increased rapidly and can have dramatic impacts on the life and health of city inhabitants, on their economy, and on jobs. Even in a relatively small city, just one or two factories dumping wastes into the only nearby river can contaminate everyone's drinking, washing, and cooking water. Many slums and shanties crowd close to hazardous industries, as this is land no one else wants. This proximity has magnified the risks for the poor, a fact demonstrated by great loss of life and human suffering in various recent industrial accidents.

The uncontrolled physical expansion of cities has also had serious implications for the urban environment and economy. Uncontrolled development makes provision of housing, roads, water supply, sewers, and public services prohibitively expensive. Cities are often built on the most productive agricultural

TABLE 10.3
Environmental Problems in Third World Cities

Out of India's 3,119 towns and cities, only 209 had partial and only 8 had full sewage and sewage treatment facilities. On the river Ganges, 114 cities each with 50,000 or more inhabitants dump untreated sewage into the river every day. DDT factories, tanneries, paper and pulp mills, petrochemical and fertilizer complexes, rubber factories, and a host of others use the river to get rid of their wastes. The Hoogly estuary (near Calcutta) is choked with untreated industrial wastes from more than 150 major factories around Calcutta. Sixty percent of Calcutta's population suffer from pneumonia, bronchitis, and other respiratory diseases related to air pollution.

Chinese industries, most of which use coal in outdated furnaces and boilers, are concentrated around 20 cities and ensure a high level of air pollution. Lung cancer mortality in Chinese cities is four to seven times higher than in the nation as a whole, and the difference is largely attributable to heavy air pollution.

In Malaysia, the highly urbanized Klang Valley (which includes the capital, Kuala Lumpur) has two to three times the pollution levels of major cities in the United States, and the Klang river system is heavily contaminated with agricultural and industrial effluents and sewage.

Sources: Centre for Science and Environment, *State of India's Environment: A Citizens' Report* (New Delhi: 1983); V. Smil, *The Bad Earth: Environmental Degradation in China* (London: Zed Press, 1986); Sahabat Alam Malaysia, *The State of Malaysian Environment 1983–84—Towards Greater Environmental Awareness* (Penang, Malaysia: 1983).

land, and unguided growth results in the unnecessary loss of this land. Such losses are most serious in nations with limited arable land, such as Egypt. Haphazard development also consumes land and natural landscapes needed for urban parks and recreation areas. Once an area is built up, it is both difficult and expensive to re-create open space.

In general, urban growth has often preceded the establishment of a solid, diversified economic base to support the build-up of housing, infrastructure, and employment. In many places, the problems are linked to inappropriate patterns of industrial development and the lack of coherence between strategies for agricultural and urban development. . . . The world economic crisis of the 1980s has not only reduced incomes, increased unemployment, and eliminated many social programs. It has also exacerbated the already low priority given to urban problems, increasing the chronic shortfall in resources needed to build, maintain, and manage urban areas.

Notes

1. This [selection] draws heavily on four background papers prepared for WCED: I. Burton, "Urbanization and Development" (1985); J. E. Hardoy and D. Satterthwaite, "Shelter, Infrastructure and Services in Third World Cities" (1985), printed in *Habitat International* 10, no. 4 (1986); J. E. Hardoy and D. Satterthwaite, "Rethinking the Third World City" (1986); and I. Sachs, "Human Settlements: Resource and Environmental Management" (1985).

2. U.N., *The Growth in the World's Urban and Rural Population 1920–1980*, Population Studies, no. 44 (New York, 1969); U.N., *Urban, Rural, and City Populations 1950–2000* (as assessed in 1978), Population Studies, no. 68 (New York, 1980).
3. UNCHS (Habitat) position paper for October 1986 DAC meeting on Urban Environment, OECD document DAC (1986), no. 47, 27 Aug. 1986.
4. U.N. Department of International Economic and Social Affairs, "Urban and Rural Population Projections, 1984" (unofficial assessment) (New York, 1986).
5. J. E. Hardoy and D. Satterthwaite, *Shelter: Need and Response: Housing, Land and Settlement Policies in Seventeen Third World Nations* (Chichester, England: John Wiley and Sons, 1981). For the situation in São Paulo, see J. Wilheim, "São Paulo: Environmental Problems of the Growing Metropolis," submitted to WCED Public Hearings, São Paulo, 1985.
6. J. E. Hardoy and Satterthwaite, "Third World Cities and the Environment of Poverty," *Geoforum* 15, no. 3 (1984).

Anup Shah, from *Ecology and the Crisis of Overpopulation: Future Prospects for Global Sustainability* (1998)

Shah's book intelligently summarizes the impact of population change and urbanization on global sustainability and day-to-day life. The selection here focuses on the problems humans have as social animals in adapting to living in large urban areas. For instance, Shah attributes higher urban crime rates in part to the effects, especially on young males, of physical crowding and the loss of traditional village social hierarchy.

LIVING CONDITIONS IN THE FIRST WORLD CITY

Compared to the village life of our ancestors, cities are full of strangers. In the primeval tribe of around 100 members, everyone knew everyone else. In the city, there are millions of strangers. It is possible that many of the problems of city life derive from the stresses impacting on sensibilities formed by the hunter-gatherer lifestyle of 10,000 years ago. Road rage, claustrophobia, and crime are cited as among the penalties of city life.

Packed in a small area, the spatial dimension of territories has been considerably reduced in a city. This loss may matter because one of the most basic ways in which animals reduce aggression (and generally co-exist) is to form well-defined territories. These are defended spaces. By keeping to separate patches, everyone has a share of the environment. Each territory provides its owners with a spatially limited form of dominance, making it possible for them to respect the territory of others.

Crowding and Behavior in Public and Private Places

Territorial boundaries become blurred with crowding. Then relationships between people can become less harmonious, increasing chances of conflict. It is worth making a distinction between interactions in public and private places since public places dominate city life. When people are in public places the territory is not marked for anyone in particular. To achieve cohesion that is necessary in the complex world of city life we need conventions of behavior when in public places. Accordingly, city dwellers need to co-operate in a subtle set of rules about spacing in public places. Conventions, of course, require the agreement of others. Points of conflict arise when tacit agreements break down or are misinterpreted.

City life greatly increases such chances of conflict.[1] With great crowding in public places, the implicit rules of personal space in public places can become overwhelmed. There is also a greater opportunity for an individual to gain some personal advantage by breaking the implicit rules since the sanction of social disapproval in the city matters less compared to village life sanction. Even if conventions are adhered to so that cohesion is maintained and thereby conflict is contained, it means that, in public places at least, city dwellers are obliged to follow strict patterns of behavior. That means some loss of freedom and individuality.

Turning to private places, where territory boundaries are expected to be observed, there is nevertheless a problem. Residents are packed together in a city and, as a result, there is residential crowding. That makes it much more difficult to find privacy or to avoid involuntary social contact.

City Crime and Violence

Large cities generally have higher levels of crime. (Criminal statistics are notoriously subject to error because many crimes are not reported to the police. This means that most figures underestimate crime.) Within countries, there seems to be a direct relationship between the size of a city and the likelihood of crimes such as drug-taking, burglary, and bodily attacks.[2]

Fear of violence and crime is a social problem because it limits the amount of voluntary social interaction that occurs. People who are most fearful of crime and violence in public areas are less likely to venture out and meet others. So fear alters behavior—people avoid going to certain places at certain times and also avoid going out alone. In addition, restrictions on children's activities are thought to have increased because the fear of violence to children has grown (an interesting point is that the large majority of parents who restrict their children were not restricted themselves when they were children). Moreover, there are direct emotional effects of the fear—tension, anxiety—on parents and children.[3]

Territorial spacing is one way in which levels of aggression can be relaxed. Another system is that of "peck order" or social hierarchy and the theory is as follows.[4] In the ancient smaller and simpler tribe, the distance between the top and the bottom of the peck order was not dramatic. However, the contrast between the status of the city's cardboard-box dwellers and the billionaire tycoons is vast. There are so many at the bottom and so few at the top that there are always some who will be driven to seeking revenge for what they see as their suppression and exploitation. Those squashed at the bottom and too cowardly to attack the powerful source of their agony turn to weaker victims—women, children, and animals—who serve as substitutes. This may explain the prevalence of "redirected aggression" or violence for the sake of violence in cities.

Stress

The pace of the lives of our ancestors was once ruled by the rhythm of the sun and the changing seasons. However, for today's city dwellers, speed has become part of their culture. We are working much harder than before and taking less leisure. In a city, a great deal has to be done and complex negotiations conducted in often difficult and rushed circumstances. So compared to their ancestors, city dwellers lead vastly accelerated lives. The increased speed required to do things, time urgency, can be distinguished from time pressure, the need to find more time to fit in all we have to do. It is time pressure that is thought to cause stress.[5]

ADAPTING TO CITY LIFE

Overall stress is likely to be reduced when conventions of behavior work in the city. However, what is special about stress in the city is that much of it is difficult to change: the city environment, which ultimately is responsible for stress,[6] has to change and engineering a re-plan of an existing city is extremely problematic.

In our distant past, we often altered the environment to suit us. Nowadays, it appears that we largely adjust to accommodate the demands of city life such as the noise, the crowds, and the information overload. The kind of adaptation that has taken place goes something like this example: we become habituated to noise. When we experience a constant level of background noise, as we do in most cities, we tend to habituate to it such that we hardly notice it at all. Another example: we have responded to time pressure by doing things quickly and by adopting a mode of behavior which makes the time we do have at our disposal more intense (for example, aerobics instead of long walks, talking on a mobile phone while driving). City life is also made up of a huge number of encounters with potential sensory and information overload. We adapt to this by tunneling our attention—for example, when walking in a crowded street we do not stop and talk but keep our faces blank and eyes straight ahead. So city dwellers have adapted to a high level of noise, doing a great many things in a short time interval, crowds, information overload, and so on.[7]

Adaptation is costly and therefore welfare reducing. Adaptation also implies restriction of freedom or additional restriction on behavior and therefore reduces welfare yet again. The problem is that we adapt behavior to suit city environment. The solution is to re-plan cities to suit us.

Notes

1. E. Krupat, *People in Cities* (Cambridge: Cambridge University Press, 1985).
2. K. Lynch, *The Image of the City* (Cambridge: MIT Press, 1960).
3. M. Apter, *The Dangerous Edge* (New York: Free Press, 1992).
4. D. Morris, *The Human Animal* (London: BBC Books, 1994).
5. D. A. Norman, *The Psychology of Everyday Things* (New York: Basic Books, 1988).
6. D. Halpern, *Mental Health and the Built Environment* (Washington, D.C.: Taylor and Francis, 1995).
7. S. P. Newman and S. Lonsdale, *Human Jungle* (London: Ebury Press, 1996).

Michael Carley and Philippe Spapens, from *Sharing the World: Sustainable Living and Global Equity in the Twenty-first Century* (1998)

Carley and Spapens provide a clear comparative framework for understanding the problem of urban sprawl as it affects both the developed and the developing world. They also list the major environmental dimensions of this phenomenon. Their solution is more stringent planning controls that encourage higher density development, more public transport, and the like.

URBANIZATION AND TRANSPORT

In terms of urban land use and its integration with transport, sustainable management presents a complex problem. There is substantial evidence that the typical modern urban sprawl of housing, shopping, and employment, triggered by economic development, needs to be contained.[1] This is in order to reduce the take-up of increasingly scarce rural land for lower-density urbanization—for example, 2.4 million acres (one million hectares) of prime farmland converted in the U.S.A. in the past decade, and 20,000 hectares lost in Java annually[2]—to reduce the emissions associated with unsustainable automobile transport; and to reduce the proportion of land paved over by settlements and transport infrastructure, which carry an ecological backpack inversely related to settlement density. . . .

Once a suburbanized land-use pattern is established, it is almost impossible to revert to a more sustainable pattern of concentrated, nodal density, and even rail-based transport is no solution as destinations spread over the landscape. As urbanization penetrates into the countryside, political pressure grows for yet more road-building. A related problem is the cost of congestion, which wastes billions of dollars annually—money which could be devoted to social or economic investment for development. For example, the annual cost of congestion in Bangkok is US$272 million, accounting for 2.1 percent of regional GDP.[3]

Higher, urban densities are more likely to generate sustainable development than are lower, suburban densities. . . . Higher densities also reduce the pressure on the diminishing rural land base and reduce the need to use private vehicles to access shopping and workplaces, thus reducing CO_2 emissions and air pollution; they also make it easier to run public transport on an economic basis. However, higher density living runs counter to suburban values, and to a propensity in many cultures for households to desire large detached houses with sprawling private yards or gardens. Control of automobility and suburbanization is thus politically sensitive, and options for managing the situation must be socially feasible.

A related problem is the fragmentation of the remaining rural landscape, which reduces the viability of ecosystems, and the subtle but relentless destruction of what has been defined as rural tranquility by the Council for the Protection of Rural England (CPRE).[4] Rural tranquility can be dissipated by mining operations, traffic, aircraft noise, power stations, touristic developments, such as large hotels, and intensive recreational uses, such as golf courses. The process described is the suburbanization of the countryside and it is occurring all over the world. For example, the Sustainable Greece team, Nea Ecologia, warns that intensive tourist developments, exceeding biophysical and social carrying capacities, is a main challenge of sustainable land management in the country. Many local, island people are aware of the dangers of misusing their limited land resources. But the current framework for land-use planning is weak and too easily manipulated to generate short-term economic gain at the expense of long-term needs. Urbanization as a result of growing prosperity at the expense of agricultural production usually means that an increasing proportion of foodstuffs are imported. This contributes to transport emissions, to global warming, as well as to the occupation of foreign land by developed countries.

Notes

1. U.K. Departments of the Environment and Transport, *Reducing the Need to Travel Through Land Use and Transport Planning* (London: HMSO, 1995); M. Carley, "Settlement Patterns and the Crisis of Automobility," *Futures* (Sept. 1992).
2. Worldwatch Institute, *Vital Signs 1996–97* (Washington, D.C.: 1996), 81.
3. S. Euisoon et al., "Valuing the Economic Impacts of Environmental Problems: Asian Cities," World Bank, Urban Management Program, Discussion Paper (1992).
4. Council for the Protection of Rural England and Countryside Commission, *Special Report: Tranquil Areas: A New Way of Seeing the Countryside*, by Ash Consulting Group (Nov. 1995).

Richard T. T. Forman, from *Land Mosaics: The Ecology of Landscapes and Regions* (1995)

Forman outlines general principles of land planning and management and characterizes the failures of urban planning heretofore.

"Think globally, act locally" is a phrase much in vogue. Yet it has two problems. First, few people will ever give primacy to the globe in decision-making. Second, local considerations overwhelmingly determine actions. Indeed, these are two roots of environmental and societal problems etched widely in the land. Typically, our best soils are thinned by erosion or covered by suburbs; biodiversity and wildlife in large wooded areas are impoverished by dispersed logging; critical riparian and hedgerow corridors are cut or razed by macroagriculture; productive land is desertified by irrigation and overuse; and parks are surrounded by the tightening noose of development.

Can this unraveling of the land, as the primary capital of a nation, be reversed? Land patterns seem logical and ecological in exceptional places within, for example, Britain, Romania, Costa Rica, Australia, and the USA, as well as in certain spatial models. In these cases efforts in design, planning, conservation, and decision-making seem both visionary and practical. They are directed at agricultural landscapes, suburban landscapes, forested landscapes, or regions. At these scales the environment, economics, and society coalesce. The landscape and region are the central linkage between global and local.

The philosophic framework for this . . . therefore might be that we better *"Think globally, plan regionally, and then act locally."*

The perennial challenges in planning, design, and management of an area are not only to take a broad spatial view and a long temporal view, but also to address all major environmental and human issues present. Water, transportation, biodiversity, aesthetics, sense of community, food production, and much more are the essential factors. All plans, all designs, and all management should address them. This requires an exceptional breadth of expertise. The usual result is that an individual or group within a discipline either develops a plan for a large area focused on one primary objective, or alternatively carves out a piece of the area where this objective is primary. . . .

Traditionally, planners and managers have successfully planned, designed, and managed areas by explicitly integrating the human world with the biological or natural world. Familiar examples are the ancient Chinese temple areas, medieval fortress towns, notable Italian villas, Olmsted's Emerald Necklace in Boston, Letchworth in Britain, and Canberra, Australia. Yet, in some planning circles public administration and economics have been substituted for the biological component. In essence, this is an experiment, doubtless of short duration, to see if natural processes, biological patterns, and the environment can be largely ignored in planning.

The result is that planning and management themselves are now in trouble. Laws, regulations, guidelines, standard practices, building codes, and planning acts are all essentially legal "standards" that govern transportation, mining, forestry, real estate development, grazing, environmental protection, water resources, and other human endeavors dear to our hearts. . . . The standards are described in manuals, law books, and the like. They were developed to protect society from human error, that is, to protect health, safety, and welfare. The unfortunate result is that the standards also protect individuals and groups who demonstrate lack of vision or make errors, as long as "one follows the code." We are stuck in the standards. Worse still, most standards were developed before the recent explosion in ecological understanding.

We straightened streams. We filled wetlands. We built levees along rivers. We tried to eliminate fire. We exterminated large predators. Today we are literally paying the price for wetland loss, soil erosion, massive floods, pest explosions, and "forestlessness." We know many of the standards are misguided, but society finds itself painted into a corner. There is no easy way out. Some standards should be rolled back or replaced, but those are only important details. Nor should a cookbook of prescriptions replace the pile of outdated standards. Rather an easier and wiser path is embedded in "vision" and an open process, and ends with a specific yet broad mandate.

Dominique Lapierre, from *The City of Joy* (1985), translated by Kathryn Spink

This selection from Lapierre's famous account of the life of migrants to Calcutta describes the natural events (hurricanes, floods, and famines) and human disputes (the 1947 Partition of British India, and subsequent wars between India and Pakistan) behind the demographic influx that has helped tip the city into seemingly irreversible decline since the late 1930s—forces and consequences broadly similar to those shaping many other cities in the developing world today.

The city that [recent migrant] Hasari [Pal] had not hesitated to describe as "inhuman" was in fact a mirage city, to which in the course of one generation six million starving people had come in the hope of feeding their families. In the 1960s, Calcutta was still, despite its decline over the previous half century, one of the most active and prosperous cities in Asia. Thanks to its harbor and its numerous industries, its metal foundries and chemical and pharmaceutical works, its flour mills and its linen, jute, and cotton factories, Calcutta boasted the third highest average wages per inhabitant of any Indian city, immediately after Delhi and Bombay. One third of the imports and nearly half of India's exports passed along the waters of the Hooghly, the branch of the Ganges on the banks of which the city had been founded three centuries earlier. Here, thirty percent of the entire country's bank transactions were undertaken and a third of its income tax was levied. Nicknamed [after the German industrial region] the "Ruhr of India," its hinterland produced twice as much coal as France and as much steel as the combines of North Korea. Calcutta drained into its factories and warehouses all the material resources of this vast territory: copper, manganese, chromium, asbestos, bauxite, graphite, and mica as well as precious timber from

Figure 10.1 A wandering holy man cleans his teeth with water from the holy Ganges River in Varanasi, India, despite the fact that it oozes with raw sewage, garbage, and human corpses. Courtesy of CleanGanga.com.

the Himalayas, tea from Assam and Darjeeling, and almost fifty percent of the world's jute.

From this hinterland also converged each day on the city's bazaars and markets an uninterrupted flow of foodstuffs: cereals and sugar from Bengal, vegetables from Bihar, fruit from Kashmir, eggs and poultry from Bangladesh, meat from Andra, fish from Orissa, shellfish and honey from the Sundarbans, tobacco and betel from Patna, cheeses from Nepal. Vast quantities of other items and materials also fed one of the most diversified and lively trading centers in Asia. No fewer than 250 different varieties of cloth were to be counted in the bazaars of Calcutta and more than 5,000 colors and shades of saris. Before reaching this mecca of industry and commerce, these goods often had to cross vast areas that were extremely poor, areas where millions of small peasants like the Pals scratched a desperate living out of infertile patches of land. How could these poor not dream, each time disaster struck, to take the same road as those goods?

The metropolis was situated at the heart of one of the world's richest yet at the same time most ill-fated regions, an area of failing or devastating monsoons causing either drought or biblical floods. This was an area of cyclones and apocalyptic earthquakes, an area of political exoduses and religious wars such as no other country's climate or history has perhaps ever engendered. The earthquake that shook Bihar on 15 January 1937, caused hundreds of thousands of deaths and catapulted entire villages in the direction of Calcutta. Six years later a famine killed 3.5 million people in Bengal alone and ousted millions of refugees. India's independence and the Partition in 1947 cast upon Calcutta some four million Muslims and Hindus fleeing from Bihar and East Pakistan [now Bangladesh]. The conflict with China in 1962, and subsequently the war against Pakistan, washed up a further several hundred thousand refugees; and in the same year, 1965, a cyclone as forceful as ten three-megaton H-bombs capable of razing to the ground a city like New York, together with a dreadful drought in Bihar, once more sent to Calcutta entire communities.

Now, it was yet another drought that was driving thousands of starving peasants like the Pals to the city.

The arrival of these successive waves of destitute people had transformed Calcutta into an enormous concentration of humanity. In a few years the city was to condemn its ten million inhabitants to living on less than twelve square feet of space per person, while the four or five million of them who squeezed into its slums had sometimes to make do with barely three square feet each. Consequently Calcutta had become one of the biggest urban disasters in the world—a city consumed with decay in which thousands of houses and many new buildings, sometimes ten floors high or even higher, threatened at any moment to crack and collapse. With their crumbling façades, tottering roofs, and walls eaten up with tropical vegetation, some neighborhoods looked as if they had just been bombed. A rash of posters, publicity and political slogans, and advertisement billboards painted on the walls defied all efforts at renovation. In the absence of an adequate garbage collection service, 1,800 tons of refuse accumulated daily in the streets, attracting a host of flies, mosquitoes, rats, cockroaches, and other creatures.

In summer the proliferation of filth brought with it the risk of epidemics. Not so very long ago it was still a common occurrence for people to die of cholera, hepatitis, encephalitis, typhoid, and rabies. Articles and reports in the local press never ceased denouncing the city as a refuse dump poisoned with fumes, nauseating gases, and discharges—a devastated landscape of broken roads, leaking sewers, burst water pipes, and torn down telephone wires. In short, Calcutta was "a dying city."

And yet, thousands, hundreds of thousands, even millions of people swarmed night and day over its squares, its avenues, and the narrowest of its alleyways. The smallest fragment of pavement was occupied, squatted upon, covered with salesmen and pedlars, with homeless families camping out, with piles of building materials or refuse, with stalls and a multitude of altars and small temples. The result of all this was an indescribable chaos on the roads, a record accident rate, nightmarish traffic jams. Furthermore, in the absence of public toilets, hundreds of thousands of the city's inhabitants were forced to attend to their bodily needs in the street.

In those years seven out of ten families had to survive on no more than one or two rupees a day, a sum that was not even sufficient to buy a pound of rice. Calcutta was indeed that "inhuman city" where the Pals had just discovered people could die on the pavements surrounded by apparent indifference. It was also a powder flask of violence and anarchy, where the masses were to turn one day to the saving myth of communism. To hunger and communal conflicts must also be added one of the world's most unbearable climates. Torrid for eight months of the year, the heat melted the asphalt on the roads and expanded the metal structure of the great Howrah Bridge to such an extent that it measured four feet more by day than by night. In many respects the city resembled the goddess Kali whom many of its inhabitants worship—Kali the terrible, the image of fear and death, depicted with a terrifying expression in her eyes and a necklace of snakes and skulls around her neck. Even slogans on the walls proclaimed the disastrous state of this city. "Here there is no more hope," said one of them. "All that is left is anger."

Manimanjari Mitra, from *Calcutta in the Twentieth Century: An Urban Disaster* (1990)

In contrast to Lapierre, Mitra (like many post-independence Indian historians) attributes Calcutta's urban miseries to the intertwined forces of capitalism and colonialism that have warped its development for the benefit of the wealthy and foreigners since the British founded the city in 1690. He argues in particular that these aspects of European imperialism have had a demographic as well as a material impact on how the city grew and what it has become today.

Calcutta's overpopulation, unplanned expansion, stagnating economy, extensive unrelenting poverty, and total inadequacy of civic amenities—all together grotesquely affected it in numerous ways. Even though it was the capital of a colonial empire and the sole metropolitan center of the eastern region, throughout the first half of the present century it manifested atrophied urban developmental processes, which thereby characterized its shortfall in growth potentials. Gerald Breese has described third-world urbanization as "subsistence urbanization" where "ordinary citizens have only bare necessities" and individuals have to live under conditions "that may be worse than the rural areas from which they have come."[1] It is debatable whether all the urban centers of the developing countries did really represent this "subsistence" level, or was Calcutta's experience singular in this respect?

Although the city failed to offer mere "subsistence urbanization" to the majority of its inhabitants, it did remain, throughout the centuries, almost the sole center of affluence and prosperity of this region, which only a small percentage of its population enjoyed. In fact, the city's appropriate geographical location enticed occidental colonialists, who converted it into their principal commercial center. The city's primary purpose was to serve as the trading outpost of the imperial power, which extracted the resources from the countryside and transferred the same to England. This systematic strangulation bereft the region of its wealth, but filled the coffers of the city, a fact which hardly made an iota of difference, as the accumulating wealth failed to invigorate Calcutta's economy. Thus, as can be rightly judged, the apparent prosperity of the city remained only superficial in nature. This blatant transfer of wealth by the foreigners destroyed the native rural economy, and under this circumstance Calcutta, even in the present century, appeared to be an "oasis amidst the vast desert of underdevelopment." Perhaps, herein lay the root of all the problems encountered by the city. . . .

Calcutta's primary concern was its demographic structure and although throughout the years the city enlarged itself, its growth was induced through the culmination of demographic pressure rather than economic advancement. Immigrant influx was a phenomenon of the twentieth century, in which people from different parts of the country or from abroad crowded into this city, and the rate of this immigration significantly increased between 1911 and 1921. It is interesting to note that during these years, [the] mortality rate appeared to be much greater than the rate of birth, but the total city population progressively increased from year to year, due to the enormous immigrant influx. The majority of them came from various districts of Bengal, as well as from the neighboring provinces such as Assam, Bihar, Orissa, U.P. [Uttar Pradesh], etc. It was they who gave the city a cosmopolitan character and it remained throughout the decades a willing receptacle of both manpower and talent. However, there was a notable increase in demographic pressure during the '40s and '50s, with the settlement of numerous uprooted families in and around Calcutta. In addition, the evident imbalance in sex-ratio was a characteristic feature of the incoming population. The prospect of gainfully earning a livelihood drew the males into this city, while their families remained in the villages. Broadly speaking, this gave rise to dual problems, namely, the flight of capital, which retarded the city's economic growth, and the adverse female ratio, which had a degenerating effect on its social life.

The demographic pressures on the city, although posing serious hindrance to its progression, ultimately retarded the process of urbanization by creating deficiencies in land and housing within the urban limits. As population density intensified, the problem of accommodating the multitudes within the limited space of the city too multiplied. Unmitigated congestion thereby vitiated the civic atmosphere of the city. With the curtailment of all options, the swelling populace was ultimately constrained to shift from the central zone of the city and move towards the periphery. Although their means of livelihood lay centered around the city proper, they lived in the surrounding villages and the peripheral areas. Under these circumstances those areas, despite their backwardness, became practically part of the city itself. This finally necessitated the amalgamation of those semi-rural environs, which pertained more towards ruralization than to the desired urban atmosphere of a city. Inappropriate coordination in the extension of urban limits ultimately nullified the existence of metropolitan amenities, thus leading to the near total suburbanization of the city region. Thereby the inhabitants of Calcutta were largely denied the benefits of a metropolitan life-style.

In the economic front also, the mounting pressure of population began to have an adverse effect. As the colonialists failed to revitalize the economy of Calcutta, the resources of the city itself remained restricted. As such, the protracted influx of sizeable sections of job-seekers, along with the presence of emigrés from neighboring countries, only accelerated the economic crisis of Calcutta. The city's economy seemingly depended on both the secondary and tertiary sectors, but it never witnessed capital-intensive industrial growth, which could have otherwise ensured actual prosperity and urban progress. On the contrary, labor-intensive as well as activities in the non-industrial sectors continued to expand and provide means of livelihood to the numerous poor. In fact, under a colonial set up, what had developed in Calcutta was actually an export-oriented economy, which provided little impetus to industrialization; thus economic activity continued to be subservient to the alien authority, instead of serving the people of the land. Had it been otherwise, it would have struck deep roots and thereby kept the city economically resilient and vibrant. However as this was not to be, the only expansion witnessed was in the quantum jump of industrial workers and petty traders. Economic stagnation, while curtailing the city's development, also restricted elevation of living standards among the citizens. Further, the drain of wealth impoverished the city, as a large number of immigrants remitted considerable portions of their income to families living outside Calcutta. Thus the wealth generated here continued to be divested, enriching numerous districts and provinces, but leaving little for the prosperity of this city.

The various retrogressive aspects manifested in Calcutta's economy influenced the propagation of innumerable destitute laboring classes, who were employed both in industrial and non-industrial sectors. Although their labors were crucial for the maintenance of all essential services, in return they were largely ostracized by society and survived in conditions pertaining to subhuman levels. Thereby, interspersed among the imperial splendors of this city, countless survived undernourished and insufficiently clothed, languishing in slums or pavement shacks, with the sole intention of earning enough to live on. Their hunger sometimes reached such extremities that many groveled along with dogs for scraps of food from dustbins. Their abject poverty ultimately influenced their character by fostering despair and the sense of alienation, which provoked a section to indulge in nefarious activities in the city. What is intriguing is that despite the existence of such abominable conditions, the rate of influx, particularly of indigents from near and distant regions, was hardly affected. In fact, their numbers rather augmented with the passing years, whereby the city's economy remained under unyielding pressure all throughout the first six decades of the present [twentieth] century.

As a result of chaotic conditions reigning in several spheres of urban life, the city authorities were incapable of providing adequate amenities to the citizens. During the early decades, the Englishmen were reluctant to provide these beyond the limits of the "white-town." Therefore the native areas, which bore the brunt of population pressure, had to make do with abysmal civic conditions. However, it should be duly stressed that besides administrative incompetence, the financial disability endured by large sections of the population too had [an] adverse effect on their living conditions. . . . The endless inflow of a poverty-stricken rural population . . . took its toll on the city's urbanization processes. Here they were restrained from upliftment of living standards by numerous ulterior forces, whereby they preferred to adhere to their habits and cultures, thus furthering the ruralization of the city. So this semi-urban, almost rural environment, most unbecoming of a major metropolitan center like Calcutta, was positively a factor to reckon with and something which required immediate solution.

But all efforts in gearing the city's resources were negated primarily due to the size of its population. With its unprecedented increase, all developmental measures proved to be inadequate within a short span of time, thus protracting defacement of the city surroundings. To solve the pertinent problems of employment, poverty, and retarded growth of urbanization, it was exigent to diminish over-crowding in the city. However, this became impossible, as the mass influx was the consequence of economic underdevelopment—both regional and national. . . .

The British in India adhered to the policy of systematic de-industrialization, whereby traditional production modes were all but destroyed. This was done deliberately to foster this nation as an agricultural hinterland of England. Their ruinous policy duly bereft this region as numerous towns, which had previously been centers of trade and manufacture, now witnessed a steady decline. As such, this waning industrial activity effectively curtailed the growth of new urban centers. Till the early decades of the present century, manufacture of jute products [remained] the principal industry of West Bengal. These jute mills were particularly situated along the banks of the Hooghly river and in 1941 gave employment to more than fifty percent [of the] workers from various provinces of India. Other industries, such as engineering and the Calcutta dockyard, employed the rest.[2] However, in the other provinces the situation was dissimilar, viz., cotton appeared to be the principal industry of U.P. while a number of factories processed indigo and lac in Orissa. But apart from these principal commodities, hardly any other processing industries existed there.[3] Therefore, as can be surmised, the unemployed rural masses of the industrially backward regions converged on the few urban-cum-industrial centers, one of which was Calcutta. The decline of traditional activities too compelled many to migrate into this city, as was evidenced in the large-scale migration from coastal Orissa, due to the diminution of the salt industry in the early years of this [the twentieth] century.[4]

The nexus between the decline of industries in Orissa and the subsequent demographic pressure upon Calcutta is quite significant, because in cases where the former was in evidence, it compelled the unemployed to seek livelihood in other provinces with an urban-cum-industrial base. However, due to the lack of an alternative urban center in West Bengal, this city, along with its surrounding areas, bore the brunt of immigrant pressure. . . .

After independence, no fundamental, sustained efforts were made to strengthen Calcutta's economy. . . . Investments in Calcutta and West Bengal were far below that of Bombay and Maharashtra. But at the same time it must be mentioned that the post-independence era registered an unparalleled influx of emigrés. Of the 4.1 million people who came to India, 3.2 million settled in West Bengal and one third of the latter settled in Calcutta, Howrah, and neighboring areas.[5] Its only outcome was the proliferation of extensive poverty in this metropolitan city.

Notes

1. G. Breese, *Urbanization in Newly Developing Countries* (New Delhi, 1978), 5.
2. S. K. Munsi, *Calcutta Metropolitan Explosion* (New Delhi, 1975), 30.
3. Department of Statistics, India, *List of Factories and Other Large Industries in India* (1911).
4. *Census of India* (1951), Orissa, vol. 11, pt. 1.
5. CMPO, *Basic Development Plan* (1966).

Mike Davis, from *Ecology of Fear: Los Angeles and the Imagination of Disaster* (1998)

Davis's book is a history of the expansion of greater Los Angeles since the early nineteenth century, detailing the environmental impacts of each stage of that growth. This selection, which focuses on the period from World War II to the 1980s, like the rest of the volume,

Figure 10.2 *Pearblossom Hwy., 11–18th April 1986, #2* by David Hockney, 1986. Photographic collage of chromogenic prints (color), 198 × 282 cm (78 × 111 in.). The J. Paul Getty Museum, Los Angeles.

emphasizes the interplay of politics, economic interests, and ignorance as developers and public officials conspired to fend off attempts to limit growth and protect the environment.

GREENING THE URBAN DESERT?

In 1958, sociologist William Whyte—author of *The Organization Man*—had a disturbing vision as he was leaving Southern California. “Flying from Los Angeles to San Bernardino—an unnerving lesson in man’s infinite capacity to mess up his environment—the traveler can see a legion of bulldozers gnawing into the last remaining tract of green between the two cities, and from San Bernardino another legion of bulldozers gnawing westward.” When he reached New York he wrote an article for *Fortune* magazine, describing the insidious new growth form that he called “urban sprawl.”[1]

After the debacle in the San Fernando Valley, there was negligible political or bureaucratic opposition to the obliteration of the rest of Southern California’s picture postcard landscapes. Although Los Angeles County paid homage in its 1941 master plan to the “major importance” of protecting choice agricultural land from subdivision, its actual landuse policies continued to encourage sprawl. In a 1956 report, for example, the Regional Planning Commission confirmed that all the remaining citrus orchards in the eastern San Gabriel Valley would soon be bulldozed and subdivided. The commission's only concern was that "this transition to urban uses should be encouraged to take place in an orderly manner" to minimize the "dead period" between land clearance and home construction.[2] But the speculative homebuilding frontier of the 1950s produced the same glut of subdivision relative to demand—"ghost towns in reverse"—that so shocked [Frederick Law] Olmsted [Jr.] and [Harland] Bartholomew during the 1920s.

For a decade, meanwhile, at least one thousand citrus trees were bulldozed and burned every week. Between 1939 and 1970, agricultural acreage in Los Angeles County south of the San Gabriel Mountains (the richest farmland in the nation according to some agronomists) fell from 300,000 to less than 10,000 acres. One of the nation's most emblematic landscapes—the visual magnet that had attracted hundreds of thousands of immigrants to Southern California—was systematically eradicated.[3]

Hillside and canyon environments fared little better. Both Olmsted and [Richard] Neutra had denounced the privatization of hillside vistas, and Olmsted had urged public ownership of key tracts in the Santa Monica Mountains. A 1945 county citizens' committee, after reminding political leaders that the quality of recreational landscape was "the goose that lays our golden eggs," proposed extensive open space conservation in the Palos Verdes, Baldwin, Montebello, Puente, San Raphael, and Verdugo Hills. But the postwar demand for "view lots" was virtually unquenchable. Within Los Angeles County as a whole, more than 60,000 house sites were carved out of the mountains and foothills during the 1950s and early 1960s.[4]

The automobile also devoured exorbitant quantities of prime land. By 1970 more than one third of the Los Angeles region was dedicated to the car: freeways, streets, parking lots, and driveways.[5] What generations of tourists and migrants had once admired as a real-life Garden of Eden was now buried under an estimated three billion tons of concrete (or 250 tons per inhabitant).[6]

Southern California sprawl eventually became a national scandal. Thanks to the crusading efforts of Whyte, federal responsibility for the "exploding metropolis" was subjected to unprecedented scrutiny and debate. Despite fierce opposition from the National Association of Home Builders, the Kennedy administration officially acknowledged the social costs of sprawl and introduced legislation in 1961 to support the conservation of urban open space.

In California, the state legislature was prodded by the Sierra Club and California Tomorrow (a regional-planning advocacy group) into authorizing a major study of the state's "open space crisis." The eminent San Francisco firm of Eckbo, Dean, Austin, and Williams (known acronymically as EDAW), which dominated environmental planning in California during the 1960s and 1970s, was hired to consult. Although Edward A. Williams wrote the final report for the State Office of Planning in 1965, the overarching influence of the firm's most prominent partner, Garrett Eckbo, in this and subsequent studies was evident. . . .

In *Landscape for Living*, his postwar manifesto for a new philosophy of environmental design, Eckbo decried the "sordid chaos" of "general commercial speculation" and argued that it was "no more than democratic Americanism to say that such forces can be analyzed, exposed, and placed under proper public control." Rejecting the reservation of the lushest areas for the rich, he advocated a "truly democratic organization" of the landscape that would replace "the sterile formality of authority" with the "tremendous tree symphony of the future." Indeed as authentic democracy began to achieve "cultural expression in the landscape . . . the present scale of landscape values will tend to reverse itself."

> Instead of moving from the ugly city toward the peak of wilderness beauty, it will be possible to move from the wilderness through constantly more magnificent and orderly rural refinements of the face of the earth, to urban communities composed of structures, paving, grass, shrubs, and trees, which are rich, sparkling, crystalline nuclei in the web of spatial relations that surrounds the earth—peak expressions of the reintegration of man and nature.[7]

The Urban Metropolitan Open Space Study that Eckbo and his partners submitted to Governor Pat Brown in 1965 as the keystone of a proposed state development plan resonated with the bold values and motifs of *Landscape for Living*. Indeed some Sacramento bureaucrats must have found little to choose between EDAW's theses and those of contemporary radicals in Berkeley's Sproul Plaza. The study warned that all of California's remaining Mediterranean valleys and foothills, including the exquisite Santa Barbara–Ventura coast, as well as the famed vineyards of Sonoma and Napa counties, were threatened with the same fate as Los Angeles's citrus belt. It condemned county governments for their "weak, timid, and unimaginative" use of zoning powers and denounced a tax system that rewarded land speculators and punished farmers. It also stressed how landscape-destroying sprawl at the urban edge devoured the bulk of regional tax resources and accelerated neighborhood decay in the center.[8]

"A clearcut crisis situation," the authors found, "exists in the Southern California urban-metropolitan area." Once again, suburbanization had entirely outpaced the production of new public space. By the most minimal standards of per capita provision of recreation space, Los Angeles County was facing a 100,000-acre shortfall of regional parks, while at the municipal level the discrepancy was frequently much worse. Indeed, the open space situation throughout the Los Angeles Basin—"1,500 square miles of low grade, monotonous suburban construction"—was so hopeless that the study focused instead on stopping sprawl at its periphery.[9]

In 1965 significant farm and foothill belts still defended Ventura-Oxnard, San Bernardino, Riverside, and San Diego from engulfment by greater Los Angeles. Although local environmentalists had targeted the Santa Monica Mountains as the most important conservation area, the study focused instead on the Puente Hills and Chino Plain, which separated the San Gabriel Valley from suburbanizing San Bernardino and Orange Counties. As "the center of the greatest population pressure within the region . . . they should become the most highly prized and zealously protected open-space resource."[10]

The second regional priority was "from Conejo to Hidden Hills, between Los Angeles, and Ventura, an area of beautiful rolling hills and valleys, peculiarly vulnerable to destruction by careless and indifferent development, yet peculiarly pregnant with possibilities for rich and imaginative design." Other environmental battlegrounds were the undeveloped parts of the Oxnard Plain, the Elsinore-Temecula corridor in southwest Riverside County, and the coastal mesas and valleys between San Diego and Vista.[11] (EDAW's prescience was subsequently borne out by the frenzy of subdivision in each of these areas during the late 1970s and 1980s.)

The study also briefly surveyed the dismal trajectory of urban overspill into the Mojave and Colorado River deserts. "The entire desert seems to be subdivided and covered with a gridiron of graded streets; such development destroys the desert as landscape and as open space, replacing [it] with nothing but the empty wasteland of ex-urbanism." Moreover, the elaboration of community designs suitable to the desert environment appeared to be simply "beyond the capability of [existing] planning processes."[12] . . .

Six years later, in 1972, EDAW produced another major open space survey: this time an exhaustive study of the Santa Monica Mountain coastline for the California legislature. Given the Santa Monicas' incalculable recreational and landscape value, EDAW expressed incredulity at the county's projected population of 405,000 for the environmentally sensitive Malibu area.[13] They pointed out that Malibu, apart from major problems with earthquakes, flooding, and landslides, also had a fire history "unique in intensity, devastating and heightened during Santa Ana wind conditions"; and echoing Olmsted and Bartholomew, they decried the ease with which developers in high-risk areas shifted the costs of fire and flood protection onto the taxpayers at large. They proposed a stringent permit system to keep new construction at a minimum while the legislature evaluated options for expanding public ownership in the Santa Monicas.[14]

The EDAW reports were seminal documents in the renaissance of regional planning and landscape conservation. Californians were suddenly forced to confront the cultural and ecological costs of their postwar wonderland, and from Eureka to San Diego, they were shocked by what they saw.

In the San Francisco Bay Area, a unique heritage of Brahmin conservationism provided Nob Hill support for successful efforts to protect wetlands and create a regional conservancy in the foothills. People for Open Space united the followers of John Muir and Lewis Mumford—environmentalists, planners, and philanthropists—in a common defense of San Francisco's great natural beauty and fashioned the first comprehensive "anti-sprawl" plan for any American metropolitan area.[15]

Southern California's counterpart movements crusaded to stop flagrant tract development in the Santa Monica Mountains and other foothill and coastal areas. Unlike Bay Area activists, however, they found little institutional support. The Los Angeles County Regional Planning Commission was theoretically the chief custodian of the regional landscape. Yet commission members, according to critics, had historically functioned as "expediters for fringe growth," producing planning documents that were seldom more than "blueprints for sprawl." As the League of Women Voters complained, the commission's efforts to preserve the Santa Monicas amounted to "color[ing] the vacant land green on its master plan."[16] After soliciting environmental development guidelines from a distinguished panel of natural scientists in 1970, the commission brazenly discarded them in order to double the area of land targeted for urbanization.[17]

At stake were the remaining fragments of the greenbelts identified by Eckbo and Williams in their 1965 study. The commission proposed to feed hungry developers another million acres of priceless agricultural and foothill landscape, while a coalition of environmental groups argued in a landmark lawsuit that population growth should be accommodated by investment within the existing urban fabric to revive declining neighborhoods and pockets of blight.[18] In 1979 the controversy over open space management suddenly erupted into full-fledged public scandal. A grand jury investigation exposed the inner workings of a regional planning system dominated and corrupted by development interests. As critics had long charged, key planning officials had advised developers in the Santa Monica Mountains and Antelope Valley on how to circumvent public hearings and environmental regulations by illegally partitioning their property among relatives and dummy corporations. An astounding *13,000* individual cases of fraudulent lot division were alleged. Further, when planning staff recommended against environmentally destructive projects in Diamond Bar and Santa Clarita, they had been summarily overruled by the commission majority.[19]

Although public outrage eventually forced the resignation of the commission's chairman, it was a modest, even a Pyrrhic victory for the environmental movement. The brief light focused on corruption within the Regional Planning Commission was never allowed to illuminate more fundamental conflicts of public and private interests within the Board of Supervisors. Once the commission reformed its most egregious practices, the steam went out of the largely legalistic battle to stop the fringe development juggernaut. To appease the court, the county was finally forced to officially designate some sixty-two "significant ecological areas" (SEAs), but no legislation was ever enacted to guarantee their preservation. To rub salt in environmentalists' wounds, the majority of appointees to the commission monitoring the SEAs have been and continue to be full- or part-time consultants to developers.[20]

As a result, suburbanization has completely devoured each of the open space buffer zones thought crucial by Eckbo and Williams. While the population of the Los Angeles region grew by forty-five percent from 1970 to 1990, its developed surface area increased by 300 percent.[21] Small environmental gains—notably by the Santa Monica Mountains Conservancy—have been parried and checked by relentless subdivision. Unlike in the San Francisco Bay Area, there have been no unqualified victories for open space preservation, just an accumulation of worthless environmental impact reports and toothless development guidelines. In part, this striking difference in outcomes must be attributed to the dissimilar political cultures and power structures in California's two major metropolitan regions. County government in Southern California is so hopelessly captive to the land development industry that sweeping electoral reforms, comparable to California's Progressive revolution of 1911, are probably the prerequisite for overthrowing the "new octopus" and transforming landuse priorities.

Yet 1970s environmentalism in Los Angeles County was also compromised by its own parochialism and historical amnesia. There was little discussion, in the spirit of New Deal modernists like Neutra and Eckbo, of the role of parks and other open areas as the "functional skeleton of the community." More often than not, environmental battles were fought piecemeal by local organizations with scant consideration of overall regional strategy and seldom in coalition with other constituencies. The class and racial dimensions of the recreational crisis were pointedly ignored. Ecology, in other words, stopped short of the more subversive but necessary politics of urban design.

THE LAST LANDSCAPE

January 1998. Time travel is becoming increasingly difficult. Traffic on California 126 is halted while a convoy of huge bottom-dump semis, each filled to the brim with the dark rich soil of the Santa Clara River Valley, lumber onto the asphalt. Farther west, even larger Caterpillar graders are leveling roadbed for the four-lane widening of Highway 126. A wrathful dust devil whirls in front of the lead grader.[22]

Eventually to be widened to eight lanes, Highway 126, once a bottleneck blocking the urbanization of the Santa Clara River Valley, will become the conduit for rapid and overwhelming growth. Transport planners already foresee its fate in the early twenty-first century when, gridlocked from Ventura to Santa Clarita, it will accommodate a staggering 360,000 trips per day. The major cause of this congestion will be Newhall Ranch, a master-planned city of 70,000 which will occupy a ten-mile corridor along the Santa Clara River. The Newhall Land

and Farming Company, one of the West's largest developers, will begin building it in the year 2000.[23] [As of 2007, building had not commenced.]

This project should be a legal impossibility. More than 25,000 housing units are designated for an agriculturally zoned floodplain of a wild river that is one of the most important SEAs in Los Angeles County and contains several endangered species protected by the state of California. Newhall Land and Farming, however, has the kind of political "juice" that makes the eyes of politicians pop. A lavish contributor to the campaigns of three of the five Los Angeles County supervisors, the company is also one of Governor Wilson's most faithful corporate supporters. By promising to preserve an 800-acre beauty strip along the Santa Clara River, and counting a large area of nearby undevelopable mountainside as integral "open space," the company purchased enough environmental legitimacy to satisfy its undemanding political allies. By the beginning of 1998, Newhall Ranch had easily cleared most of the principal zoning and regulatory hurdles.[24]

The stakes for the company are immense. Over its thirty-year development schedule, each completed phase of the Ranch will further the value of remaining raw land within the nineteen-square-mile parcel. Similarly, the growth of Newhall Ranch and the addition of more workers and consumers will consolidate the role of adjacent Valencia—an earlier Newhall Land and Farming Company community—as the industrial, retail, and cultural hub of northern Los Angeles County. The Ranch will also become an Archimedean lever inflating land values and ensuring urban development in the nearly 16,000 acres of orange and lemon groves owned by the company west of the Ventura County line.

Opposing the Ranch is a small band of ranchers, artists, and environmentalists known as the Friends of the Santa Clara River. They are acutely aware that northern Los Angeles County is evolving into a second Orange County, fueled by white flight from the aging suburbs of the San Fernando Valley. The population of the region, which has tripled since 1980, is expected to reach one million sometime around 2010. As the Friends have argued cogently but futilely to the Regional Planning Commission, unrestrained development on this scale will soon destroy the landscape "amenities"—including clean air, beautiful vistas, abundant open spaces, and relatively uncongested highways—that have attracted so many thousands to northern Los Angeles County in the first place.

Likewise, the continuing exodus from Los Angeles of so many middle-class taxpayers (led by several thousand police officers and sheriffs), inevitably followed by businesses and jobs, will only accelerate the decay of older suburban neighborhoods. Taxpayers in these areas subsidize their own decline through gas taxes for new peripheral freeways and higher water rates to amortize the costly California Aqueduct that irrigates the new edge cities. Public action mitigates the environmental crisis primarily for the top ten percent of the population who benefit from the conversion of wetlands into marinas, and from hidden subsidies for hillside living. For the other ninety percent, Southern California remains radically unplanned, undesigned, and out of control.

The Newhall Land and Farming Company, of course, would vehemently disagree. Newhall Ranch, like the dozens of other master-planned communities that offer "prestige lifestyles" in the white-flight belts of northern Los Angeles, eastern Ventura, western Riverside, southern Orange, and northern San Diego Counties, is meticulously planned down to the last cul-de-sac and red-tiled roof. The company claims, moreover, to have designed Newhall Ranch in conformity with the deepest longings of Southern Californians for an idealized "hometown America." Using "psychographic" surveys and focus groups, the company found that ninety-eight percent of potential homebuyers want to live near nature and that "vast majorities said they wanted entertainment opportunities, to feel safe, and to avoid the homogeneity of many of today's suburban tracts."[25]

Newhall Ranch, the company promises, will be this utopia. Like Disney Corporation's model community of Celebration in Florida, it has been planned to the fairytale standards of the "New Urbanism," with all homes located in traditional villages ("Riverwood," "Oak Valley," "The Mesas," and so on), a short walking distance away from shops, open areas, and sports facilities. Billed as "A Community by Nature," it will supposedly "re-create elements of the lifestyle prevalent in Southern California before World War Two."[26]

The Friends of the Santa Clara River are quick to point out the dark irony in the nostalgic trappings of Newhall Ranch. The last authentic landscape of that prewar way of life—Eden's last garden—will be destroyed in order to build its suburban simulation. The classic tenets of good planning advocated by Olmsted, Alexander, and Eckbo—repackaged as "New Urbanism"—will become sales points for the next generation of supersprawl.

The widening of Highway 126 is then a primal scene—the familiar tremor heralding an eruption of growth that will wipe away human and natural history. Suburbanization is like another one of Southern California's natural disasters—recurrent, inexorable. The big machines shifting dirt along the Santa Clara River today are the direct descendants of the bulldozers that ripped freeway corridors through the San Gabriel Valley and Orange County in the 1950s. The cast is the same, but the play is now in its final act. After the Santa Clara River Valley dies, there is only desert to feed the developers' insatiable hunger.

Notes

1. William Whyte, "Urban Sprawl," *Fortune* 57 (Jan. 1958): 302.
2. Los Angeles County, Regional Planning Commission, *Master Plan of Land Use* (Los Angeles, 1941); and *East San Gabriel Valley* (Los Angeles, 1956).
3. See Mark Northcross, "Los Angeles County: Biting the Land that Feeds Us," *California Tomorrow* 36; and Raymond Dassmann, *California's Changing Environment* (San Francisco, 1981), 81.
4. Testimony of Frederick Law Olmsted, Jr., to Citizens' Committee on Parks, Playgrounds and Beaches, *Los Angeles Times* (22 Feb. 1928); and Richard Jahns, "Seventeen Years of Response by the City of Los Angeles to Geologic Hazards," *Geologic Hazards and Public Problems: Conference Proceedings* (Santa Rosa, 1970), 266.
5. Donald Coates, *Environmental Geomorphology and Landscape Conservation. Vol. 2: Urban Areas* (Stroudsberg, 1974), 273.

6. Calculated by architect Christopher Wegscheid (Southern California Institute of Architecture, 1994) using data that I supplied on aggregate (sand and gravel) production in Los Angeles County.
7. Garrett Eckbo, *Landscape for Living* (New York, 1949), 45, 111–12.
8. EDAW [Eckbo, Dean, Austin, and Williams], *Open Space, the Choices before California: The Urban Metropolitan Open Space Study* (San Francisco, 1969), 22–23. See also EDAW, *State Open Space and Resource Conservation Program for California* (Sacramento, 1972).
9. EDAW, *Open Space*, 15, 24.
10. Ibid., 41.
11. Ibid., 42.
12. Ibid., 45.
13. Ventura–Los Angeles Mountain and Coastal Study Commission, *Final Report to the Legislature*, prepared by EDAW (6 Mar. 1972) [6.1], [12B15.1].
14. Ibid. [9.2], [12B3.1].
15. Thomas Kent, Jr., *Open Space for the California Bay Area: Organizing to Guide Metropolitan Growth* (Berkeley, 1970); and Alfred Heller, ed., *The California Tomorrow Plan* (Los Altos, 1972).
16. League of Women Voters of Los Angeles County, *Open Space in Los Angeles County* (Los Angeles, 1972), 33.
17. For a concise account, see W. David Conn, *Environmental Management in the Malibu Watershed: Institutional Framework* (Washington, D.C., 1975).
18. See *Coalition for Los Angeles County Planning in the Public Interest v Board of Supervisors, Los Angeles County*, Superior Court: C-63218 (12 Mar. 1975).
19. *Los Angeles Times* (3 Apr., 16 May, and 9 July 1979).
20. For bleak assessments, cf. ibid. (2 Dec. 1990); and Betsey Landis, "Significant Ecological Areas: The Skeleton in Los Angeles County's Closet?" in J. E. Keeley, ed., *Interface Between Ecology and Land Use in California* (Los Angeles, 1993), 112–13, 116.
21. Data from Southern California Association of Governments (1996).
22. What follows is largely based on Jim Churchill's detailed critique of the Newhall Ranch Environmental Impact Report in 1996 and subsequent interviews with Barbara Wample of the Friends of the Santa Clara River in fall 1997.
23. *Newhall Land Fact Sheet.*
24. *Los Angeles Times* (3 and 24 Nov. 1996). Newhall Land and Farming Company's campaign contributions, 1992–96, furnished by Jim Churchill.
25. Ibid. (24 Nov. 1996); and Newhall Ranch brochures.
26. Newhall Ranch brochures.

Marc Reisner, from *Cadillac Desert: The American West and Its Disappearing Water* (1993 [1986])

The two excerpts in the selection from Reisner depict the dilemma Californians face over their water supply, as well as their counterproductive attempts to address the problem. The first passage discusses the state's human-constructed landscape and economy based on unsound use of water resources. The second takes a representative moment in the struggle to provide water for Los Angeles (a history portrayed in the film Chinatown*), and predicts a catastrophically dry future for the city.*

Everyone knows there is a desert somewhere in California, but many people believe it is off in some remote corner of the state—the Mojave Desert, Palm

Figure 10.3 The banks of Lake Oroville in California reveal its recession. Courtesy of G. Donald Bain / The Geo-Images Project, Dept. of Geography, UC–Berkeley.

Springs, the eastern side of the Sierra Nevada. But inhabited California, most of it, is, by strict definition, a semidesert. Los Angeles is drier than Beirut; Sacramento is as dry as the Sahel; San Francisco is just slightly rainier than Chihuahua. About sixty-five percent of the state receives under twenty inches of precipitation a year. California, which fools visitors into believing it is "lush," is a beautiful fraud.

California is the only state in America with a truly seasonal rainfall pattern—stone dry for a good part of the year, wet during the rest. Arizona is much drier overall, but has two distinct rainy seasons. Nevada is the driest state, but rain may come at any time of year. If you had to choose among three places to try to grow a tomato relying on rainfall alone, South Dakota, West Texas, or California, you would be wise to choose South Dakota or West Texas, because it rains in the summer there. California summers are mercilessly dry. In San Francisco, average rainfall in May is four-tenths of an inch. In June, a tenth of an inch. In July, none. In August, none. In September, a fifth of an inch. In October, an inch. Then it receives eighteen inches between November and March, and for half the year looks splendidly green. The reason for all this is the Pacific high, one of the most bewildering and yet persistent meteorological phenomena on earth—a huge immobile zone of high pressure that shoves virtually all precipitation toward the north, until it begins slipping southward to Mexico in October, only to move back up the coast in late March. More than any other thing, the Pacific high has written the social and economic history of California.

Actually, San Francisco looks green all year long, if one ignores the rain-starved hills that lie disturbingly behind its emerald-and-white summer splendor, but this is the second part of the fraud, the part perpetrated by man. There was not a single tree growing in San Francisco when the first Spanish arrived; it was too dry and windblown for trees to take hold. Today, Golden Gate Park looks as if Virginia had mated with Borneo, thanks to water brought nearly 200 miles by tunnel. The same applies to Bel Air, to Pacific Palisades, to the manicured lawns of La Jolla, where the water comes from three directions and from a quarter of a continent away.

The whole state thrives, even survives, by moving water from where it is, and presumably isn't needed, to where it isn't, and presumably is needed. No other state has done as much to fructify its deserts, make over its flora and fauna, and rearrange the hydrology God gave it. No other place has put as many people where they probably have no business being. There is no place like it anywhere on Earth. Thirty-one million people (more than the population of Canada), an economy richer than all but seven nations' in the world, one third of the table food grown in the United States—and none of it remotely conceivable within the preexisting natural order.

For all its seasonal drought, its huge southern deserts, and its climatic extremes, there is plenty of water in California for all the people who live there today. If, God forbid, another twenty-five million arrive, there will still be plenty for them. The only limiting factor will be energy: to get to where the people are likely to settle, a lot of the water has to be lifted over mountains. Take any ten of the largest reservoirs—Shasta, Bullard's Bar, Pine Flat, Don Pedro, New Melones, Trinity, a few others—and you have enough water for the reasonable needs of twenty-five million people; enough for their homes, their schools, their offices, their industries, even (in all but the driest times) their swimming pools and lawns. As for the other 1,190 California dams and reservoirs, their purposes are threefold: power, flood control, and, above all, water for irrigation. What few people, including some Californians, know is that agriculture uses eighty-one percent of all the water in this most populous and industrialized of states.

California's $18 billion agricultural industry—and it is a gigantic, complex, integrated industry—is the largest and still the most important in the state, Silicon Valley notwithstanding. That figure, $18 billion, only begins to convey what agriculture really means to California. A great proportion of its freight traffic is agricultural produce. A disproportionate amount of the oil and gas mined in the state is used by agriculture. California agriculture supports a giant chemicals industry (it uses about thirty percent of all the pesticides produced in the United States), a giant agricultural-implements industry, an unrivaled amount of export trade. Because it relies on irrigation—and therefore on dams, aqueducts, and canals—there is a close symbiotic relationship with the construction industry, which is why politicians who lobby hard on behalf of new dams can count on great infusions of campaign cash from the likes of the Operating Engineers Local No. 3 and the AFL-CIO. And, more than any other state, California has been a source of opportunities for the Bureau of Reclamation and the [Army] Corps of Engineers.

All of this production, all of these jobs, all of these concentric rings of income-earning activity nourish California's awesome $485 billion GNP. It is a gross *state* product, obviously, but everyone seems to refer to it as the "California GNP," as if the state were a nation unto itself—which it really is, and nowhere more so than in the example of water. California has preached and practiced water imperialism against its neighbor states in a manner that would have done Napoleon proud, and, in the 1960s, it undertook, by itself, what was then the most expensive public-works project in history. That project, the State Water Project, more than anything else, is *the* symbol of California's immense wealth, determination, and grandiose vision—a demonstration that it can take its rightful place in the company of nations rather than mere states. It has also offered one of the country's foremost examples of socialism for the rich.

* * *

In California, when the issue is water, the ironies seem to string out in seamless succession. Bill Warne, the man who built the California Water Project, was in government service nearly all his life, and never made a great deal of money. In his mid-seventies, Warne was still doing consulting work; he also owned a small almond orchard outside of Sacramento. The consulting work was lucrative, but unpredictable. The almonds, on the other hand, were a good, reliable source of income. Or they were until Tenneco, by far the largest almond grower in the state, made a bid in 1981 to control the market—the same kind of power play that Prudential made with olives. "The bastards really went for our throats," Warne admitted ruefully during an interview early in 1982. "They beat the hell out of the rest of us in the market, and that includes me." Of course, one could just as well have said that Warne beat the hell out of himself. It was *his* project that irrigated Tenneco's almond orchards; it was *his* aqueduct that flowed practically within view of his small almond ranch, destined for the huge factory farms in the desolate southern reaches of the valley. Because of the hot climate down there, the crops grown on irrigation water have always been, in large part, specialty crops: almonds, pistachios, grapes, olives, kiwis, melons, canning tomatoes. And because the national acreage given over to such crops is comparatively small (California accounts for most of it), a single big grower who doesn't mind being a little ruthless can whiphand the market pretty much as he pleases.

Bill Warne's project had become a Frankenstein's monster. But its maker still refused to turn against his creation. "The moment we began settling California, we overran our water supply," he said. "We've never gotten to the point where you could just stop. And we never will."

Whether or not that is true, it is hard to imagine, by 1985, how the State

Water Project would ever be completed. The old warhorses, the Bill Warnes and Pat Browns, might still be talking about the "unconscionable waste" of water flooding down the Eel River each winter (as Warne did, to whomever would listen), or saying that "the Columbia doesn't need all that water that flows down there—it's ridiculous, between you and me" (as Pat Brown did during an interview in 1979), but those who followed them in public office and were faced with the nitty-gritty problem of diverting the Eel, or the Columbia, or any so-called "surplus" water that could be found, discovered that it was like uncovering a nest of killer bees. Jerry Brown's successor, George Deukmejian, was elected with large infusions of cash from the growers in the San Joaquin Valley, where he is from. As expected, Deukmejian, a deeply conservative Republican, proved himself ideologically double-jointed on the issue of water development; while wading through the state budget with a machete, he made a wide circle around the Peripheral Canal, which he wanted to build but call something else, and he spoke approvingly of plans to send a lot more water southward. The reaction from northern California politicians, who, in the meantime, had managed to seize control of the speaker's chair in the legislature, and, through Congressman George Miller (who represents the Delta) of a key committee in Congress that can probably thwart much of what Deukmejian hopes to build, was so intemperate that the governor, after a year in office, was hardly mentioning the canal anymore.

Deukmejian may merely have decided to lie low, but by 1985 the people who will feel the impending shortages most acutely—the growers and the cities of the South Coast—appeared to have given up on the idea; either that, or they were mollifying their opposition while they stealthily plotted some hydrologic equivalent of Pearl Harbor. In June, the State Water Contractors, an organization representing all the customers of the State Water Project, issued a report predicting a shortfall, by the year 2010, of 4.9 million acre-feet state-wide—the domestic consumption of twenty million people. The deficit within the State Project service area alone would be about 1.9 million acre-feet. Without more construction, the San Joaquin Valley would receive 733,000 fewer acre-feet than it was counting on. The South Coast cities and irrigation districts, which signed contracts to buy 2,497,500 acre-feet from the State Water Project, could be guaranteed a firm yield of only 1,120,000 acre-feet. Only in wet years could each region hope for more; during extended droughts they would receive even less. Meanwhile, a state report on groundwater pumping was describing the overdraft as "potentially critical" in eleven subregions of the Central Valley, most of which were in the service area of the State Water Project. What made things worse was that the valley's ancient saltwater aquifer, lying below the fresh water, could eventually rise to take its place.

Those figures, if they were accurate, bespoke calamity from both regions' points of view. What was startling, therefore, was the fact that the report said virtually nothing about sending more water from northern California southward. Its solutions—which it admitted were only halfway solutions—were for the most part the same ones that had been proposed by the environmental lobby, and which the water lobby had scorned just a few years earlier. The Imperial Valley farmers, according to the report, could conserve about 250,000 acre-feet if they lined their earthen canals and improved their irrigation practices; the water could then be sold to Los Angeles. The occasional surplus Colorado River flows below Parker Dam, as long as they lasted, could be stored in groundwater basins near Los Angeles and San Diego. Reusing treated sewage water (the report didn't go so far as to advocate drinking it) could save a few tens of thousands of acre-feet. Delta channels could be widened and levees rebuilt to allow slightly greater flows. The state could buy the surplus water in the Central Valley Project, for as long as that lasted. It was nickel-and-dime stuff, no heroics; the water savings might amount to 1.6 million acre-feet, which would only make up a third of the projected statewide shortfall. Only two new reservoirs, both off-stream and judiciously located south of San Francisco, were even mentioned, and the report didn't even advocate that they be built; it merely called for "investigations." (Initial investigations by the Department of Water Resources suggested a per acre-foot price range of $310 to $400 from one of the reservoirs, Los Banos Grande; since that was fifteen to twenty times the cost of Oroville water, it was hard to imagine who in his right mind would buy it, at least as long as there was groundwater to overdraft.) Not a word was said about the Peripheral Canal.

Ironically, the State Water Contractors' report was accompanied by a rather lengthy history of the State Water Project, written by the first head of the Department of Water Resources, Harvey Banks, which called the project "a high water mark symbolizing the results of the collective efforts of people of many points of view to resolve their ward with a program of statewide benefit." Reviewing the history of the project, it was hard for some to see how Banks managed to arrive at such a conclusion. To begin with, Californians had been sold a pig in a poke: a project whose cost was deliberately and extravagantly understated, and whose delivery capability was much less than they had been led to believe. Completing just the first phase of construction had required federal cost-sharing at San Luis Dam, nearly half a billion dollars in tidelands oil subsidies, and several hundred million dollars in scavenged new revenue bonds. Then, when spectacular agricultural and urban growth had occurred on the promise of water the project couldn't deliver, a new leadership of "new age" politicians had tried to sell the voters an even bigger and more expensive pig, which they had spurned. Los Angeles had fought with the growers, then formed an alliance with them, then fought again, then formed another alliance; two of the biggest growers had been instrumental in launching the project, then played an indispensable role in the defeat of the Peripheral Canal; and, all the while, the state had remained bitterly divided along the geographic and climatologic lines the project was supposed to supersede. This was "cooperation"?

As for the "statewide benefit" Banks wrote about, the California Water Project may have been necessary if the state was to continue to grow at its historical, breathtaking rate. But that was the point. The growth it created was not "orderly" growth, to use that buzzword of which the water developers are so fond. It was giantism. It was chaotic growth. In southern California, project water is allowing hundreds of acres to be subdivided, malled, and paved over each week, transforming what could have been a Mediterranean paradise into one of the twentieth century's urban nightmares. In Kern County, it created, solidified, and enriched land monopolies that are waging economic war against the small farmers who are so important to the state's economic stability, and who give its agricultural regions what little charm they have. To drive from east to west across the San Joaquin Valley, from a pretty little palm-colonnaded city such as Chowchilla, made prosperous by the Central Valley Project and surrounding small farms, to a shabby town such as Huron, surrounded by endless tracts of irrigated land farmed by distant corporate owners, is to fathom the sorry social impact agricultural monopoly can have.

And what is worse, the State Water Project fostered growth in the desert, willy-nilly, without a secure foundation of water. Twenty million people may live between Santa Barbara and San Diego in 2010; the current outlook, according to the State Water Contractors, is that five million of them won't have water unless some drastic conservation steps are taken and occasional surpluses are scavenged from every available source. Even if the groundwater overdraft in the San Joaquin Valley continues to increase—and the chairman of the California Water Commission said recently that it may become "intolerable" by the year 2000—a shortfall of nearly a million acre-feet looms ahead there. The likeliest "solution" to the shortages, as things now stand, will be a lot of land going out of production. The farmers who are apt to give up first are those who are wholly dependent on farming for their livelihood. The ones likely to continue are those to whom farming is a sideline to oil refining or banking or running a railroad, or a tax writeoff—a way to accumulate a little judicious financial loss.

Gary Snyder, "Night Song of the Los Angeles Basin" (1986)

Snyder poetically confronts modern Los Angeles, the epitome of our civilization of machines, with the persistent and almost magical survival of endangered nature. In doing so, he refutes the post-Cartesian claims of Western culture, which assert the disenchantment of the world, but also imagines the power of that culture to destroy the environment.

Night Song of the Los Angeles Basin

Owl

calls,

pollen dust blows

Swirl of light strokes writhing

Knot-tying light paths,

calligraphy of cars.

Los Angeles basin and hill slopes
Checkered with streetways. Floral loops
Of the freeway express and exchange.

Dragons of light in the dark
sweep going both ways
in the night city belly.
The passage of light end to end and rebound,
—ride drivers all heading somewhere—
etch in their traces to night's eye-mind

calligraphy of cars.

Vole paths. Mouse trails worn in
On meadow grass;
Winding pocket-gopher tunnels,
Marmot lookout rocks.
Houses with green watered gardens
Slip under the ghost of the dry chaparral,

Ghost
shrine to the L.A. River.
The *jinja* that never was there
is there.
Where the river debouches
the place of the moment
of trembling and gathering and giving
so that lizards clap hands there
—just lizards
come pray, saying
"please give us health and long life."

A hawk,
a mouse.

Slash of calligraphy of freeways of cars.

Into the pools of the channelized river
the Goddess in tall rain dress
tosses a handful of meal.

Gold bellies roil
mouth-bubbles, frenzy of feeding,
the common ones, the bright-colored rare ones
show up, they tangle and tumble,

godlings ride by in Rolls Royce
wide-eyed in brokers' halls
lifted in hotels
being presented to, platters
of tidbit and wine,
snatch of fame,

churn and roil,

meal gone the water subsides.

A mouse,
a hawk.

The calligraphy of lights on the night
freeways of Los Angeles

will long be remembered.

Owl
calls;
late-rising moon.

Figure 10.4 Suburban sprawl. Courtesy of USDA.

PART TWO

Foundational Disciplines and Topics

I. BIOLOGICAL INTERACTIONS

The study of the natural world or "the environment" has long been the province of what is commonly called the natural sciences, usually distinguished from two other areas, the humanities and the social sciences. The three sections in Part Two of *Environment* mirror these three areas, and by doing so make the claim that all three are connected to environmental issues, and that each one contributes something important to the understanding of those issues.

The natural sciences are usually delineated to include physics, "the queen of the sciences," chemistry, biology, astronomy, geology, air and atmospheric science, oceanography, and sometimes mathematics. In this section we elevate biology to its place of honor as the "queen" of the environmental sciences for a simple, yet controversial reason: study of the environment takes on urgency in direct proportion to its effect on biological systems and life.

In other words, our concerns focus first on how environmental change affects living organisms: if air is polluted, it is not the air, for its own sake, that we worry about—we worry about the negative consequences for human health or for the health of species or whole ecosystems that depend on that air. Such interdependency guides the structure of this section on Biological Interactions. The first chapter, "Biodiversity and Conservation Biology," focuses on interactions within biological systems. Subsequent chapters expand the focus outward in two directions, to include interactions with non-living components of the environment and with human beings. The second chapter adds a partly living, partly non-living, interacting entity—soil—to the mix, and enriches our understanding of the fundamental human undertaking of agriculture. The third chapter adds two non-living yet vital constituents—air and water—whose cycles sustain yet are simultaneously threatened by human activities. The fourth chapter on energy, incorporating important aspects of chemistry and physics, underscores how crucial human energy use is to the well-being of every organism on the planet. Finally, the fifth chapter on toxicology documents the impacts of industrial economies on the chemistry and biology of our total environment.

The endangerment and extinction of species and the degradation of ecosystems have strong interactive effects. The loss of a species can have cascading effects through chains of species that once interacted with it. For example, if a pollinator is lost, it is likely that the plant species that relied on the pollination service will eventually be lost as well. If a predator is lost, the former prey population may grow at such an exponential rate that it will entirely degrade its own food source. The known examples of such interactions are endless. The science of ecology has attempted to codify interactions, and we might use that codification as a first pass in trying to understand how things interact in nature. If the interaction between two species results in a benefit to both species, as in pollination, where the insect benefits nutritionally and the plant benefits reproductively, ecologists call this *mutualism* and often designate it as +/+. In cases in which one species benefits to the detriment of another (+/–), as when a lion eats a wildebeest or a parasite weakens a human, ecologists call this *exploitation* (predation and parasitism are subcategories of exploitation). *Antagonism* (–/–) is the term used for cases in which both species are harmed by the interaction, as when blue jays and crows compete for the same food sources, like berries and nuts. Although ecologists use the scheme of "mutualism, exploitation, and antagonism" only to describe interactions between species, anthropologists have applied the scheme to human cultures, too. It is worth considering here how it applies to *all* interactions between different entities in the environment. This is possible to do because even the non-living components of environmental systems have many characteristics of living systems. Consider, when reading about interactions between human groups, plant and animal species, and non-living components of the environment: who gains and who loses? Is the answer always clear-cut? To what extent is it valid to include non-living entities in the scheme?

The definition of "life" and "living system" has always been, and will always be, problematic and controversial. Are viruses alive? Are growing crystals? The answer will depend upon the suite of characteristics one chooses to include—there is no right or wrong answer. Under most definitions, water in a glass is not alive, nor is it a living system. However, moving water in a stream, though chemically identical to the water in the glass, is alive in many ways, as ecosystems are alive—the stream persists through time and maintains its identity though constantly changing; it responds dynamically to changes in its own environment, sometimes cyclically and predictably, sometimes in a highly idiosyncratic manner. As a corollary, only something that is alive, or at least shares many properties of living systems, has the potential to be killed. In this sense, all the subjects of the five chapters in this section hold together: species can be extirpated and ecosystems can be degraded; soil can be washed away or otherwise eroded so that it no longer can support forests, grasslands, or agricultural fields; aquatic systems can be so polluted that they no longer can sustain the species and ecosystems that once inhabited them; water sources for human sustenance can be poisoned or depleted to the point of uselessness; and air chemistry can be altered in ways that negatively affect every terrestrial organism and ecosystem. Finally, it is now *our* use of the resources of the Earth, mostly for our energy needs, and our transmogrification of those resources into poisons and waste, that has become the single greatest driving force for the alteration of every one of these living systems.

Although unscientific by modern standards, the ancients grasped a deep truth when they reckoned that there are four elements: earth, air, water, and fire (energy). The largest, most basic environmental systems, those that constitute and sustain all life, are directly interconnected and interact: soil, the atmosphere and climate, water, and energy use.

Our greatest hope to curtail the degradation of these systems is to understand them. We need to understand the biology, chemistry, and physics of the systems; to understand the interactions between their components; to know when they are acting mutualistically, exploitatively, or antagonistically; and, finally, to examine our motives in the degradation of these systems. Then we can identify options and choices for changing our actions in order to achieve a sustainable, healthy balance of all systems. Such an understanding is an interactive, interdisciplinary one, and it runs through every selection in this book.

Keystone Essay: Thomas Berry, from "The Ecozoic Era" (1997)

The changes presently taking place in human and earthly affairs are beyond any parallel with historical change or cultural modification as these have occurred in the past. This is not like the transition from the classical period to the medieval period or from the medieval to the modern period. These changes reach far beyond the civilizational process, beyond even the human process, into the biosystems and even the geological structures of the Earth itself.

There are only two other moments in the history of this planet that offer us some sense of what is happening. These two moments are the end of the Paleozoic Era 220 million years ago, when some ninety percent of all species living at that time were extinguished, and the terminal phase of the Mesozoic Era sixty-five million years ago, when there was also very extensive extinction.

Then, in the emerging Cenozoic Era the story of life on this planet flowed over into what could be called the lyric period of Earth history. The trees had come before this, the mammals already existed in a rudimentary form, the flowers had appeared perhaps thirty million years earlier. But in the Cenozoic Era, there was wave upon wave of life development, with the flowers, the birds, the trees, and the mammalian species particularly all leading to that luxuriant display of life upon Earth such as we have known it.

In more recent times, during the past million years this region of New England went through its different phases of glaciation, also its various phases of life development. New England's trees especially developed a unique grandeur. Possibly no other place on Earth has such color in its fall foliage as this region. It was all worked out during these past sixty-five million years. The songbirds we hear also came about in this long period.

Then we, the human inhabitants of the Earth, came into this region with all the ambivalences we bring with us. Not only here but throughout the planet we have become a profoundly disturbing presence. In this region and to the north in southern Quebec, the native maple trees are dying out in great numbers due to pollutants we have put into the atmosphere, the soil, and the water.

Their demise is largely a result of the carbon compounds we have loosed into the atmosphere through the use of fossil fuels, especially of petroleum, for our fuel and energy. Carbon is, as you know, the magical element. The whole life

Spring by Giuseppe Arcimboldo, 1573.

structure of the planet is based upon the element carbon. So long as the life process is guided by its natural patterns, the integral functioning of the Earth takes place. The wonderful variety expressed in marine life and land life, the splendor of the flowers and the birds and animals—all these could expand in their gorgeous coloration, in their fantastic forms, in their dancing movements, and in their songs and calls that echo over the world.

To accomplish all this, however, nature must find a way of storing immense quantities of carbon in the petroleum and coal deposits, also in the great forests. This process was worked out over some hundreds of millions of years. A balance was achieved, and the life systems of the planet were secure in the interaction of the air and the water and the soil with the inflowing energy from the sun.

But then we discovered that petroleum could produce such wonderful effects. It can be made into fertilizer to nourish crops; it can be spun into fabrics; it can fuel our internal combustion engines for transportation over the vast highway system we have built; it can produce an unlimited variety of plastic implements; it can run gigantic generators and produce power for lighting and heating of our buildings.

It was all so simple. We had no awareness of the deadly consequences that would result from the residue from our use of petroleum for all these purposes. Nor did we know how profoundly we would affect the organisms in the soil with our insistence that the patterns of plant growth be governed by artificial human demands met by petroleum-based fertilizers rather than by the spontaneous rhythms within the living world. Nor did we understand that biological systems are not that adaptable to the mechanistic processes we imposed upon them.

I do not wish to dwell on the devastation we have brought upon the Earth but only to make sure we understand the nature and the extent of what is happening. While we seem to be achieving magnificent things at the microphase level of our functioning, we are devastating the entire range of living beings at the macrophase level. The natural world is more sensitive than we have realized. Unaware of what we have done or its order of magnitude, we have thought our achievements to be of enormous benefit for the human process, but we now find that by disturbing the biosystems of the planet at the most basic level of their functioning, we have endangered all that makes the planet Earth a suitable place for the integral development of human life itself.

Our problems are primarily problems of macrophase biology. Macrophase biology, the integral functioning of the entire complex of biosystems of the planet, is something biologists have given almost no attention. Only with James Lovelock and some other more recent scientists have we even begun to think about this larger scale of life functioning. The delay is not surprising, for we are caught in the microphase dimensions of every phase of our human endeavor. This is true in law and medicine and in the other professions as well as in biology.

Macrophase biology is concerned with five basic spheres: land, water, air, life—and how these interact with one another to enable the planet Earth to be what it is—and a very powerful sphere: the human mind. Consciousness is certainly not limited to humans. Every living being has its own mode of consciousness. We must be aware, however, that consciousness is an analogous concept. It is qualitatively different in its various modes of expression. Consciousness can be regarded as the capacity for intimate presence of things to one another through knowledge and sensitive identity. But obviously the consciousness of a plant and the consciousness of an animal are qualitatively different, as are the consciousness of insects and the consciousness of birds or fish. Similarly, there is a difference in consciousness between fish and human: for the purposes of the fish, human modes of consciousness would be more a defect than an advantage. So too, tiger consciousness would be inappropriate for the bird.

It is also clear that the human mode of consciousness is capable of unique intrusion into the larger functioning of the planetary life systems. So powerful is this intrusion that the human has established an additional sphere that might be referred to as a technosphere, a way of controlling the functioning of the planet for the benefit of the human at the expense of the other modes of being. We might even consider that the technosphere in its subservience to industrial-commercial uses has become incompatible with the other spheres that constitute the basic functional context of the planet. . . .

Our present system, based on the plundering of the Earth's resources, is certainly coming to an end. It cannot continue. The industrial world on a global scale, as it functions presently, can be considered definitively bankrupt. There is no way out of the present recession [of 1991] within the context of our existing commercial-industrial processes. This recession is not only a financial recession or a human recession even. It is a recession of the planet itself. The Earth cannot sustain such an industrial system or its devastating technologies. In the future the industrial system will have its moments of apparent recovery, but these will be minor and momentary. The larger movement is toward dissolution. The impact of our present technologies is beyond what the Earth can endure.

Nature has its own technologies. The entire hydrological cycle can even be regarded as a huge engineering project, a project vastly greater than anything humans could devise with such beneficent consequences throughout the life systems of the planet. We can differentiate between an acceptable human technology and an unacceptable human technology quite simply: an acceptable one is compatible with the integral functioning of the technologies governing the natural systems; an unacceptable one is incompatible with the technologies of the natural world. . . .

It is awesome to consider how quickly events of such catastrophic proportions are happening. When I was born in 1914, there were only one and a half billion people in the world. Children of the present will likely live to see ten billion. The petrochemical age had hardly begun in my early decades. Now the planet is saturated with the residue from spent oil products. There were fewer than a million automobiles in the world when I was born. In my childhood the tropical rain forests were substantially intact; now they are devastated on an immense scale. The biological diversity of life forms was not yet threatened on an extensive scale. The ozone layer was still intact.

In evaluating our present situation, I submit that we have already terminated the Cenozoic Era of the geo-biological systems of the planet. Sixty-five million years of life development are terminated. Extinction is taking place throughout the life systems on a scale unequaled since the terminal phase of the Mesozoic Era.

A renewal of life in some creative context requires that a new biological period come into being, a period when humans would dwell upon the Earth in a mutually enhancing manner. This new mode of being of the planet I describe as the Ecozoic Era, the fourth in the succession of life eras thus far identified as the Paleozoic, the Mesozoic, and the Cenozoic. But when we propose that an Ecozoic Era is succeeding the Cenozoic, we must define the unique character of this emergent era.

I suggest the name "Ecozoic" as a better designation than "Ecological." Eco-logos refers to an *understanding* of the interaction of things. Eco-zoic is a more biological term that can be used to indicate the integral *functioning* of life systems in their mutually enhancing relations.

The Ecozoic Era can be brought into being only by the integral life community itself. If other periods have been designated by such names as "Reptilian" or "Mammalian," this Ecozoic period must be identified as the Era of the Integral Life Community. For this to emerge, there are special conditions required on the part of the human, for although this era cannot be an anthropocentric life period, it can come into being only under certain conditions that dominantly concern human understanding, choice, and action.

When we consider the conditions required of humans for the emergence of such an Ecozoic Era in Earth history, we might list these as follows:

The first condition is to understand that the universe is a communion of subjects, not a collection of objects. Every being has its own inner form, its own spontaneity, its own voice, its ability to declare itself and to be present to other components of the universe in a subject-to-subject relationship. Whereas this is true of every being in the universe, it is especially true of each component member of the Earth community. Each component of the Earth is integral with every other component. This is also true of the living beings of the Earth in their relations with one another. . . .

The second condition for entering the Ecozoic Era is a realization that the Earth exists, and can survive, only in its integral functioning. It cannot survive in fragments any more than any organism can survive in fragments. Yet the Earth is not a global sameness. It is a differentiated unity and must be sustained in the integrity and interrelations of its many bioregional contexts. This inner coherence of natural systems requires an immediacy of any human settlement with the life dynamics of the region. Within this region the human right to habitat must respect the right to habitat possessed by the other members of the life community. Only the full complex of life expression can sustain the vigor of any bioregion.

A third condition for entering the Ecozoic Era is recognition that the Earth is a onetime endowment. We do not know the quantum of energy contained in the Earth, its possibilities or its limitations. We must reasonably suppose that the Earth is subject to irreversible damage in the major patterns of its functioning and even to distortions in its possibilities of development. Although there was survival and further development after the great extinctions at the end of the Paleozoic and the Mesozoic Eras, life was not so highly developed as it is now. Nor were the very conditions of life at those times negated by such changes as we have wrought through our toxification of the planet. . . .

A fourth condition for entering the Ecozoic Era is a realization that the Earth is primary and humans are derivative. The present distorted view is that humans are primary and the Earth and its integral functioning only a secondary consideration—thus the pathology manifest in our various human institutions. The only acceptable way for humans to function effectively is by giving first consideration to the Earth community and then dealing with humans as integral members of that community. The Earth must become the primary concern of every human institution, profession, program, and activity, including economics. In economics the first consideration cannot be the human economy, because the human economy does not even exist prior to the Earth economy. Only if the Earth economy is functioning in some integral manner can the human economy be in any way effective. The Earth economy can survive the loss of its human component, but there is no way for the human economy to survive or prosper apart from the Earth economy. The absurdity has been to seek out a rising Gross National Product in the face of a declining Gross Earth Product. . . .

A fifth condition for the rise of the Ecozoic Era is to realize that there is a single Earth community. There is no such thing as a human community in any manner separate from the Earth community. The human community and the natural world will go into the future as a single integral community, or we will both experience disaster on the way. However differentiated in its modes of expression, there is only one Earth community—one economic order, one health system, one moral order, one world of the sacred. . . .

Earlier I mentioned five conditions for the integral emergence of the Ecozoic Era. Here I would continue with a sixth condition: that we understand fully and respond effectively to our own human role in this new era. For while the Cenozoic Era unfolded in its full splendor entirely apart from any role fulfilled by the human, almost nothing of major significance is likely to happen in the Ecozoic Era that humans will not be involved in. The entire pattern of Earth's functioning is being altered in this transition from the Cenozoic to the Ecozoic. We did not even exist until the major developments of the Cenozoic were complete. In the Ecozoic, however, the human will have a pervasive influence on almost everything that happens. We are approaching a critical watershed in the entire modality of Earth's functioning. Our positive power of creativity in the natural life systems is minimal; our power of negating is immense. Whereas we cannot make a blade of grass, there is liable not to be a blade of grass unless it is accepted, fostered, and protected by the human. Protected mainly from ourselves so that Earth can function from within its own dynamism.

11 Biodiversity and Conservation Biology

Every rock has its own environment, as does every molecule of water. In the rock's case, interaction with its environment influences its size, form, and location. But even though every non-living thing on Earth has an environment, when we think of the word "environment," we most often think of it in relation to living organisms, or populations or communities of living organisms. The most pressing aspect of the study of these organisms and their ecosystems is their endangerment and loss. Species are becoming endangered and suffering extinction, ecosystems are being degraded, and genetic diversity is diminishing at a historically accelerated rate, mostly as a result of direct or indirect human activity. The discipline that concerns itself with this loss is conservation biology.

The term "biodiversity" usually suggests counting species, frequently measured either as the number of different species present in a particular ecosystem, as in the claim that "the tropical rainforest has more biodiversity than any other kind of ecosystem on Earth," or as the number of species in a larger, more inclusive taxonomic group, as in the statement that "there is greater biodiversity of insects than mammals."

However, biodiversity is not that simple. Ecosystems can be degraded and ecosystem types can be lost without the extinction or endangerment of any particular species. The loss of diversity of ecosystems themselves must also be considered a loss of biodiversity. Moreover, genetic variability within one species (or population) may be diminished without the immediate loss or endangerment of that species. This too is loss of biodiversity. Thus, biodiversity can be threatened at many different levels. In fact, its most common definition, best phrased by the United States Office of Technology Assessment (OTA), is:

> Biological diversity refers to the variety and variability among living organisms and the ecological complexes in which they occur. Diversity can be defined as the number of different items and their relative frequency. For biological diversity, these items are organized at many levels, ranging from complete ecosystems to the chemical structures that are the molecular basis of heredity. Thus, the term encompasses different ecosystems, species, genes, and their relative abundance.

Figure 11.1 *The Entry of the Animals into Noah's Ark* by Jan Brueghel the Elder, 1613. Oil on panel, 54.6 × 83.8 cm (21.5 × 33 in.). The J. Paul Getty Museum, Los Angeles.

Yes, all that and yet more. Biodiversity, in theory and in practice, is even broader than this OTA definition. For between ecosystems, species, and genes, there are other entities: populations, chromosomes, ecological guilds, and higher taxonomic groups, such as genera and families. And then there are all the interactions between all of the entities at all of the levels. The section on biotechnology and genetically manipulated organisms (see Chapter 4) presents the monarch migration, an endangered *biological phenomenon.*

Both "ecosystem" and "species" are wonderfully rich concepts, and both are susceptible to a range of meanings. In order to conserve biodiversity, we must understand as much about ecosystems and species as we can, including the broad set of responses that people have to them and the different definitions that they engender. For example, despite efforts spanning more than two centuries, biologists still can disagree about what, exactly, constitutes a species. "Diversity begets diversity" is not just an ecological law—it is a psychological response of humans to nature, and the following selections present a starting point for exploring diversity in both senses.

FURTHER READING

Botkin, Daniel. *Discordant Harmonies.* Oxford: Oxford University Press, 1990.

Mayden, R. L. "A Hierarchy of Species Concepts: The Denouement in the Saga of the Species Problem," pages 381–424 in M. F. Claridge, H. A. Dawah, and M. R. Wilson, ed., *Species: The Units of Biodiversity.* London: Chapman & Hall, 1997.

Perlman, Dan, and Glenn Adelson. *Biodiversity: Exploring Values and Priorities in Conservation.* Malden, Mass.: Blackwell Science, 1997.
Peters, R. H. *A Critique for Ecology.* Cambridge: Cambridge University Press, 1991.
Shrader-Frechette, K. S., and E. D. McCoy. *Method in Ecology.* Cambridge: Cambridge University Press, 1993.

The Bible, Genesis 1:20–31

Genesis 1:20–28 epitomizes the theology of human separation from and dominion over nature often blamed as a root cause of the West's sorry ecological record. Verses 29–31 then arguably define this dominion as a stewardship resembling God's own love for creation. Regardless of one's propositional belief in the following story, the lesson that can be drawn from these verses is that our planet is the reservoir of a remarkable amount of biodiversity and "it is very good."

[20] And God said, "Let the waters bring forth swarms of living creatures, and let birds fly above the earth across the firmament of the heavens."
[21] So God created the great sea monsters and every living creature that moves, with which the waters swarm, according to their kinds, and every winged bird according to its kind. And God saw that it was good.
[22] And God blessed them, saying, "Be fruitful and multiply and fill the waters in the seas, and let birds multiply on the earth."
[23] And there was evening and there was morning, a fifth day.
[24] And God said, "Let the earth bring forth living creatures according to their kinds: cattle and creeping things and beasts of the earth according to their kinds." And it was so.
[25] And God made the beasts of the earth according to their kinds and the cattle according to their kinds, and everything that creeps upon the ground according to its kind. And God saw that it was good.
[26] Then God said, "Let us make man in our image, after our likeness; and let them have dominion over the fish of the sea, and over the birds of the air, and over the cattle, and over all the earth, and over every creeping thing that creeps upon the earth."
[27] So God created man in his own image, in the image of God he created him; male and female he created them.
[28] And God blessed them, and God said to them, "Be fruitful and multiply, and fill the earth and subdue it; and have dominion over the fish of the sea and over the birds of the air and over every living thing that moves upon the earth."
[29] And God said, "Behold, I have given you every plant yielding seed which is upon the face of all the earth, and every tree with seed in its fruit; you shall have them for food.

Figure 11.2 *Adam and Eve* by Lucas Cranach the Elder, 1530. Courtauld Institute of Art Gallery, London.

[30] And to every beast of the earth, and to every bird of the air, and to everything that creeps on the earth, everything that has the breath of life, I have given every green plant for food." And it was so.
[31] And God saw everything that he had made, and behold, it was very good. And there was evening and there was morning, a sixth day.

Peter M. Vitousek, Harold A. Mooney, Jane Lubchenco, and Jerry M. Mellilo, from "Human Domination of Earth's Ecosystems" (1997)

The most important part of ecosystem conservation is knowing how people have changed the world's ecosystems. Agriculture and urbanization have modified, degraded, and often destroyed ecosystems directly, but human enterprise has had an additional, indirect effect

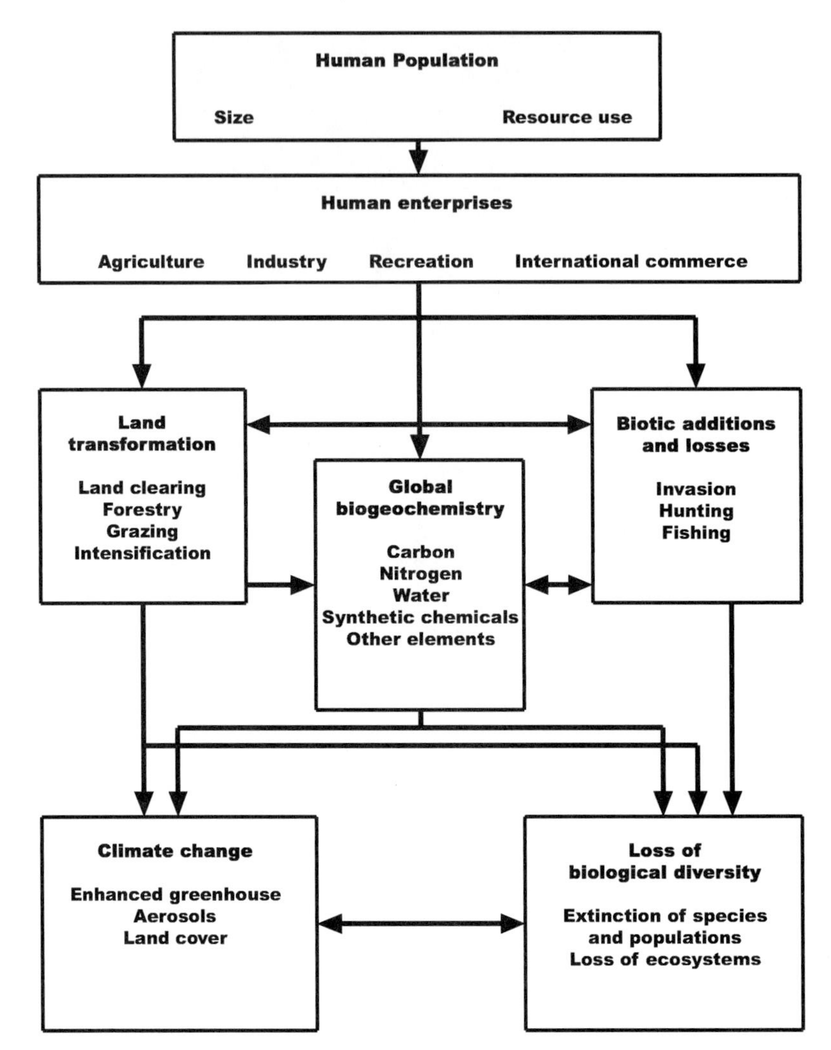

Figure 11.3 Schematic of human domination of the Earth system.

on all of the Earth's ecosystems through pollution, water use, and species movement. "Human Domination of Earth's Ecosystems" introduced a special issue of the journal Science *devoted to the world's ecosystems.*

All organisms modify their environment, and humans are no exception. As the human population has grown and the power of technology has expanded, the scope and nature of this modification has changed drastically. Until recently, the term "human-dominated ecosystems" would have elicited images of agricultural fields, pastures, or urban landscapes; now it applies with greater or lesser force to all of Earth. Many ecosystems are dominated directly by humanity, and no ecosystem on Earth's surface is free of pervasive human influence.

This article . . . explore[s] how large humanity looms as a presence on the globe—how, even on the grandest scale, most aspects of the structure and functioning of Earth's ecosystems cannot be understood without accounting for the strong, often dominant influence of humanity.

We view human alterations to the Earth system as operating through the interacting processes summarized in Figure 11.3. The growth of the human population, and growth in the resource base used by humanity, is maintained by a suite of human enterprises such as agriculture, industry, fishing, and international commerce. These enterprises transform the land surface (through cropping, forestry, and urbanization), alter the major biogeochemical cycles, and add or remove species and genetically distinct populations in most of Earth's ecosystems. Many of these changes are substantial and reasonably well quantified; all are ongoing. These relatively well-documented changes in turn entrain further alterations to the functioning of the Earth system, most notably by driving global climatic change and causing irreversible losses of biological diversity.

LAND TRANSFORMATION

The use of land to yield goods and services represents the most substantial human alteration of the Earth system. Human use of land alters the structure and functioning of ecosystems, and it alters how ecosystems interact with the atmosphere, with aquatic systems, and with surrounding land. Moreover, land transformation interacts strongly with most other components of global environmental change.

The measurement of land transformation on a global scale is challenging; changes can be measured more or less straightforwardly at a given site, but it is difficult to aggregate these changes regionally and globally. In contrast to analyses of human alteration of the global carbon cycle, we cannot install instruments on a tropical mountain to collect evidence of land transformation. Remote sensing is a most useful technique, but only recently has there been a serious scientific effort to use high-resolution civilian satellite imagery to evaluate even the more visible forms of land transformation, such as deforestation, on continental to global scales.

Land transformation encompasses a wide variety of activities that vary substantially in their intensity and consequences. At one extreme, 10 to 15 percent of Earth's land surface is occupied by row-crop agriculture or by urban-industrial areas, and another 6 to 8 percent has been converted to pastureland;[1] these systems are wholly changed by human activity. At the other extreme, every terrestrial ecosystem is affected by increased atmospheric carbon dioxide (CO_2), and most ecosystems have a history of hunting and other low-intensity resource extraction. Between these extremes lie grassland and semiarid ecosystems that are

Figure 11.4 Human domination or alteration of several major components of the Earth system, expressed as (from left to right) percentage of the land surface transformed; percentage of the current atmospheric carbon dioxide concentration that results from human action; percentage of accessible surface fresh water used; percentage of terrestrial nitrogen fixation that is human-caused; percentage of plant species in Canada that humanity has introduced from elsewhere; percentage of bird species on Earth that have become extinct in the past two millennia, almost all of them as a consequence of human activity; and percentage of major marine fisheries that are fully exploited, overexploited, or depleted.

grazed (and sometimes degraded) by domestic animals, and forests and woodlands from which wood products have been harvested; together, these represent the majority of Earth's vegetated surface.

The variety of human effects on land makes any attempt to summarize land transformations globally a matter of semantics as well as substantial uncertainty. Estimates of the fraction of land transformed or degraded by humanity (or its corollary, the fraction of the land's biological production that is used or dominated) fall in the range of 39 to 50 percent[2] (Figure 11.4). These numbers have large uncertainties, but the fact that they are large is not at all uncertain. Moreover, if anything these estimates understate the global impact of land transformation, in that land that has not been transformed often has been divided into fragments by human alteration of the surrounding areas. This fragmentation affects the species composition and functioning of otherwise little modified ecosystems.

Overall, land transformation represents the primary driving force in the loss of biological diversity worldwide. Moreover, the effects of land transformation extend far beyond the boundaries of transformed lands. Land transformation can affect climate directly at local and even regional scales. It contributes ~20 percent to current anthropogenic [human-caused] CO_2 emissions, and more substantially to the increasing concentrations of the greenhouse gases methane and nitrous oxide; fires associated with it alter the reactive chemistry of the troposphere, bringing elevated carbon monoxide concentrations and episodes of urban-like photochemical air pollution to remote tropical areas of Africa and South America; and it causes runoff of sediment and nutrients that drive substantial changes in stream, lake, estuarine, and coral reef ecosystems.

The central importance of land transformation is well recognized within the community of researchers concerned with global environmental change. Several research programs are focused on aspects of it; recent and substantial progress toward understanding these aspects has been made, and much more progress can be anticipated. Understanding land transformation is a difficult challenge; it requires integrating the social, economic, and cultural causes of land transformation with evaluations of its biophysical nature and consequences. This interdisciplinary approach is essential to predicting the course, and to any hope of affecting the consequences, of human-caused land transformation. . . .

BIOTIC CHANGES

Human modification of Earth's biological resources—its species and genetically distinct populations—is substantial and growing. Extinction is a natural process, but the current rate of loss of genetic variability, of populations, and of species is far above background rates; it is ongoing and it represents a wholly irreversible global change. At the same time, human transport of species around Earth is homogenizing Earth's biota, introducing many species into new areas where they can disrupt both natural and human systems.

Losses. Rates of extinction are difficult to determine globally, in part because the majority of species on Earth have not yet been identified. Nevertheless, recent calculations suggest that rates of species extinction are now on the order of 100 to 1000 times those before humanity's dominance of Earth.[3] For particular well-known groups, rates of loss are even greater; as many as one-quarter of Earth's bird species have been driven to extinction by human activities over the past two millennia, particularly on oceanic islands[4] (Figure 11.4). At present, 11 percent of the remaining birds, 18 percent of the mammals, 5 percent of fish, and 8 percent of plant species on Earth are threatened with extinction.[5] There has been a disproportionate loss of large mammal species because of hunting; these species played a dominant role in many ecosystems, and their loss has resulted in a fundamental change in the dynamics of those systems, one that could lead to further extinctions. The largest organisms in marine systems have been affected similarly, by fishing and whaling. Land transformation is the single most important cause of extinction, and current rates of land transformation eventually will drive many more species to extinction, although with a time lag that masks the true dimensions of the crisis. Moreover, the effects of other components of global environmental change—of altered carbon and nitrogen cycles, and of anthropogenic climate change—are just beginning.

As high as they are, these losses of species understate the magnitude of loss of genetic variation. The loss to land transformation of locally adapted populations within species, and of genetic material within populations, is a human-caused change that reduces the resilience of species and ecosystems while precluding human use of the library of natural products and genetic material that they represent.

Although conservation efforts focused on individual endangered species have yielded some successes, they are expensive—and the protection or restoration of whole ecosystems often represents the most effective way to sustain genetic, population, and species diversity [see Noss et al., below]. Moreover, ecosystems themselves may play important roles in both natural and human-dominated landscapes. For example, mangrove ecosystems protect coastal areas from erosion and provide nurseries for offshore fisheries, but they are threatened by land transformation in many areas.

Invasions. In addition to extinction, humanity has caused a rearrangement of Earth's biotic systems, through the mixing of floras and faunas that had long been isolated geographically. The magnitude of transport of species, termed "biological invasion," is enormous; invading species are present almost everywhere. On many islands, more than half of the plant species are nonindigenous, and in many continental areas the figure is 20 percent or more (see Figure 11.4).[6]

As with extinction, biological invasion occurs naturally—and as with extinction, human activity has accelerated its rate by orders of magnitude [multiples of ten]. Land transformation interacts strongly with biological invasion, in that human-altered ecosystems generally provide the primary foci for invasions, while in some cases land transformation itself is driven by biological invasions. International commerce is also a primary cause of the breakdown of biogeographic barriers; trade in live organisms is massive and global, and many other organisms are inadvertently taken along for the ride. In freshwater systems, the combination of upstream land transformation, altered hydrology, and numerous deliberate and accidental species introductions has led to particularly widespread invasion, in continental as well as island ecosystems.

In some regions, invasions are becoming more frequent. For example, in the San Francisco Bay of California, an average of one new species has been established every 36 weeks since 1850, every 24 weeks since 1970, and every 12 weeks for the last decade.[7] Some introduced species quickly become invasive over large areas (for example, the Asian clam in the San Francisco Bay), whereas others become widespread only after a lag of decades, or even over a century.[8]

Many biological invasions are effectively irreversible; once replicating biological material is released into the environment and becomes successful there, calling it back is difficult and expensive at best. Moreover, some species introductions have consequences. Some degrade human health and that of other species; after all, most infectious diseases are invaders over most of their range. Others have caused economic losses amounting to billions of dollars; the recent invasion of North America by the zebra mussel is a well-publicized example. Some disrupt ecosystem processes, altering the structure and functioning of whole ecosystems. Finally, some invasions drive losses in the biological diversity of native species and populations; after land transformation, they are the next most important cause of extinction.

CONCLUSIONS

The global consequences of human activity are not something to face in the future—as Figure 11.4 illustrates, they are with us now. All of these changes are ongoing, and in many cases accelerating; many of them were entrained long before their importance was recognized. Moreover, all of these seemingly disparate phenomena trace to a single cause—the growing scale of the human enterprise. The rates, scales, kinds, and combinations of changes occurring now are fundamentally different from those at any other time in history; we are changing Earth more rapidly than we are understanding it. We live on a human-dominated planet—and the momentum of human population growth, together with the imperative for further economic development in most of the world, ensures that our dominance will increase. . . .

Recognition of the global consequences of the human enterprise suggests three complementary directions. First, we can work to reduce the rate at which we alter the Earth system. Humans and human-dominated systems may be able to adapt to slower change, and ecosystems and the species they support may cope more effectively with the changes we impose, if those changes are slow. Our footprint on the planet might then be stabilized at a point where enough space and resources remain to sustain most of the other species on Earth, for their sake and our own [see Chapter 5]. Reducing the rate of growth in human effects on Earth involves slowing human population growth and using resources as efficiently as is practical. Often it is the waste products and by-products of human activity that drive global environmental change.

Second, we can accelerate our efforts to understand Earth's ecosystems and how they interact with the numerous components of human-caused global change. Ecological research is inherently complex and demanding: it requires measurement and monitoring of populations and ecosystems; experimental studies to elucidate the regulation of ecological processes; the development, testing, and validation of regional and global models; and integration with a broad range of biological, earth, atmospheric, and marine sciences. The challenge of understanding a human-dominated planet further requires that the human dimensions of global change—the social, economic, cultural, and other drivers of human actions—be included within our analyses.

Finally, humanity's dominance of Earth means that we cannot escape responsibility for managing the planet. Our activities are causing rapid, novel, and substantial changes to Earth's ecosystems. Maintaining populations, species, and ecosystems in the face of those changes, and maintaining the flow of goods and services they provide humanity, will require active management for the foreseeable future. There is no clearer illustration of the extent of human dominance of Earth than the fact that maintaining the diversity of "wild" species and the functioning of "wild" ecosystems will require increasing human involvement.

Notes

1. Olson, J. S., J. A. Watts, and L. J. Allison, *Carbon in Live Vegetation of Major World Ecosystems* (Washington, D.C.: Office of Energy Research, U.S. Department of Energy, 1983).
2. Vitousek, P. M., P. R. Ehrlich, A. H. Ehrlich, and P. A. Matson, *Bioscience* 36 (1986): 368; Kater, R. W., B. L. Turner, and W. C. Clark, in Turner B. L. et al., *The Earth as Transformed by Human Action* (Cambridge: Cambridge University Press, 1990); Daily, G. C., *Science* 269 (1995): 350.
3. Lawton, J. H. and R. M. May, ed., *Extinction Rates* (Oxford: Oxford University Press, 1995); Pimm, S. L., G. J. Russell, J. T. Gittleman, and T. Brooks, *Science* 269 (1995): 347.
4. Olson, S. L., in Western, D. and M. C. Pearl, ed., *Conservation for the Twenty-First Century* (Oxford: Oxford University Press, 1989): 50; Steadman, D. W., *Science* 267 (1995): 1123.
5. Barbault, R. and S. Sastrapradja, in Heywood, V. H., ed., *Global Biodiversity Assessment* (Cambridge: Cambridge University Press, 1995).
6. Rejmánek, M. and J. Randall, *Madrono* 41 (1994): 161.
7. Cohen, A. N. and J. T. Carlton, *Biological Study: Nonindigenous Aquatic Species in a United States Estuary* (Washington, D.C.: U.S. Fish and Wildlife Service, 1995).
8. Kowarik, I. in Pysek, P., K. Prach, M. Rejmánek, and M. Wade, ed., *Plant Invasions—General Aspects and Special Problems* (Amsterdam: SPB Academic, 1995): 15.

Charles Darwin, from "Galapagos Archipelago" in *Voyage of the Beagle* (1845)

During his voyage on the Beagle, *Charles Darwin first observed the diversity that initiated ideas leading to his theory of evolution by natural selection. These observations are recorded in this selection on the Galapagos Islands, where he first noticed the variation in finch (*Geospiza*) species and reflected on the similarity of the Galapagos tortoise to the tortoise known on the Indian Ocean island of Maritius. Only because Darwin had such a vast knowledge of the whole of the diversity of life was he able to see such similarities and differences—the generation of an idea as powerful as natural selection is dependent on the accumulated knowledge of particulars.*

The natural history of [the Galapagos] islands is eminently curious, and well deserves attention. Most of the organic productions are aboriginal creations, found nowhere else; there is even a difference between the inhabitants of the different islands; yet all show a marked relationship with those of America, though separated from that continent by an open space of ocean, between 500 and 600

Figure 11.5 Charles Darwin, pictured on the British ten-pound note.
Courtesy of the Bank of England.

miles in width. The archipelago is a little world within itself, or rather a satellite attached to America, whence it has derived a few stray colonists, and has received the general character of its indigenous productions. Considering the small size of these islands, we feel the more astonished at the number of their aboriginal beings, and at their confined range. Seeing every height crowned with its crater, and the boundaries of most of the lava-streams still distinct, we are led to believe that within a period, geologically recent, the unbroken ocean was here spread out. Hence, both in space and time, we seem to be brought somewhat near to that great fact—that mystery of mysteries—the first appearance of new beings on this Earth. . . .

Of land-birds I obtained twenty-six kinds, all peculiar to the group and found nowhere else, with the exception of one lark-like finch from North America (*Dolichonyx oryzivorous*), which ranges on that continent as far north as 54°, and generally frequents marshes. The other twenty-five birds consist, firstly, of a hawk, curiously intermediate in structure between a buzzard and the American group of carrion-feeding Polybori; and with these latter birds it agrees most closely in every habit and even tone of voice. Secondly, there are two owls, representing the short-eared and white barn-owls of Europe. Thirdly, a wren, three tyrant fly-catchers (two of them species of *Pyrocephalus*, one or both of which would be ranked by some ornithologists as only varieties), and a dove—all analogous to, but distinct from, American species. Fourthly, a swallow, which though differing from the *Progne purpurea* of both Americas, only in being rather duller colored, smaller, and slenderer, is considered by Mr. Gould as specifically distinct. Fifthly, there are three species of mocking-thrush—a form highly characteristic of America. The remaining land-birds form a most singular group of finches, related to each other in the structure of their beaks, short tails, form of body, and plumage: there are thirteen species, which Mr. Gould has divided into four subgroups. All these species are peculiar to this archipelago; and so is the whole group, with the exception of one species of the subgroup Cactornis, lately brought from Bow Island, in the Low Archipelago. Of Cactornis, the two species may be often seen climbing about the flowers of the great cactus-trees; but all the other species of this group of finches, mingled together in flocks, feed on the dry and sterile ground of the lower districts. The males of all, or certainly of the greater number, are jet black; and the females (with perhaps one or two exceptions) are brown. The most curious fact is the perfect gradation in the size of the beaks in the different species of *Geospiza*, from one as large as that of a hawfinch to that of a chaffinch, and . . . even to that of a warbler. The beak of Cactornis is somewhat like that of a starling; and that of fourth subgroup, Camarhynchus, is slightly parrot-shaped. Seeing this gradation and diversity of structure in one small, intimately related group of birds, one might really fancy that from an original paucity of birds in this archipelago, one species had been taken and modified for different ends. In a like manner it might be fancied that a bird originally a buzzard, had been induced here to undertake the office of the carrion-feeding Polybori of the American continent. . . .

But it is the circumstance, that several of the islands possess their own species of the tortoise, mocking-thrush, finches, and numerous plants, these species having the same general habits, occupying analogous situations, and obviously filling the same place in the economy of this archipelago, that strikes me with wonder. It may be suspected that some of these representative species, at least in the case of the tortoise and of some of the birds, may hereafter prove to be only well-marked races; but this would be of equally great interest to the philosophical naturalist. . . . I must repeat, that neither the nature of the soil, nor height of the land, nor the climate, nor the general character of the associated beings, and therefore their action one on another, can differ much in the different islands. If there be any sensible difference in their climates, it must be between the windward group (namely Charles and Chatham Islands), and that to leeward; but there seems to be no corresponding difference in the productions of these two halves of the archipelago.

The only light which I can throw on this remarkable difference in the inhabitants of the different islands is that very strong currents of the sea running in a westerly and WNW direction must separate, as far as transportal by the sea is concerned, the southern islands from the northern ones; and between these northern islands a strong NW current was observed, which must effectually separate James and Albemarle Islands. As the archipelago is free to a most remarkable degree from gales of wind, neither the birds, insects, nor lighter seeds would be blown from island to island. And lastly, the profound depth of the ocean between the islands, and their apparently recent (in a geological sense) volcanic origin, render it highly unlikely that they were ever united; and this, probably, is a far more important consideration than any other with respect to the geographical distribution of their inhabitants. Reviewing the facts here given, one is astonished at the amount of creative force, if such an expression may be used, displayed on these small, barren, and rocky islands; and still more so at its diverse yet analogous action on points so near each other. I have said that the Galapagos Archipelago might be called a satellite attached to America, but it should rather be called a group of satellites, physically similar, organically distinct, yet intimately related to each other, and all related in a marked, though much lesser degree, to the great American continent.

Edward O. Wilson, from "Biodiversity Reaches the Peak" in *The Diversity of Life* (1992)

Edward O. Wilson presents a narrative tour of the last 540 million years, speculating on which events in which eras provided the raw material for biological diversification. His book The Diversity of Life, *from which this selection is taken, provides both an evolutionary biologist's description of the generation of diversity and a conservationist's concern about its protection. Wilson's conclusion that over hundreds of millions of years "the survival rate of species . . . is 1 in 2000," means that each species today is infinitely valuable, as the sole survivor of that vast original stock.*

Approximately 540 million years ago, near the beginning of the Cambrian period, earliest of the time segments of the Phanerozoic eon in which we now live, a seminal event occurred in the history of life. Animals increased in size and diversified explosively. The supply of free oxygen in the atmosphere was by this time near the 21 percent level of today. The two trends are probably linked, for the simple reason that large, active animals need aerobic respiration and a rich supply of oxygen. Within a few million years, the fossil record held almost every modern phylum of invertebrate animals a millimeter or more in length and possessed of skeletal structures, hence easily preserved and detectable later. A large portion of present-day classes and orders had also come on stage. Thus occurred the Cambrian explosion, the big bang of animal evolution. Bacteria and single-celled organisms had long since attained comparable levels of biochemical sophistication. Now, in a dramatic new radiation, they augmented their niches to include life on the bodies and waste materials of the newly evolved animals. They created a new, microscopic suzerainty of pathogens, symbionts, and decomposers. In broad outline at least, life in the sea attained an essentially modern aspect no later than 500 million years ago.

By this time a strong ozone layer existed as well, screening out lethal short-wave radiation. The intertidal reaches and dry land were safe for life. By the late Ordovician period, 450 million years ago, the first plants, probably derived from multicellular algae, invaded the land. The terrain was generally flat, lacking mountains, and mild in climate. Animals soon followed: invertebrates of still unknown nature burrowed and tunneled through the primitive soil. Paleontologists have found their trails but still no bodies. Within 50 or 60 million years, early into the Devonian period, the pioneer plants had formed thick mats and low shrubbery widely distributed over the continents. The first spiders, mites, centipedes, and insects swarmed there, small animals truly engineered for life on the land. They were followed by the amphibians, evolved from lobe-finned fishes, and a burst of land vertebrates, relative giants among land animals, to inaugurate the Age of Reptiles. Next came the Age of Mammals and finally the Age of Man, amid continuing tumultuous change at the level of class and order.

By 340 million years before the present, the pioneer vegetation had given way to the coal forests, dominated by towering lycophyte trees, seed ferns, tree horsetails, and a great variety of ferns. Life was close to the attainment of its maximal biomass. More organic matter was invested in organisms than ever before. The forests swarmed with insects, including dragonflies, beetles, and cockroaches. By late Paleozoic and early Mesozoic times, close to 240 million years ago, most of the coal vegetation had died out, with the exception of the ferns. Dinosaurs arose among a newly constituted, mostly tropical vegetation of ferns, conifers, cycads, and cycadeoids. From 100 million years on, the flowering plants swept to domination of the land vegetation, reconstituting the forests and grasslands of the world. The dinosaurs died out during the hegemony of this essentially modern vegetation, at a time when tropical rain forests were assembling the greatest concentration of biodiversity of all time.

For the past 600 million years, the thrust of biodiversity, mass extinction episodes notwithstanding, has been generally upward. In the sea the orders of marine animals climbed slightly above a hundred during the Cambrian and Ordovician periods and stayed in place for the remaining 450 million years. Families, genera, and species closely traced the same pattern to the end of the Paleozoic era, 245 million years ago. They were knocked down sharply by the extinction catastrophe that closed the Paleozoic era, followed in only 50 million years by a smaller spasm in the Triassic period. Thereafter they climbed steeply, with a dip at the end of the Mesozoic, reaching unprecedented levels during the past several million years. Plant and animal diversity on the land, after a delay of 100 million years during which colonization took place, followed the same trajectory to our time.

Each extinction spasm reduced the numbers of species most and the numbers of classes and phyla least. The lower the taxonomic category, the more it was diminished. At the end of the Paleozoic era, as many as 96 percent of the species of marine animals and foraminiferans vanished, compared with 78 to 84 percent of the genera and 54 percent of the families. Apparently no phyla came to an end.

This descending vulnerability by taxonomic rank is an artifact, a straightforward consequence of the hierarchical manner in which biologists classify organisms. But it is an interesting and technically useful artifact, for reasons most clearly expressed by what has been called the Field of Bullets Scenario. Imagine advance infantrymen walking forward in eighteenth-century manner, posting arms in one serried rank into a field of fire. Each man represents a species, belonging to a platoon (genus), which is a unit of a company (family), in turn a unit of a battalion (order), and so on up to the corps (phylum). Each man has the same chance each moment of being hit by a bullet. When he falls the species he represents is extinguished, but other members of the platoon-genus march on, so that even though diminished in size the genus survives. In time all the members of the genus may perish, yet remnants of other genera are still standing down the line, and so the company-family presses forward. At the end of the long lethal march, the vast majority of species, genera, families, orders, and even classes may be gone, but so long as one species out of the multitude remains alive the phylum-corps survives.

Across 600 million years of Phanerozoic evolution, the turnover of species was nearly total. More than 99 percent of all species that ever lived in each period perished, to be replaced by even larger numbers drawn from the descendants of the survivors. Such is the nature of dynastic successions through the history of life, often initiated by the extinction spasms that bring down entire company-families and battalion-orders. Ninety-nine percent is not a surprising figure. Imagine a group such as the archaic amphibians of the Paleozoic. A thousand species die, and one survives to produce the primitive reptiles. A thousand of those also die, but one survivor carries on to become the ancestor of the Mesozoic dinosaurs. The survival rate of species in this sequence is 1 in 2,000. In other words, only a single line persists out of the 2,000 created, and yet life flourishes as diversely as before.

William Cronon, from "The View from Walden" in *Changes in the Land* (1983)

"Ecosystem" can be defined as the dynamic collection of living organisms interacting among themselves and the abiotic (non-living) environment in which they exist. Ecosystem boundaries are usually both fuzzy and variable: an ecosystem can be as small as the intestines of a deer or the hollow of a dead tree and as large as the planet. In "The View from Walden," William Cronon takes on two other questions concerning ecosystems boundaries: are ecosystems bounded in time as well as in space, and do they include people within them?

When one asks how much an ecosystem has been changed by human influence, the inevitable next question must be: "changed in relation to what?" There is no simple answer to this. Before we can analyze the ways people alter their environments, we must first consider how those environments change in the absence of human activity, and that in turn requires us to reflect on what we mean by an ecological "community." Ecology as a biological science has had to deal with this problem from its outset. The first generation of academic ecologists, led by Frederic Clements, defined the communities they studied literally as superorganisms which experienced birth, growth, maturity, and sometimes death much as individual plants and animals did. Under this model, the central dynamic of community change could be expressed in the concept of "succession." Depending on its region, a biotic community might begin as a pond, which was then gradually transformed by its own internal dynamics into a marsh, a meadow, a forest of pioneer trees, and finally to a forest of dominant trees. This last stage was assumed to be stable and was known as the "climax," a more or less permanent community that would reproduce itself indefinitely if left undisturbed. Its equilibrium state defined the mature forest "organism," so that all members of the community could be interpreted as functioning to maintain the stability of the whole. Here was an apparently objective point of reference: any actual community could be compared with the theoretical climax, and differences between them could then usually be attributed to "disturbance." Often the source of disturbance was human, implying that humanity was somehow outside of the ideal climax community.

This functionalist emphasis on equilibrium and climax had important consequences, for it tended to remove ecological communities from history. If all ecological change was either self-equilibrating (moving toward climax) or nonexistent (remaining in the static condition of climax), then history was more or less absent except in the very long time frame of climatic change or Darwinian evolution. The result was a paradox. Ecologists trying to define climax and succession for a region like New England were faced with an environment massively altered by human beings, yet their research program demanded that they determine what that environment would have been like without a human presence. By peeling away the corrupting influences of man and woman, they could discover the original ideal community of the climax. One detects here a certain resemblance to Thoreau's reading of William Wood: historical change was defined as an aberration rather than the norm.

In time, the analogy comparing biotic communities to organisms came to be criticized for being both too monolithic and too teleological [oriented to one final end]. The model forced one to assume that any given community was gradually working either to become or to remain a climax, with the result that the dynamics of nonclimax communities were too easily ignored. For this reason, ecology by the mid-twentieth century had abandoned the organism metaphor in favor of a less teleological "ecosystem." Now individual species could simply be described in terms of their associations with other species along a continuous range of environments; there was no longer any need to resort to functional analysis in describing such associations. Actual relationships rather than mystical superorganisms could become the focus of study, although an infusion of theory from cybernetics encouraged ecologists to continue their interest in the self-regulating, equilibrating characteristics of plant and animal populations.

With the imperatives of the climax concept no longer so strong, ecology was prepared to become at least in part a historical science, for which change was less

Figure 11.6 Walden Pond. Courtesy of Bonnie McGrath.

the result of "disturbance" than of the ordinary processes whereby communities maintained and transformed themselves. Ecologists began to express a stronger interest in the effects of human beings on their environment. What investigators had earlier seen as an inconvenient block to the discovery of ideal climax communities could become an object of research in its own right. But accepting the effects of human beings was only part of this shift toward a more historical ecology. Just as ecosystems have been changed by the historical activities of human beings, so too have they had their own less-recorded history: forests have been transformed by disease, drought, and fire, species have become extinct, and landscapes have been drastically altered by climatic change without any human intervention at all. . . . The period of human occupation in postglacial New England has seen environmental changes on an enormous scale, many of them wholly apart from human influence. There has been no timeless wilderness in a state of perfect changelessness, no climax forest in permanent stasis.

But admitting that ecosystems have histories of their own still leaves us with the problem of how to view the people who inhabit them. Are human beings inside or outside their systems? In trying to answer this question, appeal is too often made to the myth of a golden age, as Thoreau sometimes seemed inclined to do. If the nature of Concord in the 1850s—a nature which many Americans now romanticize as the idyllic world of Thoreau's own Walden—was as "maimed" and "imperfect" as he said, what are we to make of the wholeness and perfection which he thought preceded it? It is tempting to believe that when the Europeans arrived in the New World they confronted Virgin Land, the Forest Primeval, a wilderness which had existed for eons uninfluenced by human hands. Nothing could be further from the truth. In Francis Jennings's telling phrase, the land was less virgin than it was widowed. Indians had lived on the continent for thousands of years, and had to a significant extent modified its environment to their purposes. The destruction of Indian communities in fact brought some of the most important ecological changes which followed the Europeans' arrival in America. The choice is not between two landscapes, one with and one without a human influence; it is between two human ways of living, two ways of belonging to an ecosystem.

The riddle . . . is to explore why these different ways of living had such different effects on New England ecosystems. A group of ecological anthropologists has tried to argue that for many non-Western societies, like those of the New England Indians, various ritual practices have served to stabilize people's relationships with their ecosystems. In effect, culture in this anthropological model becomes a homeostatic, self-regulating system much like the larger ecosystem itself. Thus have come the now famous analyses designed to show that the slaughter of pigs in New Guinea, the keeping of sacred cows in India, and any number of other ritual activities, all function to keep human populations in balance with their ecosystems. Such a view would describe precolonial New England not as a virgin landscape of natural harmony but as a landscape whose essential characteristics were kept in equilibrium by the cultural practices of its human community.

Unfortunately, this functional approach to culture has the same penchant for teleology as does the organism model of ecological climax. Saying that a community's rituals and social institutions "function" unconsciously to stabilize its ecological relationships can lead all too quickly into a static and ahistorical view of both cultural agency and ecological change. If we assume *a priori* that cultures are systems which tend toward ecological stability, we may overlook the evidence from many cultures—even preindustrial ones—that human groups often have significantly *unstable* interactions with their environments. When we say, for instance, that the New England Indians burned forests to clear land for agriculture and to improve hunting, we describe only what they themselves thought the purpose of burning to be. But to go further than this and assert its unconscious "function" in stabilizing Indian relationships with the ecosystem is to deny the evidence from places like Boston and Narragansett Bay that the practice could sometimes go so far as to remove the forest altogether, with deleterious effects for trees and Indians alike.

All human groups consciously change their environments to some extent—one might even argue that this, in combination with language, is the crucial trait distinguishing people from other animals—and the best measure of a culture's ecological stability may well be how successfully its environmental changes maintain its ability to reproduce itself. But if we avoid assumptions about environmental equilibrium, the *instability* of human relations with the environment can be used to explain both cultural and ecological transformations. An ecological history begins by assuming a dynamic and changing relationship between environment and culture, one as apt to produce contradictions as continuities. Moreover, it assumes that the interactions of the two are dialectical. Environment may initially shape the range of choices available to a people at a given moment, but then culture reshapes environment in responding to those choices. The reshaped environment presents a new set of possibilities for cultural reproduction, thus setting up a new cycle of mutual determination. Changes in the way people create and re-create their livelihood must be analyzed in terms of changes not only in their *social* relations but in their *ecological* ones as well. . . .

The view from Walden in reality contained far more than Thoreau saw that January morning in 1855. Its relationships stretched beyond the horizons of Concord to include vistas of towns and markets and landscapes that were not in Thoreau's field of vision. If only for this reason, we must beware of following him too closely as our guide in these matters. However we may respect his passion, we must also recognize its limits:

> I take infinite pains to know all the phenomena of the spring, for instance, thinking that I have here the entire poem, and then, to my chagrin, I hear that it is but an imperfect copy that I possess and have read, that my ancestors have torn out many of the first leaves and grandest passages, and mutilated it in many places. I should not like to think that some demigod had come before me and picked out some of the best of the stars. I wish to know an entire heaven and an entire earth.

We may or may not finally agree with Thoreau in regretting the changes which European settlers wrought in the New World, but we can never share his certainty about the possibility of knowing an entire heaven and an entire earth. Human and natural worlds are too entangled for us, and our historical landscape does not allow us to guess what the "entire poem" of which he spoke might look like. To search for that poem would in fact be a mistake. Our project must be to locate a nature which is within rather than without history, for only by so doing can we find human communities which are inside rather than outside nature.[1]

Note

1. Thoreau, H. D., *Journal, VIII* (23 March 1856): 221.

Donald Worster, from "Thinking Like a River" in *The Wealth of Nature* (1993)

The oceans contain over 300 million cubic miles of water, the fresh-water lakes contain 30,000, and the rivers of the world contain only 300. Yet, despite this relative inconsequentiality, rivers, as channeled flowing water, have tremendous importance, both for human use and as ecosystems in their own right. Environmental historian Donald Worster describes the various ways people have thought about rivers and then narrows his focus to the problems of water use in the American West.

Throughout history, the water cycle [see Chapter 13, below] has served humans as a model of the natural world. Early civilizations saw in it a figure of the basic pattern of life, the cycle of birth, death, and return to the source of being. More recently, science has added to that ancient religious metaphor a new perception: the movement of water in an unending, undiminished loop can stand as a model for understanding the entire economy of nature. Looking for a way to make the principles of ecology clear and vivid, Aldo Leopold suggested that nature is a "round river," like a stream flowing into itself, going round and round in an unceasing circuit, going through all the soils, the flora, and the fauna of Earth.[1] Another scientist, Robert Curry, has argued that the watershed (the area the river drains, its body, as it were) is the most appropriate unit for thinking about and dealing with nature.[2] The watershed is a complex whole, uniting biota, geochemistry, and energy in a single, interdependent system, in a dynamically balanced set of countervailing forces—erosion and construction, productivity and grace. Each watershed has its own peculiar shape and its own way of moving toward an elegant equilibrium of forces. The language of these scientists may be novel, but the insight is old and familiar. In water we see all of nature reflected. And in our use of that water, that nature, we see much of our past and future mirrored.

The first commandment for living successfully in nature—living for the long term at the highest possible level of moral development—is to understand how that round river and its watershed work together and to adapt our behavior accordingly. Taking a purely economic attitude toward water, on the other hand, is the surest way to fail in that understanding. In a strictly economic appraisal, water becomes merely the commodity H_2O, bulked here as capital to invest some day, spent freely when the market is high. It comes to be seen as a "cash flow," no longer as the lifeblood of the land. And then we begin to do foolish things with our streams and rivers. We fail to see that the meanderings of a creek across a meadow exemplify not chaotic or wasted motion but a fundamental rationality, that those meanderings make sense. Government engineers, confident that they know better, straighten the creek with bulldozers so that it will carry off floodwater faster, and in the process they destroy all the wild riparian edges that express a rationality that is different from economics, one we have not yet fully understood. The elementary need in learning how to farm water effectively, Leopold would have said, is to stop thinking about the problem exclusively as economists and engineers and to begin learning the logic of the river. . . .

It is time we began to rethink our agricultural relation to water. The problems are so numerous and complex that a flurry of quick-fix solutions is no longer ad-

Figure 11.7 New River irrigation ditch, Brawley, California. Courtesy of G. Donald Bain/the Geo-Images Project, Dept. of Geography, UC–Berkeley.

equate. What is needed is a fundamentally new approach to the challenge of how to extract a farm living from the hydrological cycle, both in humid and in arid regions. That requires vision more than technique: a way of perceiving, a set of mental images, an ethic controlling agricultural policy and practice. It demands, as I have said before, learning to think like a river.

For a long while this country [the United States] has been perfecting a strategy of comprehensive river planning. We have called this approach "conservation," though it has drastically remade rather than conserved what nature has given us. It has always been based on technological instead of ecological thinking; planners have defined their task as taking entire river systems apart, putting them back together in more "useful" arrangements, and delivering the goods wholesale to the farmer. A more reasonable strategy would be to focus on the individual farm, asking what its particular needs are and how they can be met with the least possible interference with the water cycle. Start with the local and specific rather than the general and grand. Develop a well-considered set of ends for public water policy. Seek then to meet those ends with the most elegantly simple means of water use possible, means that are economical, appropriately scaled to place and need, and capable of enduring indefinitely.

No farmer needs to grow corn in the desert or to use the Colorado River to do so. He may be inclined to choose to raise corn because he grew up in Indiana where corn was the traditional crop and then took his habits west, or because a county agent has told him that the world price of corn will be soaring after harvest. But that is not an analysis of need; it is a selection of methods. What the farmer really needs is a comfortable living for himself and his family, along with a chance to use intelligence and initiative to gain a measure of satisfaction from his work. The American consumer needs something to eat—nutrition and taste on the table at a not outlandish price. Can those genuine needs be met without turning the Colorado or Snake River into an elaborate artificial plumbing system, plagued by mounting costs and environmental destruction at every joint and faucet? Of course they can, if we are willing to put our ingenuity to work inventing, disciplining, and adapting; they can be met, that is, if we will undertake a radical reconstruction of methods.

The first specific step toward a new water consciousness is to end all federal subsidies of irrigation projects in the West. The subsidies should not be halted abruptly, but gradually, reversing with care and sensitivity the existing policy that has been in effect nearly a century now. Americans have no reason to fear such a change. The greatest portion of artificially watered acreage in the West raises crops that can be grown more cheaply elsewhere: 37 percent of all federal reclamation land, for example, is used for hay and forage; 21 percent for corn, barley, and wheat; 10 percent for cotton.[3] The United States will hardly starve if we do not subsidize those crops, for farmers in the East will raise them instead, and they can do so in ways far less disturbing ecologically. But since western farmers have long been induced by government incentives to move west and set up their irrigation plans, they should not be made to suffer by this policy reversal. What is needed is a new "homestead program," equivalent to the one devised in the mid-nineteenth century, that will encourage many western farmers to relocate in the more humid areas and learn the best practices for those places. For most of our national history we have assumed that to go forward was to go west. Now a sustainable agriculture requires a redirection of progress: Go east, young man or woman, and grow up with the country.

The West is now overpopulated, grossly exceeding its natural river capacity, and a new sense of water limits should stimulate the region's city residents as well as farmers to resettle eastward where they can be supported more easily. Those who remain, who constitute a permanent population in equilibrium with the environment, must have new water technologies that will enable them to survive, enjoy a modest prosperity, and grow some of their food close at hand. Unfortunately, almost no official thought has been devoted to alternative technologies that can provide for that population, although we know that at some point the great reservoirs will be filled with silt. It is time, perhaps past time, to begin the process of reinventing the West. The farsighted desert or plains farmer will start now to work out his own salvation, not wait for the planners, although he can use the advice of hydrologists, geneticists, and engineers. He will study the art of adaptive dry farming. He will demand some new crop varieties that can survive in places of little rain, and perhaps he will convince legislators to grant some aid to ease that changeover, much as they have subsidized home solar-energy conversions. With his neighbors, he will devise ways of diverting floodwaters without appropriating the entire river, ways of guaranteeing a minimum flow in the channel to support its ecology while making use of the river for crops. Where local markets require fresh fruits, vegetables, and milk, he will install drip irrigation, which uses far less water than furrow methods. Confronting the decline of fossil fuels, he will let the sun and gravity do most of the work of lifting and circulating water through his farm. These are a few of the ways in which agriculture can begin to adjust to an arid setting. Rather than insisting that drylands be made over into a version of Missouri, producing crops for which they are ill suited, we should begin imagining a future West that is finely tuned to its unique environment.

In more humid areas, farmers face the challenge of not letting a wealth of rain wash away their common sense and with it their soil. The first principle of good water management, it still bears repeating, is the maintenance of soil cover. In many places this means restoring the natural forest vegetation or planting the strong defense of a tree crop. In other places a thick sponge of grass will be enough to absorb the impact of falling water, slow its race to the sea, and keep the rivers clean and sweet. New perennial crops that make plowing unnecessary can cut erosion losses to natural replacement levels. Diminished use of pesticides and chemicals can reduce groundwater contamination. These familiar therapies are all parts of a larger vision: an agriculture in which every farm fits harmoniously within the dynamics of its own local watershed, rather than an agriculture in which every farm seeks to maximize its share of the money economy.

By now it should be evident that no market will ever pay farmers for accommodating themselves to their watershed. To be sure, the marketplace will reward long-range calculation more handsomely than many farmers are aware. But finally the marketplace is an institution that teaches self-advancement, private acquisition, and the domination of nature. Its way of thinking is incompatible with the round river. Ecological harmony is a non-market value that takes a collective will to achieve. It requires that farmers living along a stream cooperate to preserve it and to pass a fertile world down to another generation. It requires that urban consumers be willing to pay farmers to use good conservation techniques as well as to produce food. Without a public willingness to bid against market pressures, there will not be a radical reconstruction of farming methods or a rapprochement between agriculture and nature.

Americans, like people in other places and times, have a history of considerable violence toward the land and its life-giving rivers. Perhaps we have done more violence than most nations—certainly we have done more damage in a shorter period than most. Violence is typically a sporadic act, ill considered and destructive to the perpetrator as well as the victim; it is never the basis for permanence. What is now required in our agriculture, if it is to be secure, is a rejection of violence. Fortunately, we still can find in this country a broad enough margin of resources that we are not forced to violent land and water use; we can make other choices and avoid frantic, draconian measures. We are in a position to think not only about self-preservation but also about generosity and peace—about ethics.

Almost forty years ago, Aldo Leopold wrote that we will never get along well with nature until we learn to regard it morally. We must develop, he maintained, a sense of belonging to the larger community of nature, a community that has many interests and claims besides our own. We must cultivate a moral sensitivity to that community's integrity and beauty. He spoke of the need for a "land ethic," including in it a moral responsiveness to all parts of the ecological whole.[4] But given the centrality of water in our lives, and given the magnitude of the problems we confront in farming our watersheds, it also makes sense to talk about a "water ethic." Water, after all, covers most of this planet's surface. Even more than land, water is the essence and the context of life, the sphere of our being and that of other creatures. It has a value that extends beyond the economic use we make of it on our farms. Preserving that value of water through a new American agriculture is an extension of ethics as well as of wisdom.

Notes

1. Aldo Leopold, *Round River* (New York: Oxford University Press, 1953).
2. Robert Curry, "Watershed Form and Process: The Elegant Balance," *Co-Evolution Quarterly* (Winter 1976–77): 14–21.
3. Bureau of Reclamation, *Water and Land Resource Accomplishments* (1977), 8.
4. Aldo Leopold, *A Sand County Almanac* (New York: Oxford University Press, 1949).

Mark Kurlansky, from "With Mouth Wide Open" in *Cod: A Biography of the Fish That Changed the World* (1997)

Cotton Mather writes that when a preacher at Plymouth Bay Colony of Massachusetts in the 1620s declaimed to the settlers that they had ventured into the wilderness for the freedom of worship, one among them bluntly remonstrated him, "Sir you are mistaken, our main end in coming here was to catch fish." A little later, John Adams had inscribed on the crest of the most accomplished family in America the words "Piscemur ut olim"*: May we fish for a good, long while. The history of the peoples of the North Atlantic is intimately intertwined with the natural history of the Atlantic cod, as Mark Kurlansky explains in this excerpt from his book* Cod: A Biography of the Fish That Changed the World.

The hero, *Gadus morhua*, is not a nice guy.

It is built to survive. Fecund, impervious to disease and cold, feeding on most any food source, traveling to shallow waters and close to shore, it was the perfect commercial fish, and the Basques had found its richest grounds. Cod should have lasted forever, and for a very long time it was assumed that it would. As late as 1885, the Canadian Ministry of Agriculture said, "Unless the order of nature is overthrown, for centuries to come our fisheries will continue to be fertile."

The cod is omnivorous, which is to say it will eat anything. It swims with its mouth open and swallows whatever will fit—including young cod. Knowing this, sports fishermen in New England and Maritime Canada jig for cod, a baitless means of fishing, where a lure by its appearance and motion imitates a favorite prey of the target fish. A cod jigger is a piece of lead, sometimes fashioned to resemble a herring, but often shaped like a young cod. . . .

The cod's greed makes it easy to catch, but the fish is not much fun for sportsmen. A cod, once caught, does not fight for freedom. It simply has to be hauled up, and it is often large and heavy. New England anglers would far rather catch a bluefish than a cod. Bluefish are active hunters and furious fighters, and once hooked, a struggle ensues to reel in the line. But the bluefish angler brings home a fish with dark and oily flesh, characteristic of a midwater fighter who uses muscles for strong swimming. The cod, on the other hand, is prized for the whiteness of its flesh, the whitest of the white-fleshed fish, belonging to the order Gadiformes. The flesh is so purely white that the large flakes almost glow on the plate. Whiteness is the nature of the sluggish muscle tissue of fish that are suspended in the near-weightless environment at the bottom of the ocean. The cod will try to swim in front of an oncoming trawler net, but after about ten minutes it falls to the back of the net, exhausted. White muscles are not for strength but for quick action—the speed with which a cod, slowly cruising, will suddenly pounce on its prey.

Cod meat has virtually no fat (0.3 percent) and is more than 18 percent protein, which is unusually high even for fish. And when cod is dried, the more than 80 percent of its flesh that is water having evaporated, it becomes concentrated protein—almost 80 percent protein.

There is almost no waste to a cod. The head is more flavorful than the body, especially the throat, called a tongue, and the small disks of flesh on either side, called cheeks. The air bladder, or sound, a long tube against the backbone that can fill or release gas to adjust swimming depth, is rendered to make isinglass, which is used industrially as a clarifying agent and in some glues. But sounds are also fried by cod-fishing peoples, or cooked in chowders or stews. The roe [egg mass] is eaten, fresh or smoked. Newfoundland fishermen also prize the female gonads, a two-pronged organ they call the britches, because its shape resembles a pair of pants. Britches are fried like sounds. Icelanders used to eat the milt, the sperm, in whey. The Japanese still eat cod milt. Stomachs, tripe, and livers are all eaten, and the liver oil is highly valued for its vitamins.

Icelanders stuff cod stomachs with cod liver and boil them until tender and eat them like sausages. This dish is also made in the Scottish Highlands, where its dubious popularity is not helped by the local names: Liver-Muggie or Crappin-Muggie. Cod tripe is eaten in the Mediterranean.

The skin is either eaten or cured as leather. Icelanders used to roast it and serve it with butter to children. What is left from the cod, the remaining organs and bones, makes an excellent fertilizer, although until the twentieth century, Icelanders softened the bones in sour milk and ate them too. . . .

Codfish include ten families with more than 200 species. Almost all live in cold salt water in the Northern Hemisphere. Cod were thought to have developed into their current forms about 120 million years ago in the Tethys Sea, a tropical sea that once ran around the Earth east-west and connected all other oceans. . . .

Despite the warm-water origins, only one tropical cod remains: the tiny bregmaceros, of no commercial value and almost unknown habits. There is also one South Atlantic species and even one freshwater cod, the burbot, whose white flesh, though not quite the quality of an Atlantic cod, is enjoyed by lake fishermen in Alaska, the Great Lakes, New England, and Scandinavia. Norwegians think the burbot has a particularly delectable liver. There are other gadiforms that are pleasant to eat but of no commercial value. Sportsmen like to jig the coastline of Long Island and New England for the small tomcod, which also has a Pacific counterpart.

But to the commercial fisherman, there have always been five kinds of gadiform: the Atlantic cod, the haddock, the pollock, the whiting, and the hake. Increasingly, a sixth gadiform must be added to the list, the Pacific cod, *Gadus macrocephalus*, a smaller version of the Atlantic cod whose flesh is judged of only slightly lesser quality.

The Atlantic cod, however, is the largest, with the whitest meat. In the water, its five fins unfurl, giving an elegant form that is streamlined by a curving white stripe up the sides. It is also recognizable by a square rather than forked tail and a curious little appendage on the chin, which biologists think is used for feeling the ocean floor.

The smaller haddock has a similar form but is charcoal-colored on the back where the cod is spotted browns and ambers; it also has a black spot on both sides above the pectoral fin. The stripe on a haddock is black instead of white. In New England, there is a traditional explanation for this difference. There, cod is sometimes referred to as "the sacred cod." In truth, this is because it has earned New Englanders so many sacred dollars. But according to New England folklore, it was the fish that Christ multiplied to feed the masses. In the legend, Satan tried to do the same thing, but since his hands were burning hot, the fish wriggled away. The burn mark of Satan's thumb and forefinger left black stripes; hence the haddock.

This story illustrates the difference, not only in stripes but in status, between cod and haddock. British and Icelandic fishermen only reluctantly catch haddock after their cod quotas are filled, because cod always brings a better price. Yet Icelanders prefer eating haddock and rarely eat cod except dried. Asked why this is so, Reykjavík chef Úlfar Eysteinsson said, "We don't eat money." . . .

Figure 11.8 The Sacred Cod presides over the Massachusetts State House in Boston. © Newscom / Index Stock Photos.

But in spite of the occasional local preference, on the world market, cod is the prize. This was true in past centuries when it was in demand as an inexpensive, long-lasting source of nutrition, and it is true today as an increasingly expensive delicacy. Even with the Grand Banks closed, worldwide more than six million tons of gadiform fish are caught in a year, and more than half are *Gadus morhua*, the Atlantic cod. For fishermen, who are extremely tradition bound, there is status in fishing cod. Proud cod fishermen are indignant, or at least saddened, by the suggestion that they should switch to what they see as lesser species.

In addition to its culinary qualities, the cod is eminently catchable. It prefers shallow water, only rarely venturing to 1,800 feet, and it is commonly found in 120 feet (twenty fathoms) or less. Cod migrate for spawning, moving into still-shallower water close to coastlines, seeking warmer spawning grounds and making it even easier to catch them.

They break off into subgroups, which adapt to specific areas, varying in size and color, from yellow to brown to green to gray, depending on local conditions. In the dark waters off of Iceland, they are brown with yellow specks, but it takes only two days in the brightly lit tank of an aquarium in the Westman Islands, off of Iceland, for a cod to turn so pale it looks almost albino. The so-called northern stock, the cod off of Newfoundland and Labrador, are smaller for their age than the cod off of Massachusetts, where the water is warmer. Though always a coldwater fish, preferring water temperatures between thirty-four and fifty degrees, cod grows faster in the warmer waters of its range. Historically, but not in recent years because of overfishing, the cod stock off of Massachusetts was the largest and meatiest in the world.

Cod manufacture a protein that functions like antifreeze and enables the fish to survive freezing temperatures. If hauled up by a fisherman from freezing water, which rarely happens since they are then underneath ice, the protein will stop functioning and the fish will instantly crystallize.

Cod feed on the sea life that clusters where warm and cold currents brush each other—where the Gulf Stream passes by the Labrador current off North America, and again where it meets arctic currents off the British Isles, Scandinavia, and Russia. The Pacific cod is found off of Alaska, where the warm Japanese current touches the arctic current. In fact, the cod follow this edge of warm and cold currents so consistently that some scientists believe the shifting of weather patterns can be monitored by noting where fishermen find cod. When cold northern waters become too cold, the cod populations move south, and in warmer years they move north.

From Newfoundland to southern New England, there is a series of shallow areas called banks, the southernmost being Georges Bank off of Massachusetts, which is larger than the state. Several large banks off of Newfoundland and Labrador are together called the Grand Banks. The largest of the Grand Banks, known as the Grand Bank, is larger than Newfoundland. These are huge shoals on the edge of the North American continental shelf. The area is rich in phytoplankton, a growth produced from the nitrates stirred up by the conflicting currents. Zooplankton, tiny sea creatures, gorge themselves on the phytoplankton. Tiny shrimp-like free-floating creatures called krill eat the zooplankton. Herring and other midwater species rise to eat the krill near the surface, and seabirds dive for both the krill and the fish. Humpback whales also feed on krill. And it is this rich environment on the banks that produces cod by the millions. In the North Sea, the cod grounds are also found on banks, but the North American banks, where the waters of the Gulf of Mexico meet the arctic Greenland waters, had a greater density of cod than anything ever seen in Europe. This was the Basques' secret.

Still more good news for the fishermen, a female cod forty inches (102 centimeters) long can produce three million eggs in a spawning. A fish ten inches longer can produce nine million eggs. A cod may live to be twenty or even thirty years old, but it is the size more than the age that determines its fecundity. [Alexander] Dumas's image of all the eggs hatching so that someone could walk across the ocean on the backs of cod is typical nineteenth-century enthusiasm about the abundance of the species. But it could never happen. In the order of nature, a cod produces such a quantity of eggs precisely because so few will reach maturity. The free-floating eggs are mostly destroyed as they are tossed around the ocean's surface, or they are eaten by other species. After a couple of weeks, the few surviving eggs hatch and hungrily feed, first on phytoplankton and soon zooplankton and then krill. That is, if they can get to those foods before the other fish, birds, and whales. The few cod larvae that are not eaten or starved in the first three weeks will grow to about an inch and a half. The little transparent fish, called juveniles, then leave the upper ocean and begin their life on the bottom, where they look for gravel and other rough surfaces in which to hide from their many predators, including hungry adult cod. A huge crop of eggs is necessary for a healthy class, as biologists call them, of juveniles. If each female cod in a lifetime of millions of eggs produces two juveniles that live to be sexually mature adults, the population is stable. The first year is the hardest to survive. After that, the cod has few predators and many prey. Because a cod will eat most anything, it adapts its diet to local conditions, eating mollusks in the Gulf of Maine, and herring, capelin, and squid in the Gulf of St. Lawrence. The Atlantic cod is particularly resistant to parasites and diseases, far more so than haddock and whiting.

If ever there was a fish made to endure, it is the Atlantic cod—the common fish. But it has among its predators man, an openmouthed species greedier than cod.

Michael E. Soulé, "What Is Conservation Biology?" (1985)

Published at the beginning of the modern era of conservation biology, this selection by Michael Soulé, one of the founders of the Society for Conservation Biology, continues to be a clear-headed exposition of the practices and principles of this applied science. The major expansion of Soulé's ideas has been in his holistic concept of multidisciplinarity, as more and more conservation biologists incorporate law, economics, and social sciences into their daily practice and teaching.

Conservation biology differs from most other biological sciences in one important way: it is often a crisis discipline. Its relation to biology, particularly ecology, is analogous to that of surgery to physiology and war to political science. In crisis disciplines, one must act before knowing all the facts; crisis disciplines are thus a mixture of science and art, and their pursuit requires intuition as well as information. A conservation biologist may have to make decisions or recommendations about design and management before he or she is completely comfortable with the theoretical and empirical bases of the analysis. Tolerating uncertainty is often necessary.

Conservation biologists are being asked for advice by government agencies and private organizations on such problems as the ecological and health consequences of chemical pollution, the introduction of exotic species and artificially produced strains of existing organisms, the sites and sizes of national parks, the definition of minimum conditions for viable populations of particular target species, the frequencies and kinds of management practices in existing refuges and managed wildlands, and the ecological effects of development. For political reasons, such decisions must often be made in haste.

For example, the rapidity and irreversibility of logging and human resettlement in Western New Guinea (Irian Jaya) prompted the Indonesian government to establish a system of national parks. Two of the largest areas recommended had never been visited by biologists, but it appeared likely that these areas harbored endemic biotas. Reconnaissance later confirmed this. The park boundaries were established in 1981, and subsequent development has already precluded all but minor adjustments. Similar crises are now facing managers of endangered habitats and species in the United States—for example, grizzly bears in the Yellowstone region, black-footed ferrets in Wyoming, old-growth Douglas-fir forests in the Pacific Northwest, red-cockaded woodpeckers in the Southeast, and condors in California.

As illustrated in Figure 11.9, conservation biology shares certain characteristics with other crisis-oriented disciplines. A comparison with cancer biology illustrates some of these characteristics, including conservation biology's synthetic, eclectic, multidisciplinary structure. Furthermore, both fields take many of their questions, techniques, and methods from a broad range of fields, not all biological. This illustration is also intended to show the artificiality of the dichotomy between pure and applied disciplines.

Finally, this figure illustrates the dependence of the biological sciences on social science disciplines. Today, for example, any recommendations about the location and size of national parks should consider the impact of the park on indigenous peoples and their cultures, on the local economy, and on opportunity costs such as forfeited logging profits.

There is much overlap between conservation biology and the natural resource fields, especially fisheries biology, forestry, and wildlife management. Nevertheless, two characteristics of these fields often distinguish them from conservation biology. The first is the dominance in the resource fields of utilitarian, economic objectives. Even though individual wildlife biologists honor Aldo Leopold's land ethic and the intrinsic value of nature, most of the financial resources for management must go to enhancing commercial and recreational values for humans. The emphasis is on *our* natural *resources*.

The second distinguishing characteristic is the nature of these resources. For the most part, they are a small number of particularly valuable target species (e.g., trees, fishes, deer, and waterfowl)—a tiny fraction of the total biota. This distinction is beginning to disappear, however, as some natural resource agencies become more "ecological" and because conservation biologists frequently focus on individual endangered, critical, or keystone species.

Conservation biology tends to be holistic, in two senses of the word. First, many conservation biologists, including many wildlife specialists, assume that ecological and evolutionary processes must be studied at their own macroscopic

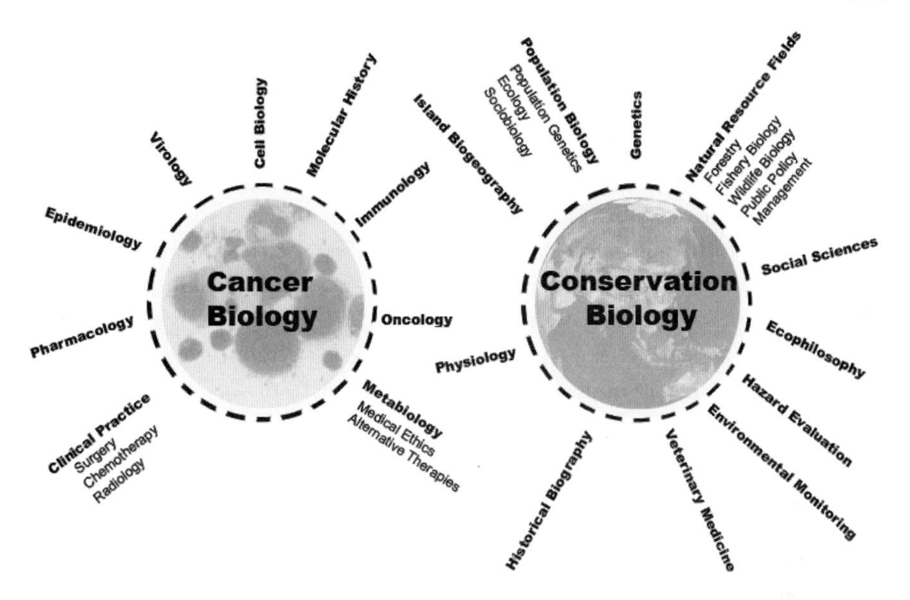

Figure 11.9 Cancer biology and conservation biology are both synthetic, multidisciplinary sciences. The dashed line indicates the artificial nature of the borders between disciplines and between "basic" and "applied" research.

levels and that reductionism alone cannot lead to explanations of community and ecosystem processes such as body-size differences among species in guilds, pollinator-plant coevolution, succession, speciation, and species-area relationships. Even ecological reductionists, however, agree that the proper objective of conservation is the protection and continuity of entire communities and ecosystems. The holistic assumption of conservation biology should not be confused with romantic notions that one can grasp the functional intricacies of complex systems without conducting scientific and technological studies of individual components. Holism is not mysticism.

The second implication of the term *holistic* is the assumption that multidisciplinary approaches will ultimately be the most fruitful. Conservation biology is certainly holistic in this sense. Modern biogeographic analysis is now being integrated into the conservation movement. Population genetics, too, is now being applied to the technology of wildlife management. Multidisciplinary research, involving government agencies and wildlife biologists, is also evident in recent efforts to illuminate the question of viable population size.

Another distinguishing characteristic of conservation biology is its time scale. Generally, its practitioners attach less weight to aesthetics, maximum yields, and profitability, and more to the long-range viability of whole systems and species, including their evolutionary potential. Long-term viability of natural communities usually implies the persistence of diversity, with little or no help from humans. But for the foreseeable future, such a passive role for managers is unrealistic, and virtually all conservation programs will need to be buttressed artificially. For example, even the largest nature reserves and national parks are affected by anthropogenic factors in the surrounding area, and such refuges are usually too small to contain viable populations of large carnivores. In addition, poaching, habitat fragmentation, and the influx of feral animals and exotic plants require extraordinary practices such as culling, eradication, wildlife immunization, habitat protection, and artificial transfers. Until benign neglect is again a possibility, conservation biology can complement natural resource fields in providing some of the theoretical and empirical foundations for coping with such management conundrums.

POSTULATES OF CONSERVATION BIOLOGY

These are working propositions based partly on evidence, partly on theory, and partly on intuition. In essence, they are a set of fundamental axioms, derived from ecology, biogeography, and population genetics, about the maintenance of both the form and function of natural biological systems. They suggest the rules for action. A necessary goal of conservation biology is the elaboration and refinement of such principles.

The first, the evolutionary postulate states: *Many of the species that constitute natural communities are the products of coevolutionary processes.* In most communities, species are a significant part of one another's environment. Therefore, their genetically based physiological and behavioral repertoires have been naturally selected to accommodate the existence and reactions of a particular biota. For example, the responses of prey to a predator's appearance or of a phytophagous insect to potential host plants are continually "tuned" by natural selection.

This postulate merely asserts that the structure, function, and stability of coevolved, natural communities differ significantly from those of unnatural or synthetic communities. It does not necessarily rely on deterministic factors like density-dependent population dynamics or the molding by competition of morphological relationships in communities over both ecological and evolutionary time. In addition, this postulate is neutral on the issue of holistic versus reductionistic analysis of community structure. (In practice, a reductionistic methodology, including autecological research, may be the best way to establish the holistic structure of communities.)

There are many "corollaries" of this postulate. Strictly speaking, most of them are empirically based generalizations. The following all assume the existence of community processes as well as a coevolutionary component in community structure.

Species are interdependent. Not only have species in communities evolved unique ways of avoiding predators, locating food, and capturing and handling prey, but mutualistic relationships are frequent. This is not to say that every species is essential for community function, but that there is always uncertainty about the interactions of species and about the biological consequences of an extinction. Partly for this reason, Aldo Leopold admonished conservationists to save all of the parts (species) of a community.[1]

Many species are highly specialized. Perhaps the majority of animal species, including phytophagous insects, parasites, and parasitoids, depend on a particular host. This means that the coattails of endangered host species can be very long, taking with them dozens or hundreds of small consumer species when they go.

Extinctions of keystone species can have long-range consequences. The extinction of major predators, large herbivores, or plants that are important as breeding or feeding sites for animals may initiate sequences of causally linked events that ultimately lead to further extinctions.

Introductions of generalists may reduce diversity. The introduction of exotic plant and animal species may reduce diversity, especially if they are large or generalist species. Apparently, the larger the land mass, the less the impact of exotics.

The evolutionary postulate and its corollaries formalize the evidence that natural communities comprise species whose genetic makeups have been mutually affected by their coexistence. An alternative theory, the null hypothesis that communities are randomly assembled, is usually restricted to "horizontal" subcommunities such as guilds, specific taxa, or trophic levels. In general, this latter thesis lacks empirical support, except that competitive structuring within guilds or trophic levels is often absent or difficult to demonstrate, and that harsh environments or the vagaries of dispersal may often be more important than biological interactions in determining local community composition.

The second functional postulate concerns the scale of ecological processes: *Many, if not all, ecological processes have thresholds below and above which they become discontinuous, chaotic, or suspended.* This postulate states that many ecological processes and patterns (including succession, nutrient cycling, and density-dependent phenomena) are interrupted or fail altogether where the system is too small. Smallness and randomness are inseparable.

Nonecological processes may also dominate at the other end of the spatial and temporal scale, in very large or very old systems. In very large systems, such as continents, climatic and physiographic phenomena often determine the major patterns of the landscape, including species distribution. In very old systems, ecological processes give way to geological and historical ones or to infrequent catastrophic events, such as inundation, volcanism, and glaciation. In other words, ecological processes belong to an intermediate scale of physical size and time, and these processes begin to fail or are overwhelmed near the extremities of these ranges.

Two major assumptions, or generalizations, underlie this postulate. First, *the temporal continuity of habitats and successional stages depends on size.* The random disappearance of resources or habitats will occur frequently in small sites but rarely, if ever, in large ones. The reasons include the inherent randomness of such processes as patch dynamics, larval settlement, or catastrophic events, as well as the dynamics of contagious phenomena such as disease, windstorm destruction, and fire. The larger an area, the less likely that all patches of a particular habitat will disappear simultaneously. Species will disappear if their habitats disappear.

Second, *outbursts reduce diversity.* If population densities of ecologically dominant species rise above sustainable levels, they can destroy local prey populations and other species sharing a resource with such species. Outbursts are most probable in small sites that lack a full array of population buffering mechanisms, including habitat sinks for dispersing individuals, sufficient predators, and alternative feeding grounds during inclement weather. The unusually high population densities that often occur in nature reserves can also increase the rate of disease transmission, frequently leading to epidemics that may affect every individual.

Taken together, the corollaries of this postulate lead to the conclusion that survival rates of species in reserves are proportional to reserve size. Even though there is now a consensus that several small sites can contain as many species as one large site (when barriers to dispersal are absent), the species extinction rate is generally higher in small sites.

The third functional postulate concerns the scale of population phenomena: *Genetic and demographic processes have thresholds below which nonadaptive, random forces begin to prevail over adaptive, deterministic forces within populations.* The stochastic factors in population extinction have been discussed extensively in the context of the minimum conditions for population viability. The main implication of this postulate for conservation is that the probability of survival of a local population is a positive function of its size. One of the corollaries of this postulate is that below a certain population size (between 10 and 30), the probability of extinction from random demographic events increases steeply.

The next three corollaries are genetic. First, populations of outbreeding organisms will suffer a chronic loss of fitness from inbreeding depression at effective population sizes of less than 50 to 100. Second, genetic drift in small populations (less than a few hundred individuals) will cause a progressive loss of genetic variation; in turn, such genetic erosion will reduce immediate fitness because multilocus heterozygosity is generally advantageous in outbreeding species. (The genetic bases of these two corollaries may be the same: homozygosity for deleterious, recessive alleles.) Finally, natural selection will be less effective in small populations because of genetic drift and the loss of potentially adaptive genetic variation.

The fourth functional postulate is that *nature reserves are inherently disequilibrial for large, rare organisms.* There are two reasons for this. First, extinctions are inevitable in habitat islands the size of nature reserves; species diversity must be artificially maintained for many taxa because natural colonization (reestablishment) from outside sources is highly unlikely. Second, speciation, the only other nonartificial means of replacing species, will not operate for rare or large organisms in nature reserves because reserves are nearly always too small to keep large or rare organisms isolated within them for long periods, and populations isolated in different reserves will have to be maintained by artificial gene flow if they are to persist. Such gene flow would preclude genetic differentiation among the colonies. . . .

CONCLUSIONS

Conservation biology is a young field, but its roots antedate science itself. Each civilization and each human generation responds differently to the forces that weaken the biological infrastructure on which society depends and from which it derives much of its spiritual, aesthetic, and intellectual life. In the past, the responses to environmental degradation were often literary, as in the Babylonian Talmud (Vol. I, Shabbath 129a, chap. xviii, p. 644), Marsh, Leopold, Carson, and others. More recently, legal and regulatory responses have been noticeable, especially in highly industrialized and democratized societies. Examples include the establishment of national parks and government policies on human population and family planning, pollution, forest management, and trade in endangered species. At this point in history, a major threat to society and nature is technology, so it is appropriate that this generation look to science and technology to complement literary and legislative responses.

Our environmental and ethical problems, however, dwarf those faced by our ancestors. The current frenzy of environmental degradation is unprecedented, with deforestation, desertification, and destruction of wetlands and coral reefs occurring at rates rivaling the major catastrophes in the fossil record and threatening to eliminate most tropical forests and millions of species in our lifetimes. The response, therefore, must also be unprecedented. It is fortunate, therefore, that conservation biology, and parallel approaches in the social sciences, provides academics and other professionals with constructive outlets for their concern.

Conservation biology and the conservation movement cannot reverse history and return the biosphere to its prelapsarian majesty. The momentum of the human population explosion, entrenched political and economic behavior, and withering technologies are propelling humankind in the opposite direction. It is, however, within our capacity to modify significantly the *rate* at which biotic diversity is destroyed, and small changes in rates can produce large effects over long periods of time. Biologists can help increase the efficacy of wildland management; biologists can improve the survival odds of species in jeopardy; biologists can help mitigate technological impacts. The intellectual challenges are fascinating, the opportunities plentiful, and the results can be personally gratifying.

Note

1. Leopold, Aldo. *The Round River* (New York: Oxford University Press, 1953).

Charles Elton, from "The Invaders" in *The Ecology of Invasions by Animals and Plants* (1958)

Charles Elton was one of the early giants of the modern discipline of ecology. Author of the influential text, Animal Ecology, *and founder of the* Journal of Animal Ecology, *he is today best known for the prescient work from which this selection is taken. In the previous selection, Michael Soulé described conservation biology as a crisis discipline. Over the last two decades we have learned, unfortunately, that invasion ecology has become one of the most urgent fields of study within conservation biology. Some of the species that have been carried around the globe by ever-expanding human transportation technology have caused the extinction and endangerment of native species and the destruction of native ecosystems. For further readings on this field, see Chapter 2, "Species in Danger"—especially the case of the brown tree snake in Guam.*

Nowadays we live in a very explosive world, and while we may not know where or when the next outburst will be, we might hope to find ways of stopping it or at any rate damping down its force. It is not just nuclear bombs and wars that threaten us, though these rank very high on the list at the moment: there are other sorts of explosions, and this book is about ecological explosions. An ecological explosion means the enormous increase in numbers of some kind of living organism—it may be an infectious virus like influenza, or a bacterium like bubonic plague, or a fungus like that of the potato disease, a green plant like the prickly pear, or an animal like the grey squirrel. I use the word "explosion" deliberately, because it means the bursting out from control of forces that were previously held in restraint by other forces. . . .

Ecological explosions differ from some of the rest by not making such a loud noise and in taking longer to happen. That is to say, they may develop slowly and they may die down slowly; but they can be very impressive in their effects, and many people have been ruined by them, or died or forced to emigrate. At the end of the First World War, pandemic influenza broke out on the Western Front, and thence rolled right round the world, eventually, not sparing even the Eskimos of Labrador and Greenland, and it is reputed to have killed 100 million human beings. Bubonic plague is still pursuing its great modern pandemic that started at the back of China in the end of last century, was carried by ship rats to India, South Africa, and other continents, and now smolders among hundreds of species of wild rodents there, as well as in its chief original home in Eastern Asia. In China it occasionally flares up on a very large scale in the pneumonic form, resembling the Black Death of medieval Europe. In 1911 about 60,000 people in Manchuria died in this way. This form of the disease, which spreads directly from one person to another without the intermediate link of a flea, has mercifully been scarce in the newly invaded continents. Wherever plague has got into natural ecological communities, it is liable to explode on a smaller or larger scale, though by a stroke of fortune for the human race, the train of contacts that starts this up is not very easily fired. In South Africa the gerbils living on the veld carry the bacteria permanently in many of their populations. Natural epidemics flare up among them frequently. From them the bacteria can pass through a flea to the multimammate mouse; this species, unlike the gerbils, lives in contact with man's domestic rat; the latter may become infected occasionally and from it isolated human cases of bubonic plague arise. These in turn may spread into a small local epidemic, but often do not. In the United States and Canada a similar underworld of plague (with different species in it) is established over an immense extent of the Western regions, though few outbreaks have happened in man. Here, then, the chain of connections is weaker even than in South Africa, though the potentiality is present. Although plague-stricken people and plague-infected rats certainly landed from ships in California early this century, it is still possible that the plague organism was already present in North America. Professor Karl Meyer, who started the chief ecological research on sylvatic plague there, says: "The only conclusion one can draw is that the original source and date of the creation of the endemic sylvatic plague area on the North American Continent, inclusive [of] Canada, must remain a matter of further investigation and critical analysis."[1]

Another kind of explosion was that of the potato fungus from Europe that partly emptied Ireland through famine a hundred years ago. Most people have had experience of some kind of invasion by a foreign species, if only on a moderate scale. Though these are silent explosions in themselves, they often make quite a loud noise in the Press, and one may come across banner headlines like "Malaria Epidemic Hits Brazil," "Forest Damage on Cannock Chase," or "Rabbit Disease in Kent." This arrival of rabbit disease—myxomatosis—and its subsequent spread have made one of the biggest ecological explosions Great Britain has had this century, and its ramifying effects will be felt for many years.

But it is not just headlines or a more efficient news service that make such events commoner in our lives than they were last century. They are really happening much more commonly; indeed they are so frequent nowadays in every continent and island, and even in the oceans, that we need to understand what is causing them and try to arrive at some general viewpoint about the whole business. Why should a comfortably placed virus living in Brazilian cotton-tail rabbits suddenly wipe out a great part of the rabbit populations of Western Europe? Why do we have to worry about the Colorado potato beetle now, more than 300 years after the introduction of the potato itself? Why should the pine looper moth break out in Staffordshire and Morayshire pine plantations two years ago? It has been doing this on the Continent for over 150 years; it is not a new introduction to this country [England].

The examples given above point to two rather different kinds of outbreaks in populations: those that occur because a foreign species successfully invades another country, and those that happen in native or long-established populations. This book is chiefly about the first kind—the invaders. But the interaction of fresh arrivals with the native fauna and flora leads to some consideration of ecological ideas and research about the balance within and between communities as a whole. In other words, the whole matter goes far wider than any technological discussion of pest control, though many of the examples are taken from applied ecology. The real thing is that we are living in a period of the world's history when the mingling of thousands of kinds of organisms from different parts of the world is setting up terrific dislocations in nature. We are seeing huge changes in the natural population balance of the world. Of course, pest control is very important, because we have to preserve our living resources and protect ourselves from diseases and the consequences of economic dislocation. But one should try to see the whole matter on a much broader canvas than that. I like the words of Dr. Johnson: "Whatever makes the past, the distant, or the future, predominate over the present, advances us in the dignity of thinking beings."[2] The larger ecological explosions have helped to alter the course of world history, and, as will be shown, can often be traced to a breakdown in the isolation of continents and islands built up during the early and middle parts of the Tertiary Period.

In order to focus the subject, here are seven case histories of species which were brought from one country and exploded into another. About 1929, a few African mosquitoes accidentally reached the northeast corner of Brazil, having probably been carried from Dakar on a fast French destroyer. They managed to get ashore and founded a small colony in a marsh near the coast—the Mosquito Fathers as it were. At first not much attention was paid to them, though there was a pretty sharp outbreak of malaria in the local town, during which practically every person was infected. For the next few years the insects spread rather quietly along the coastal region, until at a spot about 200 miles farther on explosive malaria blazed up and continued in 1938 and 1939, by which time the mosquitoes were found to have moved a further 200 miles inland up the Jaguaribe River valley. It was one of the worst epidemics that Brazil had ever known, hundreds of thousands of people were ill, some twenty thousand are believed to have died, and the life of the countryside was partially paralyzed. The biological reasons for this disaster were horribly simple: there had always been malaria-carrying mosquitoes in the country, but none that regularly flew into houses like the African species, and could also breed so successfully in open sunny pools outside the shade of the forest. Fortunately both these habits made control possible, and the Rockefeller Foundation combined with the Brazil government to wage a really astounding campaign, so thorough and drastic was it, using a staff of over three thousand people who dealt with all the breeding sites and sprayed the inside of houses. This prodigious enterprise succeeded, at a cost of over two million dollars, in completely exterminating *Anopheles gambiae* on the South American continent within three years.[3]

Here we can see three chief elements that recur in this sort of situation. First there is the historical one; this species of mosquito was confined to tropical Africa but got carried to South America by man. Secondly, the ecological features—its method of breeding, and its choice of place to rest and to feed on man. It is quite certain that the campaign could never have succeeded without the intense ecological surveys and study that lay behind the inspection and control methods. The third thing is the disastrous consequences of the introduction. One further consequence was that quarantine inspection of aircraft was started, and in one of these they discovered a tsetse fly, *Glossina palpalis*, the African carrier of sleeping sickness in man, and at the present day not found outside Africa.

The second example is a plant disease. At the beginning of this century sweet chestnut trees in the eastern United States began to be infected by a killing disease caused by a fungus, *Endothia parasitica*, that came to be known as the chestnut blight. It was brought from Asia on nursery plants. In 1913 the parasitic fungus was found on its natural host in Asia, where it does no harm to the chestnuts. But the eastern American species, *Castanea dentata*, is so susceptible that it has almost died out over most of its range. This species carries two native species of *Endothia* that do not harm it, occurring also harmlessly on some other trees like oak; one of these two species also comes on the chestnut, *C. sativa*, in Europe. Even by 1911 the outbreak, being through wind-borne spores, had spread to at least ten states, and the losses were calculated to be at least twenty-five million dollars up to that date.[4] In 1926 it was still spreading southwards, and by 1950 most of the chestnuts were dead except in the extreme south; and it is now on the Pacific coast too. So far, the only answer to the invasion has been to introduce the Chinese chestnut, *C. mollissima*, which is highly though not completely immune through having evolved into the same sort of balance with its parasite, as had the American trees with theirs; much as the big game animals of Africa can support trypanosomes in their blood that kill the introduced domestic animals like cattle and horses. The biological dislocation that occurs in this trypanosomiasis is the kind of thing that presumably would have happened also if the American chestnut had been introduced into Asia. The Chinese chestnut is immune both in Asia and America. Already by 1911 the European chestnuts grown in America had been found susceptible. In 1938 the blight appeared in Italy where it has exploded fast and threatens the chestnut groves that there are grown in pure stands for harvesting the nuts; it has also reached Spain and will very likely reach Britain in the long or short run. Unfortunately the Chinese chestnut will not flourish in Italy, and hopes are placed solely on the eventual breeding of a resistant variety of hybrid.

The third example is the European starling, *Sturnus vulgaris*, which has spread over the United States and Canada within a period of sixty years. (It has also become established in two other continents—South Africa and Australia, as well as in New Zealand.) This subspecies of starling has a natural range extending into Siberia, and from the north of Norway and Russia down to the Mediterranean. We should therefore expect it to be adaptable to a wide variety of continental habitats and climate. Nevertheless, the first few attempts to establish it in the United States were unsuccessful. Then from a stock of about eighty birds put into Central Park, New York, several pairs began to breed in 1891. After this the increase and spread went on steadily, apart from a severe mortality in the very cold winter of 1917–18. But up to 1916 the populations had not established beyond the Allegheny Mountains. The breeding range had extended concentrically, with outlying records of non-breeding birds far beyond the outer breeding limits, which had moved beyond the Alleghenies but nowhere westward of a line running about southwards from Lake Michigan.[5] By 1954 the process was nearly reaching its end, and the starling was to be found, at any rate on migration outside its breeding season, almost all over the United States, though it was not fully entrenched yet in parts of the West Coast states. It was penetrating northern Mexico during migration, and in 1953 one starling was seen in Alaska.[6] This was an ecological explosion indeed, starting from a few pairs breeding in a city park; just as the spread of the North American muskrat, *Ondatra zibethica*, over Europe was started from only five individuals kept by a landowner in Czechoslovakia in 1905. The muskrat now inhabits Europe in many millions, and its range has been augmented by subsidiary introductions for fur-breeding, with subsequent establishment of new centers of escaped animals and their progeny. Since 1922, over 200 transplantations of muskrats have been started in Finland, some originally from Czechoslovakia in 1922, and the annual catch is now between 100,000 and 240,000.[7] Independent Soviet introductions have also made the muskrat an important fur animal in most of the great river systems of Siberia and northern Russia, as well as in Kazakstan. In zoogeographical terminology, a purely Palaearctic species (the starling) and a purely Nearctic species (the muskrat) have both become Holarctic within half a century.

[Elton here provides his fifth, sixth, and seventh examples: rice-grass in Britain, sea lampreys in North America, and mitten crabs in China.]

These seven examples alone illustrate what man has done in deliberate and accidental introductions, especially across the oceans. Between them all they cover the waters of sea, estuary, river, and lake; the shores of sea and estuary; tropical and temperate forest country, farm land, and towns. In the eighteenth century there were few ocean-going vessels of more than 300 tons. Today there are thousands. A Government map made for one day, 7 March 1936, shows the position of every British Empire ocean-going vessel all over the world. There are 1,462 at sea and 852 in port; and this map does not include purely coasting vessels. Some idea of what this can mean for the spread of animals can be got from the results of an ecological survey done by Myers, a noted tropical entomologist, while traveling on a Rangoon rice ship from Trinidad to Manila in 1929. He amused himself by making a list of every kind of animal on board, from cockroaches and rice beetles to fleas and pet animals.[8] Altogether he found forty-one species of these travelers, mostly insects. And when he unpacked his clothes in the hotel in Manila, he saw some beetles walk out of them. They were *Tribolium castaneum*, a well-known pest of stored flour and grains, which was one of the species living among the rice on the ship.

A hundred years of faster and bigger transport has kept up and intensified this bombardment of every country by foreign species, brought accidentally or on purpose, by vessel and by air, and also overland from places that used to be isolated. Of course, not all the plants and animals carried around the world manage to establish themselves in the places they get to; and not all that do are harmful to man, though they must change the balance among native species in some way. But this world-wide process, gathering momentum every year, is gradually breaking down the sort of distribution that species had even a hundred years ago.

To see the full significance of what is happening, one needs to look back much further still, in fact many millions of years by the geological time-record. It was Alfred Russel Wallace who drew general public attention to the existence of great faunal realms in different parts of the world, corresponding in the main to the continents. These came to be known as Wallace's Realms. . . . He supposed these realms to have been left isolated for such long periods that they had kept or evolved many special groups of animals. When one was a child, this circumstance was very simply summed up in books about animals. The tiger lives in India. The wallaby lives in Australia. The hippopotamus lives in Africa. One might have learned that the coypu or nutria lives in South America. A very advanced book might have speculated that this big water rodent was evolved inside South America, which we now know to be so. But nowadays, it would have to add a footnote to later editions, saying that the coypu is also doing quite well in the States of Washington, Oregon, California, and New Mexico; also in Louisiana (where 374,000 were trapped in one year recently); in south-east U.S.S.R.; in France; and in the Norfolk Broads of East Anglia.[9] In the Broads it carries a special kind of fur parasite, *Pitrufquenia coypus*, belonging to a family (Gyropidae) that also evolved in South America.[10] These fur lice have antennae shaped like monkey-wrenches, which perhaps explains how they managed to hang on so well all the way from South America.

But in very early times, say 100 million years ago in the Cretaceous Period, the world's fauna was much more truly cosmopolitan, not so much separated off by oceans, deserts, and mountains. If there had been a Cretaceous child living at the time the chalk was deposited in the warm shallow seas at Marlborough or Dover, he would have read in his book, or slate perhaps: "Very large dinosaurs occur all over the world except in New Zealand; keep out of their way." Or that water monsters occurred in more than one loch in the world. In fact, zoogeographically, it would have been rather a dull book, though the illustrations and accounts of the habits of animals would have been terrifically interesting. There would have been much less use for zoos: you just went out, with suitable precautions, and did dinosaur-watching wherever you were, and made punch-card records of their egg clutch-sizes. But the significance of these dinosaurs for the serious historical evidence is that you couldn't then get an animal the size of a lorry from one continent to another except by land; therefore the continents must have been joined together, at any rate fairly frequently, as geological time is counted.

This early period of more or less cosmopolitan land and fresh-water life was about three times longer than that between the Cretaceous Period and the present day. It was in the later period that Wallace's Realms were formed, because the sea, and later on great obstructions like the Himalaya and the Central Asian deserts, made impassable barriers to so many species. In fact the world had not one, but five or six great faunas, besides innumerable smaller ones evolved on isolated islands like Hawaii or New Zealand or New Caledonia, and in enormous remote lakes like Lake Baikal or Tanganyika. Man was not the first influence to start breaking up this world pattern. A considerable amount of re-mixing has taken place in the few million years before the Ice Age and since then: two big factors in this were the emergence of the Panama Isthmus from the sea, and the passage at various times across what is now Bering Strait. But we are artificially stepping up the whole business, and feeling the manifold consequences.

For thirty years I have read publications about this spate of invasions; and many of them preserve the atmosphere of first-hand reporting by people who have actually seen them happening, and give a feeling of urgency and scale that is absent from the drier summaries of text-books. We must make no mistake: we are seeing one of the great historical convulsions in the world's fauna and flora. We might say, with Professor Challenger, standing on Conan Doyle's "Lost World," with his black beard jutting out: "We have been privileged to be present at one of the typical decisive battles of history—the battles which have determined the fate of the world." But how will it be decisive? Will it be a Lost World? These are questions that ecologists ought to try to answer.

Notes

1. Meyer, Karl, "The known and unknown in the plague." *American Journal of Tropical Medicine* 22 (1942): 9–36.
2. Johnson, Samuel, *A Journey to the Western Islands of Scotland* (London: 1775).
3. Soper, F. L., and D. B. Wilson, Anopheles gambiae *in Brazil, 1930 to 1940* (New York: 1943).
4. Metcalfe, H., and J. F. Collins, "The control of the chestnut bark disease." *Farmers' Bulletin U.S. Department of Agriculture* 467 (1911): 1–24.
5. Cooke, M. T., "The spread of the European starling in North America (to 1928)." *Circular U.S. Department of Agriculture* 40 (1928): 1–9.
6. Kalmbach, E. R., "Pigeon, sparrow, and starling control." *Pest Control* 22 (1954).

7. Artimo, A., "Finland a profitable muskrat land. Preliminary report." *Suom. Riista* 4 (1949): 7–61.
8. Myers, J. G., "The arthropod fauna of a rice-ship, trading from Burma to the West Indies." *Journal of Animal Ecology* 3 (1934): 146–149.
9. Lauries, E. M. O., "The coypu (*Myocastor coypus*) in Great Britain." *Journal of Animal Ecology* 15 (1946): 22–34.
10. Freeman, R. B., "*Pitrofquenia coypus* Marelli (Mallophaga, Gyropidae), an ectoparasite on *Myocastor coypus* Mol." *Ent. Mon. Mag.* 82 (1946): 226–227.

William R. Jordan III, from *The Sunflower Forest* (2003)

Conservation biology can serve both to protect and to restore. Threatened and endangered species, through monitoring and, later, captive breeding, were the first conservation targets to receive restorative treatment. But threatened and endangered species can thrive only if their habitats are intact. More important than individual species restoration is the larger effort of restoration ecology. William Jordan provides examples of the types of disturbed ecosystems that restoration ecology targets and then asks some profound questions about the nature of the restorative enterprise.

Ecological restoration is the attempt, sometimes breathtakingly successful, sometimes less so, to make nature whole. To do this the restorationist does everything possible to heal the scars and erase the signs of disturbance or disruption. He or—just as often—she removes exotic plants and animals that have invaded an area. He reintroduces species that have been eliminated. In some cases—in grasslands, for example—this may mean replanting hundreds of species. In others, such as some kinds of wetlands, it means planting just one or two, by hand, rice-paddy fashion, at densities of many thousands per acre. To restore seagrass beds in the shallow water along coastlines, restorationists go underwater, hovering over the ocean floor with scuba equipment, setting plants and fastening them in place with metal staples. When they can, they enlist the help of the natural forces they seek to reinstate. Restorationists working in arid ecosystems in the West have relied on kangaroo rats to collect and disperse seed; others, working on landfills in New Jersey, have set up perches to attract starlings and robins, counting on the seed they deposit in their droppings to help establish a new plant community. When necessary, restorationists rehabilitate soil, recontouring it, adding nutrients to promote growth of native plants, or in some cases finding ways to remove nutrients in order to discourage the growth of fast-growing, weedy species. They reintroduce natural drainage patterns and seasonal patterns of flood and drought, closing ditches, putting the bends back into channelized streams, probing soil with metal rods to find—and then remove or break—drainage tiles put in place generations earlier, in order to nudge a mowing meadow back toward wet prairie, or a field that had been producing soybeans back toward bottomland forest. Working on creeks and rivers, they go upstream to eliminate or abate sources of stress: road salt; toxic chemicals; acid draining from a mine; or altered drainage patterns. When appropriate, as on the prairies, they reintroduce fire, insects, bison, rocks. Restorationists working on Curtis Prairie at the University of Wisconsin–Madison Arboretum in the 1930s disassembled stone fences, rescattering the stones where glaciers had left them ten thousand years earlier, before replanting prairie around them. In some cases restorationists have recontoured entire landscapes at sites disturbed by mining or other forms of land remodeling. For the past decade the Army Corps of Engineers has been putting the meanders back into a stretch of the Kissimmee River in southern Florida, replumbing the landscape to specifications provided by nature and history.

On a smaller scale, restorationist Ed Collins is supervising a project to do the same thing on Nippersink Creek in northern Illinois. There a series of kames, or "gumdrop" hills left by glaciers, were mined decades ago for material that was used to channelize the creek. Collins is taking this "filler" out again, restoring the old meanders and using the recovered glacial till to rebuild the kames. Today, the view from the top of one of Ed's new kames is beginning to resemble that of two hundred years ago. Farm fields are gone, and a program is underway to begin easing out exotic species and replace them with species that once made up the prairies and savannas of the area. As in all restoration projects, the idea—or ideal—is ultimately to replace everything, not just trees, flowers, and free-flowing rivers. If rattlesnakes, mosquitoes, poison ivy, fire, or flood belong in the system, they are brought in too or are encouraged to return. The outcome, when all this is successful, is paradoxical. The aim of the restorationist is to erase the mark of his own kind from the landscape. Yet through the process of restoration he enters into a peculiarly profound and intimate relationship with it.

Restoration is not a new idea. In a general sense humans have been rehabilitating ecosystems altered or degraded by activities such as agriculture or tree cutting for millennia, through practices such as tree planting and the fallowing of land. And projects that were restoration efforts in the fully modern sense—that is, active attempts to recreate whole, ecologically accurate examples of historic landscapes or ecosystems—were undertaken in the United States as early as the 1920s. These were an initiative of conservationists that in certain respects anticipated concerns of the environmentalists of more recent generations. Yet, while restoration has been available, both as an idea and as a practice for most of the past century, until very recently its role in conservation has been negligible. Only since the mid or late 1980s have environmentalists and conservation practitioners begun to take restoration seriously as a conservation strategy.

Today thousands of projects are underway in virtually every kind of ecosystem, from tallgrass prairies and alpine meadows to the coral reefs and tropical forests that have traditionally served as tropes for the irreplaceability of natural ecosystems. Canadian restorationists have done extensive work to restore vegetation on cliff faces of the Niagara Escarpment damaged by climbers, and at least one project has been undertaken to restore the ecosystem inside a cave. Others are working on the restoration of woodland in downtown Manhattan. Restoration projects range in scale from backyard projects of a few square yards to projects like the multi-billion-dollar CalFed program being carried out in the watershed of the Sacramento River in California, or the restoration of Everglades National Park, a project that will eventually encompass tens of thousands of acres. Some of these projects have been planned and are being carried out entirely by professionals, though most involve amateurs, participating as volunteers. Many, especially in major cities like Chicago, San Francisco, Seattle, and New York, have been carried out almost entirely by volunteers. Some have been remarkably successful, summoning back bits and pieces of historic landscapes in places where, surrounded by suburbs, shopping malls, or cornfields, they have reappeared as biological Rip Van Winkles, visitors from the past and rare islands of biological diversity. Others have been less successful. For many, perhaps most, the outcome is not yet clear. In a basic sense, all are works in progress and always will be—a circumstance that many find unsettling.

For generations environmentalists have assumed that the loss or degradation of areas generally described as "natural" is an irreversible process—that we can only subtract from or degrade the natural landscape, never add to or improve it. Yet the best work of restorationists shows that this is not always true. It is possible to recreate reasonably accurate versions of some landscapes or ecosystems. Besides this—and even more important—restoration, properly understood, turns out to be the key to the survival—or preservation—of *all* natural landscapes, not just those that have obviously been degraded or abused.

Whatever we may prefer to think, indulging fantasies of untrammeled wilderness or Edenic landscapes unviolated by human influence, the hard logic of ecology—the principle that everything interacts with (and therefore influences) everything else—makes the conclusion unavoidable: preservation in the strict sense is impossible. Protection from damage or "outside" influence may be a crucial step in the conservation of a classic landscape, but it is only a first step because it is never complete. However high the wall we put around a preserve, we cannot protect it from all novel or outside influences. And since an ecosystem will always respond to these influences, however subtle or indirect, in the long run, which is the only run that counts in natural-area conservation, the best natural—or "classic"—ecosystems will not be those that have been "preserved" from human influences but those that have been subject to management aimed at identifying those influences that do exist and then compensating for them in an ecologically effective way—that is, in an ongoing program of restoration.

While the intensity of restoration will of course vary enormously, reflecting the coefficient of interaction between a particular classic landscape and the novel ecologies created by humans, some kind of restoration is always necessary to keep a classic landscape on its historical course because the coefficient of interaction is never zero. In any event, however small it is, we must never ignore it. As the ecological term in the equation defining our relationship with the landscape, it must be the basis for any program of management, any idea of our relationship with the classic landscape, and any philosophy of nature grounded in ecology. For this reason we are not free to "neglect" it, as a mathematician may drop a small term from an equation to make it easier—or possible—to solve. And if that makes the equation troublingly complex, or even impossible to solve, there is a lesson in that: our relationship with the rest of nature, as with anything, is not really an equation or a problem to be solved. It is a mystery.

So far as wilderness areas are concerned, we will, as environmental managers have long insisted, want to make them as large and as remote as possible, to reduce "outside" influences on them, and to maximize their self-organizing abilities. And we will naturally take pains to minimize our own influence on them. But we must never regard or treat them simply as *preserves*, since to do that is merely to abandon them to drift in the variable breezes of ecological influence toward a future shaped as much by "outside" influences as by the "internal" forces of succession and natural selection. Of course we may choose to allow this on rare occasions, as a kind of experiment. But the landscapes that result from this kind of neglect will not be "preserves" in the old sense, but zones of sacrifice, entailing changes in ecological character and historic quality, and justified not by their effects on the landscape, but by the information they produce or the feelings they generate.

To say this is not to deny the value of wilderness or the importance of preservation as a conservation goal. It is, rather, to acknowledge that restoration, understood in this way, is simply the best way we have of reaching that goal. Understood not merely as an emergency measure, to be deployed in response to dramatic or acute insults to ecosystems such as oil spills or surface mining, but as an ongoing process of ecological compensation, restoration quite simply defines the terms of our relationship with natural landscapes, and the terms on which they will survive through the twenty-first century and beyond.

Americans first learned this lesson on the tallgrass prairies of the Midwest during the early decades of the twentieth century as conservationists sought ways to reverse what was by then the near extinction of this once vast ecosystem. But that lesson has implications for every kind of ecosystem influenced by humans, which is to say, ultimately, for every ecosystem on the planet. It is relevant, for example, to riparian and arid areas in the West, altered by decades of grazing, alterations of hydrological cycles, and introduction of exotic species such as tamarisk. It is relevant to wetlands affected by filling, alterations in hydrology, or the introduction of exotic species. It is relevant to the forests of the Northeast, where in many areas forests have recovered fairly well from clearing during the colonial period but often lack "minor" species such as ants or salamanders or the spring flowers that lend these forests so much of their charm, and which in many instances do not reappear unless reintroduced.

More dramatically, it is urgently needed in the ponderosa pine forests of the Southwest, where years of fire suppression have resulted in accumulations of fuel and closing of canopies that have made thousands of square miles dangerously combustible. This is bad enough, as the fires of 2002 dramatically illustrated. What makes it worse is that, in contrast to fires in other, moister forests farther north, crown fires in ponderosa pine forests are ecologically terminal events. Sterilizing the landscape, they are followed on steep slopes by catastrophic erosion, not only killing the ecological community but destroying the infrastructure of soil and seed that sustains it. Once burned, many such areas are left a moonscape where vegetation of any kind will not recover for centuries.

The situation here is dramatically different from that on the prairies, which die if deprived of fire, but die quietly, leaving the ecosystem intact as the prairie is gradually replaced by another kind of community, often oak forest. The ponderosa pine forests, in contrast, die violently, striking back viciously and self-destructively, demonstrating in a way the prairies do not the worst consequences of human neglect. If it was on the prairies that conservation learned the value of restoration, it may be in the ponderosa pine forests that the rest of us will at last learn its importance.

Importance for the landscape, but also—and inseparable from that—for our relationship with it, since restoration provides, as other paradigms of relationship between our own species and the rest of nature do not, a basis for a relationship that is active, positive, and, in a psychological sense, comprehensive. And this, of course, brings us to the root of our problem. Ultimately, the future of a natural ecosystem depends not on protection from humans but on its relationship with the people who inhabit it or share the landscape with it. This relationship must not only be respectful, it must also be ecologically robust, economically productive, and psychologically rewarding. This being the case, the central task of natural-area conservation is to provide a role for people inside the old, "natural" system that is not only both active and constructive, but that engages and challenges all the human interests and abilities, including those for manipulation and invention as well as for observation, description, and caretaking. Even more challenging, it must provide a way of confronting and coming to terms with the negative and troubling, as well as the more attractive, positive aspects of nature and our relationship with it.

And so far as the classic landscape is concerned, restoration provides the only way of doing this. In effect creating an ecologically and psychologically satisfactory niche for humans in that landscape, it actually solves the "dilemma of use" that has troubled environmentalism for more than a century. Besides this, as projects in cities, suburbs, and rural and wilderness areas are showing, restoration offers a way to bring the human community together and to strengthen the relationship between human and nonhuman nature, on which the fate of the classic landscape ultimately depends.

Despite this, environmentalism has been extremely slow to accept restoration or take it seriously as a conservation strategy, and I have come to believe that the reasons for this have as much to do with the ideas and values of environmentalism as they have to do with the nature and short-comings of restoration. There are many reasons for skepticism about restoration. Some are political and reflect the concern that the promise of restoration might be used—as in fact it has been used—to undermine arguments for the preservation of existing historic landscapes. Others are economic, restoration being, in many cases, a costly and uncertain process that is often seen as a kind of fool tax—the pound of cure that is demanded of those who neglect the ounce of prevention. Still others are conceptual. Sixty years ago conservationists took little interest in restoration because the restorationist's emphasis on historical accuracy and ecological completeness seemed silly and self-indulgent—a kind of boutique conservation. More recently, environmentalists have seen it as presumptuous. Many have been uneasy with restoration because it violates in a particularly troubling way the categories of "nature" and "culture" that play so large a role in environmental thinking and rhetoric. Overall, restoration, with all its complexity and uncertainty, is simply not the way we have expected a healthy relationship with nature to *feel*.

Only recently has that expectation begun to change as environmentalists and managers have begun to discover the value of restoration as a strategy for conserving classic landscapes and developing a vital, satisfying relationship with them. As this has happened it has become clear that the neglect of restoration has been a mistake—indeed, one of the defining mistakes of twentieth-century environmentalism. . . .

Restoration raises many questions. Some of these, like those that led . . . me to the notion of restoration ecology, were merely technical or ecological questions about the process of restoration and the ecology of the ecosystems being restored. But others were more fundamental. What, for example, does the metaphor of restoration—or "standing up again"—mean when we apply it to a dynamic, living system such as an ecological community or ecosystem? If such a system is a moving target, with a complex history and an unpredictable future, what does it mean to restore it? How should we choose the historic ecosystem or landscape we want to use as a model, and how can we describe or define it so that we will know whether we have succeeded in our attempt to reproduce it? Even assuming that we can define an objective in some meaningful way, is restoration possible? Given the complexity of the work and the uncertainty of the results, wouldn't it be better simply to let a disturbed landscape alone, removing obvious sources of disruption and relying on natural processes to bring about recovery? What is the value of a restored landscape? And what sort of thing is it? Is it "nature," or merely a human artifact misrepresented as nature—at best a poor substitute and at worst a deception and a fake, as some have argued? What is the proper role of humans in the restoration—and by extension the management, and ultimately the ecology—of natural or historic landscapes? Is restoration the epitome of a responsible, nurturing relationship between ourselves and our fellow species and with the rest of nature generally? Or is it, as some have insisted, simply another arena for the exercise and aggrandizement of the human ego?

What, more generally, is the proper relationship between ourselves and the rest of nature? If restoration, understood as compensation for "outside" influences on a landscape, is the indispensable paradigm for the conservation of historic ecosystems, what are we to make of the fact that it is difficult, uncertain, and in some ways emotionally challenging work?

All these questions apply to all aspects of our relationship with nature, but many of them are easy to ignore in a landscape that is regarded as a "preserve" or found object. They are almost impossible to ignore in a restored landscape. . . . Thinking them over, I gradually realized that these questions were perhaps the most valuable thing about restoration, but that they were also the reason why several generations of environmentalists had either ignored or resisted it. Such questions point to our ignorance, which is troubling enough, but they also bring us face to face with our deepest uncertainties, and in doing so they challenge ideas about nature, about relationship, and about the nature of relationship that environmental thinkers have taken for granted since the time of Emerson and Thoreau.

12 Soil and Agriculture

All terrestrial life depends on the sun, water, and air. Although we rarely think of "dirt" as essential, such life depends equally on soil. Ninety-seven percent of human food comes ultimately from the soil, a general term denoting all loose covering on land. As population increases and arable land is used for buildings and roads, more marginal soils are farmed. Unfortunately, marginal land when cultivated is especially susceptible to degradation and increased erosion. The natural phenomenon of erosion moves soil and even creates it by breaking down rock. But soil erosion becomes problematic when wind and water remove disproportionate amounts of organic material. Soil and its erosion are crucial considerations in all terrestrial ecosystems, including forests and uncultivated areas. Erosion is the enemy of cultivation and, unfortunately, cultivation speeds erosion.

Individual soils exhibit varied microbial ecologies and chemical compositions. The interactions of plants in the wild, farming, fertilizers, crop selection, climate, water, and wind are complex. In order to sustain food production over long periods of time, these interactions must be carefully monitored and managed. Poor farming methods and climatic extremes create a recipe for catastrophe, as the 1930s Dust Bowl in the American Midwest showed. Repeated planting of one crop depletes vital soil nutrients. Slash and burn agriculture and inadequate water threaten soil preservation. Short-term policies and practices that temporarily boost crop yields or profits often cause long-term degradation, even permanent loss. In 2004 the U.N. estimated that by 2025 two-thirds of arable land in Africa may disappear, along with one-third in Asia and one-fifth in South America.

The introduction of new strains of seeds developed in the later twentieth century, the so-called Green Revolution, dramatically increased crop yields for rice, maize (corn), soy, and other staples. Genetically manipulated seeds (see Chapter 4, above; note especially the Mangelsdorf selection on the history of corn improvement) provide benefits such as pest resistance; however, they can cause problematic changes in ecological systems, some known, others unforeseen. Diets rich in meat, particularly red meat, heap added burdens on soil and

Figure 12.1 Egyptian farming approximately 4,000 years ago, from original Middle Kingdom wall-painting near Luxor.

Figure 12.2 Egyptian farming today.

water resources. Moreover, it is unclear if heavy use of artificial nitrogen fertilizers can continue indefinitely without causing significant environmental stress. The nitrogen cycle, which carries nitrogen and energy through the atmosphere, plants, animals, water, oceans, and nitrogen-fixing bacteria, is too pervasive to control easily yet too delicate to be immune from damage. Pesticides can degrade soils. Soil contamination by toxins often forces costly and protracted cleanups.

Enough food exists to feed everyone, but poverty forces two billion people to be malnourished. The chief question remains: given increasing population pressures, what agricultural practices can we sustain over long periods of time? These practices must sustain soil itself. They include preservation of the best soils for cultivation, carefully planned irrigation, reconsideration of fertilizers and pesticides, tilling methods to reduce erosion, selection and even manipulation of seeds advantageous to local conditions and diets, and the reconsideration of diets themselves. Older farming methods, although shaped by centuries of trial and error, can often be improved. As with current water use, human exploitation of the soil as it is now occurring across the planet cannot be sustained.

The soil also supports the world's forests and native grasslands. These are vital to human life and the planet's ecological well-being. In addition to providing material for housing, paper, and fuel, forests serve as CO_2 sinks, recreation areas, and habitat for countless species of birds, other animals, and plants (see Chapter 6, above). Our economic prosperity and our psychological health, as well as the biodiversity of the planet, rest on the preservation, good management, and sustainability of the Earth's woodlands and grasslands.

Both the past and the present are replete with examples not only of outright exploitation and greed, but also of institutional and intellectual failures to come to grips with the ecology of forests and grasslands and human impacts upon them. Skewed legislation and deficient management practices on U.S. lands indicate that we are not yet sufficiently attuned to what this irreplaceable dimension of our ecology and our economy will require in order to survive into the twenty-second century. Finding answers involves something like a revival of the interdisciplinary study of geography—long in decline in America, though less neglected in Canada and Britain—under the rubric of environmental studies. For it is only when biology and ecology meet political science, economics, and history (material, intellectual, and cultural) that we can understand how to live with, and for, our forests.

The Earth's surface is three-fourths water, but human interaction with the environment takes place disproportionately on land. Urban areas are responsible for concentrated environmental degradation, but it is the conversion of forests and grasslands to agriculture that has always left the largest overall human footprint on the land. Even in a time when the occupation "farmer" has been removed from the U.S. census, agriculture in its industrial, corporate form has unprecedented influence on the health of ecosystems, and the greatest steps we can take for environmental protection remain those that work toward making agriculture more ecologically sound.

FURTHER READING

Barrow, C. J. *Land Degradation: Development and Breakdown of Terrestrial Environments.* Cambridge: Cambridge University Press, 1991.

———. *Regenerating Agriculture: Policies and Practice for Sustainability and Self-Reliance.* London: Earthscan, 1995.

Clay, Jason. *World Agriculture and the Environment.* Washington, D.C.: Island Press, 2004.

Kimbrell, Andrew, ed. *Fatal Harvest: The Tragedy of Industrial Agriculture.* Washington, D.C.: Island Press, 2002.

McCann, James. *Maize and Grace: Africa's Encounter with a New World Crop, 1500–2000.* Cambridge, Mass.: Harvard University Press, 2005.

Pollan, Michael. *The Omnivore's Dilemma: A Natural History of Four Meals.* New York: Penguin Press, 2006.

Pretty, Jules N. *Agri-Culture: Reconnecting People, Land and Nature.* London: Earthscan, 2002.

Steiner, Frederick R. *Soil Conservation in the United States: Policy and Planning.* Baltimore: John Hopkins University Press, 1990.

Worster, Donald. *Dust Bowl: The Southern Plains in the 1930s.* New York: Oxford University Press, 1979.

Duane L. Winegardner, "The Fundamental Concept of Soil" in *An Introduction to Soils for Environmental Professionals* (1996)

This selection is drawn from a book entitled An Introduction to Soils for the Environmental Professionals. *As the title suggests, the nature of soil is integral to environmental concerns, and overall environmental concerns are essential to the specialized study of soil. For centuries, care of soil followed farming techniques refined over generations. Since about 1850 "scientific" farming and chemical analyses have enhanced and further explained the qualities of soil. Land-grant universities and agricultural colleges were founded to improve and spread this knowledge.*

From a purely pragmatic viewpoint, soil can be considered everything that is included in the superficial covering of the Earth's land area. Based on environmental perspectives, soil is an aggregate of unconsolidated mineral and organic particles produced by the combined physical, chemical, and biological processes of water, wind, and life activity.

In common usage, the definition of soil varies according to the user. For agriculturists (and related scientists), the important factors in soil focus on the upper few feet of the soil column, which are important to plant growth. Civil engineers consider soil a structural material with definable physical and chemical properties that can be manipulated (or tolerated) for construction purposes.

Microbiologists are interested in the interaction of microbes in the life cycle of the soil. Soil chemists study the detailed chemical reactions that result from the continuously changing underground laboratory. Questions asked by geoscientists include: Where did it originate? How did it get here? What is the next stage of its development? and Of what is it composed?

The science of soil physics, which evolved from agricultural studies, considers the mechanical functions of soil such as fluid flow, interparticle relationships, and the complex interaction between the soil atmosphere, water, mineral surfaces, and organic matter.

Scientists in the individual disciplines of soil science identified in the preceding paragraphs tend to focus their efforts in specific directions toward the solution of well-defined problems. The environmental specialist is expected to be a "jack-of-all-trades" or (in medical terms) a "general practitioner." Environmental science recognizes that all terrestrial life depends on soil for its existence. It is important that environmental specialists have at least a deep appreciation of the complex interactions of all of the various happenings occurring within the "soil sphere."

The chemical composition and physical structure of the soil at any given location are determined by: (a) the type of geological material from which it originated, (b) the vegetative cover, (c) length of time that the soil has been weathered, (d) topography, and (e) the artificial changes caused by human activities. Land surfaces almost everywhere are covered by this unconsolidated debris (soil) sometimes called the *regolith.* This blanket above the bedrock may be very thin, or it may extend to depths of hundreds of feet. Its physical and chemical composition may vary, not only horizontally, but vertically, and its geological origin is not always the same, even within the local area. Some soils are formed from the bedrock (or other geological material) which immediately underlies it. Other soils develop as the result of transported materials being deposited in their current location by the action of water, ice, or wind.

Regardless of the origin, most soils consist of four basic components: mineral matter, water, air, and organic matter. These materials are present in a fine state of subdivision (individual particles) and intimately mixed. In fact, the mixture is so completely blended that separation is difficult. The more solid part of the soil, and naturally the most noticeable, is composed of mineral fragments in various stages of decomposition and disintegration. A variable amount of organic matter, depending on the horizon observed, is blended with inorganic substances. Normally, the largest amount of organic matter is in the surface layer.

The mineral material has its genesis in the parent material. Some soil minerals persist almost intact, while others are quickly transformed into new minerals in response to the current soil environment. Various sizes of mineral particles occur, ranging from coarse sand (2 mm in diameter) to finely divided clay particles (<0.002 mm in diameter).

Organic matter represents the accumulation of plant and animal residues (generally in an active state of decay), especially in shallow horizons. Most organic matter is distributed in finely divided clay-sized particles. The organic content of soil ranges from less than 0.05 percent to greater than eighty percent (in highly organic soil such as peat), but is most often found between two and five percent.

All soil, even the most dense, has significant pore volume between soil grains. Soil pores are variable as to continuity, local dimensions, and total volume. These void spaces may be filled with air or water. The proportion of water-filled or air-filled pores is dependent upon the character of the soil, the conditions of formation, and when the last water was added to the system. Soft clay may have void spaces of over sixty percent by volume, while a well-compacted, uniformly sized sand can have pore volumes in the twenty-five percent range.

Water in the soil structure can be either transitory or fixed (at least temporarily). The part of the water that is retained is held with varying degrees of tenacity. Surface contact between the water and soil particles is the key to many chemical and physical reactions. Soil water is never without solutes (dissolved components) as its solvent properties, along with varying acidic and oxidation potential, causing it to be a major player in the dynamic nature of soil.

The general components of the soil atmosphere are nitrogen, oxygen, and carbon dioxide. Nitrogen is relatively inert unless fixed by soil bacteria. Free oxygen is present unless it has reacted with mineral matter or organic matter (especially living tissue).

Carbon dioxide can result from respiration of microbes, or abiotic chemical reactions. The concentration of carbon dioxide in the soil air is usually greater than the outside atmosphere. The proportion of oxygen, nitrogen, and carbon dioxide not only varies inversely with the amount of water present in the soil, but is extremely variable as to the ratio of the gases present.

Soils have significant variations in appearance, fertility, and chemical characteristics, depending on the mineral and plant materials from which they were formed and continue to be transformed. The soil realm has been compared to an elaborate chemical laboratory where a large number of reactions occur simultaneously. A few of the reactions are relatively simple and well understood, while the vast majority are not yet completely explained. The reactions range from simple solution and substitution to complex biologically mitigated multistep processes. Many reactions depend on the participation of water, mineral, and biological factors in a dynamic setting.

Soil, then, is a dynamic environment; almost a living structure. Continuous processes (although relatively slow) are active in even the most remote settings. Soil has been called "the bridge between life and the mineral world." All life owes its existence to a few elements that must be ultimately derived from the

Earth's crust. After weathering and other processes create soil, plants (including microbes) perform the intermediary role of assimilating these necessary elements, making them available to animals and humans.

All mineral energy sources on Earth, as we know it, come from plants that have grown in the soil while obtaining their energy from the sun. Most "natural" materials used by man are derived in some way from the soil.

As concern for the environment increases, it is comforting to recognize that soil, when properly used, can offer an unlimited potential for disposal and recycling of waste materials. Knowledge of physical, chemical, and biological reactions is more important to us today than ever before.

Alfredo Sfeir-Younis and Andrew K. Dragun, from *Land and Soil Management: Technology, Economics, and Institutions* (1993)

These authors represent the newer breed of environmental professionals examining one issue in its larger contexts: how are problems of soil conservation related to technological advances and larger structures of society—markets, taxes, government policies, incentives, subsidies, and education? These structures influence and at times dictate what people, including farmers, decide to do. How can lessons learned in one country or locale be best applied in another?

Roots require a proper environment for growth as they expand into pores or cracks in the soil. This proper environment is characterized by appropriate temperature, degree of moisture, and aeration capacity. Soil texture and structure determine the working properties of a soil, especially water relationships and the availability of nutrients to growing plants.

Soils contain organic matter and nutrients. One gram of soil contains millions of bacteria, their quantity, type, and role depending on their soil environment. Nutrients come in varying quantities and in degree of solubility and availability to plants. Major (or macro-) nutrients include nitrogen, phosphorus, and potassium, while trace elements (micronutrients) include iron, zinc, and boron. . . .

SOIL EROSION

It should be remembered that soil is also being continuously created by natural processes. As bare rock is broken down by the action of water, temperature, and wind, soil is created and the top layer is mixed with organic matter. The process is very slow, however; a study in Indiana found rates of soil formation from different sources to vary one inch every 25 years to one inch every 500 years. When soil erosion begins however, it usually proceeds at a much faster rate than soil creation, thereby leading to a long-term net loss of soil.

Figure 12.3 Soil erosion exposes tree roots in western Kenya.
Courtesy of Dino J. Martins.

Soil erosion is a physical phenomenon of the soil surface which has economic effects, both on upstream soil quality (and thus on yields) and on amounts of waterway sediment (and thus on water and habitat quality). Erosion is a process that includes three steps, the *detachment* of particles of soil by wind and water from the surface, the *transportation* of these particles, and the *deposition* of these particles in another place.

In water erosion, detachment occurs when raindrops strike the soil, or when running water picks up loosened soil and transports it, or when water acts as a lubricant and causes soil to slip down slopes.

The amount of detachment by rainfall is closely related to the intensity or rate of rainfall. The amount of soil that can be washed away by surface runoff depends on the volume of runoff and the velocity of flow. Some of the rain is absorbed by plants and soil, the surplus becoming surface runoff. The amount of runoff is increased where there is a large amount of rain and low absorption, because of poor plant cover or shallow soil.

Soils damaged by erosion thus become more vulnerable to further erosion. The velocity of surface runoff affects its power to cause erosion, so steep slopes or long slopes are more at risk. One of the chief control measures is to decrease steep slopes by bench terracing, or to break up long slopes by horizontal banks that intercept the runoff.

Water erosion has been classified in many different ways. Here the following four classifications are used:[1]

1. *Sheet erosion*—the removal of more or less uniform layers of soil. It is caused by raindrop splash and the flow of sheets of water over entire sur-

faces. Sheet erosion may appear to be a slow process because only a few millimeters of soil are removed annually, but relative to total soil removed and geographical extent, it is the most serious form of erosion. One of the worst aspects of sheet erosion is its insidiousness [i.e., it is not easily observed or measured].

2. *Rill erosion*—erosion occurring along small channels or rills small enough to be obliterated by normal tillage. Rill erosion is common in newly tilled land, on steep slopes, and in such places as building sites and road embankments. Both detachment and transport of soil occur along the rills.
3. *Gully erosion*—erosion occurring in channels too large to be eliminated by tillage. Gullies extend by headward recession and waterfall action, as well as by the undermining of their sides by flowing water. Gullies act as major lines of transport out of a catchment for sheet-eroded, rill-eroded, and gully-eroded soil—a high velocity flash flood may do this.
4. *Stream bank erosion*—erosion of soil from banks by water in permanently flowing streams. Stream bank erosion is not triggered directly by rainfall, but is increased by higher rainfall. . . .

Humans are an important agent of change, affecting the rate at which soils are being eroded around the world. The most important factors are: deforestation, driven most often by demand for wood-fuel energy as well as traditional forestry products; certain forms of intensive agriculture; inappropriate agricultural practices, such as plowing up-down rather than across the slope of a hill; shifting cultivation, when land is left fallow only for short periods of time; overgrazing; and forest fires.

Other factors that deserve mention are poverty (creating pressures to use land continuously to satisfy subsistence needs) and high population density compared with the carrying capacity of existing soils. Important institutional factors are insecurity of land tenure, which makes farmers unwilling to invest in land seen not to be theirs, and small holdings that make it impossible to carry out conventional soil conservation practices.

Economic factors are also important. . . . Changes in prices, interest rates, taxes, subsidies, and marketing costs are all significant determinants of farmers' attitudes toward the conservation of soil. These factors are often called "incentives," whether they operate through markets or outside market forces (that is, as state regulations). . . .

The Process of Erosion

Economic analysis of soil erosion programs often begins with the establishment of some notion of productivity, commonly crop yields. Erosion alters the characteristics of soils as a medium for plant growth by changing productive capacity. Erosion reduces productivity mainly through loss of water retention capacity in the soil, causing water stress in plants. Erosion may reduce the depth of the root zone, the water-holding capacity of the remaining topsoil, or both.

The primary effect of erosion is loss of topsoil available for agricultural production. This type of damage occurs when the rate of soil erosion is greater than the rate of topsoil formation in any given parcel of land. Generally, the topsoil is the most workable and permeable soil, providing a hospitable medium for root growth, water retention, and nutrient storage. Losses in topsoil affect these vital functions. Above and beyond the simple removal of topsoil, erosion leads to changes in soil structure, usually for the worse, causing problems of surface crusting, cultivation difficulties, and uneven crop growth. The combination of loss of topsoil, tighter soil texture, and exposure of subsoil can result in lower infiltration and hence higher rates of runoff.

One of the most insidious characteristics of erosion is the difficulty of detecting it by eyesight. Erosion often reduces productivity so slowly that the change may not be recognized until the land is no longer economically suitable for growing crops.

The concomitant removal of plant nutrients is also part of the erosion process. Subsoils generally contain fewer plant nutrients than topsoil, so sediments that run off often contain more nutrients. This has often been expressed by saying that the erosion process is "selective" in that erosion removes the finer and more fertile particles, leaving behind coarse and infertile sands. Additionally, the impact of raindrops on surface can effectively seal the soil, rendering it impervious to water.

Soil Erosion As an Economic Problem

Soil conservation projects are not cost-free. They require resources that could be allocated to other economic activities. Planners have to balance the current costs of soil conservation practices with the future values resulting from those practices. In assessing these tradeoffs there are several complicating factors. First, most of the off-site effects are not considered in the farmers' decision-making process and the amount of erosion produced by one rational individual may be much greater than what society would find optimal. Second, planning occurs under extreme uncertainty with regard to the value of agricultural productivity in the future, coupled with a high probability of irreversible damages. And third, processes to combat soil erosion have long-term returns that farmers may regard as far beyond their planning horizon.

Other Forms of Land Degradation

Erosion is only one form of land degradation. Land is also degraded by accumulations of salts and alkali, organic wastes, infectious organisms, industrial inorganic wastes, pesticides, radioactive wastes, heavy metals, fertilizers, and detergents. Saline and alkali soil problems are usually associated with irrigated lands. Organic wastes are often associated with the disposal of industrial, domestic, and municipal wastes, [as] are the addition of salts and the specific ion effects of boron. Degradation due to infectious organisms is associated with disease and insects. Industrial wastes affect soils unfavorably because they release stack gases (for example, sulfur dioxide, fluorides) with inorganic residues. Pesticides are very persistent in the soil, and accumulations of chlorinated hydrocarbons have become a major concern. Radioactive effects on soils, though not now a major problem in LDCs [lesser-developed countries], may become an important source of land degradation in the future. Traces of heavy metals have shown up in many foods because of dumping of industrial wastes. Although fertilizers are a source of soil nutrients, they are also associated with contaminants—radioactive elements, for example, have been detected in raw rock phosphates. . . .

SOIL CONSERVATION METHODS

To remove or alleviate the major causes of soil erosion, policy makers may design and implement investment programs, or make institutional changes, or do both. Most investment programs include methods that may be classified into two groups—engineering or mechanical protection methods and biological protection methods. Although this classification is rather arbitrary, it helps to identify the sources of benefits and costs of soil conservation projects. Without such methods, the costs incurred—in the form of lost agricultural production, for example—can be formidable.

Many *mechanical* methods are used to control erosion. Stormwater diversion drains in ditches intercept stormwater that would otherwise flow down from higher ground onto arable land. Channel terraces, a form of earthwork at right angles to steep slopes, intercept surface runoff. Artificial watercourses, which include grass waterways, sod waterways, or meadow strips, can be used to drain runoff into water channels. Bench terraces, which convert a steep slope into a series of steps having horizontal (or nearly horizontal) ledges and vertical (or almost vertical) walls between the ledges, can be constructed on the contour to minimize runoff, or with a slight gradient like graded terraces. Irrigation terraces, planted in raised strips between irrigation channels, are extensively used to grow rice, tea, fruit trees, and other high-value crops. Orchard terraces are small, level or reverse-slope terraces, each carrying one line of trees. Contour bands, which are low banks thrown up by hand-digging approximately on the contour, are designed to conserve both soil and water. Pasture furrows are small shallow drains that spread out surface water as evenly as possible by not allowing it to concentrate in depressions or across watercourses. Their primary purpose is to conserve water by increasing infiltration, a technique often used on grasslands. Tied ridging consists of covering a whole surface with closely spaced ridges in two directions at right angles so that the ground becomes a series of rectangular depressions. Contour cultivation and grass strips involve tillage cultivation on contours between strips of grass or other close-growing vegetation. One last method is ridge and furrow, where the ground is tilled into wide parallel ridges interspersed with water-saving furrows. These combine erosion control with surface drainage. . . .

Most of the *biological* methods of soil conservation involve crop rotation, pasture use, forest planting and management, strip cropping, planting of wind breaks, and sand dune stabilization. One can classify different land use groups by the relative efficiencies of crop cover that protect the soil from erosion. A classification of this kind must be region and soil-type specific. In some regions, for example, permanent vegetation like trees may be more effective in protecting existing soil than certain row crops. With regard to cropping practices, different forms of tillage, planting methods, fertilizer use, and harvesting methods greatly affect the productivity of soils.

The distinction between mechanical and biological forms of soil conservation is not just a matter of technique. They conserve the soil in essentially different ways. Mechanical methods *control* erosion—that is, after the soil has started moving, prevent it from moving offsite. Biological methods *prevent* erosion—that is, they intercept raindrops, thus not allowing the erosion process to start. Prevention is better than cure and certainly better than control.

Any of these methods generate both costs and benefits for the farmer. The scientist can help to assess the actual benefits of each method in saving soil. The task of the economist is to quantify in monetary terms the value of the potential net incremental benefits from these soil conservation practices.

NATURAL EROSION

Erosion has always taken place and always will. The surface of the Earth is constantly changing under the forces of nature, and erosion is one aspect of this constant process of change. It is the starting point for the formation of sedimentary rocks and alluvial soils. The rate of erosion depends on climatic, topographic, geologic, and other physical parameters. Hence, it varies greatly across the world.

Human related erosion is sometimes defined as geological erosion accelerated by human activities, but it is usually difficult to distinguish the two. Only in areas completely unsettled by humans can one be certain that humans have had no influence. . . .

Humans are active agents of change and actions by humans usually accelerate the soil erosion process. However, there is a background erosion rate, known as "normal" or "geological" erosion. . . .

There are many forms of geological erosion—leaching, oxidation, landslides, and surface erosion. The speed at which geological erosion takes place varies from place to place. In desert areas, for example, the wind causes high rates of erosion. . . .

Erosion therefore exists without human intervention in all parts of the world. However, the scale of the difference between human-induced and natural erosion can best be appreciated by referring to a classic experiment conducted in Africa.[2]

Over a period of ten years a bare soil site lost an average of 127 tons/ha [hectare], while a fully protected adjacent site lost less than one ton/ha/year. Differences in the amount of water runoff (and hence soil moisture to sustain plant growth) were no less spectacular. The difference between the sort of erosion focused on [here] and geological erosion can be at least 100:1, a ratio which emphasizes just how critical the task of soil conservation is. . . .

THE MEANING OF CONSERVATION

Conservation, or the countering of depletion, is a dynamic process which reflects changes in the rate at which a resource is used over time. When any given natural resource is exploited, a redistribution of use rates toward greater use in the future leads to conservation (a decrease in present use). By the same token, a redistribution of use rates toward the present leads to depletion (an increase in present use rate). Thus, the study of conservation with respect to existing natural resources consists of comprehensive analysis and comparison of redistribution—that is, the direction of change of use rates over time. Conservation is a dynamic concept and does not mean nonuse of resources. . . .

The nature and magnitude of the erosion problem in many countries is such that action is urgently needed. Erosion is a natural phenomenon, but imbalances have been created by human actions. Worsening erosion and sedimentation will impair the achievement of long-term development objectives. . . .

We got where we are now partly because of lack of integration between sectoral policies and macroeconomic policies. Macroeconomic policies on such things as prices, export incentives, foreign exchange earnings, and the like, that might appear to have no connection with land degradation are indirectly having a substantial impact on soil quantity and quality.

. . . Land degradation is not only an economic problem. It is a social problem as well, and policy makers should understand that the effects extend beyond this generation and that its causes are not simply the actions of individual farmers.

Land degradation is also a social, demographic, physical, and, perhaps most important of all, a political problem. Difficult decisions must be made to avoid irreversible damages to the basic foundations of development, people, and national resources. Preventative actions are certainly more appropriate than curative actions. Thus, strategies to deal with land degradation require firm political commitments now. . . .

WHERE DO WE GO FROM HERE?

. . . Several determinants of success can be clearly identified:

1. In many instances, very minor changes (accompanied by a favorable policy environment) can check the rate at which land is being degraded. In other words, *soil conservation practices are not necessarily expensive.*
2. *Farmers must participate in the decision-making process.* Clearly, a lack of farmer participation is responsible for the poor sustainability of conservation decisions and practices.
3. *Management objectives and policies must be developed* and these must be accompanied by practical *guidelines and standards.*
4. Income and food-related policies must accompany conservation policies. In other words, conservation policies *cannot be enacted or put into place in isolation.*
5. Effectiveness in policy or program implementation depends upon *knowledge and information.* Lack of these two ingredients will result in decisions which, in many instances, lack even common sense.
6. Investments and policies should be judged not only on their economic merits but on other grounds, sometimes including *ethical or moral* grounds.
7. *Replicability and area coverage* are two basic determinants of project success. But the implications of this are several. Government and financial institutions must understand that in order to control land degradation, it is imperative to deal with vast areas of a country, even at the risk of supporting economically "unprofitable" projects.
8. Investments and policies should be oriented toward *internalizing the negative effects* of human actions, whether of farmers or other actors in the economy. Those responsible for the damages should somehow be made aware of the costs of correcting the damages—in some cases, by helping to pay the costs.

HOW DO WE GET WHERE WE WANT TO GO?

To achieve success in a reasonable time, it is imperative to take an integrated approach across sectors, across decision-making levels, and across private and public sector boundaries.

A move toward rational exploitation of the natural resources available to agriculture and other sectors must be guided by a clear set of management objectives. Without these objectives, investment and policy changes will be carried out in a vacuum. Several objectives can be singled out: (1) the need to look into soil and water quality as well as quantity changes; (2) the recognition of demographic problems; (3) assessment of conservation/development tradeoffs; and (4) definition of an action program to lead the way into the future.

With regard to soil quantity and quality, a basic set of management objectives must be followed to avoid irreversible damages, to reverse degradation problems in the medium term, to reduce erosion to tolerable levels, and to encourage the retention of prime agricultural land. With regard to water quantity and quality, investments and policies should be designed to minimize the adverse effects of organic wastes, to reduce pollution from excessive nutrients and salinity, to minimize levels of toxic pollutants, and to reduce sedimentation effects.

It is important to identify areas where the population places stress on the natural resource base. In many instances it may be necessary to protect fragile land while using it to produce food or fiber, while living with certain negative impacts—at least in the short term.

Coping with the challenges ahead will not be an easy task. Countries as well as financial institutions will need to establish a comprehensive conservation agenda, set reasonable targets, and focus on achievements.

Let us remember that poor land makes people poor.

Notes

1. Trolh, F., J. Hobbs, and R. Donahue, *Soil and Water Conservation for Productivity and Environmental Protection* (Englewood Cliffs, N.J.: Prentice-Hall, 1980).
2. Hudson, N., *Soil Conservation*, 2nd ed. (Ithaca: Cornell University Press, 1981).

Richard Manning, from "The Oil We Eat" (2004)

The consumption of food provides people with the energy, measured in calories, that allows their metabolic activity, and consequently their life, to continue. It also takes energy to produce the food that gets to our tables. In the past, a great deal of this energy was supplied by the muscles of farmers and oxen, but today most of it is delivered by machines. Richard Manning details the energy inputs needed to grow and distribute food in the twenty-first century and traces those inputs back to their source, which is, more often than not, oil. His investigation raises the question whether our food production energy budgets are properly balanced.

The secret of great wealth with no obvious source is some forgotten crime, forgotten because it was done neatly.—Balzac

The journalist's rule says: follow the money. This rule, however, is not really axiomatic but derivative, in that money, as even our vice president will tell you, is really a way of tracking energy. We'll follow the energy.

. . . As James Prescott Joule discovered in the nineteenth century, there is only so much energy. You can change it from motion to heat, from heat to light, but there will never be more of it and there will never be less of it. The conservation of energy is not an option, it is a fact. This is the first law of thermodynamics.

Special as we humans are, we get no exemptions from the rules. All animals eat plants or eat animals that eat plants. This is the food chain, and pulling it is the unique ability of plants to turn sunlight into stored energy in the form of carbohydrates, the basic fuel of all animals. Solar-powered photosynthesis is the only way to make this fuel. There is no alternative to plant energy, just as there is no alternative to oxygen. The results of taking away our plant energy may not be as sudden as cutting off oxygen, but they are as sure.

Scientists have a name for the total amount of plant mass created by Earth in a given year, the total budget for life. They call it the planet's "primary productivity." There have been two efforts to figure out how that productivity is spent, one by a group at Stanford University, the other an independent accounting by the biologist Stuart Pimm. Both conclude that we humans, a single species among millions, consume about forty percent of Earth's primary productivity, forty percent of all there is. This simple number may explain why the current extinction rate is 1,000 times that which existed before human domination of the planet. We six billion have simply stolen the food, the rich among us a lot more than others.

Energy cannot be created or canceled, but it can be concentrated. This is the larger and profoundly explanatory context of a national-security memo George Kennan wrote in 1948 as the head of a State Department planning committee, ostensibly about Asian policy but really about how the United States was to deal with its newfound role as the dominant force on Earth. "We have about 50 percent of the world's wealth but only 6.3 percent of its population," Kennan wrote. "In this situation, we cannot fail to be the object of envy and resentment. Our real task in the coming period is to devise a pattern of relationships which will permit us to maintain this position of disparity without positive detriment to our national security. To do so, we will have to dispense with all sentimentality and day-dreaming; and our attention will have to be concentrated everywhere on our immediate national objectives. We need not deceive ourselves that we can afford today the luxury of altruism and world-benefaction."

"The day is not far off," Kennan concluded, "when we are going to have to deal in straight power concepts."

If you follow the energy, eventually you will end up in a field somewhere. Humans engage in a dizzying array of artifice and industry. Nonetheless, more than two-thirds of humanity's cut of primary productivity results from agriculture, two-thirds of which in turn consists of three plants: rice, wheat, and corn [maize]. In the 10,000 years since humans domesticated these grains, their status has remained undiminished, most likely because they are able to store solar energy in uniquely dense, transportable bundles of carbohydrates. [In the last few centuries, sugarcane has joined their rank.] They are to the plant world what a barrel of refined oil is to the hydrocarbon world. Indeed, aside from hydrocarbons they are the most concentrated form of true wealth—sun energy—to be found on the planet.

As Kennan recognized, however, the maintenance of such a concentration of wealth often requires violent action. Agriculture is a recent human experiment. For most of human history, we lived by gathering or killing a broad variety of nature's offerings. Why humans might have traded this approach for the complexities of agriculture is an interesting and long-debated question, especially because the skeletal evidence clearly indicates that early farmers were more poorly nourished, more disease-ridden and deformed, than their hunter-gatherer contemporaries. Farming did not improve most lives. The evidence that best points to the answer, I think, lies in the difference between early agricultural villages and their pre-agricultural counterparts—the presence not just of grain but of granaries and, more tellingly, of just a few houses significantly larger and more ornate than all the others attached to those granaries. Agriculture was not so much about food as it was about the accumulation of wealth. It benefited some humans, and those people have been in charge ever since.

. . . It is no accident that no matter where agriculture sprouted on the globe, it always happened near rivers. You might assume, as many have, that this is because the plants needed the water or nutrients. Mostly this is not true. They needed the power of flooding, which scoured landscapes and stripped out competitors. Nor is it an accident, I think, that agriculture arose independently and simultaneously around the globe just as the last ice age ended, a time of enormous upheaval when glacial melt let loose sea-size lakes to create tidal waves of erosion. It was a time of catastrophe.

Corn, rice, and wheat are especially adapted to catastrophe. It is their niche. In the natural scheme of things, a catastrophe would create a blank slate, bare soil, that was good for them. Then, under normal circumstances, succession would quickly close that niche. The annuals would colonize. Their roots would stabilize the soil, accumulate organic matter, provide cover. Eventually the catastrophic niche would close. Farming is the process of ripping that niche open again and again. It is an annual artificial catastrophe, and it requires the equivalent of three or four tons of TNT per acre for a modern American farm. Iowa's fields require the energy of 4,000 Nagasaki bombs every year.

Iowa is almost all fields now. Little prairie remains, and if you can find what Iowans call a "postage stamp" remnant of some, it most likely will abut a cornfield. This allows an observation. Walk from the prairie to the field, and you probably will step down about six feet, as if the land had been stolen from beneath you. Settlers' accounts of the prairie conquest mention a sound, a series of pops, like pistol shots, the sound of stout grass roots breaking before a moldboard plow. A robbery was in progress.

When we say the soil is rich, it is not a metaphor. It is as rich in energy as an oil well. A prairie converts that energy to flowers and roots and stems, which in turn pass back into the ground as dead organic matter. The layers of topsoil build up into a rich repository of energy, a bank. A farm field appropriates that energy, puts it into seeds we can eat. Much of the energy moves from the earth to the rings of fat around our necks and waists. And much of the energy is simply wasted, a trail of dollars billowing from the burglar's satchel.

I've already mentioned that we humans take forty percent of the globe's primary productivity every year. You might have assumed we and our livestock eat our way through that volume, but this is not the case. Part of that total—almost a third of it—is the *potential* plant mass lost when forests are cleared for farming or when tropical rain forests are cut for grazing or when plows destroy the deep mat of prairie roots that held the whole business together, triggering erosion. The Dust Bowl was no accident of nature. A functioning grassland prairie produces more biomass each year than does even the most technologically advanced wheat field. The problem is, it's mostly a form of grass and grass roots that humans can't eat. So we replace the prairie with our own preferred grass, wheat. Never mind that we feed most of our grain to livestock, and that livestock is perfectly content to eat native grass. And never mind that there likely were more bison produced naturally on the Great Plains before farming than all of beef farming raises in the same area today. Our ancestors found it preferable to pluck the energy from the ground and when it ran out move on.

Today we do the same, only now when the vault is empty we fill it again with new energy in the form of oil-rich fertilizers. Oil is annual primary productivity stored as hydrocarbons, a trust fund of sorts, built up over many thousands of years. On average, it takes 5.5 gallons of fossil energy to restore a year's worth of lost fertility to an acre of eroded land—in 1997 we burned through more than 400 years' worth of ancient fossilized productivity, most of it from someplace else. Even as the earth beneath Iowa shrinks, it is being globalized. . . .

Wheat is temperate and prefers plowed-up grasslands. The globe has a limited stock of temperate grasslands, just as it has a limited stock of all other biomes. On average, about ten percent of all other biomes remain in something like their native state today. Only one percent of temperate grasslands remains undestroyed. Wheat takes what it needs.

The supply of temperate grasslands lies in what are today the United States, Canada, the South American pampas, New Zealand, Australia, South Africa, Europe, and the Asiatic extension of the European plain into the sub-Siberian steppes. This area largely describes the First World, the developed world. Temperate grasslands make up not only the habitat of wheat and beef but also the globe's islands of Caucasians, of European surnames and languages. In 2000 the countries of the temperate grasslands, the neo-Europes, accounted for about 80 percent of all wheat exports in the world, and about 86 percent of all corn. That is to say, the neo-Europes drive the world's agriculture. The dominance does not stop with grain. These countries, plus the mothership—Europe—accounted for three-fourths of all agricultural exports of all crops in the world in 1999.

. . . By the fifth century . . . wheat's strategy of depleting and moving on ran up against the Atlantic Ocean. Fenced-in wheat agriculture is like rice agriculture. It balances its equations with famine. In the millennium between 500 and

1500, Britain suffered a major "corrective" famine about every ten years; there were seventy-five in France during the same period. The incidence, however, dropped sharply when colonization brought an influx of new food to Europe.

The new lands had an even greater effect on the colonists themselves. Thomas Jefferson, after enduring a lecture on the rustic nature by his hosts at a dinner party in Paris, pointed out that all of the Americans present were a good head taller than all of the French. Indeed, colonists in all of the neo-Europes enjoyed greater stature and longevity, as well as a lower infant-mortality rate—all indicators of the better nutrition afforded by the onetime spend down of the accumulated capital of virgin soil.

The precolonial famines of Europe raised the question: What would happen when the planet's supply of arable land ran out? We have a clear answer. In about 1960 expansion hit its limits and the supply of unfarmed, arable lands came to an end. There was nothing left to plow. What happened was grain yields tripled.

The accepted term for this strange turn of events is the green revolution, though it would be more properly labeled the amber revolution, because it applied exclusively to grain—wheat, rice, and corn. Plant breeders tinkered with the architecture of these three grains so that they could be hypercharged with irrigation water and chemical fertilizers, especially nitrogen. This innovation meshed nicely with the increased "efficiency" of the industrialized factory-farm system. With the possible exception of the domestication of wheat, the green revolution is the worst thing that has ever happened to the planet.

For openers, it disrupted long-standing patterns of rural life worldwide, moving a lot of no-longer-needed people off the land and into the world's most severe poverty. The experience in population control in the developing world is by now clear: It is not that people make more people so much as it is that they make more poor people. In the forty-year period beginning about 1960, the world's population doubled, adding virtually the entire increase of three billion to the world's poorest classes, the most fecund classes. The way in which the green revolution raised that grain contributed hugely to the population boom, and it is the weight of the population that leaves humanity in its present untenable position.

Discussion of these, the most poor, however, is largely irrelevant to the American situation. We say we have poor people here, but almost no one in this country lives on less than one dollar a day, the global benchmark for poverty. It marks off a class of about 1.3 billion people, the hard core of the larger group of two billion chronically malnourished people—that is, one third of humanity. We may forget about them, as most Americans do.

More relevant here are the methods of the green revolution, which added orders of magnitude to the devastation. By mining the iron for tractors, drilling the new oil to fuel them and to make nitrogen fertilizers, and by taking the water that rain and rivers had meant for other lands, farming had extended its boundaries, its dominion, to lands that were not farmable. At the same time, it extended its boundaries across time, tapping fossil energy, stripping past assets.

The common assumption these days is that we muster our weapons to secure oil, not food. There's a little joke in this. Ever since we ran out of arable land, food *is* oil. Every single calorie we eat is backed by at least a calorie of oil, more like ten. In 1940 the average farm in the United States produced 2.3 calories of food energy for every calorie of fossil energy it used. By 1974 (the last year in which anyone looked closely at this issue), that ratio was 1:1. And this understates the problem, because at the same time that there is more oil in our food there is less oil in our oil. A couple of generations ago we spent a lot less energy drilling, pumping, and distributing than we do now. In the 1940s we got about 100 barrels of oil back for every barrel of oil we spent getting it. Today each barrel invested in the process returns only ten, a calculation that no doubt fails to include the fuel burned by the Hummers and Blackhawks we use to maintain access to the oil in Iraq.

David Pimentel, an expert on food and energy at Cornell University, has estimated that if all of the world ate the way the United States eats, humanity would exhaust all known global fossil-fuel reserves in just over seven years. Pimentel has his detractors. Some have accused him of being off on other calculations by as much as thirty percent. Fine. Make it ten years. . . .

Nitrogen can be released from its "fixed" state as a solid in the soil by natural processes that allow it to circulate freely in the atmosphere. This also can be done artificially. Indeed, humans now contribute more nitrogen to the nitrogen cycle than the planet itself does. That is, humans have doubled the amount of nitrogen in play.

This has led to an imbalance. It is easier to create nitrogen fertilizer than it is to apply it evenly to fields. When farmers dump nitrogen on a crop, much is wasted. It runs into the water and soil, where it either reacts chemically with its surroundings to form new compounds or flows off to fertilize something else, somewhere else.

That chemical reaction, called acidification, is noxious and contributes significantly to acid rain. One of the compounds produced by acidification is nitrous oxide, which aggravates the greenhouse effect. Green growing things normally offset global warming by sucking up carbon dioxide, but nitrogen on farm fields plus methane from decomposing vegetation make every farmed acre, like every acre of Los Angeles freeway, a net contributor to global warming. Fertilization is equally worrisome. Rainfall and irrigation water inevitably wash the nitrogen from fields to creeks and streams, which flow into rivers, which flood into the ocean. This explains why the Mississippi River, which drains the nation's Corn Belt, is an environmental catastrophe. The nitrogen fertilizes artificially large blooms of algae that in growing suck all the oxygen from the water, a condition biologists call anoxia, which means "oxygen-depleted." Here there's no need to calculate long-term effects, because life in such places has no long term: everything dies immediately. The Mississippi River's heavily fertilized effluvia has created a dead zone in the Gulf of Mexico the size of New Jersey.

America's biggest crop, grain corn, is completely unpalatable. It is raw material for an industry that manufactures food substitutes. Likewise, you can't eat unprocessed wheat. You certainly can't eat hay. You can eat unprocessed soybeans, but mostly we don't. These four crops cover eighty-two percent of American cropland. Agriculture in this country is not about food; it's about commodities that require the outlay of still more energy to *become* food.

About two-thirds of U.S. grain corn is labeled "processed," meaning it is milled and otherwise refined for food or industrial uses. More than forty-five percent of that becomes sugar, especially high-fructose corn sweeteners, the keystone ingredient in three-quarters of all processed foods, especially soft drinks, the food of America's poor and working classes. It is not a coincidence that the American pandemic of obesity tracks rather nicely with the fivefold increase in corn-syrup production since Archer Daniels Midland developed a high-fructose version of the stuff in the early seventies. Nor is it a coincidence that the plague selects the poor, who eat the most processed food.

It began with the industrialization of Victorian England. The empire was then flush with sugar from plantations in the colonies. Meantime the cities were flush with factory workers. There was no good way to feed them. And thus was born the afternoon tea break, the tea consisting primarily of warm water and sugar. If the workers were well off, they could also afford bread with heavily sugared jam—sugar-powered industrialization. There was a 500 percent increase in per capita sugar consumption in Britain between 1860 and 1890, around the time when the life expectancy of a male factory worker was seventeen years. By the end of the century the average Brit was getting about one-sixth of his total nutrition from sugar, exactly the same percentage Americans get today—double what nutritionists recommend.

There is another energy matter to consider here, though. The grinding, milling, wetting, drying, and baking of a breakfast cereal requires about four calories of energy for every calorie of food energy it produces. A two-pound bag of breakfast cereal burns the energy of a half-gallon of gasoline in its making. All together the food-processing industry in the United States uses about ten calories of fossil-fuel energy for every calorie of food energy it produces.

That number does not include the fuel used in transporting the food from the factory to a store near you, or the fuel used by millions of people driving to thousands of super discount stores on the edge of town, where the land is cheap. It appears, however, that the corn cycle is about to come full circle. If a bipartisan coalition of farm-state lawmakers has their way—and it appears they will—we will soon buy gasoline containing twice as much fuel alcohol as it does now. Fuel alcohol already ranks second as a use for processed corn in the United States, just behind corn sweeteners. According to one set of calculations, we spend more calories of fossil-fuel energy making ethanol than we gain from it. The Department of Agriculture says the ratio is closer to a gallon and a quart of ethanol for every gallon of fossil fuel we invest. The USDA calls this a bargain, because gaso-

Figure 12.4 A combine harvesting corn near Stockton, Kansas. Courtesy of USDA.

hol is a "clean fuel." This claim to cleanness is in dispute at the tailpipe level, and it certainly ignores the dead zone in the Gulf of Mexico, pesticide pollution, and the haze of global gases gathering over every farm field. Nor does this claim cover clean conscience; some still might be unsettled knowing that our SUVs' demands for fuel compete with the poor's demand for grain.

Green eaters, especially vegetarians, advocate eating low on the food chain, a simple matter of energy flow. Eating a carrot gives the diner all that carrot's energy, but feeding carrots to a chicken, then eating the chicken, reduces the energy by a factor of ten. The chicken wastes some energy, stores some as feathers, bones, and other inedibles, and uses most of it just to live long enough to be eaten. As a rough rule of thumb, that factor of ten applies to each level up the food chain, which is why some fish, such as tuna, can be a horror in all of this. Tuna is a secondary predator, meaning it not only doesn't eat plants but eats other fish that themselves eat other fish, adding a zero to the multiplier each notch up, easily a hundred times, more like a thousand times less efficient than eating a plant.

This is fine as far as it goes, but the vegetarian's case can break down on some details. On the moral issues, vegetarians claim their habits are kinder to animals, though it is difficult to see how wiping out 99 percent of wildlife's habitat, as farming has done in Iowa, is a kindness. In rural Michigan, for example, the potato farmers have a peculiar tactic for dealing with the predations of whitetail deer. They gut-shoot them with small-bore rifles, in hopes the deer will limp off to the woods and die where they won't stink up the potato fields.

Animal rights aside, vegetarians can lose the edge in the energy argument by eating processed food, with its ten calories of fossil energy for every calorie of food energy produced. The question, then, is: Does eating processed food such as soy burger or soy milk cancel the *energy* benefits of vegetarianism, which is to say, can I eat my lamb chops in peace? Maybe. If I've done my due diligence, I will have found out that the particular lamb I am eating was both local and grass-fed, two factors that of course greatly reduce the embedded energy in a meal. I know of ranches here in Montana, for instance, where sheep eat native grass under closely controlled circumstances—no farming, no plows, no corn, no nitrogen. Assets have not been stripped. I can't eat the grass directly. This can go on. There are little niches like this in the system. Each person's individual charge is to find such niches.

Richard Levins, "Science and Progress: Seven Developmentalist Myths in Agriculture" (1986)

Agriculture has become more mechanized, more consolidated, and more engaged with modern scientific techniques, but, in many cases, neither the quality nor distribution of food has improved. Richard Levins lays out a combination of seven "myths" that have supported these unidirectional shifts in agriculture and explains why these myths are both untrue and detrimental to the main goals: feeding people efficiently and avoiding the back-breaking labor that historically has characterized the planting, tending, and harvesting of crops.

In the third world, the view of "science" as unqualified progress is expressed in developmentalism, the view that progress takes place along a single axis from less developed to more developed, and therefore that the task of the revolutionary society is to proceed as quickly as possible along that axis of progress to overtake the advanced countries. [For Levins, a revolutionary society is one that attempts rapid development without the social inequalities that he associates with industrial capitalism.] The consequence of this view is the rapid reproduction of the worst features of world (capitalist) science and technology, the uncritical acceptance of the "modern." It fails to recognize that the pattern of modern technology is not dictated by nature but is developed through the interaction of the capitalist need to control the labor force, the desired outcome of research in the form of commodities, the intellectual climate in which the scientists work, the pattern of knowledge and ignorance coming from previous work, and the nature of the scientific problems to be solved.

In what follows, I examine the developmentalist ideology as it applies to agricultural technology, and contrast it with a more dialectical, political, and ecologically based approach. In the field of agricultural technology, developmentalism is supported by seven myths about what is "modern":

(1) *Backward is labor-intensive, modern is capital-intensive agriculture.* From this it follows that criticism of high-tech agriculture is misinterpreted as an appeal to return to primitive hand cultivation, and the critics are accused of "trying to deny to our people the advance that you have achieved." This view reflects first of all a misunderstanding of development. Its view of technology is a very nineteenth-century one, a thermodynamic model in which vast amounts of energy are used to move vast amounts of material. It has proven to be the least efficient production system in terms of energy; harmful to its productive base through erosion, water depletion, salinization, destruction of organic matter and soil organisms; and it endangers the health of people and wildlife.

The alternative that some of us have been advocating is more analogous to physiology and electronics: hormones are tiny quantities of matter that produce big effects; the nerve impulse is an insignificant amount of energy compared to the motion it is capable of initiating. The strategy of ecological agriculture is not the invention of a bigger bag of tricks allowing more kinds of interventions in crop production, but the design of systems that require minimum intervention. This is achieved by detailed knowledge of the processes affecting soil fertility, the population dynamics of insects (both pests and useful), and microclimatology. It is not an anti-technology stance. My own experience as a farmer in the mountains of Puerto Rico preparing land with a hoe left me with no nostalgia for that most burdensome of tasks. But rather than seek to remove the physical toil of plowing from human muscle to giant machines, we look for ways of loosening soil structure and reducing the tillage requirements. Our claim is that *the evolution of agricultural technology should be from labor-intensive through capital-intensive to knowledge- and thought-intensive.*

(2) *Diversity is backward, uniform monoculture is modern.* Once again, this myth is based on experience: the diversity of the minifundia [small farms] has in many places been replaced by the uniformity of agribusiness's monoculture. But monoculture inevitably creates new and serious pest problems, prevents us from using the variability of soils and climate to our advantage, depletes the soil, and makes necessary the heavy use of costly inputs.

Agro-ecologists see the possibilities of using patterns of diversity to manipulate the microclimate, for example, a shelter belt of trees on a hillside can act as a dam holding back the flow of cooler air and creating a belt of warmer air for a distance of about ten times the height of the trees. This belt, broad enough to farm with suitable mechanization, allows cultivation of a crop that requires higher temperatures. Diversity helps pest control, for instance by one crop's providing the nectar needed by wasps which parasitize pests of another crop, or by interrupting the spread of an epidemic. Crop diversity would also allow better use of labor in the face of the uncertainty of nature since some crops such as tomatoes require rapid harvest as soon as ripening occurs, while others such as cassava can be left in the ground until needed. . . . *The evolutionary sequence should be from the random heterogeneity of the peasant minifundia, through the homogeneity of the capitalist agribusiness, to the planned heterogeneity of an ecologically rational agriculture.*

(3) *Small scale is backward, large scale is modern.* The economies of large scale are recognized, but the disadvantages must also be acknowledged. For example, in dairying it helps to get to know the individual cows in a herd in order to adjust nutrition and detect the first signs of disease. This usually means that herds larger than 50 or 100 animals are not as productive as smaller herds. In field crops, large scale prevents utilization of each piece of ground for its most suitable cropping pattern. Therefore, there is an optimum size of plot which is not the greatest possible but rather large enough to make use of the necessary mechanization, and small enough to permit the use of edge effects, mosaic patterns of diversity, and adaptation of crops to topography. The unit of planning must be large in order to take into account regional patterns of hydrology, pest migrations, labor supply, and consumption needs. But the unit of planning is not the same as the unit of production, which can be much smaller. This is not a "small is beautiful" position but rather the search for an optimal geographical scale for the needs of a revolutionary society.

(4) *Backward is subjection to nature, modern implies increasingly complete control over everything that happens in the field or orchard or pasture.* However, nature is inherently variable. We can override part of the natural variability by major, costly inputs. But these often create a new vulnerability that replaces the old. For instance, irrigation works reduce the immediate dependence on rainfall but increase the vulnerability to variations in the price of oil. High yielding crop varieties often depend on a complete technical package under optimal conditions, and lose their advantage when there is unusually severe weather or the package is not available. Small differences in the weather can drastically change the synchrony between crops and their pests or pollinators, and between pests and their predators. And every change we make in nature changes the direction or intensity of natural selection, causing new directions of evolution in the many species that coexist with us. New pests adapt to our new crops or technologies, old ones acquire resistance to our control measures, beneficial predators may lose interest in the prey we had in mind for them.

An ecologically rational strategy would not pretend to set up a final, fully controlled production system but would acknowledge and utilize the variability of nature in several ways: climatic monitoring so that planting schemes could take into account the trends in temperature and moisture on a scale of years or decades (the years of poor crops in the USSR have been years of reduced rainfall roughly ten years apart that have some predictability); the growing together of crops with somewhat different environmental requirements to guarantee the food supply no matter what the weather (wet years in the United States Midwest are better for corn but worse for wheat, which suffers more from fungus diseases in wet years); the selection of varieties and of mixtures of varieties for their broad tolerance of the unexpected; the dependence on an array of natural pest control measures, with predators and parasites that have their own patterns of response to the environment so that we do not have to know how each one is doing to be sure that some populations of beneficial insects are always available; and systems of redistribution so that local failures and bumper crops can compensate for each other without major impacts on human welfare.

(5) *Folk knowledge is backward, scientific knowledge is modern.* The struggle against superstition has been and continues to be an important part of the process of liberation. However, in recent times there has been a growing appreciation of folk knowledge, particularly in the healing arts and agriculture. The aggressive claims that science is the only way to knowledge have been used to justify a chauvinist, class-based, and sexist contempt for the intellectual achievements of third world peoples, workers, and women of all countries. These claims are false. All knowledge does indeed come from experience, direct or indirect, and the reflection on that experience with intellectual tools derived from previous knowledge and experience. Modern science is one way in which that experience has been organized and used consciously for the purpose of acquiring knowledge. But all peoples learn, experiment, and analyze. In agriculture as in health, in industrial production, and indeed in all spheres, the best condition for the creation of new scientific knowledge requires the combination of the detailed, intimate, local, and particular understanding that people have of their own circumstances with the more general, theoretical, but abstract knowledge that science acquires only by distancing itself from the particular. . . .

(6) *Specialists are modern, generalists backward.* The rational kernel in this view is that there is too much to know within every discipline for anyone to know everything. The history of European thought has been an increasing subdivision of knowledge from the days of the philosopher-scholar-theologian, through general "scientists," to the present multiplication of specialties within previously coherent fields of study. For example, genetics, a part of biology, now includes molecular genetics, cytogenetics, population genetics, quantitative genetics (for plant and animal breeding), as well as further breakdowns by kinds of organisms studied. Uncritical admirers of specialization see the solution to the problem of integration in the teamwork of groups of specialists, and developing countries speak with pride of the numbers of specialists that graduate from their schools.

However, specialization prevents the researchers from seeing the whole picture, both because of the narrowness of their training and because the ideology of expertise makes it a matter of pride to consider only precise, quantitative information as real science while the rest is "philosophy" (a bad word among positivistic scientists) or "not my department." The training of specialists rather than the education of scientists encourages the combination of micro-creativity and docility that permits scientists to work on the most monstrous projects of destruction without attention to their consequences. The great failings in the application of science to human well-being have come about not because of the failure to know the details of the structure or workings of something but rather because of the failure to examine the system in its complexity. The strategy of the "green revolution" is solving many and difficult technical problems of plant breeding, but the geneticists did not anticipate problems of pest ecology, land tenure, or political economy, and as a result increases in production are sometimes associated with increases in misery. The Aswan Dam was an engineering success in that it retained the water it was intended to retain. But by stopping the seasonal flooding [of the Nile] that provided renewed soil fertility, the dam made farmers dependent on imported chemical fertilizers; the reduced flow of water into the Mediterranean Sea increased salinity and adversely affected fisheries; the outflow of the Nile was reduced to the point that it could no longer offset the erosion of the coastline; the irrigation ditches became the habitat for snails that transmit liver flukes.

It is a common experience that in large programs of development the ministries of health and agriculture do not talk to each other so that it comes as a surprise when the expansion of cotton production increases malaria. Cotton is very heavily sprayed. The natural enemies of the mosquitoes are killed, allowing the mosquitoes that transmit malaria to thrive in habitats created for them by the clearing of forest. The immigration of a labor force not previously adapted to malaria allows the parasites ideal susceptible hosts. There is a vast oral tradition of such cautionary tales. The point is that most of these "unexpected" outcomes are predictable, at least in principle. There is no longer any excuse for planners not to ask the obvious questions about a program, such as: what will it do to the position of women? New technologies are usually handed over to men, and traditional women's occupations are displaced. For instance, the use of herbicides displaces women from weeding. How will vegetational changes alter the biology of potential disease vectors? Will the new productive activity be compatible with the water needs of the people? Will the production of export crops make the food supply more vulnerable?

The outcome of short-sighted specialization is that each department takes as its starting point the products of the department next door. Crops are bred for their performance in monoculture because the machinery was designed for operations in pure stands of a single crop. The engineers design machinery for monoculture because the agronomists inform them that it can replicate what farmers do. The farmers plant monoculture because their varieties and machinery are suitable to monoculture. Each party is making rational decisions given the constraints imposed by the others, giving the whole trajectory of technological development the appearance of inevitability and necessity, while nobody looks out for the process as a whole.

(7) *The smaller the object of study, the more modern.* This is a continuation of the old pecking order of the sciences introduced by Comte [nineteenth-century

philosopher of science Auguste Comte], which sees the study of atoms as superior to the study of molecules, which is superior to the study of cells, which is superior to the study of organisms, which is superior to the study of populations, etc. It also elevates the laboratory experiment above the field study, investigations of collections of specimens, and theoretical work done in libraries. In its present insidious form, it uses the term "modern biology" to refer exclusively to molecular genetics, genetic engineering, and biotechnology. It implicitly relegates other branches of biology to lower ranks, and has even threatened the existence of museum collections which take decades to develop. [See "The Impending Extinction of Natural History" in Chapter 26, below.] It is important to recognize that a modern systematic biology, modern population genetics, modern community ecology or biogeography or epidemiology also exist: The reductionist bias implicit in this very narrow view of "modern" is especially harmful in developing countries which often give overwhelming priority to the more expensive branches of modern biology while ignoring the other areas which are of equal practical and theoretical importance. A healthy biological science requires the combination of study at all levels of organization and in the laboratory, field, library, and museum.

The struggle against these developmentalist myths about modernization is not anti-scientific. Rather, it is a program for a different kind of science. We have to insist that existing "modern" high-technology agriculture is not generic progress but a particular form of technological development under capitalist political or intellectual domination. The alternative is the development of new technologies designed not to create new input commodities or control a reluctant labor force but to provide high, sustainable yields of necessary products with minimum use of resources and damage to the environment and to people, and in a work process that promotes health and creativity.

Robert Stock, from "Agrarian Development and Change" in *Africa South of the Sahara* (2004)

The relationship between agriculture and land conservation is nowhere more rife with tension than in sub-Saharan Africa, with its rapidly increasing human population, generally poor soils, and density of large mammalian wildlife that is likely to lay waste to the farmers' fields. However, the greatest challenges to African agriculture, as Robert Stock recounts, are the changes brought about by its colonial history, especially the cash crop economy that was imposed in the first half of the twentieth century.

Africa has more unexploited arable land than any other continent. Development planners have often expressed optimism about Africa's potential and have made many proposals over the years for the transformation of African agriculture and pastoralism. Attempts at major structural change, such as the establishment of plantations or state farms, have usually achieved poor results. The alternate approach, presently favored by many development agencies, focuses on the productive capacity of small farmers.

COLONIAL EFFECTS ON INDIGENOUS PRODUCTION SYSTEMS

Colonial rule brought many changes to African agriculture. Prior to the 1970s, colonial impacts on agrarian development were usually viewed in a positive light. Emphasis was given to the apparent benefits to Africa of the introduction of new crops, new market opportunities, and new technologies. Although coercion was often used to bring about desired changes, such measures were seen to be fully justified. Scholars from the dependency school dismissed this notion of colonial altruism; rather, the motivation behind colonial policies was seen as blatant self-interest. These policies undermined the viability of indigenous systems by squeezing resources from small producers and upsetting ecological balances that, it was argued, Africans had maintained for centuries.

Large areas of land were expropriated and reserved for European farms, most notably in Kenya, Southern Rhodesia (now Zimbabwe), and South Africa. Africans expelled from their lands were relocated in confined reserves, where less fertile soils and necessarily shorter fallow periods caused declining productivity. The tax burden in settler-dominated colonies was placed squarely on the shoulders of Africans. Dual pricing systems were established that granted significantly higher prices for European crops and livestock. Forest and game reserves were created, with often serious implications for farmers and pastoralists who had traditionally used these territories. Pastoralists lost vital seasonal pastures and had their normal migration routes severed. The increasingly dense vegetation and growing wildlife populations in these reserves created ideal conditions for the proliferation of tsetse [the fly that carries sleeping sickness], which in turn made the surrounding countryside increasingly dangerous for humans and livestock.

Rural labor was appropriated to serve the interests of the colonial state. Every colonial power relied at times on forced labor to recruit troops, porters, construction workers, and plantation laborers. For the peasantry, time lost from agriculture meant smaller harvests and often hunger. Other forms of labor appropriation—such as the "voluntary" seasonal migration of peasants to earn money for taxes—had similar implications for peasant agriculture, despite their more benign appearance.

Cash crops were promoted as a means of involving Africans in the commercial economy and ensuring a supply of tropical products for European industry. Some of these crops, including cotton, groundnuts, and oil palm, had been widely cultivated in pre-colonial Africa; others, such as tea, were introduced for the first time. Africans frequently resisted the introduction of cash crops, because it diverted scarce resources away from food crop production. Some cash crops were more compatible than others with existing farming systems. For example, groundnuts could be intercropped with millet and actually increased grain yields by fixing atmospheric nitrogen in the soil, whereas cotton was viewed as a "soil robber" and poorly suited to intercropping.

When fairly attractive prices were offered for cash crops that were compatible with indigenous farming systems, African farmers could become enthusiastic adopters. Migrant farmers planted vast areas of unfarmed forest in the Gold Coast (now Ghana) with cocoa. However, the prices offered to African producers were seldom attractive, and cash cropping often clashed with food cropping. In such cases, peasant resistance to cash cropping made sense. The appropriation of profits through taxation and monopoly pricing was justified as a means of securing capital for development, but little of what was actually spent on development was of much benefit to rural areas.

Colonial agricultural agents attempted to introduce changes in indigenous agricultural practice that reflected European views of how farming should be done. For example, mixed farming integrating cropping and raising livestock was often promoted, but with little success, because few African farmers had sufficient land and wealth to engage in mixed farming. Monoculture was also promoted, despite evidence that traditional farming practices such as shifting cultivation and intercropping were actually superior. African farmers' disinclination to change seemed to confirm their backwardness. The myth of the unprogressive peasant helped to sustain the view that Africa's land and labor could be most effectively exploited by developing European-style farms and plantations.

POSTCOLONIAL POLICIES AND THE NEGLECT OF AGRICULTURE

During the 1960s, the expansion and diversification of cash cropping were seen as obvious vehicles to finance modern development. Farmers paid a "hidden tax" in the form of the low prices that they received from state agencies for cash crops. These prices were typically set well below prevailing world market prices. But increased production often brought reduced prices in world markets, because the demand for products such as coffee and cocoa is relatively inelastic (i.e., consumption is unlikely to change much in response to changes in price). For other commodities, competition from other sources (e.g., U.S. mechanized peanut farm vs. African groundnut growers) or from substitute products (e.g., synthetic products used to replace sisal) depressed both demand and price. Finally, relatively little attention was paid to the possible role of locally important cash crops.

Prior to the opening of the International Institute of Tropical Agriculture (IITA) in Ibadan, Nigeria, in 1968, food crops were virtually ignored in agricultural research and development. Little was known about the logic of indigenous production systems or the potential utility of improved food crop varieties and related innovations. When the need to actively encourage greater food production was finally recognized, many African nations were already suffering from recurrent food shortages.

The most common response to food deficits was to increase food imports, especially wheat and rice, which had become an important component of urban diets. Imported foods were often sold at less than the market prices of locally produced foods. Although food subsidies reduced the likelihood of unrest among the urban population, they also undercut indigenous food producers. Depressed prices for both food crops and cash crops created a massive disincentive for peasants to invest in their farms.

The overall structure of post-independence development has hastened the decline of agriculture. Schools and health clinics were constructed in many rural districts, but social services in urban areas remain far superior. The probability of migration was also increased by educational systems that seldom promoted agriculture or provided skills designed to foster rural development. Rather, the "hidden curriculum" tended to extol the values of modern urban life. It is little wonder that more and more youth aspired to an urban future and had little enthusiasm for agriculture or rural life.

ATTEMPTS TO RESTRUCTURE AFRICAN AGRICULTURE

Most African countries, eager to boost food production and export earnings, went through phases of importing modern agricultural machinery and establishing or promoting the development of large-scale farm enterprises. There were a few qualified successes, but these approaches typically had minimal effects on national production. In contrast, they typically had adverse effects on indigenous production systems—often through alienating land normally used by peasants and herders, and through diverting developmental resources away from small-scale producers.

State Farms

The first experiments with state farms occurred in the early 1960s in Ghana and Guinea. The collective farms of Eastern Europe provided the model for agrarian transformation. The experiments failed badly, largely because of poor management. Heavy investments in farm machinery were wasted because of parts shortages, poor maintenance, and adverse environmental conditions. Moreover, peasants were reluctant to commit themselves fully to collective work on state farms, preferring instead to farm their own land. Mozambique, Angola, and Ethiopia later tried to establish state farms as a part of a broader socialist transformation. In the former Portuguese colonies, farms formerly owned by white settlers were consolidated to form state enterprises. Mozambique invested heavily in mechanization, but was unable to mobilize adequate management. The results were so poor that most of the farms were abandoned by the mid-1980s.

Although state farms were most often associated with Marxist regimes, moderate socialist and capitalist regimes also experimented with them. For example, Tanzania established large-scale mechanized wheat farms with Canadian assistance. Zambia's experiment with state farms was unusual; each of several huge model farms was set up and managed as a foreign aid project by a different country (the United States, the Soviet Union, China, and Canada, among others). Zambia derived few benefits from this experiment, in which the quest for foreign aid took precedence over the development of a coherent national strategy of development built around an assessment of the country's own needs and resources.

Land Reform and Planned Resettlement

Planned resettlement has been initiated in several countries, but few of these initiatives achieved the anticipated benefits. One of the most common reasons for resettlement programs has been to accommodate people displaced by dams and other major projects. Resettlement has been used in several countries as a strategy for political and economic transformation. The settlement of Africans on farms reclaimed from European settlers in Kenya and Zimbabwe, the creation of *ujamaa* villages in Tanzania, and the removal of Ethiopians from drought-stricken plateaus to the sparsely populated periphery provide varied examples of politically informed resettlement.

One of the largest and most successful resettlement programs was undertaken in Kenya during the early 1960s. After the Mau Mau revolts, with independence imminent, many white farmers wished to leave the country. A series of resettlement projects was developed in which the government purchased European farms, subdivided them into smaller units, and sold them to Kenyans. The preparations involved careful evaluation of soil fertility and studies to determine optimum sizes for viable farms. The initial schemes focused on middle-class farmers who could afford to invest in relatively large farms of 6–20 ha. The resettlement program was later expanded to accommodate some 35,000 landless families in a new project, known as the Million-Acre Scheme. Colonial officials had little confidence in these poor, apparently inexperienced farmers. However, the performance of Kenya's resettlement schemes confounded the expert predictions. Total agricultural output increased four percent per year during the 1960s and early 1970s. Despite having inferior resources, smallholders were soon obtaining higher yields and higher profits than the middle-class owners. The extensive promotion of commercial farming innovations, including hybrid maize [see Mangelsdorf, Chapter 4, above], various cash crops, and dairy cattle, was one of the keys to the success of the Kenyan program.

A similar approach to land reform was planned for Zimbabwe's transition to independence in 1980. However, few of the international funds pledged to purchase white-owned farms for redistribution to small farmers materialized. Most of the white farmers remained, and large-scale commercial farming continued to dominate the agricultural sector. Disparities in wealth and access to land were causes of resentment, and in 2000 land ownership became the country's paramount political issue. President Robert Mugabe declared that most of the land occupied by white farmers would be seized and allocated to landless Africans. This politically motivated land seizure resulted in violence as well as major economic and political instability in Zimbabwe, and was widely condemned internationally.

Tanzania attempted to achieve rural transformation by encouraging its scattered population to form new *ujamaa* villages. *Ujamaa* villagers were to devote a significant amount of their time to communal efforts, including work in collective village farms. The early promise of this scheme waned, especially when increasing levels of coercion were applied in what had begun as a voluntary program. Moreover, with its total economy in decline, Tanzania was less and less able to provide material support for rural development.

Ethiopia's socialist government began a program to resettle families from the densely populated heartland of the country to the sparsely populated western and southwestern peripheries. This program was greatly expanded at the time of the 1984–1985 drought; over half a million people were moved in just over one year. The government perceived resettlement as a cure-all for the famine and its deeper environmental roots, and as a way to lessen its dependence on foreign aid by opening up fertile, underutilized territory. However, the program was implemented hastily without careful planning and with inadequate resources to ensure an orderly and successful transition.

The claims of some international organizations that the Ethiopian government was forcing people to migrate, and that tens of thousands had perished because of inadequate planning, seem to have been overstated. Although many resettled persons (e.g., unemployed urban youths) were coerced, most migrated voluntarily, albeit because of their desperate circumstances during the famine. Many migrants later returned to their former homes, but most have remained behind and become successful farmers in their new environment. After 1986, the Ethiopian government curtailed the program, primarily because of the lack of resources to support successful resettlement.

Large-Scale Capitalist Agriculture

In most of Africa, large-scale capitalist agricultural projects, whether initiated by domestic or foreign companies, play a quite minor role. However, large-scale agriculture is well developed in some countries, such as Côte d'Ivoire, and is growing rapidly in others. Nigeria allows transnational companies to establish large estates to produce more of their own raw materials (e.g., cotton for textile mills and grain for breweries) and to grow food for the domestic market. This approach conforms to the long-standing view of those in power—namely, that future prosperity depends on modernizing agriculture, rather than on developing fully the potential of indigenous farming systems. According to this vision, more and more Africans will work as wage laborers rather than as independent producers. It implies that spatial patterns of development will become increasingly uneven. In most cases, it also implies that non-Africans will exert control over scarce agricultural resources.

In South Africa and Zimbabwe, agriculture was dominated by large commercial farms owned by white farmers. These farms produced a wide variety of commodities—livestock, grains, vegetables and fruits, and other specialty crops—using modern technologies and large amounts of black labor. Their output included high-value-added products, such as the wines produced in the Cape region. Legislation, particularly the South African Native Lands Act of 1913, and the expansion of white commercial farms during the twentieth century left virtually no space for small-scale African farmers producing for the market. Africans who were not working on commercial farms were removed into tiny designated reserves, where overcrowding and adverse environmental conditions limited agriculture to very basic subsistence production.

The structure of the South African agricultural sector has not changed substantially since the end of apartheid. Indeed, the removal of international sanctions has expanded market opportunities for commercial farmers to increase exports and make new investments in land and technology. Land reform, first promised by the African National Congress in its Freedom Charter of 1955, remains a major political challenge. The government has moved slowly and deliberately since 1994 to implement a land reform program to redress past injustice and alleviate rural poverty. However, redistribution has fallen far short of the promised thirty percent of agricultural land within five years.

Production Under Contract

During the 1980s and 1990s, increasing numbers of Africans became involved in the production under contract of higher-value horticultural export crops. These crops include fresh vegetables and fruit and fresh cut flowers, most of them bound for European markets. Kenya, Zimbabwe, South Africa, Zambia, Gambia, and Côte d'Ivoire have been among the major participants in this new trade. Although African producers have yet to overtake the leading sources in Latin America (e.g., Mexico and Argentina) and Asia (e.g., China), Africa's market share is increasing rapidly. Kenya has become Europe's largest supplier of fresh horticultural produce, and South Africa is the world's third largest exporter of citrus fruits.

Production and marketing arrangements for higher-value horticultural goods differ from those associated with lower-value, bulky, less perishable "traditional" exports such as coffee. Assured product quality, flexible and reliable supply, and the speediest possible delivery to market are essential for highly perishable horticultural products. In contrast, traditional commodity exports required little or no special handling. In the horticultural sector, producer contracts govern the marketing chain that links growers to retailers, located continents apart. These contracts specify not only the quantity and timing of product delivery, but also production methods such as the use of pesticides.

Africa has several attractions as a source of produce for Europe. It offers diverse climate conditions, permitting the growth of most tropical and many temperate products. The climate permits year-round production wherever water supply is assured, while Southern Hemisphere locations offer counterseasonal supplies of produce. Cheap labor and land keep production prices low, and shipping time and costs may be lower because Africa is closer to Europe than some of its competitors are.

Produce exports offer new opportunities for producers in several African countries. Nevertheless, these opportunities come at a cost, because of the ways that importing firms control production and marketing arrangements. Smaller producers often find it difficult to meet contract specifications, especially when these force them to give priority to the export crop over all other considerations. It has increased conflict over land and labor between male and female members of households. Heavy use of pesticides, especially for flower production, is a health threat that is especially serious for children and women. On the slopes of Mount Kenya, the rapid growth of horticultural production is resulting in water shortages and struggles over water rights.

Pablo Neruda, "Ode to Wine" (1954–59), translated by James Engell

From Homer's "the bewitching wine, which sets even a wise man to singing and to laughing gently and rouses him up to dance and brings forth words that were better unspoken" to Keats's "beaker full of the warm South . . . that I might drink and leave the world unseen," poets have equated wine with their own poetic inspiration. Moreover, the quality of wine, at least in the eyes of its purists, has a greater correlation with the soil and other native conditions in which the grapes grow — collectively known as terroir *— than is true of any other crop.*

Wine is sacred to many religions, a national symbol to France and Italy, and one of the forms of agriculture most rapidly encroaching on native ecosystems. Australia, New Zealand, Neruda's own Chile, Argentina, South Africa, and the United States have all increased their percentage of lands devoted to vineyards exponentially in the last two decades. Can we afford to devote such a high percentage of the world's lands to a crop that provides no essential nutrition? On the other hand, can we afford to become such steely-eyed rationalists that we no longer appreciate the beauty of agricultural lands and the romance of the non-essential crops they often provide?

Ode to Wine

Wine colored of day, wine colored of night,
wine with purple feet or of ruby blood,
wine, starry child of earth,
wine, with heft like a sword of gold,
or soft, like sensuous velvet,
wine, spiral-swirled like a seashell
and held suspended, loving, of the sea—
never has one glass, song, or man held you:
you are choral, social,
and should at least be shared.

Sometimes you draw from mortal memories,
and on your liquid wave
we move from tomb to tomb,
you, stonecutter of cold sepulchers,
and we weep passing tears;
but your gorgeous spring dress differs:
heart's sap rises in limbs,
wind excites the day, and nothing's left behind
in your quiet soul.
Wine arouses the spring;
like a shoot, joy pushes aside the earth,
walls and rocks tumble down, the chasms close,
and song is born.

A flask of wine and thou beside me
in the wilderness, sang the old poet—
Let the pitcher of wine
double the kiss of love.

My love, so suddenly,
your hip's the curved fullness
of the wine glass, your breast the cluster,
your aureole the fruit,
the light of intoxication your hair,
your navel a pure seal
set on the vessel of your waist:
your love cascades in never-ending wine,
a clarity that heightens my senses—
earthy splendor of life.

And yet not only love or burning kiss,
or kindled heart, are you, the wine of life;
you're deeper intimacy,
transparency,
a chorus of learning,
abundance of flowers.

I love, on the table,
when there is talk,
the light of a bottle of thoughtful wine,
which they drink, so that in each drop of gold,
in every glass incarnadine,
or crimson ladle, they recall
how autumn worked to fill the casks with wine;
and, so, may the common man learn
in rituals of daily work
to remember the soil, his debt to earth,
and to voice the canticle of the vine.

13 Air and Water

Imagine a model of Earth 40 centimeters or about 16 inches in diameter, where one inch equals five hundred miles. On that model, the atmospheric shell around the globe, roughly twenty miles high except for its thinnest upper reaches, which extend to sixty miles, would span only one millimeter, barely more than one thirty-second of an inch. And most life forms on land and in the sea exist within *one-tenth* of that thickness. The atmospheric biozone of Earth is a remarkably thin, fragile envelope, yet its circulation connects all aerobic life forms. The atmosphere carries water vapor and is necessary for the water cycle. It interacts with the oceans and every body of water. Cosmic rays and all forms of energy from the sun, including UV rays, bombard it. Different cloud forms and colors reflect back into space varying amounts of energy. Small meteors constantly burn up in the atmosphere, though some survive the heat of atmospheric friction and plummet to Earth. The motion and heat differentials of a globe spinning at one thousand miles per hour at the hot equator but comparatively still at its frigid poles create massive air currents of varying velocities and temperatures.

For thousands of years, humans regarded air as a single element. Early modern scientists hypothesized the existence of an ether or "phlogiston" that permitted air to support combustion. Only two to three hundred years ago, pioneering chemists such as Robert Boyle, Antoine Lavoisier, Joseph Priestley, and Humphry Davy began to identify and study gases that make up the atmosphere. In 1802 the English chemist Luke Howard, using the model of Latin nomenclature for plants and animals introduced by Carolus Linnaeus fifty years earlier, devised the names still used for cloud types (e.g., cirrus, cumulonimbus). During the time when humans regarded the atmosphere as a single element, their activity had little impact on it. Local air pollution and smoke—for example, clouds of noxious coal dust and sulfur dioxide that hung over eighteenth-century London—were not uncommon. But the extent of the atmosphere meant that the amount of combustion controlled by humans exerted minimal influence.

The Industrial Revolution changed all that. In the nineteenth century manufacturing and domestic use of wood and especially coal brought massive pollution to urban areas. Toxic clouds proved deadly in cities such as Pittsburgh. In the twentieth century, the automobile and the generation of electricity by burning coal and oil exacerbated air pollution. In the 1950s "killer fogs" in London left scores of people dead in a few days. During the Cold War, nuclear tests above ground increased levels of atmospheric radiation. By the 1970s in Athens, air pollution ravaged statues that for 2,000 years had survived well. They were moved to air-conditioned quarters. After a generation of anti-pollution regulations and emission standards for vehicles, industrial plants, and power stations, conditions improved in many locales in the United States and other industrial countries, but they also worsened in others, and greatly deteriorated in many developing nations.

Today, everyone knows that some localities enjoy cleaner air than others. Stargazing in the Canadian Rockies or South Seas is very different than stargazing in Houston or Hong Kong. However, airline pilots now routinely report a yellow haze of pollution covering not only big cities but also blanketing areas such as Southeast Asia and the Indian Ocean. Air pollution (like ocean pollution) has gone global. Its variety is staggering. It ranges from tiny particulate matter that diesel and other exhausts contain and that can lodge permanently in the smallest air sacs of the lungs, to toxic, even deadly chemicals released from industrial plants, to the human habit of smoking. Air pollution causes acid rain, which damages forests and crops, sometimes severely. This rain acidifies lakes, effectively killing them. The sulfuric and nitric acid of such rain has ravaged South America and Asia as well as North America and Europe. Affected areas are often "wilderness," hundreds or thousands of miles from the sources of pollution. The law of ecology that says everything is connected to everything else often finds its most telling confirmation and swiftest connection in the atmosphere.

Like that of Venus and Mars, Earth's atmosphere has changed over long periods of time. Acting in concert, many factors and processes have altered its temperature and composition: lush plant life, massive volcanic activity with enormous releases of gases, constant interaction with the oceans, and numerous meteor impacts, to name a few. Earth has experienced significantly different climates over long periods of time, and the chemical composition of the atmosphere has changed across millions of years.

Now, however, in the span of one century or less, human activity is having an unprecedented effect on the atmosphere. Two dramatic though basically unrelated examples are the ozone hole (covered in this chapter) and global warming or climate change (see "Climate Shock," Chapter 1).

At low altitudes, ozone (O_3) is considered a pollutant. But in the upper atmosphere, ozone is vital to life as we know it. Stratospheric ozone helps block the UV (ultraviolet) rays that damage living tissue.

In 1974 F. Sherwood Rowland and Mario J. Molina hypothesized that various chemicals containing chlorine and used as refrigerants, pesticides, and aerosol propellants had risen to the upper atmosphere and, in a series of chemical reactions catalyzed by chlorine, were destroying ozone molecules. This hypothesis was in the next decade confirmed by measurements taken at high altitudes. The destructive effect of these chlorofluorocarbons is concentrated near the poles. When first measured in the 1980s, the stratospheric ozone layer at high latitudes was vanishing over a limited area seasonally. By September 2000 the ozone hole surrounding the South Pole covered a record 11.5 million square miles, three times the area of the United States, and had become present the entire year. Effects of these stratospheric changes began to be seen on the ground: for example, in increased blindness among New Zealand sheep and increased skin cancer rates in Australia and Scandinavian countries.

In 1987, scores of countries signed a treaty, the Montreal Protocol, which provides a comprehensive framework for the reduction and, in many cases, the elimination of ozone-destroying chemicals. Atmospheric chemists anticipate that the ozone layer might be fully restored in the second half of this century. The Montreal Protocol is a model of successful international cooperation, but it remains fragile. Methyl bromide was scheduled to be eliminated in developed countries by 2005, for example, but the U.S. demanded and received an exemption that allowed national methyl bromide use to *increase* in 2005.

So crucial is the human impact on the atmosphere that textbooks in atmospheric science now introduce this fact immediately. Yet the intricate chemistry and complex behavior of the atmosphere is still not completely understood. The stresses placed on it by human activities only increase the challenge of understanding the atmosphere and formulating appropriate responses to the changes and damage inflicted on it. The atmosphere is something akin to the skin of the living Earth; it exists as a kind of extensive covering organ that protects what is underneath, retains warmth, allows for respiration, contains moisture, regulates temperature, renews itself, and consists of multiple layers with elaborate transport systems between them. The atmosphere may be regarded like the skin in another way. If too much of it is severely damaged or altered, the life it envelops cannot survive.

Like air, water is a key ingredient of life. The quality of water can be threatened by disease-spreading microorganisms and chemical pollutants. Quantity matters too, and even more dramatically: we can survive no more than a few days without water, or a few minutes immersed in it. With a well-regulated supply of fresh water, we can build cities like Las Vegas or Los Angeles in the desert (see Chapter 10). Without it, we would be forced to abandon the most modest village.

Water is ironically both scarce and abundant. Most of the earth's surface is covered in water, and the atmosphere is saturated with it. But only two and one-half percent of the water is fresh, and only one-fifth of that is available for human use. The demands made on that scarce resource are staggeringly large and often wasteful, and they are increasing. On every continent, aquifers—large, natural underground reservoirs, our savings account of clean fresh water accumulated over millions of years—are being drawn down. Land subsides over depleted aquifers, and salt water intrudes. Some rivers are so exploited and diverted that they never reach their natural end. Large dams not only change the speed, temperature, and ecology of rivers, they also send much of the still water into the air as evaporation. At the same time, they silt up, hastening the end of their useful life as sources of hydroelectric power and (most famously on the Nile) depriving downstream soil of much-needed nutrients.

Agriculture has an insatiable thirst. The biggest limiting factor in food production is not seed, climate, or soil, but water. As one Turkish agricultural official remarked, "Without irrigation, we can't do anything." Eating meat raises water requirements markedly. In impoverished countries, as farmers till more marginal land, water needs soar. Still, with factories, golf courses, hotels, and car washes, affluent countries remain the gluttons of water consumption. The EPA and the National Wildlife Federation rank world per capita water use. If the U.S., the heaviest user, ranks 100, then the U.K. is 58, Asia 15, and Africa a meager 7.8. From 1930 to 2000, world population tripled but water demand increased sixfold. At the start of the twenty-first century, U.N. Secretary-General Kofi Annan warned, "Unless we take swift and decisive action, by 2025, two-thirds of the world's population may be living in countries that face serious water shortages."

What can rectify this imbalance between scarce supply and out-of-control demand? The economist's answer is correct pricing. When water is free for the taking, there is no economic incentive to conserve (compare with Hardin's "Tragedy of the Commons" in Chapter 23). The merit of the argument for prices that encourage conservation has, unfortunately, largely been overshadowed in the public eye by egregious examples of price gouging by international corporations that have gained control of privatized municipal supplies.

Another part of the answer is technological. We could be using water much more efficiently than we do. Drip irrigation, pioneered in Israel, not only uses much less water than conventional agriculture, but also helps prevent erosion and salinization of soil. Graywater systems reuse domestic water, for example by flushing toilets and watering lawns with the outflow from sinks and dishwashers. Porous pavements help rainwater re-infiltrate aquifers rather than being washed into sewers or out to sea. Innovations like desalinization of sea water, though too expensive and energy-intensive to be used on a large scale at present, can even help to augment the supply of fresh water.

A third part of the answer is ethical. The water crisis requires an attitudinal change, a renewed conservation ethic. We need to learn—or relearn—to regard water as precious, some would say even as sacred. (See, for example, Cronon in Chapter 11.)

If we continue to waste and foul our fresh water supplies, conflicts are inevitable (see Chapter 7). Water disputes caused wars in the ancient Tigris and Euphrates valleys, where humans first practiced large-scale irrigation more than 4,000 years ago. Rivers form boundaries; rivers and aquifers also cross boundaries. Today major rivers are in dispute, the Nile, Danube, Colorado, and Rio Grande among them. Most water laws and treaties are old; many are effectively outdated. Acute water scarcity could lead to mass illegal immigration and violent conflict in coming decades and centuries; some observers think water scarcity could be as explosive as petroleum scarcity.

Many poorer nations suffer from chronic shortages of clean and safe water today. Water-borne diseases—fatal diarrheas, worms, and schistosomiasis (a blood parasite)—kill millions each year in Africa, Asia, and Latin America. If you drink six glasses of water daily, in the average time between two of them, more than 1,500 children will die from water-borne diseases and lack of clean water. In 1980, international organizations targeted 1990 as the year to provide sufficient water globally for everyone. The target was abandoned, and even more modest goals are continually postponed and rescheduled. As the Global Water Supply and Sanitation Assessment points out (see selection below), efforts to provide safe water to all are in a race against population growth. But the greater irony, as Rodney White notes (see selection below), is that the scarcity of clean, safe water is not really a scarcity of water at all. Water is abundant in these indigent regions, or at least it is more than adequate for the modest household uses (drinking, cooking, cleaning, sewerage) that would eliminate water-borne diseases. What is scarce is the institutional capacity and political will to pump, pipe, and treat the water, the commitment to provide services to the poorest of the poor.

FURTHER READING

Baumert, K., ed. *Options for Protecting the Climate.* Washington, D.C.: World Resources Institute, 2002.

Benedick, R. E. *Ozone Diplomacy.* Cambridge: Harvard University Press, 1998.

Carson, Rachel. *The Sea Around Us.* New York: Oxford University Press, 1951.

Donahue, John M., and Barbara Rose, ed. *Water, Culture, and Power: Local Struggles in a Global Context.* Wahington, D.C.: Island Press, 1998.

Ellis, Richard. *The Empty Ocean: Plundering the World's Marine Life.* Washington, D.C.: Island Press, 2003.

Postel, Sandra. *Last Oasis: Facing Water Scarcity.* New York: W. W. Norton, 1997.

———. *Pillar of Sand: Can the Irrigation Miracle Last?* New York: W. W. Norton, 1999.

Wayne, Richard. *Chemistry of Atmospheres.* New York: Oxford University Press, 2000.

FILMS

Chinatown (1974)
A Civil Action (1998)
Erin Brockovich (2000)

John Seinfeld and Spyros Pandis, from *Atmospheric Chemistry and Physics* (1998)

The authors of a comprehensive text do something here that would have been unusual prior to 1995 and unthinkable—as well as scorned by many—before 1985. At the start *of their treatment of atmospheric science and chemistry they alert readers to changes in the atmosphere, climate, and weather produced by human activities over the last two centuries. These activities, particularly the combustion of fossil fuels, continue to intensify. To the somewhat older, still urgent issues of air pollution, acid rain, and smog must now be added yet more fundamental effects on atmospheric composition and behavior, effects global rather than local. Even if* all *human activity stopped now, these effects would last decades or centuries.*

HISTORY AND EVOLUTION OF THE EARTH'S ATMOSPHERE

It is generally believed that the solar system condensed out of an interstellar cloud of gas and dust, referred to as the "primordial solar nebula," about 4.6 billion years ago. The atmospheres of the Earth and the other terrestrial planets, Venus and Mars, are thought to have formed as a result of the release of trapped volatile compounds from the planet itself. The early atmosphere of the Earth is believed to have been a mixture of carbon dioxide (CO_2), nitrogen (N_2), and water vapor (H_2O), with trace amounts of hydrogen (H_2), a mixture similar to that emitted by present day volcanoes.

The composition of the present atmosphere bears little resemblance to the composition of the early atmosphere. Most of the water vapor that outgassed from the Earth's interior condensed out of the atmosphere to form the oceans. The predominance of the CO_2 that outgassed formed sedimentary carbonate rocks after dissolution in the ocean. It is estimated that for each molecule of CO_2 presently in the atmosphere, there are about 10^5 CO_2 molecules incorporated as carbonates in sedimentary rocks. Since N_2 is chemically inert, nonwater soluble, and noncondensable, most of the outgassed N_2 accumulated in the atmosphere over geologic time to become the atmosphere's most abundant constituent.

The early atmosphere of the Earth was a mildly reducing chemical mixture, whereas the present atmosphere is strongly oxidizing. The dramatic rise of oxygen (O_2) as an atmospheric constituent over time was the result of the production of O_2 as a byproduct of photosynthetic activity. It has been estimated that the current level of O_2 in the atmosphere was achieved approximately 400 million years ago.[1] The present level of O_2 is maintained by a balance between production from photosynthesis and removal through respiration and decay of organic carbon. If O_2 were not replenished by photosynthesis, the reservoir of surface organic carbon would be completely oxidized in about twenty years, at which time the amount of O_2 in the atmosphere would have decreased by less than 1%.[2] In the absence of surface organic carbon to be oxidized, weathering of sedimentary rocks would consume the remaining O_2 in the atmosphere, but it would take approximately four million years to do so.

The Earth's atmosphere is composed primarily of the gases N_2 (78%), O_2 (21%), and Ar (1%) whose abundances are controlled over geologic time scales by the biosphere, uptake and release from crustal material, and degassing of the interior. Water vapor is the next most abundant constituent; it is found mainly in the lower atmosphere and its concentration is highly variable, reaching concentrations as high as 3%. Evaporation and precipitation control its abundance. The remaining gaseous constituents, the *trace gases*, comprise less than 1% of the atmosphere. These trace gases play a crucial role in the Earth's radiative balance and in the chemical properties of the atmosphere. The trace gas abundances have changed rapidly and remarkably over the last two centuries. . . .

Spectacular innovations in instrumentation in the last quarter of the twentieth century have enabled identification of atmospheric trace species down to levels of almost 10^{-12} parts per part of air, 1 part per trillion (ppt) by volume. Observations have shown that the composition of the atmosphere is changing on the global scale. Present-day measurements coupled with analyses of ancient air trapped in bubbles in ice cores provide a record of dramatic, global increases in the concentrations of gases such as CO_2, methane (CH_4), nitrous oxide (N_2O), and various halogen-containing compounds. These "greenhouse gases" act as atmospheric thermal insulators. They absorb infrared radiation from the Earth's surface and reradiate a portion of this radiation back to the surface. These gases include CO_2, O_3, CH_4, N_2O, and halogen-containing compounds. The emergence of the Antarctic ozone hole provides striking evidence of the ability of emissions of trace species to perturb large-scale atmospheric chemistry. [See the following three selections for more on ozone.] Observations have documented the essentially complete disappearance of ozone in the Antarctic stratosphere during the austral spring, a phenomenon that has been termed the "Antarctic ozone hole." Observations have also documented less dramatic decreases over the Arctic and over the northern and southern midlatitudes. Whereas stratospheric ozone levels have been eroding, those at ground level in the Northern Hemisphere have, over the past century, been increasing. Paradoxically, whereas ozone in the stratosphere protects living organisms from harmful solar ultraviolet radiation, ozone in the lower atmosphere can have adverse effects on human health and plants. Ozone also has effects on climate. Climate is most sensitive to changes in ozone near the tropopause [the upper boundary of the troposphere—see below]; information on ozone levels in this region, and for the rest of the troposphere above ground level as well, is limited to a few locations globally where balloon-borne measurements have been made for about thirty years.

Quantities of airborne particles in industrialized regions of the Northern Hemisphere have increased markedly since the Industrial Revolution. Atmospheric particles (aerosols) arise both from direct emissions and from gas-to-particle conversion of vapor precursors. Aerosols can affect climate and stratospheric ozone concentrations and have been implicated in human morbidity and mortality in urban areas. The climatic role of atmospheric aerosols arises from their ability to reflect solar radiation back to space and from their role as cloud condensation nuclei. Estimates of the cooling effect resulting from the reflection of solar radiation back to space by aerosols indicate that the cooling effect may be sufficiently large to mask the warming effect of greenhouse gas increases over industrialized regions of the Northern Hemisphere.

The atmosphere is the recipient of many of the products of our technological society. These effluents include products of combustion of fossil fuels and of the development of new synthetic chemicals. Historically these emissions can lead to unforeseen consequences in the atmosphere. Classical examples include the realization in the 1950s that motor vehicle emissions could lead to urban smog and the realization in the 1970s that emissions of chlorofluorocarbons from aerosol spray cans and refrigerators could cause the depletion of stratospheric ozone.

The chemical fates of trace atmospheric species are often intertwined. The life cycles of the trace species are inextricably coupled through the complex array of chemical and physical processes in the atmosphere. As a result of these couplings, a perturbation in the concentration of one species can lead to significant changes in the concentrations and lifetimes of other trace species and to feedbacks that can either amplify or damp the original perturbation. An example of this coupling is provided by methane. Methane is the predominant organic molecule in the troposphere and it is the second most important greenhouse gas after CO_2. Methane sources such as rice paddies and cattle can be estimated and are increasing. Methane is removed from the atmosphere by reaction with the hydroxyl (OH) radical, at a rate that depends on the atmospheric concentration of OH. But, the OH concentration depends on the amount of carbon monoxide (CO), which itself is a product of CH_4 oxidation as well as a result of fossil fuel combustion and biomass burning. The hydroxyl concentration also depends on the concentration of ozone and oxides of nitrogen. Change in CH_4 can affect the total amount of ozone in the troposphere, so methane itself affects the concentration of the species, OH, that governs its removal. . . .

The extraordinary pace of the recent increases in atmospheric trace gases can be seen when current levels are compared with those of the distant past. Such comparisons can be made for CO_2 and CH_4, whose histories can be reconstructed from their concentrations in bubbles of air trapped in ice in such perpetually cold places as Antarctica and Greenland. With gases that are long-lived in the atmosphere and therefore distributed rather uniformly over the globe, such as CO_2 and CH_4, polar ice core samples reveal global average concentra-

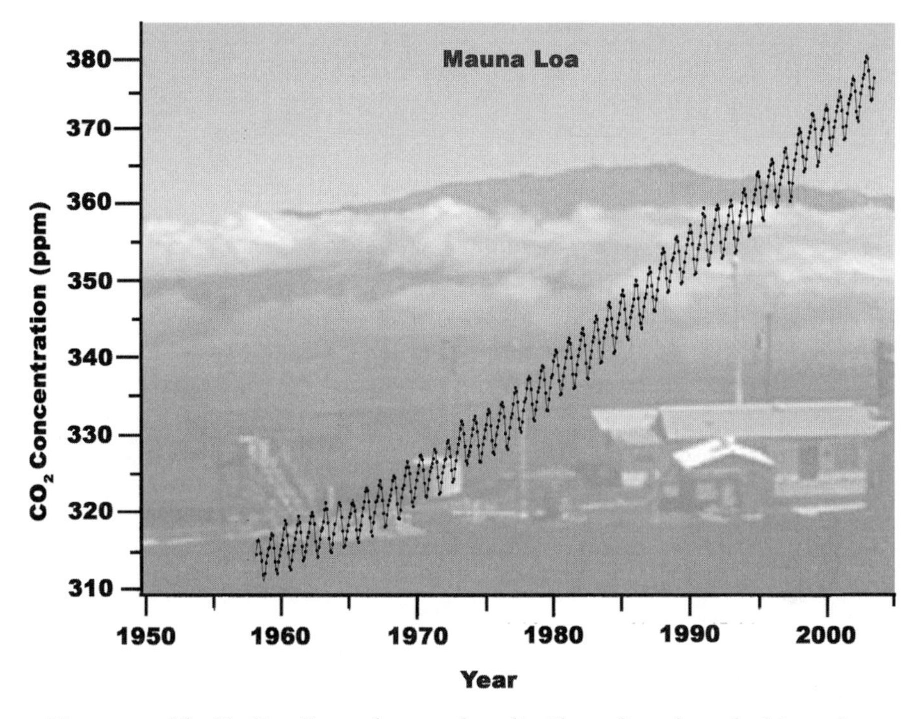

Figure 13.1 The Keeling Curve shows carbon dioxide readings from the Mauna Loa Observatory. C. D. Keeling and T. P. Whorf. 2004. Atmospheric CO_2 records from sites in the SIO air sampling network, *Trends: A Compendium of Data on Global Change*. Carbon Dioxide Information Analysis Center, Oak Ridge National Laboratory, U.S. Department of Energy.

tions of previous eras. Analyses of bubbles in ice cores show that CO_2 and CH_4 concentrations remained essentially unchanged from the end of the last ice age some 10,000 years ago until roughly 300 years ago, at mixing ratios close to 260 ppm by volume and 0.7 ppm by volume, respectively. [See Gribbin in Chapter 1 for more on evidence of carbon dioxide fluctuations from ice cores.] . . . About 300 years ago methane levels began to climb, and about 100 years ago levels of both gases began to increase markedly. Before the large-scale production of chlorofluorocarbons the natural level of chlorine in the stratosphere was 0.6 part per billion (ppb) by volume; now it is 3 ppb, a factor of 5 increase.

Activities of humans account for most of the rapid changes in the trace gases over the past 200 years—combustion of fossil fuels (coal and oil) for energy and transportation, industrial and agricultural activities, biomass burning (the burning of vegetation), and deforestation. For CO_2, for example, the sources are mainly fossil-fuel combustion and deforestation in the tropics; for CH_4, mainly rice cultivation, cattle breeding, biomass burning, microbial activity in municipal landfills, and leakage of gas during the recovery of coal, oil, and natural gas.

Figure 13.2 Atmospheric carbon dioxide concentrations over more than forty years near the South Pole. Keeling and Whorf, 2004 (see Figure 13.1).

CLIMATE

Viewed from space, the Earth is a multicolored marble; clouds and snow-covered regions of white, blue oceans, and brown continents. The white areas make Earth a bright planet; about 30% of the Sun's radiation is reflected immediately back to space. Solar energy that does not reflect off clouds and snow is absorbed by the atmosphere and the Earth's surface. As the surface warms, it sends infrared radiation back to space. The atmosphere, however, absorbs much of the energy radiated by the surface and reemits its own energy, but at much lower temperatures. Aside from gases in the atmosphere, clouds play a major climatic role. Some clouds cool the planet by reflecting solar radiation back to space; others warm the Earth by trapping energy near the surface. On balance, clouds exert a significant cooling effect on Earth, although in some areas, such as the tropics, heavy clouds can markedly warm the regional climate.

The temperature of the Earth adjusts so that solar energy reaching the Earth is balanced by that leaving the planet. Whereas the radiation budget must balance for the entire Earth, it does not balance at each particular point on the globe. Very little solar energy reaches the white, ice-covered polar regions, especially during the winter months. The Earth absorbs most solar radiation near its equator. Over time, though, energy absorbed near the equator spreads to the colder regions of the globe, carried by winds in the atmosphere and by currents in the oceans. This global heat engine, in its attempt to equalize temperatures, generates the climate with which we are all familiar. It pumps energy into storm fronts and powers hurricanes. In the colder seasons, low-pressure and high-pressure cells push each other back and forth every few days. Energy is also transported over the globe by masses of wet and dry air. Through evaporation, air over the warm oceans absorbs water vapor and then travels to colder regions and continental interiors where water vapor condenses as rain or snow, a process that releases heat into the atmosphere. In the oceans, salt helps drive the heat engine. Over some areas, like the arid Mediterranean, water evaporates from the sea faster than rain or river flows can replace it. As seawater becomes increasingly salty, it grows denser. In the North Atlantic, cool air temperatures and excess salt cause the surface water to sink, creating a current of heavy water that spreads throughout the world's oceans. By redistributing energy in this way, the oceans act to smooth out differences in temperature and salinity. Whereas the atmosphere can respond in a few days to a warming or cooling in the ocean, it takes the sea surface months or longer to adjust to changes in energy coming from the atmosphere.

Solar radiation, clouds, ocean currents, and the atmospheric circulation weave together in a complex and chaotic way to produce our climate. Until recently, climate was assumed to change on a time scale much much longer than our lifetimes and those of our children. Evidence is mounting, however, that the release of trace gases to the atmosphere, the "greenhouse gases," has the potential to lead to an increase of the Earth's temperature by several degrees Celsius. The Earth's average temperature rose about 0.5°C over the past century. It has been estimated that a doubling of CO_2 from its pre-Industrial Revolution mixing ratio of 280 ppm by volume could lead to a rise in average global temperature of 1.5 to 4.5°C. A 2°C warming would produce the warmest climate seen on Earth in 6,000 years. A 4.5°C rise would place the world in a temperature regime last experienced in the Mesozoic Era—the age of dinosaurs.

Although an average global warming of a few degrees does not sound like much, it could create dramatic changes in climatic extremes. It has been estimated, for example, that, in the event of an average global warming of 1.7°C, the frequency of periods of 5 days or more exceeding 35°C (95°F) in the Corn Belt of the United States would increase threefold. Such conditions at critical stages of the growing season are known to harm corn and lead to reduced yields. With a doubling of CO_2, the number of days exceeding 38°C (100°F) and nights above 27°C (80°F) have been estimated to rise dramatically in many major American cities. Changes in the timing and amount of precipitation would almost certainly occur with a warmer climate. Soil moisture, critical during planting and early growth periods, will change. Some regions would probably become more productive, others less so. The North American Grain Belt, according to at least one climate model, will shift northward into Canada as warming produces hotter, drier conditions in the American Midwest.

Of all the effects of a global warming, perhaps none has captured more attention than the prospect of rising sea levels. This would result from the melting of land-based glaciers and volume expansion of ocean water as it warms. Prevailing opinion is that a sea level rise of about 0.5 m could occur by 2100. In the most dramatic scenario, the West Antarctic ice sheet, which rests on land that is below sea level, could slide into the sea if the buttress of floating ice separating it from the ocean were to melt. This would raise the average sea level 5 to 6 m.[3] Even a 0.3 m (1 foot) rise would have major effects on the erosion of coastlines, salt water intrusion into the water supply of coastal areas, flooding of marshes, and inland extent of surges from large storms.

To systematically approach the complex subject of climate, the scientific community has divided the problem into two major parts, climate *forcings* and climate *responses*. Climate *forcings* are changes in the energy balance of the Earth that are imposed upon it; forcings are measured in units of heat flux—watts per square meter (Wm^{-2}). An example of a forcing is a change in energy output from the sun. *Responses* are the meteorological results of these forcings, reflected in temperatures, rainfall, extremes of weather, sea level height, and so on.

Much of the variation in the predicted magnitude of potential climate effects resulting from the increase in greenhouse gas levels hinges on estimates of the size and direction of various feedbacks that may occur in response to an initial perturbation of the climate. Negative feedbacks have an effect that damps the warming trend; positive feedbacks reinforce the initial warming. One example of a greenhouse warming feedback mechanism involves water vapor. As air warms, each cubic meter of air can hold more water vapor. Since water vapor is a greenhouse gas, this increased concentration of water vapor further enhances greenhouse warming. In turn, the warmer air can hold more water, and so on. This is an example of a positive feedback, providing a physical mechanism for multiplying the original impetus for change beyond its initial amount.

Some mechanisms provide a negative feedback, which decreases the initial impetus. For example, increasing the amount of water vapor in the air may lead to forming more clouds. Low-level, white clouds reflect sunlight, thereby preventing sunlight from reaching the Earth and warming the surface. Increasing the geographical coverage of low-level clouds would reduce greenhouse warming, whereas increasing the amount of high, convective clouds could enhance greenhouse warming. This is because high, convective clouds absorb energy from below at higher temperatures than they radiate energy into space from their tops, thereby effectively trapping energy. It is not known with certainty whether increased temperatures would lead to more low-level clouds or more high, convective clouds.

Another feedback uncertainty involves the response of plants to rising CO_2 levels. Some studies indicate that agricultural crops grow faster in a high-CO_2 environment, pulling more carbon out of the atmosphere and storing it in plant tissues. Little is known about the response of the forests, grasslands, and tundra that cover much of the Earth.

Probably of even greater significance than the feedback involving terrestrial plants and soils are the immense and very complex ocean-atmosphere interactions. Without human influence, the flows of carbon between the atmosphere, plants, and ocean would be roughly balanced. Fossil-fuel combustion today adds about 5×10^9 metric tons per year (5 Gt yr^{-1}) of carbon to the atmosphere. (1 metric ton, Mt, equals 10^6 g.) About half of the 5 Gt yr^{-1} remains in the atmosphere as rising CO_2 levels. The rest is absorbed by plants and oceans but uncertainty exists as to just how much goes to each reservoir. [Research reported in 2004 indicates that most goes to the oceans.] Deforestation is adding up to 4 Gt of carbon to the atmosphere annually. Thus, of the 5 to 9 Gt yr^{-1} added to the atmosphere by humans, 2.5 Gt remains in the atmosphere and 2.5 to 6.5 Gt yr^{-1} is absorbed by plants and the oceans. Accepting the higher end of the range of estimates for deforestation and the lower end of the range of estimates for plant uptake, plants can serve as a net source, rather than a sink, for CO_2.

The oceans contain 55 times as much carbon as does the atmosphere and 20 times as much as do land plants. Thus small changes in the oceans' capacity to store carbon can have a large effect on atmospheric concentrations. In most areas, atmospheric CO_2 interacts only with the top 100 m or so of seawater and moves downward slowly because of thermal gradients that separate the surface layer from deeper waters. However, the only effective way for the oceans to buffer the atmospheric increase in CO_2 is to pump the carbon into the deep ocean, either as dissolved gaseous CO_2 or as carbonate particles that settle in the sediments at the ocean floor. CO_2 may be drawn into the deep ocean in a few locations, like the North Atlantic, where cold surface waters sink to the bottom. Some have speculated that global warming could cause the oceans' currents to become more sluggish, reducing their ability to take up CO_2 and exacerbating the warming.

THE LAYERS OF THE ATMOSPHERE

In the most general terms, the atmosphere is divided into lower and upper regions. The lower atmosphere is generally considered to extend to the top of the stratosphere, an altitude of about 50 kilometers (km). Study of the lower atmosphere is known as *meteorology;* study of the upper atmosphere is called *aeronomy.*

The Earth's atmosphere is characterized by variations of temperature and pressure with height. In fact, the variation of the average temperature profile with altitude is the basis for distinguishing the layers of the atmosphere. The regions of the atmosphere are (Figure 13.3):

Troposphere. The lowest layer of the atmosphere, extending from the Earth's surface up to the tropopause, which is at 10 to 15 km altitude depending on latitude

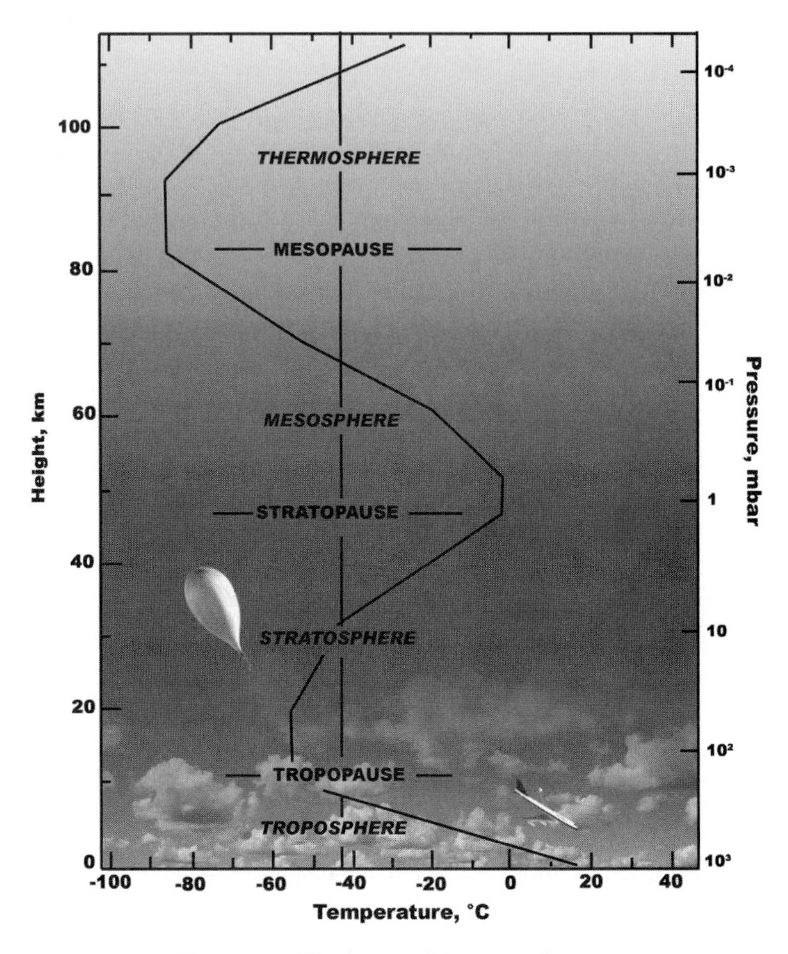

Figure 13.3 The layers of the atmosphere.

and time of year; characterized by decreasing temperature with height; rapid vertical mixing.

Stratosphere. Extends from the tropopause to the stratopause (~45 to 55 km altitude); temperature increases with altitude, leading to a layer in which vertical mixing is slow.

Mesosphere. Extends from the stratopause to the mesopause (~80 to 90 km altitude); temperature decreases with altitude to the mesopause, which is the coldest point in the atmosphere; rapid vertical mixing.

Thermosphere. The region above the mesopause; characterized by high temperatures as a result of absorption of short wavelength radiation by N_2 and O_2; rapid vertical mixing. The *ionosphere* is a region of the upper mesosphere and lower thermosphere where ions are produced by photoionization.

Exosphere. The outermost region of the atmosphere (~500 km altitude) where gas molecules with sufficient energy can escape from the Earth's gravitational attraction.

. . . The tropopause is at a maximum height over the tropics, sloping downward moving toward the poles. The name coined by British meteorologist Sir Napier Shaw, from the Greek word *tropos*, meaning turning, the troposphere is a region of ceaseless turbulence and mixing. The caldron of all weather, the troposphere contains almost all of the atmosphere's water vapor. Although the troposphere accounts for only a small fraction of the atmosphere's total height, it contains about 80% of its total mass. In the troposphere, the temperature decreases almost linearly with height. For dry air the lapse rate is 9.7 K km^{-1} [i.e., 9.7 degrees Kelvin per kilometer]. The reason for this progressive decline is the increasing distance from the sun-warmed Earth. At the tropopause, the temperature has fallen to an average of about 217 K (–56°C). The troposphere can be divided into the *planetary boundary layer*, extending from the Earth's surface up to about 1 km, and the *free troposphere*, extending from about 1 km to the tropopause.

As air moves vertically, its temperature changes in response to the local pressure. For dry air, this rate of change is substantial, about 1°C per 100 m. . . . An air parcel that is transported from the surface to 1 km can decrease in temperature from 5 to 10°C depending on its water content. Because of the strong dependence of the saturation vapor pressure on temperature, this decrease of temperature of a rising air parcel can be accompanied by a substantial increase in relative humidity (RH) in the parcel. As a result, upward air motions of a few hundreds of meters can cause the air to reach saturation (RH = 100%) and even supersaturation. The result is the formation of clouds.

Vertical motions in the atmosphere result from (1) convection from solar heating of the Earth's surface, (2) convergence or divergence of horizontal flows, (3) horizontal flow over topographic features at the Earth's surface, and (4) buoyancy caused by the release of latent heat as water condenses. Interestingly, even though an upward moving parcel of air cools, condensation of water vapor can provide sufficient heating of the parcel to maintain the temperature of the air parcel above that of the surrounding air. When this occurs, the parcel is buoyant and accelerates upward even more, leading to more condensation. Cumulus clouds are produced in this fashion, and updraft velocities of meters per second can be reached in such clouds. Vertical convection associated with cumulus clouds is, in fact, a principal mechanism for transporting air from close to the Earth's surface to the mid- and upper-troposphere.

The stratosphere, extending from about 11 km to about 50 km, was discovered at the turn of the twentieth century by the French meteorologist, Léon Philippe Teisserenc de Bort. Sending up temperature-measuring devices in balloons, he found that, contrary to the popular belief of the day, the temperature in the atmosphere did not steadily decrease to absolute zero with increasing altitude, but stopped falling and remained constant at 11 km or so. He named the region the stratosphere from the Latin word *stratum* meaning layer. Although an isothermal region does exist from about 11 to 20 km at midlatitudes, temperature progressively increases from 20 to 50 km, reaching 271 K at the stratopause, a temperature not much lower than the average of 288 K at the Earth's surface. The vertical thermal structure of the stratosphere is a result of the presence of ozone in the stratosphere. Absorption of solar ultraviolet radiation by O_3 causes the temperature in the stratosphere to be much higher than expected, based on simply extending the troposphere's lapse rate into the stratosphere. . . .

LARGE-SCALE MOTION OF THE ATMOSPHERE: GENERAL CIRCULATION

The large-scale motion of the atmosphere comprises the winds, the global structure of which is referred to as the general circulation. Moisture, momentum, and energy are exchanged between the atmosphere and the underlying ocean and land. Little mass and momentum are exchanged with space, although the atmosphere absorbs directly a portion of the solar radiation. The total dry mass of the atmosphere, calculated as an annual mean, is estimated to be 5.13×10^{18} kg.[4]

Even though the total input and output of radiant energy to and from the Earth are essentially in balance, they are not in balance at every point on the Earth. The amount of energy reaching the Earth's surface depends, in part, on the nature of the surface (e.g., land versus sea) and the degree of cloudiness, as well as on the latitude of the point. For example, at lower solar angles in the polar regions, the same amount of solar energy as radiated to the tropics must pass through more atmosphere and intercept a larger surface area. The uneven distribution of energy resulting from latitudinal variations in insolation [rate of exposure to sun's rays] and from differences in absorptivity of the Earth's surface leads to the large-scale air motions of the Earth. In particular, the tendency to transport energy from the tropics toward the polar regions, thereby redistributing energy inequalities on the Earth, is the overall factor governing the general circulation of the atmosphere.

In order to visualize the nature of the general circulation of the atmosphere, we can think of the atmosphere over either hemisphere as a fluid enclosed within a long, shallow container, heated at one end and cooled at the other. Because the horizontal dimension of the "container" is so much greater than its vertical dimension, the curvature of the earth can be neglected, and the container can be considered to be rectangular. If such a container were constructed in the laboratory and the ends differentially heated as described above, one would observe a circulation of the fluid, consisting of rising motion along the heated wall and descending motion along the cooled wall, flow in the direction of warm to cold at the top of the box, and flow in the direction of cold to warm along the bottom of the box. In the atmosphere, then, the tendency is for warm tropical air to rise and cold polar air to sink, with poleward and equatorward flows to complete the circulation.

However, the general circulation of the atmosphere is not as simple as just described. Another force arises because of the motion of the Earth, the Coriolis force. At the Earth's surface an object at the equator has a greater tangential velocity than one in the temperate zones. Air moving toward the south cannot acquire an increased eastward (the Earth rotates from west to east) tangential velocity as it moves south and thus, *to an observer on the Earth*, appears to acquire a velocity component in the westward direction. Thus air moving south in the Northern Hemisphere appears to lag behind the Earth. To an observer on the Earth it appears that the air has been influenced by a force in the westward direction. To an observer in space, it would be clear that the air is merely trying to maintain straight-line motion while the Earth turns below it. Friction between the wind and the ground diminishes this effect in the lower atmosphere.

From the standpoint of air motion, the atmosphere can be segmented vertically into two layers. Extending from the ground up to about 1000 m is the *planetary boundary layer*, the zone in which the effect of the surface is felt and in which the wind speed and direction are governed by horizontal pressure gradients, shear stresses, and Coriolis forces. Above the planetary boundary layer is the *geostrophic layer*, in which only horizontal pressure gradients and Coriolis forces influence the flow.

To predict the general pattern of macroscale air circulation on the Earth we must consider both the tendency for thermal circulation and the influence of Coriolis forces. Figure 13.4 shows the nature of the general circulation of the atmosphere. At either side of the equator is a thermal circulation, in which warm tropical air rises and cool northern air flows toward the equator. The circulation does not extend all the way to the poles because radiative cooling of the upper northward flow causes it to subside (fall) at about 30° N and S latitude. The Coriolis force acting on these cells leads to easterly winds, called the trade winds. The same situation occurs in the polar regions, in which warm air from the temperate zones moves northward in the upper levels, eventually cooling by radiation and subsiding at the poles. The result is the polar easterlies.

In the temperate regions, between 40° and 55° latitude, influences of both tropical and polar regions are felt. The major feature of the temperate regions is large-scale weather systems, which results in the circulation shown in Figure 13.4. The surface winds in the Northern Hemisphere are westerlies because of the Coriolis force.

At the boundaries between thermal circulation at the equator, 30°, and 55° N and S latitude, there are regions of calm. The observed net precipitation near the equator and the polar front is explained by rising moist air that cools. At 30° N and S latitude a strong subsidence of dry air occurs, since the air loses its moisture upon ascension in the equatorial zone. As a result, net evaporation of the oceans occurs from 10° to 40° N and S latitude.

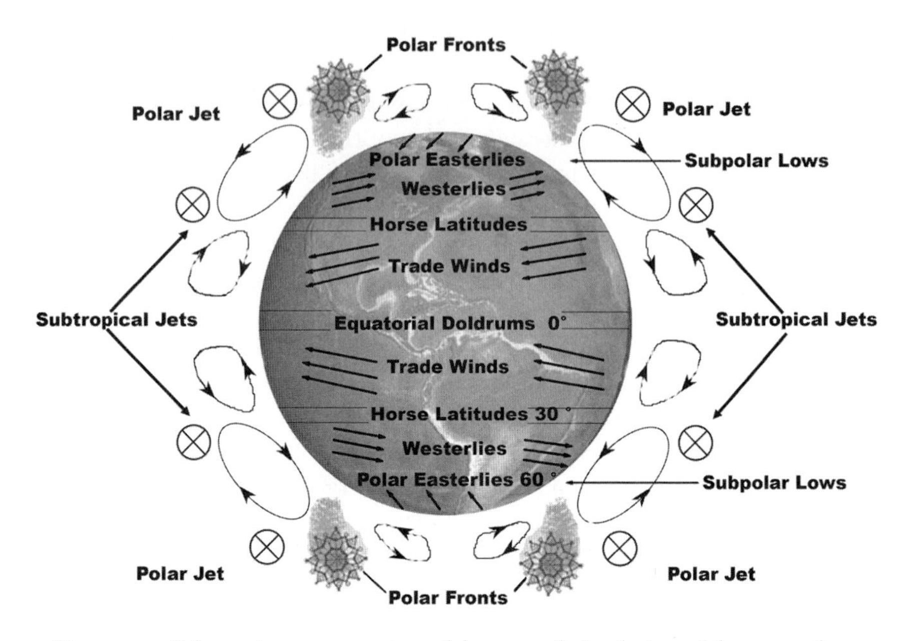

Figure 13.4 Schematic representation of the general circulation of the atmosphere.

The pattern of general circulation shown in Figure 13.4 does not represent the actual state of atmospheric circulation on a given day. The irregularities of land masses and their surface temperatures tend to disrupt the smooth global circulation patterns we have described. Another influence that tends to break up zonal patterns is the Coriolis force. Air that converges at low levels toward regions of low pressure must also execute a circular motion because of Coriolis forces. The effect of friction at the surface is to direct the winds at low levels in part toward the region of low pressure, producing an inward spiraling motion. This vortex-like motion is given the name *cyclone*. The center of a cyclone is usually a rising column of warm air. Similarly, a low-level diverging flow from a high-pressure region will spiral outward. Such a region is called an *anticyclone*. In the Northern Hemisphere the motion of a cyclone is counterclockwise and that of an anticyclone, clockwise. The dimensions of commonly occurring cyclones and anticyclones are from 100 to 1000 km. Most cyclones and anticyclones are born in one part of the world and migrate to another. These are not to be confused with hurricanes or typhoons, which, although they consist of the same type of air motion, are of a smaller scale.

An element of the cyclone-anticyclone phenomenon that has particular importance for air pollution in several parts of the world is the semipermanent subtropical anticyclone, high-pressure regions centered over the major oceans. They are called semipermanent because they shift position only slightly in summer and winter. The key feature of the subtropical anticyclone is that the cold subsiding air aloft, which results in the high pressure observed at sea level, is warmed by compression as it descends, often establishing an elevated temperature inversion. The inversion layer generally approaches closer to the ground as the distance from the center of the high pressure increases.

On the eastern side of the subtropical anticyclones the inversion is strengthened by the southerly flow of cool, dry air (recall that in the Northern Hemisphere the rotation in an anticyclone is clockwise). Particularly in coastal areas the low-level air is cooled by contact with the cold ocean, an exchange that tends to strengthen the inversion. Since the air aloft, as well as the southbound low-level flow, is warming, there is little precipitation in these regions. Thus on the west coasts of continents it is common to find arid, desert-like conditions, such as the deserts of southern California, the Sahara in North Africa, the desert in western Australia, and the coastal plains of South America.

On the other hand, on the western side of the semipermanent anticyclone, inversions are less frequent and the low-level air from the tropics is warm and moist. As it cools on its path to the north, precipitation is heavy. Thus the eastern coasts of continents in the subtropics are warm and humid, such as the eastern coasts of South America and Africa.

We can now see one of the reasons why Los Angeles is afflicted with air pollution problems. Its location on the west coast of North America in the subtropical region and on the eastern side of the Pacific anticyclone is one in which elevated inversions are frequent and strong. The lowest layer of air (the marine layer) is cooled because of its contact with the ocean. Air pollutants are trapped in the marine layer and prevented from vertically exchanging with upper-level air. Such a situation can lead to serious air pollution problems. The base of a subsidence inversion lies typically at an elevation of about 500 m, with the inversion layer extending another 500 to 1000 m upward.

In addition to the semipermanent anticyclones, there are many migratory cyclones and anticyclones in the temperate zones. Formed by confrontations between arctic and tropical air, they have a lifetime of a few weeks and drift with the westerly winds at about 800 km day^{-1}. Precipitation is often associated with the rising air over the low-pressure center of a cyclone. Cyclones are thus usually accompanied by cloudy skies and inclement weather. On the other hand, anticyclones are characterized by clear skies, light winds, and fair weather.

Notes

1. P. Cloud, "The Biosphere," *Scientific American* 249 (1983): 176–89.
2. J. C. G. Walker, *Evolution of the Atmosphere* (New York: Macmillan, 1977).
3. C. R. Bentley, "Rapid Sea-Level Rise Soon from West Antarctic Ice Sheet Collapse?" *Science* 275 (1997): 1077–78.
4. K. E. Trenberth and C. J. Guillemot, "The Total Mass of the Atmosphere," *Journal of Geophysical Research* 99 (1994): 23079–88.

S. George Philander, from "The Ozone Hole, A Cautionary Tale" in *Is the Temperature Rising?* (1998)

Philander recounts the story of the ozone hole—how ozone depletion was predicted, confirmed by observation, and addressed in a public-spirited way by the international community. Although it is a story with a happy ending, it is also a cautionary tale. It took decades for scientists to recognize the problem, and another decade or so passed before the Montreal Protocol was agreed upon. Other global environmental crises—Philander has global warming in particular in mind—might not permit such a generous timetable.

It is a curious story, about an invention that at first was viewed as a boon to mankind, then was suspected of being harmful to the atmosphere, and now is seen as a threat to life on Earth. The indictment of this invention, a family of chemicals known as chlorofluorocarbons or CFCs, came after laboratory experiments revealed that CFCs are potential vandals of our atmosphere's ozone shield. The nations of the world took swift action to prevent damage to the shield that protects us from dangerous ultraviolet rays in sunlight. In Montreal, in 1987, they agreed to limit the production of CFCs. This happened before there was any direct evidence that these extraordinarily versatile and valuable chemicals were damaging the ozone shield. When evidence of the harmful effects of CFCs did appear, it was in a dramatic and unexpected form. Whereas scientists predicted that CFCs would cause a gradual, uniform weakening of the ozone layer over many decades, a gaping hole suddenly appeared in that layer over Antarctica, where it can now be found every October. The explanation for the ozone hole involves such an unusual combination of factors that a prediction of that hole before it had been observed would have been dismissed as too improbable.

A few vocal people still refuse to accept that CFCs cause ozone depletion. They regard attempts to ban the production of CFCs as conspiracies by the chemical cartel, which has new products it wishes to market. The story of the ozone hole not only illustrates how difficult it is to predict precisely how our planet will respond when we interfere with the processes that maintain habitability, but also demonstrates how emotional the discussion of scientific matters can be.

The reason for the abundance of ozone in the stratosphere is related to the decrease in the availability of molecular oxygen with altitude—gravity keeps most oxygen molecules near Earth's surface—and the increase in the availability of atomic oxygen with altitude. Atomic oxygen (O) is created high in the atmosphere where energetic ultraviolet photons photodissociate molecular oxygen (O_2):

$$O_2 + \text{ultraviolet photon} \rightarrow O + O$$

An oxygen molecule (O_2) is converted into two oxygen atoms (O) so that a community of couples (O_2) and singles (O) evolves. Inevitably, threesomes emerge. Such an arrangement is known as ozone (O_3), an unusual molecule with three, not two atoms. The creation of ozone generates so much heat that it requires the presence of a third molecule, to be called M, to carry heat away.

$$O_2 + O + M = O_3 + M^*$$

This equation, in which M* is a more energetic version of M, describes how ozone emerges from the collision of three particles: two molecules (O_2 and M) and an atom (O). In general, the collision of three particles is a rare event at great altitudes where the density of air is very low, but it becomes more common with decreasing height (and increasing density). Not all particles become more abundant with decreasing altitude; atomic oxygen does not because it requires the presence of energetic photons to split molecular oxygen. The elevation at which ozone can be created most efficiently is therefore a compromise between low elevations where the density of air is so high that three particles collide frequently, and great heights where one of the required particles, atomic oxygen, is abundant. This compromise is reached in the stratosphere, which therefore is the location of the ozone layer.

A threesome is not always a stable arrangement, especially not in the presence of energetic ultraviolet photons.

$$O_3 + \text{photon} \rightarrow O_2 + O$$

A photon can separate the ozone into a single oxygen atom plus an oxygen molecule. This reaction drains ozone from the stratosphere, but the reaction described by the previous equation restores ozone. The continual creation and destruction of ozone in the stratosphere, by means of the reactions described by the last two equations, maintain a stable community of singles, doubles, and triplets. The result is no net chemical change—it is as if ozone were water in a bathtub that has a running faucet to compensate for a drain that is unplugged—except for an absorption of ultraviolet radiation that leads to high temperatures in the stratosphere and safe conditions for life at Earth's surface.

This theory for the ozone layer, developed by the Englishman Sydney Chapman in 1930, explains why the highest concentration of ozone is at altitudes between 15 and 50 km, but gives values for the concentration that are larger than the observed ones. There must be additional chemical reactions that destroy ozone. In 1970, the Dutch chemist Paul Crutzen proposed that those reactions involve nitrous oxide, a chemical produced by microorganisms in the soil, and was able to explain the observed concentration of ozone in the stratosphere.

The chemicals known as CFCs, which have proven the nemesis of ozone, were invented in the 1920s. At first they were regarded as a triumph of technology because they are stable, nonreactive, long-lived, nonflammable, and safe, far more so than ammonia, which they replaced in refrigerators and air conditioners. CFCs turned out to have a great many additional uses: they are effective propellants in spray cans for household products and pharmaceuticals; they are good insulators and therefore standard ingredients of plastic-foam materials; and they are excellent solvents for cleaning electronic equipment. These versatile chemicals are furthermore cheap to manufacture so that production grew rapidly during the 1950s and '60s.

In the 1970s Rowland and Molina predicted that the nonreactiveness of CFCs, considered a virtue, could prove to be a major flaw. [See selection below.] Because CFCs do not react with other chemicals, they remain in the atmosphere for a very long time, gradually dispersing upward to considerable elevations, and infiltrating the stratosphere, where they fall prey to dangerous beams in sunlight that break up the complex molecules into simpler ones. Rowland and Molina anticipated that one of the simpler chemicals, chlorine, would prove disastrous to the ozone layer. Chlorine acts as a catalyst, and a single chlorine molecule can eliminate tens of thousands of ozone molecules by participating in catalytic cycles such as the following,

$$\mathrm{Cl} + \mathrm{O_3} \rightarrow \mathrm{ClO} + \mathrm{O_2}$$

$$\mathrm{ClO} + \mathrm{O} \rightarrow \mathrm{Cl} + \mathrm{O_2}.$$

The net result is

$$\mathrm{O_3} + \mathrm{O} \rightarrow 2\mathrm{O_2}.$$

At the end of this cycle, an ozone molecule and an oxygen atom have combined to form two oxygen molecules. A chlorine atom arranges such marriages, which remove ozone from the atmosphere, but then withdraws, remaining free to arrange more marriages, hundreds of thousands of them. The result is a thinning of the ozone shield.

Scientists projected that serious damage to the ozone layer could affect all life on Earth. It could cause a significant reduction in food production and fisheries. They also feared that an increase in ultraviolet radiation could cause a suppression of the human immune system, an increase in eye cataracts and blindness, and, most disturbingly, a significant increase in the number of deaths attributable to skin cancer.

These dire predictions persuaded the nations of the world to act swiftly. Even though, in the mid-1980s, there was no firm evidence that CFCs were causing a weakening of the ozone layer, twenty-three nations, under United Nations auspices, started to negotiate a protocol to limit production of CFCs. Surprisingly, the sponsors of the protocol included both environmental groups and the main chemical companies that produced CFCs in the United States. The producers had an opportunity to gauge the strength of the environmental movement in the United States when the sales of spray propellants plummeted because of the possibility that the propellants could damage the ozone layer. The companies that produce CFCs anticipated that those chemicals would be banned in the United

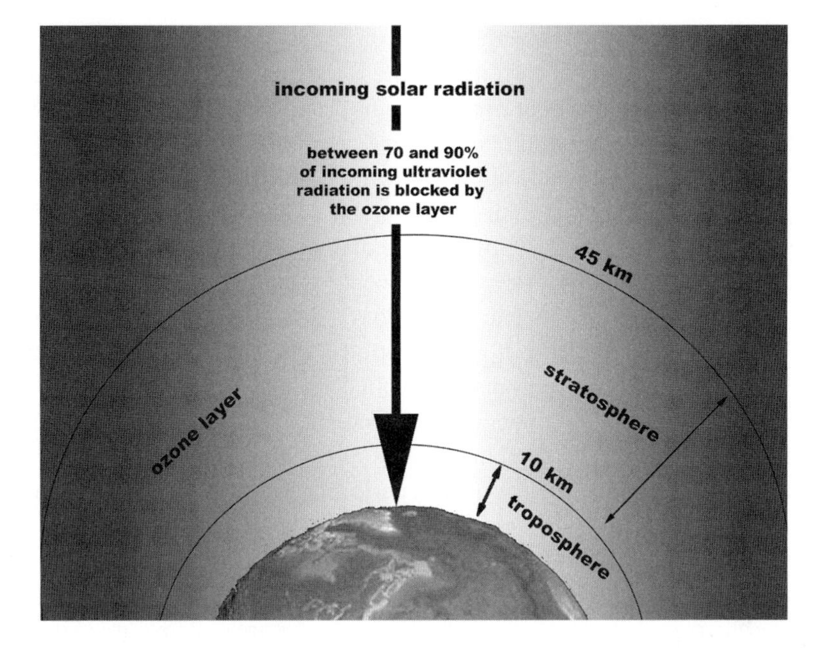

Figure 13.5 The positive effects of ozone in the atmosphere. Adapted from figure 4.1 in *North, South, and the Environmental Crisis* by Rodney R. White (University of Toronto Press, 1993).

States and realized that such unilateral action on the part of the United States would give foreign competitors an advantage. Those companies therefore favored an international agreement to ban CFCs. The first agreement, the Montreal Protocol of 1987, was for a reduction in the production of CFCs but the nations wisely agreed to permit revisions should new scientific evidence make that desirable. During the negotiations that led to the protocol, the story of ozone depletion took a curious turn.

THE OZONE HOLE OVER ANTARCTICA

Rowland and Molina predicted a gradual, global depletion of the ozone layer over the next several decades, but the first observations of ozone depletion, published in 1985, showed it to be massive and local, over Antarctica where it appears during the month of October. (The hole is absent during earlier months and disappears in November.) [The hole is now evident all year.] The measurements were so unexpected that, for a while, the cause of the ozone hole was a mystery. A field program, including flights into the stratosphere over Antarctica and laboratory experiments, established the reasons for the ozone hole. It is a consequence of chemical reactions that scientists at first ignored. Initially scientists paid attention primarily to reactions in mixtures of gases and disregarded heterogeneous reactions, those that occur on the surfaces of particles. (For two molecules to react, they have to meet. The frequency with which this happens can be low if the molecules are flying around randomly in the rarefied stratosphere, but increases should the molecules linger on the surfaces of solid particles.) Suitable particles include minute ice crystals in clouds that form high above Antarctica when temperatures are extremely low, during the long polar night. On these surfaces, certain forms of chlorine that do not react with ozone are converted into forms that do, after photodissociation by photons. The stage is set for a massive destruction of ozone when two requirements are met: temperatures must be so low that clouds composed of solid particles can form in the stratosphere, and photons must be available to split certain chlorine molecules. One of these requirements, low temperatures, is satisfied during the long Antarctic night, during July and August, but there are no energetic photons at that time. The other requirement, plentiful energetic photons, is satisfied during the southern summer, November and December, but, because of the relatively high temperatures once the Sun is above the horizon, the stratosphere at that time has no clouds composed of ice particles. Only in September and October, when the Sun is barely over Antarctica and when temperatures are still low, are conditions favorable for an ozone hole. These arguments imply that an ozone hole should also appear over the Arctic Ocean in March and April. At first, none was observed, but a hole did appear in March 1997. . . .

The story of the ozone hole is a cautionary tale about the limited abilities of scientists to anticipate how Earth is likely to respond to perturbations. Nature's response to minuscule changes in the chemical composition of the atmosphere can be in the form of bizarre phenomena (such as the ozone hole) that are difficult, if not impossible, to predict. It is to their credit that scientists succeeded in anticipating that CFCs could harm the ozone layer, and that diplomats succeeded in negotiating a treaty limiting the production of CFCs.

Mario J. Molina and F. Sherwood Rowland, "Stratospheric Sink for Chlorofluoromethanes: Chlorine Atom-Catalysed Destruction of Ozone" (1974)

This short work is a classic. Published in a premier journal of science, Nature, *it explored the fate of chlorofluoromethanes in the atmosphere, including (and especially) their role in the destruction of stratospheric ozone. In ensuing decades, atmospheric measurements and satellite data filled in the picture sketched here, and additional ozone-depleting chemicals were identified. Intricate aspects of the chemical reactions first identified by these two scientists remain under study even today.*

Abstract: Chlorofluoromethanes are being added to the environment in steadily increasing amounts. These compounds are chemically inert and may remain in the atmosphere for 40–150 years, and concentrations can be expected to reach 10 to 30 times present levels. Photodissociation of the chlorofluoromethanes in the stratosphere produces significant amounts of chlorine atom, and leads to the destruction of atmospheric ozone.

Halogenated aliphatic hydrocarbons [carbon-chain molecules (such as methanes or ethanes) that include one or more halogen atoms (such as chlorine or fluorine) in place of hydrogen] have been added to the natural environment in steadily increasing amounts over several decades as a consequence of their growing use, chiefly as aerosol propellants and as refrigerants. Two chlorofluoromethanes, CF_2Cl_2 and $CFCl_3$, have been detected throughout the troposphere in amounts (about 10 and 6 parts per 10^{11} by volume, respectively) roughly corresponding to the integrated world industrial production to date. The chemical inertness and high volatility which make these materials suitable for technological use also mean that they remain in the atmosphere for a long time. There are no obvious rapid sinks [natural destructive chemical processes] for their removal, and they may be useful as inert tracers of atmospheric motions. We have attempted to calculate the probable sinks and lifetimes for these molecules. The most important sink for atmospheric $CFCl_3$ and CF_2Cl_2 seems to be stratospheric photolytic dissociation [chemical decomposition caused by absorption of light energy] to $CFCl_2$ + Cl and to CF_2Cl + Cl, respectively, at altitudes of 20–40 km. Each of the reactions creates two odd-electron species—one Cl atom and one free radical [an atom or group of atoms with an unpaired electron]. The dissociated chlorofluoromethanes can be traced to their ultimate sinks. An extensive catalytic chain reaction leading to the net destruction of O_3 and O occurs in the stratosphere:

$$Cl + O_3 \rightarrow ClO + O_2 \quad (1)$$

$$ClO + O \rightarrow Cl + O_2 \quad (2)$$

This has important chemical consequences. Under most conditions in the earth's atmospheric ozone layer, (2) is the slower of the reactions because there is a much lower concentration of O than of O_3. The odd chlorine chain (Cl, ClO) can be compared with the odd nitrogen chain (NO, NO_2) which is believed to be intimately involved with the regulation of the present level of O_3 in the atmosphere. At stratospheric temperatures, ClO reacts with O six times faster than NO_2 reacts with O. Consequently, the Cl-ClO chain can be considerably more efficient than the NO-NO_2 chain in the catalytic conversion of $O_3 + O \rightarrow 2O_2$ per unit time per reacting chain.

PHOTOLYTIC SINK

Both $CFCl_3$ and CF_2Cl_2 absorb radiation in the far ultraviolet, and stratospheric photolysis will occur mainly in the "window" at 1,750–2,200 Å [Angstrom, a measure of wavelength equal to 10^{-10} meters] between the more intense absorptions of the Schumann-Runge regions of O_2 and the Hartley bands of O_3. [Oxygen strongly absorbs all wavelengths between 1350 and 1760 Å (the Schumann-Runge continuum), and additional wavelengths scattered between 1760 and 1926 Å (the Schumann-Runge bands). Ozone absorbs wavelengths between 2000 and 3000 Å (the Hartley bands).] We have extended measurements of absorption coefficients for the chlorofluoromethanes to cover the range 2,000–2,270 Å. Calculations of the rate of photolysis of molecules at a given altitude at these wavelengths is complicated by the intense narrow band structure in the Schumann-Runge region, and the effective rates of vertical diffusion of molecules at these altitudes are also subject to substantial uncertainties. Vertical mixing is frequently modeled through the use of "eddy" diffusion coefficients, which are presumably relatively insensitive to the molecular weight of the diffusing species. Calculated using a time-independent one-dimensional vertical diffusion model with eddy diffusion coefficients of magnitude $K \sim (3 \times 10^3) - 10^4 \text{ cm}^2 \text{ s}^{-1}$ [i.e., K falls within the range of 3,000 to 10,000 square centimeters per second] at altitudes 20–40 km, the atmospheric lifetimes of $CFCl_3$ and CF_2Cl_2 fall into the range of 40–150 yr. The time required for approach toward a steady state is thus measured in decades, and the concentrations of chlorofluoromethanes in the atmosphere can be expected to reach saturation values of 10–30 times the present levels, assuming constant injection at current rates, and no other major sinks. (The atmospheric content is now equivalent to about five years world production at current rates.) Lifetimes in excess of > 10 and > 30 yr can already be estimated from the known industrial production rates and atmospheric concentrations, and so the stratospheric photochemical sink will be important even if other sinks are discovered.

Our calculation of photodissociation rates is modeled after those of Kockarts and Brinkmann, and is globally averaged for diurnal and zenith angle effects. The photodissociation rates at an altitude of 30 km are estimated to be $3 \times 10^{-7} \text{ s}^{-1}$ for $CFCl_3$ and $3 \times 10^{-8} \text{ s}^{-1}$ for CF_2Cl_2, decreasing for each by about a factor of 10^{-2} at 20 km. The appropriate solar ultraviolet intensities at an altitude of 30 km may be uncertain by a factor of 2 or 3 and we have therefore calculated lifetimes for photodissociation rates differing from the above by factors of 3 or more. The competition between photodissociation and upward diffusion reduces the relative concentration of chlorofluoromethane at higher attitudes and the concentrations should be very low above 50 km. The peak rate of destruction, and formation of Cl atoms, occurs at 25–35 km, in the region of high ozone concentration. The rates of formation of Cl atoms at different altitudes and the chlorofluoromethane atmospheric lifetimes are sensitive to the assumed eddy diffusion coefficients, as well as to the photodissociation rates. . . .

The initial photolytic reaction produces one Cl atom from each of the parent molecules, plus a CX_3 radical (X may be F or Cl). The detailed chemistry of CX_3 radicals in O_2 or air is not completely known, but in the laboratory a phosgene-type molecule, CX_2O, is rapidly produced and another X atom—probably Cl (or ClO)—is released from CF_2Cl_2 or CF_2Cl. CX_2O may also photolyze in the atmosphere to give a third and fourth free halogen atom. Thus, each molecule of $CFCl_3$ initially photolyzed probably leads to [either two or] three Cl atom chains, and $CFCl_2$ probably produces two Cl atom chains when it is photolyzed. . . .

PRODUCTION RATES

The 1972 world production rates for $CFCl_3$ and CF_2Cl_2 are about 0.3 and 0.5 Mton yr^{-1} respectively, and [production rates] are steadily increasing (by 8.7 percent per year for total fluorocarbons in the United States from 1961–71). We have not included any estimates for other chlorinated aliphatic hydrocarbons also found in the atmosphere, such as CCl_4, $CHCl_3$, C_2Cl_4, and C_2HCl_3, for which there is no evidence for long residence times in the atmosphere. If the stratospheric photolytic sink is the only major sink for $CFCl_3$ and CF_2Cl_2, then the 1972 production rates correspond at steady state to globally averaged destruction rates of about 0.8×10^7 and 1.5×10^7 molecules cm^{-2} s^{-1} and formation rates of Cl atoms of about 2×10^7 and 3×10^7 atoms cm^{-2} s^{-1}, respectively. The total rate of production of 5×10^7 Cl atoms cm^{-2} s^{-1} from the two processes is of the order of the estimated natural flux of NO molecules ($2.5-15 \times 10^7$ NO molecules cm^{-2} s^{-1}) involved in the natural ozone cycle, and of the 5×10^7 NO molecules cm^{-2} s^{-1} whose introduction around 25 km from stratospheric aviation is estimated would cause a 6 percent reduction in the total O_3 column.

Photolysis of these chlorofluoromethanes does not occur in the troposphere because the molecules are transparent to wavelengths longer than 2,900 Å [i.e., those longer wavelengths (including the visible spectrum) that penetrate to the surface]. In fact the measured absorption coefficients for $CFCl_3$ and CF_2Cl_2 [fall] rapidly at wavelengths longer than 2,000 Å. The reaction between OH and CH_4 is believed to be important in the troposphere, but the corresponding Cl atom abstraction reaction (for example, $OH + CFCl_3 \rightarrow HOCl + CFCl_2$) is highly endothermic [requires a significant amount of energy] and is negligible under all atmospheric conditions. Neither $CFCl_3$ nor CF_2Cl_2 is very soluble in water, and they are not removed by rainout in the troposphere. Details of biological interactions of these molecules in the environment are very scarce because they do not occur naturally (except possibly in minute quantities from volcanic eruptions), but rapid biological removal seems unlikely. The relative insolubility in water together with their chemical stability (especially toward hydrolysis [water-induced chemical dissociation]) indicates that these molecules will not be rapidly removed by dissolution in the ocean, and the few measurements made so far indicate equilibrium between the ocean surface and air, and therefore a major oceanic sink cannot be inferred.

It seems quite clear that the atmosphere has only a finite capacity for absorbing Cl atoms produced in the stratosphere, and that important consequences may result. This capacity is probably not sufficient in steady state even for the present rate of introduction of chlorofluoromethanes. More accurate estimates of this absorptive capacity need to be made in the immediate future in order to ascertain the levels of possible onset of environmental problems.

As with most NO_x calculations, our calculations have been based entirely on reactions in the gas phase, and essentially nothing is known of possible heterogeneous reactions of Cl atoms with particulate matter in the stratosphere. [Subsequent research showed that nitric acid trihydrate (NAT) crystals play a catalytic role—see Philander, above, on the role of particles.] One important corollary of these calculations is that the full impact of the photodissociation of CF_2Cl_2 and $CFCl_3$ is not immediately felt after their introduction at ground level because of the delay required for upward diffusion up to and above 25 km. If any Cl atom effect on atmospheric O_3 concentration were to be observed from this source, the effect could be expected to intensify for some time thereafter. A lengthy period (of the order of calculated atmospheric lifetimes) may thus be required for natural moderation, even if the amount of Cl introduced into the stratosphere is reduced in the future.

Kathryn S. Brown, "The Ozone Layer: Burnt by the Sun Down Under" (1999)

One immediate result of stratospheric ozone depletion is increased risk of sunburn and potentially life-threatening cancers, especially of the skin and eye. Brown's charting of these risks and results first appeared in Science, *a leading international scientific journal. The danger from UV rays coming through a depleted ozone layer is higher in southern latitudes, but ozone depletion occurs in the north, too.*

When it's winter in the north and summer in the south, many cold-weary tourists from Europe and North America flock to New Zealand for its wild backcountry and radiant sunshine. They may be getting more than they bargained for.

. . . Scientists at the National Institute of Water and Atmospheric Research (NIWA) in Lauder, New Zealand, report that over the past ten years [1989–99] peak levels of skin-frying and DNA-damaging ultraviolet (UV) rays have gradually been increasing in New Zealand, just as concentrations of protective stratospheric ozone have decreased. By the summer of 1998–99, peak sunburning UV levels were about 12 percent higher than they were during similar periods earlier this decade. Experts say that the NIWA study provides the strongest evidence yet that a degraded stratospheric ozone layer causes more hazardous conditions for life on the planet's surface. "They have done about as careful a study as you can do," says atmospheric physicist Paul Newman of Goddard Space Flight Center in Greenbelt, Maryland.

Atmospheric scientists first detected the notorious "ozone hole" over the South Pole fourteen years ago [in 1985], the apparent result of chemical reactions caused by chlorofluorocarbons and other pollutants in the stratosphere. Ever since, their calculations have predicted that loss of stratospheric ozone—which acts like a protective sheath around the planet, absorbing much of the harmful UV-B radiation (290 to 315 nanometers)—would let through more of the rays. And not just in the sparsely populated polar regions: Researchers soon began to realize that stratospheric ozone was also thinning above populous midlatitude regions such as northern Europe, Canada, New Zealand, and Australia.

But nailing the expected relationship between ozone loss and increased UV-B radiation has proven to be anything but simple, says atmospheric physicist William Randel of the National Center for Atmospheric Research in Boulder, Colorado. Efforts to find a definitive link have been complicated by the fact that transient environmental features—such as clouds, snow cover, volcanic ash, or pollution—can filter or reflect UV-B. In 1993, for example, James Kerr and Thomas McElroy of Canada's Atmospheric Environment Service reported that winter levels of UV-B radiation reaching Toronto had risen more than 5 percent a year over the previous four years, a rate in step with declining peak ozone levels. But that study came under fire for being too short to detect a trend.

Now, NIWA atmospheric scientists Richard McKenzie, Brian Connor, and Greg Bodeker have come up with data that appear to clinch the connection between ozone and UV-B in the midlatitudes. They began their study in 1989, positioning their spectroradiometers and other equipment on the ground at Lauder, a rural region on New Zealand's South Island that enjoys unpolluted, cloudless days much of the year. In measurements taken each year since, the team has found that the maximum summertime UV-B levels crested higher and higher until they are now at least 12 percent above what they were at the beginning of the study. That agrees remarkably well with the roughly 15 percent increase the researchers had predicted based on the known decline in stratospheric ozone levels measured since 1978 in Lauder. Meanwhile, the longer wavelength UV-A radiation (315 to 400 nanometers), which is unimpeded by ozone, remained relatively constant.

According to meteorologist Jim Miller at the National Oceanic and Atmospheric Administration's National Centers for Environmental Prediction in Camp Springs, Maryland, New Zealand's peak UV-B levels, which are about 20 percent higher than those that bathe Toronto, could put inhabitants at increased risk of skin cancer, cataracts, and perhaps immune problems. What's more, elevated UV-B levels may perturb marine ecology, killing important algae and bacteria, says Ottawa University ecologist David Lean. Despite the increases, McKenzie notes that UV levels in New Zealand are still lower than levels in unpolluted, low-latitude regions of Australia, Africa, and South America.

Researchers should have plenty of time to study possible effects in New Zealand and elsewhere. The 1987 Montreal Protocol and its amendments, which restrict the use of ozone-destroying chemicals, have stemmed the flood of damaging pollutants reaching the stratosphere. But it will take decades for the ozone layer to recover, says McKenzie, because chlorine and bromine compounds can hang around in the atmosphere for years. "The problem isn't going to go away until the middle of the next [twenty-first] century, at the earliest," he says.

Joshua I. Barzilay, Winkler G. Weinberg, and J. William Eley, from *The Water We Drink: Water Quality and Its Effects on Health* (1999)

Not only human health but also the fate of civilizations has hinged on the availability of clean water. The history of water and health contains triumphs and object lessons.

The concept that there is a connection between water quality, sanitation, and the presence of disease was known in biblical and classical times. Efforts at safekeeping water purity, maintaining access to waters of high quality, and providing sewage disposal were widely practiced. With the demise of the Roman Empire and the beginning of the Middle Ages, these precepts were largely forgotten, and infectious illnesses secondary to polluted waters became commonplace. Only with the ascendancy of the scientific method and the discoveries of microbiology in the last one hundred years has the connection between water quality, sanitation, and health once again been discovered.

In more recent times, interest in water quality has shifted from an emphasis on infectious diseases to an emphasis on chemical pollution. This is a direct result of the industrialization of Western societies. As will become clear, this pollution is no less threatening to the health and welfare of the community than were the waterborne outbreaks of infectious diseases in the past. . . .

It was under the Romans that public health and water quality reached their acmes in the ancient world. Diligence in seeking and choosing water sources was widely practiced. By the first century C.E. the city of Rome was supplied by nine aqueducts from sources in the surrounding upland countryside, where groundwater contamination was uncommon. The aqueducts had basins at regular intervals to allow for settling of sediments, as well as sand and stones that acted as filters (much as is done today). In the city of Rome itself there were large cemented reservoirs from which water was brought by pipes to the homes of the rich or to public fountains for the masses. By the end of the third century C.E. there were more than 1,300 such public fountains. With regard to public sanitation, there was an extensive system for controlling sewage and storm

Figure 13.6 Marine panel from *The Apotheosis of Washington* by Constantino Brumidi. The Rotunda of the United States Capitol Building. By permission of the Architect of the Capitol.

water runoff in Rome, as well as a system for lowering water levels in the marshes. . . .

Despite all of these measures, the sanitary conditions in biblical and ancient times were not adequate by today's standards. Numerous plagues are listed in the Bible and in the writings of the Greek historian Thucydides and the Roman author Pliny. Still, certain rudimentary standards were observed. Although these standards would today appear self-evident, they were largely forgotten until as recently as the end of the nineteenth century. During the intervening centuries the Western world took a step backward with regard to water quality and public sanitation. . . .

THE NINETEENTH CENTURY

The concepts of public health and sanitation had their origins in the nineteenth century. They arose from two different sources. The first was the growing realization, in the first half of the century, that government had a moral and ethical responsibility to protect the welfare of its citizens. Much of this realization developed as a result of the Enlightenment and the democratic movements in the United States, England, and France during the previous century. For example, Jeremy Bentham, an English philosopher, advocated numerous governmental reforms to address social and health problems. The French philosopher Octave Mirabeau proclaimed that the health of the citizenry was the responsibility of the state. As a result of these new concepts, a greater number of legislative deliberations and official inquiries regarding sanitation were conducted. During these deliberations emphasis was placed on ensuring an adequate supply of wholesome water to the poor, since it was increasingly recognized that epidemics and poverty were closely linked. In London settling reservoirs and filters were introduced to remove gross particulate matter from the water derived from the Thames River.

The other source from which the concepts of public health and sanitation arose in this period was the development of organic chemistry and microbiology. Public health and sanitation were now supported by a scientific method that was quantitative and reproducible and presented a cohesive framework for inquiry. Scientists such as Louis Pasteur and Robert Koch showed that the organic compounds in water were microorganisms called bacteria, that these bacteria were derived from human or animal waste, that these microorganisms caused disease even in low concentrations, that water was a temporary habitat for many microorganisms and served as a medium of transfer from an infected person to another who was healthy, and that removal of microorganisms from water supplies could lead to the control and prevention of epidemics. These discoveries allowed for a major leap forward for the health of the community.

Armed with this knowledge, cities undertook sanitary measures. The first milestone in this regard followed the demonstration in 1854 by the English physician John Snow that a cholera epidemic in London could be traced back to a single source of water. His evidence was compelling, and when he removed the handle of the water pump on Broad Street, thereby shutting down the source of water infection to the public, the epidemic abated and was stopped. Soon public water commissions were set up and uniform standards for water purification were developed. Public cesspools and latrines, as well as the dumping of untreated human and animal waste and of animal entrails from slaughterhouses into rivers from which the community derived its water, were stopped. For the first time in nearly two thousand years, Europe enjoyed water of a decent quality. Consequently, by the end of the nineteenth century many waterborne infectious outbreaks began to decline in prevalence.

THE MODERN ERA OF DRINKING WATER REGULATION

The modern era of drinking water protection and regulation is most strongly reflected in developments in the United States. It is therefore appropriate to relate the history of drinking water regulation in the twentieth century by describing the events and legislation that have affected drinking water in this country.

The First Half of the Twentieth Century

In the United States in the nineteenth century, as in Europe, there were numerous outbreaks of diarrheal disease due in part to contaminated water. In New Orleans there was a large cholera epidemic in 1833. In Chicago and Philadelphia, typhoid was endemic. Indeed, diarrheal diseases were the third leading cause of death in the United States at the turn of the twentieth century.

The discovery that these diseases were infectious and transmissible through water served as an impetus to the development of methods of water purification. In Lawrence, Massachusetts, the first U.S. water filtration system was established in 1887 to sift the waters coming from the Merrimack River. Sand was the filtering medium. Within a short time the incidence of typhoid fever dropped precipitously. This success led to the introduction of several other types of mechanical and sand filters for municipal water supplies throughout the United States. By 1907 thirty-three cities in the United States had installed mechanical filters and thirteen others were using sand filters. Concurrent with these developments, in 1899 the Congress enacted the Rivers and Harbors Act, prohibiting the discharge of refuse into navigable waters.

Several years after the introduction of filtration, another method of water purification was introduced—chlorination. Chlorine had recently been found to reduce the number of microorganisms in water by diffusing through the cell walls of the infecting agents and poisoning them. Chlorination was first used in Jersey City, New Jersey, in 1909. The city's reservoir would at times become contaminated from river sewage. Rather than introduce new filtration systems, the water company decided to chlorinate the water supply. It soon became clear that the bacterial counts in the water had dropped to negligible levels and that chlorination was an effective method of ensuring water purity. Soon chlorination became an additional and commonly used method of water purification. In the United States today, chlorine is used in 75 percent of the larger municipal systems and 95 percent of the smaller systems.

In 1948 the Federal Pollution Control Act was passed into law. With subsequent amendments, it became the Clean Water Act. It enabled the government to more forcefully implement the prohibition of pollutant discharge into navigable waterways. In particular, the act regulated producers of pollution, such as factories and sewage treatment plants, and required them to pre-treat their effluents to minimize water pollution at its source.

As a result of these measures, as well as the establishment by the government of bacteriologic standards for water quality, the common infectious causes of waterborne illness became well controlled. Major outbreaks of typhoid, cholera, and other diarrheal diseases became rare occurrences or were quickly contained. It could be stated that by the middle of the twentieth century the U.S. public was enjoying a high quality of drinking water as never before.

The Second Half of the Twentieth Century

The second half of the twentieth century saw two new developments with regard to drinking water. The first was the use of drinking water as a means to promote public health. This was done through the introduction of fluoride into public drinking water supplies for protection against dental caries (cavities). This was the first time that drinking water was used in a proactive manner.

The other major development of the latter half of this century was the realization that drinking water sources were being contaminated by chemicals and that this contamination was injurious to health. The industrialization of the United States following World War II had led to a marked increase in the use of synthetic chemicals. The problems associated with disposal of these chemicals soon became noticeable when the levels of chemical pollutants in water sources rose. The threat posed by these chemicals was first noticed in the fish and mammals that lived in or near these water sources. Finding dead fish with disfiguring tumors in inland waters had a profound effect on the public consciousness and on legislative bodies. Dying sea mammals near ocean coasts also caught the public's attention. The growing scientific awareness of the adverse effects of toxic substances, the fear of cancer, the enhanced ability to measure chemical levels through improved technologies, and the rise of the environmental movement made it clear that the regulations of the 1940s were no longer adequate to protect the drinking water supply in the United States.

In response to these challenges, Congress enacted the Safe Drinking Water Act (SDWA) of 1974. This act was the most comprehensive in history to protect water sources and drinking water. It set up government oversight, through the Environmental Protection Agency (EPA), of surface and groundwater sources, programs for the development of standards and regulations, and funding for state water systems. Ongoing monitoring to ensure compliance was made an integral part of these programs. National standards for protection from harmful contaminants were authorized so as to "protect to the extent feasible, using technology, treatment techniques, and other means, which the Administrator determines are generally available (taking costs into consideration)."

To attain these goals, the EPA set up two types of regulations: (1) mandatory, enforceable maximum contaminant levels (MCLs), to be set as close to the recommended health-based goals "as is feasible"; these levels would pose no significant health risk if drunk over a lifetime and (2) non-mandatory, health-based maximum contaminant level goals (MCLGs) established for each toxic substance that "may have an adverse effect on the health of persons . . . allowing for an adequate margin of safety." . . .

The problem with this approach is twofold. First, it is very expensive. The cost may not be so strongly felt by large municipal water suppliers. On the other hand, the cost of detection, treatment, and monitoring is relatively much more expensive for small water suppliers. As of 1990 there were more than 51,000 small water suppliers in this country serving populations of fewer than 3,000 people.

The second problem is that the emergence of newer illnesses and medical concerns has made the older standards less relevant. . . . Problems as diverse as human infertility, heart disease, and deteriorating memory (dementia) may be related to the contaminant content of drinking water. These effects are subtle and take longer to come to the fore than does the development of cancer. For that reason, newer regulations for water quality are and will be necessary.

World Health Organization and UNICEF, from *Global Water Supply and Sanitation Assessment 2000 Report*

Despite modest progress in the late twentieth century, new pressures, mainly from growing populations in developing nations in the South, threaten to overwhelm earlier advances in the effort to provide safe drinking water and sanitation to all.

HEALTH HAZARDS OF POOR WATER SUPPLY AND SANITATION

Approximately four billion cases of diarrhea each year cause 2.2 million deaths, mostly among children under the age of five. This is equivalent to one child dying every fifteen seconds, or twenty jumbo jets crashing every day. These deaths represent approximately fifteen percent of all child deaths under the age of five in developing countries. Water, sanitation, and hygiene interventions reduce diarrheal disease on average by between one-quarter and one-third.

Intestinal worms infect about ten percent of the population of the developing world. These can be controlled through better sanitation, hygiene, and water supply. Intestinal parasitic infections can lead to malnutrition, anemia, and retarded growth, depending upon the severity of the infection.

It is estimated that six million people are blind from trachoma and the population at risk from this disease is approximately 500 million. . . .

200 million people in the world are infected with schistosomiasis, of whom

twenty million suffer severe consequences. The disease is still found in seventy-four countries of the world. . . .

Cholera Epidemics

Cholera is a worldwide problem that can be prevented by ensuring that everyone has access to safe drinking water, adequate excreta disposal systems, and good hygiene behaviors.

Major health risks arise where there are large concentrations of people and hygiene is poor. These conditions often occur in refugee camps, and special vigilance is needed to avoid outbreaks of disease.

Most of the 58,057 cases of cholera reported in Zaire [Democratic Republic of Congo] in 1994 occurred in refugee camps near the Rwandan border. A decrease to 553 cases in Zaire in 1995 reflected the stabilization of refugee movement.

A cholera epidemic that began in Peru in 1990 spread to sixteen other countries in Latin America. A total of 378,488 cases were reported in Latin America in 1991. Ten years later, cholera remains endemic following its absence from the continent for nearly a century.

WATER SUPPLY AND HEALTH

Lack of improved domestic water supply leads to disease through two principal transmission routes:

Waterborne disease transmission occurs by drinking contaminated water. This has taken place in many dramatic outbreaks of fecal-oral diseases such as cholera and typhoid. Outbreaks of waterborne disease continue to occur across the developed and developing world. Evidence suggests that waterborne disease contributes to background rates of disease not detected as outbreaks. The waterborne diseases include those transmitted by the fecal-oral route (including diarrhea, typhoid, viral hepatitis A, cholera, dysentery) and dracunculiasis. International efforts focus on the permanent eradication of dracunculiasis (guinea worm disease).

Water-washed disease occurs when there is a lack of sufficient quantities of water for washing and personal hygiene. When there is not enough water, people cannot keep their hands, bodies, and domestic environments clean and hygienic. Without enough water, skin and eye infections (including trachoma) are easily spread, as are the fecal-oral diseases.

Diarrhea is the most important public health problem affected by water and sanitation and can be both waterborne and water-washed.

Adequate quantities of safe water for consumption and its use to promote hygiene are complementary measures for protecting health. The quantity of water people use depends upon their ease of access to it. If water is available through a house or yard connection people will use large quantities for hygiene, but consumption drops significantly when water must be carried for more than a few minutes from a source to the household. . . .

* * *

MAIN FINDINGS OF THE *GLOBAL WATER SUPPLY AND SANITATION ASSESSMENT 2000*

. . . The percentage of people served with some form of improved water supply rose from 79 percent (4.1 billion) in 1990 to 82 percent (4.9 billion) in 2000. Over the same period the proportion of the world's population with access to excreta disposal facilities increased from 55 percent (2.9 billion people served) to 60 percent (3.6 billion). At the beginning of 2000 one-sixth (1.1 billion people) of the world's population was without access to improved water supply and two-fifths (2.4 billion people) lacked access to improved sanitation. The majority of these people live in Asia and Africa, where fewer than one-half of all Asians have access to improved sanitation and two out of five Africans lack improved water supply. Moreover, rural services still lag far behind urban services. Sanitation coverage in rural areas, for example, is less than half that in urban settings, even though eighty percent of those lacking adequate sanitation (two billion people) live in rural areas—some 1.3 billion in China and India alone. These figures are all the more shocking because they reflect the results of at least twenty years of concerted effort and publicity to improve coverage. . . .

Although an enormous number of additional people gained access to services between 1990 and 2000, with approximately 816 million additional people gaining access to water supplies and 747 million additional people gaining access to sanitation facilities, the percentage increases in coverage appear modest because of global population growth during that time. Unlike urban and rural sanitation and rural water supply, for which the percentage coverage has increased, the percentage coverage for urban water supply appears to have decreased over the 1990s. Furthermore, the numbers of people who lack access to water supply and sanitation services remained practically the same throughout the decade.

The water supply and sanitation sector will face enormous challenges over the coming decades. The urban populations of Africa, Asia, and Latin America and the Caribbean are expected to increase dramatically. The African urban population is expected to more than double over the next twenty-five years, while that of Asia will almost double. The urban population of Latin America and the Caribbean is expected to increase by almost fifty percent over the same period. . . .

Projected urban population growth, especially in Africa and Asia, suggests that urban services will face great challenges over the coming decades to meet fast-growing needs. At the same time, rural areas also face the daunting task of meeting the existing large service gap. To reach universal coverage by the year

2025, almost three billion people will need to be served with water supply and more than four billion with sanitation. . . .

INTERNATIONAL DEVELOPMENT TARGETS FOR WATER SUPPLY AND SANITATION COVERAGE . . .

Targets to be achieved are:

By 2015 to reduce by one-half the proportion of people without access to hygienic sanitation facilities, which was endorsed by the Second World Water Forum, The Hague, March 2000.

By 2015 to reduce by one-half the proportion of people without sustainable access to adequate quantities of affordable and safe water, which was endorsed by the Second World Water Forum and in the United Nations Millennium Declaration.

By 2025 to provide water, sanitation, and hygiene for all.

. . . These targets build upon the target of universal coverage established for the International Drinking Water Supply and Sanitation Decade 1981–1990, which was readopted as the target for the year 2000 at the World Summit for Children in 1990. [In other words, the original 1990 target of water and sanitation for all was postponed to 2000, and then again in 2000 to 2025.]

Rodney R. White, from "Water Supply" in *North, South and the Environmental Crisis* (1993)

While water pollution remains a serious problem almost everywhere in the world, the lack of clean water for all does not chiefly stem from an overall scarcity or technical deficiencies but—as with soil erosion and many other environmental concerns—from economic, political, and educational barriers. Despite higher population and increased demand, obstacles to useable water are more human than natural.

SOURCES OF WATER AND USES OF WATER

Globally, there is no shortage of water per se. The problem is the cost of making water available at a particular place. Figure 13.7 illustrates the difference between absolute water supplies and available (or potentially available) water for domestic and agricultural use. Roughly, of the world's total water only three percent is fresh, and of that three percent less than one percent is available at the surface; a further twenty-two percent is groundwater (i.e., underground water), some of which can be drawn up or pumped up for human use. Of the water that is available at the surface, about half is cycling through the atmosphere, through the soil, and through biological systems. The other half is on the surface in rivers, lakes, and swamps.

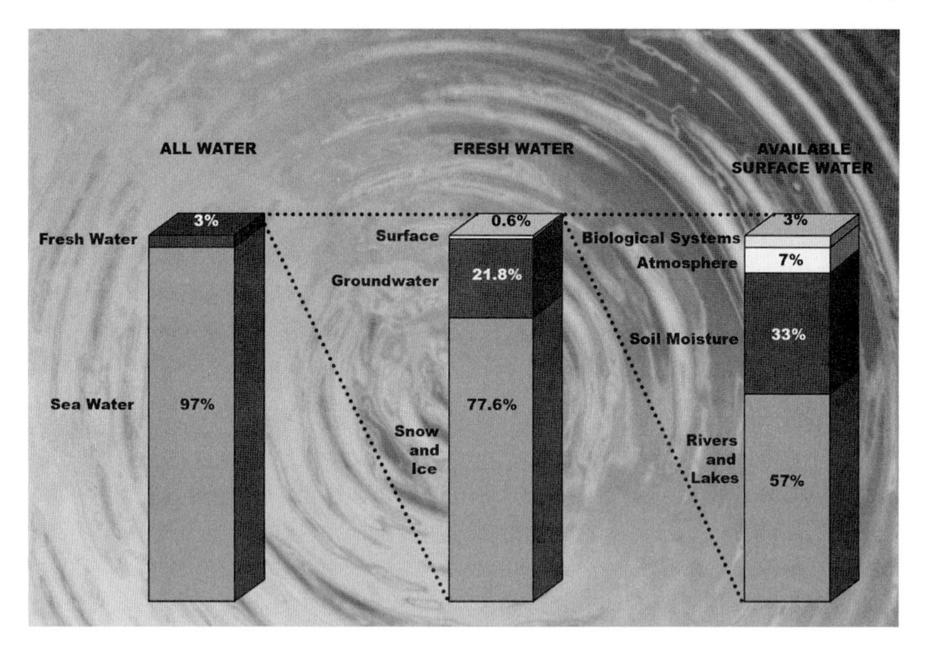

Figure 13.7 The global distribution of water (adapted from *Safeguarding the World's Water*, UNEP Environmental Brief 6, 1988, Nairobi, p. 2).

Through all of these locations water continually cycles, moving quickly through some places, such as rivers, more slowly through others, such as the atmosphere, and resting without movement for very long periods in others, such as fossil (or unreplenished) groundwater aquifers. These movements are known collectively as the hydrological cycle (Figure 13.8). The surface water is usually the most available source, meaning deliverable at reasonable cost. However, in theory, people may avail themselves of water from almost any point in the cycle. For example, they may desalinize salt water, collect rainwater, melt glaciers, or even collect moisture from low-lying clouds (as is being done experimentally with nylon nets in Chile). In practice, costs vary enormously. The *Economist* recently quoted the following costs per acre-foot, delivered to Los Angeles: from Mono Lake in the Sierra Nevada, $70; from California's Central Valley, $250; from desalinized ocean water, $2,000.

Generally people measure only part of what is used, principally for irrigation, industrial use, and urban water supply. The equally important usage by livestock and rain-fed plants can be estimated, as so much per livestock unit or so much per ton of crop produced. However, these unmeasured requirements are often ignored when an urban water supply company or an irrigated farm puts in a bid for available water. When water supply becomes constrained the various users quickly come into conflict. Such conflicts are increasingly common now as pop-

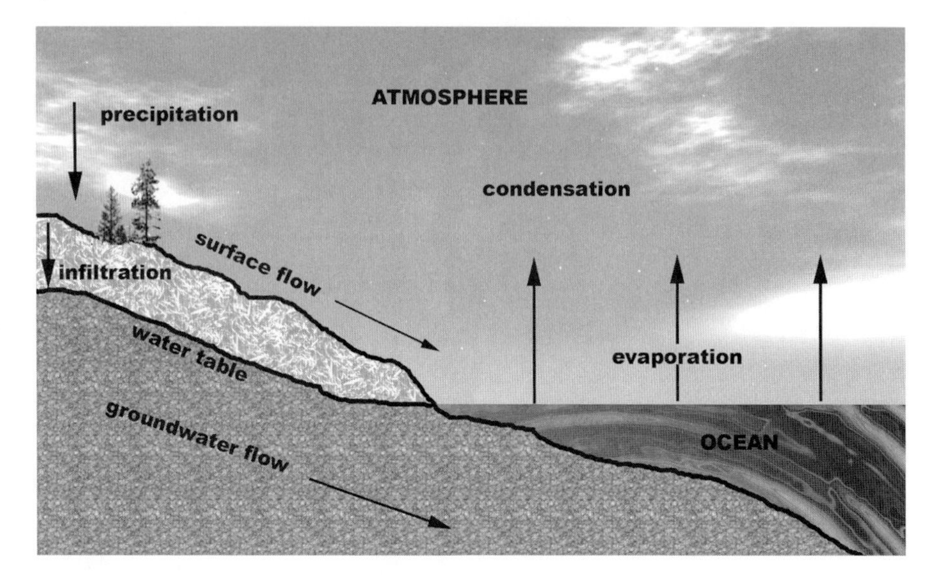

Figure 13.8 The hydrological cycle.

ulation increases; also environmentalists in the North question whether we really need new water supplies when we make such wasteful use of what is already available. The most controversial projects are those that involve dams for water retention or for water diversion and those that pump the groundwater in excess of the recharge rate.

In the North it was in the drier parts of North America and the Soviet Union that the largest dams were built, especially in the immediate postwar period in the 1950s and 1960s when there existed a veritable competition between the superpowers to build the biggest dams. This was at an age when the size of a project was considered synonymous with efficiency. The Americans could boast that the Colorado River was "fully allocated"—that is, all the water in the stream was used for hydroelectric power generation, irrigation, and urban and industrial purposes, so that by the time the river reached the coast, in Mexico, it was reduced to a muddy trickle. . . .

When these projects are viewed simply in terms of the delivery and use of water they often make economic sense: "wasted" water is being made available for productive human use. In the case of the Aswan Dam, the Sudan was able to grow more cotton, while Egypt boosted its electricity supply for its growing industrial base. As northern Sudan had no rainfall and Egypt had no coal the advantages appeared incontrovertible. Now the situation is seen to be much more complicated as various hidden, or ignored, costs have been revealed. Costs upstream of the dam include the immediate loss of agricultural, grazing, and hunting land that would be flooded and the relocation of the inhabitants. Ghana's

Volta River Dam put nearly one-sixth of the country under water, and the new lake entailed major cuts in the national transportation network. If areas around the new lake are not well drained there is increased risk from malaria. In general the upstream costs might be expected to be justified by downstream benefits. However, there are costs downstream as well, such as the disruption of the existing economy, which might include fishing and flood-recession agriculture. For the workers in the downstream irrigated fields, in the tropics, there is increased risk from schistosomaisis and other water-related diseases [see *Global Water Supply and Sanitation Assessment* selection above]. Apart from the physical costs there are distributional impacts on income as such schemes generally disrupt the most marginal members of society, while potentially bringing benefits to those with capital to invest in more intensive forms of agriculture or in industrial and commercial activities that can benefit from hydroelectric power. Projects that divert water from one drainage basin to another bring the additional problem of disruption of the ecosystem by introducing different fauna and flora. . . .

WATER AND HUMAN HEALTH

The consequences of having too little water for . . . domestic purposes are quite well known. Where water supply and water treatment are deficient, human health status is low, life expectancy is low, and infant mortality is very high. . . .

Despite widespread agreement on the importance of improved water supply and water treatment, progress is slow, and in some places conditions are deteriorating. [See *Global Water Supply and Sanitation Assessment* selection, above.] . . .

[Critics] rightly put the emphasis for failure on social and political factors, because the technical problems of delivering water to human settlements are rarely insurmountable. . . . What is needed to rectify water-supply deficiencies is very well known. The World Bank and other development agencies have rooms full of manuals on project design. This is not a scientific mystery like discovering a cure for AIDS; nor does it pose baffling technical problems like removing the CFCs that have drifted up into the stratosphere. Yet progress is imperceptible in most of the world's poorest countries, because, despite the rhetoric, the political commitment is not there. Universal water supply is simply not a priority.

Why has progress been so disappointing, if the technical requirements for solution are understood? Part of the answer lies in the fact that public health policy cannot be changed without changing almost everything else in a society, from the physical means of communication (like the roads and the telephone) to the distribution of power between rich and poor, between female and male, and between all other divisions in society. Current efforts to improve urban and rural water supply in low-income countries have been highly dependent on development-agency funding and on pretending that the main dimensions of the problem are technical. It was assumed that these technical problems could be solved by good engineering, "getting the prices right," and good management, and that it could all be done in the five-year time horizon which is convenient for development-agency projects. Later, when it was seen that most projects did not work well in rural areas, a budget was added for "training for maintenance." . . .

The failure to apply so much well-known and important knowledge about basic health care is the responsibility of both the North and the South. The rich countries, through the development agencies, persist in the convenient (but highly inaccurate) assumption that they need only provide the five-year effort to install the system, so long as a "training component" is now included in the project. The governments of poor countries, for their part, accept the projects on these terms although they resist the social transformation that is required in order to provide "adequate water for all."

Yet history shows that rapid results can be achieved. At the time of the 1890 census, malaria was the leading cause of death in the United States. The southern states reported 7,000 malaria deaths per 100,000 people, compared with 1,000 deaths in the northern states. By 1930, even in the South, the rates were down to twenty-five malaria deaths per 100,000 people. Improved drainage, land-use controls, screened houses, antimalarial drugs, and the managed fluctuation of water levels in reservoirs brought about this rapid decline. [See Barzilay, above, for more on the elimination of water-borne disease in the developed world.] It is true that tropical ecosystems provide a more difficult and varied environment for control and eradication, and therefore such rapid results should not be expected. However, the ecological differences should not obscure the fact that the main problems are social and political—and there is nothing as political as water. A Ghanaian engineer who supervised the installation of 2,000 village boreholes in northern Ghana in 1979 told me that what he wanted to do next was earn a doctoral degree in political science, using as data the files from the project. Even he, a citizen of the country, had been amazed by the intense conflicts that had been aroused by the installation of a limited number of boreholes in a place of very great need: a need that could nowhere near be met by one five-year project, however well designed and implemented.

WATER FOR ALL?

Water of adequate quality is an increasingly scarce resource. As the world's population and the world's economy continue to grow, that scarcity will become critical. In some cases it has already led to widespread violence. In Mexico one state is dumping waste into a lake which flows into another state which depends on it for its principal source of water. In Senegal a large, modern hotel complex on the coast was designed with its own water supply, pumped from its own grounds; unfortunately it is pumping in excess of the recharge rate and is drawing down the water-table underneath the nearby villages. If it continues this activity it will bring in sea water, which will seep into the fresh groundwater on which the hotel and the villages depend. In Morocco there is some regret that

Figure 13.9 Sprinklers drench a lawn in suburban Clark County, Nevada. Courtesy of USDA Natural Resources Conservation Service.

the country's tourism strategy concentrated on five-star hotels for the wealthiest travelers; it has now been found that there is a linear correlation between the number of stars earned by the hotel and the number of liters of water used per bed of hotel capacity. It is about 100 liters per star. So five-star luxury hotels are using 500 liters per bed per day, whether occupied or not, mostly to fill the swimming pools and water the gardens. Around Marrakesh the hotel owners are now in competition with farmers who are growing irrigated wheat.

The examples are countless, but the conclusions are fairly clear. Even with existing knowledge, drinking water could be provided for all, and the present irrigated land, industries, hotels, and hydroelectric-power stations could be supplied. But the social and political problems are very complex. Identification of the already existing improvements in the technology has been done; avenues for research are already mapped out. What is needed is a joint recognition by the North and the South that the commitment must be long term and comprehensive. This is not just a matter of more ribbon cutting for new dams; it is, literally, a matter of life and death.

14 Energy

Every use of energy always creates environmental consequences and impacts. The dramatic increase of human use of energy during the past two centuries rests at the foundation of many if not most environmental issues and problems.

The ability to make energy pass through human hands in a useful, applied way began to grow explosively only four lifetimes ago, at the start of the Industrial Revolution. During the previous thousands of lifetimes, wood, coal, wind, charcoal, draft animals, falling water, and sheer muscle power had propelled whatever projects and mills people could devise. The Romans had the metallurgy and knowledge needed to harness steam, but despite their engineering prowess they never put the pieces together, demonstrating that available technology does not always produce successful application. In the eighteenth century, steam engines began to drive looms, saws, grindstones, and lifts. By 1850 many ships ran on steam power. The railroad, even today the most efficient land transportation (next to the bicycle), was the child of steam. With development of the internal combustion engine, the Age of Oil began. Almost simultaneously, the dynamo, turbine, and related technologies permitted heat produced from *any* source to be transformed into electricity, transported almost instantaneously on transmission lines and modified into light, heat, and, through motors, mechanical energy. All these developments happened in less than 150 years, creating a new and completely exceptional phase for humanity, nature, and all of life.

The availability of steam power and electricity first increased at a rate far greater than population growth, putting more energy into the hands of ordinary people and enriching their lives with newfound freedoms and conveniences, from superior transportation to labor-saving appliances. In the nineteenth and twentieth centuries, cheap energy fostered dramatic increases in agricultural productivity (and thus in population), the construction and support of dense urban centers, a world tightly knit by communications and trade, the exploration of outer space, and an information revolution. (National and per capita economic productivity have almost always been correlated to energy use.) Abundant energy for human use also made possible a population burden of global scale, climate change caused by carbon dioxide emissions, new destructive powers in war-making, and the degradation of air, water, and soil on scales entirely unprecedented.

Energy from carbon-based fuels (e.g., coal, oil, gasoline, and natural gas) produces carbon dioxide, now a major cause of climate change. Lowering the use of such fuels and applying technologies to their combustion and exhaust are important steps in reducing CO_2 emissions. Imposing a carbon tax and establishing carbon credits that can be traded might seem to crimp economic growth, but in the end may not only slow climate change but also reduce long-term social costs.

It is important to make the nature of energy clear. In a strict and pertinent sense, energy is not produced on Earth. Like the shape-shifting Greek god Proteus, it simply changes forms, with some forms easier to use than others. Earth receives huge amounts of energy each second from the sun, some of which is retained and converted into forms available for human exploitation. Carbon-based fuels are compressed plants whose original energy comes from the sun. Winds are driven by Earth's rotation and by temperature differentials created by the sun. Geothermal energy originates from Earth's primordial formation and the continued compressions of gravity and movements of Earth's crust. Hydroelectric power derives from gravity and from the sun: water must evaporate before it falls and flows. Tidal energy depends on the moon's gravitational pull. Hurricanes concentrate energy in one system for a few days, then dissipate it back into the wider atmosphere. Energy moves from one form to another (mechanical to electrical, electrical to heat, chemical to heat, heat to mechanical, and so forth), often with relative ease. Yet, no conversion of energy is 100 percent efficient. Inanimate means of converting energy (automobiles, power stations, and furnaces) are far less efficient than animate ones (the metabolisms of living creatures). Thus, because energy can convert from one form to another, yet also because the inanimate means for that conversion are not very efficient, energy is supremely easy to waste.

The *work* that energy does is a measure of how much mass can be moved over how great a distance. *Power* is a measure of how much work can be performed in how short a period of time. Before 1760, few individuals could access more power than that of a team of horses or oxen. This explains why we still commonly speak of "harnessing" energy, and why "horsepower" became a common unit of power. Today, billions of people each command anywhere from five to five hundred horsepower at their fingertips. Ships, planes, trains, and turbines put out tens of thousands of horsepower.

The United States consumes more energy per capita than any other nation. Its 5 percent of the world's population accounts for 25 percent of all energy used. Not long after the "energy crisis" of the early 1970s, when OPEC (Organization of Petroleum Exporting Countries) reduced oil production, the U.S. government created the Department of Energy, in part to reduce dependence on

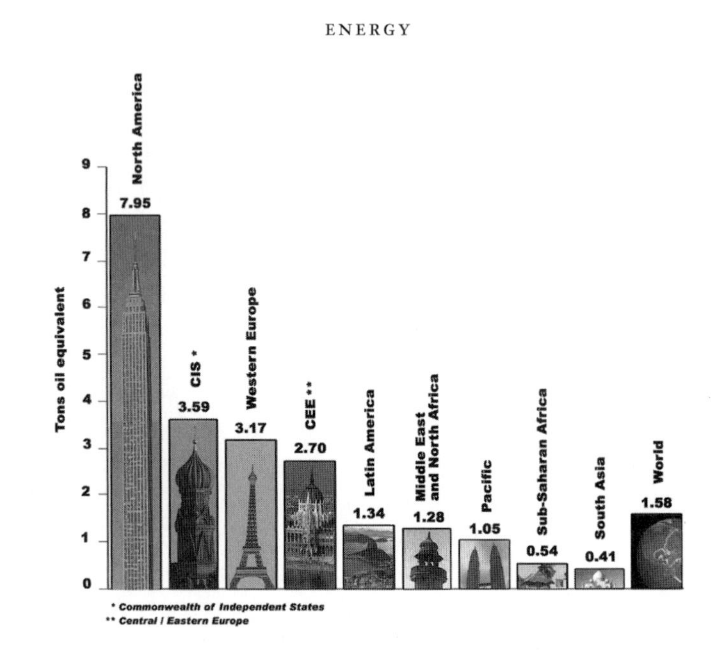

Figure 14.1 World per capita energy use by region. Based upon 2001 data provided in table 7 of *World Resources 2005: The Wealth of the Poor—Managing Ecosystems to Fight Poverty* (Washington, D.C.: World Resources Institute, 2005).

imported oil. Yet private industry and consumer practices still largely dictate energy use. There has been little in the way of an effective national policy. In all other developed nations, government policies and heavier taxes restrict use. This encourages conservation but can stifle economic growth. Proponents of low energy taxes point out that the United States is geographically large and its whole economy, including middle and lower income workers, relies heavily on vehicular transportation. (Transportation accounts for one quarter of world energy use.) Shockingly, despite new technologies, lighter materials, and computer-aided efficiencies in engine performance, the overall fuel economy of all U.S. vehicles was *worse* in 2005 than in 1985. In 2003, for the first time, the number of vehicles registered in the U.S. exceeded the number of licensed drivers.

At projected rates of use, the world's known petroleum and natural gas reserves will be eliminated in little more than a century. Experts disagree but in general estimate that oil production will peak between 2010 and 2040. Oil extraction outstrips discovery now by 3:1. Oil pumping in the United States has declined since 1970. There exist large coal deposits, oil shale fields, and tar sands in North America, Russia, and China that might be used to satisfy growing energy demands. These sources pollute more than petroleum and produce more greenhouse gases, though technologies are improving. Nuclear power may be a viable option, especially because it produces little CO_2 (see Chapter 3). Using hydrogen to replace gasoline or other fuels holds promise, but any "hydrogen

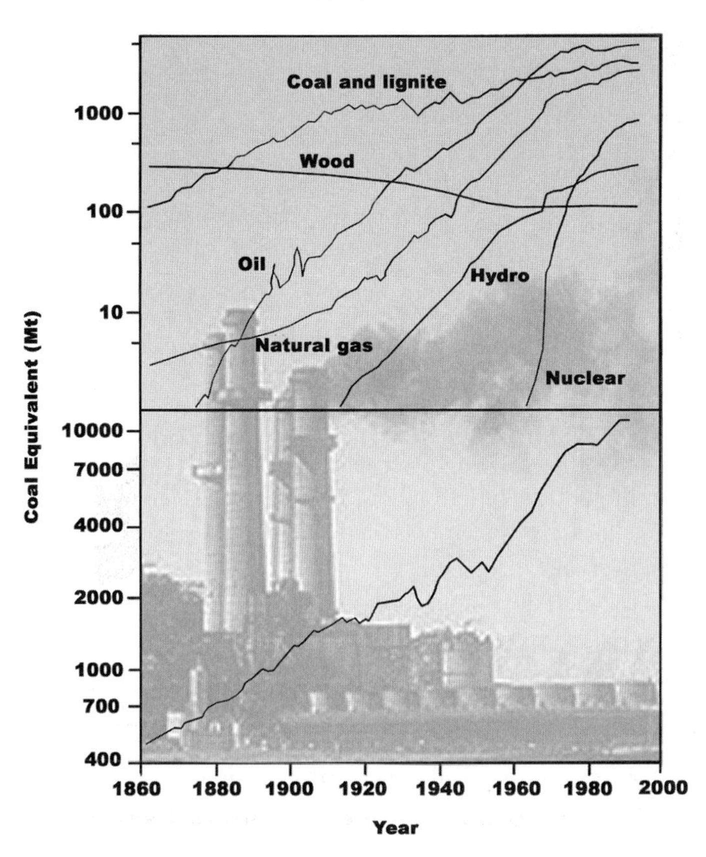

Figure 14.2 Changes in pattern of fuel usage (upper) and graph of world energy consumption (lower) since 1860; scale on left axis is logarithmic. Courtesy of Thomson Corporation. Adapted from figure 6.2, page 63 of "Sustainable Energy Development," by A. Williams and M. A. Uqaili, in *Energy Demand and Planning*, ed. J. C. McVeigh and J. G. Mordue (New York: Routledge, 1999).

economy" will likely depend on renewable energy sources to produce hydrogen in the first place; at some point, it will become cheaper to produce hydrogen from renewable sources than from natural gas.

In 2007, renewables are working many places. Denmark meets sixty percent of its electrical needs with wind generators. India, the United States, United Kingdom, and other nations are building more wind farms. The cost of generating electricity from photovoltaic cells has fallen steadily since their invention and for consumers now runs only twice the cost incurred by older means. In some markets, it actually costs less than electricity bought from a power company. Ethanol, extracted mostly from corn (maize) and sugarcane, provides energy in the American Midwest and in Brazil, where it is a vital fuel. By the middle of the twenty-first century, one-third to one-half of the world's energy appetite will probably be met with renewable sources. Off the coasts of Portugal and

Scotland, large mechanical snakes are converting wave motion to electricity. Trials are scheduled for U.S. coastal waters. Each snake, about 150 meters long, produces electricity for about 500 households.

Some large energy companies such as Royal Dutch Shell and British Petroleum realize that the Age of Oil will recede. Some automakers do, too, and several produce popular hybrids with high gas mileage. Fuel cell technology is another area of increased investment and research. In some countries, solar cookers can replace ovens that pollute and contribute to deforestation. The jigsaw puzzle of the world's energy future is hard to solve. Its pieces will be more diverse, and more of them will be renewable. In 2006, a *Fortune* 500 company, Whole Foods, became the largest buyer of wind energy credits in North America. It bought 100 percent of its projected energy use for that year in such credits. The EPA reported that this was roughly equal to removing 60,000 cars from the road. Other corporations are likely to follow. Businesses are getting interested in renewables—for business as well as for environmental reasons. The two sets of reasons often overlap.

"Distributed generation" holds promise. In this practice, multiple local small operations generate energy from hydro, wind, sun, and geothermal sources. (Large hydropower necessitates large dams, with all their ecological drawbacks.) Distributed generation can replace or supplement, and thus make less vulnerable, the power grids fed by large generating stations. Subject to accident and human error, these massive grids, each serving millions of people, have failed several times.

In 2007 and likely for decades to come, the single greatest untapped "source" of energy, and one available immediately with little or no capital investment, is not increased oil exploration, "clean" coal technology, hydrogen, nuclear power, or even renewable sources such as biomass, wind, hydroelectric, or geothermal. It is energy conservation. Since energy is always conserved in some form, even if dissipated into the atmosphere inefficiently as heat, "conservation" is more aptly called "lower overall use," applying just enough to perform work that needs to be done, and no more. This can mean throwing off a switch, turning down a dial, properly matching a source of power to work performed, or realizing that the work is not required at all. Conservation could decrease energy use in the U.S. by twenty percent while still accomplishing the same work. In 2004, despite a larger population and a bigger economy, the U.S. used no more water than it did in 1984. If similar conservation measures were applied to energy, the savings and environmental benefits would be enormous. Some have been; more are needed. Vaclav Smil points out in *Energy at the Crossroads* (2003) that increased efficiency alone, however, has never yet led to lower overall use. Lowering per capita energy use overall is a trend that over any appreciable span of time has not yet become evident in any advanced or developing nation.

FURTHER READING

See also Chapter 1 on Climate Shock and Chapter 3 on Nuclear Power in Part One.
Alliance to Save Energy and the Tellus Institute. *America's Energy Choices: Investing in a Strong Economy and a Clean Environment.* Cambridge, Mass.: Union of Concerned Scientists, 1991–92.
Freese, Barbara. *Coal, A Human History.* New York: Perseus Books, 2003.
Geller, Howard S. *Energy Revolution: Policies for a Sustainable Future.* Washington, D.C.: Island Press, 2003.
Rifkin, Jeremy. *The Hydrogen Economy: The Creation of the Worldwide Energy Web and the Redistribution of Power on Earth.* New York: Penguin Putnam, 2002.
Roberts, Paul. *The End of Oil: On the Edge of a Perilous New World.* Boston: Houghton Mifflin, 2004.
Smil, Vaclav. *Energy at the Crossroads: Global Perspectives and Uncertainties.* Cambridge: MIT Press, 2003.
Yergin, Daniel. *The Prize: The Epic Quest for Oil, Money, and Power.* New York: Simon and Schuster, 1991.

National Commission on Energy Policy, from "Ending the Energy Stalemate: A Bipartisan Strategy to Meet America's Energy Challenges" (2004)

In late 2004 a bipartisan commission published detailed recommendations for future energy use, security, conservation, and innovation in the U.S. While largely dealing with the availability of energy and with developing a diversity of resources, the report of the commission speaks directly to environmental issues, too. It is instructive to compare its recommendations with those of scientists such as Pacala and Socolow (see next selection) who advocate specific energy uses in order to control CO_2 *emissions. The two sets of recommendations overlap significantly.*

IMPROVING OIL SECURITY

To enhance the nation's energy security and reduce its vulnerability to oil supply disruptions and price shocks, the Commission recommends:

- Increasing and diversifying world oil production while expanding the global network of strategic petroleum reserves.
- Significantly raising federal fuel economy standards for cars and light trucks while reforming the 30-year-old Corporate Average Fuel Economy (CAFE) program to allow more flexibility and reduce compliance costs. New standards should be phased in over a five-year period beginning no later than 2010.

- Providing $3 billion over ten years in manufacturer and consumer incentives to encourage domestic production and boost sales of efficient hybrid and advanced diesel vehicles.

Today's combination of tight oil supplies and high and volatile prices is likely to continue, given trends in global consumption (expected to grow by more than 50 percent over the next two decades), continuing instability in the Middle East and other major oil-producing regions, and a global decline in spare production capacity.

Oil production in the United States peaked in the 1970s and has been flat or declining since. Although highly important to the nation's economy and energy security, it cannot compensate for anticipated growth in domestic demand, which is expected to reach 29 million barrels per day by 2025—a more than 40 percent increase over current consumption levels.

Improving the nation's energy security and reducing its vulnerability to high oil prices and supply disruptions are more meaningful and ultimately achievable policy goals than a misplaced focus on energy independence per se. Achieving these goals requires focusing in equal measure on expanding and diversifying oil supplies and improving efficiency, especially in the transportation sector. Additional Commission recommendations aim to expand transportation fuel supplies by enabling production of unconventional oil and alternative fuels.

The Commission's recommendations for improving passenger vehicle fuel economy, increasing the contribution from alternative fuels, and improving the efficiency of the heavy-duty truck fleet and passenger vehicle replacement tires, could reduce U.S. oil consumption in 2025 by 10–15 percent or 3–5 million barrels per day. These demand reductions, in concert with increased oil production, would significantly improve domestic oil security.

REDUCING RISKS FROM CLIMATE CHANGE

To address the risks of climate change resulting from energy-related greenhouse gas emissions, without disrupting the nation's economy, the Commission recommends:

- Implementing in 2010 a mandatory, economy-wide tradable-permits system designed to curb future growth in the nation's emissions of greenhouse gases while capping initial costs to the U.S. economy at $7 per metric ton of carbon dioxide-equivalent.
- Linking subsequent action to reduce U.S. emissions with comparable efforts by other developed and developing nations to achieve emissions reductions via a review of program efficacy and international progress in 2015.

The Commission believes the United States must take responsibility for addressing its contribution to the risks of climate change, but must do so in a manner that recognizes the global nature of this challenge and does not harm the competitive position of U.S. businesses internationally.

The Commission proposes a flexible, market-based strategy designed to slow projected growth in domestic greenhouse gas emissions as a first step toward later stabilizing and ultimately reversing current emissions trends if comparable actions by other countries are forthcoming and as scientific understanding warrants.

Under the Commission's proposal, the U.S. government in 2010 would begin issuing permits for greenhouse gas emissions based on an annual emissions target that reflects a 2.4 percent per year reduction in the average greenhouse gas emissions intensity of the economy (where intensity is measured in tons of emissions per dollar of GDP).

Most permits would be issued at no cost to existing emitters, but a small pool, 5 percent at the outset, would be auctioned to accommodate new entrants, stimulate the market in emission permits, and fund research and development of new technologies. Starting in 2013, the amount of permits auctioned would increase by one-half of one percent each year (i.e., to 5.5 percent in 2013; 6 percent in 2014, and so on) up to a limit of 10 percent of the total permit pool.

The Commission's proposal also includes a safety valve mechanism that allows additional permits to be purchased from the government at an initial price of $7 per metric ton of carbon dioxide (CO_2)-equivalent. The safety valve price would increase by 5 percent per year in nominal terms to generate a gradually stronger market signal for reducing emissions without prematurely displacing existing energy infrastructure.

In 2015, and every five years thereafter, Congress would review the tradable-permits program and evaluate whether emissions control progress by major trading partners and competitors (including developing countries such as China and India) supports its continuation. If not, the United States would suspend further escalation of program requirements. Conversely, international progress, together with relevant environmental, scientific, or technological considerations, could lead Congress to strengthen U.S. efforts.

Absent policy action, annual U.S. greenhouse gas emissions are expected to grow from 7.8 billion metric tons of CO_2-equivalent in 2010 to 9.1 billion metric tons by 2020—a roughly 1.3 billion metric ton increase. Modeling analyses suggest that the Commission's proposal would reduce emissions in 2020 by approximately 540 million metric tons. If the technological innovations and efficiency initiatives proposed elsewhere in this report further reduce abatement costs, then fewer permits will be purchased under the safety valve mechanism and actual reductions could roughly double to as much as 1.0 billion metric tons in 2020, and prices could fall below the $7 safety valve level.

The impact of the Commission's proposed greenhouse gas tradeable-permits program on future energy prices would be modest. Modeling indicates that relative to business-as-usual projections for 2020, average electricity prices would be expected to rise by 5–8 percent (or half a cent per kilowatt-hour); natural gas prices would rise by about 7 percent (or $0.40 per mmBtu); and gasoline prices would increase 4 percent (or 6 cents per gallon). Coal use would decline by 9 percent below current forecasts, yet would still increase in absolute terms by 16 percent relative to today's levels, while renewable energy production would grow more substantially; natural gas use and overall energy consumption, meanwhile, would change only minimally (1.5 percent or less) relative to business-as-usual projections.

Overall, the Commission's greenhouse gas recommendations are estimated to cost the typical U.S. household the welfare equivalent of $33 per year in 2020 (2004 dollars) and to result in a slight reduction in expected GDP growth, from 63.5 percent to 63.2 percent, between 2005 and 2020.

IMPROVING ENERGY EFFICIENCY

To improve the energy efficiency of the U.S. economy, the Commission—in addition to an increase in vehicle fuel economy standards—recommends:

- Updating and expanding efficiency standards for new appliances, equipment, and buildings to capture additional cost-effective energy-saving opportunities.
- Integrating improvements in efficiency standards with targeted technology incentives, R&D, consumer information, and programs sponsored by electric and gas utilities.
- Pursuing cost-effective efficiency improvements in the industrial sector.

In addition, efforts should be made to address efficiency opportunities in the heavy-duty truck fleet, which is responsible for roughly 20 percent of transportation energy consumption, but is not subject to fuel economy regulation, and in the existing vehicle fleet where a substantial opportunity exists to improve efficiency by, for example, mandating that replacement tires have rolling-resistance characteristics equivalent to the original equipment tires used on new vehicles.

In updating and implementing efficiency standards, policy makers should seek to exploit potentially productive synergies with targeted technology incentives, research and development initiatives, information programs (such as the federal "Energy Star" label), and efficiency programs sponsored by both electricity and natural gas utilities.

Energy efficiency advances all of the critical policy objectives identified elsewhere in this report and is therefore essential to successfully managing the nation's, and the world's, short- and long-term energy challenges. Absent substantial gains in the energy efficiency of motor vehicles, buildings, appliances, and equipment, it becomes difficult to construct credible scenarios in which secure, low-carbon energy supplies can keep pace with increased demand. As a nation that consumes more energy than any other in the world, improving domestic energy efficiency can have a notable effect on global energy demand.

EXPANDING ENERGY SUPPLIES

The United States and the world will require substantially increased quantities of electricity, natural gas, and transportation fuels over the next 20 years. In addition to the measures discussed previously for improving oil security, the Commission's recommendations for assuring ample, secure, clean, and affordable supplies of energy address established fuels and technologies (such as natural gas and nuclear power), as well as not-yet-commercialized options, such as coal gasification and advanced biomass (including waste-derived) alternative transportation fuels.

Natural Gas

To diversify and expand the nation's access to natural gas supplies, the Commission recommends:

- Adopting effective public incentives for the construction of an Alaska natural gas pipeline.
- Addressing obstacles to the siting and construction of infrastructure needed to support increased imports of liquefied natural gas (LNG).

Other Commission recommendations aim to: (1) improve the ability of agencies like the Bureau of Land Management to evaluate and manage access to natural gas resources on public lands and (2) increase R&D efforts to develop technologies for tapping nonconventional natural gas supplies, such as natural gas hydrates, which hold tremendous promise.

The above recommendations are intended to address growing stresses on North American natural gas markets that have already resulted in sharply higher and more volatile gas prices, and created substantial costs for consumers and gas-intensive industries. Construction of a pipeline would provide access to significant natural gas resources in Alaska's already-developed oilfields (potentially lowering gas prices by at least 10 percent over the pipeline's first decade). Support for a pipeline in the form of loan guarantees, accelerated depreciation, and tax credits was included in legislation passed by Congress late in 2004, but the Commission believes that additional incentives are likely to be necessary given the high cost, lengthy construction period, uncertainty about future gas prices, and other siting and financing hurdles associated with the project.

In addition to the Alaska pipeline, expanded LNG infrastructure would further increase the nation's ability to access abundant global supplies of natural gas, providing important benefits in terms of lower and less volatile gas prices and more reliable supplies for electricity generators and for other gas-intensive industries. Accordingly, the Commission recommends concerted efforts to overcome current siting obstacles, including improved federal-state cooperation in reviewing and approving new LNG facilities and efforts to educate the public regarding related safety issues.

Advanced Coal Technologies

To enable the nation to continue to rely upon secure, domestic supplies of coal to meet future energy needs while addressing the risks of global climate change due to energy-related greenhouse gas emissions, the Commission recommends:

- Providing $4 billion over ten years in early deployment incentives for integrated gasification combined cycle (IGCC) coal technology.
- Providing $3 billion over ten years in public incentives to demonstrate commercial-scale carbon capture and geologic sequestration at a variety of sites.

Coal is an abundant and relatively inexpensive fuel that is widely used to produce electricity in the United States and around the world. Finding ways to use coal in a manner that is both cost-effective and compatible with sound environmental stewardship is imperative to ensure a continued role for this important resource.

IGCC technology—in which coal is first gasified using a chemical process and the resulting synthetic gas is used to fuel a combustion turbine—has the potential to be significantly cleaner and more efficient than today's conventional steam boilers. Moreover, it can assist in effectively controlling pollutants such as mercury and can open the door to economic carbon capture and storage. The gasification process itself is already commonly used in the manufacture of chemicals, but—with the exception of a handful of demonstration facilities—has not yet been widely applied to producing power on a commercial scale.

Nuclear Power

To help enable nuclear power to continue to play a meaningful role in meeting future energy needs, the Commission recommends:

- Fulfilling existing federal commitments on nuclear waste management.
- Providing $2 billion over ten years from federal research, development, demonstration, and deployment (RDD&D) budgets for the demonstration of one to two new advanced nuclear power plants.
- Significantly strengthening the international nonproliferation regime.

Worldwide, some 440 nuclear power plants account for about one-sixth of total electricity supplies and about half of all non-carbon electricity generation. In the United States, 103 operating nuclear power plants supply about 20 percent of the nation's electricity and almost 70 percent of its non-carbon electricity. The contribution of nuclear energy to the nation's power needs will decline in the future absent concerted efforts to address concerns about cost, susceptibility to accidents and terrorist attacks, management of radioactive wastes, and proliferation risks. [See Chapter 3.]

Government intervention to address these issues and to improve prospects for an expanded, rather than diminished, role for nuclear energy is warranted by several important policy objectives, including reducing greenhouse gas emissions, enhancing energy security, and alleviating pressure on natural gas supplies from the electric-generation sector.

Renewable Energy

To expand the contribution of clean, domestic, renewable energy sources to meeting future energy needs, the Commission recommends:

- Increasing federal funding for renewable technology research and development by $360 million annually. Federal efforts should be targeted at overcoming key hurdles in cost competitiveness and early deployment.
- Extending the federal production tax credit for a further four years (i.e., from 2006 through 2009), and expanding eligibility to all non-carbon energy sources, including solar, geothermal, new hydropower generation, next generation nuclear, and advanced fossil fuel generation with carbon capture and sequestration. (This is in addition to the extension recently passed by Congress for 2004–2005.)
- Supporting ongoing efforts by the Federal Energy Regulatory Commission (FERC) to promote market-based approaches to integrating intermittent resources into the interstate grid system, while ensuring that costs are allocated appropriately and arbitrary penalties for over- and under-production are eliminated.
- Establishing a $1.5 billion program over ten years to increase domestic production of advanced nonpetroleum transportation fuels from biomass (including waste).

Renewable energy already plays an important role in the nation's energy supply, primarily in the form of hydropower for electricity production and corn-based ethanol as a transportation fuel. Other renewable options—including wind, solar, and advanced biomass technologies for power generation together with alternative transportation fuels from woody or fibrous (cellulosic) biomass and organic wastes—have made considerable progress in recent years, but still face substantial cost or technology hurdles as well as, in some cases, siting challenges.

Figure 14.3 Wind farm near Whitewater, California. Courtesy of G. Donald Bain/The Geo-Images Project, Dept. of Geography, UC–Berkeley.

The Commission's recommendations aim to improve the performance and cost-competitiveness of renewable energy technologies while also addressing deployment hurdles by providing more planning certainty in terms of federal tax credits, boosting R&D investments, and addressing issues related to the integration of renewable resources with the interstate transmission grid.

STRENGTHENING ENERGY SUPPLY INFRASTRUCTURE

To sustain access to the essential energy supplies and services on which the economy depends, the Commission recommends:

- Reducing barriers to the siting of critical energy infrastructure.
- Protecting critical infrastructure from accidental failure and terrorist threats.
- Supporting a variety of generation resources—including both large scale power plants and small scale "distributed" and/or renewable generation—and demand reduction (for both electricity and natural gas), to ensure affordable and reliable energy service for consumers.

- Encouraging increased transmission investment and deployment of new technologies to enhance the availability and reliability of the grid, in part by clarifying rules for cost-recovery.
- Enhancing consumer protections in the electricity sector and establishing an integrated, multi-pollutant program to reduce power plant emissions.

The Commission believes there is a national imperative to strengthen the systems that deliver secure, reliable, and affordable energy. Priorities include: siting reforms to enable the expansion and construction of needed energy facilities; greater efforts to protect the nation's energy systems from terrorist attack; and reforms to improve the reliability and performance of the electricity sector.

DEVELOPING ENERGY TECHNOLOGIES FOR THE FUTURE

To ensure that technologies capable of providing clean, secure, and affordable energy become available in the timeframe and on the scale needed, the Commission recommends:

- Doubling federal government funding for energy research and development, while improving the management of these efforts and promoting effective public-private partnerships.
- Increasing incentives for private sector energy research, development, demonstration, and early deployment (ERD^3).
- Expanding investment in cooperative international ERD^3 initiatives and improving coordination among relevant federal agencies.
- Providing incentives for early deployment of (1) coal gasification and carbon sequestration; (2) domestically produced efficient vehicles; (3) domestically produced alternative transportation fuels; and (4) advanced nuclear reactors.

Overcoming the energy challenges faced by the United States and the rest of the world requires technologies superior to those available today. To accelerate the development of these technologies, the federal government must increase its collaboration with the private sector, with states, and with other nations to develop and deploy technologies that will not be pursued absent greater federal support.

Investments by both the private and public sectors in energy research, development, demonstration, and early deployment have been falling short of what is likely to be needed to meet the energy challenges confronting the nation and the world in the twenty-first century. This insufficiency of investment is compounded by shortcomings in the government's management of its energy-technology-innovation portfolio and in the coordination and cooperation among relevant efforts in state and federal government, industry, and academia.

TABLE 14.1
A Revenue Neutral Strategy for Investing in Energy Technology Development

The Commission proposes to double current federal spending on energy innovation, substantially expand early deployment efforts for advanced energy technologies, and triple investment in cooperative international energy research. To offset additional costs to the Treasury, the Commission proposes that the federal government each year auction a small percentage of greenhouse gas emissions permits.

Additional expenditures		*Annual*	*10-year total*
RD&D	Double current investment	$1.7 billion	$17 billion
Incentives for early deployment	Coal IGCC, biofuels, advanced nuclear, non-carbon production tax credit (PTC), manufacturer and consumer auto efficiency incentives, Alaska pipeline	$1.4 billion	$14 billion
International cooperation	Triple current investment	$500 million	$5 billion
Total			**$36 billion**
Additional revenues			*10-year total*
Greenhouse gas permit sales	5 percent permit auction in 2010 with 0.5 percent annual increase starting in 2013		$26 billion
	Revenue from expected permit sales under the safety valve		$10 billion
Total			**$36 billion**

The Commission proposes that the nation devote the resources generated by the sale of greenhouse gas emissions permits to enhance the development and deployment of improved energy technologies. The approximately $36 billion that Commission analysis indicates will be generated over ten years by the proposed greenhouse gas tradeable-permits program—most of which will come from auctioning a small portion of the overall permit pool—will offset the specific additional public investments summarized [in Table 14.1].

Stephen Pacala and Robert Socolow, "Stabilization Wedges: Solving the Climate Problem for the Next Fifty Years with Current Technologies" (2004)

To ease the disruption and damage of climate shock caused by increased atmospheric levels of CO_2 from the burning of carbon-based fuels, these two Princeton scientists propose a series of actions that will produce "wedges." Each wedge will reduce CO_2 emissions very modestly compared with taking no new action and continuing current practices ("business as usual"). But the sum of all or even several of these wedges will permit a leveling off of atmospheric CO_2. In other words, there is no magic bullet, but a series of concerted actions together can yield dramatic results.

Abstract: Humanity already possesses the fundamental scientific, technical, and industrial know-how to solve the carbon and climate problem for the next half-century. A portfolio of technologies now exists to meet the world's energy needs over the next 50 years and limit atmospheric CO_2 to a trajectory that avoids a doubling of the preindustrial concentration. Every element in this portfolio has passed beyond the laboratory bench and demonstration project; many are already implemented somewhere at full industrial scale. Although no element is a credible candidate for doing the entire job (or even half the job) by itself, the portfolio as a whole is large enough that not every element has to be used.

The debate in the current literature about stabilizing atmospheric CO_2 at less than a doubling of the preindustrial concentration has led to needless confusion about current options for mitigation. On one side, the Intergovernmental Panel on Climate Change (IPCC) has claimed that "technologies that exist in operation or pilot stage today" are sufficient to follow a less-than-doubling trajectory "over the next hundred years or more."[1] On the other side, a recent review in *Science* asserts that the IPCC claim demonstrates "misperceptions of technological readiness" and calls for "revolutionary changes" in mitigation technology, such as fusion, space-based solar electricity, and artificial photosynthesis.[2] We agree that fundamental research is vital to develop the revolutionary mitigation strategies needed in the second half of this century and beyond. But it is important not to become beguiled by the possibility of revolutionary technology. Humanity can solve the carbon and climate problem in the first half of this century simply by scaling up what we already know how to do.

WHAT DO WE MEAN BY "SOLVING THE CARBON AND CLIMATE PROBLEM FOR THE NEXT HALF-CENTURY"?

Proposals to limit atmospheric CO_2 to a concentration that would prevent most damaging climate change have focused on a goal of 500 ± 50 parts per mil-

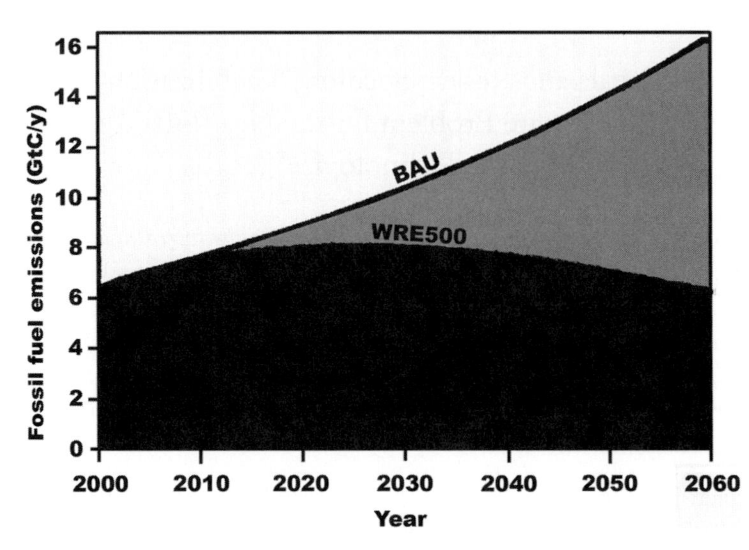

Figure 14.4 The top curve is a representative BAU emissions path for global carbon emissions as CO_2 from fossil fuel combustion and cement manufacture: 1.5% per year growth starting from 7.0 GtC/year in 2004. The bottom curve is a CO_2 emissions path consistent with atmospheric CO_2 stabilization at 500 ppm by 2125.[11] [See also SOM text.] The bottom curve assumes an ocean uptake calculated with the High-Latitude Exchange Interior Diffusion Advection (HILDA) ocean model[12] and a constant net land uptake of 0.5 GtC/year. The area between the two curves represents the avoided carbon emissions required for stabilization.

lion (ppm), or less than double the preindustrial concentration of 280 ppm.[3–7] The current concentration is ~375 ppm. The CO_2 emissions reductions necessary to achieve any such target depend on the emissions judged likely to occur in the absence of a focus on carbon [called a business-as-usual (BAU) trajectory], the quantitative details of the stabilization target, and the future behavior of natural sinks for atmospheric CO_2 (i.e., the oceans and terrestrial biosphere). We focus exclusively on CO_2, because it is the dominant anthropogenic greenhouse gas; industrial-scale mitigation options also exist for subordinate gases, such as methane and N_2O.

Very roughly, stabilization at 500 ppm requires that emissions be held near the present level of 7 billion tons of carbon per year (GtC/year) for the next 50 years, even though they are currently on course to more than double (Figure 14.4). The next 50 years is a sensible horizon from several perspectives. It is the length of a career, the lifetime of a power plant, and an interval for which the technology is close enough to envision. The calculations behind Figure 14.4 are explained in Section 1 of the supporting online material (SOM) text [see notes]. The BAU and stabilization emissions in Figure 14.4 are near the center of the cloud of variation in the large published literature.[8]

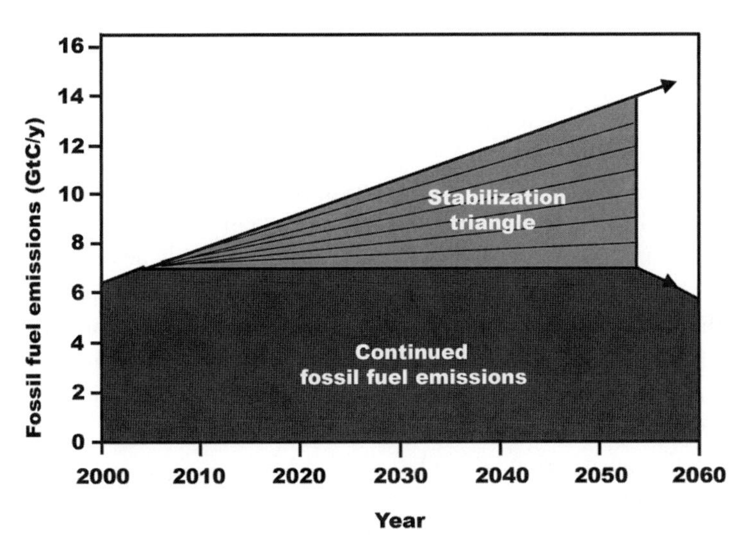

Figure 14.5 Idealization of Figure 14.4: A stabilization triangle of avoided emissions (gray) and allowed emissions (black). The allowed emissions are fixed at 7 GtC/year beginning in 2004. The stabilization triangle is divided into seven wedges, each of which reaches 1 GtC/year in 2054. With linear growth, the total avoided emissions per wedge is 25 GtC, and the total area of the stabilization triangle is 175 GtC. The arrow at the bottom right of the stabilization triangle points downward to emphasize that fossil fuel emissions must decline substantially below 7 GtC/year after 2054 to achieve stabilization at 500 ppm.

THE STABILIZATION TRIANGLE

We idealize the 50-year emissions reductions as a perfect triangle in Figure 14.5. Stabilization is represented by a "flat" trajectory of fossil fuel emissions at 7 GtC/year, and BAU is represented by a straight-line "ramp" trajectory rising to 14 GtC/year in 2054. The "stabilization triangle," located between the flat trajectory and BAU, removes exactly one-third of BAU emissions.

To keep the focus on technologies that have the potential to produce a material difference by 2054, we divide the stabilization triangle into seven equal "wedges." A wedge represents an activity that reduces emissions to the atmosphere that starts at zero today and increases linearly until it accounts for 1 GtC/year of reduced carbon emissions in 50 years. It thus represents a cumulative total of 25 GtC of reduced emissions over 50 years. In this paper, to "solve the carbon and climate problem over the next half-century" means to deploy the technologies and/or lifestyle changes necessary to fill all seven wedges of the stabilization triangle.

Stabilization at any level requires that net emissions do not simply remain constant, but eventually drop to zero. For example, in one simple model[9] that begins with the stabilization triangle but looks beyond 2054, 500-ppm stabilization is achieved by 50 years of flat emissions, followed by a linear decline of about two-thirds in the following 50 years, and a very slow decline thereafter that matches the declining ocean sink. To develop the revolutionary technologies required for such large emissions reductions in the second half of the century, enhanced research and development would have to begin immediately.

Policies designed to stabilize at 500 ppm would inevitably be renegotiated periodically to take into account the results of research and development, experience with specific wedges, and revised estimates of the size of the stabilization triangle. But not filling the stabilization triangle will put 500-ppm stabilization out of reach. In that same simple model,[9] 50 years of BAU emissions followed by 50 years of a flat trajectory at 14 GtC/year leads to more than a tripling of the preindustrial concentration.

It is important to understand that each of the seven wedges represents an effort beyond what would occur under BAU. Our BAU simply continues the 1.5% annual carbon emissions growth of the past 30 years. This historic trend in emissions has been accompanied by 2% growth in primary energy consumption and 3% growth in gross world product (GWP) (Section 1 of SOM text). If carbon emissions were to grow 2% per year, then ~10 wedges would be needed instead of 7, and if carbon emissions were to grow at 3% per year, then ~18 wedges would be required (Section 1 of SOM text). Thus, a continuation of the historical rate of decarbonization of the fuel mix prevents the need for three additional wedges, and ongoing improvements in energy efficiency prevent the need for eight additional wedges. Most readers will reject at least one of the wedges listed here, believing that the corresponding deployment is certain to occur in BAU, but readers will disagree about which to reject on such grounds. On the other hand, our list of mitigation options is not exhaustive.

WHAT CURRENT OPTIONS COULD BE SCALED UP TO PRODUCE AT LEAST ONE WEDGE?

Wedges can be achieved from energy efficiency, from the decarbonization of the supply of electricity and fuels (by means of fuel shifting, carbon capture and storage, nuclear energy, and renewable energy), and from biological storage in forests and agricultural soils. Below, we discuss 15 different examples of options that are already deployed at an industrial scale and that could be scaled up further to produce at least one wedge (summarized in Table 14.2). Although several options could be scaled up to two or more wedges, we doubt that any could fill the stabilization triangle, or even half of it, alone.

Because the same BAU carbon emissions cannot be displaced twice, achieving one wedge often interacts with achieving another. The more the electricity system becomes decarbonized, for example, the less the available savings from greater efficiency of electricity use, and vice versa. Interactions among wedges are discussed in the SOM text. Also, our focus is not on costs. In general, the achievement of a wedge will require some price trajectory for carbon, the details of which depend on many assumptions, including future fuels prices, public acceptance, and cost reductions by means of learning. Instead, our analysis is intended to complement the comprehensive but complex "integrated assessments"[1] of carbon mitigation by letting the full-scale examples that are already in the marketplace make a simple case for technological readiness.

Category I: Efficiency and Conservation

Improvements in efficiency and conservation probably offer the greatest potential to provide wedges. For example, in 2002, the United States announced the goal of decreasing its carbon intensity (carbon emissions per unit GDP) by 18% over the next decade, a decrease of 1.96% per year. An entire wedge would be created if the United States were to reset its carbon intensity goal to a decrease of 2.11% per year and extend it to 50 years, and if every country were to follow suit by adding the same 0.15% per year increment to its own carbon intensity goal. However, efficiency and conservation options are less tangible than those from the other categories. Improvements in energy efficiency will come from literally hundreds of innovations that range from new catalysts and chemical processes, to more efficient lighting and insulation for buildings, to the growth of the service economy and telecommuting. Here, we provide four of many possible comparisons of greater and less efficiency in 2054. (See references and details in Section 2 of the SOM text.)

Option 1: Improved fuel economy. Suppose that in 2054, 2 billion cars (roughly four times as many as today) average 10,000 miles per year (as they do today). One wedge would be achieved if, instead of averaging 30 miles per gallon (mpg) on conventional fuel, cars in 2054 averaged 60 mpg, with fuel type and distance traveled unchanged.

Option 2: Reduced reliance on cars. A wedge would also be achieved if the average fuel economy of the 2 billion 2054 cars were 30 mpg, but the annual distance traveled were 5,000 miles instead of 10,000 miles.

Option 3: More efficient buildings. According to a 1996 study by the IPCC, a wedge is the difference between pursuing and not pursuing "known and established approaches" to energy-efficient space heating and cooling, water heating, lighting, and refrigeration in residential and commercial buildings. These approaches reduce mid-century emissions from buildings by about one-fourth. About half of potential savings are in the buildings in developing countries.[1]

Option 4: Improved power plant efficiency. In 2000, coal power plants, operating on average at 32% efficiency, produced about one-fourth of all carbon emissions: 1.7 GtC/year out of 6.2 GtC/year. A wedge would be created if twice today's quantity of coal-based electricity in 2054 were produced at 60% instead of 40% efficiency.

TABLE 14.2

Potential Wedges: Strategies Available to Reduce the Carbon Emission Rate in 2054 by 1 GtC/year or to Reduce Carbon Emissions from 2004 to 2054 by 25 GtC

	Option	*Effort by 2054 for one wedge, relative to 14 GtC/year BAU*	*Comments, issues*
Energy efficiency and conservation	Economy-wide carbon-intensity reduction (emissions/$GDP)	Increase reduction by additional 0.15% per year (e.g., increase U.S. goal of 1.96% reduction per year to 2.11% per year)	Can be tuned by carbon policy
	1. Efficient vehicles	Increase fuel economy for 2 billion cars from 30 to 60 mpg	Car size, power
	2. Reduced use of vehicles	Decrease car travel for 2 billion 30-mpg cars from 10,000 to 5,000 miles per year	Urban design, mass transit, telecommuting
	3. Efficient buildings	Cut carbon emissions by one-fourth in buildings and appliances projected for 2054	Weak incentives
	4. Efficient baseload coal plants	Produce twice today's coal power output at 60% instead of 40% efficiency (compared with 32% today)	Advanced high-temperature materials
Fuel Shift	5. Gas baseload power for coal baseload power	Replace 1400 GW 50%-efficient coal plants with gas plants (four times the current production of gas-based power)	Competing demands for natural gas
CO_2 Capture and Storage (CCS)	6. Capture CO_2 at baseload power plant	Introduce CCS at 800 GW coal or 1600 GW natural gas (compared with 1060 GW coal in 1999)	Technology already in use for H_2 production
	7. Capture CO_2 at H_2 plant	Introduce CCS at plants producing 250 MtH_2/year from coal or 500 MtH_2/year from natural gas (compared with 40 MtH_2/year today from all sources)	H_2 safety, infrastructure

	8. Capture CO_2 at coal-to-synfuels plant	Introduce CCS at synfuels plants producing 30 million barrels a day from coal (200 times Sasol), if half of feedstock carbon is available for capture	Increased CO_2 emissions, if synfuels are produced without CCS
	Geological storage	Create 3,500 Sleipners	Durable storage, successful permitting
Nuclear fission	9. Nuclear power for coal power	Add 700 GW (twice the current capacity)	Nuclear proliferation, terrorism, waste
Renewable electricity and fuels	10. Wind power for coal power	Add 2 million 1-MW-peak windmills (50 times the current capacity) "occupying" 30×10^6 ha, on land or offshore	Multiple uses of land because windmills are widely spaced
	11. PV power for coal power	Add 2000 GW-peak PV (700 times the current capacity) on 2×10^6 ha	PV production cost
	12. Wind H_2 in fuel-cell car for gasoline in hybrid car	Add 4 million 1-MW-peak windmills (100 times the current capacity)	H_2 safety, infrastructure
	13. Biomass fuel for fossil fuel	Add 100 times the current Brazil or U.S. ethanol production, with the use of 250×10^6 ha (one-sixth of world cropland)	Biodiversity, competing land use
Forests and agricultural soils	14. Reduced deforestation, plus reforestation, afforestation, and new plantations	Decrease tropical deforestation to zero instead of 0.5 GtC/year, and establish 300 Mha of new tree plantations (twice the current rate)	Land demands of agriculture, benefits to biodiversity from reduced deforestation
	15. Conservation tillage	Apply to all cropland (10 times the current usage)	Reversibility, verification

Category II: Decarbonization of Electricity and Fuels

(See references and details in Section 3 of the SOM text.)

Option 5: Substituting natural gas for coal. Carbon emissions per unit of electricity are about half as large from natural gas power plants as from coal plants. Assume that the capacity factor of the average baseload coal plant in 2054 has increased to 90% and that its efficiency has improved to 50%. Because 700 GW of such plants emit carbon at a rate of 1 GtC/year, a wedge would be achieved by displacing 1400 GW of baseload coal with baseload gas by 2054. The power shifted to gas for this wedge is four times as large as the total current gas-based power.

Option 6: Storage of carbon captured in power plants. Carbon capture and storage (CCS) technology prevents about 90% of the fossil carbon from reaching the atmosphere, so a wedge would be provided by the installation of CCS at 800 GW of baseload coal plants by 2054 or 1600 GW of baseload natural gas plants. The most likely approach has two steps: (i) precombustion capture of CO_2, in which hydrogen and CO_2 are produced and the hydrogen is then burned to produce electricity, followed by (ii) geologic storage, in which the waste CO_2 is injected into subsurface geologic reservoirs. Hydrogen production from fossil fuels is already a very large business. Globally, hydrogen plants consume about 2% of primary energy and emit 0.1 GtC/year of CO_2. The capture part of a wedge of CCS electricity would thus require only a tenfold expansion of plants resembling today's large hydrogen plants over the next 50 years.

The scale of the storage part of this wedge can be expressed as a multiple of the scale of current enhanced oil recovery, or current seasonal storage of natural gas, or the first geological storage demonstration project. Today, about 0.01 GtC/year of carbon as CO_2 is injected into geologic reservoirs to spur enhanced oil recovery, so a wedge of geologic storage requires that CO_2 injection be scaled up by a factor of 100 over the next 50 years. To smooth out seasonal demand in the United States, the natural gas industry annually draws roughly 4,000 billion standard cubic feet (Bscf) into and out of geologic storage, and a carbon flow of 1 GtC/year (whether as methane or CO_2) is a flow of 69,000 Bscf/year (190 Bscf per day), so a wedge would be a flow to storage 15 and 20 times as large as the current flow. Norway's Sleipner project in the North Sea strips CO_2 from natural gas offshore and reinjects 0.3 million tons of carbon a year (MtC/year) into a non-fossil-fuel-bearing formation, so a wedge would be 3,500 Sleipner-sized projects (or fewer, larger projects) over the next 50 years.

A worldwide effort is under way to assess the capacity available for multicentury storage and to assess risks of leaks large enough to endanger human or environmental health.

Option 7: Storage of carbon captured in hydrogen plants. The hydrogen resulting from precombustion capture of CO_2 can be sent offsite to displace the consumption of conventional fuels rather than being consumed onsite to produce electricity. The capture part of a wedge would require the installation of CCS, by 2054, at coal plants producing 250 MtH_2/year, or at natural gas plants producing 500 MtH_2/year. The former is six times the current rate of hydrogen production. The storage part of this option is the same as in Option 6.

Option 8: Storage of carbon captured in synfuels plants. Looming over carbon management in 2054 is the possibility of large-scale production of synthetic fuel (synfuel) from coal. Carbon emissions, however, need not exceed those associated with fuel refined from crude oil if synfuels production is accompanied by CCS. Assuming that half of the carbon entering a 2054 synfuels plant leaves as fuel but the other half can be captured as CO_2, the capture part of a wedge in 2054 would be the difference between capturing and venting the CO_2 from coal synfuels plants producing 30 million barrels of synfuels per day. (The flow of carbon in 24 million barrels per day of crude oil is 1 GtC/year; we assume the same value for the flow in synfuels and allow for imperfect capture.) Currently, the Sasol plants in South Africa, the world's largest synfuels facility, produce 165,000 barrels per day from coal. Thus, a wedge requires 200 Sasol-scale coal-to-synfuels facilities with CCS in 2054. The storage part of this option is again the same as in Option 6.

Option 9: Nuclear fission. On the basis of the Option 5 estimates, a wedge of nuclear electricity would displace 700 GW of efficient baseload coal capacity in 2054. This would require 700 GW of nuclear power with the same 90% capacity factor assumed for the coal plants, or about twice the nuclear capacity currently deployed. The global pace of nuclear power plant construction from 1975 to 1990 would yield a wedge, if it continued for 50 years.[10] Substantial expansion in nuclear power requires restoration of public confidence in safety and waste disposal [see Chapter 3], and international security agreements governing uranium enrichment and plutonium recycling.

Option 10: Wind electricity. We account for the intermittent output of windmills by equating 3 GW of nominal peak capacity (3 GW_p) with 1 GW of baseload capacity. Thus, a wedge of wind electricity would require the deployment of 2,000 GW_p that displaces coal electricity in 2054 (or 2 million 1-MW_p wind turbines). Installed wind capacity has been growing at about 30% per year for more than 10 years and is currently about 40 GW_p. A wedge of wind electricity would thus require 50 times today's deployment. The wind turbines would "occupy" about 30 million hectares (about 3% of the area of the United States), some on land and some offshore. Because windmills are widely spaced, land with windmills can have multiple uses.

Option 11: Photovoltaic electricity. Similar to a wedge of wind electricity, a wedge from photovoltaic (PV) electricity would require 2,000 GW_p of installed capacity that displaces coal electricity in 2054. Although only 3 GW_p of PV are currently installed, PV electricity has been growing at a rate of 30% per year. A wedge of PV electricity would require 700 times today's deployment, and about 2 million hectares of land in 2054, or 2 to 3 m^2 per person.

Option 12: Renewable hydrogen. Renewable electricity can produce carbon-free hydrogen for vehicle fuel by the electrolysis of water. The hydrogen produced by 4 million 1-MW_p windmills in 2054, if used in high-efficiency fuel-cell cars, would achieve a wedge of displaced gasoline or diesel fuel. Compared with Option 10, this is twice as many 1-MW_p windmills as would be required to produce the electricity that achieves a wedge by displacing high-efficiency baseload coal. This interesting factor-of-two carbon-saving advantage of wind-electricity over wind-hydrogen is still larger if the coal plant is less efficient or the fuel-cell vehicle is less spectacular.

Option 13: Biofuels. Fossil-carbon fuels can also be replaced by biofuels such as ethanol. A wedge of biofuel would be achieved by the production of about 34 million barrels per day of ethanol in 2054 that could displace gasoline, provided the ethanol itself were fossil-carbon free. This ethanol production rate would be about 50 times larger than today's global production rate, almost all of which can be attributed to Brazilian sugarcane and United States corn. An ethanol wedge would require 250 million hectares committed to high-yield (15 dry tons/hectare) plantations by 2054, an area equal to about one-sixth of the world's cropland. An even larger area would be required to the extent that the biofuels require fossil-carbon inputs. Because land suitable for annually harvested biofuels crops is also often suitable for conventional agriculture, biofuels production could compromise agricultural productivity.

Category III: Natural Sinks

Although the literature on biological sequestration includes a diverse array of options and some very large estimates of the global potential, here we restrict our attention to the pair of options that are already implemented at large scale and that could be scaled up to a wedge or more without a lot of new research. (See Section 4 of the SOM text for references and details.)

Option 14: Forest management. Conservative assumptions lead to the conclusion that at least one wedge would be available from reduced tropical deforestation and the management of temperate and tropical forests. At least one half-wedge would be created if the current rate of clear-cutting of primary tropical forest were reduced to zero over 50 years instead of being halved. A second half-wedge would be created by reforesting or afforesting approximately 250 million hectares in the tropics or 400 million hectares in the temperate zone (current areas of tropical and temperate forests are 1500 and 700 million hectares, respectively). A third half-wedge would be created by establishing approximately 300 million hectares of plantations on nonforested land.

Option 15: Agricultural soils management. When forest or natural grassland is converted to cropland, up to one-half of the soil carbon is lost, primarily because annual tilling increases the rate of decomposition by aerating undecomposed organic matter. About 55 GtC, or two wedges' worth, has been lost historically in this way. Practices such as conservation tillage (e.g., seeds are drilled into the soil without plowing), the use of cover crops, and erosion control can reverse the losses. By 1995, conservation tillage practices had been adopted on 110 million hectares of the world's 1,600 million hectares of cropland. If conservation tillage could be extended to all cropland, accompanied by a verification program that enforces the adoption of soil conservation practices that actually work as advertised, a good case could be made for the IPCC's estimate that an additional half to one wedge could be stored in this way.

CONCLUSIONS

In confronting the problem of greenhouse warming, the choice today is between action and delay. Here, we presented a part of the case for action by identifying a set of options that have the capacity to provide the seven stabilization wedges and solve the climate problem for the next half-century. None of the options is a pipe dream or an unproven idea. Today, one can buy electricity from a wind turbine, PV array, gas turbine, or nuclear power plant. One can buy hydrogen produced with the chemistry of carbon capture, biofuel to power one's car, and hundreds of devices that improve energy efficiency. One can visit tropical forests where clear-cutting has ceased, farms practicing conservation tillage, and facilities that inject carbon into geologic reservoirs. Every one of these options is already implemented at an industrial scale and could be scaled up further over 50 years to provide at least one wedge.

Notes

1. IPCC, *Climate Change 2001: Mitigation*, B. Metz et al., ed. (Geneva: IPCC Secretariat, 2001) available at www.grida.no/climate/ipcc_tar/wg3/index.htm.
2. M. I. Hoffert et al., *Science* 298 (2002): 981.
3. R. T. Watson et al., *Climate Change 2001: Synthesis Report. Contribution to the Third Assessment Report of the Intergovernmental Panel on Climate Change* (Cambridge: Cambridge University Press, 2001).
4. B. C. O'Neill and M. Oppenheimer, *Science* 296 (2002): 1971.
5. Royal Commission on Environmental Pollution, *Energy: The Changing Climate* (2000); available at www.rcep.org.uk/energy.htm.
6. Environmental Defense, *Adequacy of Commitments—Avoiding "Dangerous" Climate Change: A Narrow Time Window for Reductions and a Steep Price for Delay* (2002); available at www.environmentaldefense.org/documents/2422_COP_time.pdf.
7. "Climate Options for the Long Term (COOL) synthesis report," *NRP Report 954281* (2002); available at www.wau.nl/cool/reports/COOLVolumeAdef.pdf.
8. IPCC, *Special Report on Emissions Scenarios* (2001); available at www.grida.no/climate/ipcc/emission/index.htm.
9. R. Socolow, S. Pacala, and J. Greenblatt, Proceedings of the *Seventh International Conference on Greenhouse Gas Control Technology* (Vancouver, B.C., September 2004).
10. BP, *Statistical Review of World Energy* (2003); available at www.bp.com/subsection.do?categoryId=95&contentId=2006480.
11. T. M. L. Wigley, in *The Carbon Cycle*, T. M. L. Wigley, D. S. Schimel, ed. (Cambridge: Cambridge University Press, 2000), 258–76.

12. G. Shaffer and J. L. Sarmiento, *Journal of Geophysical Research* 100 (1995): 2659.
13. The authors thank J. Greenblatt, R. Hotinski, and R. Williams at Princeton; K. Keller at Penn State; and C. Mottershead at BP. This paper is a product of the Carbon Mitigation Initiative (CMI) of the Princeton Environmental Institute at Princeton University. CMI (www.princeton.edu/~cmi) is sponsored by BP and Ford.
Supporting Online Material (SOM):
www.sciencemag.org/cgi/content/full/305/5686/968/DC1
SOM Text; Figs. S1 and S2; Tables S1 to S5; References

Michael B. McElroy, from "Industrial Growth, Air Pollution, and Environmental Damage: Complex Challenges for China" (1998)

China, the world's fastest growing major economy, holds one-fifth its population. Much of its energy comes from significant coal reserves, burned with scant regulation and less enforcement. Coal is one of the more environmentally disruptive means of harnessing energy. Mining and burning it create air and water pollution, acid rain, greenhouse gases, and toxins. China's increasing affluence means higher energy use. Auto manufacturers are determined to capture a huge market by building assembly plants there. What the large populations in China and India do will greatly affect everyone. With energy use, as with so much else, there is no "other side of the world." McElroy's work here presents issues for China but also provides an excellent overall picture of the interaction of human energy use with atmospheric chemistry and climate change in general.

INTRODUCTION

This article addresses the environmental challenges faced by China as a consequence of its ongoing rapid industrialization. We emphasize problems in the atmosphere associated with the use of fossil fuel, primarily coal. We distinguish between issues that are mainly local, for example particulate air pollution, and problems that are intrinsically global, primarily the risk of climate change caused by an increasing burden of greenhouse gases such as carbon dioxide released by the consumption of fossil fuel anywhere on earth.

There are important problems which arise on spatial scales intermediate between local and global, often referred to as regional hazards. Oxides of nitrogen and sulfur, emitted as byproducts of fossil fuel consumption, can acidify rain and snow falling thousands of kilometers from where the fuel is used. Photochemical processes triggered by industrial emissions of nitrogen oxides, in combination with hydrocarbons and carbon monoxide of both natural and anthropogenic origin, can raise levels of ozone in the lower atmosphere, damaging public health and agricultural productivity over extensive areas. Acid precipitation and elevated levels of tropospheric (lower atmospheric) ozone exemplify issues in which actions taken in one country can go beyond local or national boundaries and have serious consequences for its neighbors; they therefore require a regional response. The climate issue, far broader still, must find a response from the global community as a whole.

The industrial revolution began in England in the latter half of the eighteenth century and spread rapidly to neighboring countries in Western Europe and to the New World. Technological innovations, notably Watt's heat pump, Edison's commercial electricity, and Daimler's internal combustion engine, ushered in a period of unprecedented, prolonged, economic growth, fueled for the most part in the early years by coal. At the end of the twentieth century oil is supplanting coal as the leading fuel of the global economy, but coal supplies more than 70 percent of the commercial energy consumed in China, and, reflecting China's natural endowments, is likely to continue to dominate the national energy economy for the foreseeable future.

Local environmental problems from the use of fossil fuel were obvious even in the early years of the industrial revolution. Accidents, black lung disease, and accumulating slag heaps took a heavy toll on coal mining communities. The connection between burning coal and the filthy, particulate-laden air in industrial cities was inescapable, but it drew relatively little attention from policymakers even in the richest western countries until recently; the pollution associated with a fossil fuel economy was accepted as an inevitable price to be paid for industrial progress—at least until it became so severe that policymakers could not escape demands for action by an aroused citizenry. It is interesting that the first laws regulating emissions of smoke in the United States were passed not by the federal government but by cities—Chicago and Cincinnati in the 1880s. Thousands of people died in a killer smog (a mixture of smoke and fog) in London in 1952, and following a similar episode in Donora, Pennsylvania, in 1948. Publicity surrounding these events was quickly translated into calls for action. While disasters such as those of Donora and London may not yet have struck in China, conditions may be ripe for them to occur in the future.

The solution to the local smog problem was clear: aggressive reduction of particulates (smoke) from domestic chimneys and industrial smokestacks. The challenge was particularly urgent for cities where large quantities of coal were used either industrially, as in Donora, or both industrially and domestically, as in London. As discussed below, a variety of strategies were available to deal with the problem of particulates, and for the most part it has been eliminated, or at least seriously mitigated, in developed countries. Particulate pollution remains a serious issue, however, for the developing world, and is recognized by policymakers as a priority for environmental action in China. The smoke burden today in major Chinese industrial cities such as Chongqing is significantly worse than it was in London in 1952.

At about the same time that countries and localities were gearing up to deal with the particulate problem, a new issue of air quality affecting public health began to receive attention in Los Angeles. Mysterious at first, it eventually was attributed to the increased presence of tropospheric (lower level) ozone, synthesized in the urban atmosphere by the interaction of sunlight with a mix of hydrocarbons and nitrogen oxides. Ozone is highly toxic when inhaled. It can result in a variety of pulmonary problems, including emphysema, bronchitis, and asthma; at high concentrations it can lead to heart failure.[1] Chronic exposure to ground ozone levels above 50 parts per billion (ppb) also causes damage to sensitive plants and can reduce agricultural productivity. As discussed below, this is potentially important for China.

The ozone issue provides an interesting contrast to the London-style smog problem. The health effects in Los Angeles were perceived before there was any understanding of their cause. Important detective work was required to identify ozone as the culprit and additional difficult research was needed to develop even a rudimentary understanding of the chemical processes responsible for high ozone levels in Los Angeles. It was correspondingly difficult to develop a realistic strategy for mitigation. Fifty years later the chemistry of tropospheric ozone remains a topic of active research. The basic elements are finally understood but the details remain elusive. It is clear that ozone pollution, in contrast to the fog-smoke mixture in London, is a secondary, or indirect, product of industrial activity. The primary emissions responsible for it are oxides of nitrogen, carbon monoxide, and hydrocarbons. These largely come from automobile use, and the ultimate solution will require an expensive redefinition of the role of the automobile in ozone-affected cities such as Los Angeles. However, the problem is not unique to Los Angeles. High levels of ozone will develop in any environment where air is trapped and subject to significant input of the primary responsible chemicals. Photochemical smog, the technical term for the Los Angeles condition, is an issue of overriding importance and a significant constraint on economic activity in cities as diverse as Paris, New York, Mexico City, São Paulo, and Tokyo. It will assuredly emerge as a future concern for China as direct emissions of toxic chemicals and health-threatening particulates are successfully reduced.

A common approach to the problem of air pollution in the 1950s and 1960s was to install high smoke stacks on large point sources of emissions. While mitigating the local impact, this distributed the pollution over much larger areas, including regions previously relatively pristine. Recourse to tall stacks in large measure explains the spread of acid precipitation. Combustion of fossil fuel, particularly coal, emits oxides of nitrogen and sulfur. In the absence of controls, coal is about three times more polluting in terms of nitrogen oxides than either oil or natural gas. It is similarly more polluting than oil in the case of sulfur while natural gas is essentially sulfur-free. Nitrogen and sulfur oxides are converted by chemical processes in the atmosphere to acids. These acids can cause serious damage to vegetation when they come into contact with leaf surfaces and the problem is exacerbated when the acids are concentrated in fog. They are removed from the atmosphere in part by contact with vegetative surfaces (dry deposition) and in part by incorporation in either rain or snow (wet deposition). The latter mechanism is responsible for the phenomenon of acid precipitation (familiarly, "acid rain"). This was recognized first as a serious problem in regions such as Scandinavia and eastern North America, where soils, lakes, and rivers are especially vulnerable to inputs of acids.

Like photochemical smog, acid rain became a public issue in advance of scientific understanding of either its cause or effects. It first came to public attention when it was noticed that fish were dying in lakes in southern Norway and that the fish kills were associated with increasing levels of acidity in lakes. It was unclear, though, why the fish were dying. Was the problem a direct consequence of acidity? Or, was the impact indirect? Was there something else in the lakes, related perhaps to acidity, that was responsible for the drop in fish population? If acid rain and snow were the primary culprits, where did the acidic compounds come from? Clearly, they were not local. The suspects in Scandinavia included high smoke stacks in the United Kingdom and emissions from industrial regions of both Eastern and Western Europe. In North America, use of high sulfur coal in the Midwest was generally alleged to play an important role in the problems identified in New England and eastern Canada, although closer sources of pollution were also implicated. To deal with the problem, it was necessary to develop at least tentative answers to contentious scientific questions. In the absence of scientific understanding, politicians had an excuse to equivocate and to postpone expensive strategies for remediation.

The acid rain problem was also important in that it was perhaps the first of the air pollution issues to call clearly for not only national but also international action. It highlighted the fact that transport of pollution proceeds without respect for national and state boundaries. Steps taken in Norway could not effectively address the problem of fish kills in Norwegian lakes in the absence of a response by its neighbors to the south. Actions to deal with the problems of acid rain and elevated ozone in New England required cooperation from other states, as well as from Canada. A bilateral international approach was clearly needed to alleviate the difficulties between the U.S. and Canada, while a communitywide response was required to address transnational transport of pollutants in Europe.

The complexity of air pollution issues is illustrated instructively by the contentious debate in the United States that eventually led to the Clean Air Act Amendments of 1990.[2] The 1990 legislation represents the most recent of a long series of laws dealing with air pollution in the United States, dating back to the Air Pollution Control Act of 1955. For the most part, the legislation addresses environmental damage caused by the use of fossil fuel. Specifically it targets problems associated with mobile sources (trucks and automobiles) and energy intensive industries, such as the generation of electric power and the production of steel and various chemicals. A number of the topics now covered by laws have emerged since the first legislation in 1955: among them are photochemical smog and acid precipitation, and the damage to the stratospheric (upper atmosphere) ozone layer from emissions of industrial chlorinated and brominated halocarbons.

Without exception, U.S. air pollution legislation had a problem-trailing character; it addressed only problems that were already apparent, many of them unanticipated when the responsible technologies were first introduced. The history of attempts to deal with air pollution in the United States provides an interesting case study for policymakers in China, and a challenge to chart a course to industrial development different from that followed by China's historical predecessors. A modern approach should guide investment so as to anticipate problems, and thus minimize the need for expensive retrospective action. It is hoped that, by outlining these problems, this chapter can contribute to this objective.

We begin with an overview of issues relating to the use of fossil fuel with an emphasis on coal, the primary fuel for the Chinese economy. We follow with more specific accounts of the health and environmental impacts of particulates and oxides of nitrogen and sulfur, before turning to the problem of carbon dioxide (CO_2) and its potential role as an agent for change in global climate. We describe the effects of emissions both in China and in a global context. We conclude with a recommendation for future research and a broad consideration on the management of industrialization.

EMISSIONS FROM COMBUSTION OF FOSSIL FUEL

Coal, a carbon-rich residue of ancient fossilized plant material, is by far the most abundant of the world's fossil fuels. It is estimated that China has 11 percent of total global coal reserves, sufficient to fuel its economy at current rates for more than a hundred years. The energy liberated in burning coal is contributed mainly by oxidation of its constituent carbon. In contrast, a significant fraction of the energy released in burning oil is derived from oxidation of hydrogen, and the hydrogen contribution from natural gas is even larger. On a per unit mass basis, the energy content of oil is approximately twice that of coal while natural gas is even more efficient than oil. On an energy yield basis, emission of CO_2 is highest for coal (approximately 60 pounds of carbon per million British thermal units [Btu]), intermediate for oil (45 pounds per million Btu), and lowest for natural gas (32 pounds per million Btu).

From an environmental viewpoint, natural gas, consisting mainly of methane (CH_4), is the cleanest of the fossil fuels. Efficient combustion of natural gas produces mainly CO_2 and water (H_2O) with variable quantities of nitrogen oxides (NO_x) generated during the combustion process when air is raised to high temperature. Crude oil is composed of a complex range of hydrocarbons, a number of which are known carcinogens. Reflecting its origin as plant material, it contains, in addition to carbon and hydrogen, variable quantities of chemically reduced nitrogen and sulfur. Combustion of oil results in emission of CO_2, H_2O, and CO (carbon monoxide, less or more depending on the efficiency of the combustion process) and a range of hydrocarbons in addition to significant quantities of nitrogen and sulfur oxides. Combustion of coal poses by far the largest environmental risk. Its products include, in addition to CO_2, CO, and H_2O, significant quantities of NO_x and sulfur oxides (SO_x) and, reflecting the geological origins of particular coal beds, variable quantities of particulate material rich in trace metals and other elements present in abundance in the Earth's crust.

Particulates

Particulate emissions from the combustion of coal are represented by a suite of finely dispersed liquid and solid materials with sizes in the range 0.05 to 100 microns. The composition of coal-derived particulate emissions varies with the nature of coal used and details of the combustion process. In addition to unsightly black soot, coal emissions contain a mix of organic and inorganic substances including sulfates, metallic oxides, and various naturally occurring radionuclides and salts. Respiration of this material can result in a range of pulmonary problems, including emphysema, asthma, bronchitis, and lung cancer. Particles in the smallest size range, with diameters less than 10 microns (PM10), can penetrate deep into the lung, where particles as small as 0.05 microns may remain for as long as the life of an individual. The smallest particles are judged most serious in terms of their impact on public health.[3]

National standards for six so-called criteria pollutants (total suspended particulates, sulfur dioxide [SO_2], CO, lead, ozone [O_3], and nitrogen dioxide [NO_2]) were set by the Environmental Protection Agency (EPA) for the United States as required by the Clean Air Act of 1970. The initial ruling for particles defined a standard only for total suspended particulate matter (TSP) but was amended in 1987 to include a separate requirement for PM10 and was further amended in 1997 to include standards for particles smaller than 2.5 microns (PM2.5). The volume of total annual particulate emissions decreased in the United States by about 60 percent between 1970 and 1990, from about 18.5 million tons to about 7.2 million tons. Despite this improvement, which was achieved despite a significant growth in economic activity, fifty-eight of 247 Air Quality Control Regions (as defined in the 1990 Clean Air Act) were in violation of national standards for TSP in the United States in 1990.

The particulate situation is much worse in China and India. Standards for TSP defined by the World Health Organization (WHO) were violated on 294 days of a typical year between 1980 and 1984 in New Delhi. Performance was only slightly better in China. . . . The data indicate that annual mean concentrations of TSP for . . . five cities exceeded maximum acceptable WHO standards, by a factor of five in the worst case (Xian), and by no less than a factor of two in even the best case (Guangzhou). For a typical year, Freedman and Jaggi (1993)

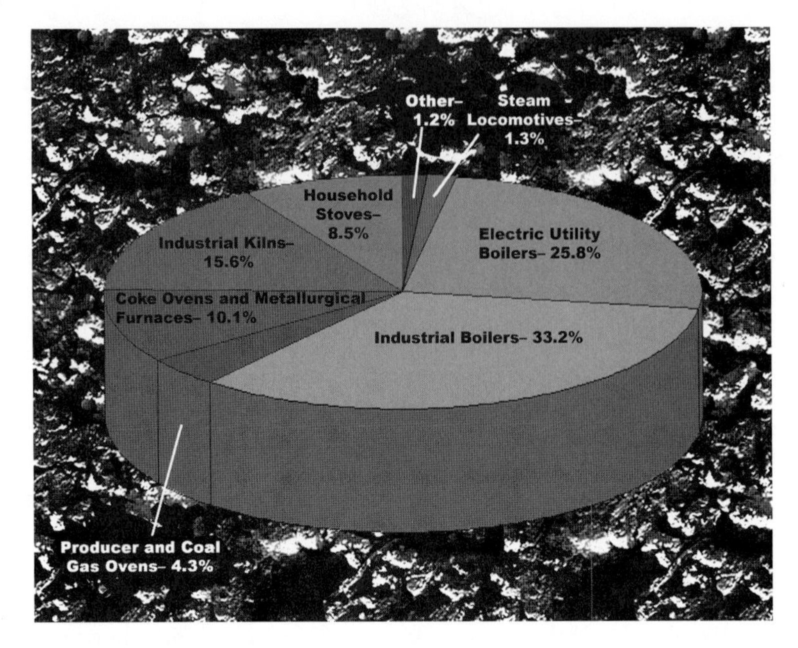

Figure 14.6 Coal use in China in 1990. *Source:* Sinton et al. 1996.

report that WHO standards for TSP are violated for an average of 274, 272, and 219 days in the cities of Xian, Beijing, and Shenyang.[4]

Approximately 26 percent of the coal used in China is consumed in the generation of electricity, with an additional 63 percent employed in a variety of industrial applications (see Figure 14.6). Only a small fraction, 8.5 percent, of China's coal is consumed domestically. Although cost-effective means for small emission sources may be more elusive, relatively straightforward technological strategies are available to reduce particulate emissions from large industrial sources, including power plants. These include the precombustion cleaning of coal and the use of electrostatic precipitators, wet scrubbers, fabric filters, and mechanical precipitators. . . . As early as 1975, it was possible to remove as much as 99.7 percent of particulate emissions from power plants in the United States using electrostatic precipitation[5] although in practice efficiencies achieved operationally were usually less than this ideal by about 10 percent. By 1987, it was possible to eliminate essentially all large particles by means of electrostatic precipitation,[6] although this method is less effective for small particles. These may be trapped, however, by fine meshed filters and removed mechanically. By 1987, a number of electric power plants in the western United States had adopted this dual approach with success. Whatever the mix of strategies China adopts to reduce particulate emissions, it is clear that this goal is technologically achievable and indeed progress has been already achieved. From 1985 to 1995, average ambient concentrations fell from 700 $\mu g/m^3$ to around 450 $\mu g/m^3$ for cities over 1 million and from 500 $\mu g/m^3$ to around 300 $\mu g/m^3$ for cities 0.1–1 million. It may be bad now, but it used to be much worse. The World Health Organization (WHO) guideline is 100 $\mu g/m^3$, however, and in recent years concentrations have stopped declining, holding around the 1995 levels mentioned above. Protecting the health of its population and reducing losses in economic productivity are strong incentives for China to take action against particulate pollution.

In addition to the particulate material emitted directly from burning fossil fuels, the atmosphere contains a number of particles formed by secondary reactions following the emission of gaseous compounds such as SO_2. Sulfur dioxide is oxidized in the atmosphere to sulfuric acid (H_2SO_4), either in the gas phase or more commonly in the liquid phase in clouds. As clouds containing dilute mixtures of H_2SO_4 and H_2O evaporate (due to a rise in temperature, for example) sulfur will be retained differentially in droplets: the vapor pressure of H_2O in equilibrium over a mixed sulfate-water condensed phase is less than that for pure water. Evaporation of cloud particles containing H_2SO_4 will lead to production of a suite of small sulfate-rich aerosols (airborne particles). When condensation resumes, these particles serve as nuclei for subsequent condensation. The larger the abundance of condensation nuclei, the larger the number of particles available to absorb newly forming liquid: a fixed quantity of condensed water is distributed over a larger number of cloud particles. The size of the average cloud particle will decrease but the integrated surface area of particles contained in a given volume will be larger. Since the reflective properties of clouds depend on integrated surface area, a high concentration of condensation nuclei will tend to favor formation of more reflective clouds.

This suggests that a high abundance of sulfate aerosols over China, maintained by emission of large and growing quantities of industrial sulfur, may be responsible for a significant drop in daytime temperatures over China. This is likely to pose an unfortunate dilemma for policymakers: a reduction in emission of industrial sulfur, as required both for public health and to reduce acid precipitation, may result in an increase in temperature exacerbating the rise in temperature expected due to higher levels of greenhouse gases such as CO_2 as discussed below. . . .

Acid Rain

Rain in equilibrium with atmospheric CO_2 has a pH of about 5.6. Addition of alkaline material such as calcium carbonate or ammonia can raise the pH of precipitation to values as high as 7 or larger. In contrast, inclusion of nitric and sulfuric acids and acidic aerosols can cause pH to drop below 4. We use the term acid rain here to refer to precipitation with pH below about 5, recognizing the significance of the limiting value imposed for pristine conditions by CO_2. The pH of cloud and fog water is typically lower than that of either rain or snow. In one extreme case, water collected from a dissipating fog near Los Angeles was measured to have a pH of 1.69, roughly similar to that of lime juice.[7]

Oxides of nitrogen and sulfur, mainly NO and SO_2, released as byproducts of combustion, are the primary precursors for formation of acid rain in polluted environments. . . .

Acid rain is a problem primarily in southeastern and south-central portions of China (the provinces of Sichuan, Guizhou, Hunan, Guangxi, and Guangdong), where soils are naturally acidic and coal is often high in sulfur. It is less serious in the north where soils are more alkaline and where high concentrations of carbonate minerals in airborne dust are effective in neutralizing acidity contributed by nitric and sulfuric acid. An annual average of 3.2 was reported for the pH of rain in Guiyang (the capital city of Guizhou) in 1993 while the average for suburban Chongqing (Sichuan Province) was only slightly higher, 4.18.[8] Acid rain has been implicated in causing serious damage to structural materials in cities and has been blamed also for the decline of forests over extensive regions of southwestern China.[9] Leaching of aluminum by acid rain percolating through soils can lead to accumulation of toxic concentrations of this element in lakes with serious consequences for fish. Despite the high priority attached to the issue of acid rain by the Chinese government in the Seventh and Eighth Five-Year Plans, it appears that the situation has deteriorated and that the area affected by acid rain in China has actually expanded in recent years. . . .

A number of approaches are available to reduce emissions of sulfur compounds from the combustion of coal. The simplest, flue gas desulfurization (FGD), involves exposure of stack gases to a spray of chemicals, usually a slurry of lime (CaO) and limestone ($CaCO_3$), which acts to convert volatile forms of sulfur to less volatile products such as calcium sulfite ($CaSO_3$) and calcium sulfate ($CaSO_4$). Additional strategies currently under study include pretreating coal to remove sulfur before combustion (coal-washing), or removing sulfur during combustion by burning the coal on a bed of limestone sorbent (fluidized bed combustion). Commercial applications of the latter technologies have been relatively limited to date but are expected to grow in the future. In the FGD process, the sludge is recovered and disposed of elsewhere. This must be done carefully if additional environmental problems are to be avoided. The cost of retrofitting power plants with FGD has dropped significantly in recent years, from more than $220 per kilowatt in 1982 to less than $170 in 1990, with prospects for further reduction.

Emission of sulfur compounds to the atmosphere can be reduced by as much as 95 percent with current FGD technology. There is a cost, however, not only in terms of capital investment in equipment but also in terms of energy used in operation of the FGD equipment itself. This energy use reduces the output of the plant and itself generates emissions of CO_2.[10] To date only one large power plant equipped with FGD has been installed in China and there are reports that the desulfurization equipment on this plant is typically unused.[11] . . .

Carbon Dioxide, the Greenhouse Effect, and Climate

Carbon dioxide produced by burning fossil fuel—coal, oil, or natural gas—is the largest single waste product of modern industrial society. Emissions to the atmosphere globally in 1997 exceeded six billion tons, more than a ton for every man, woman, and child on the planet. The United States, with approximately 5 percent of the world's population, was responsible for 22 percent of global emissions [some other sources state as high as 25 percent]. China, which ranked number ten in 1950, is now number two, having recently surpassed Russia. Given current trends, there is little doubt that China will eventually be number one and that it will set the pace globally for CO_2 at some point in the not too distant future.

Per unit of energy delivered, coal is the largest source of CO_2, followed by oil and gas. In 1991, oil accounted for 37.1 percent of global consumption of commercial energy, coal 29.2 percent, and gas 23.7 percent, with the balance attributable to nuclear (7 percent) and hydroelectric (3 percent). The relative contributions of oil, coal, and gas to commercial emissions of CO_2 globally in the same year were 42 percent, 38 percent, and 17 percent respectively with the balance associated with gas flaring and the manufacture of cement.

We have an excellent account of variations in the concentration of atmospheric CO_2 over the past 160,000 years. [See Chapter 1, Climate Shock, particularly Figure 1.10.] The modern instrumental record, based on sampling of air at a variety of remote sites around the world, dates back to 1958. Studies of gases trapped in polar ice extend the measurements back in time and indicate that the modern rise in CO_2 began in the early part of the eighteenth century. Over the past 200 years, the period of most rapid industrial development in the west, the concentration of CO_2 in the atmosphere has risen from about 280 parts per million by volume (ppmv) to close to 360 ppmv. Burning of fossil fuel is primarily responsible, though deforestation—primarily in the tropics—also contributes as the plant matter is burned, releasing stored carbon to the atmosphere. The latter effect is offset to some extent by regrowth of vegetation, which re-absorbs the gas from the atmosphere, accumulating it in the wood of trunks and limbs, primarily at midlatitudes of the northern hemisphere. . . .

There is . . . general agreement that an increase in CO_2 will result in an increase in surface temperature. The Intergovernmental Panel on Climate Change concluded that the build-up of greenhouse gases anticipated to occur over the next century may be expected to cause a rise in global average temperature of between 1 and 3.5°C with an associated increase in global average sea level of between 15 and 95 cm. [These 1996 estimates were increased in 2001.] The panel concluded that the temperature increase that has taken place already in this [twentieth] century (about 0.5°C) "is unlikely to be entirely natural in origin" and that "the balance of the evidence suggests that there is discernible human influence on climate." [See Chapter 1, Climate Shock.]

An increase in sea level of less than a meter seems small, but an increase in sea level by as little as 50 centimeters would be sufficient to inundate an area of more than 40,000 square kilometers in China[12] and would have even more serious consequences for densely populated, low-lying countries such as Bangladesh. It would be associated also, most likely, with intrusion of saltwater into freshwater aquifers worldwide and could exacerbate damage from coastal storms.

A warming of the atmosphere may be associated also with greater weather turbulence. Floods in south and central China were responsible for more than a thousand deaths in each of the summers of 1994, 1995, and 1996, while at the same time drought in northern China caused the Yellow River to dry up for a cumulative total of 333 days between 1990 and 1995.[13] Between 1990 and 1995 weather-related problems in the United States, such as unseasonable rains in California, floods in the Midwest, and hurricanes in Florida, resulted in economic losses of more than $150 billion ($38 billion in 1995 alone). Weather events cost the insurance industry $50 billion, almost three times the total paid out over the decade of the '80s.[14]

It is not our intention to argue that the anomalies in weather experienced recently should be considered necessarily a harbinger of what is to be expected in the future. But they do emphasize our vulnerability to unexpected changes in climate and the need for great caution about any human steps which could provoke such changes. At great expense, we have built roads and bridges and sited human habitations, relying on what weather did in the past as a guide to what it will do in the future. If weather patterns change, and we are unprepared, the consequences for human beings could be very intense. Rich societies such as the United States may be able to adjust, at some expense, but responses for densely populated developing countries such as China may be significantly more difficult.

CONCLUDING REMARKS

China faces a baffling array of challenges as it confronts the complex range of issues relating to the use of fossil fuel, including impacts on air quality, climate, public health, and agriculture. To the extent that China's leaders perceive the issues of the environment and economic growth as interconnected, and see economic advance as depending on a sustainable healthy environment, they are more likely to integrate policies for development with policies for environmental protection. Environment and energy policies adopted in China have implications not only for China but also for its neighbors—and vice versa.

The ability to feed its expanding population remains an overriding imperative for China's government; memories of the horrendous famine of the late 1950s and early '60s are still very much alive in the consciousness of the Chinese people and their leaders. Beyond that imperative, China's priorities are to improve the material quality of life for its citizens, and to provide jobs and the reliable supply of food and energy essential for sustained economic growth and political stability.

It is clear that China's policies with respect to environmental protection place larger emphasis on issues perceived as immediate and local, as distinct from problems such as climate change, seen as global and futuristic, for which others may be expected to take the lead. But environmental stability, whether local or global, cannot be taken for granted as industrialization and agricultural modernization proceed. Changes in climate may complicate China's plans to expand its domestic production of grain, and an increase in tropospheric ozone could hamper crop productivity, as discussed above.

China's aspirations for development and its desire to provide a better and more sustainable environment for its citizens merit our most serious respect and support. But for this respect and support to be made concrete and effective, China must cooperate. There is an urgent need to engage the resources of the developed world to reduce its current reliance on coal and to diversify its energy economy. There may well be rewards from searching for sources of fossil energy more environmentally friendly than coal, such as oil and gas. Nonfossil sources of energy such as nuclear, solar, geothermal, wind, and hydroelectricity can and should play an important role in China's energy future. Where coal must be used, incentives for using it as efficiently as possible and in a fashion designed to minimize damage to the environment must be put in place.

China faces a daunting choice. It can opt to pursue an independent path seeking essentially one goal, the most rapid possible economic growth, risking serious long-term, potentially irreparable, damage to her environment in the interest of short-term returns. Or it can learn from the historical experience of other countries—that it is less costly in the long run to avoid present environmental damage, eliminating the need to pay for expensive remediation in the future. . . . [This second option] is a challenge that neither we nor China can afford to forego.

Notes

1. American Lung Association, *Danger Zones: Air Pollution and Our Children* (New York: American Lung Association, 1995). [McElroy cites other sources as well.]
2. Gary C. Bryner, *Blue Skies, Green Politics: The Clean Air Act of 1990 and Its Implementation* (Washington, D.C.: CQ Press, 1995).
3. Richard Wilson and John Spengler, *Particles in Our Air: Concentrations and Health Effects* (Cambridge: Harvard School of Public Health, 1996).
4. Martin Freedman and Bikki Jaggi, *Air and Water Pollution Regulation: Accomplishments and Economic Consequences* (Westport, Conn.: Quorum Books, 1993).
5. R. H. White, *The Price of Power, Update: Electric Utilities and the Environment* (New York: Council on Economic Priorities, 1977).
6. Freedman and Jaggi.
7. D. J. Jacob, J. M. Waldman, J. W. Munger, and M. R. Hoffmann, "Chemical Composition of Fogwater Collected Along the California Coast," *Environmental Science and Technology* 19 (1985): 730–36.

8. J. E. Sinton, D. G. Fridley, Z. P. Jiang, M. D. Levine, X. Zhuang, F. Q. Yang, K. J. Jiang, and X. F. Liu, ed., *China Energy Data Book*, Report no. LBL-32822 Rev. 4 (Berkeley, Calif.: Lawrence Berkeley National Laboratory, 1996).
9. Dianwu Zhao and Jiling Xiong, "Acidification in Southwestern China," in *Acidification in Tropical Countries*, ed. H. Rodhe and R. Herrera (Chichester: John Wiley and Sons, 1988).
10. Fiona E. Murray and Peter P. Rogers, "Living with Coal: Coal-based Technology Options for China's Electric Power Generating Sector," in *Energizing China: Reconciling Environmental Protection and Economic Growth*, ed. Michael B. McElroy, Chris P. Nielsen, and Peter Lydon (Cambridge: Harvard University Committee on Environment, 1998), 167–200.
11. Michael B. McElroy and Chris Nielsen, "Energy, Agriculture, and the Environment: Prospects for Sino-American Cooperation," *Living with China: U.S.-China Relations in the Twenty-first Century*, ed. Ezra Vogel (New York: W. W. Norton, 1997).
12. Mike Hulme et al., *Climate Change Due to the Greenhouse Effect and Its Implications for China* (Gland, Switzerland: World Wide Fund for Nature, 1992).
13. *China News Digest* (electronic news service), News Briefs of 16 June, 6 Aug. 1994; 18 Aug. 1995; 27 May, 15 July, 29 July 1996.
14. Christopher Flavin and Odil Tunali, "Climate of Hope: New Strategies for Stabilizing the World's Atmosphere," *World Watch Paper 130* (World Watch Institute, June 1996).

15 Toxicology

"The dose makes the poison." With that insight, the medieval physician Paracelsus laid the groundwork for the modern discipline of toxicology. Living tissues operate within certain chemical parameters; too much (or in some cases too little) of a substance will damage or destroy them. Even water can be harmful in excess, as marathon runners who monitor their blood salt levels know.

Normally our interest in toxicology is in substances that are harmful in small doses. Nature abounds with these; plants, animals, and fungi have developed a wide variety of toxins to deter, stun, and kill both predators and prey. Nature often puts antidotes close at hand as well, enabling indigenous peoples, farmers, experienced hikers, and others fluent in the chemistry of their surroundings to navigate them without fear.

Ancient civilizations marveled over the power of acute poisons such as arsenic and cyanide, but were less commonly aware of the chronic effects of lead in water pipes and kitchenware. With the industrial revolution and the advent of synthetic chemistry, the number of toxic substances in the environment has exploded.

Of these new synthetic chemicals, pesticides have attracted the greatest concern and regulatory attention—and rightly so, since they are designed to be toxic. It was the development of military applications, especially government investment in basic research and productive capacity in the U.S. during World War II, that launched the pesticide industry as we know it today. Chloropicrin was a tear gas before it was a soil fumigant. DDT was a makeshift military weapon against malarial mosquitoes before it was marketed for use on civilian food crops. Overuse and misuse of DDT in the postwar United States was a central theme in Rachel Carson's *Silent Spring*, which raised public consciousness about environmental toxins and spurred governmental action. (Ironically, indiscriminate pesticide use has caused many pests to develop resistance. According to the research of David Pimentel, while U.S. pesticide use increased 33-fold between 1945 and 1985, the percentage of crops lost to pests actually increased.)

Figure 15.1 Toxic chemicals can have a devastating impact on wildlife. Over 100 tons of dead fish had to be pulled from the Tisza River in Hungary and former Yugoslavia following an accidental release of cyanide-laced tailings from the Romanian Baia Mare gold mine in 2000, the largest industrial disaster in Eastern Europe since Chernobyl. Although the river has partially recovered from the initial cyanide poisoning, toxic heavy metals lodged in sediment will be a part of the river ecosystem for generations to come. Courtesy of Tibor Kocsis, Flóra Film International.

But there are many other types of environmental toxins in the modern world. Thanks to mining and other industrial activities, metal compounds and other minerals normally locked in the Earth's crust now circulate in significant quantities in air, water, and living tissue. Mercury released into the atmosphere from the smokestacks of coal-fired power plants, for example, settles on rivers and lakes and accumulates in the aquatic food chain. Human activity concentrates naturally occurring radioactive isotopes and produces new forms of radioactive waste that require careful handling and disposal.

A good deal of petroleum is used not for energy, but as a base material for producing plastics. Synthetic plastics have revolutionized modern life, and in the span of a few generations have become ubiquitous. But because they are so new, their long-term environmental effects remain largely unknown. On the one hand, many plastics are notoriously resistant to biodegradation, leading to a variety of waste-related problems. Images of sea animals choking on discarded plastic wrappers have become iconic of the litter problem. On the other hand, some plastics, like the PVCs (polyvinylchlorides) used in piping, food packaging, and children's toys, are known to change chemically over time and leach toxic degradates and additives.

The effect of a toxin depends on the dose, as Paracelsus recognized, but also on other factors, including the route of exposure (e.g., oral, inhalation, or dermal), the frequency of exposure, the speed at which the compound accumulates in tissue or the body flushes it out, and the presence of enabling or inhibiting compounds. Toxins harm living tissue in a variety of ways. Some attack particular organs or organ systems; the neurotoxicity of lead is a well-known example. Carcinogens interfere with the normal process of cellular reproduction. Teratogens interfere with normal development of fetuses *in utero*. Endocrine disruptors block or mimic hormones; these chemicals can affect fertility and even change the sexual morphology or behavior of animals in the wild. Recent research has focused concern on pharmaceuticals and personal care pollutants, or PPCPs, a class that includes thousands of chemicals from the ingredients of cosmetics, fragrances, and sunscreen to caffeine and nicotine, to over-the-counter and prescription drugs. Not surprisingly, given the ubiquity of these products—and pharmacists' instructions to flush unwanted medicines down the toilet—PPCPs are now found at low levels in many environmental waters. The long-term effects of minute but regular doses of this "cocktail" in drinking water are unknown.

The science of toxicology is, if no longer in its infancy, still in its youth. We have a large amount of data on the effects of individual chemicals on rats, mice, and other animals at moderate to high doses in laboratory settings. But only a fraction of known chemicals have been tested—and at the same time, the models used to extrapolate low-dose effects in humans from those laboratory studies are largely based on conjecture. (Some toxicologists argue compellingly that current models are unrealistically overprotective—see the article by Ames and Gold listed in Further Reading.) We still have much to learn about the effects of chemicals in real-world settings—occurring not alone but in combination, degrading and metabolizing to form new compounds, biomagnifying in the food chain. We need to know more about how toxins work, about what causes differences in susceptibility among species, among individuals, and even within individuals over time—increased susceptibility of the young and the immune-deficient, for example, and the process of sensitization, in which an exposure at one time may make a person abnormally sensitive, or allergic, at a later time.

As scientists fill gaps in our knowledge, policy-makers often face dilemmas. The introduction of chlorine as a disinfectant in drinking water treatment is one of the great success stories in the history of public health, credited with eradicating typhus and other water-borne diseases in many parts of the world—but recent research shows that chlorine and other disinfectants can react with natural organic matter in water to form carcinogens. When the Clean Air Act Amendments of 1990 accelerated the phasing out of lead in gasoline, air quality

Figure 15.2 Controlled experiments on rabbits and other animals help expand our knowledge of the properties of toxic chemicals, but they create an ethical dilemma: does the knowledge gained to protect the health of humans and wildlife justify inflicting deliberate harm on animals in the laboratory? Courtesy of the British Union for the Abolition of Vivisection.

improved in many parts of the United States—but it now appears that the most popular lead substitute, a highly soluble petroleum-derived chemical known as MTBE, is nearly ubiquitous in the nation's ground water.

Toxins are a window on our hubris. We create new chemicals with particular purposes in mind, and we think we are their masters, but they obey their own laws. We are surprised when hand-washing does not prevent lead poisoning, when an insect poison kills birds, when a drug intended for the mother crosses the placenta and affects the fetus. We should not be surprised at any of these things, of course, but people at first resisted believing each of them. Gregory Bateson, in *Steps to an Ecology of Mind*, identifies this partial blindness to unintended consequences as an essential quality of the human mind, one we will never master, but one that we must recognize and learn caution from if we are to survive on the planet.

Today, policy-makers in many countries are adopting the precautionary principle, which places the burden on the manufacturer of a new chemical to demonstrate that it is not dangerous before introducing it in mass quantities into the workplace, the marketplace, and the environment. Despite such promising developments, the sheer number, variety, and complexity of chemicals already in the environment make it likely that toxicologists and ecotoxicologists will be studying the chemical soup we have already cooked up for generations to come.

FURTHER READING

Ames, Bruce N., and Lois Swirsky Gold. "Paracelsus to Parascience: The Environmental Cancer Distraction." *Mutation Research* 447 (2000): 3–13.
Colborn, Theo, Dianne Dumanoski, and John Peterson Myers. *Our Stolen Future.* New York: Plume, 1997.
Klaasen, Curtis, Mary Amdur, and John Doull, ed. *Casarett and Doull's Toxicology: The Basic Science of Poisons.* 5th ed. New York: McGraw-Hill, 1996.
Steingraber, Sandra. *Living Downstream: An Ecologist Looks at Cancer and the Environment.* Reading, Mass.: Perseus, 1997.
Wargo, John. *Our Children's Toxic Legacy: How Science and Law Fail to Protect Us from Pesticides*, 2nd ed. New Haven: Yale University Press, 1998.

Alice Hamilton, from *Exploring the Dangerous Trades* (1943)

From the earliest days of mining and metalwork, exposure to toxic chemicals has occurred in the workplace. As workplace conditions evolve, toxicology, humanitarian aid, and government regulation struggle to keep pace. In the nineteenth century, milliners risked becoming "mad as hatters" because the felt they handled was laced with mercury. Today, office workers succumb to "sick building syndrome," a chemical sensitization caused by a combination of poor ventilation and the "off-gassing" of volatile chemicals like formaldehyde from carpeting, upholstery, plastics, and electronics. This selection is from the autobiography of Alice Hamilton, a founding figure of toxicology and workplace safety in the United States. The work she began in 1910 is carried on today by the Occupational Safety and Health Administration (OSHA), created in 1970.

Thirty-two years ago, in 1910, I went as a pioneer into a new, unexplored field of American medicine, the field of industrial disease. . . .

It was while I was living in Hull-House [in Chicago] and working in bacteriological research that the opportunity came to me to investigate the dangerous trades of Illinois[1]—not those where violent accidents occurred, but those with the less spectacular hazard of sickness from some industrial poison. It was a voyage of exploration that we undertook, our little group of physicians and student assistants, for nobody in Illinois knew even where we should make our investigations, beyond a few notorious lead trades. American medical authorities had never taken industrial diseases seriously, the American Medical Associations had never held a meeting on the subject, and while European journals were full of articles on industrial poisoning, the number published in American medical journals up to 1910 could be counted on one's fingers.

Figure 15.3 Alice Hamilton. Courtesy of the Schlesinger Library, Radcliffe Institute, Harvard University.

. . . No young doctor nowadays can hope for work as exciting and rewarding. Everything I discovered was new and most of it was really valuable. I knew nothing of manufacturing processes, but I learned them on the spot, and before long every detail of the Old Dutch Process and the Carter Process of white-lead production was familiar to me, also the roasting of red lead and litharge and the smelting of lead ore and refining of lead scrap. From the first I became convinced that what I must look for was lead dust and lead fumes, that men were poisoned by breathing poisoned air, not by handling their food with unwashed hands. Nowadays that fact has been so strongly established by experimental proof that nobody would think of disputing it. But in 1910 and for many years after, the firm (and comforting) belief of foremen and employers was that if a man was poisoned by lead it was because he did not wash his hands and scrub his nails, although a little intelligent observation would have been enough to show its absurdity.

This fact, that lead poisoning is brought about far more rapidly and intensely by the breathing of lead-laden air than by the swallowing of lead, is of the greatest practical importance. There can be no intelligent control of the lead danger in industry unless it is based on the principle of keeping the air clear from dust and fumes. The English authority, Sir Thomas Legge, after some thirty years' experience in the prevention of industrial disease, reached the conclusion that the air is the only important source of occupational lead poisoning and that the only efficient measures for its prevention are those directed toward the prevention of dust and fumes. A hundred years ago Tanquerel des Planches, who is called the Columbus of lead poisoning, noted that severe plumbism [lead poisoning] never followed the handling of solid lead but only exposure to dust and "emanations." [Hamilton and other experts opposed the introduction of lead into automotive fuel—and automotive exhaust—in the 1920s, when lead was one of several competing "anti-knock" additives. Leaded gasoline nevertheless became the standard and remained so until a phase-out began in the 1970s in the United States and the 1990s in Europe.]

Lead is the oldest of the industrial poisons except carbon monoxide, which must have begun to take its toll soon after Prometheus made the gift of fire to man. In Roman days lead poisoning was known, for Pliny the Elder includes it among the "diseases of slaves," which were potters' and knife grinders' phthisis, lead, and mercurial poisoning. Throughout all the centuries since then men have used this valuable metal in many ways, and from time to time an observant physician has seen the results and described them, notably Ramazzini in the eighteenth century, and early in the nineteenth century the great Frenchman, Tanquerel des Planches. It is a poison which can act in many different ways, some of them so unusual and outside the experience of the ordinary physician that he fails to recognize the cause. I could never feel that I had uncovered all the cases in any community, no matter how small, even after I had talked with all the doctors and gone through the hospital records, for some doctors would not pronounce a case to be due to lead poisoning unless there was either colic or palsy, which is as if he refused to recognize alcoholism unless there were an attack of delirium tremens.

It is true that a severe attack of colic is the most characteristic symptom of lead poisoning, and palsy—usually in the form of wristdrop—is the one most easily recognized, but there are many other manifestations of this protean malady, as every physician knows today. Thirty years ago it was not hard to find extremely severe forms, such as could come only from an exposure so great as to seem criminal to us now, but which then attracted no attention. Here are four histories, picked at random, from my notes of 1910.

A Bohemian, an enameler of bathtubs, had worked eighteen months at his trade, without apparently becoming poisoned, though his health had suffered. One day, while at the furnace, he fainted away and for four days he lay in coma, then passed into delirium during which it was found that both forearms and both ankles were palsied. He made a partial recovery during the following six months but when he left for his home in Bohemia he was still partly paralyzed.

A Hungarian, thirty-six years old, worked for seven years grinding lead paint. During this time he had three attacks of colic, with vomiting and headache. I saw him in the hospital, a skeleton of a man, looking almost twice his age, his limbs soft and flabby, his muscles wasted. He was extremely emaciated, his color was a dirty grayish yellow, his eyes dull and expressionless. He lay in an apathetic condition, rousing when spoken to and answering rationally but slowly, with often an appreciable delay, then sinking back into apathy.

A Polish laborer worked only three weeks in a very dusty white-lead plant at an unusually dusty emergency job, at the end of which he was sent to the hospital with severe lead colic and palsy of both wrists.

A young Italian, who spoke no English, worked for a month in a white-lead plant but without any idea that the harmless-looking stuff was poisonous. There was a great deal of dust in his work. One day he was seized with an agonizing pain in his head which came on him so suddenly that he fell to the ground. He was sent to the hospital, semiconscious, with convulsive attacks, and was there for two weeks; when he came home, he had a relapse and had to go back to the hospital. Three months later he was still in poor health and could not do a full day's work.

Every article I wrote in those days, every speech I made, is full of pleading for the recognition of lead poisoning as a real and serious medical problem. It was easy to present figures demonstrating the contrast between lead work in the United States under conditions of neglect and ignorance, and comparable work in England and Germany, under intelligent control. For instance, when I went to England in 1910 I found that a factory which produced white and red lead, employing ninety men, had not had a case of lead poisoning in five successive years. And I compared it with one in the United States, employing eighty-five men, where the doctor's records showed thirty-five men "leaded" in six months. . . .

Life at Hull-House had accustomed me to going straight to the homes of people about whom I wished to learn something and talking to them in their own surroundings, where they have courage to speak out what is in their minds. [The same technique yielded notable success to Erin Brockovich also, eighty years later.] They were almost always foreigners, Bulgarians, Serbs, Poles, Italians, Hungarians, who had come to this country in the search for a better life for themselves and their children. Sometimes they thought they had found it, then when sickness struck down the father things grew very black and there were no old friends and neighbors and cousins to fall back on as there had been in the old country. . . .

It sometimes seemed to me that industry was exploiting the finest and best in these men—their love of their children, their sense of family responsibility. I think of an enameler of bathtubs whom I traced to his squalid little cottage. He was a young Slav who used to be so strong he could run up the hill on which his cottage stood and spend all the evening digging in his garden. Now, he told me, he climbed up like an old man and sank exhausted in a chair, he was so weary, and if he tried to hoe or rake he had to give it up. His digestion had failed, he had a foul mouth, he couldn't eat, he had lost much weight. He had had many attacks of colic and the doctor told him if he did not quit he would soon be a wreck. "Why did you keep on," I asked, "when you knew the lead was getting you?"

"Well, there were the payments on the house," he said, "and the two kids." The house was a bare, ugly, frame shack, the children were little, underfed things, badly in need of a handkerchief, but for them a man had sacrificed his health and his joy in life. When employers tell me they prefer married men, and encourage their men to have homes of their own, because it makes them so much steadier, I wonder if they have any idea of all that that implies. . . .

What was the prevalence of lead poisoning among enamel grinders and enamelers? That was a question I had the opportunity to explore before I left Chicago on my Federal survey. A strike was declared in a large sanitary-ware enameling works and it occurred to me that here was a chance to meet the usual working force, not only the invalided men, and to see if any of them were leaded. I went to A. F. of L. [American Federation of Labor] headquarters and there learned that the strikers were meeting in a Polish saloon and John Fitzpatrick was trying to organize them. He was willing to take me along, so while he harangued the men in the front room I interviewed them one by one in the back room.

Today the making of a diagnosis of lead poisoning is an elaborate affair, including examination of the blood, both chemical and microscopic, quantitative chemical tests of excreta, delicate tests for nerve response, as well as the usual physical examination. None of such aids were available to me, my methods were as crude as those of Tanquerel des Planches. I adopted a rigid standard. I would not accept a case as positive unless there were a clear "lead line" in addition to a typical history, or, if there were no line, a diagnosis of lead poisoning made by a doctor on the occasion of an acute attack. The "lead line" nowadays is very rarely seen. It is a deposit of black lead sulphide in the cells of the lining of the mouth, usually clearest on the gum along the margin of the front teeth, and it is caused by the action of sulphureted hydrogen on the lead in these cells, the sulphureted hydrogen coming from decaying protein food in the mouth. Of course if a lead worker has a clean, healthy mouth, he will not have a lead line, and I am sure I have not seen one in over ten years, but in 1910 I looked on it as a common sign in lead workers.

Among the 148 men I examined—all Slavs, many of them powerfully built peasants—there were 54 whom I accepted as leaded but I refrained from adding 38 others who in these days would be classed as "probable cases." It would be simply impossible to find nowadays in this country a group of lead workers revealing such a condition, and after an exposure averaging less than six years. Fifty-four out of 148 is more than one third.

Because enameling tubs was notoriously hard, hot, and dangerous, most American men shunned it and I found in Pittsburgh and the surrounding towns, in Trenton and in Chicago, foreign-born workmen, Russians, Bohemians, Slovaks, Croatians, Poles. I remember a foreman saying to me, as we watched the enamelers at work, "They don't last long at it. Four years at the most, I should say, then they quit and go home to the Old Country." "To die?" I asked. "Well, I suppose that is about the size of it," he answered. It was not the lead, as I discovered then, that did the greatest harm, but the silica dust. Many of the doctors I talked to told me that there was where my attention should be turned, to the pulmonary consumption; lead poisoning was a minor evil. But lead dust was my job, not silica.

The lead dust was bad. In the enameling rooms of the plants, this would be the picture. In front of the great furnaces stood the enameler and his helper. The door swung open and, with the aid of a mechanism which required strength to operate, a redhot bathtub was lifted out. The enameler then dredged as quickly as possible powdered enamel over the hot surface, where it melted and flowed to form an even coating. His helper stood beside him working the turntable on which the tub stood so as to present all its inner surface to the enameler. The dredge was big and so heavy that part of its weight had to be taken by a chain from the roof. The men during this procedure were in a thick cloud of enamel dust, and were breathing rapidly and deeply because of the exertion and the extreme heat. I found that I could not stand the heat any nearer than twelve feet but the workmen had to come much closer. They protected their faces and eyes by various devices, a light tin pan with eyeholes and a hoop to go around the head, or a piece of wood with eyeholes and a stick nailed at a right angle so that it could be held between the teeth. . . .

Seventeen years later, in 1929, I made a second survey of this industry, this time with special regard to the dust hazard. The risk of lead poisoning had been moderated; silicosis had by then become the most important of the industrial diseases and I knew that in making porcelain-enameled sanitary ware not only the enamelers but many other workmen were exposed to silica dust. I found the work of the enamelers not nearly so bad as it had been. In the first place no plant was using as much lead as it had in 1912 and three out of ten used none at all for ordinary enamel. Then there was far less dust both in the grinding room and in enameling. . . .

But the department that interested me most in this second study was the one where the ironware is prepared for coating with enamel. As the tubs come from the foundry their surface is smooth and enamel will not stick to it, so in order to roughen it a blast of fine sand is turned on and the million particles of sand make millions of tiny dents in the metal so that it is frosted all over. This, every industrial physician knew to be one of the most dangerous jobs in all industry and the National Safety Council had appointed a committee, of which I was a member, to study sandblasting and suggest how the danger might be controlled. So I had seen a good deal of this process in other industries but never anything approaching what I saw in sanitary-ware manufacture.

Imagine a great room filled with men cleaning mold sand off the tubs from the foundry. At one end through a thick fog of dust can dimly be seen eight lamplit little rooms open to the big room, in each of which a grotesque figure is manipulating a sandblast which, with a deafening roar, shoots sand at something, one cannot see what. Great clouds of sand come eddying out into the room and filling the air the cleaners must breathe. Some attempt is made to protect the sandblaster, none at all to protect the cleaner.

This was the worst plant I saw, but three others were almost as bad, all in states where no compensation law for occupational disease existed. The others were all much less dangerous and one was excellent in every respect. This was the Kohler plant in Sheboygan, Wisconsin, or rather in the charming village of Kohler. I remembered it from 1912, for Walter Kohler was the only employer I met then who was seriously concerned with the problems of lead poisoning and of dust. In the intervening years he had done away with the lead and now he had brought the dust under control. In some of the other plants I saw excellent conditions, but Kohler's stood at the head in 1929. What Mr. Kohler had done was to make the sandblast chambers dust-proof, so that no man outside was endangered, and to provide the sandblaster with pure, dust-free air, fed to him through a pipe which led to his respirator. And of course a physician kept close watch of all these men.

Enameling sanitary ware is still a dangerous trade, for only the greatest precautions can prevent silicosis in sandblasters and enamelers, and even if there is no lead in the enamel the excessive heat is harmful. But compared with the situation in 1912 the trade has made enormous progress, aided, no doubt, by the passage in 1939 of a law in Pennsylvania awarding compensation for occupational disease.

Note

1. A State Commission was appointed by Governor Deneen in 1910 to report on "occupational diseases in Illinois."

Rachel Carson, from "And No Birds Sing" in *Silent Spring* (1962)

Rachel Carson's Silent Spring *was one of those rare books, like* Uncle Tom's Cabin *and* The Jungle, *that capture the imagination of an era and catalyze history-making events and reforms. Supreme Court Justice William O. Douglas called* Silent Spring *"the most important chronicle of this century for the human race." Carson's analysis of the effects of indiscriminate pesticide use — especially DDT — on wildlife in the United States was initially dismissed by some high-placed scientists as female hysteria, but subsequent investigations confirmed that her science was sound and her concerns well-founded. Carson's book, along with contemporaneous events like the revelation of birth defects caused by the widely prescribed drug thalidomide and the ongoing debate over nuclear power, shook Americans' faith in the forward march of technological progress and in the wisdom and beneficence of government and industry "experts." In the years after* Silent Spring, *the modern environmental movement was born; also, the U.S. Environmental Protection Agency was founded, and the first pesticide it banned was DDT.*

The survival of the robin, and indeed of many other species as well, seems fatefully linked with the American elm, a tree that is part of the history of thousands of towns from the Atlantic to the Rockies, gracing their streets and their village squares and college campuses with majestic archways of green. Now the elms are stricken with a disease that afflicts them throughout their range, a disease so serious that many experts believe all efforts to save the elms will in the end be futile. It would be tragic to lose the elms, but it would be doubly tragic if, in vain efforts to save them, we plunge vast segments of our bird populations into the night of extinction. Yet this is precisely what is threatened.

The so-called Dutch elm disease entered the United States from Europe about 1930 in elm burl logs imported for the veneer industry. It is a fungus disease; the organism invades the water-conducting vessels of the tree, spreads by spores carried in the flow of sap, and by its poisonous secretions as well as by mechanical clogging causes the branches to wilt and the tree to die. The disease is spread from diseased to healthy trees by elm bark beetles. The galleries which the insects have tunneled out under the bark of dead trees become contaminated with spores of the invading fungus, and the spores adhere to the insect body and are carried wherever the beetle flies. Efforts to control the fungus disease of the elms have been directed largely toward control of the carrier insect. In community after community, especially throughout the strongholds of the American elm, the Midwest and New England, intensive spraying has become a routine procedure.

What this spraying could mean to bird life, and especially to the robin, was first made clear by the work of two ornithologists at Michigan State University, Professor George Wallace and one of his graduate students, John Mehner. When Mr. Mehner began work for the doctorate in 1954, he chose a research project that had to do with robin populations. This was quite by chance, for at that time no one suspected that the robins were in danger. But even as he undertook the work, events occurred that were to change its character and indeed to deprive him of his material.

Spraying for Dutch elm disease began in a small way on the university campus in 1954. The following year the city of East Lansing (where the university is located) joined in, spraying on the campus was expanded, and, with local programs for gypsy moth and mosquito control also under way, the rain of chemicals increased to a downpour.

During 1954, the year of the first light spraying, all seemed well. The following spring the migrating robins began to return to the campus as usual. Like the bluebells in Tomlinson's haunting essay "The Lost Wood," they were "expecting no evil" as they reoccupied their familiar territories. But soon it became evident that something was wrong. Dead and dying robins began to appear on the campus. Few birds were seen in their normal foraging activities or assembling in their usual roosts. Few nests were built; few young appeared. The pattern was repeated with monotonous regularity in succeeding springs. The sprayed area had become a lethal trap in which each wave of migrating robins would be eliminated in about a week. Then new arrivals would come in, only to add to the numbers of doomed birds seen on the campus in the agonized tremors that precede death.

"The campus is serving as a graveyard for most of the robins that attempt to take up residence in the spring," said Dr. Wallace. But why? At first he suspected some disease of the nervous system, but soon it became evident that "in spite of the assurances of the insecticide people that their sprays were 'harmless to birds' the robins were really dying of insecticidal poisoning; they exhibited the well-known symptoms of loss of balance, followed by tremors, convulsions, and death."

Several facts suggested that the robins were being poisoned, not so much by direct contact with the insecticides as indirectly, by eating earthworms. Campus earthworms had been fed inadvertently to crayfish in a research project and all the crayfish had promptly died. A snake kept in a laboratory cage had gone into violent tremors after being fed such worms. And earthworms are the principal food of robins in the spring.

A key piece in the jigsaw puzzle of the doomed robins was soon to be supplied by Dr. Roy Barker of the Illinois Natural History Survey at Urbana. Dr. Barker's work, published in 1958, traced the intricate cycle of events by which the robins' fate is linked to the elm trees by way of the earthworms. The trees are sprayed in the spring (usually at the rate of two to five pounds of DDT per 50-foot tree, which may be the equivalent of as much as *23 pounds per acre* where elms are numerous) and often again in July, at about half this concentration. Powerful sprayers direct a stream of poison to all parts of the tallest trees, killing directly not only the target organism, the bark beetle, but other insects, including pollinating species and predatory spiders and beetles. The poison forms a tenacious film over the leaves and bark. Rains do not wash it away. In the autumn the leaves fall to the ground, accumulate in sodden layers, and begin the slow process of becoming one with the soil. In this they are aided by the toil of the earthworms, who feed in the leaf litter, for elm leaves are among their favorite foods. In feeding on the leaves the worms also swallow the insecticide, accumulating and concentrating it in their bodies. Dr. Barker found deposits of DDT throughout the digestive tracts of the worms, their blood vessels, nerves, and body wall. Undoubtedly some of the earthworms themselves succumb, but others survive to become "biological magnifiers" of the poison. In the spring the robins return to provide another link in the cycle. As few as 11 large earthworms can transfer a lethal dose of DDT to a robin. And 11 worms form a small part of a day's rations to a bird that eats 10 to 12 earthworms in as many minutes.

Not all robins receive a lethal dose, but another consequence may lead to the extinction of their kind as surely as fatal poisoning. The shadow of sterility lies over all the bird studies and indeed lengthens to include all living things within its potential range. There are now only two or three dozen robins to be found each spring on the entire 185-acre campus of Michigan State University, compared with a conservatively estimated 370 adults in this area before spraying. In 1954 every robin nest under observation by Mehner produced young. Toward the end of June, 1957, when at least 370 young birds (the normal replacement of the adult population) would have been foraging over the campus in the years before spraying began, Mehner could find *only one young robin.* A year later Dr. Wallace was to report: "At no time during the spring or summer [of 1958] did I see a fledgling robin anywhere on the main campus, and so far I have failed to find anyone else who has seen one there."

Part of this failure to produce young is due, of course, to the fact that one or more of a pair of robins dies before the nesting cycle is completed. But Wallace has significant records which point to something more sinister—the actual destruction of the birds' capacity to reproduce. He has, for example, "records of robins and other birds building nests but laying no eggs, and others laying eggs and incubating them but not hatching them. We have one record of a robin that sat on its eggs faithfully for 21 days and they did not hatch. The normal incubation period is 13 days. . . . Our analyses are showing high concentrations of DDT in the testes and ovaries of breeding birds," he told a congressional committee in 1960. "Ten males had amounts ranging from 30 to 109 parts per million in the testes, and two females had 151 and 211 parts per million respectively in the egg follicles in their ovaries."

Soon studies in other areas began to develop findings equally dismal. Professor Joseph Hickey and his students at the University of Wisconsin, after careful comparative studies of sprayed and unsprayed areas, reported the robin mortality to be at least 86 to 88 percent. The Cranbrook Institute of Science at Bloomfield Hills, Michigan, in an effort to assess the extent of bird loss caused by the spraying of the elms, asked in 1956 that all birds thought to be victims of DDT poisoning be turned in to the institute for examination. The request had a response beyond all expectations. Within a few weeks the deep-freeze facilities of the institute were taxed to capacity, so that other specimens had to be refused. By 1959 a thousand poisoned birds from this single community had been turned in or reported. Although the robin was the chief victim (one woman calling the institute reported 12 robins lying dead on her lawn as she spoke), 63 different species were included among the specimens examined at the institute.

The robins, then, are only one part of the chain of devastation linked to the spraying of the elms, even as the elm program is only one of the multitudinous spray programs that cover our land with poisons. Heavy mortality has occurred among about 90 species of birds, including those most familiar to suburbanites and amateur naturalists. The populations of nesting birds in general have declined as much as 90 percent in some of the sprayed towns. As we shall see, all the various types of birds are affected—ground feeders, treetop feeders, bark feeders, predators.

It is only reasonable to suppose that all birds and mammals heavily dependent on earthworms or other soil organisms for food are threatened by the robins' fate. Some 45 species of birds include earthworms in their diet. Among them is the woodcock, a species that winters in southern areas recently heavily sprayed with heptachlor. Two significant discoveries have now been made about the woodcock. Production of young birds on the New Brunswick breeding grounds is definitely reduced, and adult birds that have been analyzed contain large residues of DDT and heptachlor.

Already there are disturbing records of heavy mortality among more than 20 other species of ground-feeding birds whose food—worms, ants, grubs, or other soil organisms—has been poisoned. These include three of the thrushes whose songs are among the most exquisite of bird voices, the olive-backed, the wood, and the hermit. And the sparrows that flit through the shrubby understory of the woodlands and forage with rustling sounds amid the fallen leaves—the song sparrow and the white-throat—these, too, have been found among the victims of the elm sprays.

Mammals, also, may easily be involved in the cycle, directly or indirectly. Earthworms are important among the various foods of the raccoon, and are eaten in the spring and fall by opossums. Such subterranean tunnelers as shrews and moles capture them in some numbers, and then perhaps pass on the poison to predators such as screech owls and barn owls. Several dying screech owls were picked up in Wisconsin following heavy rains in spring, perhaps poisoned by feeding on earthworms. Hawks and owls have been found in convulsions—great horned owls, screech owls, red-shouldered hawks, sparrow hawks, marsh hawks. These may be cases of secondary poisoning, caused by eating birds or mice that have accumulated insecticides in their livers or other organs.

Nor is it only the creatures that forage on the ground or those who prey on them that are endangered by the foliar [leaf] spraying of the elms. All of the treetop feeders, the birds that glean their insect food from the leaves, have disappeared from heavily sprayed areas, among them those woodland sprites the kinglets, both ruby-crowned and golden-crowned, the tiny gnatcatchers, and many of the warblers, whose migrating hordes flow through the trees in spring in a multicolored tide of life. In 1956, a late spring delayed spraying so that it coincided with the arrival of an exceptionally heavy wave of warbler migration. Nearly all species of warblers present in the area were represented in the heavy kill that followed. In Whitefish Bay, Wisconsin, at least a thousand myrtle warblers could be seen in migration during former years; in 1958, after the spraying of the elms, observers could find only two. So, with additions from other communities, the list grows, and the warblers killed by the spray include those that most charm and fascinate all who are aware of them: the black-and-white, the yellow, the magnolia, and the Cape May; the ovenbird, whose call throbs in the Maytime woods; the Blackburnian, whose wings are touched with flame; the chestnut-sided, the Canadian, and the black-throated green. These treetop feeders are affected either directly by eating poisoned insects or indirectly by a shortage of food.

The loss of food has also struck hard at the swallows that cruise the skies, straining out the aerial insects as herring strain the plankton of the sea. A Wisconsin naturalist reported: "Swallows have been hard hit. Everyone complains of how few they have compared to four or five years ago. Our sky overhead was full of them only four years ago. Now we seldom see any. . . . This could be both lack of insects because of spray, or poisoned insects."

Of other birds this same observer wrote: "Another striking loss is the phoebe. Flycatchers are scarce everywhere but the early hardy common phoebe is no more. I've seen one this spring and only one last spring. Other birders in Wisconsin make the same complaint. I have had five or six pair of cardinals in the past, none now. Wrens, robins, catbirds, and screech owls have nested each year in our garden. There are none now. Summer mornings are without bird song. Only pest birds, pigeons, starlings, and English sparrows remain. It is tragic and I can't bear it."

The dormant sprays applied to the elms in the fall, sending the poison into every little crevice in the bark, are probably responsible for the severe reduction observed in the number of chickadees, nuthatches, titmice, woodpeckers, and brown creepers. During the winter of 1957–58, Dr. Wallace saw no chickadees or nuthatches at his home feeding station for the first time in many years. Three nuthatches he found later provided a sorry little step-by-step lesson in cause and effect: one was feeding on an elm, another was found dying of typical DDT symptoms, the third was dead. The dying nuthatch was later found to have 226 parts per million of DDT in its tissues.

The feeding habits of all these birds not only make them especially vulnerable to insect sprays but also make their loss a deplorable one for economic as well as less tangible reasons. The summer food of the white-breasted nuthatch and the brown creeper, for example, includes the eggs, larvae, and adults of a very large number of insects injurious to trees. About three quarters of the food of the chickadee is animal, including all stages of the life cycle of many insects. The chickadee's method of feeding is described in Bent's monumental *Life Histories* of North American birds: "As the flock moves along each bird examines minutely bark, twigs, and branches, searching for tiny bits of food (spiders' eggs, cocoons, or other dormant insect life)."

Various scientific studies have established the critical role of birds in insect control in various situations. Thus, woodpeckers are the primary control of the Engelmann spruce beetle, reducing its populations from 45 to 98 percent and are important in the control of the codling moth in apple orchards. Chickadees and other winter-resident birds can protect orchards against the cankerworm.

But what happens in nature is not allowed to happen in the modern, chemical-drenched world, where spraying destroys not only the insects but their principal enemy, the birds. When later there is a resurgence of the insect population, as almost always happens, the birds are not there to keep their numbers in check. As the Curator of Birds at the Milwaukee Public Museum, Owen J. Gromme, wrote to the *Milwaukee Journal:* "The greatest enemy of insect life is other predatory insects, birds, and some small mammals, but DDT kills indiscriminately, including nature's own safeguards or policemen. . . . In the name of progress are we to become victims of our own diabolical means of insect control to provide temporary comfort, only to lose out to destroying insects later on? By what means will we control new pests, which will attack remaining tree species after the elms are gone, when nature's safeguards (the birds) have been wiped out by poison?"

Mr. Gromme reported that calls and letters about dead and dying birds had been increasing steadily during the years since spraying began in Wisconsin. Questioning always revealed that spraying or fogging had been done in the area where the birds were dying.

Mr. Gromme's experience has been shared by ornithologists and conservationists at most of the research centers of the Midwest such as the Cranbrook Institute in Michigan, the Illinois Natural History Survey, and the University of Wisconsin. A glance at the Letters-from-Readers column of newspapers almost anywhere that spraying is being done makes clear the fact that citizens are not only becoming aroused and indignant but that often they show a keener understanding of the dangers and inconsistencies of spraying than do the officials who order it done. "I am dreading the days to come soon now when many beautiful birds will be dying in our back yard," wrote a Milwaukee woman. "This is a pitiful, heartbreaking experience. . . . It is, moreover, frustrating and exasperating, for it evidently does not serve the purpose this slaughter was intended to serve. . . . Taking a long look, can you save trees without also saving birds? Do they not, in the economy of nature, save each other? Isn't it possible to help the balance of nature without destroying it?"

The idea that the elms, majestic shade trees though they are, are not "sacred cows" and do not justify an "open end" campaign of destruction against all other forms of life is expressed in other letters. "I have always loved our elm trees which seemed like trademarks on our landscape," wrote another Wisconsin woman. "But there are many kinds of trees. . . . We must save our birds, too. Can anyone imagine anything so cheerless and dreary as a springtime without a robin's song?"

To the public the choice may easily appear to be one of stark black-or-white simplicity: Shall we have birds or shall we have elms? But it is not as simple as that, and by one of the ironies that abound throughout the field of chemical control we may very well end by having neither if we continue on our present, well-traveled road. Spraying is killing the birds but it is not saving the elms. The illusion that salvation of the elms lies at the end of a spray nozzle is a dangerous will-o'-the-wisp that is leading one community after another into a morass of heavy expenditures, without producing lasting results. Greenwich, Connecticut, sprayed regularly for ten years. Then a drought year brought conditions especially favorable to the beetle and the mortality of elms went up 1000 percent. In Urbana, Illinois, where the University of Illinois is located, Dutch elm disease first appeared in 1951. Spraying was undertaken in 1953. By 1959, in spite of six years' spraying, the university campus had lost 86 percent of its elms, half of them victims of Dutch elm disease.

In Toledo, Ohio, a similar experience caused the Superintendent of Forestry, Joseph A. Sweeney, to take a realistic look at the results of spraying. Spraying was begun there in 1953 and continued through 1959. Meanwhile, however, Mr. Sweeney had noticed that a city-wide infestation of the cottony maple scale was worse after the spraying recommended by "the books and the authorities" than it had been before. He decided to review the results of spraying for Dutch elm disease for himself. His findings shocked him. In the city of Toledo, he found, "the only areas under any control were the areas where we used some promptness in removing the diseased or brood trees. Where we depended on spraying the disease was out of control. In the country where nothing has been done the disease has not spread as fast as it has in the city. This indicates that spraying destroys any natural enemies.

"We are abandoning spraying for the Dutch elm disease. This has brought me into conflict with the people who back any recommendations by the United States Department of Agriculture but I have the facts and will stick with them."

It is difficult to understand why these midwestern towns, to which the elm disease spread only rather recently, have so unquestioningly embarked on ambitious and expensive spraying programs, apparently without waiting to inquire into the experience of other areas that have had longer acquaintance with the problem. New York State, for example, has certainly had the longest history of continuous experience with Dutch elm disease, for it was via the Port of New York that diseased elm wood is thought to have entered the United States about 1930. And New York State today has a most impressive record of containing and suppressing the disease. Yet it has not relied upon spraying. In fact, its agricultural extension service does not recommend spraying as a community method of control.

How, then, has New York achieved its fine record? From the early years of the battle for the elms to the present time, it has relied upon rigorous sanitation, or the prompt removal and destruction of all diseased or infected wood. In the beginning some of the results were disappointing, but this was because it was not at first understood that not only diseased trees but all elm wood in which the beetles might breed must be destroyed. Infected elm wood, after being cut and stored for firewood, will release a crop of fungus-carrying beetles unless burned before spring. It is the adult beetles, emerging from hibernation to feed in late April and May, that transmit Dutch elm disease. New York entomologists have learned by experience what kinds of beetle-breeding material have real importance in the spread of the disease. By concentrating on this dangerous material, it has been possible not only to get good results, but to keep the cost of the sanitation program within reasonable limits. By 1950 the incidence of Dutch elm disease in New York City had been reduced to 2/10 of 1 percent of the city's

Figure 15.4 *Dead Blackbird,* after Albrecht Dürer. © The British Museum.

55,000 elms. A sanitation program was launched in Westchester County in 1942. During the next 14 years the average annual loss of elms was only 2/10 of 1 percent a year. Buffalo, with 185,000 elms, has an excellent record of containing the disease by sanitation, with recent annual losses amounting to only 3/10 of 1 percent. In other words, at this rate of loss it would take about 300 years to eliminate Buffalo's elms.

What has happened in Syracuse is especially impressive. There no effective program was in operation before 1957. Between 1951 and 1956 Syracuse lost nearly 3,000 elms. Then, under the direction of Howard C. Miller of the New York State University College of Forestry, an intensive drive was made to remove all diseased elm trees and all possible sources of beetle-breeding elm wood. The rate of loss is now well below 1 percent a year.

Tina Rosenberg, from "What the World Needs Now Is DDT" (2004)

Toxins in the environment pose a clear danger to the health of humans and wildlife. But, like nuclear power, chemicals sometimes inspire public dread out of proportion to the actual risks. As Rosenberg reports, there are places and situations where DDT is the most suitable anti-malarial measure—but the stigma attached to the chemical since Silent Spring *was published prevents its effective use.*

The year 2000 was a time of plague for the South African town of Ndumo, on the border of Mozambique. That March, while the world was focused on AIDS, more than 7,000 people came to the local health clinic with malaria. The South African Defense Force was called in, and soldiers set up tents outside the clinic to treat the sick. At the district hospital thirty miles away in Mosvold, the wards filled with patients suffering with the headache, weakness, and fever of malaria—2,303 patients that month. "I thought we were going to get buried in malaria," said Hervey Vaughan Williams, the hospital's medical manager.

Today, malaria has all but vanished in Ndumo. In March 2003, the clinic treated nine malaria cases; Mosvold Hospital, only three.

As malaria surges once again in Africa, victories are few. But South Africa is beating the disease with a simple remedy: spraying the inside walls of houses in affected regions once a year. Several insecticides can be used, but South Africa has chosen the most effective one. It lasts twice as long as the alternatives. It repels mosquitoes in addition to killing them, which delays the onset of pesticide-resistance. It costs a quarter as much as the next cheapest insecticide. It is DDT. . . .

To Americans, DDT is simply a killer. Ask Americans over forty to name the most dangerous chemical they know, and chances are that they will say DDT. Dichloro-diphenyl-trichloroethane was banned in the United States in 1972. The chemical was once sprayed in huge quantities over cities and fields of cotton and other crops. Its persistence in the ecosystem, where it builds up to kill birds and fish, has become a symbol of the dangers of playing God with nature, an icon of human arrogance. Countries throughout the world have signed a treaty promising to phase out its use.

Yet what really merits outrage about DDT today is not that South Africa still uses it, as do about five other countries for routine malaria control and about ten more for emergencies. It is that dozens more do not. Malaria is a disease Westerners no longer have to think about. Independent malariologists believe it kills two million people a year, mainly children under five and 90 percent of them in Africa. Until it was overtaken by AIDS in 1999, it was Africa's leading killer. One in twenty African children dies of malaria, and many of those who survive are brain-damaged. Each year, 300 to 500 million people worldwide get malaria. During the rainy season in some parts of Africa, entire villages of people lie in bed, shivering with fever, too weak to stand or eat. Many spend a good part of the year incapacitated, which cripples African economies. A commission of the World Health Organization found that malaria alone shrinks the economy in countries where it is most endemic by 20 percent over fifteen years. There is currently no vaccine. While travelers to malarial regions can take prophylactic medicines, these drugs are too toxic for long-term use for residents.

Yet DDT, the very insecticide that eradicated malaria in developed nations, has been essentially deactivated as a malaria-control tool today. The paradox is that sprayed in tiny quantities inside houses—the only way anyone proposes to use it today—DDT is most likely not harmful to people or the environment. Certainly, the possible harm from DDT is vastly outweighed by its ability to save children's lives.

No one concerned about the environmental damage of DDT set out to kill African children. But various factors, chiefly the persistence of DDT's toxic image in the West and the disproportionate weight that American decisions carry worldwide, have conspired to make it essentially unavailable to most malarial nations. With the exception of South Africa and a few others, African countries depend heavily on donors to pay for malaria control. But at the moment, there is only one country in the world getting donor money to finance the use of DDT: Eritrea, which gets money for its program from the World Bank with the understanding that it will look for alternatives. Major donors, including the United States Agency for International Development, or USAID, have not financed any use of DDT, and global health institutions like WHO and its malaria program, Roll Back Malaria, actively discourage countries from using it.

. . . "For us to be buying and using in another country something we don't allow in our own country raises the specter of preferential treatment," said E. Anne Peterson, the assistant administrator for global health at USAID. "We certainly have to think about 'What would the American people think and want?' and 'What would Africans think if we're going to do to them what we wouldn't do to our own people?'"

Given the malignant history of American companies employing dangerous drugs and pesticides overseas that they would not or could not use at home, it is understandable why Washington officials say it would be hypocritical to finance DDT in poor nations. But children sick with malaria might perceive a more deadly hypocrisy in our failure to do so: America and Europe used DDT irresponsibly to wipe out malaria. Once we discovered it was harming the ecosystem, we made even its safe use impossible for far poorer and sicker nations.

Sharon Guynup, "Arctic Life Threatened by Toxic Chemicals, Groups Say" (2002)

Some contaminants stay put, but others travel widely. A variety of chemicals are now found in the Arctic and Antarctic at levels harmful to both wildlife and human populations. Top predators, including humans, are often at the highest risk because many chemicals bioaccumulate in the tissue of simpler organisms and biomagnify in the food chain.

There's something seriously wrong in the Great White North. Polar bears are birthing fewer cubs. Seals that swim in northern seas carry high levels of mercury and cadmium in the body fat that insulates them from the cold—and animals from reindeer to whales to sea birds also carry industrial chemicals in their bodies. Some Inuit newborns are born with high blood pressure that persists into elementary school.

The reason, according to a new study, is that the Arctic has become a repository for some of the world's most toxic chemicals, and at higher concentrations than previously thought.

Although the brilliant white snow and clear blue Arctic seas appear pristine, small concentrations of industrial chemicals are carried here on air, river, and ocean currents from as far away as Asia and gradually build up.

This is why "the Arctic is a very important area to take the pulse of the globe," said Lars Otto Reiersen, leader of the Norway-based Arctic Monitoring and Assessment Program (AMAP), who co-produced the new report *Arctic Pollution 2002* in collaboration with the World Wildlife Federation (WWF).

NERVOUS SYSTEM DAMAGE, WEAKENED IMMUNITY

"[These chemicals] come from us," said Samantha Smith, director of the WWF Arctic Program. "They come from people in industrialized countries, from the factories that make our products and the way that we grow our food."

The Inuit Circumpolar Conference, an organization representing Inuit people in Alaska, Canada, Greenland, and Russia, expressed concern over the report. The group's chair, Sheila Watt-Cloutier, called for expanded research on threats from toxic industrial chemicals, and asked for international cooperation to protect Arctic indigenous people.

The study showed that levels of some heavy metals like mercury, lead, and cadmium; and persistent organic pollutants (POPs)—toxins like Polychlorinated Biphenyls (PCBs), the insecticide DDT, and dioxins—exceeded previous estimates or hadn't dissipated over time. POPs are chemicals that break down slowly in the environment. They damage the nervous system and interfere with development. They also weaken immunity: fur seals and polar bears with high PCB levels had increased rates of infection.

MERCURY RISING

Mercury has risen to dangerous levels. Among some indigenous people, levels are high enough to affect childhood development, causing nerve and brain damage. It may also be affecting the reproduction of peregrine falcons.

"The increase in levels of organic mercury in some parts of the Arctic is primarily due to increased burning of coal for energy production in Southeast Asia, showing once again the tight links between the Arctic—as recipient of pollutants—and the rest of the world," said Reiersen.

Lake sediments in Greenland show mercury concentrations are three times higher than in pre-industrial times. Globally, 5,000 tons of mercury are present in the air at any time.

In addition to known pollutants, newly detected toxins were added to the list. Among those were flame retardants that affect brain development and weaken immunity, and perfluorooctane sulfonate (PFOS), a stain repellant. PFOS, which was recently found in the livers of northern Alaskan bears, is of particular concern because of its "extreme persistence." "It does not seem to break down under any circumstance," the authors said.

Many of these chemicals persist longer here than in other regions because of the frigid climate and the lack of soil and vegetation to absorb pollution. Even small amounts of toxins go a long way since northern animals accumulate them over a lifetime in the fat they store to survive the extreme cold.

CHEMICALS IN BREAST MILK

Inuit people are particularly at risk because the staples of their diet include animals that sit high on the food chain, like seal, whale, and fish, that have absorbed large quantities of contaminants. Chemicals have also been found in breast milk.

There was some good news. Since the introduction of non-leaded gasoline in North America in the 1970s, lead levels have dropped steadily in Greenland ice core samples. But tests on animals, from moose in the Yukon Territory to Swedish reindeer, show little change in the amount of lead stored in body tissues.

Steps are being taken to address the problem. In 2001, the United Nations Environment Program identified the most dangerous pollutants and initiated global negotiations, a move that created the Stockholm Convention on Persistent Organic Pollutants, an international treaty to ban these chemicals. As of July [2002], Canada, Iceland, Norway, and Sweden had ratified the agreement.

WWF says that toxic chemicals are slowly poisoning some of Earth's most unique residents, and is urging the United States and Russia to act. "Without a global ban, we can't protect indigenous communities and wildlife in the Arctic," said Smith. "The U.S. and Russia need to stop ignoring the scientific evidence and ratify the Stockholm Convention." [As of early 2007, the treaty had still not been ratified by either country.]

II. HUMAN DIMENSIONS

The creative and intellectual endeavors of literature, art, ethics, history, philosophy, the spiritual life, and gender studies reveal human life in its widest context: nature and the non-human. The humanities ask how women and men should value what they know about the natural world, and how such evaluation should guide actions and change habits. This means that science must inform the humanities, yet science does not teach us what to value, care for, or commit ourselves to save and strengthen. Without humanistic guidance, those choices usually lead to unsustainable practices and environmental destruction.

Similarly, if one could regard all environmental values as economic, or all environmental action through only a legal or political lens, then the humanities might be dismissed. Yet personal consciousness, group beliefs, and individual conviction have wellsprings deeper than material interests and express themselves in artistic and symbolic ways that supplement and subtly guide politics and social action. Art, literature, music, and spiritual practices affect and can even transcend socioeconomic differences and political borders. Many figures who promote environmental awareness and knowledge also practice the humanities or turn to humanistic perspectives: Francis of Assisi, William Wordsworth, Henry David Thoreau, Albert Bierstadt, Rachel Carson, Aldo Leopold, Michel Serres, Ansel Adams, and Annie Dillard are representative.

Humans exert an impact on local and global environments greater than any other species. It seems imperative to ask how we have regarded—and how we ought to regard—our connections with life that is not our life and with environmental systems we did not create but now influence or dominate. So much depends on human judgment, informed by scientific and social knowledge, and on a commitment to values that shape policy, practice, and law. For instance, national parks serve both preservation and recreation. Yet, if recreation were valued too highly, soon preservation would lose out and then, eventually, so too would recreation of the kind originally envisioned.

Artistic culture, from ancient paintings in the Dordogne caves of France to contemporary visual arts and sculpture, largely represents both human life and the natural world. But whenever artistic representations of nature present a flatly objective copy, they seem dead. One could obtain that with a tape recorder, microscope, or camera—in short, with a scientific or recording instrument. Instead, art re-creates and inflects nature with uniquely human perceptions, emotions, and history. Works of art amalgamate the natural and human; in art the two can fuse as personally experienced and perceived, complementing scientific examination and social analyses. Art imitates a total, *felt* relationship with nature. Art also becomes formative. It breaks old habits and opens new perceptions. Once the eye has witnessed the work of an impressionist painter such as Monet, it will never look on a fog, sunrise, haystack, cathedral door, or pond of water lilies in the same way again. Art is or should be the mediator between human experience and the natural world.

Many people regard their relationship to nature as spiritual or holy. Nature may be read as a sacred "book" to be decoded, encouraging contemplation or inspiring sanctity. To take the opposite case, religious beliefs prompting blunt human mastery and exploitation of nature rather than reverence and co-existence produced deeply problematic results.

For rapidly evolving environmental situations, new responsibilities demand new modes of ethical inquiry and evaluation. However indispensable, increased knowledge and technological sophistication do *not*, in and of themselves, guarantee environmental preservation, clean water, fresh air, or biodiversity. Technology must be directed by values. Often, mutually exclusive choices among different technologies and policies present themselves.

Most creatures more complex than single-cell organisms rely on sexual difference for reproduction, genetic variation, and adaptability to changing conditions. Sexual behavior and roles vary across species, as they do across human cultures and within one culture over time and across segments of its population. Yet, because of specific biological differences between the two human sexes, it is plausible that their general outlook and attitude to nature, nurturing, and the environment, while overlapping, at points differ significantly. This does not predict attitudes of specific individuals but suggests that, for example, in any given culture, women as a group may present a different profile of environmental awareness and consciousness than do men. The study of gender informs any attitude to the environment.

Predominant social attitudes often disempower women and impoverish the relationship between human cultures and the environment. As Rosalind Miles notes in *The Women's History of the World*, it is easier and cheaper for most women globally to buy a Coke than to obtain clean water, and to buy cigarettes than contraceptives. However, women's environmental roles have been historically central. Women probably originated the first tools for gathering and containing food. The menstrual cycle in human females permitted a phenomenal population expansion (great apes and other higher primates have much lower rates of fertility). Until a few thousand years ago, most cultures revered a mother goddess or female figure of nature instead of a father god in the sky. As Miles notes, the Kayaba Indians of Colombia sing, "The Mother of songs, the Mother of our whole seed, bore us in the beginning. She is the Mother of all races of men, and all tribes. She is the Mother of the thunder, of the rivers, of the trees, and of the grain. She is the only Mother we have, and She alone is the Mother of all things. She alone."

"Not even the gods can change what has been done." So runs the proverb. In environmental study, therefore, it may seem acceptable and even efficient to ignore the genesis and evolution of environmental history and the human actions that shaped it. Why not take current situations simply as givens and work from there? Why is history worth it?

History explains and unfolds; it has predictive power. It teaches cause and effect, as well as the impact of accidents and the unintended consequences of rational decisions. Its study sharpens judgment about the probable results of policies and actions. Environmental education without historical awareness would be like a struggle for human rights ignorant of its earlier champions and achievements.

As William Faulkner said, "The past isn't dead. It isn't even past." To grasp the flowing, not the fixed, character of natural processes and human impact on them, it would be unwise to settle for today's snapshot when longer films, taken from different angles, can be freeze-framed to capture any instant. The explanatory power of such a narrative is greater. So is its predictive force. The Nobel chemist Roald Hoffmann asserts that scientific discovery itself constructs and follows explanations, hypotheses, or hunches that are basically narrative explanations. The ancient manner of telling these interconnected stories to explain nature, the gods, and human life, was called mythology. Our modern manner is similar, but our complex mythology entails historical research. It includes the history of science. It demands that we determine what we value in our relationships to all of life and to the non-human. And it re-imagines and discloses the natural world in color, word, song, and tale.

Keystone Essay: Rachel Carson, from "The Real World Around Us" (1954)

The pleasures, the values of contact with the natural world are not reserved for the scientists. They are available to anyone who will place himself under the influence of a lonely mountain top—or the sea—or the stillness of a forest; or who will stop to think about so small a thing as the mystery of a growing seed.

I am not afraid of being thought a sentimentalist when I stand here tonight and tell you that I believe natural beauty has a necessary place in the spiritual development of any individual or any society. I believe that whenever we destroy beauty, or whenever we substitute something man-made and artificial for a natural feature of the Earth, we have retarded some part of man's spiritual growth.

I believe this affinity of the human spirit for the Earth and its beauties is deeply and logically rooted. As human beings, we are part of the whole stream of life. We have been human beings for perhaps a million years. But life itself passes on something of itself to other life—that mysterious entity that moves and is aware of itself and its surroundings, and so is distinguished from rocks or sense-

Rachel Carson. Courtesy of the Lear/Carson Collection, Connecticut College.

less clay [from which] life arose many hundreds of millions of years ago. Since then it has developed, struggled, adapted itself to its surroundings, evolved an infinite number of forms. But its living protoplasm is built of the same elements as air, water, and rock. To these the mysterious spark of life was added. Our origins are of the Earth. And so there is in us a deeply seated response to the natural universe, which is part of our humanity.

. . . The world of today threatens to destroy much of that beauty that has immense power to bring us a healing release from tension. Women have a greater intuitive understanding of such things. They want for their children not only physical health but mental and spiritual health as well. I bring these things to your attention tonight because I think your awareness of them will help, whether you are practicing journalists, or teachers, or librarians, or housewives and mothers. [Carson is addressing an audience of women journalists.]

What are these threats of which I speak? What is this destruction of beauty—this substitution of man-made ugliness—this trend toward a perilously artificial world. . . .

We see it in small ways in our own communities, and in larger ways in the community of the state of the nation. We see the destruction of beauty and the suppression of human individuality in hundreds of suburban real estate developments where the first act is to cut down all the trees and the next is to build an infinitude of little houses, each like its neighbor.

We see it in distressing form in the nation's capital, where I live. There in the heart of the city, we have a small but beautiful woodland area—Rock Creek Park. It is a place where one can go, away from the noise of traffic and of man-made confusions, for a little interval of refreshing and restoring quiet—where one can hear the soft water sounds of a stream on its way to river and sea, where the wind flows through the trees, and a veery sings in the green twilight. Now they propose to run a six-lane arterial highway through the heart of that narrow woodland valley, destroying forever its true and immeasurable value to the city and the nation.

Those who place so great a value on a highway apparently do not think the thoughts of an editorial writer for the *New York Times* who said, "But a little lonesome space, where nature has her own way, where it is quiet enough at night to hear the patter of small paws on leaves and the murmuring of birds, can still be afforded. The gift of tranquility, wherever found, is beyond price."

We see the destructive trend on a national scale in proposals to invade the national parks with commercial schemes such as the building of power dams. The parks were placed in trust for all the people, to preserve for them just such recreational and spiritual values as I have mentioned. Is it the right of this, our generation, in its selfish materialism, to destroy these things because we are blinded by the dollar sign? Beauty—and all the values that derive from beauty—are not measured and evaluated in terms of the dollar.

Years ago I discovered in the writings of the British naturalist Richard Jefferies a few lines that so impressed themselves upon my mind that I have never forgotten them. May I quote them for you now? "The exceeding beauty of the Earth, in her splendor of life, yields a new thought with every petal. The hours when the mind is absorbed by beauty are the only hours when we really live. All else is illusion, or mere endurance." Those lines are, in a way, a statement of the creed I have lived by, for, as perhaps you have seen tonight, a preoccupation with the wonder and beauty of the Earth has strongly influenced the course of my life.

Since *The Sea Around Us* was published, I have had the privilege of receiving many letters from people who, like myself, have been steadied and reassured by contemplating the long history of the earth and sea, and the deeper meanings of the world of nature. These letters have come from all sorts of people. There have been hairdressers and fishermen and musicians; there have been classical scholars and scientists. So many of them have said, in one phrasing or another: "We have been troubled about the world and had almost lost faith in man; it helps to think about the long history of the Earth, and of how life came to be. And when we think in terms of millions of years, we are not so impatient that our own problems be solved tomorrow."

In contemplating "the exceeding beauty of the Earth" these people have found calmness and courage. For there is symbolic as well as actual beauty in the migration of birds; in the ebb and flow of the tides; in the folded bud ready for the spring. There is something infinitely healing in these repeated refrains of nature—the assurance that dawn comes after night, and spring after winter.

Mankind has gone very far into an artificial world of his own creation. He has sought to insulate himself, with steel and concrete, from the realities of earth and water. Perhaps he is intoxicated with his own power, as he goes farther and farther into experiments for the destruction of himself and his world. For this unhappy trend there is no single remedy—no panacea. But I believe that the more clearly we can focus our attention on the wonders and realities of the universe about us, the less taste we shall have for destruction.

16 The Inner Life

Religion is sometimes considered to be a regressive cultural force of diminishing relevance, especially to progressive causes like the environment. Likewise, raising the issue of the psychology of environmentalism might appear to undermine the objectivity and rationalism that many environmentalists believe characterize their scientific, economic, and political endeavors on behalf of the planet. Yet understanding the dimensions of the inner life is of great importance for anyone studying the environment. Historically, for instance, there is the tremendous impact since the Romantic period of what might be called the religion of nature: the belief that human spiritual awareness is most acutely activated not by contemplating a transcendent deity or consulting sacred texts, but by directly witnessing and reflecting on the complexity, order, wonder, and regenerative organicism of the natural world, which predates and will outlast the human species. Similarly, sociologists confirm that today, even in highly secularized societies, religion continues to shape cultural values: secularization typically consists in a decline in formal religious participation rather than in abandoning deeply rooted societal belief structures or moral codes.

Attitudes toward the environment reflect this persistence. Lynn White, Jr., argues, for instance, that Judaism and Christianity, both through sacred texts like the Book of Genesis and the theologies and ethical structures developed from them, have generally caused the West to treat the environment poorly. The rise of modern science has only extended a religious sanction for human exploitation of nature under a different guise. Wendell Berry, in contrast, admits the Judeo-Christian tradition's ultimate responsibility for many environmental problems, but points to texts like Genesis 1:20–31 (see Chapter 11, Biodiversity and Conservation Biology) or Matthew 6:24–34, and the theology of the Promised Land, which create an ethical imperative for stewardship and loving care of the environment. Other strands in the faith of Israel and of Jesus, he says, provide a basis for Jewish and Christian environmental activism today—a view seconded by scholars who point out that ancient Israel was an agricultural society highly attuned to natural cycles. In this context, the opposition between humanity and nature from which White proceeds seems anachronistic, since it misses the intimacy and mutuality felt by the Jewish people toward their environment.

Other traditions and cultures manifest a similarly persistent religious influence on environmental attitudes. Seyyed Hossein Nasr sees the frequent environmental depredations in Islamic countries as stemming from their continuing cultural, economic, and political domination by the secularized West. Differing with White, he opposes this pattern to what he regards as true Christian attitudes toward the natural world as well as to the traditional belief of Islam. For both emphasize human responsibility toward other creatures and the need to use resources justly. In doing so, Muslims also recognize Allah's lordship over and Presence in all created things. Islamic environmentalism thus marks not an innovation, but a return to orthodoxy and a rejection of false secular values, playing a role in moderate Islam's nuanced response to modernity and its temptations.

Similarly, but less confrontationally, Tu Wei-Ming's "The Continuity of Being: Chinese Visions of Nature" illustrates how Hindu, Buddhist, Taoist, and Confucian religious and philosophical assumptions about the cosmos have led to fundamentally different views than in the West about the human relationship to the natural world, and hence, our ethical responsibility toward the environment. So extensive are these differences that it suggests that the environmental movement must adapt itself to specific cultures if it is to energize local populations by appealing to their heritage and values. The *Mundaka Upanishad* and the story of "The Bull Elephant" reinforce this lesson from Hindu and Buddhist perspectives; as do Chief Seattle's famous reply and David Cusick's version of the Iroquois Creation Story in the context of the confrontation between Protestant American and Native American cultures in the nineteenth century. Historically we have failed to be open to cultural difference concerning both spirituality and the environment.

Phyllis Windle's "The Ecology of Grief" challenges us to look at another dimension of the inner life in relation to the environment: our personal psychological response. In particular, she shows how the grieving process manifests itself in the way those committed to the environment suffer in the face of species loss, habitat destruction, and other environmental degradations. Instead of repressing these feelings, she suggests that we understand and even ritualize the ways in which we process them in order to achieve healing. A healthy psychological life and a healthy environment complement each other, a fact that encourages further work in the relatively neglected field of the psychology of the environment.

FURTHER READING

Boff, Leonardo. *Ecology and Liberation: A New Paradigm.* Trans. John Cumming. Ecology and Justice Series, ed. Mary Evelyn Tucker et al. Maryknoll, N.Y.: Orbis Books, 1995.

Conroy, Donald B., and Rodney L. Peterson, ed. *Earth at Risk: An Environmental Dialogue Between Religion and Science.* Amherst, N.Y.: Humanity Books/Prometheus Books, 2000.

Elder, John C., and Steven C. Rockefeller, ed. *Spirit and Nature. Why the Environment Is a Religious Issue: An Interfaith Dialogue.* Boston: Beacon Press, 1992.

Gosling, David L. *Religion and Ecology in India and Southeast Asia.* Foreword by Ninian Smart. London: Routledge, 2001.

Mabey, Richard. *Nature Cure.* London: Chatto and Windus, 2005.

Maguire, Daniel C., and Larry L. Rasmussen. *Ethics for a Small Planet: New Horizons on Population, Consumption, and Ecology.* Intro. Rosemary Radford Ruether. SUNY Series in Religious Studies, ed. Harold Coward. Albany: State University of New York Press, 1998.

McFague, Sallie. *The Body of God: An Ecological Theology.* Minneapolis: Fortress Press, 1993.

Roszak, Theodore, Mary E. Gomes, and Allen D. Kramer, ed. *Ecopsychology: Restoring the Earth, Healing the Mind.* San Francisco: Sierra Club Books, 1995.

Scharper, Stephen Bede. *Redeeming the Time: A Political Theology of the Environment.* New York: Continuum, 1998.

Tucker, Mary Evelyn, and John Grim, ed. *Religions of the World and Ecology Series.* Cambridge: Harvard University Press for the Harvard University Center for the Study of World Religions, Harvard Divinity School, 1997.

From *Mundaka Upanishad*, translated by Sanderson Beck

This late Vedic Hindu scripture opens with a creation account typical of many of the Upanishads, identifying mind and matter pantheistically as manifestations of universal being. It also affirms the existence of two kinds of knowledge, higher and lower. The former provides insight into the unity of all in an all-pervading, creative life spirit. Mundaka thus contrasts with mainstream Judeo-Christian thought, which emphasizes the division of creator from creation, spiritual from physical, and human from natural, with negative environmental consequences.

. . . "Two kinds of knowledge are to be known
which the knowers of God speak of, the higher and lower.
Of these the lower is the Vedas: Rig, Yajur, Sama, Atharva,
phonetics, ritual, grammar, definition, metrics, astrology.
The higher is that by which the imperishable is apprehended.

"That which is invisible, intangible,
without family, without class,
without sight or hearing, without hands or feet,
eternal, all-pervading, omnipresent, most subtle,
that is the imperishable
which the wise perceive as the source of creation.

Figure 16.1 *Krishna and Radha by the Jumna River*, attributed to the Family of Nainsukh, about 1800. Kangra style, Punjab Hills, Northern India. Ink and opaque watercolor on paper. 26.6 × 32 cm (10.5 × 12.375 in.). Museum of Fine Arts, Boston. Ross-Coomaraswamy Collection. Photograph © Museum of Fine Arts, Boston.

"As the spider puts out and gathers in,
as plants grow on the earth,
as hair on the head and body of a living person,
so from the imperishable arises everything here.

"By discipline God expands.
From that, matter is produced;
from matter, life, mind, reality,
the worlds, and in works immortality.
Whoever is all-knowing and all-wise,
whose discipline consists of knowledge,
from this is produced what is God here,
name and form and matter.
This is that truth. . . .

"This is the truth:
as from a blazing fire
thousands of flaming sparks come forth,
so from the imperishable, my friend,
various beings come forth and return there also.
Divine and formless is the Spirit,
which is outside and inside, unborn, not breath, not mind,
pure, higher than the high imperishable.

"From this is produced breath, mind, and all the senses,
space, air, light, water, and earth supporting all.
Fire is its head, its eyes the sun and moon,
the regions of space its ears, the revealed Vedas its speech,
air its breath, its heart the world.
The earth is its footstool.

"It is the inner soul of all beings.
From it comes fire whose fuel is the sun,
from the moon, rain, plants on the earth;
the male pours seed in the female;
thus creatures are produced from the Spirit.

"From it come the hymns, the chants,
the formulas, the rites of initiation,
and all the sacrifices, ceremonies, and offerings,
the year too, and the sacrificer,
and the worlds where the moon shines and the sun.

"From it also are born various gods, the celestials,
people, cattle, birds, the in-breath and the out-breath,
rice and barley, discipline,
faith, truth, chastity, and the law.

"From it come forth the seven life-breaths,
the seven flames, their fuel, the seven oblations,
these seven worlds in which move the life-breaths
set within the secret place, seven and seven.

"From it the seas and mountains all;
from it flow the rivers of all kinds;
from it come all plants and the essence
by which the inner soul lives in the elements.

"The Spirit itself is all this here:
works and discipline and God, beyond death.
Whoever knows that which is set in the secret place,
that one here on earth, my friend,
cuts apart the knot of ignorance.
Manifest, hidden, moving in the secret place, the great home.
In it lives all that moves and breathes and sees."

The Bible, Matthew 6:24–34

Jesus's parable of "the lilies of the field" picks up the theme from Genesis 1:29–31 (see Chapter 11) that human dominion over nature involves a kind of stewardship resembling God's own love for creation. The parable also reflects ancient Israel's closeness as an agricultural community to the land.

[24] "No one can serve two masters; for either he will hate the one and love the other, or he will be devoted to the one and despise the other. You cannot serve God and mammon.
[25] Therefore I tell you, do not be anxious about your life, what you shall eat or what you shall drink, nor about your body, what you shall put on. Is not life more than food, and the body more than clothing?
[26] Look at the birds of the air: they neither sow nor reap nor gather into barns, and yet your heavenly Father feeds them. Are you not of more value than they?
[27] And which of you by being anxious can add one cubit to his span of life?
[28] And why are you anxious about clothing? Consider the lilies of the field, how they grow; they neither toil nor spin;
[29] yet I tell you, even Solomon in all his glory was not arrayed like one of these.
[30] But if God so clothes the grass of the field, which today is alive and tomorrow is thrown into the oven, will he not much more clothe you, O men of little faith?
[31] Therefore do not be anxious, saying, 'What shall we eat?' or 'What shall we drink?' or 'What shall we wear?'
[32] For the Gentiles seek all these things; and your heavenly Father knows that you need them all.
[33] But seek first his kingdom and his righteousness, and all these things shall be yours as well.
[34] Therefore do not be anxious about tomorrow, for tomorrow will be anxious for itself. Let the day's own trouble be sufficient for the day."

Udana IV.5, "Naga Sutta" ("The Bull Elephant"), translated by John D. Ireland

In depicting the parallel needs of the bull elephant and the Lord Buddha for solitude, this selection suggests a number of environmentally relevant Buddhist concepts. These include the unity of consciousness across species, the oneness of all life, and the ethical and spiritual imperative to respect all forms of being.

Thus have I heard. At one time the Lord was staying near Kosambi at the Ghosita monastery. At that time the Lord was living hemmed in by bhikkhus and bhikkhunis, by male and female lay followers, by kings and royal ministers, by sectarian teachers and their disciples, and he lived in discomfort and not at ease. Then the Lord thought: "At present I am living hemmed in by bhikkhus and bhikkhunis . . . by sectarian teachers and their disciples, and I live in discomfort and not at ease. Suppose I were to live alone, secluded from the crowd?"

Then the Lord, having put on his robe in the forenoon and taken his bowl and outer cloak, entered Kosambi for almsfood. Having walked for almsfood in Kosambi and returned after the meal, he set his lodging in order by himself, took his bowl and cloak, and without informing his attendant or taking leave of the Order of bhikkhus, he set off alone, without a companion, for Parileyyaka. Walking on tour by stages, he arrived at Parileyyaka and stayed near Parileyyaka in a protected forest at the foot of an auspicious sal-tree.

Now a certain bull elephant was living hemmed in by elephants and she-elephants, by elephant calves and sucklings; he ate grass with the tips pulled off and they ate the branches he had broken down. He drank muddied water and on going down and coming out of the water he was jostled by she-elephants; and he lived in discomfort and not at ease. Then that bull elephant thought: "At present I am living hemmed in by elephants and she-elephants, by elephant calves and sucklings; I eat grass with the tips pulled off and they eat the branches which I break down. I drink muddied water and on going down and coming out of the water I am jostled by she-elephants; and I live in discomfort and not at ease. Suppose I were to live alone, secluded from the crowd?"

So that bull elephant left the herd and went to Parileyyaka, to the protected forest, and approached the Lord at the foot of the auspicious sal-tree. On reaching the place where the Lord was staying that bull elephant kept the place free of grass and brought water with his trunk for the Lord's use.

Then, while the Lord was in solitude and seclusion, this thought arose in his mind: "Formerly I was living hemmed in by bhikkhus and bhikkhunis . . . and I was living in discomfort and not at ease. But now I live not hemmed in by bhikkhus and bhikkhunis . . . in comfort and at ease." And also this thought arose in that bull elephant's mind: "Formerly I was living hemmed in by elephants and she-elephants . . . and I was living in discomfort and not at ease, but now I live not hemmed in by elephants and she-elephants . . . I eat unbroken grass and (others) do not eat the branches which I break down. I drink clear water and on going down and coming out of the water I am not jostled by she-elephants, and I live in comfort and at ease."

Then the Lord, on observing his own solitude, understood with his mind the thought in the mind of that bull elephant, and uttered on that occasion this inspired utterance:

This unites mind with mind,
The perfected one and the bull elephant
With tusks as long as chariot-poles:
That each delights in being alone in the forest.

From the Iroquois Creation Story, as told in David Cusick, *Sketches of Ancient History of the Six Nations* (1827)

This dualistic creation account tells of the birth of two primordial twins: the good mind, who is responsible for creation, positive changes in the environment, and the provision of animals friendly or useful to humans; and the bad mind, who would have left the world in its natural state, and who made animals that are hostile to humanity, like snakes. The eventual dominance of the former inscribes a hope-filled, harmonious view of the relationship between humankind and the environment.

Among the ancients there were two worlds in existence. The lower world was in a great darkness, the possession of the great monster; but the upper world was inhabited by mankind; and there was a woman conceived and would have the twin born.

. . . The woman remained in a state of unlimited darkness, and she was overtaken by her travail to which she was subject. While she was in the limits of distress one of the infants was moved by an evil opinion, and he was determined to pass out under the side of the parent's arm, and the other infant in vain endeavored to prevent his design. The woman was in a painful condition during the time of their disputes, and the infants entered the dark world by compulsion, and their parent expired in a few moments. They had the power of sustenance without a nurse, and remained in the dark regions. After a time . . . the infants were grown up, and one of them [was] possessed with a gentle disposition and named Enigorio, i.e., the good mind. The other youth possessed an insolence of character, and was named Enigonhahetgea, i.e., the bad mind. The good mind was not contented to remain in a dark situation, and he was anxious to create a great light in the dark world; but the bad mind was desirous that the world should remain in a natural state. The good mind determined to prosecute his designs, and therefore commences the work of creation. At first he took the parent's head (the deceased), of which he created an orb, and established it in the center of the firmament, and it became of a very superior nature to bestow light in the new world (now the sun); and again he took the remnant of the body, and formed another orb, which was inferior to the light (now the moon). In the orb a cloud of legs appeared to prove it was the body of the good mind (parent). The former was to give light to the day, and the latter to the night; and he also created numerous spots of light (now stars); these were to regulate the days, nights, seasons, years, etc. Whenever the light extended to the dark world the monsters were displeased and immediately concealed themselves in the deep places, lest they should be discovered by some human beings. The good mind continued the works of creation, and he formed numerous creeks and rivers on the Great Island, and then created numerous species of animals of the smallest

Figure 16.2 *The Tree of Peace* by Ron Henry.
Courtesy of the artist/Long Ago & Far Away.

and greatest to inhabit the forests, and fish of all kinds to inhabit the waters. When he had made the universe he was in doubt respecting some being to possess the Great Island; and he formed two images of the dust of the ground in his own likeness, male and female, and by his breathing into their nostrils he gave them the living souls, and named them Ea-gwe-howe, i.e., a real people; and he gave the Great Island all the animals of game for their maintenance; and he appointed thunder to water the earth by frequent rains, agreeable to the nature of the system; after this the Island became fruitful, and vegetation afforded the animals subsistence. The bad mind, while his brother was making the universe, went throughout the Island and made numerous high mountains and falls of water, and great steeps, and also created various reptiles which would be injurious to mankind; but the good mind restored the Island to its former condition. The bad mind proceeded further in his motives, and he made two images of clay in the form of mankind; but while he was giving them existence they became apes; and when he had not the power to create mankind he was envious against his brother, and again he made two of clay. The good mind discovered his brother's contrivances, and aided in giving them living souls.[1] (It is said these had the most knowledge of good and evil.) The good mind now accomplishes the works of creation, notwithstanding the imaginations of the bad mind were continually evil; and he attempted to enclose all the animals of game in the earth, so as to deprive them from mankind; but the good mind released them from confinement (the animals were dispersed, and traces of them were made on the rocks near the cave where it was closed). The good mind experiences that his brother was at variance with the works of creation, and feels not disposed to favor any of his proceedings, but gives admonitions of his future state. Afterwards the good mind requested his brother to accompany him, as he was proposed to inspect the game, etc., but when a short distance from their nominal residence, the bad mind became so unmanly that he could not conduct his brother any more. The bad mind offered a challenge to his brother and resolved that who gains the victory should govern the universe; and appointed a day to meet the contest. The good mind was willing to submit to the offer, and he enters the reconciliation with his brother; which he falsely mentions that by whipping with flags would destroy his temporal life; and he earnestly solicits his brother also to notice the instrument of death, which he manifestly relates by the use of deer horns, beating his body he would expire. On the day appointed the engagement commenced, which lasted for two days; after pulling up the trees and mountains as the track of a terrible whirlwind, at last the good mind gains the victory by using the horns, as mentioned the instrument of death, [by] which he succeeded in deceiving his brother, and he crushed him in the earth; and the last words uttered from the bad mind were, that he would have equal power over the souls of mankind after death; and he sinks down to eternal doom, and became the Evil Spirit. After this tumult the good mind repaired to the battle ground, and then visited the people and retires from the earth.

Note

1. It appears by the fictitious accounts that the said beings become civilized people, and made their residence in the southern parts of the Island; but afterwards they were destroyed by the barbarous nations, and their fortifications were ruined unto this day.

Francis of Assisi, "The Canticle of Brother Sun" (twelfth–thirteenth century C.E.), translated by Benen Fahy, O.F.M.

Francis of Assisi's vision of all creatures and aspects of existence united in praise of their Creator reflects his mystical theology, in which nature mirrors God's excellence and so allows humans access to the divine. This breaks down all barriers and hierarchies among species, as illustrated by Francis's sermon to the birds and his treaty with the Wolf of Gubbio, stories collected contemporaneously with this canticle in the posthumous fourteenth-century Little Flowers of Saint Francis.

Most high, all-powerful, all good, Lord!
  All praise is yours, all glory, all honor
  And all blessing.

Figure 16.3 Sermon to the birds from *The Life of St. Francis* by Giotto, 1297–1300. Basilica of St. Francis of Assisi. Courtesy of the Franciscan Friars of Australia.

To you, alone, Most High, do they belong.
  No mortal lips are worthy
  To pronounce your name.
All praise be yours, my Lord, through all that you have made,
  And first my lord Brother Sun,
  Who brings the day; and light you give to us through him.
How beautiful is he, how radiant in all his splendor!
  Of you, Most High, he bears the likeness.
All praise be yours, my Lord, through Sister Moon and Stars;
  In the heavens you have made them, bright
  And precious and fair.
All praise be yours, my Lord, through Brothers Wind and Air,
  And fair and stormy, all the weather's moods,
  By which you cherish all that you have made.
All praise be yours, my Lord, through Sister Water,
  So useful, lowly, precious and pure.
All praise be yours, my Lord, through Brother Fire,

Through whom you brighten up the night.
How beautiful is he, how gay! Full of power and strength.
All praise be yours, my Lord, through Sister Earth, our mother,
Who feeds us in her sovereignty and produces
Various fruits with colored flowers and herbs.
All praise be yours, my Lord, through those who grant pardon
For love of you; through those who endure
Sickness and trial.
Happy those who endure in peace,
By you, Most High, they will be crowned.
All praise be yours, my Lord, through Sister Death,
From whose embrace no mortal can escape.
Woe to those who die in mortal sin!
Happy those She finds doing your will!
The second death can do no harm to them.
Praise and bless my Lord, and give him thanks,
And serve him with great humility.

Chief Seattle, from "Chief Seattle's Speech" (reconstructed 1887 [1854])

Replying to a proposal that his tribe move to a reservation, Chief Seattle contrasts the European settlers' apparent destiny with that of his own people. He also compares Western culture's instrumentalization and exploitation of nature with Native American respect for the land, attributing this difference in attitude to the West's individualism, lack of a sense of historical continuity, and Protestant belief that nature is material, disenchanted, and opposed to the realm of spirit. (Bracketed additions are as in the source, which is one of several versions of this speech, reconstructed from memory by a Euro-American journalist.)

There was a time when our people covered the whole land, as the waves of a wind-ruffled sea cover its shell-paved floor. But that time has long since passed away with the greatness of tribes now almost forgotten.

... Your God loves your people and hates min[e]; he [folds his strong] arms [lovingly around the white man and] leads him as a father leads his infant son, but he has forsaken his red children; he makes your people wax strong every day, and soon they will fill the land; while my people are ebbing away like a fast-receding tide, that will never flow again.

... Your God seems to us to be partial. He came to the white man. We never saw Him; never even heard His voice; He gave the white man laws but He had no word for His red children whose teeming millions filled this vast continent as the stars fill the firmament. No, we are two distinct races and must ever remain so. There is little in common between us. The ashes of our ancestors are sacred and their final resting place is hallowed ground, while you wander away from the tombs of your fathers seemingly without regret.

Your religion was written on tables of stone by the iron finger of an angry God, lest you might forget it. The red man could never remember nor comprehend it.

Our religion is the traditions of our ancestors, the dreams of our old men, given them by the great Spirit, and the visions of our sachems, and is written in the hearts of our people.

Your dead cease to love you and the homes of their nativity as soon as they pass the portals of the tomb. They wander far off beyond the stars, are soon forgotten, and never return. Our dead never forget the beautiful world that gave them being. They still love its winding rivers, its great mountains and its sequestered vales, and they ever yearn in tenderest affection over the lonely hearted living and often return to visit and comfort them.

. . . But why should we repine? Why should [I murmur at the] fate of my people? Tribes are made up of individuals and are no better than [th]ey. Men come and go like the waves of the sea. A tear, a tamanawus, a dirge, and they are gone from our longing eyes forever. Even the white man, whose God walked and talked with him, as friend to friend, is not exempt from the common destiny. We *may* be brothers after [all.] We shall see.

We will ponder your proposition, and when we have decided we will tell you. But should we accept it, I here and now make this the first condition: That we will not be denied the privilege, without molestation, of visiting at will the graves of our ancestors and friends. Every part of this country is sacred to my people. Every hill-side, every valley, every plain and grove has been hallowed by some fond memory or some sad experience of my tribe.

Even the rocks that seem to lie dumb as they swelter in the sun along the silent seashore in solemn grandeur thrill with memories of past events connected with the fate of my people, and the very dust under your feet responds more lovingly to our footsteps than to yours, because it is the ashes of our ancestors, and our bare feet are conscious of the sympathetic touch, for the soil is rich with the life of our kindred.

The sable braves, and fond mothers, and glad-hearted maidens, and the little children who lived and rejoiced here, and whose very names are now forgotten, still love these solitudes, and their deep fastnesses at eventide grow shadowy with the presence of dusky spirits. And when the last red man shall have perished from the earth and his memory among white men shall have become a myth, these shores shall swarm with the invisible dead of my tribe, and when your children's children shall think themselves alone in the field, the store, the shop, upon the highway or in the silence of the woods they will not be alone. In all the earth there is no place dedicated to solitude. At night, when the streets of your cities and villages shall be silent, and you think them deserted, they will throng with

Figure 16.4 Chief Seattle. Courtesy of the Museum of History and Industry, Seattle, Wash.

the returning hosts that once filled and still love this beautiful land. The white man will never be alone. Let him be just and deal kindly with my people, for the dead are not altogether powerless.

Lynn White, Jr., from "The Historical Roots of Our Ecologic Crisis" in *Machina Ex Deo: Essays in the Dynamism of Western Culture* (1968)

Historian Lynn White, Jr., goes back beyond the scientific revolution and the rise of democracy to locate the ancient and medieval roots of our current ecological crisis. With exceptions like Francis of Assisi, he holds Western Christian theologians to blame for beliefs (e.g., in a divinely ordained human separation from and domination over nature) that developed into "the presuppositions that underlie modern technology and science," with negative results for the environment.

What did Christianity tell people about their relations with the environment? . . . Christianity inherited from Judaism not only a concept of time as nonrepetitive and linear but also a striking story of creation. By gradual stages a loving and all-powerful God had created light and darkness, the heavenly bodies, the earth and all its plants, animals, birds, and fishes. Finally, God had created Adam and, as an afterthought, Eve to keep man from being lonely. Man named all the animals, thus establishing his dominance over them. God planned all of this explicitly for man's benefit and rule: no item in the physical creation had any purpose save to serve man's purposes. And, although man's body is made of clay, he is not simply part of nature: he is made in God's image.

. . . Man shares, in great measure, God's transcendence of nature. Christianity, in absolute contrast to ancient paganism and Asia's religions (except, perhaps, Zoroastrianism), not only established a dualism of man and nature but also insisted that it is God's will that man exploit nature for his proper ends.

At the level of the common people this worked out in an interesting way. In Antiquity every tree, every spring, every stream, every hill had its own *genius loci*, its guardian spirit. These spirits were accessible to men, but were very unlike men; centaurs, fauns, and mermaids show their ambivalence. Before one cut a tree, mined a mountain, or dammed a brook, it was important to placate the spirit in charge of that particular situation, and to keep it placated. By destroying pagan animism, Christianity made it possible to exploit nature in a mood of indifference to the feelings of natural objects.

. . . The spirits *in* natural objects, which formerly had protected nature from man, evaporated. Man's effective monopoly on spirit in this world was confirmed, and the old inhibitions to the exploitation of nature crumbled. . . .

The Christian dogma of creation, which is found in the first clause of the Creeds, has another meaning for our comprehension of today's ecologic crisis. By revelation, God had given man the Bible, the Book of Scripture. But since God had made nature, nature also must reveal the divine mentality. The religious study of nature for the better understanding of God was known as natural theology. In the early Church, and always in the Greek East, nature was conceived primarily as a symbolic system through which God speaks to men: the ant is a sermon to sluggards; rising flames are the symbol of the soul's aspiration. This view of nature was essentially artistic rather than scientific. . . .

However, in the Latin West by the early thirteenth century natural theology was following a very different bent. It was ceasing to be the decoding of the physical symbols of God's communication with man and was becoming the effort to understand God's mind by discovering how his creation operates. . . . From the thirteenth century onward into the eighteenth, every major scientist, in effect, explained his motivations in religious terms. Indeed, if Galileo had not been so expert an amateur theologian he would have got into far less trouble: the professionals resented his intrusion. It was not until the late eighteenth century that the hypothesis of God became unnecessary to many scientists.

. . . Since both *science* and *technology* are blessed words in our contemporary vocabulary, some may be happy at the notions, first, that, viewed historically, modern science is an extrapolation of natural theology and, second, that modern technology is at least partly to be explained as an Occidental, voluntarist realization of the Christian dogma of man's transcendence of, and rightful mastery over, nature. But, as we now recognize, somewhat over a century ago science and technology, hitherto quite separate activities, joined to give mankind powers which, to judge by many of the ecologic effects, are out of control. If so, Christianity bears a huge burden of guilt.

Wendell Berry, from "The Gift of Good Land" in *The Gift of Good Land* (1981)

Berry attempts to refute Lynn White's argument that due to their scriptures and theology, Judaism and Christianity have historically been and today remain predominantly harmful influences on Western attitudes toward the environment. He claims that concepts such as that of stewardship (not lordship) in Genesis, the later theology of the Promised Land, and the New Testament's injunctions to do charity all form the basis for a practical, even decidedly activist Judeo-Christian ecological ethic.

Some of the reluctance to make a forthright Biblical argument against the industrial rape of the natural world seems to come from the suspicion that this rape originates with the Bible, that Christianity cannot cure what, in effect, it has caused. Judging from conversations I have had, the best known spokesman for this view is Professor Lynn White, Jr., whose essay, "The Historical Roots of Our Ecologic Crisis," has been widely published.

Professor White asserts that it is a "Christian axiom that nature has no reason for existence save to serve man." He seems to base his argument on one Biblical passage, Genesis 1:28, in which Adam and Eve are instructed to "subdue" the earth. "Man," says Professor White, "named all the animals, thus establishing his dominance over them." There is no doubt that Adam's superiority over the rest of Creation was represented, if not established, by this act of naming; he *was* given dominance. But that this dominance was meant to be tyrannical, or that "subdue" meant to destroy, is by no means a necessary inference. Indeed, it might be argued that the correct understanding of this "dominance" is given in Genesis 2:15, which says that Adam and Eve were put into the Garden "to dress it and to keep it."

But these early verses of Genesis can give us only limited help. The instruction in Genesis 1:28 was, after all, given to Adam and Eve in the time of their innocence, and it seems certain that the word "subdue" would have had a different intent and sense for them at that time than it could have for them, or for us, after the Fall.

It is tempting to quarrel at length with various statements in Professor White's essay, but he has made that unnecessary by giving us two sentences that define both his problem and my task. He writes, first, that "God planned all of this [the Creation] explicitly for man's benefit and rule: no item in the physical creation had any purpose save to serve man's purposes." And a few sentences later he says: "Christianity . . . insisted that it is God's will that man exploit nature for his *proper* ends" [my emphasis].

It is certainly possible that there might be a critical difference between "man's purposes" and "man's *proper* ends." And one's belief or disbelief in that difference, and one's seriousness about the issue of propriety, will tell a great deal about one's understanding of the Judeo-Christian tradition.

I do not mean to imply that I see no involvement between that tradition and the abuse of nature. I know very well that Christians have not only been often indifferent to such abuse, but have often condoned it and often perpetrated it. That is not the issue. The issue is whether or not the Bible explicitly or implicitly defines a *proper* human use of Creation or the natural world. Proper use, as opposed to improper use, or abuse, is a matter of great complexity, and to find it adequately treated it is necessary to turn to a more complex story than that of Adam and Eve.

The story of the giving of the Promised Land to the Israelites is more serviceable than the story of the giving of the Garden of Eden, because the Promised Land is a divine gift to a *fallen* people. For that reason the giving is more problematical, and the receiving is more conditional and more difficult. In the Bible's long working out of the understanding of this gift, we may find the beginning—and, by implication, the end—of the definition of an ecological discipline.

The effort to make sense of this story involves considerable difficulty because the tribes of Israel, though they see the Promised Land as a gift to them from God, are also obliged to take it by force from its established inhabitants. And so a lot of the "divine sanction" by which they act sounds like the sort of rationalization that invariably accompanies nationalistic aggression and theft. It is impossible to ignore the similarities to the westward movement of the American frontier. The Israelites were following their own doctrine of "manifest destiny," which for them, as for us, disallowed any human standing to their opponents. In Canaan, as in America, the conquerors acted upon the broadest possible definition of idolatry and the narrowest possible definition of justice. They conquered with the same ferocity and with the same genocidal intent.

But for all these similarities, there is a significant difference. Whereas the greed and violence of the American frontier produced an ethic of greed and violence that justified American industrialization, the ferocity of the conquest of Canaan was accompanied from the beginning by the working out of an ethical system antithetical to it—and antithetical, for that matter, to the American conquest with which I have compared it. The difficulty but also the wonder of the story of the Promised Land is that, there, the primordial and still continuing dark story of human rapaciousness begins to be accompanied by a vein of light which, however improbably and uncertainly, still accompanies us. This light originates in the idea of the land as a gift—not a free or a deserved gift, but a gift given upon certain rigorous conditions.

It is a gift because the people who are to possess it did not create it. It is accompanied by careful warnings and demonstrations of the folly of saying that "My power and the might of mine hand hath gotten me this wealth" (Deuteronomy 8:17). Thus, deeply implicated in the very definition of this gift is a specific warning against *hubris* which is the great ecological sin, just as it is the great sin of politics. People are not gods. They must not act like gods or assume godly authority. If they do, terrible retributions are in store. In this warning we have the root of the idea of propriety, of *proper* human purposes and ends. We must not use the world as though we created it ourselves.

The Promised Land is not a permanent gift. It is "given," but only for a time, and only for so long as it is properly used. It is stated unequivocally, and repeated again and again, that "the heaven and the heaven of heavens is the Lord's thy God, the earth also, with all that therein is" (Deuteronomy 10:14). What is given is not ownership, but a sort of tenancy, the right of habitation and use: "The land shall not be sold forever: for the land is mine; for ye are strangers and sojourners with me" (Leviticus 25:23). . . .

How are they to prove worthy?

First of all, they must be faithful, grateful, and humble; they must remember that the land is a gift: "When thou hast eaten and art full, then thou shalt bless the Lord thy God for the good land which he hath given thee" (Deuteronomy 8:10).

Second, they must be neighborly. They must be just, kind to one another, generous to strangers, honest in trading, etc. These are social virtues, but, as they invariably do, they have ecological and agricultural implications. For the land is described as an "inheritance"; the community is understood to exist not just in space, but also in time. One lives in the neighborhood, not just of those who now live "next door," but of the dead who have bequeathed the land to the living, and of the unborn to whom the living will in turn bequeath it. But we can have no direct behavioral connection to those who are not yet alive. The only neighborly thing we can do for them is to preserve their inheritance: we must take care, among other things, of the land, which is never a possession, but an inheritance to the living, as it will be to the unborn.

And so the third thing the possessors of the land must do to be worthy of it is to practice good husbandry. The story of the Promised Land has a good deal to say on this subject, and yet its account is rather fragmentary. We must depend heavily on implication. For sake of brevity, let us consider just two verses (Deuteronomy 22:6–7):

> If a bird's nest chance to be before thee in the way in any tree, or on the ground, whether they be young ones, or eggs, and the dam sitting upon the young, or upon the eggs, thou shalt not take the dam with the young:
>
> But thou shalt in any wise let the dam go, and take the young to thee; that it may be well with thee, and that thou mayest prolong thy days.

This, obviously, is a perfect paradigm of ecological and agricultural discipline, in which the idea of inheritance is necessarily paramount. The inflexible rule is that the source must be preserved. You may take the young, but you must save the breeding stock. You may eat the harvest, but you must save seed, and you must preserve the fertility of the fields.

What we are talking about is an elaborate understanding of charity. It is so elaborate because of the perception, implicit here, explicit in the New Testament, that charity by its nature cannot be selective—that it is, so to speak, out of human control. It cannot be selective because between any two humans, or any two creatures, all Creation exists as a bond. Charity cannot be just human, any more than it can be just Jewish or just Samaritan. Once begun, wherever it begins, it cannot stop until it includes all Creation, for all creatures are parts of a whole upon which each is dependent, and it is a contradiction to love your neighbor and despise the great inheritance on which his life depends. Charity even for one person does not make sense except in terms of an effort to love all Creation in response to the Creator's love for it.

And how is this charity answerable to "man's purposes"? It is not, any more than the Creation itself is. Professor White's contention that the Bible proposes any such thing is, so far as I can see, simply wrong. It is not allowable to love the Creation according to the purposes one has for it, any more than it is allowable to love one's neighbor in order to borrow his tools. The wild ass and the unicorn are said in the Book of Job (39:5–12) to be "free," precisely in the sense that they are not subject or serviceable to human purposes. The same point—though it is not the main point of that passage—is made in the Sermon on the Mount in reference to "the fowls of the air" and "the lilies of the field." Faced with this problem in Book VIII of *Paradise Lost,* Milton scrupulously observes the same reticence. Adam asks about "celestial Motions," and Raphael refuses to explain, making the ultimate mysteriousness of Creation a test of intellectual propriety and humility:

. . . for the Heav'n's wide Circuit, let it speak
The Maker's high magnificence, who built
So spacious, and his Line stretcht out so far;
That Man may know he dwells not in his own;
An Edifice too large for him to fill,
Lodg'd in a small partition, and the rest
Ordain'd for uses to his Lord best known.
(lines 100–106)

The Creator's love for the Creation is mysterious precisely because it does not conform to human purposes. The wild ass and the wild lilies are loved by God for their own sake and yet they are part of a pattern that we must love because it includes us. This is a pattern that humans can understand well enough to respect and preserve, though they cannot "control" it or hope to understand it completely. The mysterious and the practical, the Heavenly and the earthly, are thus joined. Charity is a theological virtue and is prompted, no doubt, by a theological emotion, but it is also a practical virtue because it must be practiced. The requirements of this complex charity cannot be fulfilled by smiling in abstract beneficence on our neighbors and on the scenery. It must come to acts, which must come from skills. Real charity calls for the study of agriculture, soil husbandry, engineering, architecture, mining, manufacturing, transportation, the making of monuments and pictures, songs and stories. It calls not just for skills but for the study and criticism of skills, because in all of them a choice must be made: they can be used either charitably or uncharitably.

How can you love your neighbor if you don't know how to build or mend a fence, how to keep your filth out of his water supply and your poison out of his air; or if you do not produce anything and so have nothing to offer, or do not take care of yourself and so become a burden? How can you be a neighbor without *applying* principle—without bringing virtue to a practical issue? How will you practice virtue without skill?

The ability to be good is not the ability to do nothing. It is not negative or passive. It is the ability to do something well—to do good work for good reasons. In order to be good you have to know how—and this knowing is vast, complex, humble and humbling; it is of the mind and of the hands, of neither alone.

The divine mandate to use the world justly and charitably, then, defines every person's moral predicament as that of a steward. But this predicament is hopeless and meaningless unless it produces an appropriate discipline: stewardship. And stewardship is hopeless and meaningless unless it involves long-term courage, perseverance, devotion, and skill. This skill is not to be confused with any accomplishment or grace of spirit or of intellect. It has to do with everyday proprieties in the practical use and care of created things—with "right livelihood." . . .

The great study of stewardship, then, is "to know / That which before us lies in daily life" and to be practiced and prepared "in things that most concern." The angel is talking about good work, which is to talk about skill. In the loss of skill we lose stewardship; in losing stewardship we lose fellowship; we become outcasts from the great neighborhood of Creation. It is possible—as our experience in *this* good land shows—to exile ourselves from Creation, and to ally ourselves with the principle of destruction—which is, ultimately, the principle of nonentity. It is to be willing *in general* for beings to not-be. And once we have allied ourselves with that principle, we are foolish to think that we can control the results. The "regulation" of abominations is a modern governmental exercise that never succeeds. If we are willing to pollute the air—to harm the elegant creature known as the atmosphere—by that token we are willing to harm all creatures that breathe, ourselves and our children among them. There is no begging off or "trading off." You cannot affirm the power plant and condemn the smokestack, or affirm the smoke and condemn the cough.

That is not to suggest that we can live harmlessly, or strictly at our own expense; we depend upon other creatures and survive by their deaths. To live, we must daily break the body and shed the blood of Creation. When we do this knowingly, lovingly, skillfully, reverently, it is a sacrament. When we do it ignorantly, greedily, clumsily, destructively, it is a desecration. In such desecration we condemn ourselves to spiritual and moral loneliness, and others to want.

Seyyed Hossein Nasr, "Sacred Science and the Environmental Crisis—An Islamic Perspective" (1993)

Seyyed Hossein Nasr attributes the frequently poor environmental records of Islamic nations to their adoption of the values of modern Western secularism, especially scientific materialism and an absolutizing of humanity and its short-term self-interest. He outlines the theological and scriptural bases by which Islam critiques modernity and its disregard for the environment, revealing the proper relationship among the divine, the human, and the natural, and the practical responsibilities that flow from it.

The Islamic view of the natural order and the environment, as everything that is Islamic, has its roots in the Quran, the very Word of God, which is the central theophany of Islam.[1] The message of the Quran is in a sense a return to the primordial message of God to man. It addresses what is primordial in the inner nature of men and women; hence Islam is called the primordial religion (*al-dīn al-ḥanīf*).[2] As the "Primordial Scripture," the Quran addresses not only men and women but the whole of the cosmos. In a sense, nature participates in the Quranic revelation. Certain verses of the Quran address natural forms as well as human beings, while God calls nonhuman members of His creation, such as plants and animals, the sun and the stars, to bear witness in certain other verses. The Quran does not draw a clear line of demarcation between the natural and the supernatural, nor between the world of man and that of nature. The soul which is nourished and sustained by the Quran does not regard the world of nature as its natural enemy to be conquered and subdued but as an integral part of man's religious universe sharing in his earthly life and in a sense even in his ultimate destiny. . . .

The Quran depicts nature as being ultimately a theophany which both veils and reveals God. The forms of nature are so many masks which hide various Divine Qualities, while also revealing these same Qualities to those whose inner eye has not become blinded by the concupiscent ego and the centrifugal tendencies of the passionate soul.

In an even deeper sense, it can be claimed that according to the Islamic perspective God Himself *is* the ultimate environment which surrounds and encompasses man. It is of the utmost significance that in the Quran God is said to be the All-Encompassing (*Muḥīṭ*), as in the verse, "But to God belong all things in the heavens and on the earth: And He it is who encompasseth (*muḥīṭ*) all things" (IV.126), and that the term *muḥīṭ* also means environment. In reality, man is immersed in the Divine *Muḥīṭ* and is only unaware of it because of his own forgetfulness and negligence (*ghaflah*), which is the underlying sin of the soul, only to be overcome by remembrance (*dhikr*). To remember God is to see Him everywhere and to experience His reality as *al-Muḥīṭ*. The environmental crisis may in fact be said to have been caused by man's refusal to see God as the real "environment" which surrounds man and nourishes his life. The destruction of the environment is the result of modern man's attempt to view the natural environment as an ontologically independent order of reality, divorced from the Divine Environment without whose liberating grace it becomes stifled and dies. To remember God as *al-Muḥīṭ* is to remain aware of the sacred quality of nature, the reality of natural phenomena as signs (*āyāt*) of God and the presence of the natural environment as an ambience permeated by the Divine Presence of that Reality which alone is the ultimate "environment" from which we issue and to which we return.

The traditional Islamic view of the natural environment is based on this inextricable and permanent relation between what is today called the human and natural environments and the Divine Environment which sustains and permeates them. . . .

This Islamic love of nature as manifesting the "signs of God" and being impregnated with the Divine Presence must not be confused with naturalism as understood in Western philosophy and theology. Christianity, having been forced to combat the cosmolatry and rationalism of the ancient Mediterranean world, branded as naturalism both the illegitimate nature worship of the decadent forms of Greek and Roman religion and the very different concern and love for nature of northern Europeans such as the Celts, a love which nevertheless survived in a marginal manner after the Christianization of Europe, as one can see in the works of a Hildegard of Bingen or in early medieval Irish poems pertaining to nature. Now, one must never forget that the Islamic love for nature has nothing to do with the naturalism anathematized by the Church Fathers. Rather, it is much closer to the nature poetry of the Irish monks and the addresses to the sun and the moon by the patron saint of ecology, St. Francis of Assisi. Or perhaps it should be said that it is he who among all the great medieval saints is the closest to the Islamic perspective as far as the love of nature is concerned. In any case the Islamic love of nature and the natural environment and the emphasis upon the role of nature as a means to gain access to God's wisdom as manifested in His creation, do not in any sense imply the negation of transcendence or the neglect of the archetypal realities. On the contrary, on the highest level it means to understand fully the Quranic verse, "Whithersoever ye turn, there is the Face of God" (II.115). It means, therefore, to see God everywhere and to be fully aware of the Divine Environment which surrounds and permeates both the world of nature and the ambience of man.

The Islamic teachings concerning nature and the environment cannot be fully understood without dealing with the Islamic conception of man, who has been always viewed in various traditional religions as the custodian of nature and who has now become its destroyer, having changed his role, thanks to modern civilization, from the being who had descended from Heaven and who lived in harmony with the earth to a creature who considers himself as having ascended from below and who has now become the earth's most deadly predator and exterminator. Islam considers man as God's vice-gerent (*al-khalīfah*) on earth and the Quran asserts explicitly, "I am setting on the earth a vice-gerent (*khalīfah*)" (II.30). This quality of vice-gerency is, moreover, complemented by that of servantship (*al-'ubūdiyyah*) towards God. Man is God's servant (*'abd Allāh*) and must obey Him accordingly. As *'abd Allāh*, he must be passive towards God and receptive to the grace that flows from the world above. As *khalīfat Allāh*, he must be active in the world, sustaining cosmic harmony and disseminating the grace for which he is the channel as a result of his being the central creature in the terrestrial order.[3]

In the same way that God sustains and cares for the world, man as His vice-gerent must nurture and care for the ambience in which he plays the central role. He cannot neglect the care of the natural world without betraying that "trust" (*al-amānah*) which he accepted when he bore witness to God's lordship in the pre-eternal covenant (*al-mīthāq*), to which the Quran refers in the famous verse, "Am I not your Lord? They [that is, members of humanity] uttered, Yea, we bear witness" (VII.172).

To be human is to be aware of the responsibility which the state of vice-gerency entails. Even when in the Quran it is stated that God has subjected (*sakhkhara*) nature to man as in the verse, "Hast thou not seen how God has subjected to you all that is in the earth" (XXII.65), this does not mean the ordinary conquest of nature, as claimed by so many modern Muslims thirsty for the power which modern science bestows upon man. Rather, it means the dominion over things which man is allowed to exercise only on the condition that it be according to God's laws and precisely because he is God's vice-gerent on earth, being therefore given a power which ultimately belongs to God alone and not to man who is merely a creature born to journey through this earthly life and to return to God at the moment of death.

That is why also nothing is more dangerous for the natural environment than the practice of the power of vice-gerency by a humanity which no longer accepts to be God's servant, obedient to His commands and laws. There is no more dangerous a creature on earth than a *khalīfat Allāh* who no longer considers himself to be *'abd Allāh* and who therefore does not see himself as owing allegiance to a being beyond himself. Such a creature is able to possess a power of destruction which is truly Satanic in the sense that "Satan is the ape of God"; for such a human type wields, at least for a short time, a godlike but destructive dominion over the earth because this dominion is devoid of the care which God displays towards all His creatures and bereft of that love which runs through the arteries of the universe.

. . . The Divine Law (*al-Sharī'ah*) is explicit in extending the religious duties of man to the natural order and the environment. One must not only feed the poor but also avoid polluting running water. It is pleasing in the eyes of God not only to be kind to one's parents, but also to plant trees and treat animals gently and with kindness. Even in the realm of the Divine Law, and without turning to the metaphysical significance of nature, one can see the close nexus created by Islam between man and the whole natural order. Nor could it be otherwise, for the primordial character of the Islamic revelation reinstates man and the cosmos in a state of unity, harmony, and complementarity, reaffirming man's inner bond to the whole of creation, which shares the Quranic revelation in the deepest sense with man.

. . . In contrast since the advent of Renaissance humanism Western civilization has absolutized earthly man. While depriving man of his center and creating a veritable centerless culture and art, Western humanism has sought to bestow upon this centerless humanity the quality of absoluteness.[4] It is this purely earthly man defined by rationalism and humanism who developed seventeenth century science based upon the domination and conquest of nature, who sees nature as his enemy and who continues to rape and destroy the natural environment, always in the name of the rights of man, which are seen by him to be absolute. It is this terrestrial man, now absolutized, who destroys vast forests in the name of immediate economic welfare, without thinking for a moment of the consequences of such action for future generations of men or for other creatures of this world. It is such a creature who, in seeing his earthly life as being absolute, tries to prolong it at all costs, creating a medicine which has produced both wonders and horrors, including the destruction of the ecological balance through human overpopulation. Neither God nor nature have any right in the eyes of a humanity which sees itself as absolute even when talking about man being an insignificant observer on a small planet on the periphery of a minor galaxy, as if all this superficial humility were not also based upon the absolutization of the sense-experience and rational powers of earthly man.

Now Islam has always stood strongly opposed to this absolutization of what one might call the Promethean and Titanic man. It has never permitted the glorification of man at the expense of either God or His creation. Nothing is more detestable to traditional Muslim sensibilities than some of the Titanic art of the Renaissance created for the glorification of a humanity in rebellion against

Heaven. If modern science and with it a civilization which gave and still gives itself absolute right of domination over the earth and even the heavens did not come into being in the Islamic world, it was not because of the lack of mathematical or astronomical knowledge. Rather, it was because the Islamic perspective excluded the possibility of the deification of earthly man or of the total secularization of nature. In Islamic eyes, only the Absolute is absolute.

Notes

1. Concerning the Islamic view of nature, see S. H. Nasr, *An Introduction to Islamic Cosmological Doctrines* (Albany, N.Y., 1993); idem, *Science and Civilization in Islam;* and idem, *Islamic Life and Thought* (Albany, N.Y., 1981), especially chapter 19.
2. On the concept of Islam as the primordial religion, see F. Schuon, *Understanding Islam,* and S. H. Nasr, *Ideals and Realities of Islam* (London, 1989). Islam is also called *dīn al-fiṭrah,* which means the religion that is in the very nature of things and engraved in man's primordial and eternal substance.
3. On the Islamic concept of man, see G. Eaton, "Man," in Nasr, ed., *Islamic Spirituality—Foundations,* vol. 19 of *World Spirituality: An Encyclopedic History of the Religious Quest* (New York, 1987), chapter 19; also Nasr, *Knowledge and the Sacred* (Edinburgh, 1981), 358–77.
4. For a profound study of this loss of center in Western man as a result of the advent of [secular] humanism, see F. Schuon, *Having a Center* (Bloomington, Ind., 1990), 160ff.

Tu Wei-Ming, "The Continuity of Being: Chinese Visions of Nature" (1984)

Tu Wei-Ming summarizes traditional Chinese cosmogony, using its belief in ch'i (the "vital force" pervading all) and the continuity, wholeness, and dynamism of the process of things to explain Chinese views on humanity's relationship to and ethical responsibilities toward nature. In contradistinction to at least some Judeo-Christian theologies, Confucianism holds that we are not "the rulers of creation" but rather, the "guardians of the universe," a distinction we must earn "through self-cultivation."

The Chinese belief in the continuity of being, a basic motif in Chinese ontology, has far-reaching implications in Chinese philosophy, religion, epistemology, aesthetics, and ethics. F. W. Mote comments:

> The basic point which outsiders have found so hard to detect is that the Chinese, among all peoples ancient and recent, primitive and modern, are apparently unique in having no creation myth; that is, they have regarded the world and man as uncreated, as constituting the central features of a spontaneously self-generating cosmos having no creator, god, ultimate cause, or will external to itself.[1]

This strong assertion has understandably generated controversy among Sinologists. Mote has identified a distinctive feature of the Chinese mode of thought.

In his words, "the genuine Chinese cosmogony is that of organismic process, meaning that all of the parts of the entire cosmos belong to one organic whole and that they all interact as participants in one spontaneously self-generating life process."[2]

However, despite Mote's insightfulness in singling out this particular dimension of Chinese cosmogony for focused investigation, his characterization of its uniqueness is problematic. For one thing, the apparent lack of a creation myth in Chinese cultural history is predicated on a more fundamental assumption about reality; namely, that all modalities of being are organically connected. Ancient Chinese thinkers were intensely interested in the creation of the world. Some of them, notably the Taoists, even speculated on the creator (*tsao-wu chu*) and the process by which the universe came into being.[3] Presumably indigenous creation myths existed although the written records transmitted by even the most culturally sophisticated historians do not contain enough information to reconstruct them. The real issue is not the presence or absence of creation myths, but the underlying assumption of the cosmos: whether it is continuous or discontinuous with its creator. . . . It was not a creation myth as such but the Judeo-Christian version of it that is absent in Chinese mythology. But the Chinese, like numerous peoples throughout human history, subscribe to the continuity of being as self-evidently true.

An obvious consequence of this basic belief is the all-embracing nature of the so-called spontaneously self-generating life process. Strictly speaking, it is not because the Chinese have no idea of God external to the created cosmos that they have no choice but to accept the cosmogony as an organismic process. Rather, it is precisely because they perceive the cosmos as the unfolding of continuous creativity that it cannot entertain "conceptions of creation *ex nihilo* by the hand of God, or through the will of God, and all other such mechanistic, teleological, and theistic cosmologies."[4] The Chinese commitment to the continuity of being, rather than the absence of a creation myth, prompts them to see nature as "the all-enfolding harmony of impersonal cosmic functions."[5]

The Chinese model of the world, "a decidedly psychophysical structure" in the Jungian sense,[6] is characterized by Joseph Needham as "an ordered harmony of wills without an ordainer."[7] What Needham describes as the organismic Chinese cosmos consists of dynamic energy fields rather than static matter-like entities. Indeed, the dichotomy of spirit and matter is not at all applicable to this psychophysical structure. The most basic stuff that makes the cosmos is neither solely spiritual nor material but both. It is a vital force. This vital force must not be conceived of either as disembodied spirit or as pure matter. Wing-tsit Chan, in his influential *Source Book of Chinese Philosophy*, notes that the distinction between energy and matter is not made in Chinese philosophy. He further notes that H. H. Dubs's rendering of the indigenous term for this basic stuff, *ch'i*, as "matter-energy" is "essentially sound but awkward and lacks an adjective form."[8]

Although Chan translates *ch'i* as "material force," he cautions that since *ch'i*, before the advent of Neo-Confucianism in the eleventh century, originally "denotes the psychophysiological power associated with blood and breath," it should be rendered as "vital force" or "vital power."[9]

. . . The organismic process as a spontaneously self-generating life process exhibits three basic motifs: continuity, wholeness, and dynamism. All modalities of being, from a rock to heaven, are integral parts of a continuum which is often referred to as the "great transformation" (*ta-hua*).[10] Since nothing is outside of this continuum, the chain of being is never broken. A linkage will always be found between any given pair of things in the universe. We may have to probe deeply to find some of the linkages, but they are there to be discovered. These are not figments of our imagination but solid foundations upon which the cosmos and our lived world therein are constructed. *Ch'i*, the psychophysiological stuff, is everywhere. It suffuses even the "great void" (*t'ai-hsü*) which is the source of all beings in Chang Tsai's philosophy.[11] The continuous presence of *ch'i* in all modalities of being makes everything flow together as the unfolding of a single process. Nothing, not even an almighty creator, is external to this process. . . .

It is important to note that continuity and wholeness in Chinese cosmological thinking must be accompanied in the third motif, dynamism, lest the idea of organismic unity imply a closed system. While Chinese thinkers are critically aware of the inertia in human culture which may eventually lead to stagnation, they perceive the "course of heaven" (*t'ien-hsing*) as "vigorous" (*chien*) and instruct people to model themselves on the ceaseless vitality of the cosmic process.[12] What they envision in the spontaneously self-generating life process is not only inner connectedness and interdependence but also infinite potential for development. . . .

The organismic life process, which Mote contends is the genuine Chinese cosmogony, is an open system. As there is no temporal beginning to specify, no closure is ever contemplated. The cosmos is forever expanding; the great transformation is unceasing. The idea of unilinear development, in this perspective, is one-sided because it fails to account for the whole range of possibility in which progress constitutes but one of several dominant configurations. By analogy, neither cyclic nor spiral movements can fully depict the varieties of cosmic transformation. Since it is open rather than closed and dynamic rather than static, no geometric design can do justice to its complex morphology.

Earlier, I followed Mote in characterizing the Chinese vision of nature as the "all-enfolding harmony of impersonal cosmic function" and remarked that this particular vision was prompted by the Chinese commitment to the continuity of being. Having discussed the three basic motifs of Chinese cosmology—wholeness, dynamism, and continuity—I can elaborate on Mote's characterization by discussing some of its implications. The idea of all-enfolding harmony involves two interrelated meanings. It means that nature is all-inclusive, the spontaneously self-generating life process which excludes nothing. The Taoist idea of *tzu-jan* ("self-so"),[13] which is used in modern Chinese to translate the English word *nature*, aptly captures this spirit. To say that *self-so* is all-inclusive is to posit a nondiscriminatory and nonjudgmental position, to allow all modalities of being to display themselves as they are. This is possible, however, only if competitiveness, domination, and aggression are thoroughly transformed. Thus, all-enfolding harmony also means that internal resonance underlies the order of things in the universe. Despite conflict and tension, which are like waves of the ocean, the deep structure of nature is always tranquil. The great transformation of which nature is the concrete manifestation is the result of concord rather than discord and convergence rather than divergence.

This vision of nature may suggest an unbridled romantic assertion about peace and love, the opposite of what Charles Darwin realistically portrayed as the rules of nature. Chinese thinkers, however, did not take the all-enfolding harmony to be the original naïveté of the innocent. Nor did they take it to be an idealist utopia attainable in a distant future. They were acutely aware that the world we live in, far from being the "great unity" (*ta-t'ung*) recommended in the *Evolution of the Rites*,[14] is laden with disruptive forces including humanly caused calamities and natural catastrophes. They also knew well that history is littered with internecine warfare, oppression, injustice, and numerous other forms of cruelty. It was not naïve romanticism that prompted them to assert that harmony is a defining characteristic of the organismic process. They believed that it is an accurate description of what the cosmos really is and how it actually works.

One advantage of rendering *ch'i* as "vital force," bearing in mind its original association with blood and breath, is its emphasis on the life process. To Chinese thinkers, nature is vital force in display. It is continuous, holistic, and dynamic. Yet, in an attempt to understand the blood and breath of nature's vitality, Chinese thinkers discovered that its enduring pattern is union rather than disunion, integration rather than disintegration, and synthesis rather than separation. The eternal flow of nature is characterized by the concord and convergence of numerous streams of vital force. It is in this sense that the organismic process is considered harmonious.

. . . We humans, therefore, do not find the impersonal cosmic function cold, alien, or distant, although we know that it is, by and large, indifferent to and disinterested in our private thoughts and whims. Actually, we are an integral part of this function; we are ourselves the result of this moving power of *ch'i.* Like mountains and rivers, we are legitimate beings in this great transformation.

. . . Forming one body with the universe can literally mean that since all modalities of being are made of *ch'i*, human life is part of a continuous flow of the blood and breath that constitutes the cosmic process. Human beings are thus organically connected with rocks, trees, and animals.

. . . In a strict sense, then, human beings are not the rulers of creation; if they intend to become guardians of the universe, they must earn this distinction through self-cultivation. There is no preordained reason for them to think otherwise.

... In the metaphorical sense, then, forming one body with the universe requires continuous effort to grow and to refine oneself. We can embody the whole universe in our sensitivity because we have enlarged and deepened our feeling and care to the fullest extent. However, there is no guarantee at the symbolic nor the experiential level that the universe is automatically embodied in us. Unless we see to it that the Mandate of Heaven is fully realized in our nature, we may not live up to the expectation that "all things are complete in us."[15]

... Despite our superior intelligence, we do not have privileged access to the great harmony. As social and cultural beings, we can never get outside ourselves to study nature from neutral ground. The process of returning to nature involves unlearning and forgetting as well as remembering. The precondition for us to participate in the internal resonance of the vital forces in nature is our own inner transformation. Unless we can first harmonize our own feelings and thoughts, we are not prepared for nature, let alone for an "interflow with the spirit of Heaven and Earth."[16] It is true that we are consanguineous with nature. But as humans, we must make ourselves worthy of such a relationship.

Notes

1. Frederick W. Mote, *Intellectual Foundations of China* (New York: Alfred A. Knopf, 1971), 17–18.
2. Ibid., 19.
3. For a thought-provoking discussion on this issue, see N. J. Griardot, *Myth and Meaning in Early Taoism* (Berkeley: University of California Press, 1983), 275–310.
4. Mote, *Intellectual Foundations of China*, 20.
5. Ibid.
6. See Jung's Foreword to the *I Ching* (*Book of Changes*), translated into English by Cary F. Baynes from the German translation of Richard Wilhelm, Bollingen Series, vol. 19 (Princeton: Princeton University Press, 1967), xxiv.
7. Joseph Needham and Wang Ling, *Science and Civilization in China*, 2 vols. (Cambridge: Cambridge University Press, 1969), 2:287.
8. Wing-tsit Chan, trans. and comp., *A Source Book in Chinese Philosophy* (Princeton: Princeton University Press, 1969), 784.
9. Ibid.
10. A paradigmatic discussion on this is to be found in the *Commentaries on the Book of Changes*. See Wing-tsit Chan, *Source Book in Chinese Philosophy*, 264.
11. See Chang Tsai's "Correcting Youthful Arrogance" in Wing-tsit Chan, *Source Book in Chinese Philosophy*, 501.
12. For this reference in the *Chou I*, see *A Concordance to Yi Ching*, Harvard-Yenching Institute Sinological Index Series Supplement No. 10 (reprint; Taipei: Chinese Materials and Research Aids Service Center, Inc., 1966), 1/1.
13. Chuang Tzu, chap. 7. See the Harvard-Yenching Index on the *Chuang Tzu*, 20/7/11.
14. See William T. de Bary, Wing-tsit Chan, and Burton Watson, comps., *Sources of Chinese Tradition* (New York: Columbia University, 1960), 191–92.

15. *Mencius,* 7A4.
16. *Chuang Tzu,* chap. 33, and *Chuang Tzu ying-te,* 93/33/66.

Phyllis Windle, from "The Ecology of Grief" (1995)

Windle reveals that while scientists and ecologists often mourn the loss of species and habitats, they often repress these feelings, unwilling to admit the non-cognitive dimensions of environmental experience. In fact, the well-known Kübler-Ross grieving process model best describes the movement from shock to denial to accommodation or negotiation to acceptance that we frequently feel at environmental loss. Knowledge of the power of this emotional circuit along with environmental rituals can help heal us.

Henry Mitchell's gardening column in the *Washington Post* for Sunday, 8 April 1990, was entitled "The Demise of the Dogwood." Mitchell wrote it shortly after the dogwood anthracnose (*Discula destructiva*) was first detected in the Washington area; it had been killing trees in New England and on the West Coast since the late 1970s and was already widespread in the Great Smoky Mountains. Mitchell interviewed Jay Stipes, a plant pathologist, who feared the disease could "annihilate the species."

. . . I have now been following news of dogwood anthracnose for several years. Why, I wonder, did this bad news for the environment hit me so hard? Why do I want to commemorate the dying trees? I am an ecologist. Also, I am a trained hospital chaplain, an expert on death, dying, and grief. Finally, I realize: I am in mourning for these beautiful trees.

This realization was slow in coming because almost all of the literature on grief pertains to the death of humans. However, a significant number of professional veterinary societies and veterinary schools now research pet loss and counsel grief-stricken pet owners. Their work shows the similarity between grieving for the human members of our families and for the animals to which we are attached. Additional research indicates that other types of loss also cause grief: reactions to the loss of an arm or leg, of a home, or of a job show similarities to the loss of someone we love.

. . . Biologists often love their organisms. Ecologists often love their field sites. Does anyone really doubt it? Read E. O. Wilson's work and imagine how he feels toward ants. Watch Jane Goodall interact with chimpanzees and ask what she feels for them. Read some of George Woodwell's essays, or Rachel Carson's, and gauge the depth of their passion.

This passion, this ability and willingness to admire and care about other species and places, may be among biologists' most admirable qualities. Our attachments may even be necessary and important. Nobel Prize–winning geneticist Barbara McClintock speaks about her "feeling for the organism," her intimate knowledge of the individual corn plants in her research projects, and her deep enjoyment of that knowledge. "Good science cannot proceed without a deep emotional investment on the part of a scientist," concludes McClintock's biographer, Evelyn Fox Keller.[1]

. . . The importance of our relationships to the natural world should surprise ecologists less than it does others. Ours is a science of relationships. Usually, though, we do not consider our personal attachments to the organisms and systems we study. Perhaps the ideal of the dispassionate observer stands in our way. Or perhaps women scientists notice these connections more readily. Maybe we needed more female peers before we could speak of these matters openly.

I no longer doubt the importance or nature of my attachment to dogwoods. Nor do I feel alone in my grief for their loss.

. . . David Norton writes about the loss of an endemic New Zealand mistletoe, subtitling his article "An Obituary for a Species." Aldo Leopold entitles an essay about a favorite place "Marshland Elegy." I ask a U.S. Forest Service pathologist about his reaction to dogwood anthracnose, and he speaks of his depression. "It's sad," *The Seattle Times* quotes Jim Litchatowitch, a biologist, when officials declare the lower Columbia coho salmon extinct.

. . . Likewise, Michael Soulé's anguish regards broad ecological losses: "As the number of exotics in most regions produces a cosmopolitanization of remnant wildlands, there will be an agonizing period of transition, especially for ecologists. . . . There are moments when the destruction of a favorite place, of entire biotas and ecosystems, seems unbearable and the future looks bleak indeed."[2]

Scientists and resource managers usually do not speak freely about this aspect of our feelings for the places and organisms that are part of our work any more than of our love for nature. "The sadness discernible in some marshes arises, perhaps, from their having once harbored cranes. Now they stand humbled, adrift in history."

I treasure the poetry of Aldo Leopold's expression here. I suspect, though, that it is on our faces, not the marsh's, that the sadness is discernible. Perhaps it is our discomfort with that sadness that sees a marsh in tears. Any chaplain would say that we do better by crying our own tears.

But mourning for ecological losses has no simple or predictable path. I suspect that ecologists, like other scientists, are prone to inhibiting the pain of grief. We are solidly attached to the life of the mind, and of the several steps experts consider essential to recovery, only the first is intellectual.

. . . People have always used rituals to help themselves mourn and recover from grief. For example, funerals usually reinforce the awareness of loss, sanction remembering, facilitate the expression of feelings, provide support, guide the needed reorganization of life, and affirm life's meaning. Funerals and memorial services serve as a rite of passage between initial shock and the longer, more private phases of grieving. Not all mourning customs are religious, though.

Figure 16.5 *Ophelia* by John Everett Millais, 1852. © Tate Britain, London, 2003.

We give gifts, eat together, show group solidarity, and protect mourners—all ways to help the grief-stricken.

The NAMES quilt—that collection of more than fourteen thousand fabric panels memorializing those dead of AIDS—is a particularly effective nonreligious ritual commemorating private and public loss. Making individual panels heals the makers; viewing the assemblage links those touched by tragedy. It has also become a powerful means to educate people and to call for political action.[3]

There are scientists among us who also think in terms of rituals, even funerals, for the species and places we are losing. The Wisconsin Society for Ornithology dedicated a monument to the passenger pigeon in a state park in 1947. Ecologists gathered on 12 October 1992, where Columbus may have landed in the Bahamas to "conduct a funeral ceremony for the natural environment of the Western hemisphere. They will mourn the demise of the New World's natural heritage and the eradication of entire groups of indigenous Caribbean people."[4]

This is dramatic stuff, perhaps too outrageous for many ecologists' tastes. The importance of rituals in helping mourners cope is undisputed, however, and I see no reason why ecologists should not tap this resource in these difficult times. We could create a quilt of our own, with panels to celebrate the species we have loved and lost. We could hold a wake for a precious piece of land, gathering to tell stories of the field trips, research, and academic degrees that one particular place provided. We could create a family album, filled with the recollections of our professional grandparents, writing about the natural areas they have loved and lost in their lifetimes. We could create a fund for memorials to invest our losses with public meaning. Our mourning rituals could celebrate, too, and affirm our faith in the processes of ecology and evolution. We could note the remaining beauty of the earth, the birth of new species or subspecies, and the grand rhythms of the biogeochemical cycles.

Most of our contemporary mourning customs are important in the first weeks and months of the grieving process. I suspect that ecologists are more likely to need support in a longer, continuing way. Environmental losses are intermittent, chronic, cumulative, and without obvious beginnings and endings. Thus, we may have to devise our own unique customs. But they might be customs much needed by a society facing many kinds of transitions.

Experts urge us to grieve not only because successful grief is beneficial, but also because failure to grieve can have such far-reaching consequences. Generally, problems originate in two ways. Mourning can become excessive and prolonged, leading to chronic grief from which recovery never seems to come. Alternatively, we can inhibit the process. It becomes distorted, and grief emerges in different forms. The results are not trivial. Unresolved grief is the underlying cause of problems for as much as twenty percent of the people treated at some substance abuse centers.[5]

Grief is not pleasant, as anyone knows who has mourned a child, a parent, a close friend, or a spouse. At the same time, it has its own bittersweet richness and intensity. Charles Darwin concluded that grieving serves us well in the long term.

. . . Perhaps the transition ahead for ecologists is just such a risky one. This makes it especially urgent that we do our grief work.

What might we get from tackling this seemingly unpleasant task? People emerge from grief with new insights about their relationship to the deceased and renewed energy for loving again. The benefits might extend far beyond our individual recovery. Aldo Leopold's work is a case study. Robert Finch describes Leopold's evolution, with the theme of environmental loss as a way station, as "a necessary, important sojourn in the wilderness of loss, ignorance, and self-education from which Leopold will finally wrest his holistic 'land ethic.'"

As ethicists and others explore the underpinnings necessary for our care-filled treatment of the Earth, they often return to this same idea: the importance of the nature and depth of our relationship to other organisms and to the Earth itself. Stephen Kellert suggests that, for an environmental ethic to succeed, nature would need to be meaningful to us on a variety of levels, including the emotional. Here again it is ecologists' deep attachment to organisms and systems that is our strength—a potential model for others to emulate.

Notes

1. Evelyn Fox Keller, *A Feeling for the Organism* (San Francisco: W. H. Freeman, 1983), 198.
2. Michael Soulé, "The Onslaught of Alien Species and Other Challenges in the Coming Decades," *Conservation Biology* 4 (1990), 234, 238.
3. NAMES Project, *The Quilt: Stories from the NAMES Project* (New York: Pocket Books, 1988).
4. See H. J. Viola and C. Margolis, ed., *Seeds of Change* (Washington, D.C.: Smithsonian Institution Press, 1991), 249.
5. G. W. Davidson, *Understanding Mourning* (Minneapolis: Augsburg Publications, 1984).

17 Ethics, Philosophy, Gender

Until recently, ethics has limited itself to one species, addressing the actions of human beings toward other humans. Before 1900, only rarely did Western ethics consider the human relationship to nature or non-human organisms. The main stumbling block to such a consideration has been lack of reciprocity. When considering whether one has an ethical obligation to another human being, it is easy to ask whether that person has a reciprocal obligation. In other words, I have a right not to be killed by you only by dint of our implicit contract (or the contract imposed upon us by the state) that you have the reciprocal right not to be killed by me. Nature requires no reciprocity. If we are to establish an ethical precept that a gorilla has rights and therefore it is wrong to kill a gorilla, it is not because we think that it is *morally* wrong for a gorilla to kill a human. We understand that it is not *wrong* for nature to be deadly, even to the extent that nature can inflict injury and death *on us* in a manner that we cannot address with our current ethics and philosophy.

The acknowledged deadliness of nature has posed a second stumbling block to applying standard ethics to the natural world. A signal human achievement is that we have largely overcome nature's imminent threat to our lives. We have killed animals that might otherwise kill us and drastically modified ecosystems that posed similar dangers. Thus now, not worrying every day whether we will survive natural threats ranging from large predators to tiny microbes, we can consider such questions as whether animals or ecosystems have rights. Yet such thinking goes against the grain of a half million years of human activity.

Ethics is the branch of human thought that determines what is *right* and what is *wrong*. A turning point in our relationship with nature came with Aldo Leopold's "The Land Ethic," which famously states, "A thing is right when it tends to preserve the integrity, stability, and beauty of the biotic community. It is wrong when it tends otherwise." These are words to live by, but not to interpret too strictly. By this logic, tornados and floods are morally wrong—but we know that cannot be true. A fruitful way to resolve this paradox is to look into the meanings *right* and *wrong* themselves.

A bit of history behind John Muir's "Hetch Hetchy Valley" illustrates the problematic concepts of *right* and *wrong* in the ethics of nature. Muir wrote this piece after he realized that President Theodore Roosevelt's administration was not going to prevent flooding of the valley. Earlier Muir had enjoyed a friendly relationship with Gifford Pinchot, the first director of the U.S. Forest Service (see Pinchot's "The Birth of Conservation," at the beginning of Part One). Muir and Pinchot had hiked together, enjoying the natural beauty and each other's companionship. On one occasion, after observing a tarantula and discussing its natural history, Pinchot expressed his intention to step on the tarantula and kill it. Muir admonished him, saying "the tarantula had just as much *right* to be here as we do."

A gloss of Muir's admonition reveals two meanings of the word *right* that must be teased apart in order to make sense of an ethic of nature. There is a difference between "it is right that the tarantula continue to live in its natural place" and "the tarantula has a right to live in its natural place." The first addresses cosmic appropriateness, the second legal status. Identifying the first with the second creates many conceptual problems in environmental ethics.

The issue of distinguishing between *right* and *wrong* is only one way to shape an ethic of the environment. In fact, the diversity of nature begets a diversity of philosophical approaches to it. Richard Lewontin's "Organism and Environment" addresses the environment's ontological status, arguing that no such thing as an environment exists without an organism living in it. Both Goethe and Emerson adjure us to understand what Emerson calls "method of nature," for only then can we assess the adequacy of our own methods.

Of all philosophical questions about the environment, the most difficult to answer is "What is Nature?" As Kate Soper notes, cultural critic Raymond Williams called "nature" perhaps the most complex word in the English language. Soper provides an intelligent catalogue of the ways this word is used, yet even her exhaustive compilation cannot embrace all the nuances.

One aspect of any philosophy of the environment is the relationship between environmental attitudes and gender. One school of thought correlates environmental destruction with a masculine approach to the world and environmental protection with a feminine approach. It is instructive to note the cataloguing of types of ecofeminism in Carolyn Merchant's selection and to compare that concept's diversity with the diverse definitions of "nature." (See also Quinby's "Ecofeminism and the Politics of Resistance" in Chapter 21.)

The present interdisciplinary anthology itself assumes a particular philosophical perspective toward environmental study. Because environment is everywhere and part of everything, it engages all categories of human thought. Moreover, a desire to protect the diversity of nature must, following Emerson, incorporate nature's method into its own. If nature's method is to create diversity, then our understanding of nature must incorporate a diversity of knowledge. Accordingly, every selection in this anthology can be read as one piece of this larger approach, and the reading of every selection concludes with the implicit question, "What other fields of thought can further illuminate the issues that this one selection presents?"

FURTHER READING

Callicott, J. Baird. *Beyond the Land Ethic: More Essays in Environmental Philosophy.* Albany: State University of New York Press, 1999.

Gould, Stephen Jay. "Non-moral Nature." In *Hen's Teeth and Horses' Toes: Further Reflections in Natural History.* New York: W. W. Norton, 1984.

Light, Andrew, and Holmes Rolston III, ed. *Environmental Ethics: An Anthology.* Oxford: Blackwell, 2003.

Plumwood, Val. *Feminism and the Mastery of Nature.* London: Routledge, 1993.

Taylor, Paul. *Respect for Nature: A Theory of Environmental Ethics.* Princeton: Princeton University Press, 1986.

Kate Soper, from *What Is Nature? Culture, Politics, and the Non-Human* (1995)

The words "nature" and "natural" are often used in environmental circles as if they had universally understood meanings. But is a breakfast cereal "natural" in the same way an old-growth forest is "natural"? "Nature" is an essential term, like freedom or love, that is both an entity and a concept, and is capable of infinite flexibility.

"Nature," as Raymond Williams has remarked, is one of the most complex words in the language.[1] Yet, as with many other problematic terms, its complexity is concealed by the ease and regularity with which we put it to use in a wide variety of contexts. It is at once both very familiar and extremely elusive: an idea we employ with such ease and regularity that it seems as if we ourselves are privileged with some "natural" access to its intelligibility; but also an idea which most of us know, in some sense, to be so various and comprehensive in its use as to defy our powers of definition. On the one hand, we are perfectly at home with it, whether the reference is to the "nature" of rocks or to rocks as a part of "nature"; to that "great nature that exists in the works of mighty poets"[2] or to the humbler stuff of "natural" fiber; to the "Nature" park or the nature encroaching on our allotment [of the common land]; to the rudeness of "nature" or to a "naturalness" of manners. On the other hand, merely to contemplate this range of usage is to sense a loss of grip on what it is that we here have in mind. For the "nature" of rocks which refers us to their essential qualities is not the

Figure 17.1 *Nature Unveiling Herself Before Science* by Louis-Ernest Barrias, 1899. Bronze, 58.4 cm (23 in.). Edwin E. Jack Fund, 1974.413. Photograph © 2003 Museum of Fine Arts, Boston.

"nature" conceived as the totality of non-human matter to which they are said to belong. Nor, it seems, is the latter quite what the poet is invoking, or the poet's nature the kind of thing we eat for breakfast. Equally, we may ask how we may so readily speak of what is clearly humanly cultivated, whether it be breakfast cereal or our own modes of comportment, as "natural" while also distinguishing so firmly between what "we" are and do, and the being and productions of "nature"; how we speak of both preserved land and wilderness as "nature," or think of our garden or allotment as both belonging to "nature" and keeping it at bay.

To attempt to disentangle these various threads of nature discourse is immediately to realize what a vast range of possible topics a work such as this one might be addressing. For nature refers us to the object of study of the natural and biological sciences; to issues in metaphysics concerning the differing modes of being of the natural and the human; and to the environment and its various non-human forms of life. The natural is both distinguished from the human and the cultural, but also the concept through which we pose questions about the more or less natural or artificial quality of our own behavior and cultural formations; about the existence and quality of human nature; and about the respective roles of nature and culture in the formation of individuals and their social milieu. Nature also carries an immensely complex and contradictory symbolic load; it is the subject of very contrary ideologies; and it has been represented in an enormous variety of differing ways. In recent times, it has come to occupy a central place on the political agenda as a result of ecological crisis, where it figures as a general concept through which we are asked to rethink our current use of resources, our relations to other forms of life, and our place within, and responsibilities towards the ecosystem. . . .

THE DISCOURSES OF NATURE

In its commonest and most fundamental sense, the term "nature" refers to everything which is not human and distinguished from the work of humanity. Thus "nature" is opposed to culture, to history, to convention, to what is artificially worked or produced, in short, to everything which is defining of the order of humanity. I speak of this conception of nature as "otherness" to humanity as fundamental because, although many would question whether we can in fact draw any such rigid divide, the conceptual distinction remains indispensable. Whether, for example, it is claimed that "nature" and "culture" are clearly differentiated realms or that no hard and fast delineation can be made between them, all such thinking is tacitly reliant on the humanity-nature antithesis itself and would have no purchase on our understanding without it. The implications of this are not always as fully appreciated as they might be either by those who would have us view "nature" as a variable and relative construct of human discourse or by those who emphasize human communality with the "rest of nature." . . . Suffice it to note here that an *a priori* discrimination between humanity and "nature" is implicit in all discussions of the relations between the two, and thus far it is correct to insist that "nature" is the idea through which we conceptualize what is "other" to ourselves.

But for the most part, when "nature" is used of the non-human, it is in a rather more concrete sense to refer to that part of the environment which we have had no hand in creating. It is used empirically to mark off that part of the material world which is given prior to any human activity, from that which is humanly shaped or contrived. This is the sort of distinction which John Passmore makes central to his work on *Man's Responsibility for Nature*, where he writes he will be using the word "nature"

> so as to include only that which, setting aside the supernatural, is human neither in itself nor in its origins. This is the sense in which neither Sir Christopher Wren nor St. Paul's Cathedral forms part of "nature" and it

may be hard to decide whether an oddly shaped flint or a landscape where the trees are evenly spaced is or is not "natural."[3]

Passmore himself admits that this is to use the term in one of its narrower senses; yet it is also, I think, to use it in the sense which corresponds most closely to ordinary intuitions about its essential meaning. The idea of "nature" as that which we are not, which we are external to, which ceases to be fully "natural" once we have mixed our labor with it, or which we have destroyed by our interventions, also propels a great deal of thought and writing about "getting back" to nature, or rescuing it from its human corruption. Ecological writing, for example, very frequently works implicitly with an idea of nature as a kind of pristine otherness to human culture, whose value is depreciated proportionately to its human admixture, and this is an idea promoted by Robert Goodin, in his attempt to supply a "green theory of value." What is crucial to a "green theory of value," argues Goodin, is that it accords value to what is created by natural processes rather than by artificial human ones; and he employs the analogy with fakes and forgeries in art to argue that replications of the environment by developers, even if absolutely exact, will never be the same, or have the same value, precisely because they will not be independent of human process:

> . . . a restored bit of nature is necessarily not as valuable as something similar that has been "untouched by human hands." Even if we simply stand back and "let nature take its course" once again, and even if after several decades most of what we see is the handiwork of nature rather than of humanity, there will almost inevitably still be human residues in its final product. Even if we subsequently "let nature take its course," *which* course it has taken will typically have been dictated by that human intervention in the causal history. To the extent that that is true, even things that are largely the product of natural regeneration are still to some (perhaps significant) degree the product of human handiwork. And they are, on the green theory of value, that much less valuable for being so.[4]

But persuasive as these approaches may seem, in some ways, there are a number of reasons to question their tendency to elide "nature" defined as that "which is human neither in itself nor its origins" with "nature" defined as that part of the environment which is humanly unaffected. Much, after all, that is "natural" in the first sense is also affected by us, including, one may argue, the building materials which have gone into the making of St. Paul's Cathedral. On the other hand, if "nature" is identified with that part of the environment that is humanly unaffected, then, as Passmore rightly notes, it is being defined in such a way as to leave us uncertain of its empirical application, at least in respect of the sort of examples he gives: the oddly shaped flint, the landscape where the trees are straight, and so on. The fact that we may not be able concretely to determine what is or is not "natural" in this sense is no objection, of course, to its conceptualization as that which is unaffected by human hand; but when we consider how much of our environment we most certainly know *not* to be "natural" in this sense, and how much of the remainder we may be rather doubtful about, we may feel that the conceptual distinction, though logically clear enough, has lost touch with the more ordinary discriminations we make through the idea of "nature"—as between the built and unbuilt environment, the "natural" and the "artificial" coloring, the "Nature" park and the opera house, and so forth.

If we consider, that is, the force of Marx's remark that: "the nature which preceded human history no longer exists anywhere (except perhaps on a few Australian coral-islands of recent origin)";[5] and if we then consider the human "contamination" to which these possible "exceptions" have been subject since he wrote, then it is difficult not to feel that in thinking of "nature" as that which is utterly unaffected by human dealings, we are thinking of a kind of being to which rather little on the planet in reality corresponds. Now this, it might be said, is precisely the force of so construing it, namely that it brings so clearly into view its actual disappearance; the extent, that is, to which humanity has destroyed, nay obliterated, "nature" as a result of its occupation of the planet. This certainly seems to be the kind of prescriptive force that Goodin, and some of those associated with a "deep ecology" approach, would wish to draw from it, insofar as they present human beings as always desecrating nature howsoever they intervene in it.

But to press this kind of case is inevitably to pose some new conceptual problems. For it is to present humanity as in its very being opposed to nature, and as necessarily destroying, or distraining on its value, even in the most minimal pursuit of its most "natural" needs. Since merely to walk in "nature," to pluck the berry, to drink the mountain stream, is, on this theory of value, necessarily to devalue it, the logical conclusion would seem to be that it would have been better by far had the species never existed. But, at this point, we might begin to wonder why the same argument could not apply to other living creatures, albeit they are said, unlike ourselves, to belong to nature, since they, too, make use of its resources, destroy each other, and in that sense corrupt its pristine paradise. In other words, we may ask what it is exactly that makes a human interaction with "nature" intrinsically devaluing, where that of other species is deemed to be unproblematic—of the order of nature itself. If humanity is thought to be an intrusion upon this "natural" order, then it is unclear why other creatures should not count as "intrusions" also, and inanimate "nature" hence as better off without them. We may begin to wonder what it is exactly that renders even the most primitive of human dwellings an "artificial" excrescence, but allows the beehive or ant-heap to count as part of nature; or, conversely, whether the humanity-nature relationship is not here being conceived along lines that might logically require us to question the "naturality" of species other than our own. Or to put the point in more political terms: we may suspect that this is an approach to the "value" of nature that is too inclined to abstract from the impact on the environment of the different historical modes of "human" interaction with it, and thus to mislocate the source of the problem—which arguably resides not in any inherently "devaluing" aspect of human activity, but in the specific forms it has taken.

But rather than pursue these issues further here, let me return to the point I earlier raised concerning ordinary parlance about "nature." For there is no doubt that any definition of nature as that untouched by human hand is belied by some of the commonest uses of the term. In other words, if we count as "nature" only that which preceded human history, or is free from the impact of human occupancy of the planet, then it might seem as if we were committed to denying the validity of much of our everyday reference to "nature." To speak of the "nature walk" or "Nature Park"; of "natural" as opposed to "artificial" additives; of the "natural" environment which we love and seek to preserve—all this, it might follow from this approach, is a muddle; and a muddle, it might be further argued, that we ought to seek to correct through an adjustment of language. But tempting as it might seem, in view of the conceptual imprecision of ordinary talk of "nature," to want to police the term in this way, there are a number of reasons to resist the move. In the first place, talk of the countryside and its "natural" flora and fauna may be loose, but it still makes discriminations that we would want to observe between different types of space and human uses of it. If ordinary discourse lacks rigor in referring to woodland or fields, the cattle grazing upon them, and so forth, as "nature," it is still marking an important distinction between the urban and industrial environment. . . . The criteria employed in such distinctions may be difficult to specify, but the distinctions are not of a kind that we can readily dispense with, or that a more stringent use of terminology can necessarily capture more adequately. Or to put the point in more Wittgensteinian terms, it may be a mistaken approach to the meaning of terms to attempt to specify *how* they should be employed as opposed to exploring the *way* in which they are actually used. The philosopher's task, suggested Wittgenstein, was not to prescribe the use of terms in the light of some supposedly "strict" or essential meaning, but to observe their usage in "ordinary" language itself; and it is certainly in that spirit that much of my pursuit here of the "meaning" of nature will be conducted, even if that only serves to expose its theoretical laxity relative to any particular definition we might insist it ought to have. Indeed, there is perhaps something inherently mistaken in the attempt to define what nature is, independently of how it is thought about, talked about, and culturally represented. There can be no adequate attempt, that is, to explore "what nature is" that is not centrally concerned with what it has been *said* to be, however much we might want to challenge that discourse in the light of our theoretical rulings.

Notes

1. Raymond Williams, *Problems in Materialism and Culture* (London: Verso, 1980), 68.
2. William Wordsworth, *The Prelude*, Book Five, lines 594–5.
3. John Passmore, *Man's Responsibility for Nature* (London: Duckworth, 1980), 207.
4. Robert Goodin, *Green Political Theory* (Oxford: Polity Press, 1992), 41, cf. 30–40.
5. Karl Marx, *The German Ideology* (Moscow: Progress Press, 1968), 59.

Johann Wolfgang von Goethe, from "Formation and Transformation" (1817–24), translated by Bertha Mueller

In this piece, Goethe establishes the science of morphology, whose principle of natural change and evolution foreshadowed Darwin. Although today, morphologists tend to look at the forms as fixed objects of study, Goethe shows that in its original conception morphology is the study of things that, in modern slang, "morph." To provide a sense that all science is not progress, this selection should be read together with the Emerson selections that follow; scientific reductionism ignores the Goethian and Emersonian views of nature, it has not disproven them.

The German language has the word "Gestalt" to designate the complex of life in an actual organism. In this expression the element of mutability is left out of consideration: it is assumed that whatever forms a composite whole is made fast, is cut off, and is fixed in its character.

However, when we study forms, the organic ones in particular, nowhere do we find permanence, repose, or termination. We find rather that everything is in ceaseless flux. This is why our language makes such frequent use of the term "Bildung" to designate what has been brought forth and likewise what is in the process of being brought forth.

In introducing a science of morphology, we must avoid speaking in terms of what is fixed. Thus, if we use the term "Gestalt" at all, we ought to have in mind only an abstract idea or concept, or something which in actuality is held fast for but an instant.

What has just been formed is instantly transformed, and if we would arrive, to some degree, at a vital intuition of Nature, we must strive to keep ourselves as flexible and pliable as the example she herself provides. . . .

Each living creature is a complex, not a unit; even when it appears to be an individual, it nevertheless remains an aggregation of living and independent parts, identical in idea and disposition, but in outward appearance identical or similar, unlike or dissimilar. These organisms are partly united by origin; partly they discover each other and unite. They separate and seek each other out again, thus bringing about endless production in all ways and in all directions.

Ralph Waldo Emerson, from *Nature* (1836) and from "The Method of Nature" (1841)

Emerson's simple observation "the sun shines also today" is echoed by Thoreau in the last sentence of Walden: "the sun is but a morning star." These statements are variations on the same Trancendentalist rallying cry: envision the world anew, do not be fettered by the old ways of seeing. For Emerson, this is accomplished by immersing oneself in Nature in order to learn her ways and make them our own. Richard Lewontin modernizes one primary lesson — that which "inhabits the organ . . . makes the organ" — in "Organism and Environment," below: the organism creates its own environment.

FROM *NATURE*

Our age is retrospective. It builds the sepulchres of the fathers. It writes biographies, histories, and criticism. The foregoing generations beheld God and nature face to face; we, through their eyes. Why should not we also enjoy an original relation to the universe? Why should not we have a poetry and philosophy of insight and not of tradition, and a religion by revelation to us, and not the history of theirs? Embosomed for a season in nature, whose floods of life stream around and through us, and invite us, by the powers they supply, to action proportioned to nature, why should we grope among the dry bones of the past, or put the living generation into masquerade out of its faded wardrobe? The sun shines today also. There is more wool and flax in the fields. There are new lands, new men, new thoughts. Let us demand our own works and laws and worship.

Undoubtedly we have no questions to ask which are unanswerable. We must trust the perfection of the creation so far as to believe that whatever curiosity the order of things has awakened in our minds, the order of things can satisfy. Every man's condition is a solution in hieroglyphic to those inquiries he would put. He acts it as life, before he apprehends it as truth. In like manner, nature is already, in its forms and tendencies, describing its own design. Let us interrogate the great apparition that shines so peacefully around us. Let us inquire, to what end is nature?

All science has one aim, namely, to find a theory of nature. We have theories of races and of functions, but scarcely yet a remote approach to an idea of creation. We are now so far from the road to truth, that religious teachers dispute and hate each other, and speculative men are esteemed unsound and frivolous. But to a sound judgment, the most abstract truth is the most practical. When-

Figure 17.2 Ralph Waldo Emerson. bMS Am 1280.235 (706.3c).
By permission of the Houghton Library, Harvard University.

ever a true theory appears, it will be its own evidence. Its test is, that it will explain all phenomena. Now many are thought not only unexplained but inexplicable: as language, sleep, madness, dreams, beasts, sex.

Philosophically considered, the universe is composed of Nature and the Soul. Strictly speaking, therefore, all that is separate from us, all which Philosophy distinguishes as the NOT ME, that is, both nature and art, all other men and my own body, must be ranked under this name, NATURE. In enumerating the values of nature and casting up their sum, I shall use the word in both senses—in its common and in its philosophical import. In inquiries so general as our present one, the inaccuracy is not material; no confusion of thought will occur. *Nature*, in the common sense, refers to essences unchanged by man: space, the air, the river, the leaf. *Art* is applied to the mixture of his will with the same things, as in a house, a canal, a statue, a picture. But his operations taken together are so insignificant, a little chipping, baking, patching, and washing, that in an impression so grand as that of the world on the human mind, they do not vary the result. . . .

At present, man applies to nature but half his force. He works on the world with his understanding alone. He lives in it and masters it by a penny-wisdom; and he that works most in it is but a half-man, and while his arms are strong and his digestion good, his mind is imbruted, and he is a selfish savage. His relation to nature, his power over it, is through the understanding, as by manure; the economic use of fire, wind, water, and the mariner's needle; steam, coal, chemical agriculture; the repairs of the human body by the dentist and the surgeon. This is such a resumption of power as if a banished king should buy his territories inch by inch, instead of vaulting at once into his throne. Meantime, in the thick darkness, there are not wanting gleams of a better light—occasional examples of the action of man upon nature with his entire force—with reason as well as understanding. . . .

The problem of restoring to the world original and eternal beauty is solved by the redemption of the soul. The ruin or the blank that we see when we look at nature, is in our own eye. The axis of vision is not coincident with the axis of things, and so they appear not transparent but opaque. The reason why the world lacks unity, and lies broken and in heaps, is because man is disunited with himself. He cannot be a naturalist until he satisfies all the demands of the spirit. Love is as much its demand as perception. Indeed, neither can be perfect without the other. In the uttermost meaning of the words, thought is devout, and devotion is thought. Deep calls unto deep. But in actual life, the marriage is not celebrated. There are innocent men who worship God after the tradition of their fathers, but their sense of duty has not yet extended to the use of all their faculties. And there are patient naturalists, but they freeze their subject under the wintry light of the understanding. Is not prayer also a study of truth—a sally of the soul into the unfound infinite? No man ever prayed heartily without learning something. But when a faithful thinker, resolute to detach every object from personal relations and see it in the light of thought, shall, at the same time, kindle science with the fire of the holiest affections, then will God go forth anew into the creation.

It will not need, when the mind is prepared for study, to search for objects. The invariable mark of wisdom is to see the miraculous in the common.

* * *

"THE METHOD OF NATURE"

The method of nature: who could ever analyze it? That rushing stream will not be observed. We can never surprise nature in a corner; never find the end of a thread; never tell where to set the first stone. The bird hastens to lay her egg: the egg hastens to be a bird. The wholeness we admire in the order of the world is the result of infinite distribution. Its smoothness is the smoothness of the pitch of the cataract. Its permanence is a perpetual inchoation. Every natural fact is an emanation, and that from which it emanates is an emanation also, and from every emanation is a new emanation. If anything could stand still, it would be crushed and dissipated by the torrent it resisted, and if it were a mind, would be crazed;

as insane persons are those who hold fast to one thought and do not flow with the course of nature. Not the cause, but an ever novel effect, nature descends always from above. It is unbroken obedience. The beauty of these fair objects is imported into them from a metaphysical and eternal spring. In all animal and vegetable forms, the physiologist concedes that no chemistry, no mechanics, can account for the facts, but a mysterious principle of life must be assumed, which not only inhabits the organ but makes the organ.

John Muir, from "Hetch Hetchy Valley" in *The Yosemite* (1912)

Raised a strict Calvinist, John Muir uses biblical analogies to make the case for saving the Hetch Hetchy Valley, "one of Nature's rarest and most precious mountain temples." Despite Muir's eloquent plea, Congress passed the Raker Act in 1913, which permitted the damming of Hetch Hetchy to provide water to the growing city of San Francisco. The "Restore Hetch Hetchy" campaign (http://www.hetchhetchy.org/) hopes to convince the U.S. Department of the Interior to breach the dam and permit the Tuolumne River to run free again.

The making of gardens and parks goes on with civilization all over the world, and they increase both in size and number as their value is recognized. Everybody needs beauty as well as bread, places to play in and pray in, where Nature may heal and cheer and give strength to body and soul alike. This natural beauty-hunger is made manifest in the little windowsill gardens of the poor, though perhaps only a geranium slip in a broken cup, as well as in the carefully tended rose and lily gardens of the rich, the thousands of spacious city parks and botanical gardens, and in our magnificent National parks—the Yellowstone, Yosemite, Sequoia, etc.—Nature's sublime wonderlands, the admiration and joy of the world. Nevertheless, like anything else worth while, from the very beginning, however well guarded, they have always been subject to attack by despoiling gain-seekers and mischief-makers of every degree from Satan to Senators, eagerly trying to make everything immediately and selfishly commercial, with schemes disguised in smug-smiling philanthropy, industriously, shampiously crying, "Conservation, conservation, panutilization," that man and beast may be fed and the dear Nation made great. Thus long ago a few enterprising merchants utilized the Jerusalem temple as a place of business instead of a place of prayer, changing money, buying and selling cattle and sheep and doves; and earlier still, the first forest reservation, including only one tree, was likewise despoiled. Ever since the establishment of the Yosemite National Park, strife has been going on around its borders and I suppose this will go on as part of the universal battle between right and wrong, however much its boundaries may be shorn, or its wild beauty destroyed. . . .

One of my later visits to the Valley was made in the autumn of 1907 with the late William Keith, the artist. The leaf-colors were then ripe, and the great god-

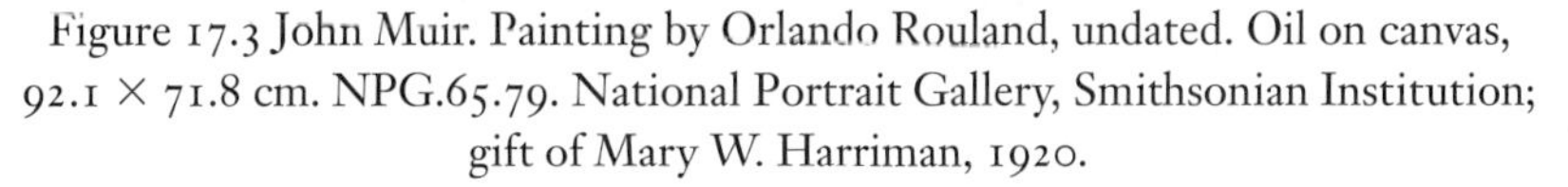

Figure 17.3 John Muir. Painting by Orlando Rouland, undated. Oil on canvas, 92.1 × 71.8 cm. NPG.65.79. National Portrait Gallery, Smithsonian Institution; gift of Mary W. Harriman, 1920.

like rocks in repose seemed to glow with life. The artist, under their spell, wandered day after day along the river and through the groves and gardens, studying the wonderful scenery; and, after making about forty sketches, declared with enthusiasm that although its walls were less sublime in height, in picturesque beauty and charm Hetch Hetchy surpassed even Yosemite.

That any one would try to destroy such a place seems incredible; but sad experience shows that there are people good enough and bad enough for anything. The proponents of the dam scheme bring forward a lot of bad arguments to prove that the only righteous thing to do with the people's parks is to destroy them bit by bit as they are able. Their arguments are curiously like those of the devil, devised for the destruction of the first garden—so much of the very best Eden fruit going to waste; so much of the best Tuolumne water and Tuolumne scenery going to waste. Few of their statements are even partly true, and all are misleading. . . .

"Damming and submerging it 175 feet deep would enhance its beauty by forming a crystal-clear lake." Landscape gardens, places of recreation and worship, are never made beautiful by destroying and burying them. The beautiful sham lake, forsooth, would be only an eyesore, a dismal blot on the landscape,

Figure 17.4 Hetch Hetchy Valley. Photographer untraced.

like many others to be seen in the Sierra. For, instead of keeping it at the same level all the year, allowing Nature centuries of time to make new shores, it would, of course, be full only a month or two in the spring, when the snow is melting fast; then it would be gradually drained, exposing the slimy sides of the basin and shallower parts of the bottom, with the gathered drift and waste, death and decay of the upper basins, caught here instead of being swept on to decent natural burial along the banks of the river or in the sea. Thus the Hetch Hetchy dam-lake would be only a rough imitation of a natural lake for a few of the spring months, an open sepulcher for the others. . . .

These temple destroyers, devotees of ravaging commercialism, seem to have a perfect contempt for Nature, and, instead of lifting their eyes to the God of the mountains, lift them to the Almighty Dollar.

Dam Hetch Hetchy! As well dam for water-tanks the people's cathedrals and churches, for no holier temple has ever been consecrated by the heart of man.

Aldo Leopold, from "The Land Ethic" in *A Sand County Almanac* (1949)

In December 1947, Aldo Leopold submitted his manuscript "Great Possessions" to Oxford University Press. Five months later he died of a heart attack after fighting a fire on his neighbor's farm. In 1949, Oxford published "Great Possessions" as A Sand County Almanac. *Providing a moral basis for human interaction with the land, "The Land Ethic," the concluding chapter of his book, sums up Leopold's philosophy.*

When godlike Odysseus returned from the wars in Troy, he hanged all on one rope a dozen slave-girls of his household whom he suspected of misbehavior during his absence.

This hanging involved no question of propriety. The girls were property. The disposal of property was then, as now, a matter of expediency, not of right and wrong.

Concepts of right and wrong were not lacking from Odysseus's Greece: witness the fidelity of his wife through the long years before at last his black-prowed galleys clove the wine-dark seas for home. The ethical structure of that day covered wives, but had not yet been extended to human chattels. During the three thousand years which have since elapsed, ethical criteria have been extended to many fields of conduct, with corresponding shrinkages in those judged by expediency only.

THE ETHICAL SEQUENCE

This extension of ethics, so far studied only by philosophers, is actually a process in ecological evolution. Its sequences may be described in ecological as well as in philosophical terms. An ethic, ecologically, is a limitation on freedom of action in the struggle for existence. An ethic, philosophically, is a differentiation of social from anti-social conduct. These are two definitions of one thing. The thing has its origin in the tendency of interdependent individuals or groups to evolve modes of co-operation. The ecologist calls these symbioses. Politics and economics are advanced symbioses in which the original free-for-all competition has been replaced, in part, by co-operative mechanisms with an ethical content.

The complexity of co-operative mechanisms has increased with population density, and with the efficiency of tools. It was simpler, for example, to define the anti-social uses of sticks and stones in the days of the mastodons than of bullets and billboards in the age of motors.

The first ethics dealt with the relation between individuals; the Mosaic Decalogue [Ten Commandments] is an example. Later accretions dealt with the relation between the individual and society. The Golden Rule tries to integrate the individual to society; democracy to integrate social organization to the individual.

There is as yet no ethic dealing with man's relation to land and to the animals and plants which grow upon it. Land, like Odysseus's slave-girls, is still property. The land-relation is still strictly economic, entailing privileges but not obligations.

The extension of ethics to this third element in human environment is, if I read the evidence correctly, an evolutionary possibility and an ecological necessity. It is the third step in a sequence. The first two have already been taken. Individual thinkers since the days of Ezekiel and Isaiah have asserted that the despoliation of land is not only inexpedient but wrong. Society, however, has not yet affirmed their belief. I regard the present conservation movement as the embryo of such an affirmation.

An ethic may be regarded as a mode of guidance for meeting ecological situations so new or intricate, or involving such deferred reactions, that the path of social expediency is not discernible to the average individual. Animal instincts are modes of guidance for the individual in meeting such situations. Ethics are possibly a kind of community instinct in-the-making.

THE COMMUNITY CONCEPT

All ethics so far evolved rest upon a single premise: that the individual is a member of a community of interdependent parts. His instincts prompt him to compete for his place in that community, but his ethics prompt him also to cooperate (perhaps in order that there may be a place to compete for).

The land ethic simply enlarges the boundaries of the community to include soils, waters, plants, and animals, or collectively: the land.

This sounds simple: do we not already sing our love for and obligation to the land of the free and the home of the brave? Yes, but just what and whom do we love? Certainly not the soil, which we are sending helter-skelter downriver. Certainly not the waters, which we assume have no function except to turn turbines, float barges, and carry off sewage. Certainly not the plants, of which we exterminate whole communities without batting an eye. Certainly not the animals, of which we have already extirpated many of the largest and most beautiful species. A land ethic of course cannot prevent the alteration, management, and use of these "resources," but it does affirm their right to continued existence, and, at least in spots, their continued existence in a natural state.

In short, a land ethic changes the role of *Homo sapiens* from conqueror of the land-community to plain member and citizen of it. It implies respect for his fellow-members, and also respect for the community as such.

In human history, we have learned (I hope) that the conqueror role is eventually self-defeating. Why? Because it is implicit in such a role that the conqueror knows, *ex cathedra*, just what makes the community clock tick, and just what and who is valuable, and what and who is worthless, in community life. It always turns out that he knows neither, and this is why his conquests eventually defeat themselves.

In the biotic community, a parallel situation exists. Abraham knew exactly what the land was for: it was to drip milk and honey into Abraham's mouth. At the present moment, the assurance with which we regard this assumption is inverse to the degree of our education.

The ordinary citizen today assumes that science knows what makes the community clock tick; the scientist is equally sure that he does not. He knows that the biotic mechanism is so complex that its workings may never be fully understood.

That man is, in fact, only a member of a biotic team is shown by an ecological interpretation of history. Many historical events, hitherto explained solely in terms of human enterprise, were actually biotic interactions between people and land. The characteristics of the land determined the facts quite as potently as the characteristics of the men who lived on it. . . .

THE OUTLOOK

It is inconceivable to me that an ethical relation to land can exist without love, respect, and admiration for land, and a high regard for its value. By value, I of course mean something far broader than mere economic value; I mean value in the philosophical sense.

Perhaps the most serious obstacle impeding the evolution of a land ethic is the fact that our educational and economic system is headed away from, rather than toward, an intense consciousness of land. Your true modern is separated from the land by many middlemen, and by innumerable physical gadgets. He has no vital relation to it; to him it is the space between cities on which crops grow. Turn him loose for a day on the land, and if the spot does not happen to be a golf links or a "scenic" area, he is bored stiff. If crops could be raised by hydroponics instead of farming, it would suit him very well. Synthetic substitutes for wood, leather, wool, and other natural land products suit him better than the originals. In short, land is something he has "outgrown." . . .

One of the requisites for an ecological comprehension of land is an understanding of ecology, and this is by no means co-extensive with "education"; in fact, much higher education seems deliberately to avoid ecological concepts. An understanding of ecology does not necessarily originate in courses bearing ecological labels; it is quite as likely to be labeled geography, botany, agronomy, history, or economics. This is as it should be, but whatever the label, ecological training is scarce. . . .

The "key-log" [the log that, moved, frees the log jam] which must be moved to release the evolutionary process for an ethic is simply this: quit thinking about decent land-use as solely an economic problem. Examine each question in terms of what is ethically and aesthetically right, as well as what is economically expedient. A thing is right when it tends to preserve the integrity, stability, and beauty of the biotic community. It is wrong when it tends otherwise.

It of course goes without saying that economic feasibility limits the tether of what can or cannot be done for land. It always has and it always will. The fallacy the economic determinists have tied around our collective neck, and which we now need to cast off, is the belief that economics determines *all* land-use. This is simply not true. An innumerable host of actions and attitudes, comprising perhaps the bulk of all land relations, is determined by the land-user's tastes and predilections, rather than by his purse. The bulk of all land relations hinges on investments of time, forethought, skill, and faith rather than on investments of cash. As a land-user thinketh, so is he.

I have purposely presented the land ethic as a product of social evolution because nothing so important as an ethic is ever "written." Only the most superficial student of history supposes that Moses "wrote" the Decalogue; it evolved in the minds of a thinking community, and Moses wrote a tentative summary of it for a "seminar." I say tentative because evolution never stops.

The evolution of a land ethic is an intellectual as well as emotional process. Conservation is paved with good intentions which prove to be futile, or even dangerous, because they are devoid of critical understanding either of the land, or of economic land-use. I think it is a truism that as the ethical frontier advances from the individual to the community, its intellectual content increases.

Richard Lewontin, from "Organism and Environment" (1982)

In most modern-day biological investigations, an organism's environment is viewed as an entity independent from the organism. Geneticist Richard Lewontin criticizes this method of investigation, noting that organisms collectively construct their environments. He argues for evolution and ecology, environment and organism, to be seen as a complex web of interdevelopment and influence, not a linear unfolding and not a pock-marked hit-or-miss set of variations.

THE INTERPENETRATION OF ORGANISM AND ENVIRONMENT

The realization that organisms convert a generalized *a priori* selection pressure into a particular "selection for," *a posteriori*, brings us to the central problem in the metaphor of trial-and-error adaptation. It leads us, moreover, to the central contradiction in evolutionary epistemologies [of unfolding]. If organisms adapt either in evolution or in learning, there must be something out there to which they are being fitted. The notion of adaptation is the notion of an independent and pre-existent circumstance to which an object is being adapted, as a key is ground to fit a lock, or my American electric shaver is fitted by an adapter to work on British voltage. There must be a challenge for there to be a response, a problem for there to be a solution. That is, the metaphor of adaptation begins with a world in which an organism's environment is somehow defined without reference to the organism itself, but as a given to which the organism adapts itself. This is the notion that the world is divided up into pre-existent ecological niches and that evolution consists of the progressive fitting of organisms into these niches. Environment begins as alienated from the organism, which must then bring itself into conformity with the given world. This view of environment as causally prior to, and ontologically independent of, organisms is the surfacing in evolutionary theory of the underlying Cartesian structure of our world view. The world is divided into causes and effects, the external and the internal, environments and the organisms they "contain." While this structure is fine for clocks, since mainsprings move the hands and not vice versa, it creates indissoluble contradictions when taken as the meta-model of the living world.

The world external to a given organism can be partitioned into *a priori* ecological niches in a non-denumerable infinity of ways. Yet only some niches are occupied by organisms. How can we know, in the absence of the organisms that already occupy them, which of the partitions of the world are niches? Unless some way exists for picking them out, the concept of niche loses all value. Moreover, quite aside from the epistemological problem of how we know an unoccupied niche when we see it, there is the ontological problem of whether such unoccupied niches have a prior existence. There are no vertebrates that lay eggs, wriggle on their abdomens, and eat grass. That is, there are no herbivorous snakes. Birds live in trees, but no flying vertebrate eats the leaves at the tops of trees either.

In fact, it is impossible to describe an environment except by reference to organisms that interact with it and define it. Organism and environment are dialectically related. There is no organism without an environment, but there is no environment without an organism. The interpenetration of organism and environment takes a number of forms, all of which need to be incorporated into any theory of how organisms evolve.

First, *organisms assemble their environments out of the bits and pieces of the world.* Indeed, an environment is nature organized by an organism. Phoebes in my garden gather bits of dead grass to make nests. It is that act which makes dead grass part of the phoebe's environment. The stones among which the grass grows are not part of the phoebe's environment, but they are part of the environment of the thrushes that use them as anvils to break snails on. Neither stones nor grass are part of the environment of the sapsucker who lives in the tree at whose base both grass and stones lie. Nor is it only behavior that determines which part of closely adjacent bits of the world belong to an organism's environment. Animals and plants are surrounded by a layer of warm air a few millimeters thick, produced by their own metabolic heat. In the case of human beings, this insulating layer is in motion, from the feet and torso up over the face and head. We do not live in the outside air, but literally in an atmospheric shell of our own manufacture. Moreover, small organisms like fleas or other ectoparasites live completely immersed in that boundary layer, provided they are small enough. Should they grow a bit larger they would suddenly find themselves only up to their knees in the warm boundary layer while their bodies and heads would emerge into the stratosphere of the outside world. It is the total morphology, physiology, and behavior of an organism that determines what bits of the world are assembled into its environment.

Second, *organisms alter their environments.* It is a fundamental feature of life that organisms both create and destroy the conditions for their own existence by physical alterations in their milieu. Small mammals burrow in the ground to make their nests, and beavers dramatically alter water levels to create the ponds on which they depend for defense and feeding. In the northeastern United States beaver activity has been the main force altering water tables and topography in forested regions since the end of the Pleistocene glaciation. Even maggots who burrow in rotting fruit increase the surface areas on which the yeasts grow that are the food of the maggots. At the same time organisms destroy the conditions that made their lives possible in the first place. They consume food, they deposit wastes. They may strangle their own offspring. White pine in New England cannot maintain itself in pure stands because the dense shade it creates prevents its own seedlings from surviving, so it gives way to shade-tolerant hardwoods. The entire phenomena of ecological succession in which fields yield to weeds, weeds to shrubs, shrubs to trees, and trees to other trees is a consequence of the alteration of soil texture, chemistry, moisture, and light created by assemblages of plant species to their own detriment. Obviously human beings are both the producers and consumers of their own environment, but they only bring to a higher pitch activities that are characteristics of all organisms.

Third, *organisms transduce physical inputs qualitatively.* The physical changes that occur in nature are not necessarily detected by organisms as changes of the same quality. A change in external temperature is not sensed by my body organs as heat flux but as a change in the concentration of various hormones, sugars, and inorganic ions that are released and consumed when I thermoregulate. The photon energy and compression waves that reach my eyes and ears when I see and hear a rattlesnake become converted by my central nervous system into chemical flows of adrenalin and the release of stored sugar. The same sight and sound will presumably have a very different effect on another rattlesnake.

Fourth, *organisms modulate signals from the environment statistically.* Variations in environmental factors are altered in amplitude and in period by physiological and behavioral mechanisms as they are transformed into effects on organisms. In this sense, animals and plants are like analogue computers that can integrate signals over time to iron out fluctuations or that can differentiate signal sequences so that the instantaneous rate of change is what matters rather than the actual level of the factor. Fluctuations in available food supply are damped by storage devices, either directly as when a squirrel stores nuts and acorns for the winter or indirectly in storage organs like body fat, glycogen, starchy roots, etc. Indeed, direct storage of food by animals depends on the indirect storage by plants. The potato tuber may have evolved as a statistical damping device for *Solanum tuberosum*, but it has been captured by *Homo sapiens.* Temperature "storage" occurs when plants initiate flower development only after a sufficient number of accumulated degree days above a certain temperature. These are all devices for damping the effect of environmental fluctuation. On the other hand, signal variation is also amplified. While homeotherms [warm-blooded animals] damp temperature fluctuations internally, poikilotherms [cold-blooded animals] can amplify temperature signals, as when butterflies orient their wings flat to the sun when it comes out from behind the clouds, to act as solar collectors. The central nervous system of vertebrates has a system of volume-control feedback circuits so that at low levels of light or sound there is internal amplification of small differences. . . .

The importance of these various forms of dialectical interaction between organism and environment is that we cannot regard evolution as the "solution" by species of some predetermined environmental "problems" because it is the life activities of the species themselves that determine both the problems and solutions simultaneously. Nothing better illustrates the error of the problem-solution model than the seemingly straightforward example of the horse's hoof given by Lorenz. He writes that the

> central nervous apparatus does not prescribe the laws of nature any more than the hoof of the horse prescribes the form of the ground. . . . But just as the hoof of the horse is adapted to the ground of the steppe which it copes with, so our central nervous apparatus for organizing the image of the world is adapted to the real world with which man has to cope.

And, further,

> . . . the hoof of the horse is already adapted to the ground of the steppe before the horse is born and the fin of the fish is adapted to the water before the fish hatches. No sensible person believes that in any of these cases the form of the organ "prescribes" its properties to the subject.

Indeed, there is a real world out there, but Lorenz makes the same mistake as Ruskin who believed in the "innocent eye." It is a long way from the "laws of nature" to the horse's hoof. Rabbits, kangaroos, snakes, and grasshoppers, all of whom traverse the same ground as the horse, do not have hooves. Hooves come not from the nature of the ground, but from an animal of certain size, with four legs, running, not hopping, over the ground at a certain speed and for certain periods of time. The small gracile ancestors of the horse had toes and toenails, not hooves, and they got along very well indeed. So, too, our central nervous systems are not fitted to some absolute laws of nature but to laws of nature operating within a framework created by our own sensuous activity. Our nervous system does not allow us to see the ultraviolet reflections from flowers, but a bee's central nervous system does. And bats "see" what night hawks do not. We do not further our understanding of evolution by general appeals to "laws of nature" to which all life must bend. Rather, we must ask how, within the general constraints of the laws of nature, organisms have constructed environments that are the conditions for their further evolution and reconstruction of nature into new environments. Organisms within their individual lifetimes and in the course of their evolution as a species do not *adapt* to environments; they *construct* them. They are not simply *objects* of the laws of nature, altering themselves to bend to the inevitable, but active *subjects* transforming nature according to its laws.

Carolyn Merchant, from "Gaia: Ecofeminism and the Earth" in *Earthcare: Women and the Environment* (1996)

Nature, as a life-giving source, has long been associated with the female. Recasting our views of nature as being gendered provides not only a set of important philosophical perspectives, but prescriptions on how humans could and should interact with the natural world. Carolyn Merchant introduces the concept of "ecofeminism" and presents diverse categories of ecofeminism, each offering a different perspective and prescription.

Ecofeminism emerged in the 1970s with an increasing consciousness of the connections between women and nature. The term, "écoféminisme," was coined in 1974 by French writer Françoise d'Eaubonne who called upon women to lead an ecological revolution to save the planet.[1] Such an ecological revolution would entail new gender relations between women and men and between humans and nature.

Developed by Ynestra King at the Institute for Social Ecology in Vermont about 1976, the concept became a movement in 1980 with a major conference on "Women and Life on Earth" held in Amherst, Massachusetts, and the ensuing Women's Pentagon Action to protest anti-life nuclear war and weapons development.[2] During the 1980s cultural feminists in the United States injected new life into ecofeminism by arguing that both women and nature could be liberated together.

Liberal, cultural, social, and socialist feminism have all been concerned with improving the human/nature relationship, and each has contributed to an ecofeminist perspective in different ways (Table 17.1).[3] Liberal feminism is consistent with the objectives of reform environmentalism to alter human relations with nature from within existing structures of governance through the passage of new laws and regulations. Cultural ecofeminism analyzes environmental problems from within its critique of patriarchy and offers alternatives that could liberate both women and nature.

Social and socialist ecofeminism, on the other hand, ground their analyses in capitalist patriarchy. They ask how patriarchal relations of reproduction reveal the domination of women by men, and how capitalist relations of production re-

Figure 17.5 *Earth Mother* by Edward Burne-Jones, 1882. Sarah C. Garver and Charlotte E. W. Buffington Funds, Worcester Art Museum, Worcester, Massachusetts. Photograph © Worcester Art Museum.

veal the domination of nature by men. They seek the total restructuring of the market economy's use of both women and nature as resources. Although cultural ecofeminism has delved more deeply into the woman-nature connection, social and socialist ecofeminism have the potential for a more thorough critique of domination and for a liberating social justice.

TABLE 17.1

Feminism and the Environment

	Nature	*Human nature*	*Feminist critique of environmentalism*	*Image of a feminist environmentalism*
Liberal feminism	Atoms Mind/body dualism Domination of nature	Rational agents Individualism Maximization of self-interest	"Man and his environment" leaves out women	Women in natural resources and environmental sciences
Marxist feminism	Transformation of Nature by science and technology for human use Domination of nature as a means to human freedom	Creation of human nature through mode of production, praxis Historically specific—not fixed	Critique of capitalist control of resources and accumulation of goods and profits	Socialist society will use resources for good of all men and women Resources will be controlled by workers
	Nature is material basis of life: food, clothing, shelter, energy	Species nature of humans		Environmental pollution could be minimal since no surpluses would be produced Environmental research by men and women

Cultural feminism	Nature is spiritual and personal Conventional science and technology problematic because of their emphasis on domination	Biology is basic Humans are sexually reproducing bodies Sexed by biology/gendered by society	Unaware of interconnectedness of male domination of nature and women Male environmentalism retains hierarchy Insufficient attention to environmental threats to woman's reproduction (chemicals, nuclear war)	Woman/Nature both valorized and celebrated Reproductive freedom Against pornographic depictions of both women and nature Cultural ecofeminism
Socialist feminism	Nature is material basis of life: food, clothing, shelter, energy Nature is socially and historically constructed Transformations of nature by production and reproduction	Human nature created through biology and praxis (sex, race, class, age) Historically specific and socially constructed	Leaves out nature as active and responsive Leaves out women's role in reproduction and reproduction as a category Systems approach is mechanistic and not dialectical	Both nature and human production are active Centrality of biological and social reproduction Dialectic between production and reproduction Multileveled structural analysis Dialectical (not mechanical) systems Socialist ecofeminism

Ecofeminist actions address the contradiction between production and reproduction. Women attempt to reverse the assaults of production on both biological and social reproduction by making problems visible and proposing solutions. When radioactivity from nuclear powerplant accidents, toxic chemicals, and hazardous wastes threaten the biological reproduction of the human species, women experience this contradiction as assaults on their own bodies and on those of their children and act to halt them. Household products, industrial pollutants, plastics, and packaging wastes invade the homes of First World women threatening the reproduction of daily life, while direct access to food, fuel, and clean water for many Third World women is imperiled by cash cropping on traditional homelands and by pesticides used in agribusiness. First World women combat these assaults by altering consumption habits, recycling wastes, and protesting production and disposal methods, while Third World women act to protect traditional ways of life and reverse ecological damage from multinational corporations and the extractive industries. Women challenge the ways in which mainstream society reproduces itself through socialization and politics by envisioning and enacting alternative gender roles, employment options, and political practices.

Many ecofeminists advocate some form of an environmental ethic that deals with the twin oppressions of the domination of women and nature through an ethic of care and nurture that arises out of women's culturally constructed experiences. As philosopher Karen Warren conceptualizes it:

> An ecofeminist ethic is both a critique of male domination of both women and nature and an attempt to frame an ethic free of male-gender bias about women and nature. It not only recognizes the multiple voices of women, located differently by race, class, age, [and] ethnic considerations, it centralizes those voices. Ecofeminism builds on the multiple perspectives of those whose perspectives are typically omitted or undervalued in dominant discourses, for example Chipko women, in developing a global perspective on the role of male domination in the exploitation of women and nature. An ecofeminist perspective is thereby . . . structurally pluralistic, inclusivist, and contextualist, emphasizing through concrete example the crucial role context plays in understanding sexist and naturist practice.[4]

An ecofeminist ethic, she argues, would constrain traditional ethics based on rights, rules, and utilities, with considerations based on care, love, and trust. Yet an ethic of care, as elaborated by some feminists, falls prey to an essentialist critique that women's nature is to nurture.

My own approach is a partnership ethic that treats humans (including male partners and female partners) as equals in personal, household, and political relations and humans as equal partners with (rather than controlled-by or dominant-over) nonhuman nature. Just as human partners, regardless of sex, race, or class, must give each other space, time, and care, allowing each other to grow and develop individually within supportive nondominating relationships, so humans must give nonhuman nature space, time, and care, allowing it to reproduce, evolve, and respond to human actions. In practice, this would mean not cutting forests and damming rivers that make people and wildlife in flood plains more vulnerable to "natural disasters"; curtailing development in areas subject to volcanos, earthquakes, hurricanes, and tornados to allow room for unpredictable, chaotic, natural surprises; and exercising ethical restraint in introducing new technologies such as pesticides, genetically engineered organisms, and biological weapons into ecosystems. Constructing nature as a partner allows for the possibility of a personal or intimate (but not necessarily spiritual) relationship with nature and for feelings of compassion for nonhumans as well as for people who are sexually, racially, or culturally different. It avoids gendering nature as a nurturing mother or a goddess and avoids the ecocentric dilemma that humans are only one of many equal parts of an ecological web and therefore morally equal to a bacterium or a mosquito.

Notes

1. Françoise d'Eaubonne, "Feminism or Death," in Elaine Marks and Isabelle de Courtivron, ed., *New French Feminisms: An Anthology* (Amherst: University of Massachusetts Press, 1980), 64–67, but see especially 25.
2. Ynestra King, "Toward an Ecological Feminism and a Feminist Ecology," in Joan Rothschild, ed., *Machina Ex Dea* (New York: Pergamon Press, 1983), 118–29.
3. Alison Jaggar, *Feminist Politics and Human Nature* (Totowa, N.J.: Rowman and Allanheld, 1983); Karen Warren, "Feminism and Ecology: Making Connections," *Environmental Ethics* 9, no. 1 (1987): 3–10.
4. Karen Warren, "Toward an Ecofeminist Ethic," *Studies in the Humanities* (December 1988): 140–56, quotation on 151.

18 Poetry

Poetry is the most distilled verbal expression of environmental awareness. Fusing perception, cognition, and the passions in a compact way, poetry conveys complex states of feeling and, through symbols, transmits ideas. Poetry calls on many faculties—vision, outward sense, an ear for music, inner visualization (the "mind's eye"), ethics, emotion, memory, reason, and the visionary or prophetic impulse—and especially on the one faculty connecting them all, the imagination. Poetry is distinguished from science, as Samuel Taylor Coleridge notes, by proposing for its immediate object pleasure, not truth. Yet poetry adumbrates truths about human relationships to the environment. Poetry therefore offers one way of knowing as well as feeling: it is knowledge felt. Poetry and science are the great complements of each other. Coleridge adds, "As Poetry is the *identity* [i.e., unity] of all other knowledge, so a Poet cannot be a great Poet but as being likewise and inclusively an Historian and Naturalist."

Environmental consciousness always includes an abiding sense of aesthetics and ethics. The chief duty for developing the aesthetic side rests with poetry and the arts. Poetry contributes to ethical awareness, too, by awakening a perception of the world previously unapprehended. It prevents experience from hardening into thoughtless habit; its alchemy makes the familiar unfamiliar. In this sense, aesthetics, ethics, and epistemology (the study of knowledge) merge. Poetry itself is interdisciplinary.

So is environmental understanding. "Everything is connected to everything else" is the first of four laws of ecology introduced by Barry Commoner in *The Closing Circle: Man, Nature and Technology*. The way to discover or understand any one thing always involves articulation in terms of other things. As with ecosystems, the parts of a poem compose a larger whole. Coleridge says that in a poem, the parts "mutually support and explain each other; all in their proportion harmonizing with" the whole. Moreover, we grasp knowledge in sets of relationships, comparisons, similarities, or parallels of structure and process; knowledge then inflects our action. This calls on the power of metaphor, a condensed relationship of words in which an idea, image, or symbol illuminates and uncovers previously hidden qualities in different ideas, images, symbols, and feelings. Metaphor presents interactive comparisons, not static or superficial similarities. Beyond embellishing ordinary realities, imaginative metaphor can enable profound cognition and open the gateway to deeper realities. In *The Triple Helix* (2000), geneticist Richard Lewontin states that it "is not possible to do the work of science without using a language that is filled with metaphors."

Poetry has been environmentally conscious for thousands of years, in part because of its subject matter. In Eastern traditions, images of nature act as abiding analogues or reflections of human activity. The Japanese lyric poet Matsuo Bashō (A.D. 1644–94) wrote this haiku: "On sweet plum blossoms / The sun rises suddenly. / Look, a mountain path!" Finely delineated images represent and elicit emotions, the natural and human become conjoined—plum blossoms, a path—and a physical experience correlates with internal delight. Many lyrics depict a nature that should not be violated or subdued, but respected, even worshipped. In Western traditions, Virgil's *Georgics* lovingly describe apiaries, plowing, winemaking, and pollination. In *The Seasons*, the first long descriptive poem of the environment in English (1726–46), James Thomson vividly pictures everything from the delicate, frost-like film covering plums, to comets and meteors. Erasmus Darwin, grandfather of the scientist, presents in poems such as *Zoönomia* an idealized evolution of the natural kingdom. In the period known as Romanticism, poets turn directly to nature as both a sensory experience and a symbolic presence pointing to a transcendent order.

Complex, interrelated forms inhabit both poetry and the environment. An aesthetically rich poem is an ecosystem in which constituent parts each play a role and whose total effect is determined not by their addition but by how they interact and interpenetrate with one another to create a whole greater than the sum of its parts. This vision of poetry is sometimes called "organic." Each poem is an environment in which grammar, syntax, word choice, rhythm, line length, imagery, and sounds participate in a greater totality. In this sense, words are, as William Wordsworth called them, "living things." They change over time and represent a living world in flux, a theme as old as the Latin poet Lucretius and summed up by the critic William Hazlitt: "Poetry describes the flowing, not the fixed." The language of poetry possesses a generative, creative power that inflects the very nature of our knowledge.

Reading the Selections: Poetry should not be studied in quite the same manner as prose. The old Japanese method was to memorize a poem and recite it throughout the day, consciously and subconsciously, until it became almost physically and psychically part of one's being. Then, one can reflect on its possible meanings and make any of them a part of one's own character, so that the poem permanently affects sensibility or conduct. Poetry at its best is possessed, almost physically taken in, so that it has a bodily feel. Its words compose a form of music. It is often better to read aloud, and slowly. Let words sink in, without rush.

There is no "secret" to understanding poetry; it is more a matter of taking time. Wherever there is a pause demanded either by sense or punctuation, give that pause its full measure of rest.

FURTHER READING

Bate, Jonathan. *The Song of the Earth.* Cambridge: Harvard University Press, 2000.
Fletcher, Angus. *A New Theory for American Poetry: Democracy, the Environment, and the Future of Imagination.* Cambridge: Harvard University Press, 2003.
Kroeber, Karl. *Ecological Literary Criticism: Romantic Imagining and the Biology of the Mind.* New York: Columbia University Press, 1994.

Lucretius, "Alma Venus" from *De rerum natura* (first century B.C.E.), translated by James Engell

For centuries, various western cultures regarded poetry as a legitimate and accurate vehicle for natural philosophy or science. In about seven thousand lines, Of the Nature of Things (De rerum natura) *treats atoms, volcanoes, ocean currents, earthquakes, magnetism, the human passions and soul, and much more. Lucretius advocates a philosophy of atomic materialism and describes an infinite universe not under complete divine guidance. However, his great opening invocation praises Venus as creative and procreative, sexual and benevolent, the driving power of life, and a guarantor of peace.*

Alma Venus

Mother of Rome's race, desire of mortal
Men and gods, You, sweet nourishing Venus,
Who, under turning constellations of the sky,
Make the sea teem with ships and Earth with fruits,
By your power all life's creatures come to be,
And, rising up, attain the sun's life-light.
From you, Goddess, from you flee the winds
And from your coming flee the clouds of heaven;
For you the daedal[1] earth sends forth her flowers,
For you the calm seascape smiles sparklingly
And the still air shines clear—with radiant light!
Now, as soon as days show signs of spring
And west winds blow creating, swift and freed,
First, the female birds proclaim you, Goddess,
Heralds of your coming, their hearts full pierced
By your ravishing power. Next, wild creatures,
And flocks and herds, leap on blessèd fields, swim swift streams:
Captivated by your allure they follow you,

Their hearts' desire, wheresoe'er you lead them on.
And then, through seas, mountains, and scouring floods,
Through leaf-fringed homes of birds, through verdant plains,
Inciting love's seduction in all their souls,
You each impel with irresistible
Desire to couple and create their kind
With lasting progeny. Since you—and you
Alone—govern the genesis and life
Of all Things, so nothing in the realms of light
Rises without you, nor sweetens, nor bears fruit,
I too desire your coupling, in these lines,
Which I strive to compose on the Nature of Things,
Dedicated to Memmius, my friend, to whom
You've always given excellence in every grace.
For his sake, therefore, now infuse my words,
Divine Goddess, with lasting power and charm.
And meanwhile, make violent acts of war,
On sea and land, cease. Yes, for only you
Can rescue humankind with tranquil peace.
All war's brutal works are Mars' dominion;
This fierce Lord of Arms, utterly undone
By wounds of Love, rests now upon your breast,
And, propping up his well-turned neck, so looks
And gazes on you with parted lips, with love
Feeding his hungry eyes. There he reclines,
Your sacred limbs enfolding him. So come,
Embrace him, Glorious One, and from your mouth
Pour forth sweet Nothings, and for your Romans
Woo the prize of peace.

Note

1. [From Daedalus; wonderfully made.]

Tu Fu, "A Traveler at Night Writes His Thoughts" (eighth century C.E.), translated by Burton Watson

Regarded as China's greatest poet, Tu Fu revitalized both the language and forms of verse. He also turned poetry to new subjects. In this lyric, human experience and the natural realm appear to blend into one picture, one set of mutually reinforcing feelings, in which neither dominates. The lines are tightly structured for layered effects. Description, actions, questions, history, the future, self-doubt, natural rhythms: all converge as music, word, and image simultaneously; by a kind of natural magic they fuse nature and human emotion.

Figure 18.1 *A Solitary Poet by a Lakeside*, formerly attributed to Ma Yuan, late thirteenth–fourteenth century. Ink and color on silk, 24.3 × 19 cm (9.56 × 7.5 in.). Museum of Fine Arts, Boston. Charles Bain Hoyt Collection, 50.1453. Photograph © 2004 Museum of Fine Arts, Boston.

A Traveler at Night Writes His Thoughts

Delicate grasses, faint wind on the bank;
stark mast, a lone night boat:
stars hang down, over broad fields sweeping;
the moon boils up, on the great river flowing.
Fame—how can my writings win me that?
Office—age and sickness have brought it to an end.
Fluttering, fluttering—where is my likeness?
Sky and earth and one sandy gull.

Wang Wei, "Stopping by the Temple of Incense Massed" (eighth century C.E.), translated by Stephen Owen

A physician and painter as well as poet, Wang Wei often portrays quiet scenes with unsurpassed delicacy. Using deft personification ("a stream's sounds choked") and synaesthesia or mixing of the senses ("hues of sunlight were chilled"), he paints a picture of natu-

ral events that match or parallel the drama of his inner life. Yet there is no anthropomorphic projection. The natural environment remains ancient, refuses human didacticism, and exists "with no one there."

Stopping By the Temple of Incense Massed

I knew not of the Temple of Incense Massed,
I went several miles into cloudy peaks.

Ancient trees, trails with no one there,
deep in hills, a bell from I knew not where.

A stream's sounds choked on steep-pitched stones,
and hues of sunlight were chilled by green pines.

Towards dusk at the bend of a deserted pool,
in meditation's calm I mastered passion's dragon.

Meng Jiao, "A Visit to the South Mountains" (eighth–ninth century C.E.), translated by Stephen Owen

Like Tu Fu (above), a poet of the late T'ang Dynasty, Meng Jiao conveys a sober, unflinching, even chilly view of the world. A later poet referred to his work as a "cold cicada's call." Yet he could also see nature as a tonic force sweeping away human illusion and the vanities of learning. His imagery, at times abrupt and bold, creates novel, sharp impressions, yet ones that seem true to sensory experience, making the familiar new and powerful, and rendering the unfamiliar as if it were a suddenly recovered memory.

A Visit to the South Mountains

The South Mountains block up earth and sky,
on its rocks are born sun and moon.

The high crests keep sunlight into the night,
deep valleys don't brighten in daytime.

In mountains men are by nature straight,
though paths are steep, the mind is even.

Steady winds drive through cypress and pine,
their sounds brush pure thousands of ravines.

Reaching this place I repent my learning,
each dawn drawn nearer to groundless renown.

Matsuo Bashō, "You Summer Grasses!" and "Into the Old Pond" (seventeenth century C.E.), translated by Daniel C. Buchanan

Bashō (see also Chapter 20) mastered the haiku, a poem of three lines with syllables arranged five, seven, five. But the haiku is much more than syllable counting. It epitomizes the suggestive genius and allusive spirit present in Japanese poetry and found also in its literary ancestor, Chinese verse. Each haiku opens itself to multiple interpretations. They offer the joy or sad poignancy that a sudden discovery of natural images can reveal. While the discovery may be in the moment, it requires time, patience, intuition, and openness to realize it. A haiku often conveys the unmarked path of Zen. A haiku read once is not read at all. It should be repeated, memorized, reconsidered, and then, to one who is aware, it will reveal, perhaps, something like the full presence of life in this palpable world.

Figure 18.2 *A Hundred Geese*, Yuan dynasty, late thirteenth–early fourteenth century. Handscroll (detail): ink and light color on silk, 22.3 × 228 cm (8.75 × 89.75 in.). Museum of Fine Arts, Boston. Charles Bain Hoyt Collection, 50.1455. Photograph © Museum of Fine Arts, Boston.

You Summer Grasses!

You summer grasses!
Glorious dreams of great warriors
Now only ruins.

Into the Old Pond

Into the old pond
A frog suddenly plunges.
The sound of water.

Navajo Songs, from "Beautyway" (traditional)

This selection presents but part of the larger myth of Beautyway, a complex story of sisters, marriage, nature, education, healing, and tribal culture. The rituals of Beautyway often physically involve the immediate presence and use of plants and animals. The ceremony invokes lightning and rain, mountains and fire and frogs; each creature and phenomenon flows into the next, for all are imbued with meaning and relation. All teach lessons. From this web of life supporting a specific ceremonial purpose emerges the realization that "All is beautiful."

BEAUTYWAY WAS PERFORMED OVER HER

Then, it seems, she had become whole again, and they returned with her to the place of the family into which she had married. After that, it seems, things were told to her as time went on. As she was learning Beautyway, one year passed in this manner. Continuing this occupation, she spent another winter, and also spent another winter, which made three. She also spent another winter, which made four. In reality, she spent four winters, they say. But she thought it had been four days, when in reality she happened to spend four years [there]. By this time, then, the songs that had been sung had all become hers, the sandpaintings and prayersticks made had all become her property, and the foot liniments employed and the incense burnt had become hers. In this way, it seems, she had learned everything without omitting a single item. "A ceremonial shall be held over you with it, over you there shall be a sing, then you shall be holy, it will be your body," she was told, they say. Then, it seems, a sing was planned for her, in the bough circle way a nine-night sing was begun over her, they say. "On that side of the wide Chokecherry Patch your older sister is like yourself, she too has finished learning," she was told. "As a matter of fact, your learning proceeded side by side with hers," she was told. "On that side, too, a ceremonial is beginning over your older sister, so that both ceremonials are starting at the same time," she was told.

Then it seems the ceremonials for them began simultaneously, exactly the same amount of days were left as they proceeded, and on exactly the same day the final day arrived. Over there, on her older sister's side, the Mountaintopway ceremonial was in progress for her, while on this side Beautyway ceremonial was being held for her. Here, it seems, after the ceremonial over her had begun, the singer entered after dark. Immediately, it seems, five unravelings were put down, the weasel [skin], snakeweed, grama grass, rock sage, prairie sagewort, wild rice grass, too, and arm attachments, they say, also the rain plume. Then, it seems, the unravelings were made. These, it seems, were undone on both soles of her feet, then along both of her palms, and along the top of her head. The Toad Man's song was used in making the unravelings:

[1] Down below I am sitting *hai-i-ye* earth, sky, Tortoise . . . stubby rainbow . . . growing things . . . falling rain . . . beads swinging . . . perfect shell disk . . . *lo-ol* call . . . down below I am sitting . . . *hai-i-yo-ye-i-iyi-ye yo-ye*. . . .

Sky . . . Mount Taylor . . . Mountain Lion . . . long rainbow . . . falling rain . . . growing things . . . swinging beads . . . perfect shell disk . . . call of Cornbeetle . . . down below I sit *hai-i-yo-ye* . . .

Badger . . . Yellow Bill . . . badger tail . . . lightning struck close by . . . *tsi-ya-ine* growing things . . . falling rain [etc., as in first stanza]. . . .

Rain is so close . . . rain passed by . . . Bear . . . Salamander . . . falling rain . . . growing things [etc., as in second stanza]. . . .

[2] Upright [straight] I am sitting. . . .

[3] Down below his call is. . . .

[4] Upward his call is. . . .

[5] All [frogs] are singing *hai-i-ye hehe*. . . .

[6] All is beautiful *wowo-hai*. . . .

With these songs, it seems, the unravelings were made, and thus one part of the ceremonial was completed. Then, it seems, on the following day the fire drill was made. The bark of a juniper which had been struck by lightning was brought home and immediately shredded. Then fuel was brought in. Dry lower limbs of piñon and the dried lower branches of juniper, to be used in the fire ceremony, were brought inside for that purpose. They were placed at both sides of the entrance with their growing ends aligned in the regular sunwise fashion, one behind the other. Scrub oak branches were broken off from all four sides [of a tree]. These were stuck into the crevices of the hogan at the four cardinal points. Four fire pokers were brought, they say.

The chokecherry [poker] cut off first was placed pointing to the fire from the east side, the next one from the south, the next one from the west, the one cut off last at the time pointed into the fire from the north side. Then, of the fuel brought in as mentioned, two short twigs were broken off and placed side by side with growing ends to the east side. On top of these the aforementioned bark was placed. Then the [block] on top of which they were to drill was placed in position, growing end sunwise.[1] A notch was then cut in. For the place where [the drill] was to be set, a hollow was made. Down into this hollow lightning-struck stone chips, after motions were made toward it from the four directions, were finally dropped in from above. The stick to be used in drilling the fire was also set in from above, after the same motions had been made over it, in the same order from the four points. A stalk of bear grass was drilled [i.e., twirled]. So, it seems, four men then sat down facing it as they twirled it. Singing was then begun, while the badger rattle was shaken:

> [1] . . . *haga yo-wo . . . yo-ye* . . . earth, sky, Tortoise . . . *awe-ne* . . . flash lightning at night *awe-ne ha-wa yo-wo . . . yo-ye* . . .
>
> Sky . . . Mount Taylor . . . Mountain Lion *awe-ne* . . . dim flashes of yellow above horizon *awe-ne* . . .
>
> Badger . . . Yellow Bill, badger tail *owene* . . . flash lightning *a-wene yo-awo hai-i-ye. . . .*
>
> Rain is close . . . rain passed by . . . Bear *owene* . . . dim yellow flashes *owene . . . ha-ayo ye-yehe. . . .*
>
> [2] Beads [which she wears] *hai-i-ye* [etc., as in preceding song].

Then, it seems, the emetic was placed on the fire. Douglas fir, chokecherry, dinas, manzanita, bearberry, ground juniper, blue sprucc, piñon, juniper, big nettle, slim nettle, cattail flag, bog rush, horsetail, all herbs that are milky, all herbs that have rattles [pods], this entire lot was chopped together. Big nettle and slim nettle are called the chief ones required for the emetic, they say.[2]

Notes

1. Cottonwood is preferred.
2. Some of the other herbs may be used, but all of them need not be gathered.

William Wordsworth, "Nutting" (1798)

At a time of unprecedented industrial and urban growth, Wordsworth (see Chapters 10, above, and 20, below) held that nature possesses strength and wisdom toward which human beings can and should turn. He believed that nature is a guide, admonisher, and comforter; that the natural world commands respect, often fear, but—above all else—love, a love that might also shape relations among men. John Stuart Mill wrote that Wordsworth's poetry saved him from severe depression and mental breakdown. One of Wordsworth's bold personifications claims: "Nature never did betray / The heart that loved her." Fearing the dehumanization of metropolitan life, he fought to preserve the natural beauty of his native Lake District.

Figure 18.3 *William Wordsworth* by Robert Hancock, 1798. National Portrait Gallery, London.

Nutting

——It seems a day
(I speak of one from many singled out)
One of those heavenly days that cannot die;
When, in the eagerness of boyish hope,
I left our cottage-threshold, sallying forth
With a huge wallet o'er my shoulders slung,
A nutting-crook in hand; and turned my steps
Tow'rd some far-distant wood, a Figure quaint,
Tricked out in proud disguise of cast-off weeds[1]
Which for that service had been husbanded,
By exhortation of my frugal Dame—
Motley accoutrement, of power to smile
At thorns, and brakes, and brambles,—and, in truth,
More raggèd than need was! O'er pathless rocks,
Through beds of matted fern, and tangled thickets,
Forcing my way, I came to one dear nook
Unvisited, where not a broken bough

Drooped with its withered leaves, ungracious sign
Of devastation; but the hazels rose
Tall and erect, with tempting clusters hung,
A virgin scene!—A little while I stood,
Breathing with such suppression of the heart
As joy delights in; and, with wise restraint
Voluptuous, fearless of a rival, eyed
The banquet;—or beneath the trees I sate
Among the flowers, and with the flowers I played;
A temper known to those, who, after long
And weary expectation, have been blest
With sudden happiness beyond all hope.
Perhaps it was a bower beneath whose leaves
The violets of five seasons re-appear
And fade, unseen by any human eye;
Where fairy water-breaks do murmur on
For ever; and I saw the sparkling foam,
And—with my cheek on one of those green stones
That, fleeced with moss, under the shady trees,
Lay round me, scattered like a flock of sheep—
I heard the murmur, and the murmuring sound,
In that sweet mood when pleasure loves to pay
Tribute to ease; and, of its joy secure,
The heart luxuriates with indifferent things,
Wasting its kindliness on stocks and stones,
And on the vacant air. Then up I rose,
And dragged to earth both branch and bough, with crash
And merciless ravage: and the shady nook
Of hazels, and the green and mossy bower,
Deformed and sullied, patiently gave up
Their quiet being: and unless I now
Confound my present feelings with the past,
Ere from the mutilated bower I turned
Exulting, rich beyond the wealth of kings,
I felt a sense of pain when I beheld
The silent trees, and saw the intruding sky—
Then, dearest Maiden, move along these shades
In gentleness of heart; with gentle hand
Touch—for there is a spirit in the woods.

Note

1. [clothes]

Samuel Taylor Coleridge, "This Lime-Tree Bower My Prison" (1797)

Coleridge wrote several poems in which he relates an event or natural scene as if he were speaking in animated fashion to a friend, in this case to Charles Lamb. The poet, remaining at home with a scalded foot, imagines his friends out walking. He sees into nature and intuits behind it an omnipotent spirit. As a result, he treasures the smallest plot and leaf as well as magnificent tracts and vistas. Everywhere nature is present and its forms symbolize bonds deeper than material ones alone. Like many writers of the Romantic era, Coleridge senses a transcendent power working in and through the fabric of life, air, light, and water, flesh and imagination. Prospero in Shakespeare's The Tempest *sat in a lime-tree "cell" too while he worked his charms. The lime tree here refers to what is usually called a linden tree.*

This Lime-Tree Bower My Prison

Well, they are gone, and here must I remain,
This lime-tree bower my prison! I have lost
Beauties and feelings, such as would have been
Most sweet to my remembrance even when age
Had dimm'd mine eyes to blindness! They, meanwhile,
Friends, whom I never more may meet again,
On springy heath, along the hill-top edge,
Wander in gladness, and wind down, perchance,
To that still roaring dell, of which I told;
The roaring dell, o'erwooded, narrow, deep,
And only speckled by the mid-day sun;
Where its slim trunk the ash from rock to rock
Flings arching like a bridge; that branchless ash,
Unsunned and damp, whose few poor yellow leaves
Ne'er tremble in the gale, yet tremble still,
Fanned by the water-fall! and there my friends
Behold the dark green file of long lank weeds,
That all at once (a most fantastic sight!)
Still nod and drip beneath the dripping edge
Of the blue clay-stone.
Now, my friends emerge
Beneath the wide wide Heaven—and view again
The many-steepled tract magnificent
Of hilly fields and meadows, and the sea,
With some fair bark, perhaps, whose sails light up
The slip of smooth clear blue betwixt two Isles

Figure 18.4 View through the leaves of a linden tree. Photo by Glenn Adelson.

Of purple shadow! Yes! they wander on
In gladness all; but thou, methinks, most glad,
My gentle-hearted Charles! for thou hast pined
And hungered after nature, many a year,
In the great city pent, winning thy way
With sad yet patient soul, through evil and pain
And strange calamity![1] Ah, slowly sink
Behind the western ridge, thou glorious Sun!
Shine in the slant beams of the sinking orb,
Ye purple heath-flowers! Richlier burn, ye clouds!
Live in the yellow light, ye distant groves!
And kindle, thou blue ocean! So my friend
Struck with deep joy may stand, as I have stood,
Silent with swimming sense; yea, gazing round
On the wide landscape, gaze till all doth seem
Less gross than bodily; a living thing
Which *acts* upon the mind—and with such hues
As veil the Almighty Spirit, when he makes
Spirits perceive his presence.

A delight
Comes sudden on my heart, and I am glad
As I myself were there! Nor in this bower,

This little lime-tree bower, have I not marked
Much that has soothed me. Pale beneath the blaze
Hung the transparent foliage; and I watched
Some broad and sunny leaf, and loved to see
The shadow of the leaf and stem above
Dappling its sunshine! And that walnut-tree
Was richly tinged, and a deep radiance lay
Full on the ancient ivy, which usurps
Those fronting elms, and now, with blackest mass
Makes their dark branches gleam a lighter hue
Through the late twilight; and though now the bat
Wheels silent by, and not a swallow twitters,
Yet still the solitary humble-bee
Sings in the bean-flower! Henceforth I shall know
That Nature ne'er deserts the wise and pure—
No plot so narrow, be but Nature there,
No waste so vacant, but may well employ
Each faculty of sense, and keep the heart
Awake to love and beauty! And sometimes
'Tis well to be bereft of promised good,
That we may lift the soul, and contemplate
With lively joy the joys we cannot share.
My gentle-hearted Charles! when the last rook
Beat its straight path across the dusky air
Homewards, I blessed it, deeming its black wing
(Now a dim speck, now vanishing in the light)
Had crossed the mighty orb's dilated glory,
While thou stoodst gazing; or, when all was still,
Flew creaking o'er thy head, and had a charm
For thee, my gentle-hearted Charles, to whom
No sound is dissonant which tells of Life.

Note

1. [Lamb's insane sister murdered her mother.]

Percy Bysshe Shelley, "Mont Blanc" (1816)

When Shelley wrote this poem, only a few climbers had ever scaled Mont Blanc, the highest peak in the Alps. It appeared as remote and alien as did the top of Everest or even the mountains of Mars a century or two later. In Shelley's poem the imposing mountain asks—as it still may ask—questions about the relation (if any) of the mind to the stupendous yet particular creations of the world. While it hardly provides one system or set

Figure 18.5 *View of Mont-Blanc* by Jean Dubois.
Centre d'Iconographie genevoise, coll. BPU.

of conclusions, the poem is philosophical, for "even these primaeval mountains / Teach the adverting mind." What they teach seems half beyond human words.

Mont Blanc

I

The everlasting universe of things
Flows through the mind, and rolls its rapid waves,
Now dark—now glittering—now reflecting gloom—
Now lending splendor, where from secret springs
The source of human thought its tribute brings
Of waters—with a sound but half its own,
Such as a feeble brook will oft assume
In the wild woods, among the mountains lone,
Where waterfalls around it leap for ever,
Where woods and winds contend, and a vast river
Over its rocks ceaselessly bursts and raves.

II

Thus thou, Ravine of Arve—dark, deep ravine—
Thou many-colored, many-voiced vale,
Over whose pines, and crags, and caverns sail
Fast cloud-shadows and sunbeams: awful scene,

Where Power in likeness of the Arve comes down
From the ice-gulfs that gird his secret throne,
Bursting through these dark mountains like the flame
Of lightning through the tempest;—thou dost lie,
Thy giant brood of pines around thee clinging,
Children of elder time, in whose devotion
The chainless winds still come and ever came
To drink their odors, and their mighty swinging
To hear—an old and solemn harmony;
Thine earthly rainbows stretched across the sweep
Of the ethereal waterfall, whose veil
Robes some unsculptured image; the strange sleep
Which when the voices of the desert fail
Wraps all in its own deep eternity;
Thy caverns echoing to the Arve's commotion,
A loud, lone sound no other sound can tame;
Thou art pervaded with that ceaseless motion,
Thou art the path of that unresting sound,
Dizzy ravine!—and when I gaze on thee
I seem as in a trance sublime and strange
To muse on my own separate fantasy,
My own, my human mind, which passively
Now renders and receives fast influencings,
Holding an unremitting interchange
With the clear universe of things around;
One legion of wild thoughts, whose wandering wings
Now float above thy darkness, and now rest
Where that or thou art no unbidden guest,
In the still cave of the witch Poesy,
Seeking among the shadows that pass by,
Ghosts of all things that are, some shade of thee,
Some phantom, some faint image; till the breast
From which they fled recalls them, thou art there!

III

Some say that gleams of a remoter world
Visit the soul in sleep, that death is slumber,
And that its shapes the busy thoughts outnumber
Of those who wake and live. I look on high;
Has some unknown omnipotence unfurled
The veil of life and death? Or do I lie
In dream, and does the mightier world of sleep
Spread far around and inaccessibly

Its circles? For the very spirit fails,
Driven like a homeless cloud from steep to steep
That vanishes among the viewless gales!
Far, far above, piercing the infinite sky,
Mont Blanc appears, still, snowy, and serene;
Its subject mountains their unearthly forms
Pile around it, ice and rock; broad vales between
Of frozen floods, unfathomable deeps,
Blue as the overhanging heaven, that spread
And wind among the accumulated steeps;
A desert peopled by the storms alone,
Save when the eagle brings some hunter's bone,
And the wolf tracks her there. How hideously
Its shapes are heaped around!—rude, bare, and high,
Ghastly, and scarred, and riven. Is this the scene
Where the old Earthquake-daemon taught her young
Ruin? Were these their toys? Or did a sea
Of fire envelop once this silent snow?
None can reply—all seems eternal now.
The wilderness has a mysterious tongue
Which teaches awful doubt, or faith so mild,
So solemn, so serene, that man may be,
But for such faith, with Nature reconciled;
Thou hast a voice, great Mountain, to repeal
Large codes of fraud and woe; not understood
By all, but which the wise, and great, and good
Interpret, or make felt, or deeply feel.

IV

The fields, the lakes, the forests, and the streams,
Ocean, and all the living things that dwell
Within the daedal earth; lightning, and rain,
Earthquake, and fiery flood, and hurricane,
The torpor of the year when feeble dreams
Visit the hidden buds, or dreamless sleep
Holds every future leaf and flower; the bound
With which from that detested trance they leap;
The works and ways of man, their death and birth,
And that of him and all that his may be;
All things that move and breathe with toil and sound
Are born and die; revolve, subside, and swell.
Power dwells apart in its tranquility,
Remote, serene, and inaccessible:

And *this*, the naked countenance of earth,
On which I gaze, even these primaeval mountains
Teach the adverting mind. The glaciers creep
Like snakes that watch their prey, from their far fountains,
Slow rolling on; there, many a precipice
Frost and the sun in scorn of mortal power
Have piled: dome, pyramid, and pinnacle,
A city of death, distinct with many a tower
And wall impregnable of beaming ice.
Yet not a city, but a flood of ruin
Is there, that from the boundaries of the sky
Rolls its perpetual stream; vast pines are strewing
Its destined path, or in the mangled soil
Branchless and shattered stand; the rocks, drawn down
From yon remotest waste, have overthrown
The limits of the dead and living world,
Never to be reclaimed. The dwelling-place
Of insects, beasts, and birds, becomes its spoil;
Their food and their retreat for ever gone,
So much of life and joy is lost. The race
Of man flies far in dread; his work and dwelling
Vanish, like smoke before the tempest's stream,
And their place is not known. Below, vast caves
Shine in the rushing torrents' restless gleam,
Which from those secret chasms in tumult welling
Meet in the vale, and one majestic river,[1]
The breath and blood of distant lands, for ever
Rolls its loud waters to the ocean waves,
Breathes its swift vapors to the circling air.

V

Mont Blanc yet gleams on high: the Power is there,
The still and solemn Power of many sights,
And many sounds, and much of life and death.
In the calm darkness of the moonless nights,
In the lone glare of day, the snows descend
Upon that mountain; none beholds them there,
Nor when the flakes burn in the sinking sun,
Or the starbeams dart through them; winds contend
Silently there, and heap the snow with breath
Rapid and strong, but silently! Its home
The voiceless lightning in these solitudes
Keeps innocently, and like vapor broods

Over the snow. The secret strength of things
Which governs thought, and to the infinite dome
Of heaven is as a law, inhabits thee!
And what were thou, and earth, and stars, and sea,
If to the human mind's imaginings
Silence and solitude were vacancy?

Note

1. [The Arve joins the Rhône.]

Robert Frost, "Spring Pools" (1928)

Born in San Francisco but a self-adopted son of New England, Frost developed a style that remains fresh. It amalgamates qualities of mind (irony, wit), invention in phrasing (rural in its roots yet beneath it all educated and well read), and first-hand contact with local nature. He seeks nature but refuses the sentimental and transcendent sides of Romanticism. Poems such as "Spring Pools" seem at first easy. Then one discovers how demanding, taut, and efficient the form is, and how closely Frost has observed what he depicts. The lines also capture complex ecological processes—forest growth and succession, wild flowers, light, shade, the change of seasons, hydrology, and the role of temporary wetlands.

Spring Pools

These pools that, though in forests, still reflect
The total sky almost without defect,

Figure 18.6 *Robert Frost* by Gardner Cox, undated. Gift of Gardner Cox, Harvard University Portrait Collection. Photograph © 2004 President and Fellows of Harvard College.

And like the flowers beside them, chill and shiver,
Will like the flowers beside them soon be gone,
And yet not out by any brook or river,
But up by roots to bring dark foliage on.

The trees that have it in their pent-up buds
To darken nature and be summer woods—
Let them think twice before they use their powers
To blot out and drink up and sweep away
These flowery waters and these watery flowers
From snow that melted only yesterday.

Wallace Stevens, "The Planet on the Table" (1953)

Full of intellectual play, cryptic allusions, and philosophical speculation anchored in apt, memorable images, Stevens's work explores the relation of the mind to the external world. Unlike Coleridge, Wordsworth, or Shelley in idiom (to whom he nonetheless owed much), Stevens offers a terse, enigmatic picture of the human and the natural. In "The Planet on the Table," poet and sun and planet seem to share a creative power or deep affinity, at least some of the time.

The Planet on the Table

Ariel was glad he had written his poems.
They were of a remembered time
Or of something seen that he liked.

Other makings of the sun
Were waste and welter
And the ripe shrub writhed.

His self and the sun were one
And his poems, although makings of his self,
Were no less makings of the sun.

It was not important that they survive.
What mattered was that they should bear
Some lineament or character,

Some affluence, if only half-perceived,
In the poverty of their words,
Of the planet of which they were part.

A. R. Ammons, "Corsons Inlet" (1965)

This poem, which owes something to the landscapes and walks of romantic poetry, presents a case that the human mind is an integral part of nature. It evolved not out of nature but as and with nature. Therefore, whatever it produces, even this poem itself, is a shared act of mind and nature. It is a not entirely pleasant experience; life involves death, often sudden, and moments of extinction, personal and near universal. But nature and the minds populating it manage to continue in some intelligible order, even if that overall order is too large and fluid for any one rational accounting, or for any one poem or walk.

Corsons Inlet

I went for a walk over the dunes again this morning
to the sea,
then turned right along
    the surf
                                        rounded a naked headland
                                        and returned

    along the inlet shore:

it was muggy sunny, the wind from the sea steady and high,
crisp in the running sand,
                some breakthroughs of sun
    but after a bit

continuous overcast:

the walk liberating, I was released from forms,
from the perpendiculars,
                straight lines, blocks, boxes, binds
of thought
into the hues, shadings, rises, flowing bends and blends
                        of sight:

                                    I allow myself eddies of meaning:
yield to a direction of significance
running
like a stream through the geography of my work:
                you can find

in my sayings
swerves of action
like the inlet's cutting edge:
there are dunes of motion,
organizations of grass, white sandy paths of remembrance
in the overall wandering of mirroring mind:

but Overall is beyond me: is the sum of these events
I cannot draw, the ledger I cannot keep, the accounting
beyond the account:

in nature there are few sharp lines: there are areas of
primrose
more or less dispersed;
disorderly orders of bayberry; between the rows
of dunes,
irregular swamps of reeds,
though not reeds alone, but grass, bayberry, yarrow, all. . .
predominantly reeds:

I have reached no conclusions, have erected no boundaries,
shutting out and shutting in, separating inside
from outside: I have
drawn no lines:
as

manifold events of sand
change the dune's shape that will not be the same shape
tomorrow,

so I am willing to go along, to accept
the becoming
thought, to stake off no beginnings or ends, establish
no walls:

by transitions the land falls from grassy dunes to creek
to undercreek: but there are no lines, though
change in that transition is clear
as any sharpness: but "sharpness" spread out,
allowed to occur over a wider range
than mental lines can keep:

the moon was full last night: today, low tide was low:
black shoals of mussels exposed to the risk

of air
and, earlier, of sun,
waved in and out with the waterline, waterline inexact,
caught always in the event of change:
            a young mottled gull stood free on the shoals
            and ate
to vomiting: another gull, squawking possession, cracked a crab,
picked out the entrails, swallowed the soft-shelled legs, a ruddy
turnstone running in to snatch leftover bits:

risk is full: every living thing in
siege: the demand is life, to keep life: the small
white blacklegged egret, how beautiful, quietly stalks and spears
                the shallows, darts to shore
                    to stab—what? I couldn't
            see against the black mudflats—a frightened
            fiddler crab?

                the news to my left over the dunes and
reeds and bayberry clumps was
                fall: thousands of tree swallows
                gathering for flight:
                an order held
                in constant change: a congregation
rich with entropy: nevertheless, separable, noticeable
                as one event,
                not chaos: preparations for
flight from winter,
cheet, cheet, cheet, cheet, wings rifling the green clumps,
beaks
at the bayberries:
            a perception full of wind, flight, curve,
            sound:
            the possibility of rule as the sum of rulelessness:
the "field" of action
with moving, incalculable center:

in the smaller view, order tight with shape:
blue tiny flowers on a leafless weed: carapace of crab:
snail shell:
                pulsations of order
                in the bellies of minnows: orders swallowed,

broken down, transferred through membranes
to strengthen larger orders: but in the large view, no
lines or changeless shapes: the working in and out, together
and against, of millions of events: this,
so that I make
no form of
formlessness:

orders as summaries, as outcomes of actions override
or in some way result, not predictably (seeing me gain
the top of a dune,
the swallows
could take flight—some other fields of bayberry
could enter fall
berryless) and there is serenity:

no arranged terror: no forcing of image, plan,
or thought:
no propaganda, no humbling of reality to precept:

terror pervades but is not arranged, all possibilities
of escape open: no route shut, except in
the sudden loss of all routes:

I see narrow orders, limited tightness, but will
not run to that easy victory:
still around the looser, wider forces work:
I will try
to fasten into order enlarging grasps of disorder, widening
scope, but enjoying the freedom that
Scope eludes my grasp, that there is no finality of vision,
that I have perceived nothing completely,
that tomorrow a new walk is a new walk.

Mary Oliver, "The Kingfisher" (1990)

With a direct, unflinching kind of honesty, Mary Oliver's poetry often seeks to be with nature on its own terms, insofar as that is possible. In doing so, she raises ethical, aesthetic, personal, and even historical questions about the environment and our relation to it. Despite her direct language, the answers she suggests are not pat statements; rather, they urge fewer preconceptions and practice clearer sight. The sympathy created is unsentimental yet does not refuse sentiment.

Figure 18.7 Illustration of the belted kingfisher from *Birds of America* by John James Audubon. The Academy of Natural Sciences of Philadelphia, Ewell Sale Stewart Library.

The Kingfisher

The kingfisher rises out of the black wave
like a blue flower, in his beak
he carries a silver leaf. I think this is
the prettiest world—so long as you don't mind
a little dying, how could there be a day in your whole life
that doesn't have its splash of happiness?
There are more fish than there are leaves
on a thousand trees, and anyway the kingfisher
wasn't born to think about it, or anything else.
When the wave snaps shut over his blue head, the water
remains water—hunger is the only story
he has ever heard in his life that he could believe.
I don't say he's right. Neither
do I say he's wrong. Religiously he swallows the silver leaf

with its broken red river, and with a rough and easy cry
I couldn't rouse out of my thoughtful body
if my life depended on it, he swings back
over the bright sea to do the same thing, to do it
(as I long to do something, anything) perfectly.

19 History and the Environment

The interactions between societies and their natural environments have histories that help explain how their current relationships originated. As Donald Worster argues in the selection included here, the widespread contemporary phenomenon of single-crop or single-species agricultural ecosystems is better understood, for example, when we consider how such systems first developed during the industrial revolution of the eighteenth century. The same market economic doctrines that shaped these ecosystems then predominate in developed and developing economies today, creating a perceived imperative that the environment be managed for maximum short-term profit. Environmental study as a discipline and environmentalism as a movement also have closely related histories that can be traced in the context of intellectual and social history. Shifts in ideas or ideology, changes in worldview, and developments in the social structure or the economy have all fundamentally affected how we perceive both individual natural phenomena and the environment as a whole. William Cronon maintains, in fact, that the most important contribution history can make toward environmental studies is to introduce us to the concept of the "social construction" of nature, i.e., how human culture shapes our scientific concepts, our views of the environment, and our patterns of interaction with it. For the dominant view among contemporary historians is that the environment is less an objective reality capable of unbiased study, and more a cultural construct shaped by such factors as the ideologies, intellectual assumptions, sociohistorical circumstances, and scientific paradigms of each succeeding generation.

Leo Marx and Lawrence Buell illustrate this, presenting two views of how, in adapting the European pastoral tradition to the New World, American culture has been shaped by and in turn has intellectually and materially shaped the environment under the influence of a variety of historical, cultural, and social factors like the circumstances of New World settlement, the economics of urbanization and industrialization, gender, and the influence of classical literature on American republican thought. Alfred W. Crosby, Jr., and Patricia Nelson Limerick likewise give us case studies of how history and historical-cultural assumptions affected both the material fate of the environment and environmental perceptions in, respectively, the wake of Columbus's discovery of America, and the nineteenth- and early-twentieth-century domination of the American environmental movement by Eastern white male scholars and writers.

Whether one accepts the "social constructivist" position or not—and there are many who object to it—its proponents' claims amply prove the relevance of history to environmental studies, with regard both to the past and to the present. For our own position is not that different from that of our ancestors, living in and interacting with an environment, but also living in and being shaped by the intellectual, social, and cultural forces and choices of the historical moment. One does not exist without the other, and so, both must be studied together. Samuel P. Hays and Barbara D. Hays demonstrate the truth of this claim in their account of the history of the conservation and environmental movement, which rounds out the chapter.

FURTHER READING

Bailes, Kendall E., ed. *Environmental History: Critical Issues in Comparative Perspective.* Lanham, Md.: University Press of America for the American Society for Environmental History, 1985.

Cronon, William. *Changes in the Land: Indians, Colonists, and the Ecology of New England.* New York: Hill and Wang, 1983.

Crosby, Alfred W., Jr. *Ecological Imperialism: The Biological Expansion of Europe, 900–1900.* Cambridge: Cambridge University Press, 1986.

Knorr-Cetina, Karin D. *The Manufacture of Knowledge: An Essay on the Constructivist and Contextual Nature of Science.* Oxford: Pergamon Press, 1981.

Krutch, Joseph Wood. "The Colloid and the Crystal." In *The Best Nature Writing of Joseph Wood Krutch.* New York: William Morrow, 1969. 309–11.

Merchant, Carolyn, ed. *Major Problems in American Environmental History: Documents and Essays.* Lexington, Mass.: D. C. Heath, 1993.

Worster, Donald, ed. *The Ends of the Earth: Perspectives on Modern Environmental History.* Cambridge: Cambridge University Press, 1988.

———. *Nature's Economy: A History of Ecological Ideas.* 1977. 2nd ed. Cambridge: Cambridge University Press, 1994.

Donald Worster, "Transformations of the Earth: Toward an Agroecological Perspective in History" (1990)

Worster maintains that environmental history should explore the material reality of past human-natural interactions, the effects of human technological interventions on nature over time, and the impact of cultural, ideological, and other mental factors on human environmental practices. He then uses the rise of capitalist monocultural agroecosystems (i.e., market-driven reconstructions of parts of nature in order to produce a single species) to illustrate the relevance of environmental history for understanding environmental problems today.

The new field of ecological or environmental history . . . rejects the common assumption that human experience has been exempt from natural constraints, that people are a separate and uniquely special species, that the ecological consequences of our past deeds can be ignored. . . .

Environmental history deals with the role and place of nature in human life. It studies all the interactions that societies in the past have had with the nonhuman world, the world we have not in any primary sense created. . . .

There are three levels on which the new history proceeds, each drawing on a range of other disciplines and requiring special methods of analysis. The first involves the discovery of the structure and distribution of natural environments of the past. Before one can write environmental history one must first understand nature itself—specifically, nature as it was organized and functioning in past times. The task is more difficult than might first appear, for although nature, like society, has a story of change to tell, there are few written records to reveal most of that story. To make such a reconstruction, consequently, the environmental historian must turn for help to a wide array of the natural sciences and must rely on their methodologies, sources, and evidence, though now and then the documentary materials with which historians work can be a valuable aid to the scientists' labors.

The second level of environmental history is more fully the responsibility of the historian and other students of society, for it focuses on productive technology as it interacts with the environment. . . . Here the focus is on understanding how technology has restructured human ecological relations, that is, with analyzing the various ways people have tried to make nature over into a system that produces resources for their consumption. In that process of transforming the earth, people have also restructured themselves and their social relations. . . . On this level of inquiry, one of the most interesting questions is who has gained and who has lost power as modes of productions have changed.

Finally, forming a third level for the environmental historian is that more intangible, purely mental type of encounter in which perceptions, ideologies, ethics, laws, and myths have become part of an individual's or group's dialogue with nature. People are continually constructing cognitive maps of the world around them, defining what a resource is, determining which sorts of behavior may be environmentally degrading and ought to be prohibited, and generally choosing the ends to which nature is put. Such patterns of human perception, ideology, and value have often been highly consequential, moving with all the power of great sheets of glacial ice, grinding and pushing, reorganizing and recreating the surface of the planet. . . .

The gathering strength of the human imagination over nature is so obvious and dramatic that it is in no danger of being neglected by historians. What has been neglected, however, or left conceptually underdeveloped, is the second level of inquiry I mentioned. And it is to that middle level, the analysis of modes of production as ecological phenomena, and particularly as they are articulated in agriculture, that the rest of this essay is devoted. . . .

To undertake this project, the historian might begin by adopting the scientist's concept of the *ecosystem* and then asking how it might be applied to the agriculture practiced in any setting or period. There is a tall pile of books and scientific papers on the complicated ways in which ecosystems are structured, work, and evolve; but in simplest terms, one might define an ecosystem as the collective entity of plants and animals interacting with one another and the nonliving (abiotic) environment in a given place. . . . Until rather recently, all those ecosystems have been understood by ecologists to have self-equilibrating powers, like automatic mechanisms that slow themselves when they get too hot or speed up when they begin to sputter and stall. Outside disturbances might affect equilibrium, throwing the system temporarily off its regular rhythm, but always (or almost always) it was supposed to return to some steady state. The number of species constituting an ecosystem was believed to fluctuate around some determinable point, the flow of nutrients and energy through the system staying more or less constant. A dominant concern among ecologists has been to explain how such systems manage to cohere, to maintain order and balance, in the midst of all the perturbations to which they are subject.[1]

But historians wanting to undertake an ecological analysis should be aware that lately the conventional ecosystem model sketched above has been coming under considerable criticism from some scientists, and there is no longer any consensus on how it functions or how resilient it is. . . .

Unquestionably, all agriculture has brought revolutionary changes to the planet's ecosystems; and, most agroecologists would agree, those changes have often been destructive to the natural order and imperfect in design and execution. Yet as they have gained understanding of how agricultural systems have interacted with nature, scientists have discovered plenty of reasons to respect the long historical achievement of billions of anonymous traditional farmers. . . .

The landscapes that resulted from such traditional practices were carefully integrated, functional mosaics that retained much of the wisdom of nature; they were based on close observation and imitation of the natural order. Here a field was selected and cleared for intensive crop production; there a forest was preserved as supply of fuel and mast; over there a patch of marginal land was used for pasturing livestock. What may have appeared scattered and happenstance in the premodern agricultural landscape always had a structure behind it—a structure that was at once the product of nonhuman factors and of human intelligence, working toward a mutual accommodation. . . .

So it was, that is, until the modern era and the rise of the capitalist mode of production. Beginning in the fifteenth century and accelerating in the eighteenth and nineteenth centuries, the structure and dynamics of agroecosystems began to change radically. . . .

First in England and then in every part of the planet, agroecosystems were rationally and systematically reshaped in order to intensify, not merely the production of food and fiber, but the accumulation of personal wealth.

Despite many variations in time and place, the capitalistic agroecosystem shows one clear tendency over the span of modern history: a movement toward the radical simplification of the natural ecological order in the number of species found in an area and the intricacy of their interconnections. As markets developed and transportation improved, farmers increasingly concentrated their energies on producing a smaller and smaller number of crops to sell for profit. They became, in short, specialists in production, even to the point of producing virtually nothing for their own direct personal consumption. But that is not all: the land itself evolved into a set of specialized instruments of production. What had once been a biological community of plants and animals so complex that scientists can hardly comprehend it, what had been changed by traditional agriculturists into a still highly diversified system for growing local foodstuffs and other materials, now increasingly became a rigidly contrived apparatus competing in widespread markets for economic success. In today's parlance we call this new kind of agroecosystem a *monoculture*, meaning a part of nature that has been reconstituted to the point that it yields a single species, which is growing on the land only because somewhere there is strong market demand for it. Although farmers in isolated rural neighborhoods may have continued to plant a broad, multispecies spectrum of crops, the trend over the past two hundred years or so has been toward the establishment of monocultures on every continent. As Adam Smith realized back in the eighteenth century, specialization is at the very heart of the capitalist mode of production. It should not be surprising then that it would eventually become the rule in agriculture and land use as it is in manufacturing.[2] . . .

The vulnerabilities inherent in modern monoculture now have a long history to be studied and understood. They include an unprecedented degree of susceptibility to disease, predation, and pest population explosions; a heightened overall instability in the system; a constant tendency of the human manager to take risks for short-term profit, including mining the soil (and in the American West mining a limited underground water resource); an increasing reliance on technological substitutes for natural plant and animal services; a reliance on chemical inputs that have often been highly toxic to humans and other organisms; a dependence on imports from distant regions to keep the local system functioning; and finally, a demand for capital and expertise that fewer and fewer individual farmers could meet.[3] This last characteristic is one of the earliest to show up and has been widely studied in rural history, though seldom from an ecological point of view. Farming communities reflect the biological systems they rest on.

A society cannot radically diminish the diversity of natural ecosystems for the sake of maximum crop production, nor keep the land regimented for profit, nor augment the flow of energy through the system by introducing fossil fuels without changing the rhythms and diversity and structure of power within its various communities. An ecological approach helps explain why capitalistic agriculture has had its peculiar social effects as well as its managerial problems.

Notes

1. The classic explication of the ecosystem concept is Eugene Odum, *Fundamentals of Ecology* (Philadelphia, 1971), 8–23.
2. Adam Smith, *An Inquiry into the Nature and Causes of the Wealth of Nations* (New York, 1937), 4–5.
3. David Pimentel et al., "Land Degradation: Effects on Food and Energy Resources," *Science* (8 Oct. 1976): 149–55.

William Cronon, from "Modes of Prophecy and Production: Placing Nature in History" (1990)

Cronon exemplifies the "social constructivist" view of history and the environment. He criticizes Worster's Marxist-influenced definition of environmental history and his account of capitalism's agroecological impact as both too materialist and too deterministic. While denying neither the material form of human-natural interactions nor the attractions of broad, teleological historical visions, he emphasizes that the environment is substantially the "cultural construct" of specific and often complex sociohistorical circumstances that defy easy systemization.

One of my chief reservations about Worster's proposed research agenda, then, is its potentially excessive materialism. . . . *Food*, like *nature*, is not simply a system of bundled calories and nutrients that sustains the life of a human community by concentrating the trophic energy flows of an ecosystem; it is also an elaborate cultural construct. How and why people choose to eat what they do depends as much on what they *think*—about themselves, their relations to each other, their work, their plants and animals, their gods—as on the organisms they actually eat.

Worster understands this, and he correctly notes that most environmental historians adopt an "interactionist" or "possibilist" philosophical stance when trying to allocate historical causation between materialist and idealist forces. But his decision to emphasize modes of production and his apparent belief that such modes generally occupy the second of his analytical levels suggest his preference for materialist styles of analysis. In fact, *mode of production* broadly construed presumably includes all the familial, social, religious, ideological, and other institutions that allow political-economic relationships to function and reproduce themselves from one generation to the next. One can ignore neither the cultural contexts in which a mode is embedded nor the mode of social reproduction that goes with it. This too Worster understands, and yet the elements of the agroecosystems he stresses in this essay are for the most part material. They thus encourage the bias against integrating ideology with political economy and environment that has been a continuing problem for environmental history. . . .

The final warning I might make about modes of production has to do with the mode that dominates Worster's argument: capitalism. It would be foolish to argue against Worster's claim that the growth of capitalism over the past half millennium has been one of the greatest forces for environmental change in human history. His critique of capitalist agriculture—its commodification of land, its drastic ecological simplification, its affection for dangerously vulnerable monocultures, its promotion of divisions of labor that in the long run can do great damage to nature and human community—is one that for the most part I share. The narrative Worster offers of a transition out of a traditional subsistence agriculture into a market-oriented capitalist agriculture has great force, and environmental historians would be foolish to ignore that great transformation. His emphasis on capitalism is crucial, and so too is his effort to remind historians that even the most urban industrial societies are ultimately agricultural at their base: environmental history without agricultural history is inconceivable.

And yet there may be danger even in so compelling an argument. The greatest attraction of Karl Marx's modes of production was their ability to fit a complex series of historical changes to a single narrative trajectory that organized both past history and future prophecy—from past feudalism to present capitalism to future communism. The modes were so encompassing that virtually any social change could be accommodated within them, giving what might otherwise have appeared incomprehensibly complex the familiar Aristotelian shape of beginning, middle, and end. The same attraction holds for environmental historians. Even though most of us agree about only one mode of production—the capitalist one—that mode allows us to narrate our stories as an endless series of transitions, out of some "traditional" predecessor and into the world we know. The ecological contradictions of capitalism, which we both discover in history and borrow from modern environmentalism, supply the basis for a powerful prophecy about the future environmental disasters that capitalism could (will?) all too easily spawn. If we follow Worster's lead by framing environmental history as a transition into and out of capitalism, we energize our historical argument with all the power of prophecy.

I would be the last to argue that doing this is intrinsically wrong. . . . The capitalist mode of production is one of his [Worster's] most powerful analytical tools as a historian—and one of his most compelling rhetorical tropes as a prophet. Therein lies its attraction and its danger. The phenomenon called capitalism—if it really is the singular thing its label suggests—has been so complicated and hydra-headed that no single analysis or narrative is likely to encompass it. Even if we can recognize certain imperatives that seem to flow from the logic of the capitalist marketplace, their implications in different cultural and environmental contexts are so complex that a metanarrative concentrating only on exploitation and despoliation is unlikely to do them full justice. Were all environmental historians to embark on an analysis of agroecosystems of the sort Worster proposes, I fear they might soon discover themselves telling the same story, albeit in different times and places, over and over again. Perhaps that oft-repeated story—of soils eroded, habitats destroyed, food crops simplified, communities dismantled, ecosystems destabilized—might in some broad sense be historically true, but it might also soon come to seem a Procrustean bed.

Leo Marx, from "Sleepy Hollow, 1844" in *The Machine in the Garden: Technology and the Pastoral Ideal in America* (1964)

Leo Marx describes how, partly because of widespread popular objections to urban modernity and partly because American literature often contrasts a rural national past with an ironically mirroring technological present, America has historically defined itself in terms of the pastoral ideal of a utopian life. An 1844 incident in which Nathaniel Hawthorne's meditation on the "Sleepy Hollow," in Concord, Massachusetts, was interrupted by a train's arrival, symbolizing the industrial revolution, illustrates this.

The pastoral ideal has been used to define the meaning of America ever since the age of discovery, and it has not yet lost its hold upon the native imagination. The reason is clear enough. The ruling motive of the good shepherd, leading figure of the classic, Virgilian mode, was to withdraw from the great world and begin a new life in a fresh, green landscape. And now here was a virgin continent! Inevitably the European mind was dazzled by the prospect. With an unspoiled hemisphere in view it seemed that mankind actually might realize what had been thought a poetic fantasy. Soon the dream of a retreat to an oasis of harmony and joy was removed from its traditional literary context. It was embodied in various utopian schemes for making America the site of a new beginning for Western society. In both forms—one literary and the other in essence political—the ideal has figured in the American view of life. . . .

At first thought the relevance of the ancient ideal to our concerns in the second half of the twentieth century is bound to seem obscure. What possible bearing can the urge to idealize a simple, rural environment have upon the lives men lead in an intricately organized, urban, industrial, nuclear-armed society? The answer to this central question must start with the distinction between two kinds of pastoralism—one that is popular and sentimental, the other imaginative and complex.

The first, or sentimental kind, is difficult to define or even to locate because it is an expression less of thought than of feeling. It is widely diffused in our culture, insinuating itself into many kinds of behavior. An obvious example is the current "flight from the city." An inchoate longing for a more "natural" envi-

Figure 19.1 *Lackawanna Valley* by George Inness, c. 1856. Gift of Mrs. Huttleston Rogers, National Gallery of Art, Washington, D.C. Image © Board of Trustees, National Gallery of Art.

ronment enters into the contemptuous attitude that many Americans adopt toward urban life (with the result that we neglect our cities and desert them for the suburbs). Wherever people turn away from the hard social and technological realities this obscure sentiment is likely to be at work. We see it in our politics, in the "localism" invoked to oppose an adequate national system of education, in the power of the farm bloc in Congress, in the special economic favor shown to "farming" through government subsidies, and in state electoral systems that allow the rural population to retain a share of political power grossly out of proportion to its size. It manifests itself in our leisure-time activities, in the piety toward the out-of-doors expressed in the wilderness cult, and in our devotion to camping, hunting, fishing, picnicking, gardening, and so on. . . .

That such desires are not peculiar to Americans goes without saying; but our experience as a nation unquestionably has invested them with peculiar intensity. The soft veil of nostalgia that hangs over our urbanized landscape is largely a vestige of the once dominant image of an undefiled, green republic, a quiet land of forests, villages, and farms dedicated to the pursuit of happiness. . . .

When we turn from the general to the "high" literary culture, however, we are struck at once by the omnipresence of the same motive. One has only to consider the titles which first come to mind from the classical canon of our literature—the American books admired most nowadays—to recognize that the theme of withdrawal from society into an idealized landscape is central to a remarkably large number of them. Again and again, the imagination of our most respected writers—one thinks of Cooper, Thoreau, Melville, Faulkner, Frost, Hemingway—has been set in motion by this impulse. But while the starting point of their work and of sentimental pastoralism may be the same, the results could hardly be more different. . . .

While in the culture at large it is the starting point for infantile wish-fulfillment dreams, a diffuse nostalgia, and a naïve, anarchic primitivism, yet it also is the source of writing that is invaluable for its power to enrich and clarify our experience. Where, then, shall we locate the point of divergence between these two modes of consciousness. . . .

Since Jefferson's time the forces of industrialism have been the chief threat to the bucolic image of America. The tension between the two systems of value had the greatest literary impact in the period between 1840 and 1860, when the nation reached that decisive stage in its economic development which W. W. Rostow calls the "take-off" . . . the "great watershed in the life of modern societies" when the old blocks and resistances to steady development are overcome and the forces of economic progress "expand and come to dominate the society." . . . Much of the singular quality of this era is conveyed by the trope of the interrupted idyll. The locomotive, associated with fire, smoke, speed, iron, and noise, is the leading symbol of the new industrial power. It appears in the woods, suddenly shattering the harmony of the green hollow, like a presentiment of history bearing down on the American asylum. . . .

Thought and feeling flow both ways. The radical change in the character of society and the sharp swing between two states of feeling, between an Arcadian vision and an anxious awareness of reality, are closely related: they illuminate each other. All of which is another way of accounting for the symbolic power of the motif: it brings the political and the psychic dissonance associated with the onset of industrialism into a single pattern of meaning. Once generated, of course, that dissonance demands to be resolved.[1] . . .

Now the great world is invading the land, transforming the sensory texture of rural life—the way it looks and sounds—and threatening, in fact, to impose a new and more complete dominion over it. . . . True, it may be said that agents of urban power had been ravaging the countryside throughout recorded history. After they had withdrawn, however, the character of rural life had remained essentially unchanged. But here the case is different: the distinctive attribute of the new order is its technological power, a power that does not remain confined to the traditional boundaries of the city. It is a centrifugal force that threatens to break down, once and for all, the conventional contrast between these two styles of life. The Sleepy Hollow episode prefigures the emergence, after 1844, of a new, distinctively American, post-romantic, industrial version of the pastoral design. And the feelings aroused by this later design will have the effect of widening the gap, already great, between the pastoralism of sentiment and the pastoralism of mind.

Note

1. Ralph Waldo Emerson, "Art," in *Essays, First Series, The Complete Works of Ralph Waldo Emerson*, 11 vols. (Boston, 1885), 2: 328.

Lawrence Buell, from *The Environmental Imagination: Thoreau, Nature Writing, and the Formation of American Culture* (1995)

Elsewhere in his book Buell builds on the writings of Leo Marx and others to develop a more descriptively comprehensive and conceptually complex account of pastoralism and its literary and cultural impact. In particular he seeks to include experiences of and writings about nature from contexts outside the United States. Here he provides a summary of his argument for the intellectual and practical relevance of pastoral as a myth and an ideology to the environmental movement today.

American Literary Naturism was a variant of a motif to be found worldwide among literary cultures in European languages generated by former colonies. Many, if not most, post-European literatures harbor traditions of envisioning their cultures as nonmetropolitan spaces set apart from the imperium for better or for worse. That provincial self-conception has given rise to latter-day versions of the insecurity Cicero long ago evinced when formulating Rome's relation to *its* fountainhead, Greece, and to varieties of cultural nationalism that try to turn the European perception of the (post)colonial periphery into a cultural asset. From this have arisen myths of the frontier, of the bush, of Africanity.

Here we find the source of much of the ideological mobility of American pastoral. For in the service of cultural self-definition, pastoral has been used by European immigrants to underwrite a program of conquest and by indigenes to decry such conquest. It has been used by corrective forces within settler culture, in the way Emerson and Thoreau criticized the hypercivilized effeteness of Boston, and self-critically by postcolonial writers [such] as V. S. Naipaul who accept metropolitan values. Sometimes it is hard to discern, as we have seen, whether a given text is being accommodationist or oppositional—whether a romantic landscape painting by Thomas Cole participates in the rhetoric of American expansionism or whether it should be seen as a protoenvironmentalist indictment of expansionism.[1] In this chapter I develop the case for new world pastoral's adaptability for ecocentric purposes in light of these complications and in turn its capacity to serve as something more than ideological theater: its capacity, in particular, to register actual physical environments as against idealized abstractions of those. Traditional pastoral, although vaguely localized, was so inclined toward the latter as to tempt one to conclude that "it thematizes the act of fictionalizing."[2] The Renaissance invention of Europe's new worlds under the sign of pastoral, however, set all the following in motion: it held out the prospect that the never-never lands of pastoral might truly be located in actual somewheres; it helped energize quests, both selfish and unselfish, to map and understand those territories; and it thereby helped ensure a future interplay between projective fantasy and responsiveness to actual environments in which pastoral thinking both energized environmental perception and organized that energy into schemas. New world pastoral thus offered both to filter the vision of those enchanted by it and to stimulate them to question metropolitan culture itself (even while participating in it).

Two considerations make new world pastoral especially significant to our study. First, it promotes the idea of vast territories of the actual globe subsisting under the sign of nature. During the era of colonization this idea remains a rudimentary albeit luminous one, unaccompanied by any conservationist impulse—indeed quite the reverse. Still, it lays the groundwork for developing the myth of the land as properly unspoiled, a myth that can give shape and impetus to more recent environmental restoration projects. Second, new world pastoral anticipates the modern would-be environmentalist's dilemma of having to come to terms with actual natural environments while participating in the institutions of a technologized culture that insulates one from the natural environment and splits one's allegiances. Modern environmentalists wishing to speak for the green world are contemporary new world pastoralists. Their challenge is also one of decolonization, insofar as they must fall back on conceptual instruments derived from metropolitan educations that have inevitably somewhat alienated them from the green world, whether they are genealogically settlers or indigenes. In order to inhabit their environment responsibly, in order even to see it, they have to perceive it as something other than just a green world, a dream, a concept. The green world myth is a start. It is the best they can perhaps do at a certain stage. It marks the beginning of the possibility of a mature conception of a heterotopic alternative to the poisoned environments that we increasingly find ourselves inhabiting.[3] But it can become productive only as people learn to use it in earth's interest as well as in humanity's, and this new responsibility cannot be assumed until one begins to look past the mythical vision as well as through it.

Notes

1. See Angela L. Miller, *The Empire of the Eye: Landscape Representation and American Cultural Politics, 1825–1875* (Ithaca: Cornell University Press, 1993), 21–64.
2. Wolfgang Iser, "Pastoralism as a Paradigm of Literary Fictionality," in *The Fictive and the Imaginary: Charting Literary Anthropology* (Baltimore: Johns Hopkins University Press, 1993), 24.
3. "Heterotopia" is a Foucaultian term, referring to actual places within society that function as countersites, where opposition to the predominant culture is located. This insight has been usefully applied to green utopianism in William Chaloupka and R. McGreggor Cawley, "The Great

Wild Hope: Nature, Environmentalism, and the Open Secret," in *In the Nature of Things: Language, Politics, and the Environment*, ed. Jane Bennett and William Chaloupka (Minneapolis: University of Minnesota Press, 1993), 6–21.

Alfred W. Crosby, Jr., from *The Columbian Exchange: Biological and Cultural Consequences of 1492* (1972)

Crosby demonstrates the relevance of historical research to environmental studies by showing that in addition to its far-reaching biological effects, the European discovery of America had a long-lasting intellectual impact. It immediately challenged early Renaissance Europe's Christian and Aristotelian worldview; began the discrediting of the Bible's scientific authority; and set the stage for a centuries-long debate between theories of single creation (monogeneticism) and multiple creation (polygeneticism) only resolved by Charles Darwin and Alfred Russel Wallace.

On the evening of October 11, 1492, Christopher Columbus, on board the *Santa Maria* in the Atlantic Ocean, thought he saw a tiny light far in the distance.

Figure 19.2 *Columbus Encounters the Indians* by Theodor De Bry, late sixteenth century. Courtesy of the Hispanic Society of America, New York.

A few hours later, Rodrigo de Triana, lookout on the *Pinta*'s forecastle, sighted land. In the morning a party went ashore. Columbus had reached the Bahamas. The connection between the Old and New Worlds, which for more than ten millennia had been no more than a tenuous thing of Viking voyages, drifting fishermen, and shadowy contacts via Polynesia, became on the twelfth day of October 1492 a bond as significant as the Bering land bridge had once been.[1]

The two worlds, which God had cast asunder, were reunited, and the two worlds, which were so very different, began on that day to become alike. That trend toward biological homogeneity is one of the most important aspects of the history of life on this planet since the retreat of the continental glaciers.

The Europeans thought they were just off the coast of Asia—back to Eurasia again—but they were struck by the strangeness of the flora and fauna of the islands they had discovered. . . .

The Europeans had emerged from the Middle Ages with intellectual systems, Christian and Aristotelian, claimed by the orthodox (and so few even guessed there was anything beyond orthodoxy) to explain everything from the first and last ticks of history to what happens in the egg prior to the hatching of the chick. These systems proved too cramped to accommodate the New World. . . .

The Bible was the source of most wisdom, and the book of Genesis told all that one needed to know about the beginning of the heavens, earth, angels, plants, animals, and men. There was one God and there had been one Creation; when mankind had offended God, God caused a great flood in which all land creatures, including men, had perished, except those preserved in Noah's ark. This explanation seemed sufficiently broad to include within its bounds all the diversity of life—plant, animal, and human—which the European was obliged to acknowledge up to the end of the fifteenth century. Then da Gama and Columbus brought whole new worlds crashing into the area of European perception.

The problems of explaining Africa and Asia were difficult but surmountable. After all, it had always been known that they were there and, if Europeans had not seen elephants, they had at least always known about them. But America, who had ever dreamed of America? The uniqueness of the New World called into question the whole Christian cosmogony. If God had created all of the life forms in one week in one place and they had then spread out from there over the whole world, then why are the life forms in the eastern and western hemispheres so different? And if all land animals and men had drowned except for those in the ark, and all that now exist are descended from those chosen few, then why the different kinds of animals and men on either side of the Atlantic? Why are there no tree sloths in the African and Asian tropics, and why do the Peruvian heathens worship Viracocha instead of Baal or some other demon familiar to the ancient Jews? The effort to maintain the Hebraic version of the origin of life and man was to "put many learned Christians upon the rack to make it out."[2]

The problem tempted a few Europeans to toy with the concept of multiple creations, but the mass of people clung to monogeneticism. They had to; it was basic to Christianity. . . .

The papacy remained undisturbed in its confidence that the book of Genesis provided all the paleontology that a Christian needed. But America was such a very square peg to fit into the round hole of Genesis. In 1520 Philippus Paracelsus, whose mind was ballasted with little dogma of any variety, is supposed to have said that no one would easily believe that "those who have been found in the out-of-the-way islands . . . are the posterity of Adam and Eve. . . . It is most probable that they are descended from another Adam."[3] Joseph de Acosta was a churchman, but the contrast between the creatures of the Old and New Worlds, which he had seen with his own eyes, also led him to the brink of heresy. There are in America, he wrote,

> a thousand different kindes of birdes and beasts of the forrest, which have never beene knowne, neither in shape nor name; and whereof there is no mention made, neither among the Latins nor Greeks, nor any other nations of the world.

He offered the explanation that "it may be God hath made a new creation of beasts."[4]

The problem of America troubled the seventeenth century, too, helping to lead some few men into unorthodoxy and at least one right into jail. If Eden and Mount Ararat were both in Asia, then how could man and animals be in America? The most influential of the men opposed to orthodox views on the subject was Isaac de La Peyrère. He was more inspired to heresy by biblical ambiguities and references in ancient documents to seemingly pre-Adamite events in Egypt and Phoenicia than by the enigma of a biologically unique America, but his theory provided explanations for all three sources of confusion. Adam was the product of a second creation and father only to the Jews. The first Creation, which preceded that of Adam by a very long time, had included the creation of the ancestors of all the non-Jews—the pre-Adamites—and the Flood had been only Palestinian in extent and had not affected them. Among the descendants of the pre-Adamites were "the Mexicans whom Columbus discovered not so long ago." La Peyrère's book was burned and he was arrested, but polygeneticism lived on.[5]

In 1857 Philip L. Sclater, one of Britain's leading zoologists . . . read before the Linnean Society a paper in which he showed himself to be one who still entertained the idea of multiple creations. This idea would explain how his birds and all other land animals, including man, were distributed as they were. Like all polygeneticists, he started with the false premise that

> every species of animal must have been created within or over the geographical area which it now occupies. Such being the case, if it can be shown

> that the areas now occupied by the primary varieties of mankind correspond with the primary zoological provinces of the globe, it would be an inevitable deduction, that these varieties of Man had their origin in the different parts of the world where they are now found, and the awkward necessity of supposing the introduction of the red man into America by Behring's Straits, and of colonizing Polynesia by stray pairs of Malays floating over the water like cocoa-nuts, and all similar hypotheses would be avoided.

It was in this paper that Sclater put forth his hypothesis that the birds of the world are distributed in six distinctive regions. These regions he divided into two groups, one for the Old World and one for the New. The titles he chose for the two prove him a brother of Acosta: *Creatio Palaeogeana* and *Creatio Neogeana*,[6] or Old World and New World Creation.

Sclater was among the last of the respectable polygeneticists. In 1858 Charles Darwin and Alfred Russel Wallace presented essays to the Linnean Society in which they put forth the modern theory of evolution. One year later, Darwin published *On the Origin of Species*, shattering the concept of multiple creations (while also knocking loose a large part of the foundation of traditional Judaism and Christianity). Once the new theory of evolution was accepted, polygeneticism lived on only as a rationale for racism, in which capacity it still serves.[7]

The real source of conflict between orthodox Christians and the tiny but stubborn number of polygeneticists was that Christians had no adequate concept of change on which to base an explanation of how the earth and the life on earth had reached their present condition. The concept of evolution had existed from at least the time of Aristotle, but it was neither popular nor orthodox: the task of the Christian philosopher and biologist was to provide man with the intellectual means to freeze reality into a stable system, and not to send it slipping and tumbling down the slope of time toward no destination in particular. The accepted belief was that all the kinds of plants and animals, plus the first two people, had been created during the first week of time and that all species were complete as of that first Sunday and were without possibility of developing into new species.

Notes

1. The theoretical basis of this chapter and this book in general is neatly summed up in George Gaylord Simpson, *The Geography of Evolution* (New York: Capricorn Books, 1965), 69–132.
2. Quoted in T. Bendyshe, "The History of Anthropology," *Memoirs of the Anthropological Society of London* 1 (1863–1864): 365.
3. Ibid., 353.
4. Joseph de Acosta, *The Natural and Moral History of the Indies*, trans. Edward Grimston (New York: Burt Franklin [n.d.]), 1: 277–78.
5. Bendyshe, "History of Anthropology," 355–66; Matthew Hale, *The Primitive Origination of Mankind, Considered and Examined According to the Light of Nature* (London: Printed by W. Godbid for W. Shrowsbery, 1677), 182ff; Lee Eldridge Huddleston, *Origins of the American Indians, European Concepts, 1492–1729* (Austin: University of Texas Press, 1967), 139–40; David Rice McKee, "Isaac

de La Peyrère, A Precursor of Eighteenth Century Critical Deists," *Publications of the Modern Language Association of America* 59 (June 1944): 456–85; Margaret T. Hodgen, *Early Anthropology in the Sixteenth and Seventeenth Century* (Philadelphia: University of Pennsylvania Press, 1964), 207–53; Isaac de La Peyrère, *Prae-Adamitae* (n.p.: 1655), 23.

6. Philip L. Sclater, "On the General Geographical Distribution of the Members of the Class Aves," *Journal of the Proceedings of the Linnaean Society (Zoological)* 2 (1858): 131, 145.
7. For an example of polygenetic racism, see Alexander Winchell, *Preadamites, or a Demonstration of the Existence of Men Before Adam* (Chicago: Scott, Foresman, 1901).

Patricia Nelson Limerick, from "Disorientation and Reorientation: The American Landscape Discovered from the West" in *Something in the Soil: Legacies and Reckonings in the New West* (2000)

Herself a historian of the American West, Limerick describes how "the old framework of the east-to-west movement of white men" has dominated accounts of the history and nature of the North American environment. She then tries to establish better standards and a less biased model by which we may analyze the developmental history of regions and landscapes.

In the month of September, in the year 1972, I traveled from the Pacific Coast to the Atlantic Coast, from southern California to southern New England. To the east of Arizona lay wilderness, and the exotic names of that wilderness both chilled and lured me: Tennessee, Virginia, the District of Columbia, Philadelphia, and most alarming of all, Manhattan. In the course of this journey, I discovered the Eastern United States, an event as consequential to me as it was insignificant to the residents.

As I drove across Oklahoma, crossing what I later learned was the ninety-eighth meridian, discovery joined up with its usual partner, disorientation. The air became humid, clammy, and unpleasant, and the landscape turned distressingly green. The Eastern United States, I learned with every mile, was badly infested by plants. Even where they had been driven back, the bushes, shrubs, and trees gave every sign of anticipating a reconquest. But the even more remarkable fact was this: millions of people lived in this muggy, congested world, and most astonishing of all, they considered it *normal*. . . .

Had I read them before my journey to the East, the standard scholarly writings on the discovery of the American landscape would have puzzled me as deeply as the landscape of the Eastern United States did and for many of the same reasons. The study of American responses to landscape ran on an east-to-west track, following the physical and mental migrations of white, English-speaking men. In the conventional view, the process of discovery reached completion when the maps had their blank spots filled in and literate white Americans had seen all the places worth seeing. What the discoverers explored was wilderness, a kind of pristine natural landscape in which Indians lived more as symbols than as three-dimensional human beings. In the same spirit, the authors of the reigning texts in this field took for granted a norm of a green, plant-filled landscape, a norm that they shared with nineteenth-century American explorers of the American West.

In 1972, with these premises locked into place and with the problem of point of view seemingly settled, one could write comfortably and complacently about this subject. In two decades, comfort and complacency have fled the field. Attention to the east-to-west process of exploration now has to compete with a recognition of Indian prior presence, as well as of northward movements from Mexico and eastward movements from Asia. The notion of a pristine wilderness is in well-deserved tatters, and the discoverers now appear as late arrivals in an already fully occupied and much-changed landscape. Thanks to various environmental messes and crises of scarce resources, the celebration of a completed process of discovery, ending in a landscape known, mastered, and put to good use, seems at best silly and at worst dangerous. The assumption that "normal" means green and well watered has lost credibility; through pollution and overallocation of what once seemed to be abundant supplies, the residents of the Eastern United States have created for themselves a number of the dilemmas in water scarcity long familiar to the West.

The new, improved, "revised standard" orthodoxy on the discovery of landscape thus does not miss a beat in bringing its charges against the old orthodoxy. The conventional studies concentrated wholeheartedly on the thinking of English-speaking, westward-moving, literate, record-keeping middle- and upper-class, pre-twentieth-century, white men. Offered as studies of American attitudes toward landscape, these standard works were in fact investigations into the minds of a minority. In the late twentieth century, such exclusivity in scholarly inquiry is no longer tenable.

And at this point, the current orthodoxy comes to an abrupt halt. Working on this essay led me to a full recognition of how the new assumptions escort one to the edge of one's ignorance and then lead one to contemplate the vacancy. Of course, I recognize that human beings discovered the American landscape from all directions: east to west, from Europe and Africa; south to north, from Mexico and South America; west to east, from Asia and the Pacific Islands; from a variety of directions, for Indian people whose arrival took place in another time frame entirely.

And yet almost all of what I know on the subject pulls me back into the old framework of the east-to-west movement of white men. . . .

How to do it better? Ranking high in the category of things easier said than done, here is a list of topics that should figure in the study of the discovery of landscape:

1. The geologic, climatic, botanical, and zoological qualities of a particular place
2. The indigenous people's worldviews and the values they invested (and may well continue to invest) in the landscapes of their home
3. The impact of the native people's actions in creating a landscape that was neither purely natural nor wholly human-made
4. The perceptions and actions of intruders, explorers, invaders, colonizers, and conquerors, responding to places that were at once new to them and familiar to the original residents
5. The second-generation experience of those born to the colonizers—the perspective of the invaders' children to whom the new and exotic area had become home
6. The continuing arrival of both new residents and impression-gathering travelers who may well have felt a sense of discovery, whether or not that sense of originality seemed legitimate to the residents who preceded them
7. The ways in which these various groups saw each other—as legitimate residents, illegitimate invaders, enrichers, despoilers, improvers, or devastators of the landscape or as quaint figures or eyesores in the view
8. Change and continuity in the physical components and arrangements of the landscape over time
9. The ongoing process of discovery and rediscovery, as people with different concerns, needs, and assumptions found and find new meaning in landscapes, even when those landscapes, by an earlier judgment composed of equal parts of cheer and gloom, seemed to have been fully discovered, known, mastered, and thereby reduced in interest
10. A reckoning, in the historian's best judgment and with subjectivity fully acknowledged, of what this whole business added up to—a kind of balance sheet of costs and benefits, gains that can be sustained and gains that prove temporary, injuries and losses where repair and compensation are imaginable and injuries where repair and compensation are beyond imagination. . . .

The landscape thus has a number of layers, all demanding the scholar's attention: rock and soil; plants and animals; humans as a physical presence, manifested in their physical works; *and* humans as an emotional and spiritual presence, manifested in the accumulated stories of their encounter with a place. Our attention and curiosity here cannot be exclusive. One can glimpse the full power of a place only in the full story of human presence there. Thus, exclusive attention to the movements, actions, and impressions of Anglo-Americans is equivalent to the arbitrary editing of a scripture, skipping entire chapters and devoting disproportionate attention to a few featured verses. The complete story of the investment of human consciousness in the American landscape requires attention to the whole set of participants—indigenous people as well as invaders, eastward-moving Asian American people as well as westward-moving Euro-American people. With anything less, the meaning of the landscape is fragmented and truncated.

Samuel P. Hays in collaboration with Barbara D. Hays, from *Beauty, Health, and Permanence: Environmental Politics in the United States, 1955–1985* (1987)

Samuel and Barbara Hays describe the cultural values and material interests that typified the early twentieth-century conservation movement in contrast to those shaping the environmental movement later in the century. In each case, the environment was encountered through a process of appropriation rather than just described and analyzed.

Accounts of the rise of environmentalism frequently have emphasized its roots in the conservation movement of the early twentieth century. But environmental differed markedly from conservation affairs. The conservation movement was an effort on the part of leaders in science, technology, and government to bring about more efficient development of physical resources. The environmental movement, on the other hand, was far more widespread and popular, involving public values that stressed the quality of human experience and hence of the human environment. Conservation was an aspect of the history of production that stressed efficiency, whereas environment was a part of the history of consumption that stressed new aspects of the American standard of living.

Environmental objectives arose out of deep-seated changes in preferences and values associated with the massive social and economic transformation in the decades after 1945. Conservation had stirred technical and political leaders and then worked its way down from the top of the political order, but environmental concerns arose later from a broader base and worked their way from the middle levels of society outward, constantly to press upon a reluctant leadership. Many of the tendencies in efficient management of material resources originating in the conservation era came into sharp conflict with newer environmental objectives. The two sets of values were continually at loggerheads.

At the outset we explore this discontinuity between conservation and environment, this transformation of aim from efficient production to better quality of life, in order to define the historical distinctiveness of environmental affairs.

RIVER DEVELOPMENT

The first clear notion about conservation as more efficient resource use developed in connection with water in the West. As settlement proceeded, water limited farming and urban development. It was not so much that water was scarce as that rainfall came unevenly throughout the year and winter snow melted quickly in the spring. Such patterns did not conform to the seasonal uses of agriculture or the more sustained needs of city dwellers.

How to conserve water? The initial thought was to construct reservoirs to hold rain and snowmelt for use later in the year. Cities began to build storage for urban water supply much as in the East. But for irrigated farmland larger engineering works were needed. The Newlands Act, which Congress passed in 1902, provided for a reclamation fund from the proceeds of the sale of western public lands that would finance irrigation works. By World War II vast projects included plans for a series of dams on the Missouri and the Colorado as well as other western rivers. . . .

The spirit of intensive management, born in these projects, was extended to rivers in other sections of the nation. As projects of lesser cost and scale were completed, the agencies moved on to those of greater and more extensive water transfer. Although the initial projects emphasized the main stems of the larger rivers, the drive to control water expanded steadily into the headwaters to include the flow of entire river basins. By the end of World War II, multipurpose river development was extending its influence to wider and wider realms of action. And it ran headlong into conflict with newer environmental interests that began to emphasize the importance of free-flowing streams unmodified by large engineering structures.

SUSTAINED-YIELD FORESTRY

The spirit of large-scale management for efficient resource development also pervaded the early forestry movement. Concern for the depletion of the nation's wood supplies grew steadily in the last quarter of the nineteenth century.

. . . A number of individuals and organizations pressed for innovations in forestland management. Foremost among these was Gifford Pinchot, who in 1905 became the first chief forester of the U.S. Forest Service. Congress in 1891 had enacted legislation that permitted the president by executive order to establish national forest reserves that would be retained permanently in public ownership as protected forestlands rather than be sold to private individuals. By 1907 Presidents Harrison, Cleveland, McKinley, and Roosevelt had established 150 million acres of timberland in the West as national forests.

. . . The major theme of forest management was "scientific forestry." This involved reforestation of cutover lands, protection from fire, and a balance of annual cut with annual growth, to produce a continuous supply of wood—known as sustained-yield forest management. Over the years foresters improved techniques for measuring yield and cut and for controlling both through a regulatory process that shaped more precisely the flow of growth and harvest.

The economic value of wood production was emphasized above all else. Pinchot was firm in his view that forestry could be promoted in the United States only if it could be profitable through the sale of wood. This spirit continued over the years through the more intensive application of science, technology, and capital to the production of more wood per acre. But this direction also ran counter to new values that were emerging in the American public as the meaning of forestland began to change. Forests were increasingly viewed as environments, aesthetic resources that provided amenities and enhanced daily life, rather than as simply sources of commodities. . . .

FROM CONSERVATION TO ENVIRONMENT

In conservation, forests and waters were closely linked. As soil conservation and game management developed they became allied with both water and forest conservation in a shared set of attitudes. Together they emphasized the scientific management of physical commodities and brought together technical specialists for a common purpose. Departments of state government dealing with such affairs were commonly called departments of natural resources. And professional training at academic institutions evolved from an initial interest in forestry to a larger set of natural resource or conservation matters.

The management of natural resources often displayed a close kinship with the entire movement for scientific management that evolved in the twentieth century and pervaded both industry and government. It emphasized large-scale systems of organization and control and increasing output through more intensive input. Professional expertise played an important role in all four facets of conservation, with strong links among them and a sense of kinship with the wider community of technical professions as a whole. Their self-respect came to be firmly connected to the desire to maintain high professional standards in resource management.

Equally important was the evolution of a common political outlook among resource specialists that professionals should be left free from political influence to determine how resources should be managed and for what purpose. This shared sense of professionalism was itself a political stance, an assertion that those with special training and expertise should determine the course of affairs. From the management of commodity resources in water, forests, soils, and wildlife emerged not just a sense of direction that stressed maximum output of physical resources but also a view about who should make decisions and how they should be made.

The coming conflicts between conservation and environment were rooted in different objectives: efficiency in the development of material commodities or amenities to enhance the quality of life. In these earlier years the national-parks movement and leaders such as John Muir had provided important beginnings for the latter. After World War II extensive changes in human values gave these intangible natural values far greater influence. To them now was added the growing view that air and water, as well as land, constituted a valuable human environment.[1] The early conservation movement had generated the first stages in shaping a "commons," a public domain of public ownership for public use and the public ownership of fish and wildlife as resources not subject to private appropriation. This sense of jointly held resources became extended in the later years to the concept of air, land, and water as an environment. Their significance as common resources shifted from a primary focus on commodities to become also meaningful as amenities that could enhance the quality of life.

THE SEARCH FOR ENVIRONMENTAL AMENITIES

The most widespread source of emerging environmental interest was the search for a better life associated with home, community, and leisure. A new emphasis on smaller families developed, allowing parents to invest their limited time and income in fewer children. Child rearing was now oriented toward a more extended period of childhood in order to nurture abilities. Parents sought to provide creative-arts instructions, summer camps, and family vacations so as to foster self-development. Within this context the phrase "environmental quality" would have considerable personal meaning.

It also had meaning for place of residence. Millions of urban Americans desired to live on the fringe of the city where life was less congested, the air cleaner, noise reduced, and there was less concentrated waste from manifold human activities. In the nineteenth century only the well-to-do could afford to live some distance from work. Although streetcars enabled white-collar workers to live in the suburbs and work downtown, blue-collar employees still could not pay the cost of daily transportation. But the automobile largely lifted this limitation, and after World War II blue-collar workers were able to escape the industrial community as a place of residence. Still, by the 1970s as many as one-third of urban Americans wished they could live farther out in the countryside.

The search for a higher quality of living involved a desire for more space both inside and outside the home. Life in the city had been intensely crowded for urban dwellers. Often the street in front of the house had constituted the only available open space. Moving to the suburbs reflected a desire to enjoy a more natural setting, but it also evidenced the search for nature beyond the metropolitan area in the parks and woodlands of the countryside. This desire increased with the ease of access to rural areas by means of the automobile. The state-parks movement of the 1920s expressed the demand by city dwellers for places in which to enjoy the countryside on the weekend or during summer vacations.

There was also the desire to obtain private lands in the countryside so as to enjoy nature not found in the city. In the 1960s and 1970s the market for vacation homesites boomed. Newspaper advertisements abounded with phrases that signaled the important values: "by a sparkling stream," "abundant wildlife," "near the edge of a forest road," "200 feet of lakefront," "on the edge of a state forest."

This pursuit of natural values by city dwellers led to a remarkable turnabout in the attitudes of Americans toward natural environments. These had long been thought of as unused wastelands that could be made valuable only if developed. But after World War II many such areas came to be thought of as valuable only if left in their natural condition. Forested land, once thought of by many as dark, forbidding, and sinister, a place to be avoided because of the dangers lurking within, now was highly esteemed.

Wetlands, formerly known as swamplands, fit only for draining so that they could become productive agricultural land, were valued as natural systems, undisturbed and undeveloped. Similar positive attitudes were expressed for the prairies of the Midwest, the swamps of the South, and the pine barrens of the East. For many years wild animals had been seen as a threat to farmers and others. Little concern had been shown for the sharp decline even in the deer population, let alone among the bear and bobcat. Yet by the 1960s and 1970s predators, as well as deer, small mammals, and wild turkey, had assumed a positive image for many Americans, and special measures were adopted to protect them and increase their numbers.

Close on the heels of these changes in attitude were new views about western deserts. The desert had long been thought of as a forbidding land where human habitation was impossible and travel was dangerous. The desert hardly figured in the debate over the Wilderness Act of 1964. But by the late 1970s this had changed. The increased popularity of nature photography had brought home the desert to the American people as a place of wonder and beauty. By 1976 western deserts had been explored and identified by many Americans as lands that should be protected in their natural condition.

ENVIRONMENTAL HEALTH AND WELL-BEING

The search for greater health and well-being constituted an equally significant element of the drive for environmental quality. Such concerns had firm roots in the earlier public health movement, which emphasized the social conditions that gave rise to health problems. . . .

The new concerns for environmental health also focused on the workplace. Occupational dangers to workers had long been thought of mainly as posed by physical factors such as machinery. Increasingly the workplace was seen as an environment in which the air itself could transmit harmful substances to cause diseases in workers. Recognition of this danger came only slowly. Much of it awaited evidence accumulated from long-term studies of the relationship between occupational exposure and disease.

The concern for environmental health was primarily an urban phenomenon.

The incidence of cancer was twice as high in cities as in the rural countryside, a difference attributed to the impact of urban pollution. The chemical products involved in manufacturing, increasing with each passing year after World War II, seemed especially to affect urban people adversely. The extensive use of the automobile in cities also posed continuing pollution threats. And studies of indoor air identified health hazards in offices and households.

Although other waterborne diseases had been controlled through chlorination and disinfection of drinking-water supplies, the rapid accumulation of newer chemical pollutants in the nation's rivers and its underground water generated new health concerns. Synthetic organic compounds, as well as heavy metals from industry, were discovered in many drinking-water sources. The disposal of industrial toxic wastes constituted an even more pervasive concern; they were often injected underground, but just as frequently they were disposed of in landfills from which they leaked into water supplies.[2]

The increasing emphasis on environmental health arose from a rising level of expectations about health and well-being. As life expectancy increased, the average American could look forward to a decade or more of active life after retirement. As the threat of infectious disease decreased, fear of sudden death or disability from polio, secondary infections from simple surgical procedures such as appendectomies, or other dangers declined sharply. All this led to a new focus in health associated more with expectations of well-being than with fear of death. There was a special interest in the quality of life of elderly people. An increasing portion of the population became concerned about preventive health care, showing interest in physical fitness, food and diet, and protection from exposure to environmental pollutants. This marked innovation in ideas about personal health was an important element in the expanding concern for one's environment as a critical element in well-being.[3]

THE ECOLOGICAL PERSPECTIVE

Ecological objectives—an emphasis on the workings of natural biological and geological systems and the pressures human actions placed on them—were a third element of environmental concern. Whereas amenities involved an aesthetic response to the environment, and environmental health concerned a choice between cleaner and dirtier technologies within the built-up environment, ecological matters dealt with imbalances between developed and natural systems that had both current and long-term implications. These questions, therefore, involved ideas about permanence.

The term "ecology" had long referred to a branch of biology that emphasized study of the interaction of living organisms with their physical and biological environment. Popular ecology in the 1960s and 1970s went beyond that scientific meaning. One heard of the impact of people on "the ecology." Professional ecologists disdained this corruption of the word as they had used it. Popular use involved both a broad meaning, the functioning of the biological and geological world, and a narrower one, the disruption of natural processes by human action, as well as the notion that the two, natural systems and human stress, needed to be brought into a better balance.

The popular ecological perspective was reflected in the ecology centers that arose in urban areas. Initially these grew out of the recycling movement—the collection of paper, glass, and tin cans for reprocessing. These centers drew together people who wished to help solve the litter problem and thus to enhance the aesthetic quality of their communities. But soon the concept of recycling seemed to spill over into larger ideas about natural cycles, a traditional ecological theme, and to human action to foster such processes. Ecology centers often expanded their activities into community organic gardens, nutrition and food for better health, and changing life-styles to reduce the human load on natural resources and natural systems.

An ecological perspective grew from the popularization of knowledge about natural processes. These were ideas significant to the study of ecology, but selected and modified by popular experience rather than as a result of formal study. An increasing number of personal or media encounters with the natural world gave rise to widely shared ideas about the functioning of biological and geological systems and the relationship of human beings to them. . . .

These varied tendencies in thought and action that constituted the strands of environmental quality and ecology often came together. The most pervasive factor in this was the emphasis on the importance of a larger role for the natural world in the advanced industrial society. Aesthetic appreciation of nature often was closely connected with intellectual understanding as many people sought both to enjoy and to comprehend the natural settings around them. The adverse impacts of pollutants on human health and one's environment emphasized that the natural world was fragile and had to be cared for. Personal responsibility in lifestyle identified the natural world as a vantage point from which one tested appropriate human behavior. It was no wonder that ideas associated with biology, ecology, and geology came to be integral parts of popular thinking about the quality and permanence of life in modern society.

Notes

1. In 1972 the statement of principles of the League of Women Voters was changed, at the request of leagues around the country. Formerly one such principle had read: "The League of Women Voters believes that responsible government should promote conservation and development of natural resources in the public interest and promote a stable and expanding economy." The words "development" and "expanding" were now dropped. This "signaled a change in emphasis from development to conservation." See statement of Ruth Clusen in *Proceedings, National Watershed Congress*, 19th National Watershed Congress (San Diego, 1972), 126.

2. Environmental Defense Fund and Robert H. Boyle, *Malignant Neglect* (New York, 1979), 82–101.
3. "It is out-of-date to think of health solely in negative terms—as the absence of disease and disability. The healthy individual is not merely unsick. He is strong, aware of his powers and eager to use them. . . . The truly healthful environment is not merely safe but stimulating." Dr. William H. Stewart, Surgeon General, quoted in *Environmental Science and Technology* 2 (1968): 21.

20 Nature Writing

Nature has long been a subject for writers of poetry and prose. Moreover, many literary works contain passages that focus on specific environmental problems and concerns. Nineteenth-century American fiction alone provides striking examples like James Fenimore Cooper's attacks on deforestation in *The Prairie* and Herman Melville's depictions of the whaling industry in *Moby-Dick.* Still other texts reflect the recent history of attitudes toward nature and the environment, as well as the history of the natural sciences and the environmental movement. As a specific kind of literature, however, nonfiction prose nature writing is relatively recent and perhaps disproportionately American in provenance.

The writings included here have been chosen because collectively they reflect all these dimensions of nature writing. First, they exemplify accomplished nonfiction prose, characterized by keen observation and acute intelligence, and provoked by personally significant encounters with the natural world. From Matsuo Bashō's crafted contrasts between nature and civilization to John Steinbeck's descriptions of marine life, they represent acknowledged excellence in such genres as the nature essay and travel writing. Second, the selections engage specific environmental issues, both in the past and from today. Some of these (e.g., those by Gilbert White, William Wordsworth, and Gretel Ehrlich) involve discrete concerns like the management of forests or water use in a semi-arid climate. Others confront broader questions raised by the likes of Annie Dillard and Aldo Leopold. These include the ethics of human living in relation to the environment, the claims of wilderness preservation versus land use for human material needs, and humanity's impact on ecological systems. But all show the creative imagination at work, coming to grips with the problems posed by human interactions with nature.

Finally, these works each reflect moments in the history of the natural sciences and of attitudes toward the environment over the past four centuries. Though his treatment of the natural world is based upon Zen principles, the seventeenth-century Japanese writer Bashō represents the pastoral impulse, in which the author retreats to a life lived in accord with nature, in part to criticize civilization. (This literary, ethical, and spiritual stance in the West stretches from Theocritus and Horace to Robert Frost and Donald Hall.) By contrast, J. Hector St. John de Crèvecoeur and Gilbert White accept the division of the human from the natural that pastoral tries to repair. Though they express it in terms of the rationalism of the eighteenth-century Enlightenment, both the form and content of their writing embody the pre-modern Western assumption—dating to the Book of Genesis—that humanity has a duty to dominate and make use of nature. They likewise exemplify the inductive methodology of the scientific tradition and literary form known as natural history. Wordsworth hints at the ways in which natural history evolved in the decades before Darwin in response to the socio-cultural and environmental impact of the industrial revolution. And the passages from Thoreau's *Walden* and *The Maine Woods* explore the division between humanity and nature, the latter tracing it to its philosophical origins.

Eliza Farnham's rejection of what many would regard as the male opposition between the human and the natural reminds us that cultural assumptions centered on gender are an important filter through which we experience the environment. Yet Annie Dillard's complaint that nature is now regarded merely materially, as disenchanted and disinspirited, underscores the persistence of the romantic tradition of Thoreau and his contemporaries, as well as its transcendence of gender lines. Aldo Leopold, in contrast, when read, say, against the selections from John Muir found elsewhere in this volume, both reminds us of the deep early twentieth-century division in American environmentalism between the advocates of wilderness protection and conservation management, and, by his rejection of quasi-religious romantic idealizations of nature like Dillard's, marks the way forward out of that dispute. Instead, Leopold strongly advocates a preservation ethic based on humanity's needs and responsibilities as one species cohabiting with others in an ecosphere subject to damage and destruction.

As this suggests, the more recent selections manifest the diversity of views found in mid- and late-twentieth-century American nature writing. One fault line continues the split in worldview, with some taking a more materialist; some, a more romantic and religious; and some, a less polarized view of the nature of nature. Similarly, some selections (for example, Steinbeck's) bespeak the influence on American writers and thinkers of intellectual positions like pre-Darwinian science, Social Darwinism, and Marxism. They also manifest the resilience and health of nature writing as a genre—a fact reflected in its popularity both inside and outside the classroom.

FURTHER READING

Finch, Robert, and John Elder, ed. *The Norton Book of Nature Writing*. New York: Norton, 1990.

Glacken, Clarence. *Traces on the Rhodian Shore: Nature and Culture in Western Thought from Ancient Times to the End of the Eighteenth Century.* Berkeley: University of California Press, 1967.
Glotfeldy, Cheryll, and Harold Fromm. *The Ecocriticism Reader: Landmarks in Literary Ecology.* Athens: University of Georgia Press, 1996.
Hildebidle, John. *Thoreau: A Naturalist's Liberty.* Cambridge: Harvard University Press, 1983.
Irmscher, Christoph. *The Poetics of Natural History from John Bartram to William James.* New Brunswick: Rutgers University Press, 1999.
Kolodny, Annette. *The Land Before Her: Fantasy and Experience of the American Frontiers, 1630–1860.* Chapel Hill: University of North Carolina Press, 1984.
Lopez, Barry H. *Crossing Open Ground.* New York: Vintage Books, 1989.
Mabey, Richard, ed. *The Oxford Book of Nature Writing.* Oxford: Oxford University Press, 1995.
McKusick, James, and Bridget Keegan, eds. *Literature and Nature: Four Centuries of Nature Writing.* Upper Saddle River, N.J.: Prentice-Hall, 2000.
Quammen, David. *The Flight of the Iguana: A Sidelong View of Science and Nature.* New York: Anchor Books, 1989.
Stewart, Frank. *A Natural History of Nature Writing.* Washington, D.C.: Island Press and Shearwater Books, 1995.
Thomas, Keith. *Man and the Natural World: Changing Attitudes in England, 1500–1800.* London: Allen Lane, 1983. [American edition subtitled *A History of the Modern Sensibility.* New York: Pantheon, 1983.]
Torrance, Robert, ed. *Encompassing Nature: A Sourcebook.* Boulder: Counterpoint Press, 1998.

Matsuo Bashō, from "Prose Poem on the Unreal Dwelling" (1691), translated by Donald Keene

This prose excerpt about his travels expresses Matsuo Bashō's desire to live a simple life in harmony with nature, whose evocative beauty and integrity contrast with civilization. This idealization of nature and association of the ascetic life with it parallel many Western nature writers such as Wordsworth. Bashō's views are specifically grounded in a Zen Buddhist belief in the interdependence of all things and the necessity of union between the seer and the seen.

My body, now close to fifty years of age, has become an old tree that bears bitter peaches, a snail which has lost its shell, a bagworm separated from its bag; it drifts with the winds and clouds that know no destination. Morning and night I have eaten traveler's fare, and have held out for alms a pilgrim's wallet. On my last journey my face was burnt by the sun of Matsushima, and I wetted my sleeve at the holy mountain. I longed to go as far as that shore where the puffins cry and the Thousand Islands of the Ainu [an indigenous people of Japan] can be seen in the distance, but my companion drew me back, telling how dangerous so long a

Figure 20.1 *Pilgrims on the Slope of Mt. Fuji* by Shibata Zeshin, 1880. © The British Museum.

journey would be with my sickness. I yielded. Then I bruised my heels along the rough coast of the northern sea, where each step in the sand dunes is painful. This year I roamed by the shores of the lake in quest of a place to stay, a single stalk of reed where the floating nest of the grebe might be borne to rest by the current. This is my Unreal Dwelling, and it stands by the mountain called Kokubu. . . . Indeed it is true that all the delusions of the senses are summed up in the one word *unreality*, and there is no way to forget even for a moment change and its swiftness.

The mountains do not extend to any great depth, but the houses are spaced well apart. Stone Mountain is before my hut, and behind stands Gorge Mountain. From the lofty peaks descends a fragrant wind from the south, and the northern wind steeped in the distant sea is cool. It was the beginning of the fourth moon when I arrived, and the azaleas were still blossoming. Mountain wisteria hung on the pines. Cuckoos frequently flew past, and there were visits from the swallows. Not a peck from a woodpecker disturbed me, and in my joy I called to the wood dove, "Come, bird of solitude, and make me melancholy!" I could not but be happy—the view would not have blushed before the loveliest scenes of China.

Between Hieda Mountain and the peak of Hira, I can see the pine of Karasaki engulfed in mist, and at times a castle glittering in the trees; when the rain clears by the bridge of Seta, sunset lingers in the pine groves. Mikami Mountain looks like Fuji, and reminds me of my old cottage at its foot. Nearby on Tanagami Mountain I have sought the traces of the men of old. Sometimes, wishing to enjoy an uninterrupted view, I climb the peak behind my hut. On the summit I have built a shelf of pine boughs, on which I spread a round straw mat: this I call the "monkey's perch." I am no follower of that eccentric who built a nest in a crab-apple tree where he drank with his friends, for that was in the city and noisy; nor would I give up my perch for the hut which Wang the Sage once tied together. . . .

But I should not have it thought from what I have said that I am devoted to solitude and seek only to hide my traces in the wilderness. Rather, I am like a sick man weary of people, or someone who is tired of the world. What is there to say? I have not led a clerical life, nor have I served in normal pursuits. Ever since I was very young I have been fond of my eccentric ways, and once I had come to make them the source of a livelihood, temporarily I thought, I discovered myself bound for life to the one line of my art, incapable and talentless as I am. I labor without results, am worn of spirit and wrinkled of brow. Now, when autumn is half over, and every morning and each evening brings changes to the scene, I wonder if that is not what is meant by dwelling in unreality. And here too I end my words.

J. Hector St. John de Crèvecoeur, from *Letters from an American Farmer* (1782–84)

J. H. St. John de Crèvecoeur's Letters from an American Farmer *No. 10 typifies Enlightenment views on humanity and nature. The hummingbird selection affirms reason's power to comprehend, analyze, depict, and so, control a natural world expressly designed by a benevolent and rational Creator for human understanding. The snake passage symbolically manifests eighteenth-century fears of natural violence and disorder, challenging the rationalist model of human intellectual domination and exploitation it otherwise embodies.*

I have amused myself a hundred times in observing the great number of hummingbirds with which our country abounds: the wild blossoms everywhere attract the attention of these birds, which, like bees, subsist by suction. From this retreat I distinctly watch them in all their various attitudes; but their flight is so rapid that you cannot distinguish the motion of their wings. On this little bird Nature has profusely lavished her most splendid colors; the most perfect azure, the most beautiful gold, the most dazzling red, are forever in contrast, and help to embellish the plumes of his majestic head. The richest pallet of the most luxuriant painter could never invent anything to be compared to the variegated tints with which this insect-bird is arrayed. Its bill is as long and as sharp as a coarse sewing-needle; like the bee, nature has taught it to find out the calyx of flowers and blossoms, those mellifluous particles that serve it for sufficient food; and yet it seems to leave them untouched, undeprived of anything that our eyes can possibly distinguish. When it feeds, it appears as if immovable, though continually on the wing; and sometimes, from what motives I know not, it will tear and lacerate flowers into a hundred pieces; for, strange to tell, they are the most irascible of the feathered tribe. Where do passions find room in so diminutive a body? They often fight with the fury of lions, until one of the combatants falls a sacrifice and dies. When fatigued, it has often perched within a few feet of me and, on such favorable opportunities, I have surveyed it with the most minute attention. Its little eyes appear like diamonds, reflecting light on every side: most elegantly finished in all parts, it is a miniature-work of our great Parent, who seems to have formed it the smallest and, at the same time, the most beautiful, of the winged species.

As I was one day sitting . . . I beheld two snakes of considerable length, the one pursuing the other, with great celerity, through a hemp-stubble field. The aggressor was of the black kind, six feet long; the fugitive was a watersnake, nearly of equal dimensions. They soon met and, in the fury of their first encounter, they appeared in an instant firmly twisted together; and, whilst their united tails beat the ground, they mutually tried with open jaws to lacerate each other. What a fell aspect did they present! Their heads were compressed to a very small size, their eyes flashed fire; and, after this conflict had lasted about five minutes, the second found means to disengage itself from the first, and hurried toward the ditch. Its antagonist instantly assumed a new posture; and, half creeping and half erect, with a majestic mien, overtook and attacked the other again, which placed itself in the same attitude and prepared to resist. The scene was uncommon and beautiful; for, thus opposed, they fought with their jaws, biting each other with the utmost rage; but, notwithstanding this appearance of mutual courage and fury, the watersnake still seemed desirous of retreating toward the ditch, its natural element. This was no sooner perceived by the keen-eyed black one than, twisting its tail twice round a stalk of hemp and seizing its adversary by the throat, not by means of its jaws, but by twisting its own neck twice round that of the watersnake, pulled it back from the ditch. To prevent a defeat the latter took hold likewise of a stalk on the bank and, by the acquisition of that point of resistance, became a match for its fierce antagonist. Strange was this to behold; two great snakes strongly adhering to the ground, mutually fastened together, by means of the writhings which lashed them to each other and, stretched at their full length, they pulled, but pulled in vain; and, in the moments of greatest exertion, that part of their bodies which was entwined seemed extremely small, while the rest appeared inflated and now and then convulsed with strong undulations rapidly following each other. Their eyes seemed on fire and ready to start out of their heads; at one time the conflict seemed decided; the watersnake bent itself into two great folds and by that operation rendered the other more than commonly outstretched; the next minute, the new struggles of the black one gained an unexpected superiority; it acquired two great folds likewise, which necessarily extended the body of its adversary in proportion as it had contracted its own. These efforts were alternate; victory seemed doubtful; inclining sometimes to the one side and sometimes to the other; until at last, the stalk, to which the black snake fastened, suddenly gave way and, in consequence of this accident, they both plunged into the ditch. The water did not extinguish their vindictive rage; for, by their agitations, I could trace, though not distinguish, their mutual attacks. They soon reappeared on the surface twisted together as in their first onset; but the black snake seemed to retain its wonted superiority, for its head was exactly fixed above that of the other, which it incessantly pressed down under the water until it was stifled and sunk. The victor no sooner perceived its enemy incapable of further resistance than, abandoning it to the current, it returned on shore and disappeared.

Gilbert White, from *The Natural History and Antiquities of Selbourne in the County of Southampton* (1789)

Gilbert White exemplifies the methods of natural history (which preceded modern biology), investigating nature not through experimentation, but, as John Hildebidle has noted, in a "sequence of steps, from observation to generalization to explanation to corroboration." Letter 7 illustrates this as it describes one way rural communities traditionally tried to manage the environment—even though the eighteenth century was the first great age of innovation in land and forestry management, plant and animal breeding, and so on.

Forests and wastes, when their allurements to irregularities are removed, are of considerable service to neighborhoods that verge upon them, by furnishing them with peat and turf for their firing; with fuel for the burning their lime, and with ashes for their grasses; and by maintaining their geese and their stock of young cattle at little or no expense.

The manor farm of the parish of Greatham has an admitted claim, I see . . . of turning all livestock on the forest at proper seasons, *bidentibus exceptis*.[1] The reason I presume why sheep[2] are excluded is because, being such close grazers, they would pick out all the finest grasses and hinder the deer from thriving.

Though (by statute 4 and 5 Wm. and Mary, c. 23) "to burn on any waste, between Candlemas and Midsummer, any grig, ling, heath and furze, goss or fern, is punishable with confinement in the House of Correction," etc., yet in this forest, about March or April, according to the dryness of the season, such vast heath-fires are lighted up that they often get to a masterless head and, catching the hedges, have sometimes been communicated to the underwoods, woods, and coppices, where great damage has ensued. The plea for these burnings is that

Figure 20.2 *Wooded Landscape with Figures and Cows at a Watering Place* by Thomas Gainsborough, 1776–77. © 2003 Tate Britain, London.

when the old coat of heath, etc., is consumed, young will sprout up and afford much tender browse for cattle; but where there is large old furze, the fire, following the roots, consumes the very ground, so that for hundreds of acres nothing is to be seen but smother and desolation, the whole circuit round looking like the cinders of a volcano; and the soil being quite exhausted, no traces of vegetation are to be found for years. These conflagrations, as they take place usually with a northeast or east wind, much annoy this village with their smoke, and often alarm the country; and once in particular I remember that a gentleman who lives beyond Andover, coming to my house, when he got on the Downs between that town and Winchester at twenty-five miles distance, was surprised much with smoke and a hot smell of fire, and concluded that Alresford was in flames; but when he came to that town, he then had apprehensions for the next village, and so on to the end of his journey.

Notes

1. For this privilege the owner of that estate used to pay the king annually seven bushels of oats.
2. In the Holt [wood], where a full stock of fallow-deer has been kept up till lately, no sheep are admitted to this day.

William Wordsworth, from *Guide to the Lakes* (1810)

William Wordsworth's description of the succession of forest trees in Guide to the Lakes, *Part III, shows how fully the poet assimilated the methods of contemporary natural history. It also affirms aesthetic values like wildness, naturalness, nativeness, and irregularity in order to criticize changes in the environment caused by urbanization, the industrial revolution, and the rise of the middle classes. It projects the author's growing social conservatism onto the landscape, politicizing nature. (See also Chapters 10 and 18.)*

If these general rules be just, what shall we say to whole acres of artificial shrubbery and exotic trees among rocks and dashing torrents, with their own wild wood in sight—where we have the whole contents of the nurseryman's catalogue jumbled together—color at war with color, and form with form—among the most peaceful subjects of Nature's kingdom, everywhere discord, distraction, and bewilderment! But this deformity, bad as it is, is not so obtrusive as the small patches and large tracts of larch-plantations that are overrunning the hillsides. To justify our condemnation of these, let us again recur to Nature. The process by which she forms woods and forests is as follows. Seeds are scattered indiscriminately by winds, brought by waters, and dropped by birds. They perish, or produce, according as the soil and situation upon which they fall are suited to them; and under the same dependence, the seedling or the sucker, if not cropped by animals (which Nature is often careful to prevent by fencing it about with brambles or other prickly shrubs), thrives, and the tree grows, sometimes single, taking its own shape without constraint, but for the most part compelled to conform itself to some law imposed upon it by its neighbors. From low and sheltered places, vegetation travels upwards to the more exposed; and the young plants are protected, and to a certain degree fashioned, by those that have preceded them. The continuous mass of foliage which would be thus produced is broken by rocks, or by glades or open places, where the browsing of animals has prevented the growth of the wood. As vegetation ascends, the winds begin also to bear their part in molding the forms of the trees; but, thus mutually protected, trees, though not of the hardiest kind, are enabled to climb high up the mountains. Gradually, however, by the quality of the ground, and by increasing exposure, a stop is put to their ascent; the hardy trees only are left; those also, by little and little, give way—and a wild and irregular boundary is established, graceful in its outline, and never contemplated without some feeling, more or less distinct, of the powers of Nature by which it is imposed.

Contrast the liberty that encourages, and the law that limits, this joint work of Nature and time, with the disheartening necessities, restrictions, and disadvantages, under which the artificial planter must proceed, even he whom long

Figure 20.3 *Langdale Pike* by John Frederick Kensett, 1858. Cornell Fine Arts Museum, Rollins College, Winter Park, Florida. Gift of Madame Charlotte Gero, 1963.011.P.

observation and fine feeling have best qualified for his task. In the first place his trees, however well chosen and adapted to their several situations, must generally start all at the same time; and this necessity would of itself prevent that fine connection of parts, that sympathy and organization, if I may so express myself, which pervades the whole of a natural wood, and appears to the eye in its single trees, its masses of foliage, and their various colors, when they are held up to view on the side of a mountain; or when, spread over a valley, they are looked down upon from an eminence. It is therefore impossible, under any circumstances, for the artificial planter to rival the beauty of Nature.

Eliza Farnham, from *Life in Prairie Land* (1846)

Eliza Farnham's memoir portrays the female experience of the frontier in more familial, domestic, and interpersonal terms than contemporary men's accounts. Greeted by nature "as a daughter welcomed by 'a strong and generous parent' and as a sibling able to turn to the trees 'as to elder brothers,'" Farnham expresses joy at her homecoming, mixed with a Wordsworthian moralization of the land and its flora and fauna typical of the time.

Summer had worn away, with its wealth of golden grains and flowers. The luxuriant harvest had disappeared from the farms in the adjacent country, the tall corn was in its sere and yellow leaf, the late fruits began to ripen, the prairies faded from their rich green, save where here and there a "late burn" showed the tender grass, like an emerald island in the vast brown ocean. Autumn in the prairie land is scarcely excelled for the richness of its charms by any other season. Coupled with the perfection of the wide vegetable world is an idea of repose which fills the soul. An immense country, whose energies have been springing all the previous months with ceaseless toil, whose rank luxuriance evinces the employment of tremendous powers, now lies all around you in the deep quiet which ushers in a truly natural death. The sun pours forth a rich, mellow light; dim and soft, as if like a tender nurse he watched over this sleep of nature. The native birds, happy in the abundance which they cannot consume, fly cheerfully but quietly about, as if, their labor done, the season of rest had come to them also. The quail whistles and dances among the brown hazel thickets; the grouse flies from field to field, dividing his depredations through the neighborhood, and bearing off, when unmolested, a full crop to the plains, which he loves better than the abodes of man. The crow calls from the wood top, or wheels his long and lazy flights above the naked prairies, seeming really more amiable than at any other season. The air is filled with the smoke of distant fires; some day they creep up into your own neighborhood, and when night comes, light the heavens and the earth as far as the eye can reach. These are magnificent spectacles. I have stood upon the roof of our large hotel in the evening, and looked into a sea of fire which appeared to be unbroken for miles. These incidents occasionally interrupt the dreamy rest to which everything tends, but they pass away in a few hours, and the next day is as quiet as before.

Henry David Thoreau, from "The Bean-Field" in *Walden* (1854) and from "Ktaadn" (1848) in *The Maine Woods* (1864)

These two selections by Thoreau reflect a common romantic pattern of alternating affirmation and skepticism about humanity's relation to nature. "The Bean-Field" chapter from Walden *investigates this relationship during the industrial revolution, asking how we may live with and in nature, as natural beings, yet ones distanced from nature by civilization. It provides a positive answer in Thoreau's ability to observe and meditate on natural phenomena, teasing out their significance. In the "Ktaadn" section of* The Maine Woods, *recording his ascent of Mount Katahdin, Thoreau realizes that mankind's hostility to nature stems from a divorce between our consciousness and the external world. This alienation from nature prevents us from finding any spiritual or metaphysical significance in our natural experience.*

FROM "THE BEAN-FIELD" IN *WALDEN* (1854)

Meanwhile my beans, the length of whose rows, added together, was seven miles already planted, were impatient to be hoed, for the earliest had grown considerably before the latest were in the ground; indeed they were not easily to be put off. What was the meaning of this so steady and self-respecting, this small Herculean labor, I knew not. I came to love my rows, my beans, though so many more than I wanted. They attached me to the earth, and so I got strength like Antæus. But why should I raise them? Only Heaven knows. This was my curious labor all summer—to make this portion of the earth's surface, which had yielded only cinquefoil, blackberries, johnswort, and the like, before, sweet wild fruits and pleasant flowers, produce instead this pulse. What shall I learn of beans or beans of me? I cherish them, I hoe them, early and late I have an eye to them; and this is my day's work. It is a fine broad leaf to look on. My auxiliaries are the dews and rains which water this dry soil, and what fertility is in the soil itself, which for the most part is lean and effete. My enemies are worms, cool days, and most of all woodchucks. The last have nibbled for me a quarter of an acre clean. But what right had I to oust johnswort and the rest, and break up their ancient herb garden? Soon, however, the remaining beans will be too tough for them, and go forward to meet new foes. . . .

This was one field not in Mr. Coleman's report [Thoreau misspells the surname of Henry Colman, author of a report on Massachusetts agriculture]. And, by the way, who estimates the value of the crop which nature yields in the still wilder fields unimproved by man? The crop of *English* hay is carefully weighed, the moisture calculated, the silicates and the potash; but in all dells and pond-holes in the woods and pastures and swamps grows a rich and various crop only unreaped by man. Mine was, as it were, the connecting link between wild and cultivated fields; as some states are civilized, and others half-civilized, and others savage or barbarous, so my field was, though not in a bad sense, a half-cultivated field. They were beans cheerfully returning to their wild and primitive state that I cultivated, and my hoe played the *Ranz des Vaches* [Swiss French, "Song of the Cows"] for them. . . .

As I drew a still fresher soil about the rows with my hoe, I disturbed the ashes of unchronicled nations who in primeval years lived under these heavens, and their small implements of war and hunting were brought to the light of this modern day. They lay mingled with other natural stones, some of which bore the marks of having been burned by Indian fires, and some by the sun, and also bits of pottery and glass brought hither by the recent cultivators of the soil. When my hoe tinkled against the stones, that music echoed to the woods and the sky, and was an accompaniment to my labor which yielded an instant and immeasurable crop. It was no longer beans that I hoed, nor I that hoed beans; and I remembered with as much pity as pride, if I remembered at all, my acquaintances who had gone to the city to attend the oratorios. The nighthawk circled overhead in the sunny afternoons—for I sometimes made a day of it—like a mote in the eye, or in heaven's eye, falling from time to time with a swoop and a sound as if the heavens were rent, torn at last to very rags and tatters, and yet a seamless cope remained; small imps that fill the air and lay their eggs on the ground on bare sand or rocks on the tops of hills, where few have found them; graceful and slender like ripples caught up from the pond, as leaves are

Figure 20.4 The summit of Mount Katahdin. Courtesy of Concord Free Public Library.

raised by the wind to float in the heavens; such kindredship is in nature. The hawk is aerial brother of the wave which he sails over and surveys, those his perfect air-inflated wings answering to the elemental unfledged pinions of the sea. Or sometimes I watched a pair of hen-hawks circling high in the sky, alternatively soaring and descending, approaching and leaving one another, as if they were the embodiment of my own thoughts. Or I was attracted by the passage of wild pigeons from this wood to that, with a slight quivering winnowing sound and carrier haste; or from under a rotten stump my hoe turned up a sluggish portentous and outlandish spotted salamander, a trace of Egypt and the Nile, yet our contemporary. When I paused to lean on my hoe, these sounds and sights I heard and saw anywhere in the row, a part of the inexhaustible entertainment which the country offers.

* * *

FROM "KTAADN" (1848) IN *THE MAINE WOODS* (1864)

Perhaps I most fully realized that this was primeval, untamed, and forever untameable *Nature*, or whatever else men call it, while coming down this part of the mountain. . . . It is difficult to conceive of a region uninhabited by man. We habitually presume his presence and influence everywhere. And yet we have not seen pure Nature, unless we have seen her thus vast and drear and inhuman, though in the midst of cities. Nature was here something savage and awful, though beautiful. I looked with awe at the ground I trod on, to see what the Powers had made there, the form and fashion and material of their work. This was that Earth of which we have heard, made out of Chaos and Old Night. Here was no man's garden, but the unhandselled globe. It was not lawn, nor pasture, nor mead, nor woodland, nor lea, nor arable, nor waste-land. It was the fresh and natural surface of the planet Earth, as it was made for ever and ever—to be the dwelling of man, we say—so Nature made it, and man may use it if he can. Man was not to be associated with it. It was Matter, vast, terrific—not his Mother Earth that we have heard of, not for him to tread on, or be buried in—no, it were being too familiar even to let his bones lie there—the home, this, of Necessity and Fate. There was there felt the presence of a force not bound to be kind to man. It was a place for heathenism and superstitious rites—to be inhabited by men nearer of kin to the rocks and to wild animals than we. . . . Here not even the surface had been scarred by man, but it was a specimen of what God saw fit to make this world. What is it to be admitted to a museum, to see a myriad of particular things, compared with being shown some star's surface, some hard matter in its home! I stand in awe of my body, this matter to which I am bound has become so strange to me. I fear not spirits, ghosts, of which I am one—*that* my body might—but I fear bodies, I tremble to meet them. What is this Titan that has possession of me? Talk of mysteries!—Think of our life in nature—daily to be shown matter, to come in contact with it—rocks, trees, wind on our cheeks! the *solid* earth! the *actual* world! the *common sense! Contact! Contact! Who* are we? *where* are we?

Aldo Leopold, from "Thinking Like a Mountain" in *A Sand County Almanac* (1949)

By uniting a concern for game management with one for ecological balance, this famous essay underscores Aldo Leopold's critical position in the transition of American environmentalism from its Progressive-era division between the wilderness preservationism of John Muir and Gifford Pinchot's "wise use" conservation policies to a modern ecosystems approach. It also illustrates Leopold's passionate advocacy of a post-Darwinian, scientifically based environmental ethic that recognizes humanity's interdependence with other species and consequent moral responsibility toward the land.

A deep chesty bawl echoes from rimrock to rimrock, rolls down the mountain, and fades into the far blackness of the night. It is an outburst of wild defiant sorrow, and of contempt for all the adversities of the world.

Every living thing (and perhaps many a dead one as well) pays heed to that call. To the deer it is a reminder of the way of all flesh, to the pine a forecast of midnight scuffles and of blood upon the snow, to the coyote a promise of gleanings to come, to the cowman a threat of red ink at the bank, to the hunter a challenge of fang against bullet. Yet behind these obvious and immediate hopes and fears there lies a deeper meaning, known only to the mountain itself. Only the mountain has lived long enough to listen objectively to the howl of a wolf.

Those unable to decipher the hidden meaning know nevertheless that it is there, for it is felt in all wolf country, and distinguishes that country from all other land. It tingles in the spine of all who hear wolves by night, or who scan their tracks by day. Even without sight or sound of wolf, it is implicit in a hundred small events: the midnight whinny of a pack horse, the rattle of rolling rocks, the bound of a fleeing deer, the way shadows lie under the spruces. Only the ineducable tyro can fail to sense the presence or absence of wolves, or the fact that mountains have a secret opinion about them.

My own conviction on this score dates from the day I saw a wolf die. We were eating lunch on a high rimrock, at the foot of which a turbulent river elbowed its way. We saw what we thought was a doe fording the torrent, her breast awash in white water. When she climbed the bank toward us and shook out her tail, we realized our error: it was a wolf. A half-dozen others, evidently grown pups, sprang from the willows and all joined in a welcoming mêlée of wagging tails and playful maulings. What was literally a pile of wolves writhed and tumbled in the center of an open flat at the foot of our rimrock.

In those days we had never heard of passing up a chance to kill a wolf. In a second we were pumping lead into the pack, but with more excitement than accuracy: how to aim a steep downhill shot is always confusing. When our rifles were empty, the old wolf was down, and a pup was dragging a leg into impassable slide-rocks.

We reached the old wolf in time to watch a fierce green fire dying in her eyes. I realized then, and have known ever since, that there was something new to me in those eyes—something known only to her and to the mountain. I was young then, and full of trigger-itch; I thought that because fewer wolves meant more deer, that no wolves would mean hunters' paradise. But after seeing the green fire die, I sensed that neither the wolf nor the mountain agreed with such a view.

Since then I have lived to see state after state extirpate its wolves. I have watched the face of many a newly wolfless mountain, and seen the south-facing slopes wrinkle with a maze of new deer trails. I have seen every edible bush and seedling browsed, first to anemic desuetude, and then to death. I have seen every edible tree defoliated to the height of a saddlehorn. Such a mountain looks as if someone had given God a new pruning shears, and forbidden Him all other exercise. In the end the starved bones of the hoped-for deer herd, dead of its own too-much, bleach with the bones of the dead sage, or molder under the high-lined junipers.

Figure 20.5 A mountain stream.

I now suspect that just as a deer herd lives in mortal fear of its wolves, so does a mountain live in mortal fear of its deer. And perhaps with better cause, for while a buck pulled down by wolves can be replaced in two or three years, a range pulled down by too many deer may fail of replacement in as many decades.

So also with cows. The cowman who cleans his range of wolves does not realize that he is taking over the wolf's job of trimming the herd to fit the range. He has not learned to think like a mountain. Hence we have dustbowls, and rivers washing the future into the sea.

We all strive for safety, prosperity, comfort, long life, and dullness. The deer strives with his supple legs, the cowman with trap and poison, the statesman with pen, the most of us with machines, votes, and dollars, but it all comes to the same thing: peace in our time. A measure of success in this is all well enough, and perhaps is a requisite to objective thinking, but too much safety seems to yield only danger in the long run. Perhaps this is behind Thoreau's dictum: In wildness is the salvation of the world. Perhaps this is the hidden meaning in the howl of the wolf, long known among mountains, but seldom perceived among men.

John Steinbeck, from *The Log from the* Sea of Cortez (1941, revised 1951)

John Steinbeck's log of a marine science expedition to Baja California argues that humankind is one species among many, within ecosystems both microcosmic and macrocosmic. He attacks notions of human superiority to or separation from the natural order, but also takes solace in the universe's "equilibrium" at a macrocosmic level. He combines older, design-affirming views of nature and human fate with concepts drawn from mid-twentieth-century natural science.

That night we intended to run across the Gulf and start for home. It was good to be running at night again, easier to sleep with the engine beating. Tiny at the wheel inveighed against the waste of fish by the Japanese. To him it was a waste complete, a loss of something. We discussed the widening and narrowing picture. To Tiny the fisherman, having as his function not only the catching of fish but the presumption that they would be eaten by humans, the Japanese were wasteful. And in that picture he was very correct. But all the fish actually were eaten; if any small parts were missed by the birds they were taken by the detritus-eaters, the worms and [sea] cucumbers. And what they missed was reduced by the bacteria. What was the fisherman's loss was a gain to another group. We tried to say that in the macrocosm nothing is wasted, the equation always balances. The elements which the fish elaborated into an individuated physical organism, a microcosm, go back again into the undifferentiated macrocosm which is the great reservoir. There is not, nor can there be, any actual waste, but simply varying forms of energy. . . . The great organism, Life, takes it all and uses it all. . . . Nothing is wasted; "no star is lost." . . .

Every living thing has its niche, *a posteriori*, and God, in a real, non-mystical sense, sees every sparrow fall and every cell utilized. . . .

Among men, it seems, historically at any rate, that processes of co-ordination and disintegration follow each other with great regularity, and the index of the co-ordination is the measure of the disintegration which follows. . . .

We think these historical waves may be plotted and the harmonic curves of human group conduct observed. Perhaps out of such observation a knowledge of the function of war and destruction might emerge. Little enough is known about the function of individual pain and suffering, although from its profound organization it is suspected of being necessary as a survival mechanism. And nothing whatever is known of the group pains of the species, although it is not unreasonable to suppose that they too are somehow functions of the surviving species. It is too bad that against even such investigation we build up a hysterical and sentimental barrier. Why do we so dread to think of our species as a species? Can it be that we are afraid of what we may find? That human self-love would suffer too much and that the image of God might prove to be a mask? This could be only partly true, for if we could cease to wear the image of a kindly, bearded, interstellar dictator we might find ourselves true images of his kingdom, our eyes the nebulae, and universes in our cells.

Annie Dillard, "Teaching a Stone to Talk" (1982)

Dillard returns to the problem addressed by Thoreau in The Maine Woods *and his later writings: how do we relate to nature now that empirical science and philosophical materialism have prevailed and the romantic vision of nature as spiritually potent has fled—a question with the deepest ethical and practical consequences for the environment.*

I

The island where I live is peopled with cranks like myself. In a cedar-shake shack on a cliff—but we all live like this—is a man in his thirties who lives alone with a stone he is trying to teach to talk.

Wisecracks on this topic abound, as you might expect, but they are made as it were perfunctorily, and mostly by the young. For in fact, almost everyone here respects what Larry is doing, as do I, which is why I am protecting his (or her) privacy, and confusing for you the details. It could be, for instance, a pinch of sand he is teaching to talk, or a prolonged northerly, or any one of a number of waves. But it is, in fact, I assure you, a stone. It is—for I have seen it—a palm-sized oval beach cobble whose dark gray is cut by a band of white which runs around and, presumably, through it; such stones we call "wishing stones," for reasons obscure but not, I think, unimaginable.

He keeps it on a shelf. Usually the stone lies protected by a square of untanned leather, like a canary asleep under its cloth. Larry removes the cover for the stone's lessons, or more accurately, I should say, for the ritual or rituals which they perform together several times a day.

No one knows what goes on at these sessions, least of all myself, for I know Larry but slightly, and that owing only to a mix-up in our mail. I assume that like any other meaningful effort, the ritual involves sacrifice, the suppression of self-consciousness, and a certain precise tilt of the will, so that the will becomes transparent and hollow, a channel for the work. I wish him well. It is a noble work, and beats, from any angle, selling shoes.

Reports differ on precisely what he expects or wants the stone to say. I do not think he expects the stone to speak as we do, and describe for us its long life and many, or few, sensations. I think instead that he is trying to teach it to say a single word, such as "cup" or "uncle." For the purpose he has not, as some have seriously suggested, carved the stone a little mouth, or furnished it in any way with a pocket of air which it might then expel. Rather—and I think he is wise in this—he plans to initiate his son, who is now an infant living with Larry's estranged wife, into the work, so that it may continue and bear fruit after his death.

II

Nature's silence is its one remark, and every flake of world is a chip off that old mute and immutable block. The Chinese say that we live in the world of the ten thousand things. Each of the ten thousand things cries out to us precisely nothing.

God used to rage at the Israelites for frequenting sacred groves. I wish I could find one. Martin Buber says: "The crisis of all primitive mankind comes with the discovery of that which is fundamentally not-holy, the a-sacramental, which withstands the methods, and which has no 'hour,' a province which steadily enlarges itself." Now we are no longer primitive; now the whole world seems not-holy. We have drained the light from the boughs in the sacred grove and snuffed it in the high places and along the banks of sacred streams. We as a people have moved from pantheism to pan-atheism. Silence is not our heritage but our destiny; we live where we want to live.

The soul may ask God for anything, and never fail. You may ask God for his presence, or for wisdom, and receive each at his hands. Or you may ask God, in the words of the shopkeeper's little gag sign, that he not go away mad, but just go away. Once, in Israel, an extended family of nomads did just that. They heard God's speech and found it too loud. The wilderness generation was at Sinai; it witnessed there the thick darkness where God was: "and all the people saw the thunderings, and the lightnings, and the noise of the trumpet, and the mountain smoking." It scared them witless. Then they asked Moses to beg God, please, never speak to them directly again. "Let not God speak with us, lest we die." Moses took the message. And God, pitying their self-consciousness, agreed. He agreed not to speak to the people anymore. And he added to Moses, "Go say to them, Get into your tents again."

III

It is difficult to undo our own damage, and to recall to our presence that which we have asked to leave. It is hard to desecrate a grove and change your mind. The very holy mountains are keeping mum. We doused the burning bush and cannot rekindle it; we are lighting matches in vain under every green tree. Did the wind use to cry, and the hills shout forth praise? Now speech has perished from among the lifeless things of earth, and living things say very little to very few. Birds may crank out sweet gibberish and monkeys howl; horses neigh and pigs say, as you recall, oink oink. But so do cobbles rumble when a wave recedes, and thunders break the air in lightning storms. I call these noises silence. It could be that wherever there is motion there is noise, as when a whale breaches and smacks the water—and wherever there is stillness there is the still small voice, God's speaking from the whirlwind, nature's old song and dance, the show we drove from town. At any rate, now it is all we can do, and among our best efforts, to try to teach a given human language, English, to chimpanzees.

In the forties an American psychologist and his wife tried to teach a chimp actually to speak. At the end of three years the creature could pronounce, in a hoarse whisper, the words "mama," "papa," and "cup." After another three years of training she could whisper, with difficulty, still only "mama," "papa," and "cup." The more recent successes at teaching chimpanzees American Sign Language are well known. Just the other day a chimp told us, if we can believe that we truly share a vocabulary, that she had been sad in the morning. I'm sorry we asked.

What have we been doing all these centuries but trying to call God back to the mountain, or, failing that, raise a peep out of anything that isn't us? What is the difference between a cathedral and a physics lab? Are not they both saying: Hello? We spy on whales and on interstellar radio objects; we starve ourselves and pray till we're blue.

IV

I have been reading comparative cosmology. At this time most cosmologists favor the picture of the evolving universe described by Lemaître and Gamow. But I prefer a suggestion made years ago by Valéry—Paul Valéry. He set forth the notion that the universe might be "head-shaped."

The mountains are great stone bells; they clang together like nuns. Who shushed the stars? There are a thousand million galaxies easily seen in the Palomar reflector; collisions between and among them do, of course, occur. But these collisions are very long and silent slides. Billions of stars sift among each other untouched, too distant even to be moved, heedless as always, hushed. The sea pronounces something, over and over, in a hoarse whisper; I cannot quite make it out. But God knows I have tried.

At a certain point you say to the woods, to the sea, to the mountains, the world, Now I am ready. Now I will stop and be wholly attentive. You empty yourself and wait, listening. After a time you hear it: there is nothing there. There is nothing but those things only, those created objects, discrete, growing or holding, or swaying, being rained on or raining, held, flooding or ebbing, standing, or spread. You feel the world's word as a tension, a hum, a single chorused note everywhere the same. This is it: this hum is the silence. Nature does utter a peep—just this one. The birds and insects, the meadows and swamps and rivers and stones and mountains and clouds: they all do it; they all don't do it. There is a vibrancy to the silence, a suppression, as if someone were gagging the world. But you wait, you give your life's length to listening, and nothing happens.

The ice rolls up, the ice rolls back, and still that single note obtains. The tension, or lack of it, is intolerable. The silence is not actually suppression; instead, it is all there is.

V

We are here to witness. There is nothing else to do with those mute materials we do not need. Until Larry teaches his stone to talk, until God changes his mind, or until the pagan gods slip back to their hilltop groves, all we can do with the whole inhuman array is watch it. We can stage our own act on the planet—build our cities on its plains, dam its rivers, plant its topsoils—but our meaningful activity scarcely covers the terrain. We do not use the songbirds, for instance. We do not eat many of them; we cannot befriend them; we cannot persuade them to eat more mosquitoes or plant fewer weed seeds. We can only witness them—whoever they are. If we were not here, they would be songbirds falling in the forest. If we were not here, material events like the passage of seasons would lack even the meager meanings we are able to muster for them. The show would play to an empty house, as do all those falling stars which fall in the daytime. That is why I take walks: to keep an eye on things. And that is why I went to the Galápagos islands.

All this becomes especially clear on the Galápagos islands. The Galápagos islands are just plain here—and little else. They blew up out of the ocean, some plants blew in on them, some animals drifted aboard and evolved weird forms—and there they all are, whoever they are, in full swing. You can go there and watch it happen, and try to figure it out. The Galápagos are a kind of metaphysics laboratory, almost wholly uncluttered by human culture or history. Whatever happens on those bare volcanic rocks happens in full view, whether anyone is watching or not.

What happens there is this, and precious little it is: clouds come and go, and the round of similar seasons; a pig eats a tortoise or doesn't eat a tortoise; Pacific waves fall up and slide back; a lichen expands; night follows day; an albatross dies and dries on a cliff; a cool current upwells from the ocean floor; fishes multiply, flies swarm, stars rise and fall, and diving birds dive. The news, in other words, breaks on the beaches. And taking it all in are the trees. The *palo santo* trees crowd the hillsides like any outdoor audience; they face the lagoons, the lava lowlands, and the shores.

I have some experience of these *palo santo* trees. They interest me as emblems of the muteness of the human stance in relation to all that is not human. I see us all as *palo santo* trees, holy sticks, together watching all that we watch, and growing in silence.

In the Galápagos, it took me a long time to notice the *palo santo* trees. Like everyone else, I specialized in sea lions. My shipmates and I liked the sea lions, and envied their lives. Their joy seemed conscious. They were engaged in full-time play. They were all either fat or dead; there was no halfway. By day they played in the shallows, alone or together, greeting each other and us with great noises of joy, or they took a turn offshore and body-surfed in the breakers, exultant. By night on the sand they lay in each other's flippers and slept. Everyone joked, often, that when he "came back," he would just as soon do it all over again as a sea lion. I concurred. The sea lion game looked unbeatable.

But a year and a half later, I returned to those unpeopled islands. In the interval my attachment to them had shifted, and my memories of them had altered, the way memories do, like particolored pebbles rolled back and forth over a grating, so that after a time those hard bright ones, the ones you thought you would never lose, have vanished, passed through the grating, and only a few big, unexpected ones remain, no longer unnoticed but now selected out for some meaning, large and unknown.

Such were the *palo santo* trees. Before, I had never given them a thought. They were just miles of half-dead trees on the red lava sea cliffs of some deserted islands. They were only a name in a notebook: "*Palo santo*—those strange white trees." Look at the sea lions! Look at the flightless cormorants, the penguins, the iguanas, the sunset! But after eighteen months the wonderful cormorants, penguins, iguanas, sunsets, and even the sea lions, had dropped from my holey heart. I returned to the Galápagos to see the *palo santo* trees.

They are thin, pale, wispy trees. You walk among them on the lowland deserts, where they grow beside the prickly pear. You see them from the water on the steeps that face the sea, hundreds together, small and thin and spread, and so much more pale than their red soils that any black-and-white photograph of them looks like a negative. The stands look like blasted orchards. At every season they all look newly dead, pale and bare as birches drowned in a beaver pond—for at every season they look leafless, paralyzed, and mute. But in fact, if you look closely, you can see during the rainy months a few meager deciduous leaves here and there on their brittle twigs. And hundreds of lichens always grow on their bark in mute, overlapping explosions which barely enlarge in the course of the decade, lichens pink and orange, lavender, yellow, and green. The *palo santo* trees bear the lichens effortlessly, unconsciously, the way they bear everything. Their multitudes, transparent as line drawings, crowd the cliffsides like whirling dancers, like empty groves, and look out over cliff-wrecked breakers toward more unpeopled islands, with their freakish lizards and birds, toward the grieving lagoons and the bays where the sea lions wander, and beyond to the clamoring seas.

Now I no longer concurred with my shipmates' joke; I no longer wanted to "come back" as a sea lion. For I thought, and I still think, that if I came back to life in the sunlight where everything changes, I would like to come back as a *palo santo* tree, one of thousands on a cliff side on those godforsaken islands, where a million events occur among the witless, where a splash of rain may drop on a yellow iguana the size of a dachshund, and ten minutes later the iguana may blink. I would like to come back as a *palo santo* tree on the weather side of an island, so that I could be, myself, a perfect witness, and look, mute, and wave my arms.

VI

The silence is all there is. It is the alpha and the omega. It is God's brooding over the face of the waters; it is the blended note of the ten thousand things, the whine of wings. You take a step in the right direction to pray to this silence, and even to address the prayer to "World." Distinctions blur. Quit your tents. Pray without ceasing.

Gretel Ehrlich, from "On Water" in *The Solace of Open Spaces* (1985)

Gretel Ehrlich's essay summarizes the history of water use and management in Wyoming, but even more important, delineates water's social and cultural status on the High Plains. Her vignettes and anecdotes of people as they interact with streams and storms and droughts strikingly illustrate just how central water or its lack is, both for life in the West, and for the human imagination generally.

In Wyoming we are supplicants, waiting all spring for the water to come down, for the snow pack to melt and fill the creeks from which we irrigate. Fall and spring rains amount to less than eight inches a year, while above our ranches, the mountains hold their snows like a secret: no one knows when they will melt or how fast. When the water does come, it floods through the state as if the peaks were silver pitchers tipped forward by mistake. . . .

A season of irrigating here lasts four months. Twenty, thirty, or as many as two hundred dams are changed every twelve hours, ditches are repaired and head gates adjusted to match the inconsistencies of water flow. By September it's over: all but the major Wyoming rivers dry up. Running water is so seasonal it's thought of as a mark on the calendar—a vague wet spot—rather than a geographical site. In May, June, July, and August, water is the sacristy at which we kneel; it equates time going by too fast.

Waiting for water is just one of the ways Wyoming ranchers find themselves at the mercy of weather. . . .

Six years ago, when I lived on a large sheep ranch, a drought threatened. Every water hole on 100,000 acres of grazing land went dry. We hauled water in clumsy beet-harvest trucks forty miles to spring range, and when we emptied them into a circle of stock tanks, the sheep ran toward us. They pushed to get at the water, trampling lambs in the process, then drank it all in one collective gulp. Other Aprils have brought too much moisture in the form of deadly storms. When a ground blizzard hit one friend's herd in the flatter, eastern part of the state, he knew he had to keep his cattle drifting. If they hit a fence line and had to face the storm, snow would blow into their noses and they'd drown. "We cut wire all the way to Nebraska," he told me. During the same storm another cowboy found his cattle too late: they were buried in a draw under a fifteen-foot drift.

High water comes in June when the runoff peaks, and it's another bugaboo for the ranchers. The otherwise amiable thirty-foot-wide creeks swell and change courses so that when we cross them with livestock, the water is belly-deep or more. Cowboys in the 1800s who rode with the trail herds from Texas often worked in the big rivers on horseback for a week just to cross a thousand head of longhorn steers, losing half of them in the process. On a less-grand scale we have drownings and near drownings here each spring. When we crossed a creek this year the swift current toppled a horse and carried the rider under a log. A cowboy who happened to look back saw her head go under, dove in from horseback, and saved her. At Trapper Creek, where Owen Wister spent several summers in the 1920s and entertained Mr. Hemingway, a cloudburst slapped down on us like a black eye. Scraps of rainbow moved in vertical sweeps of rain that broke apart and disappeared behind a ridge. The creek flooded, taking out a house and a field of corn. We saw one resident walking in a flattened alfalfa field where the river had flowed briefly. "Want to go fishing?" he yelled to us as we rode by. The fish he was throwing into a white bucket were trout that had been "beached" by the flood.

Westerners are ambivalent about water because they've never seen what it can create except havoc and mud. They've never walked through a forest of wild orchids or witnessed the unfurling of five-foot-high ferns. "The only way I like my water is if there's whiskey in it," one rancher told me as we weaned calves in a driving rainstorm. That day we spent twelve hours on horseback in the rain. Despite protective layers of clothing: wool union suits, chaps, ankle-length yellow slickers, neck scarves and hats, we were drenched. Water drips off hat brims into your crotch; boots and gloves soak through. But to stay home out of the storm is deemed by some as a worse fate: "Hell, my wife had me cannin' beans for a week," one cowboy complained. "I'd rather drown like a muskrat out there."

Dryness is the common denominator in Wyoming. We're drenched more often in dust than in water; it is the scalpel and the suit of armor that make Westerners what they are. Dry air presses a stockman's insides outward. The secret, inner self is worn not on the sleeve but in the skin. It's an unlubricated condition: there's not enough moisture in the air to keep the whole emotional machinery oiled and working. "What you see is what you get, but you have to learn to look to see all that's there," one young rancher told me. He was physically reckless when coming to see me or leaving. That was his way of saying he had and would miss me, and in the clean, broad sweeps of passion between us, there was no heaviness, no muddy residue. Cowboys have learned not to waste words from not having wasted water, as if verbosity would create a thirst too extreme to bear. If voices are raspy, it's because vocal cords are coated with dust. . . .

To follow the water courses in Wyoming—seven rivers and a network of good-sized creeks—is to trace the history of settlement here. After a few bad winters the early ranchers quickly discovered the necessity of raising feed for livestock. Long strips of land on both sides of the creeks and rivers were grabbed up in the 1870s and '80s before Wyoming was a state. Land was cheap and relatively easy to accumulate, but control of water was crucial. The early ranches such as the Swan Land & Cattle Company, the Budd Ranch, the M-L, the Bug Ranch, and the Pitchfork took up land along the Chugwater, Green, Greybull, Big Horn, and Shoshone rivers. It was not long before feuds over water began. The old law of "full and undiminished flow" to those who owned land along a creek was changed to one that adjudicated and allocated water by the acre-foot to specified pieces of land. By 1890 residents had to file claims for the right to use the water that flowed through their ranches. These rights were, and still are, awarded according to the date a ranch was established regardless of ownership changes. This solved the increasing problem of upstream-downstream disputes, enabling the first ranch established on a creek to maintain the first water right, regardless of how many newer settlements occurred upstream. . . .

Everything in nature invites us constantly to be what we are. We are often like rivers: careless and forceful, timid and dangerous, lucid and muddied, eddying, gleaming, still. Lovers, farmers, and artists have one thing in common, at least—a fear of "dry spells," dormant periods in which we do no blooming, internal droughts only the waters of imagination and psychic release can civilize. All such matters are delicate of course. But a good irrigator knows this: too little water brings on the weeds while too much degrades the soil the way too much easy money can trivialize a person's initiative. In his journal Thoreau wrote, "A man's life should be as fresh as a river. It should be the same channel but a new water every instant."

John Elder, from "Succession" in *Reading the Mountains of Home* (1998)

Inspired by Robert Frost's revision of romantic views of nature, John Elder uses the succession of forest trees to declare our need for a perspective broader than a single human life if we are to comprehend the processes by which nature creates and sustains itself. Like Frost and Thoreau, he faces the implications of living with both post-romantic spiritual and emotional expectations of nature, and a generally anti-romantic, empirical worldview.

No American poet has shown more familiarity than [Robert] Frost with the cycles of plant succession in his own region. All of the fields around the Homer Noble farm would have long ago . . . reverted to thick woods, except for the fact

Figure 20.6 *Ravine* by Vincent van Gogh, 1889. Oil on canvas, 73 × 91.7 cm (28.75 × 36.125 in.). Bequest of Keith McLeod, 52.1524. Photograph © 2003, Museum of Fine Arts, Boston.

that they are annually mowed, hayed, or burned by the Bread Loaf School of English or the Forest Service. These human disturbances are intended to maintain the particular kind of dynamic edge that prevailed in the Green Mountains during most of Frost's life.

Meadowsweet and steeplebush lift their blossoms above the tall grass throughout much of the summer. These two species of *Spiraea*, the former a lovely pale pink, the latter tending toward purple, rise waist-high on their sturdy stalks. Steeplebush calls for special attention, since it gives the entire volume of poetry its name. The little individual flowers, each with five petals, cluster into tall, fuzzy-looking arms of blossom that angle up sharply and form the spire. They bloom from the top down, with the upper clusters turning brown while the lower ones are still mellowing into a dark purple. The texture and color of the steeplebush are further enhanced as the season progresses by the fact that the tops of the serrated leaves remain green, while the undersides become a pronounced yellow. These striking flowers, mentioned in several poems by Frost, are the midway point in a cycle whereby a farmer's meadow is transformed, in the natural course of things, into a realm of trees. This hundred-year cycle is too long to be of economic advantage to any human individual, as the poet specifies. But it does offer a redemptive vision of inclusiveness, with each plant, from grass to maples, as well as the human onlooker, bound together in a dynamic whole. In Frost's landscape, things are always changing, but the change is never random. There is a grand logic of transformation, meaningful to the thoughtful observer, but always transcending the limited human purposes with which we might identify one phase or another of the whole.

The woods around this January beaver pond are a laboratory of succession—the pattern of continuity within change which was so central to Frost's perception of nature. Next to the pond, where logging seems to have occurred most recently, stands a dense grove of red maples. These large, attractive hardwoods often precede sugar maples in the reforesting of moist, disturbed ground. They are quite similar to those more familiar cousins, except for the V's between the sections of their leaves as opposed to the U's dividing the foliage of sugar maples. Yellow birches, smelling of wintergreen where little curls of bark peel off their trunks, are also mixed into these woods, as are a few ancient, slow-growing hemlocks.

On the other side of the pond, although there are few signs of logging, the ground is broken into a pronounced alternation of humps and hollows. Foresters call such a pattern of disturbed ground "pit and mound" or "pillow and cradle." It registers blow-down. When trees have toppled over, a half-circle of their roots tilts up into the air, exposing a hollow in the ground that remains even after the fallen log has begun to rot and the exposed roots have subsided into a softly contoured mound covered with soil and leaves. Our County Forester, David Brynn, counsels me always to think of trees from the root-collar down. Remember, David says, that at least as much of a tree's biomass will generally be below the ground as is above it, with the root-circle of an old hemlock often spreading out for forty or fifty feet. Remember that even for the white pines which sink such a mighty taproot, an equal amount of root by weight is in the tiny filaments that spread their interwoven mat just below the topsoil's nutrient sink. Blow-down raises the hidden life of trees into the light.

I appreciate "pillow and cradle" as a phrase to describe this juxtaposition of a depression in the woods and a nearby hump of raised ground. It expresses the perpetual renewal within the rise and fall of individual trees. Another eloquent term is "nurse log," describing the way in which a fallen tree nourishes seedlings along its length, making its stored energies available to the ongoing life of the forest. A fallen evergreen may decay, and nurse new life, over a period of sixty to seventy years, Tom Wessels informs me, while a hardwood log may decompose and make itself available within a cycle of thirty to fifty years. In his essay "Ancient Forests of the Far West," from *The Practice of the Wild*, Gary Snyder celebrates the nourishing persistence of fallen elders: "And then there are some long subtle hummocks that are the last trace of an old gone log. The straight line of mushrooms sprouting along a smooth ground surface is the final sign, the last ghost, of a tree that 'died' centuries ago."

A fallen log is something for hope. Not a hope for personal immortality, and not an assurance of prosperity or any other form of individual security. A hope, rather, for involvement in the grand pattern that connects. The southward orientation of many pillow-and-cradle formations in the rising ground just west of here tells of another event that occurred in 1938, along with the death of Elinor Frost and Robert's move into the little cabin at the woods' edge. A hurricane swept through this part of New England, absolutely leveling thousands of acres. Pillows and cradles in identical alignment show me its tracks amid the trees of Bristol Cliffs. Seedlings quickly establish themselves in the full light and nutrient-rich litter of such a vast blow-down. In disaster, they have discovered grounds for hope. . . .

Death and life are both embodied in a nurse log. Nature is always unified for one who can let go and identify with a life that transcends individuality. But certain propositions are much easier to affirm in the abstract, and such letting go sometimes requires first being overturned and uprooted. . . .

Decomposition at the bacterial level may be easier to contemplate if one leaves one's own body out of the picture. But only with an inclusive perspective on the universal breakdown of organisms can one look past it to new life. Only by adopting a time-line that comprehends the visitations of glaciers and the rise and fall of whole forests may one draw the lesson home. This is the tough but liberating view of life taught by the memory-tangled, ever-new New England woods.

III. SOCIAL CONNECTIONS

The natural sciences identify and define environmental processes; the humanities foster in us a sympathy with the environment and living nature. Both deepen our understanding and motivate us to reflect upon and change our personal values. But large-scale action is primarily understood and prompted by the social disciplines. Laws, public policy, market mechanisms—these are the principal tools in any effort to better the environment.

In the 1960s, when environmentalism was a marginal countercultural movement, environmentalists tended to see economics, law, politics, and all other human-created systems as the enemy. To some, environmental problems were so urgent that they trumped all other concerns and institutions. The maturation of the environmental movement in many countries has brought with it an understanding that human systems can have both intrinsic and utilitarian value; for example, that a just and equitable legal system is both an object of inherent value in itself and an instrument for effective environmental protection. Today most environmentalists see the environment as a primary value but one balanced against others such as individual freedoms, the democratic process, the rule of law, and economic stability. In recent years, as environmentalists have made peace with economics, the result has been the first steps in the appropriation of market mechanisms for the protection of the environment. Conversely, many individuals in business now see that environmentalism is often itself a force for economic growth and long-term efficiency.

In countries, such as those in Eastern Europe, in Latin America, and in the Middle East and Central Asia, where the environmental movement emerged or is emerging in opposition to military, communist, or religious dictatorships, it is often aligned at its core with democratic values, economic freedoms, human rights, individual rights, religious freedoms, rights of minorities, and the rule of law. Hungary is an interesting case study: scientific issues were among the few areas where the Communist regime tolerated limited dissent in the 1980s. When an environmental issue arose that galvanized the public (opposition to the proposed Gabčíkovo-Nagymaros dam on the Danube), there were demonstrations and petitions against government policy on an unprecedented scale. It was in this context of semi-legal environmental protest that opposition parties were able to organize themselves, hastening the collapse of the regime in 1989. All of Hungary's opposition parties were initially "green" parties. What the 1956 revolution and three decades of the Cold War could not change, environmental concern finally catalyzed.

As well as a sphere of action, we look to the social disciplines for a more nuanced and complete understanding of the causes of environmental problems—and their solutions. Social disciplines provide insight into market failures, unsound policies, and demographic trends that shape our relationship with the environment. Just as the understanding of natural ecological and evolutionary processes helps to reveal human social and psychological relations, so the understanding of human systems illuminates the ways that natural systems work.

Most broadly, anthropology provides us with perspectives on the range of social possibility. Besides specific lessons on ecologically friendly technologies and practices, anthropology teaches that social institutions, which often seem so solid from within, are to a certain degree malleable and arbitrary, and this gives us a sense of multiple ways to relate to one another and to the environment. The study of other cultures teaches us something about our possible origins, and our possible futures.

Many milestones in the history of the environmental movement came with the codification of environmental principles into law: the Clean Air, Clean Water, and Endangered Species Acts in the United States; and the Montreal Protocol and the Convention on International Trade in Endangered Species worldwide. Will the next legal watershed expand legal protection to include environmental justice, mandating a more equitable distribution of the risks of environmental degradation among social groups, both domestically and internationally?

The study of human systems and the environment could not be complete without lessons from the one phenomenon that most impinges on the environment—human population growth itself. The rise and redistribution of human populations has not only put an ever-increasing pressure on environmental resources, it initially produced the various cultures that we study in anthropology, the markets that we examine and regulate in economics, and the political and legal systems we fashion to govern our actions.

Overspecialization in a single discipline can be dangerous and blinding. This can be especially true of the social sciences. For decades, economists neglected the external costs imposed on the environment by human actions, and jurists assumed that land lying unused was unproductive. There has been a movement in both these disciplines, and in the other social sciences, to incorporate a more environmental worldview into the fabric of the discipline itself. But even where this has failed, the failure is often the result of a lack of connection, a lack of interdisciplinary thinking. In the discourse of environmental protection, the value of engaging in any social discipline, be it law, economics, anthropology, or political science, is great. Without including these disciplines, any view of the environment and the human relationship to it is incomplete, and our chances for effectuating positive change vanishingly small.

Keystone Essay: William O. Douglas, Dissenting Opinion in *Sierra Club v. Morton* (Mineral King), U.S. Supreme Court (1972)

The critical question of "standing" [the court's recognition of a party as a proper litigant] would be simplified and also put neatly in focus if we fashioned a federal rule that allowed environmental issues to be litigated before federal agencies or federal courts in the name of the inanimate object about to be despoiled, defaced, or invaded by roads and bulldozers and where injury is the subject of public outrage. Contemporary public concern for protecting nature's ecological equilibrium should lead to the conferral of standing upon environmental

Mineral King Valley. Courtesy of G. Donald Bain / the Geo-Images Project, Dept. of Geography, UC–Berkeley.

objects to sue for their own preservation. This suit would therefore be more properly labeled as *Mineral King v. Morton.*

Inanimate objects are sometimes parties in litigation. A ship has a legal personality, a fiction found useful for maritime purposes. The corporation sole—a creature of ecclesiastical law—is an acceptable adversary and large fortunes ride on its cases. The ordinary corporation is a "person" for purposes of the adjudicatory processes, whether it represents proprietary, spiritual, aesthetic, or charitable causes.

So it should be as respects valleys, alpine meadows, rivers, lakes, estuaries, beaches, ridges, groves of trees, swampland, or even air that feels the destructive pressures of modern technology and modern life. The river, for example, is the living symbol of all the life it sustains or nourishes—fish, aquatic insects, water ouzels, otter, fisher, deer, elk, bear, and all other animals, including man, who are dependent on it or who enjoy it for its sight, its sound, or its life. The river as plaintiff speaks for the ecological unit of life that is part of it. Those people who have a meaningful relation to that body of water—whether it be a fisherman, a canoeist, a zoologist, or a logger—must be able to speak for the values which the river represents and which are threatened with destruction. . . .

Mineral King is doubtless like other wonders of the Sierra Nevada such as Tuolumne Meadows and the John Muir Trail. Those who hike it, fish it, hunt it, camp in it, frequent it, or visit it merely to sit in solitude and wonderment are legitimate spokesmen for it, whether they may be few or many. Those who have that intimate relation with the inanimate object about to be injured, polluted, or otherwise despoiled are its legitimate spokesmen.

The Solicitor General . . . takes a wholly different approach. He considers the problem in terms of "government by the Judiciary." With all respect, the problem is to make certain that the inanimate objects, which are the very core of America's beauty, have spokesmen before they are destroyed. It is, of course, true that most of them are under the control of a federal or state agency. The standards given those agencies are usually expressed in terms of the "public interest." Yet "public interest" has so many differing shades of meaning as to be quite meaningless on the environmental front. Congress accordingly has adopted ecological standards in the National Environmental Policy Act of 1969, and guidelines for agency action have been provided by the Council on Environmental Quality of which Russell E. Train is Chairman.

Yet the pressures on agencies for favorable action one way or the other are enormous. The suggestion that Congress can stop action which is undesirable is true in theory; yet even Congress is too remote to give meaningful direction and its machinery is too ponderous to use very often. The federal agencies of which I speak are not venal or corrupt. But they are notoriously under the control of powerful interests who manipulate them through advisory committees, or friendly working relations, or who have that natural affinity with the agency which in time develops between the regulator and the regulated. As early as 1894, Attorney General Olney predicted that regulatory agencies might become "industry-minded," as illustrated by his forecast concerning the Interstate Commerce Commission:

"The Commission . . . is, or can be made, of great use to the railroads. It satisfies the popular clamor for a government supervision of railroads, at the same time that that supervision is almost entirely nominal. Further, the older such a commission gets to be, the more inclined it will be found to take the business and railroad view of things."

Years later a court of appeals observed, "the recurring question which has plagued public regulation of industry [is] whether the regulatory agency is unduly oriented toward the interests of the industry it is designed to regulate, rather than the public interest it is designed to protect."

The Forest Service—one of the federal agencies behind the scheme to despoil Mineral King—has been notorious for its alignment with lumber companies, although its mandate from Congress directs it to consider the various aspects of multiple use in its supervision of the national forests.

The voice of the inanimate object, therefore, should not be stilled. That does not mean that the judiciary takes over the managerial functions from the federal agency. It merely means that before these priceless bits of Americana (such as a valley, an alpine meadow, a river, or a lake) are forever lost or are so transformed as to be reduced to the eventual rubble of our urban environment, the voice of the existing beneficiaries of these environmental wonders should be heard.

Perhaps they will not win. Perhaps the bulldozers of "progress" will plow under all the aesthetic wonders of this beautiful land. That is not the present question. The sole question is, who has standing to be heard?

Those who hike the Appalachian Trail into Sunfish Pond, New Jersey, and camp or sleep there, or run the Allagash in Maine, or climb the Guadalupes in West Texas, or who canoe and portage the Quetico Superior in Minnesota, certainly should have standing to defend those natural wonders before courts or agencies, though they live 3,000 miles away. Those who merely are caught up in environmental news or propaganda and flock to defend these waters or areas may be treated differently. That is why these environmental issues should be tendered by the inanimate object itself. Then there will be assurances that all of the forms of life which it represents will stand before the court—the pileated woodpecker as well as the coyote and bear, the lemmings as well as the trout in the streams. Those inarticulate members of the ecological group cannot speak. But those people who have so frequented the place as to know its values and wonders will be able to speak for the entire ecological community.

Ecology reflects the land ethic; and Aldo Leopold wrote in *A Sand County Almanac* (1949), "The land ethic simply enlarges the boundaries of the community to include soils, waters, plants, and animals, or collectively: the land."

That, as I see it, is the issue of "standing" in the present case and controversy.

21 Politics and Public Policy

The late Speaker of the United States House of Representatives Thomas P. ("Tip") O'Neill famously claimed, "All politics are local." Individual situations in which politics and public policy have had an impact on the environment would seem in some ways to bear O'Neill out. For instance, in explaining the ecological and social problems resulting from the exploitation of oil resources in *Genocide in Nigeria: The Ogoni Tragedy*, playwright and novelist Ken Saro-Wiwa (himself later executed after a trial widely thought to be politically inspired) gives full credit to the ethnic and religious rivalries that led to civil war and that still divide his native land. He likewise presents the issue of who benefits from the oil wealth of Nigeria as one significantly shaped by local politics and personalities. Yet Saro-Wiwa never forgets that there are broader historical forces and a wider set of players at work here too: from the legacy of British colonialism (including the tension between the nation-state and older loyalties), to international competition during the Cold War and after for political and economic influence, to the corporate interests of multinational energy companies and banks. If all politics are local, they are also, at least when it comes to clean air, toxic waste sites, nuclear energy, and the like, national and global as well.

Indeed, like environmental laws enacted on the national or regional levels, international agreements or statements of public policy concern, such as the one produced by the 1992 United Nations Earth Summit held in Rio de Janeiro, are often the product of long and torturous negotiations aimed at assuaging an array of conflicting governments, public interest groups, and concerned individual or corporate parties. This renders generalizations about the opposing perspectives and claims of the "rich and industrial North" versus the "poor or modernizing South" problematic at best, since nations on both sides of the global divide have aspirations, values, and interests that are at times shared to a surprising extent. Each situation is perhaps best understood in its own political and environmental terms, and any patterns that emerge from a series of politically affected environmental questions are most prudently regarded as provisional.

Similar attention to the complexity of specific environmental and political situations is important when addressing the practical politics of environmentalism as a movement. John Barry, in *Rethinking Green Politics*, reminds us that there is a fundamental polarity between those who see environmentalism as a "moral crusade" seeking to win hearts and minds and change the culture, and those who see it as being about the business of institutional transformation. Yoking these two groups together and exploiting the possibilities of the dynamic between them, as it expresses itself in a particular context, is the key to success. More provocatively, Lee Quinby attributes "the splits that have taken place within the ecology and feminist movements" (e.g., in the German Green Party) to the influence of those advocating a coherent and comprehensive but (ironically) hegemonically "male" ecofeminist ideology. She instead urges a more pragmatic, situation-orientated approach, one alive to complexity and contradiction, one that will combat "ecological destruction and patriarchal domination without succumbing to the totalizing impulses of masculinist politics." Cynthia Hamilton, by contrast, returns us to the local and the pragmatic in describing how poor women of color effected environmental change in Los Angeles, thereby illustrating the truth of Aristotle's claim that we are political animals, and so, politics become necessary.

A similar openness to complexity characterizes the best intellectual treatments of the politics and public policy dimension of the environment as well. The influential "Three Economies" model developed by Zygmunt Plater and his colleagues accounts for "the dynamics of modern life" by recognizing that as humans we exist within "three different intersecting economies—comprised not only of marketplace economics and natural systems economics, but also of a civic-societal economics incorporating the society's overall cumulative well-being." So too, as David Pepper shows, rigorous analyses of taken-for-granted concepts like scientific objectivity undercut our assumptions by revealing how science is at times culturally, institutionally, intellectually, and politically manipulated in reference to environmental issues and the public policy decisions they create. As asserted at greater length by many of the scholars included in "History and the Environment," Chapter 19, the interpretation of science like so much else is in part a cultural construct. No less than individual environmental situations or the practical imperatives of environmental organizing, both intellectual life and particular academic disciplines are by-products of politics—in the broadest sense—and contributors to the politics of their time and place.

FURTHER READING

Icke, David. *It Doesn't Have to Be Like This: Green Politics Explained.* London: Green Print, 1990.

Kemp, Penny, et al. *Europe's Green Alternative.* Trans. Julia Sallabank. London: Green Print, 1992.

Maddock, Su. *Challenging Women: Gender, Culture, and Organization.* London: Sage Publications, 1999.

Wall, Derek. *Green History: A Reader in Environmental Literature, Philosophy, and Politics.* London: Routledge, 1994.

Ken Saro-Wiwa, from *Genocide in Nigeria: The Ogoni Tragedy* (1992)

Saro-Wiwa describes the ecological and human effects on the Ogoni people in the Niger delta of the Bomu Oil Disaster and the energy exploration that caused it. He characterizes this ongoing catastrophe as genocide rooted in the politics of Nigeria, an artificial state created in colonial times, whose resources and institutions are still controlled by three main ethnic groups for the benefit of the British and American governments and multinational companies, but to the detriment of minorities like the Ogoni.

Today, twenty-two years after [the disaster of 1971], the wasteland remains—barren and useless, another reminder of the road to the extinction of the Ogoni people charted by the greed and racism of Shell and the complicity of the Federal Government of Nigeria.

Since then, the offending oil well has been capped, new wells dug, and gas from them has been burning methane and other hydrocarbons into the lungs of Ogoni villagers every day of the year for twenty-one years. The noise of burning gas has made the people of Dere half-deaf—they have to shout when they speak to each other; burning gas continues to turn their nights into day. They have no electricity, no pipe-borne water, and no hospital. Respiratory diseases are common in the area. Only last year, the entire village was flooded because the access roads built by Shell to its numerous locations have made a valley of the village. Studies have indicated that the level of lead in the blood of the inhabitants is at a dangerously high level.

Shell must bear full responsibility for the genocide of the Ogoni which is going on even now. The record of the Company in environmental issues in Nigeria has been most appalling. When Chevron began to prospect for oil in Ogoni twelve years ago in 1980, it had the example of Shell to go by.

The most notorious action of both companies has been the flaring of gas, sometimes in the middle of villages, as in Dere (Bomu Oilfield), or very close to human habitation as in the Yorla and Korokoro oilfields in Ogoni. This action has destroyed ALL wildlife, and plant life, poisoned the atmosphere and therefore the inhabitants in the surrounding areas, and made the residents half-deaf and prone to respiratory diseases. Whenever it rains in Ogoni, all we have is acid rain which further poisons water courses, streams, creeks, and agricultural land.

Next to the flaring of gas comes the frequency of oil spills. Shell and Chevron use the most outdated equipment and technology in Ogoni, leading to innumerable oil spills which destroy farmlands, streams and water courses, and the

Figure 21.1 Ogoni men attempt to clean up waste. © Newscom / Reuters.

creeks. One of the greatest casualties of oil spills has been the mangrove trees in the swamps which near-surround Ogoni. These trees which were a source of firewood, of seafood such as oysters, mussels, crabs, and cockles, have been unable to survive the toxicity of oil. They have now been replaced by strange, valueless palms.

Additionally, oil has poisoned the mudbanks which were formerly the home of mudskippers, clams, crabs, and periwinkles. These rich sources of protein for the Ogoni people no longer exist. The result is that the fishermen of Ogoni have lost their occupation and Ogoni people no longer have protein in their food. Children are the main sufferers at the hands of Shell and Chevron.

Similarly, the pollution of water courses, streams, and creeks by oil spillage has led to the death of another source of protein—fish. . . . These streams and creeks formerly brimmed with fish. Today, the Ogoni who normally fished in these waters have no alternatives. If they must fish, they need to go into deeper and offshore waters for which they have not been equipped or trained.

Pipelines criss-cross Ogoni territory, using up valuable land and ringing the people round with the danger of oil being pumped under very high pressure. These pipes metaphorically drain the very life-blood of the Ogoni people. As the oil mining activities of Shell and Chevron proceed apace, huge company trucks and other vehicles thunder past villages and towns day and night, an additional hazard to the peace and quiet of rural life.

As a final mark of their genocidal intent and insensitivity to human suffering, Shell and Chevron refuse to obey a Nigerian law which requires all oil companies to re-inject gas into the earth rather than flare it. Shell and Chevron think it cheaper to poison the atmosphere and the Ogoni and pay the paltry penalty imposed by the government of Nigeria than re-inject the gas as stipulated by the regulations.

To this charge of genocidal intent and insensitivity to human suffering must be added another of racism. Shell has won prizes for environmental protection in Europe where it also prospects for oil. So it cannot be that it does not know what to do. Now, why has it visited the Ogoni people with such horror as I have merely outlined here? The answer must lie in racism. In Shell's racist mind, what is good for the whites must not be good for blacks.

Shell has used its financial might to promote European culture in Nigeria. It regularly sponsors the tours of British artists to Nigeria, but spends nothing on the promotion of Ogoni culture. It is, at this moment, sponsoring the construction of a theater for chamber music in Lagos at a cost of twenty million naira (one million pounds). But it has not spent a tenth of that amount in Ogoni in thirty-four years of exploitation of Ogoni resources estimated at a total value of 100 billion U.S. dollars.

Added to this charge is that of ethnocentrism which Shell has promoted in Nigeria. Shell's best residential and office blocks in Nigeria are situated in Lagos among the Yoruba ethnic majority where there is not a drop of crude oil. It does not have a single modern building in Ogoni. Chevron pays more in rent in one year for ONE two-bedroom flat for one of its middle-level employees in Lagos among the Yorubas than it has paid in a total of ten years to the Ogoni landlords whose land it is expropriating. Both Shell and Chevron ensure that their best jobs and contracts go first to foreigners and then to the ethnic majority in Nigeria.

Shell has stated that its sole responsibility is to produce hydrocarbons efficiently. This is manifestly false. Shell and other companies routinely contribute to the social and economic well-being of the people among whom they operate. Indeed, it is a datum of management wisdom to create a conducive atmosphere for business operations. Why Shell and Chevron have decided pig-headedly to oppress, suppress, and eventually destroy the Ogoni people cannot be understood. How they can take away millions of dollars from a people who have no electricity, no pipe-borne water, no hospitals, no schools, and no future and feel no pangs of conscience whatsoever defies the imagination.

The result of the foregoing has been the total destruction of Ogoni life, human, social, cultural, and economic. [There is also] the intrinsic psychic relationship which exists between the Ogoni and their environment. What Shell and Chevron have done to Ogoni people, land, streams, creeks, and the atmosphere amounts to genocide. The soul of the Ogoni people is dying and I am witness to the fact.

I hear the plaintive cry of the Ogoni plains mourning the birds that no longer sing at dawn; I hear the dirge for trees whose branches wither in the blaze of gas flares, whose roots lie in infertile graves. The brimming streams gurgle no more, their harvest floats on waters poisoned by oil spillages.

Where are the antelopes, the squirrels, the sacred tortoises, the snails, the lions, and tigers which roamed this land? Where are the crabs, periwinkles, mudskippers, cockles, shrimps, and all which found sanctuary in mudbanks, under the protective roots of mangrove trees?

I hear in my heart the howls of death in the polluted air of my beloved homeland; I sing a dirge for my children, my compatriots, and their progeny.

Richard N. L. Andrews, from *Managing the Environment, Managing Ourselves: A History of American Environmental Policy* (1999)

Andrews concentrates on institutional transformation as a means to environmental change. In particular, he defends the role of government in shaping the law in this area, in the face of attacks from libertarians, free-market advocates, and others. This selection also comprehensively introduces his model of state involvement in environmental public policy, describing some of the problems and dynamics peculiar to this aspect of national and local politics.

Human uses of the environment are matters of governance, not merely of individual choice or economic markets.

Some may disagree with this assertion. Strict libertarians might argue, for instance, that uses of the environment should be decided by autonomous individuals acting on their own preferences and values, without compulsion by governments. Free-market economists might argue that they should be decided by the sum of individual preferences expressed in market choices, and that government should intervene only to correct "market failures." Still others might argue that the only role of government should be to protect and enforce private property rights, leaving environmental choices to those who thus own the environment.

For at least seven reasons, however, government involvement in environmental issues is both necessary and inevitable.

First, *governments assign and enforce property rights, determining who has what rights to use or transform the environment and what duties to protect it.* When someone buys an acre of land, does she own the minerals underneath it and the water and wildlife that cross it as well, or may these belong to someone else or to the public? Does a commercial fisherman have the right to sell as many fish as he can catch, destroying an entire fishery and simply moving on to another, or is that right limited by a legitimate public interest in maintaining the fishery's survival? Does an upstream industry or city have the right to use a river for waste disposal, or does a downstream homeowner or community have the right to water free of contaminants?

Governments do not merely intervene in markets. They establish the basic operating conditions for them, affirming and enforcing principles regarding who actually holds the rights to produce or sell environmental assets—or to exclude others from them—in the first place, and who is liable for the costs of environmental damage. Even libertarians need governments to enforce their rights against environmental damage caused by their neighbors.

Second, *governments define and enforce the rules of markets themselves.* Governments enforce contracts and other rules of market honesty, for instance, so that both buyers and sellers can protect themselves against cheaters. Some of these rules involve environmental conditions: adulterated foods, contaminated or toxic products, businesses with unacknowledged environmental liabilities, false advertising claims for "green" products, and other misrepresentations. For that matter, modern corporations would not even exist without government statutes that allow people to pool capital assets in such organizations, to limit their personal liability for the corporations' actions, and to operate such corporations as "legal persons" with essentially the same legal rights as individuals. Markets do not operate without government enforcement of the rules of fair transactions, and environmental conditions are often elements of such fairness.

Third, *governments protect public health and safety.* Both infectious diseases and toxic agents are spread through environmental exposures against which people cannot fully protect themselves by individual actions or market choices: through air, water, and food contamination, insect vectors, and unwitting contact with infected individuals, for instance. Historically such hazards have killed and sickened vast numbers of people. Environmental hazards are also increased by the cumulative effects of individually logical choices: disposing of wastes as cheaply as possible, building more and more homes with septic systems around lakes, paving and building along rivers and thus increasing downstream floods. Governments are needed to enforce reasonable restrictions on individual behavior patterns that create hazards to public health and safety.

Fourth, *governments protect environmental assets from "tragedies of the commons."* Many valuable environmental resources are "open access" or "common pool" resources that can easily be captured and sold by competing businesses. Examples include ocean fisheries and large underground oil pools. Each commercial fisherman has an incentive to catch and sell as many fish as possible, and each oil company to pump as rapidly as possible, since otherwise the resource will be captured and sold by their competitors. National parks and public highways experience similar problems: the cumulative effects of individual choices can lead to overcrowding or even destruction of the common resource. Garrett Hardin described such problems as tragedies of the commons. Such patterns of incentives and behavior, he argued, must be restrained by some governance mechanism—"mutual coercion, mutually agreed upon"—or the common resource will be destroyed by the cumulative results of individually rational choices.[1] [See Hardin selection in Chapter 23.]

Fifth, *governments provide collective goods that markets do not.* Many valuable environmental conditions are collective goods. Like national defense, clean air and public parklands benefit whole communities rather than just individual purchasers, and arable soils and sustainable fisheries benefit generations who are not yet present to make market choices. Markets undervalue or even fail to provide such goods: first, because it is impossible to organize payment by all the people who benefit from them, and second, because those who don't pay cannot be excluded from enjoying them as well. Acting as self-interested individuals, both businesses and consumers are tempted to "free-ride": to pay as little as possible for their share of collective goods from which everyone benefits.

Free-riding aside, individuals and markets also depend on governments to pool enough resources to finance projects that may be widely beneficial but require large investments for benefits that accrue over long periods of time. Examples include large multipurpose water resource development projects, building of transportation infrastructure, and space exploration. Not all such projects have been well justified in the past, but many clearly have been.

[Sixth and] more generally, *governments provide environmental services that people prefer to have provided collectively*, acting as voting citizens of a community, state, or nation rather than as individual consumers. Examples include public water supplies and waste management services, community amenities, and parks and recreation areas. Governments also have the unique power to redistribute access to environmental amenities and economic resources based on votes rather than purchases. In societies more complex than small face-to-face communities, only governments can redistribute resources so as to provide at least minimal access to decent living conditions regardless of wealth. Also, only governments can moderate the inequities that markets tend to produce, both in general economic opportunity and in access to environmental resources and amenities. Governments do not always produce these results—often they are influenced by powerful economic interests to redistribute from the poor to the rich—but only governments *can* accomplish redistribution.

Finally [and seventh], environmental issues are governance issues because *governments' actions have environmental impacts themselves.* Governments tax some uses and subsidize others, and they invest public funds in projects to transform the landscape for resource extraction, human settlements, transportation, food and energy production, and other uses. They preserve and manage some landscapes directly, regulate and restrict the use of others, and support research and professional expertise in areas of environmental knowledge that markets alone might not. Government actions both cause and correct environmental problems: government regulations and subsidies have been vital and effective tools for cleaning up pollution, for instance, but government incentives for Cold War military and industrial production also caused widespread contamination in the first place. Dealing with environmental issues therefore means dealing with the environmental effects of government actions themselves, not simply those of individuals and businesses.

In environmental policy issues, therefore, questions about the proper management of the environment are fundamentally intertwined with questions about the proper ends and means of governments themselves.

ENVIRONMENTAL POLICY

Environmental *protection* policy includes three elements of environmental policy that are explicitly intended to protect public health and ecological processes from adverse effects of human activities. The first element is *pollution control*, including prevention, safe management, and cleanup of waste discharges, accidental spills, and deliberate environmental dispersion of toxic materials such as pesticides. The second is *sustainable natural resource management*, including maintenance of naturally renewable resources (groundwater, forests, fish and wildlife, arable soils), regulation of rates of extraction and use of other resources (water, fuels, strategic minerals), and management of conflicting uses of landscapes for commodity production and recreation. The third is *preservation of natural and cultural heritage*, including areas of special beauty, historical and cultural significance, ecological functions, and landscape character.

Environmental policy as a whole, however, includes all government actions that alter natural environmental conditions and processes, for whatever purpose and under whatever label. Policies promoting transformation of the environment for mineral extraction, for agriculture or forestry or outdoor recreation, for urban or industrial development, or for transportation infrastructure are in their effects just as much elements of environmental policy as are pollution control regulations or habitat protection programs—whether or not they are called by that name. So are military operations, international trade agreements, and other policies with environmental impacts.

This broad definition of environmental policy has important practical implications. First, the "real" environmental policy of a government is not necessarily what its officials say their policy is, nor what the statutes and regulations say, but the cumulative effect of what government actions actually *do* to the natural environment. Many official statements of environmental policy, and even statutory mandates, are undercut by conflicting mandates or underfunded budgets, and others are ineffective in achieving their stated purposes.

Second, many of the most powerful instruments of environmental policy are lodged not in environmental protection agencies but in agencies that transform environmental conditions for other purposes. The policies of the Agriculture,

Energy, and Transportation Departments, for instance, may affect the environment at least as much as those of the Environmental Protection Agency (EPA), and their budgets are far larger. Some of the most effective policy strategies for reducing pollution, therefore, might involve not adding new EPA regulatory programs, but changing or eliminating environmentally damaging programs administered by other departments.[2] More generally, the best way to achieve an environmental policy goal, such as reducing pollution or preserving ecosystems or landscapes, might often be to improve coordination of conflicting policies across multiple sectors, rather than merely to add a single new law involving one agency.

ENVIRONMENTAL PROBLEMS AS PUBLIC POLICY ISSUES

Environmental problems share with other public policy issues a set of questions about what governments should and should not do, and how such decisions should be made. These questions are just as much a part of the debates about environmental problems as are the questions of environmental science, technology, and economics that often dominate such debates. They include the following:

Individual versus collective purposes. Which purposes should be pursued through collective decisions, and which should be left to individual choices? Some functions are intrinsically governmental and cannot be accomplished in any meaningful way by individuals or economic markets alone. Protecting public health has long been recognized as a government function, and sustaining the "commons" of environmental conditions necessary for the society's continuation and economic welfare is also such a purpose. Other functions, such as transportation, water and energy supply, waste management infrastructure, and other services, can be provided either collectively or individually.

Tradeoffs among public purposes. How should conflicting public values for uses of the environment be balanced? Even among legitimate public purposes, conflicts and tradeoffs are inevitable: among competing uses of environmental resources themselves, among competing allocations of limited budgets and staff expertise, and among different beneficiaries and victims. Which should be considered most important, by what criteria, and who should decide?

Proof versus prudence. How much evidence should be required to justify government actions? Government policies are collective actions, in which the good of society is asserted to override the rights or preferences of individuals. They must therefore be justifiable rather than arbitrary or capricious. In the United States in particular, this principle has led to elaborate requirements for scientific and economic justification of policy proposals. Uncertainty and conflicting judgments remain inescapable, however, so prudence must always substitute to some extent for proof.

As a result, policy decisions are not only decisions about substantive environmental problems but also about how much proof is necessary to justify a government action to correct them. This issue is important in itself, especially when costly or intrusive remedies are proposed for problems whose solutions are highly uncertain. It is also used by some advocates, however, as a tactic to slow or derail prudent policy proposals that they oppose for self-interested reasons: "paralysis by analysis."

Central versus local governance. At what level of governance—local, regional, state, federal, international—should environmental policy decisions be made? The principle of "subsidiarity" holds that policy decisions should be made at the lowest possible level: local decision-making provides the best venue for developing solutions tailored to specific conditions and communities. It is most accountable to those most directly affected, and most appropriate for maintaining the diversity of human cultures and communities.

By the same token, local governments also have more limited revenues and resources than those with broader jurisdictions, leaving poor communities unable to benefit from a broader base of economic support. Local decisions are also most likely to run counter to more general public values, whether pertaining to trade, environmental protection, or human rights. Granting autonomy to local governments also creates the risk that they will displace adverse effects onto other jurisdictions, and may pit local or even national governments against one another in using tax breaks, lower wages, and weaker environmental protection policies to attract or retain businesses.

More centralized governance, conversely, provides greater opportunities for setting general standards of acceptable behavior and competition, and can amass greater revenues with which to realize public purposes. Its risks, however, include greater bureaucratization, lessened accountability, dominance by powerful centralized interests, and standardization at the expense of local diversity. Policies must be designed to use the most appropriate combination of tools and levels of government to solve each problem.

Organization of government institutions. What sorts of government agencies, at any level, should be responsible for environmental protection? Should they be specialized independent agencies (for environmental regulation, for instance), or multipurpose departments encompassing both regulation and natural resource management? How can such agencies be coordinated with other agencies whose actions affect environmental conditions? There is no simple way of organizing all government purposes under one super-agency, but separate agencies have inherent tendencies to pursue their own missions at the expense of others.

Even individual units such as the EPA must be broken down into subunits that tend to focus narrowly on their own missions. Organizing by problem types such as air or water pollution, toxic chemicals, and waste management will lead to different results than organizing by problem sources such as industry, agriculture, and households or by administrative functions such as standard-setting, enforcement, and research. Conflicting perspectives and priorities must constantly be resolved.

Collective choice procedures. Who is to decide, and by what process, what environmental policies and priorities should be? Representative legislatures or appointed administrators? Experts or "the people"? Experts may be wise protectors of the society's future, or self-interested and arrogant elites. Members of Congress may be statesmen seeking the long-term good of the society, decent but parochial representatives of their constituents' wants, or merely self-interested incumbents selling themselves to interest groups to finance their own reelections. Ordinary people may be ideal citizens seeking the good of their society, or they may be just as self-serving or short-sighted as anyone else.

Each procedure for collective decision-making has strengths and weaknesses, both in principle and as a mechanism for determining what government should do.

Policy tools. What kinds of government actions are the best tools for achieving public policy goals? Regulations with civil or criminal penalties? Public expenditures, subsidies, and investments? Taxes and other economic incentives? Information disclosure requirements? Providing public services, or contracting for them? Some policy tools may be far more effective than others, either in general or for particular purposes. All have impacts on other goals as well, such as fairness, economic efficiency, and equitable distribution of the benefits and costs of the policy to particular communities. Policy choices must therefore be based on careful evaluation of their full consequences, and on experimentation and correction over time.

Intrinsic hazards of governance processes ("government failures"). Finally, governmental actions always involve complications intrinsic to collective decision-making.[3] "Free-riding" describes each participant's temptation to try to avoid paying a fair share of the cost of collective services. "Rent-seeking" reflects the equally human tendency to seek excessive compensation for one's own property or services, or even one's vote. "Pork-barreling" describes the tendency of elected representatives to collude in allocating general public revenues to benefit their own constituencies.

Other complications stem from the *transaction costs* of reaching agreements. Collective decision-making is costly and time consuming. . . .

Governments also tend to externalize the social and environmental costs of their decisions, just as businesses and individuals do. Government decisions are routinely designed to promote the short-term self-interests of public officials and powerful organized interests by providing concentrated and visible benefits while making costs and harms as widely dispersed and invisible as possible. The result is often that social and environmental impacts are displaced onto other agencies, onto other communities or countries, onto other levels of government, onto less-organized constituencies, or onto later legislatures and administrations and future generations. Examples include locating incinerators on downwind borders, imposing unfunded federal mandates on state and local governments, and subsidizing the extraction and use of natural resources at rates faster than can be sustained for future users.

Such jurisdictional externalities are in principle no different from the externalities sometimes produced by economic behavior, except that they represent government failure rather than market failure. Such problems arise whenever a government's jurisdiction and process do not include representation of all affected constituencies and responsibility for the full range of causes, consequences, and potential solutions. They are particularly common in environmental policy issues, for reasons discussed below.

In environmental policy no less than in other public policy debates, therefore, the fundamental issues include questions not only about technical and economic matters but also about the role of government, the costs and risks of its actions, and measures to minimize and correct harmful effects.

SPECIAL CHARACTERISTICS OF ENVIRONMENTAL ISSUES

Environmental issues, however, also have characteristics that differentiate them from other policy issues. For example:

Environmental values, preferences, and power relationships. Environmental issues involve particular places with distinctive natural features and histories. People identify with such places and develop strong opinions about how they should look and be used: whether they should be kept as they are, used for established economic purposes, or altered to achieve some new vision. Moreover, the uses of particular places are interdependent. Unlike budget allocations, entitlement programs, and many other policy issues in which each constituency can lobby for a share of the outcome, each participant's use of the environment affects those of the other participants: hunters and hikers and loggers, fishermen and farmers and users of municipal wastewater treatment plants. Proposed changes, therefore, are often simultaneously good to some groups and bad to others.

Creating environmental policy, therefore, often involves negotiating conflicts among mutually exclusive preferences for the use of indivisible resources. Such conflicts are far less amenable to political compromise or compensation than other policy issues. Examples include conflicts over proposals for construction of mines, landfills, and other major facilities, logging of old-growth forests, damming of free-flowing streams, and development of beaches and lakeshores.

The physical and biological realities of environmental conditions also create one-sided relationships of economic and political power. Rights to use natural resources, such as forests and minerals, confer windfall economic benefits but also create resource-dependent interests and constituencies. Upwind or upstream users can always impose externalities on their downwind or downstream neighbors, but the latter have no inherent countervailing power to negotiate fair outcomes with the former. Hunters and fishermen benefit from capturing migratory animal species, but the reproduction and growth of those species depend on restraining people who use the ecosystems where the animals spend earlier stages of their life cycles.

All these conditions shape environmental policy debates in ways that make them distinct from political controversies based only on ideology, political party, social class, or other factors.

Public attitudes toward environmental risks. Environmental risks evoke strong public attitudes and preferences. Many people demand government action to prevent environmental risks that are far more remote than risks they voluntarily incur in their daily lives. Examples include risks of exposure to trace residues of man-made chemicals in comparison to such risks as driving a car, crossing a street, smoking, or even eating foods whose natural properties pose greater health risks than do their man-made additives (high fat, cholesterol, or natural toxins, for instance). This pattern appears to reflect greater aversion to risks that are perceived as more uncontrollable and more dreadful in their consequences than others. It may also reflect other values, such as a willingness to impose greater costs of risk prevention on other parties than on themselves (on "big business" or "big government," for instance). Whatever the reasons, fear of environmental risks represents a powerful and distinct force that is different from those motivating other policy advocates.

Tragedies of the commons. Many environmental conditions are by their nature open-access resources: available to everyone, and therefore difficult to protect from the cumulative effects of overuse. Examples include the atmosphere, water bodies (lakes, rivers, estuaries, and seas), underground aquifers and oil deposits, fisheries, and unmanaged public forests and grazing lands. Garrett Hardin's classic article "The Tragedy of the Commons" described as a model of such problems the case of self-interested sheepherders, each of whom adds animals to a common pasture until the cumulative effects of their individual decisions destroy it. Some open-access resources can be privatized, to be managed (though not necessarily protected) by a single owner. Some can be converted to government property or "common property" resources, in which either government or an association of the users manages and protects it. But others are more difficult to protect. The users may be too numerous, too diverse, or too separated in space or time to create a viable regime, or the values to be protected may be too divorced from the interests of those causing the damage: as, for instance, when fisheries are destroyed not by other fishermen but by land developers or farmers. . . .

Scientific and technical premises. Environmental policy decisions are often framed by scientific and technical claims, including assertions about what is known and what options are technically possible, as well as assumptions, predictions, and uncertainties. This raises several problematic issues. First, it creates barriers to meaningful participation by people who do not understand the scientific and technical claims being made. Second, it makes the burden of proof a key issue in its own right: should governments be required to show strong scientific proof before acting to correct environmental problems, or should they act based on reasonable judgments about the risks or opportunities at stake even when significant uncertainties remain? Third, it raises questions about how much deference should be given to scientists in the policy process. Each discipline addresses only pieces of any issue, and individual scientists reach different conclusions based on the different bodies of evidence and criteria they use. Scientists are often overconfident of their judgments, and sometimes as self-interested as other policy advocates. Many scientific claims in policy debates—though by no means all—may therefore be just as political among conflicting groups of scientists as the policy decisions themselves are among conflicting public constituencies.

Irreversible damage to public interests. Finally, some environmental issues involve consequences far broader than the self-interests of particular advocates, some of which are potentially irreversible on any meaningful time scale. Examples include species extinctions, the exhaustion of nonrenewable resources and arable soils, and particularly the destruction of whole ecological systems that support living communities, such as forests, fisheries, wetlands, and estuaries. The potential for irreversible damage to irreplaceable natural conditions and processes—and more generally, to a healthful, productive, and attractive natural environment—is an important consideration in environmental policy. To the extent that claims of such damage are well founded, they deserve serious consideration.[4]

For all these reasons, environmental policy is worth studying not only as an example of public policy generally, but as an important and distinctive topic in its own right.

AMERICAN ENVIRONMENTAL POLICY

Today's American environmental policies are the legacy of a long history. The label "environmental policy" was coined only in the 1960s,[5] and with it have come important changes in both understanding and policies. The existence of environmental policies, however, dates back not just the thirty years since the first Earth Day [1970], but the two-hundred-plus years since the establishment of the current constitutional regime, and the nearly four hundred years since European empires colonized North America. Today's environmental problems are shaped by the policies that previous generations created to address earlier environmental problems and opportunities. Environmental policy choices today, in turn, will shape the problems and opportunities of the future.

Figure 21.2 Protest banner on smokestack. Courtesy of Greenpeace.

American environmental policy reflects distinctive American attitudes toward the environment. Examples include the nineteenth-century "cornucopian" perception of virtually infinite natural resources, and the more recent perception of industrial chemicals as insidious and ubiquitous cancer risks. It also reflects distinctive American attitudes toward governance, such as distrust of centralized power and authority, a preference for adversarial over authoritative decision-making, and shifting preferences for legislative, administrative, or direct popular decision-making. Both the goals and tools of environmental policy have changed greatly over the course of American history, as have the political processes by which it is made and implemented.

To the extent that the United States has a national environmental policy today, it consists not in any integrated or coherent whole, but in a heterogeneous patchwork of statutes, purposes, instruments, agencies, and levels of government. It resides in no single department comparable to the ministries of the environment in other countries. It lies in a multiplicity of agencies implementing a growing number of largely uncoordinated statutory mandates that affect the environment in conflicting ways. The Environmental Protection Agency, despite its name, has no single overall statute authorizing it to protect the environment. Even for a specific environmental policy issue such as pollution control, pollution of the air, water, and land are addressed by separate statutes, programs, policy incentives, and decision criteria. The United States in 1970 adopted a National Environmental Policy Act, but it has never translated this into any overall plan or strategy to guide its agencies toward common goals. Both the strengths and the weaknesses of U.S. environmental policy thus derive from a policy-making structure fragmented among diverse, mission-oriented programs and agencies.

Despite this lack of coherence, however, U.S. environmental policy has distinctive features that shape its results. One feature is the expansive deference it accords to private rights to transform the environment for economic gain, and the correspondingly weak powers it accords to public agencies to protect broader societal values. A second is the pervasive influence of federalism, in the form of constant renegotiation of the tension among national, state, and local governments. A third is the active role of an independent judicial branch, not only in resolving environmental disputes among individuals and businesses but in challenging the environmental actions of government agencies themselves.

In fact, perhaps the most distinctive difference between American environmental policy-making processes and those of many other governments is the broad rights of access and redress which U.S. laws and recent judicial precedents accord not only to business and labor organizations but to citizens in general. This vulnerability to judicial review has created a heavy burden of proof on public agencies to document and justify their decisions, both through elaborate environmental, economic, and risk analyses and through increasingly detailed documentation of consultation procedures.

Finally, American environmental policy has been overwhelmingly concerned with domestic issues, especially the environmental impacts of federal resource exploitation and development projects and, since 1970, the federal regulation of pollution and toxic chemicals. This preoccupation may become increasingly problematic in a twenty-first-century world in which both environmental impacts and the economic forces that cause them are increasingly global.

Notes

1. Garrett Hardin, "The Tragedy of the Commons," *Science* 162 (1968): 1243–48.
2. Many such programs, such as agricultural subsidies, below-cost logging and mining concessions, and underpriced use of water, fuels, and grazing lands, have been criticized on economic levels as well.
3. Charles Wolf, "A Theory of Non-Market Failures," *The Public Interest* 55 (1979): 114–33.
4. Two enduring problems complicate such claims, however. One is the existence of legitimate conflict among visions of the overall public interest itself, and of priorities within it, that are not merely differences in self-interest and that are not resolvable by increases in scientific knowledge. The other is the inevitable intermixture of private benefits with public goods. Even actions that

serve an overall public interest benefit some private interests more than others, including not only business or property owners but the continued organizational success of advocacy groups.

5. Lynton K. Caldwell, "Environment: A New Focus for Public Policy?" *Public Administration Review* 22 (1963): 132–39.

Zygmunt J. B. Plater, Robert H. Abrams, William Goldfarb, and Robert L. Graham, from "The Three Economies" in *Environmental Law and Policy: Nature, Law, and Society* (1998)

These four legal and environmental studies scholars analyze environmental problems in terms of the interrelationship of what they call the marketplace economy, the natural economy, and the civic-societal economy. They do so first to redress the hitherto insufficiently comprehensive "overall economic and legal accounting" of the "externalized costs" of environmental degradation; and second, to sharpen the focus of environmental law in a broader range of settings, thereby making it more effective as a public policy tool.

THE "THREE ECONOMIES"

A fundamental theme and force in virtually every environmental case, as in most of modern life, is Economics. In trying to understand the economic processes of environmental and landuse issues, Professor Joseph Sax has argued that we should recognize not one but two economic systems—an economics of the natural economy as well as of the marketplace.[1]

It seems to us, however, that a comprehensive analytic structure for understanding the dynamics of modern life requires recognition of three different intersecting economies—comprised not only of marketplace economics and natural systems economics, but also of a civic-societal economics incorporating the society's overall cumulative well-being. This three-economies construct provides a useful way to talk about an array of long-established concepts of environmental law (and public policy generally), integrating an analysis of societal necessities and ecological realities with the powerful machinery of market dynamics.[2] . . .

The Marketplace Economy is the well-known mechanism of everyday economic and political behavior. It sits at the center—the dynamic, driving force of human society, dynamically churning out economic and political power, interlocking networks of motivations and institutions, property rights, production, politics, and wastes. It dominates daily life, and is undoubtedly the most intricate and sophisticated social mechanism ever devised to manage the extraordinary onrolling complexity of human society. The marketplace economy, however, is systematically blind to many important elements of reality; for the most part it deals only with things and services which can be bought and sold. It takes for granted

Figure 21.3 The three economies. The rim banding the marketplace economy represents the governmental role. Relative scale and proportions depend upon observer's perspective.

many natural and public values that are priceless. It generates waste and externalities. These problems are massive and real, and need to be part of a societal accounting.

The Natural Economy, as noted by Sax and others, is what happens in the natural physical world, the intricate system of living and geophysical systems that sustain dynamic planetary processes. It partly overlaps the two human economies, supplying vital resources and services to both, but in part lies outside them. The economy of nature processes everything, adapts to everything (though often with altered and diminished qualities of ecosystem health and diversity), absorbs wastes and other externalities from the marketplace, and passes the effects of many of these externality impacts onward to the civic economy. . . .

Recognition of a natural "economics" forces us to recognize that nature is not just a charming sidenote to human life, but a complex and utilitarian system in its own right, superbly integrated, diversified, and efficient in processing and recycling energy and nutrients, integrating millions of ongoing and evolving life cycles, adapting to natural change, ultimately important to humans (which links it to the civic-societal economy), and vulnerable.

The Civic-Societal Economy is the ultimate "economic" forum in human terms. It incorporates the dynamic marketplace economy, but its terms and elements extend much further beyond just those things that can be bought and sold. Societal economics are the comprehensive reality of the full actual social costs, benefits, resources, energy, inputs, outputs, values, qualities, and consequences over time that carry the life and welfare of a human society from day to day and into the future, whether the marketplace acknowledges them or not. Where do externalities go? Costs externalized by marketplace enterprises—by factories, porkbarrel government agency projects and programs, or undertakings combining private and public entrepreneurs—exit the marketplace economy, but they do not thereby drop into oblivion: by the laws of physics, ecology, and logic *they are internalized into one or both of the other economies*, into the fabric of the natural and societal economics in which we and our society will continue to live. Widespread pesticide use, or overgrazing, or dumping pollutants like Kepone wastes, or highway salting, or massive layoffs of corporate workers, or a host of other cost externalizations may make powerfully good sense to market players in the narrow market terms of individual gains, but be quite irrational in societal economics terms. Regulatory agencies have mandates designed to serve the civic economy, but they spend most if not all of their daily political lives dominated by the pressures and constraints of marketplace politics. The natural economy's complex elements are often directly and substantially linked to the utilitarian concerns of societal economics. Healthy societal economic systems are founded upon healthy and sustainable ecological system cycles of soil, water, air, and living communities. When a resource system is derogated or destroyed, some enterprises may prosper greatly, but the society is likely to be far less well off. The societal public economy needs the market economy, but for the sake of its own short- and long-term interests a society must somehow contrive to value and incorporate into its governance important elements of civic economic reality that the daily dominant marketplace resists. . . .

The economy of nature often impacts critically upon the civic-societal economy. In positive terms the civic-societal economy receives the values and benefits humans derive from natural systems, like the multi-trillion-dollar "natural capital" values that resource economists say we annually receive "free" from nature, values unacknowledged by the marketplace economy.[3] The civic-societal economy also absorbs negatives from the economy of nature's disruptions from market economy impacts. Some marketplace externalities may only affect natural systems, with no human consequences, but many others pass through the environment into the human welfare context of civic-societal economics. Thus toxic spills into watercourses, global warming, lost fisheries, eroded soils, wetlands destruction, pesticide loading, and a host of other environmental issues demonstrate important linkages between all three economies. The logical connection between natural systems and societal interests provides the compelling utilitarian linkage that should force hardnosed economics-based legal and policy analyses to integrate environmental considerations.

Both natural and societal economics, however, are often ignored and overwhelmed by the marketplace economy's narrowed perspectives and power. Scratch an environmental law controversy anywhere in the world and you are likely to find a natural value or a public value being destructively disregarded by the market's political and entrepreneurial forces. Nevertheless the three economies are interdependent, and the competent practice of environmental law, and rational democratic governance, implicitly incorporate and require an understanding of the reality of all three economies.

Notes

1. Joseph Sax, "Property Rights and the Economy of Nature," *Stanford Law Review* 45 (1993): 1433.
2. See "The Three Economies," *Ecology Law Quarterly* 25, no. 3 (1988). All three can be called "economies" because each is a complex system of interconnecting subsystems processing inputs and outputs, causes and effects, and uses of resources over time. No other modern word captures the concept of a dynamic system of systems quite so well.
3. See R. Costanza et al., "The Value of the World's Ecosystem Services and Natural Capital," *Science* 387 (1997): 253; R. Costanza and H. Daly, "Natural Capital and Sustainable Development," *Conservation Biology* 6 (1992): 37–46.

David Pepper, from "Science and Society: Influencing the Answers Obtained" in *Modern Environmentalism: An Introduction* (1996)

Pepper describes how science is used in order to influence environmental public policy. Exploiting the discipline's post-Enlightenment prestige, and association with learning and objectivity, both environmentalists and their foes can package, manipulate, and ritualize data, simplify the scientific method, and even arrange for the appearance of scientists for partisan ends. This contributes to a growing skepticism about science, though Pepper admits that this suspicion originates in an increased awareness that the results of science have often been used for specific political and cultural ends.

Despite disillusionment with some products of science, scientific knowledge and method have commanded respect and authority which is slow to be shaken. Scientists are often perceived as *detached* observers (after Descartes) of nature and society. Therefore they are regarded as ethically and politically neutral, with no ax to grind about what their results should be used for, that being for politicians and the public to decide. . . . But *because* of this perceived detachment the notion is also popular that scientifically literate people should have a special place in decision making, as experts promoting rational, "common-sense" solutions according to objective truth. As S. Richards observes, science and scientists here become "truth institutionalized"; usurping the territory formerly held by religion and priests.[1] Scientific laws describe "relations and regularities that are invariable" (like "God's law" used to be). Empirical science therefore surely must be uninfluenced by cultural factors such as dogma or mere opinion: it emphasizes what can be agreed by *all* observers. . . .

This legitimating authority of science is today sought by all sorts of interests, religious, paranormal, flat earth, and creationist enthusiasts. And it is used to further commercial ends, for instance as a tool in selling. We are familiar with this use in advertising. Products are dubbed "scientifically tested." They are said to contain impressive-sounding synthetic substances mystifyingly identified by formulae—"The wonder ingredient, WM7" and other such gobbledegook (although nowadays concerns about chemical additives make this a ploy to be judiciously used). Products may actually be promoted by actors wearing the priestly robes of the white laboratory coat. The imagined intellectual integrity and detachment from interest groups of such characters may be amplified by making them appear super-intelligent (e.g., with ludicrously high foreheads) and/or eccentric or "mad" (i.e., separate from the concerns of ordinary life). The general effect is that "if they say that this product is good it *must* be true." Any product which is the "appliance of science" must be worth buying.

. . . This can happen in environmental matters. . . .

Ecocentrics have been quick to point out such perversions. For instance they have made accusations about the scientific method used by such a "respectable" body as the U.S. World Resources Institute. This Institute's 1990–1 annual report made the surprising claim that industrialized and non-industrialized countries share equal responsibility for global greenhouse gas emissions. However, said Patrick McCully in *The Ecologist*, "a close look at the raw data used by WRI, and the way in which they have interpreted it, reveals that the institute has used highly questionable estimates for the release of greenhouse gases from developing countries and that their methodology contains some very dubious science."[2] He cites Anil Agarwal, Director of the Centre for Science and Environment in New Delhi, who says "The WRI conclusions are based on patently unfair mathematical jugglery, where politics are masquerading in the name of science."

It should not be imagined that ecocentrics are immune to this sort of perversion either. For instance, a BBC Radio Four program alleged in 1989 ("The Zero Option") that Greenpeace was so anxious to claim scientific legitimacy for its view that toxic waste dumping might kill all life in the North Sea that it took the results of a Dutch scientific study on the matter and represented them in their publicity material in exactly the opposite way to the author's intentions. . . .

To leave the argument here would be simplistic. For while the findings of science are often influenced and appropriated by groups who want science's powerful legitimating authority, that authority is not infallibly present. For example, after the Chernobyl incident in 1986, Welsh farmers began to mistrust government scientists who claimed to have an objective perspective on the effects of the discharge. R. Grove-White and M. Michael cite this in discussing how the Nature Conservancy Council in England also failed to gain the confidence of certain people (in this case, the government) by projecting an image of itself as a group practicing value-free, empirical, and universalistic science.[3] The public authority of science, they believe, depends on social circumstances, and those circumstances may be changing. In the postmodern period science in the classical mode is perhaps being relegated from its position as a great arbiter and purveyor of truth.

Notes

1. S. Richards, *Philosophy and Sociology of Science: An Introduction* (Oxford: Blackwell, 1983), n.p.
2. Patrick McCully, "Discord in the Greenhouse: How WRI Is Attempting to Shift the Blame for Global Warming," *The Ecologist* 21, no. 4 (July–August 1991): 157.
3. R. Grove-White and M. Michael, "Nature Conservation: Culture, Ethics and Science," in J. Burgess, ed., *People, Economies and Nature Conservation*, University College [London] Ecology and Conservation Unit Discussion Paper 60 (London, 1993), 139–52.

Ranee K. L. Panjabi, from *The Earth Summit at Rio: Politics, Economics, and the Environment* (1997)

Panjabi captures the mixture of idealism and self-interest, internationalism and nationalism, global fear and politics as usual characterizing the 1992 United Nations Earth Summit in Rio de Janeiro. In addressing the fate of the Earth, practical considerations (like the cost of environmental cleanup) weigh against the long-term economic benefits of a clean environment; and finger-pointing vies with the solemn facts about the future we all face if steps are not taken now.

The Rio Summit has to be assessed from a variety of perspectives, and any fair analysis must include consideration of the very real and compelling conflict between idealism and self-interest evident throughout the Conference.

There was a definite theatricality to the entire Rio event, an atmosphere akin to that of a gala performance or a Hollywood opening night. The vast numbers of the immediate audience, approximately 30,000, makes it one of the largest extravaganzas ever. Innumerable world leaders—potentates and prime ministers, governmental delegations and representatives of non-governmental organizations—all vied for media attention and for a voice in the formulations being presented to the Conference. The entire world was watching on television and reading about the event in its newspapers as some 6,000 journalists from every corner of the Earth covered the Summit. Though the entire world had gathered to avert a tragedy, the atmosphere at Rio was almost that of an elaborate carnival, and one wonders whether this was really the ideal way to formulate principles of environmental law. Though it is true that similar spectacles have often resulted in significant treaties, as at the Congress of Vienna in 1814–15 following the Napoleonic Wars, the sheer size and impressive aura of the gathering at Rio may have led nations to posture more, to harden their positions, and to exaggerate their differences, factors which could have had a significant negative impact on the achievements of the Conference. One would like to think that legal principles, particularly when these involve several nations, will not be hammered out in the glare of global publicity and worldwide interest. It was perhaps fortunate that the basic elements of the Treaties and of the Declaration were formulated in a series of meetings preceding the Summit. It was also prudent of most supporters of the compromise Treaties which emerged from those earlier sessions to resist the urge to renegotiate to any great extent at Rio what had already been agreed to in earlier meetings. If Rio was largely an event to bring global attention to the tragedy of environmental degradation, then it can be deemed an unqualified success.

It was also successful in exposing the positions of various governments and in revealing which world leaders were sincere in their commitment to the environment. Rio was ultimately an exercise in global idealism, the vision of its creator, Maurice Strong, who firmly believes that the planet must be revitalized if the human species is to survive. This great and noble vision collided headlong with the national self-interest of a variety of political players at the Conference. Ironically, the positions of the United States of America and the more assertive of the developing nations like Malaysia were remarkably similar in terms of the primacy of self-interest over environmental concerns. The American Government's refusal to risk jobs in an economy battered by recession was interpreted at Rio as selfishness and callousness. That President [George H. W.] Bush was standing for re-election that year was frequently alluded to in an explanation of the American reluctance to play the role of environmental world leader.

Some developing nations also presented a position based largely on self-interest. Malaysia, rich in forests, a resource it is logging at a rate which alarms most environmentalists, resisted efforts to deem this natural treasure of wood and biodiversity part of the global heritage of mankind. Malaysia was both strident and defensive in its stance at Rio, and the country's delegates argued very strongly that if the North wants the world's forests saved then the North must be willing to subsidize the preservation of this resource. A Malaysian diplomat, Ting Wen-Lian, commented assertively that if developed countries want developing countries to conserve their forests, they should attend to "the poverty, famine, and crushing burden of external debt" which compel the poorer nations to fell their trees in order to survive.[1] It is significant that though the United States of America and Malaysia seemed to represent opposite poles in the environmental dialogue at Rio, both nations, in emphasizing the primacy of national self-interest over global idealism, were on a remarkably similar plane of thought. Both countries are committed to environmental improvement, but both resisted the economic sacrifice inherent in its implementation. Though the rhetoric highlighted the North-South divergence, the primacy of immediate national priorities in the policies of both nations demonstrates how akin they really are in their underlying attitudes. Both Malaysia and the United States of America figured in the list of the five worst environmental nations.[2] The list was prepared and widely publicized by the various environmental activists gathered to share ideas in the unofficial people's summit, labeled the Forum, which also convened in Rio at the same time as the more formal Earth Summit.

Although the nations on the extremes of the North-South drama as played out at Rio were the major hindrance to the passage of strong measures to clean up and preserve the environment, most nations in both North and South overcame their nationalistic apprehensions to sign the compromise agreements which were passed at the Conference. The endorsement by so many nations of each of the Treaties and the unanimous acceptance of the Rio Declaration presage a new global consciousness on many levels, indicating that the pre-eminence of national self-interest must now give way to the ideals of sustainable development. It is as if the majority of the nations of the world have finally realized that there is a clear difference between short-term self-interest which would resist environmental legislation in order to safeguard economic concerns and long-term self-interest in which the entire planet survives and perhaps even thrives because timely measures have been taken to implement the ideals expounded by environmentalists. Among the nations of the North, the American position was regarded as extreme. It was in marked contrast to that of the European nations, some of which argued emphatically for firm timetables for curbing carbon dioxide emissions in the Climate Change Treaty—a provision not acceptable to the United States and therefore dropped in the final document. Austria, the Netherlands, and Switzerland initiated a proposal to commit signatory nations to immediate stabilization of emission levels, a move which, according to one Austrian delegate, resulted in a threatening letter from the U.S. Government.[3] Clearly, the American position had led to its isolation in the sphere of environmental diplomacy; regrettably, an opportunity to assume the mantle of world leadership in this important facet of international relations was forsaken. It is impossible to speculate on the probable consequences of a vigorous American assumption of leadership in this attempt to expand the parameters of environmental law. Certainly, the inclusion of firm timetables in the Climate Change Treaty and U.S. endorsement of the Biodiversity Treaty would have made a profound difference. [Similarly, the United States has not nor is it likely to endorse or ratify the Kyoto Protocol.]

Notes

1. *Globe and Mail* [Toronto], 6 June 1992.
2. Ibid., 9 June 1992.
3. *New York Times*, 8 June 1992.

John Barry, from *Rethinking Green Politics: Nature, Virtue, and Progress* (1999)

Through a theoretical model for the environmental movement, Barry tries to resolve the competing claims of those who say that individual change is the key to resolving environmental questions versus those who believe that the answer rests in institutional transformation. While favoring the former, he suggests that changes in social policy and the reform of institutions can have a decisive influence on how individuals behave toward the environment.

From the point of view of green political theory, the resolution of ecological problems is not simply a matter of structural reorganization, either of the economy and the scale of technology and production, or of changing the level of legislative or policy-making power. For many of the issues that green political theory deals with, particularly those related to its moral claims, attention needs to be focused on changing people's attitudes, interests, and modes of acting. To put it simply, many of the questions raised by greens are matters of individual and collective will as much as institutional transformation. They are of course related, and the specific manner in which agents are related to structures will constitute a particular understanding of green theory. For example, we can understand deep ecology as a theory which emphasizes agent-level change (consciousness raising) which works on the assumption that the appropriate structures will "follow" from this deeper-level change. The argument here is that, for green politics, institutional change (politically, economically, socially) must be placed within a wider cultural context.

As we have seen in previous chapters, a non-anthropocentric green politics aims to alter the prevailing *attitudes* to nature. On this account, green politics is a moral crusade seeking to win "converts" away from anthropocentrism and its worldview. These approaches have much in common with what [A.] Dobson calls the "religious approach" to green change which holds that "the changes that need to take place are *too profound to be dealt with in the political arena*, and that the proper territory for action is the psyche rather than the parliamentary chamber."[1] On the other hand, so-called "reformist environmentalism" focuses mainly on "greening" existing structures, rather than reflecting on the structures themselves in the light of ecological considerations, and the relationship between structures and the behavior and attitudes of agents.

The conception of green political theory being developed here seeks to combine both agency and structural approaches. . . . The values, principles, and goals that are central to the green political project depend upon combining both agent- and structural-level change. Collective ecological management therefore has to do with both preferences and policies, agents and structures. In this respect, unlike market approaches to ecological issues, collective ecological management is a problem-solving rather than a preference-aggregating process. For green politics, understood as a form of collective ecological management, resolving environmental problems requires *cultural* and not just institutional change. Because the roots of ecological problems do not lie exclusively in either cultural norms or institutional structures, neither do the solutions. From the point of view of green politics defended here, the long-run resolution of social-environmental problems requires a politics based on an *immanent critique* of the prevailing cultural as well as institutional order. . . .

Green politics in the last analysis is not simply about macro-level changes, but is also about choosing to live in a different manner at the micro-level of individuals and communities. Building on the discussion of virtue earlier, within collective ecological management, individuals are faced not just with the question of "what ought I to do?" but more importantly have to ask themselves "what sort of person do I wish to be?" and ultimately "what sort of society do I wish to be in?" In response to [David] Hume's adage that we ought to "design institutions for knaves," one desired aim of green politics is to encourage people to be less knavish in the first place. This requires designing institutional structures to sustain "ecologically rational" modes of behavior by supporting rather than undermining ecological social practices. It is my contention that ecologically rational modes of interaction involve, in part, cultivating ecological virtue at the individual level in the various roles people play and identities they have as consumers, producers, citizens, and parents. At the social level, ecologically rational social-environmental relations ultimately require the creation of an ecologically adapted and adaptive culture, supported by an institutional structure in which the state and market are restructured so that they are instrumental to social life and democracy as popular sovereignty.

Note

1. A. Dobson, *Green Political Thought* (London: Unwin Hyman, 1990), 143 (emphasis added).

Lee Quinby, from "Ecofeminism and the Politics of Resistance" (1990)

Quinby uses French philosopher Michel Foucault's objections to hegemony (the desire for intellectual and practical domination) to critique what she sees as a tendency toward the imposition of political and ideological orthodoxies in both the ecological and feminist movements. She sees ecofeminism as an alliance instead of those who resist the patriarchy of power, institutions, and intellectual conformity as part of the interrelated struggle for "women's freedom" and "planetary well-being."

As the French theorist Michel Foucault has observed, "things never happen as we expect from a political program," for "a political program has always, or nearly always, led to abuse or political domination from a bloc, be it from technicians or bureaucrats or other people."[1] In other words, the move toward orthodoxy is complicitous with the tendency of power to totalize, to demand consensus, to authorize certain alliances and to exclude others—in short, to limit political creativity.

This movement toward totalization has already left its mark on the West German Green party. As Michael Hoexter writes in *New Politics*, within the West German Green party the philosophical and political hostilities between the "fundamentalists" and the "realists" over the power of the state "seem irreconcilable."[2] Yet Hoexter . . . calls for "a comprehensive, and comprehensible theory of the state," arguing that without such a program, "greens and socialists are doomed to repeat the disappointing, sometimes catastrophic, history of left parties and groups in government."[3] My argument, following Foucault, is that the demand for comprehensive theory and coherent practice leads to rather than avoids the kind of polarization that has occurred in the West German Green party. The call for comprehensive theory is not a new politics at all, but rather an old story. And this story—with its all-too-predictable ending—is now being told by the U.S. ecology movement, as attested to in the polarized debates between deep ecologists and social ecologists. The lines of argument between these two camps, made visible at the July 1987 gathering of Greens at Amherst, Massachusetts, have only widened and deepened.[4]

The radical feminist movement in the United States has also suffered from precisely this tendency of theory to become dogma and political practices of resistance to accordingly become less diverse. Over the last twenty years, we have witnessed a shift away from a wide-based, multiple-issues movement concerned with *women's* bodies toward a narrow-based, single-issue one, focused primarily on pornography and *woman's* body. . . .

In light of the splits that have taken place within the ecology and feminist movements, I want to argue against these calls for coherence, comprehensiveness, and formalized agendas and to cite ecofeminism as an example of theory and practice that has combated ecological destruction and patriarchal domination without succumbing to the totalizing impulses of masculinist politics. Ecofeminism as a politics of resistance operates against power understood, as Foucault puts it, as a "multiplicity of force relations," decentered and continually "produced from one moment to the next." Against such power, coherence in theory and centralization of practice make a social movement irrelevant or, worse, vulnerable, or—even more dangerous—participatory with forces of domination. As Foucault explains, decentered power requires decentered political struggle, for "there is no single locus of great Refusal, no soul of revolt, source of all rebellions, no pure law of the revolutionary. Instead there is a plurality of resistance, each of them a special case: resistances that are possible, necessary, improbable; others that are spontaneous, savage, solitary, concerted, rampant, or violent; still others that are quick to compromise, interested, or sacrificial."[5] And, indeed, the strength thus far of ecofeminism has been to target abuses of power at the local level, in a multiplicity of places. . . .

The arrogance of speaking for others, so integral to the desire for hegemonic theory, is compounded by a diminished capacity to hear what others have to say about our circumstances as well as their own. Ecofeminism as a politics of resistance forces us to question the categories of experience that order the world and the truths we have come to know, even the truths of our radical politics, by confronting us with the truths of other women and men, differently acculturated, fighting against specific threads to their particular lands and bodies. This questioning must also extend to the anthropocentric assumption that only human beings have truths to tell about their and our experiences. The cries of factory farm animals, the suffocation of fish in poisoned waters, the sounds of flood waters rushing over deforested land—these are also voices we need to heed. Listening to all voices of subjugation and hearing their insurrectionary truths make us better able to question our own political and personal practices. This questioning my well risk the end of ecofeminism as currently constituted, for, like any social movement, ecofeminism is inevitably a provisional politics, one that has struck a chord of resistance in this era of ecological destruction and patriarchal power. And if another term and a different politics emerge from this questioning, it will be in the service of new local actions, new creative energies, and new alliances against power.

Notes

1. "Michel Foucault: An Interview," *Edinburgh Review* (1986): 59.
2. Michael Hoexter, "It's Not Easy Being Green," *New Politics* 5 (Summer 1998): 106–18.
3. Ibid., 118.
4. Jay Walljasper, "The Prospects for Green Politics in the U.S.," *Utne Reader* (September–October 1987): 37–39. For the social ecology position on deep ecology, see Murray Bookchin, "Social Ecology Versus Deep Ecology," *Socialist Review* 18 (July–September 1988): 9–29. For a critique of social ecology from a deep ecology stance, see Kirkpatrick Sale, "Deep Ecology and Its Critics," *The Nation* (14 May 1988): 670–75.
5. Michel Foucault, *The History of Sexuality*, trans. Robert Hurley (New York: Vintage, 1980), 92–96.

Cynthia Hamilton, from "Women, Home, and Community: The Struggle in an Urban Environment" (1990)

Hamilton describes a success in the struggle to preserve the environment: the tale of how overwhelmingly female, working-class, minority residents of South Central Los Angeles defeated an attempt by the city council and its allies to locate a waste incinerator in their neighborhood. She draws broader conclusions about the economics and politics behind locating pollution sources in minority areas, and the power of community action in the context of women's liberation. (See Chapter 10, above, and Chapter 22, below.)

In 1956, women in South Africa began an organized protest against the pass laws. As they stood in front of the office of the prime minister, they began a new freedom song with the refrain "now you have touched the women, you have struck a rock." This refrain provides a description of the personal commitment and intensity women bring to social change. Women's actions have been characterized as "spontaneous and dramatic," women in action portrayed as "intractable and uncompromising."[1] Society has summarily dismissed these as negative attributes. When in 1986 the City Council of Los Angeles decided that a thirteen-acre incinerator called LANCER (for Los Angeles City Energy Recovery Project), burning 2,000 tons a day of municipal waste, should be built in a poor residential, Black, and Hispanic community, the women there said "No." Officials had indeed dislodged a boulder of opposition. According to Charlotte Bullock, one of the protestors, "I noticed when we first started fighting the issue how the men would laugh at the women . . . they would say, 'Don't pay no attention to them, that's only one or two women . . . they won't make a difference.' But now since we've been fighting for about a year the smiles have gone."[2]

Minority communities shoulder a disproportionately high share of the by-products of industrial development: waste, abandoned factories and warehouses, leftover chemicals and debris. These communities are also asked to house the waste and pollution no longer acceptable in White communities, such as hazardous landfills or dump sites. In 1987, the Commission for Racial Justice of the United Church of Christ published *Toxic Wastes and Race.* The commission concluded that race is a major factor related to the presence of hazardous wastes in residential communities throughout the United States. Three out of every five Black and Hispanic Americans live in communities with uncontrolled toxic sites; seventy-five percent of the residents in rural areas in the Southwest, mainly Hispanics, are drinking pesticide-contaminated water; more than two million tons of uranium tailings are dumped on Native-American reservations each year, resulting in Navajo teenagers having seventeen times the national average of organ cancers; more than 700,000 inner city children, fifty percent of them Black, are said to be suffering from lead poisoning, resulting in learning disorders. Working-class minority women are therefore motivated to organize around very pragmatic environmental issues, rather than those associated with more middle-class organizations. According to Charlotte Bullock, "I did not come to the fight against environmental problems as an intellectual but rather as a concerned mother. . . . People say, 'But you're not a scientist, how do you know it's not safe?' I have common sense. I know if dioxin and mercury are going to come out of an incinerator stack, somebody's going to be affected."

When Concerned Citizens of South Central Los Angeles came together in 1986 to oppose the solid waste incinerator planned for the community, no one thought much about environmentalism or feminism. These were just words in a community with a seventy-eight percent unemployment rate, an average income ($8,158) less than half that of the general Los Angeles population, and a residential density more than twice that of the whole city. In the first stages of organization, what motivated and directed individual actions was the need to protect home and children; for the group this individual orientation emerged as a community-centered battle. What was left in this deteriorating district on the periphery of the central business and commercial district had to be defended—a "garbage dump" was the final insult after years of neglect, watching downtown flourish while residents were prevented from borrowing enough to even build a new roof.

The organization was never gender restricted but it became apparent after a while that women were the majority. The particular kind of organization the group assumed, the actions engaged in, even the content of what was said, were all a product not only of the issue itself, the waste incinerator, but also a function of the particular nature of women's oppression and what happens as the process of consciousness begins.

Women often play a primary part in community action because it is about things they know best. Minority women in several urban areas have found themselves part of a new radical core as the new wave of environmental action, precipitated by the irrationalities of capital-intensive growth, has catapulted them forward. These individuals are responding not to "nature" in the abstract but to the threat to their homes and to the health of their children. Robin Cannon, another activist in the fight against the Los Angeles incinerator, says, "I have asthma, my children have asthma, my brothers and sisters have asthma, there are a lot of health problems that people living around an incinerator might be subjected to and I said, 'They can't do this to me and my family.'"

Women are more likely than men to take on these issues precisely because the home has been defined and prescribed as a woman's domain. According to British sociologist Cynthia Cockburn, "In a housing situation that is a health hazard, the woman is more likely to act than the man because she lives there all day and because she is impelled by fear for her children. Community action of this kind is a significant phase of class struggle, but it is also an element of women's liberation."[3]

This phenomenon was most apparent in the battle over the Los Angeles incinerator. Women who had had no history of organizing responded as protectors of their children. Many were single parents, others were older women who had raised families. While the experts were convinced that their smug dismissal of the validity of the health concerns these women raised would send them away, their smugness only reinforced the women's determination. According to Charlotte Bullock:

> People's jobs were threatened, ministers were threatened . . . but I said, "I'm not going to be intimidated." My child's health comes first . . . that's more important than my job.

> In the 1950s the city banned small incinerators in the yard and yet they want to build a big incinerator . . . the Council is going to build something in my community which might kill my child. . . . I don't need a scientist to tell me that's wrong.

None of the officials were prepared for the intensity of concern or the consistency of agitation. In fact, the consultants they hired had concluded that these women did not fit the prototype of opposition. The consultants had concluded:

> Certain types of people are likely to participate in politics, either by virtue of their issue awareness or their financial resources, or both. Members of middle or higher socioeconomic strata (a composite index of level of education, occupational prestige, and income) are more likely to organize into effective groups to express their political interests and views. All socioeconomic groupings tend to resent the nearby siting of major facilities, but the middle and upper socioeconomic strata possess better resources to effectuate their opposition. Middle and higher socioeconomic strata neighborhoods should not fall at least within the one-mile and five-mile radii of the proposed site.
>
> . . . Although environmental concerns cut across all subgroups, people with a college education, young or middle aged, and liberal in philosophy are most likely to organize opposition to the siting of a major facility. Older people, with a high school education or less, and those who adhere to a free market orientation are least likely to oppose a facility.[4]

The organizers against the incinerator in South Central Los Angeles are the antithesis of the prototype: they are high school educated or less, above middle age and young, nonprofessionals and unemployed and low-income, without previous political experience. The consultants and politicians thus found it easy to believe that opposition from this group could not be serious.

The intransigence of the City Council intensified the agitation, and the women became less willing to compromise as time passed. Each passing month gave them greater strength, knowledge, and perseverance. The council and its consultants had a more formidable enemy than they had expected, and in the end they have had to compromise. The politicians have backed away from their previous embrace of incineration as a solution to the trash crisis, and they have backed away from this particular site in a poor, Black and Hispanic, residential area. While the issues are far from resolved, it is important that the willingness to compromise has become the official position of the city as a result of the determination of "a few women."

The women in South Central Los Angeles were not alone in their battle. They were joined by women from across the city, White, middle-class, and professional women. As Robin Cannon puts it, "I didn't know we had so many things in common . . . millions of people in the city had something in common with us—the environment." These two groups of women, together, have created something previously unknown in Los Angeles—unity of purpose across neighborhood and racial lines. According to Charlotte Bullock, "We are making a difference . . . when we come together as a whole and stick with it, we can win because we are right."

This unity has been accomplished by informality, respect, tolerance of spontaneity, and decentralization. All of the activities that we have been told destroy organizations have instead worked to sustain this movement. For example, for a year and a half the group functioned without a formal leadership structure. The unconscious acceptance of equality and democratic process resulted practically in rotating the chair's position at meetings. Newspeople were disoriented when they asked for the spokesperson and the group responded that everyone could speak for the neighborhood.

It may be the case that women, unlike men, are less conditioned to see the value of small advances.[5] These women were all guided by their vision of the possible: that it *was* possible to completely stop the construction of the incinerator, that it is possible in a city like Los Angeles to have reasonable growth, that it is possible to humanize community structures and services. As Robin Cannon says, "My neighbors said, 'You can't fight City Hall . . . and besides, you work there.' I told them I would fight anyway."

None of these women were convinced by the consultants and their traditional justifications for capital-intensive growth: that it increases property values by intensifying land use, that it draws new businesses and investment to the area, that it removes blight and deterioration—and the key argument used to persuade the working class—that growth creates jobs. Again, to quote Robin Cannon, "They're not bringing real development to our community. . . . They're going to bring this incinerator to us, and then say 'We're going to *give* you fifty jobs when you get this plant.' Meanwhile they're going to shut down another factory [in Riverside] and eliminate two hundred jobs to buy more pollution rights. . . . They may close more shops."

Ironically, the consultants' advice backfired. They had suggested that emphasizing employment and a gift to the community (of two million dollars for a community development fund for park improvement) would persuade the opponents. But promises of heated swimming pools, air-conditioned basketball courts, and fifty jobs at the facility were more insulting than encouraging. Similarly, at a public hearing, an expert witness's assurance that health risks associated with dioxin exposure were less than those associated with "eating peanut butter" unleashed a flurry of derision.

The experts' insistence on referring to congenital deformities and cancers as "acceptable risks" cut to the hearts of women who rose to speak of a child's asthma, or a parent's influenza, or the high rate of cancer, heart disease, and pneumonia in this poverty-stricken community. The callous disregard of human concerns brought the women closer together. They came to rely on each other as they were subjected to the sarcastic rebuffs of men who referred to their concerns as "irrational, uninformed, and disruptive." The contempt of the male experts was directed at professionals and the unemployed, at Whites and Blacks—all the women were castigated as irrational and uncompromising. As a result, new levels of consciousness were sparked in these women.

The reactions of the men backing the incinerator provided a very serious learning experience for the women, both professionals and nonprofessionals, who came to the movement without a critique of patriarchy. They developed their critique in practice. In confronting the need for equality, these women forced the men to a new level of recognition—that working-class women's concerns cannot be simply dismissed.

Individual transformations accompanied the group process. As the struggle against the incinerator proceeded to take on some elements of class struggle, individual consciousness matured and developed. Women began to recognize something of their own oppression as women. This led to new forms of action not only against institutions but to the transformation of social relations in the home as well. As Robin Cannon explains:

> My husband didn't take me seriously at first either. . . . He just saw a whole lot of women meeting and assumed we wouldn't get anything done. . . . I had to split my time . . . I'm the one who usually comes home from work, cooks, helps the kids with their homework, then I watch a little TV and go to bed to get ready for the next morning. Now I would rush home, cook, read my materials on LANCER . . . now the kids were on their own . . . I had my own homework. . . . My husband still wasn't taking me seriously. . . . After about six months everyone finally took me seriously. My husband had to learn to allocate more time for baby sitting. Now on Saturdays, if they went to the show or to the park, I couldn't attend . . . in the evening there were hearings . . . I was using my vacation time to go to hearings during the workday.

As parents, particularly single parents, time in the home was strained for these women. Children and husbands complained that meetings and public hearings had taken priority over the family and relations in the home. According to Charlotte Bullock, "My children understand, but then they don't want to understand. . . . They say, 'You're not spending time with me.'" Ironically, it was the concern for family, their love of their families, that had catapulted these women into action to begin with. But, in a pragmatic sense, the home did have to come second in order for health and safety to be preserved. These were hard learning experiences. But meetings in individual homes ultimately involved children and spouses alike—everyone worked and everyone listened. The transformation of relations continued as women spoke up at hearings and demonstrations and husbands transported children, made signs, and looked on with pride and support at public forums.

The critical perspective of women in the battle against LANCER went far beyond what the women themselves had intended. For these women, the political issues were personal and in that sense they became feminist issues. These women, in the end, were fighting for what they felt was "right" rather than what men argued might be reasonable. The coincidence of the principles of feminism and ecology that Carolyn Merchant explains in *The Death of Nature* (San Francisco: Harper & Row, 1981) found expression and developed in the consciousness of these women: the concern for Earth as a home, the recognition that all parts of a system have equal value, the acknowledgement of process, and, finally, that capitalist growth has social costs. As Robin Cannon says, "This fight has really turned me around, things are intertwined in ways I hadn't realized. . . . All these social issues as well as political and economic issues are really intertwined. Before, I was concerned only about health and then I began to get into the politics, decision making, and so many things."

In two years, what started as the outrage of a small group of mothers has transformed the political climate of a major metropolitan area. What these women have aimed for is a greater level of democracy, a greater level of involvement, not only in their organization but in the development process of the city generally. They have demanded accountability regarding land use and ownership, very subversive concerns in a capitalist society. In their organizing, the group process, collectivism, was of primary importance. It allowed the women to see their own power and potential and therefore allowed them to consolidate effective opposition. The movement underscored the role of principles. In fact, we citizens have lived so long with an unquestioning acceptance of profit and expediency that sometimes we forget that our objective is to do "what's right." Women are beginning to raise moral concerns in a very forthright manner, emphasizing that experts have left us no other choice but to follow our own moral convictions rather than accept neutrality and capitulate in the face of crisis.

The environmental crisis will escalate in this decade and women are sure to play pivotal roles in the struggle to save our planet. If women are able to sustain for longer periods some of the qualities and behavioral forms they have displayed in crisis situations (such as direct participatory democracy and the critique of patriarchal bureaucracy), they may be able to reintroduce equality and democracy into progressive action. They may also reintroduce the value of being moved by principle and morality. Pragmatism has come to dominate all forms of political behavior and the results have often been disastrous. If women resist the "normal" organizational thrust to barter, bargain, and fragment ideas and issues, they may help set new standards for action in the new environmental movement.

Notes

1. See Cynthia Cockburn, "When Women Get Involved in Community Action," in Marjorie Mayo, ed., *Women in the Community* (London: Routledge and Kegan Paul, 1971).

2. All of the quotations from Charlotte Bullock and Robin Cannon are personal communications, 1986.
3. Cockburn 62.
4. Cerrel Associates, *Political Difficulties Facing Waste to Energy Conversion Plant Siting* (Los Angeles: California Waste Management Board, 1984), 42–43.
5. See Cockburn 63.

22 Law and Environmental Justice

United States Chief Justice Warren Burger concluded his opinion in the landmark case of *TVA v. Hill*, below, with this quotation ascribed to Sir Thomas More:

> In the thickets of the law, I'm a forester. What would you do, cut a great road through the law to get after the Devil? And when the last law was down, and the Devil turned round on you—where would you hide, the laws all being flat?

Burger used this apt environmental metaphor of the forester and the trees to defend his upholding of the laws of Congress even when he personally did not believe them justified. Since 1970, environmentalists have been using these trees of the law to defend the trees of the forest—as well as the fish of the sea and the birds of the sky. No matter what the field—biology, chemistry, public policy, economics—in order to be an effective advocate for the environment, you must know the law. This statement is not meant to steer all students of the environment into a legal career, but it does point out a great shortcoming in the undergraduate curriculum: most colleges offer no systematic training in the legal systems that govern all citizens of all countries.

We use the plural "legal systems" intentionally. There are systems of international law, national law, state or other regional law, and local or municipal law in almost every location in the world.

International law is governed almost exclusively by treaties, which can be bilateral or multilateral. For example, Canada and the United States have a bilateral treaty guaranteeing the environmental stewardship of the Great Lakes. Multilateral environmental treaties include the Convention on International Trade in Endangered Species, signed by 162 parties, and the United Nations Convention on Biodiversity, signed by 165. The most important feature of international law is that enforceability is more difficult than in national law. In international law, the parties are usually sovereign nations, and there is a limited amount that the international community can force a sovereign nation to do against its will. Often resolution to an international dispute can be brought about only by negotiation or by force, as opposed to the decision of a legal tribunal.

National law carries far more coercive power than international law. Resident violators of national laws are subject to forfeiture of their life, liberty, or property. In the United States, the laws are enacted by the legislature, enforced by the executive, and interpreted by the judiciary. This happens in parallel on the federal and state levels. For example, in *Lucas v. South Carolina*, the state legislature passed the Beachfront Management Act, which applied only within South Carolina. The executive, in the capacity of the South Carolina Coastal Council, denied Lucas a permit to build a house on his beachfront property. Lucas turned to the judicial system, in which he had a trial in the lower state court and an appeal in the South Carolina Supreme Court. After the state supreme court ruled in the state's favor, the too-simple distinction between federal and state law began to dissolve when Lucas appealed to the United States Supreme Court.

Lucas based his appeal on the most important lack of parallelism between the state and federal legal systems—the sovereignty clause of the United States Constitution, which forbids any state law from violating any provision of the federal Constitution. The Supreme Court held that South Carolina's enforcement of the state law, the denial of a building permit, deprived Lucas of his property without just compensation, thereby violating the Fifth Amendment to the United States Constitution.

Many legal issues involving the environment are resolved at levels lower than the state executive or judiciary. Local zoning laws can be vital in environmental protection, and many of the land-use decisions in the United States are initiated in front of county or municipal commissions, which act as a local executive. The major difference between county and municipal systems, on the one hand, and state and federal systems, on the other, is that local systems have legislative and executive branches, but not an independent judiciary. Of course, if any decision made on the local level violates state or federal law, executive or judicial proceedings can be brought at those higher levels.

A corollary to environmental law is the concept of environmental justice, which is not subject to the same degree of formalization and is often without systemic remedy. Exposure to environmentally dangerous or unpleasant conditions is not distributed equally among social classes: minority and impoverished citizens live near hazardous landfills and polluting industries at a much higher rate than the general population. Congress has not passed any law that particularly addresses this problem; the only direct federal action has been a 1994 executive order requiring "each Federal agency [to] make achieving environmental justice part of its mission by identifying and addressing, as appropriate, disproportionately high and adverse human health or environmental effects of its programs, policies, and activities on minority populations and low-income populations in the United States." However, no federal agency has yet found an environmental justice case that could be prosecuted under the Civil Rights Acts or any other federal statute.

Bringing together issues of environmental health and social equity, the discipline of environmental justice raises hope for the environmental movement. It promises to bring into the fold a segment of the population—the urban poor—traditionally unlikely to be a constituency of environmentalism. However, like international law, environmental justice is difficult to enforce, although for different reasons, as pointed out by Christopher H. Foreman, Jr., in "Political and Policy Limitations to Environmental Justice" (see selection below).

The law is an academic discipline of great richness and complexity, a foundation of responsible citizenship, and a necessary tool for anyone working on environmental issues.

FURTHER READING

Nagle, John Copeland, and J. B. Ruhl. *The Law of Biodiversity and Ecosystem Management.* New York: Foundation Press, 2002.

Percival, Robert V., and Dorothy C. Alevizatos. *Law and the Environment: A Mutidisciplinary Reader.* Philadelphia: Temple University Press, 1997.

Plater, Zygmunt J. B., Robert H. Abrams, William Goldfarb, and Robert L. Graham. *Environmental Law and Policy*, 3rd ed. St. Paul: West Publishing Group, 2004.

Sax, Joseph. "Property Rights and the Economy of Nature: Understanding *Lucas.*" 45 *Stanford Law Review* 1433, 1993.

Wright, J. Skelley. *Calvert Cliffs Coordinating Committee v. United States Atomic Energy Commission*, D.C. Circuit Court, 1971.

Lynton Caldwell, "Environmental Aspects of International Law" in *International Environmental Policy: Emergence and Dimensions* (1990)

From fisheries in international waters to trade in endangered species, from the ozone hole to global warming to nuclear testing, many of the most difficult environmental problems are international in scope. As recognition of the international aspects of environmental issues becomes more evident, the procedures for effecting international agreements about the environment have become more formalized. In this selection, Lynton Caldwell details some of the more prominent of these procedures.

International environmental law is not exclusively or even primarily a field of legal practice. There are, of course, international lawyers and litigants, but lawsuits primarily pertaining to environmental issues have been relatively infrequent. In the perspective of international policy, environmental law is perhaps best understood as the collective body of agreements among states [i.e., nations] regarding mutual rights and obligations affecting the environment. It is embodied in conventions among states (treaties) and, to lesser effect, in international declarations, collective principles, opinions of jurists, and generally accepted practices among states. Enforcement of its provisions, customary or specified by treaty, are usually sought through negotiation (e.g., diplomacy) rather than through adjudication. Its boundaries are definable only in broad terms because new scientific findings of international significance and enlarging perceptions of man-biosphere relationships have continually if unevenly expanded its frontiers.

Emergent aspects of international environmental law include those extensions, codifications, or reinterpretations of rights and obligations among nations that have been long accepted as customary international law. An example is the identification of transboundary air pollution as entailing an obligation of a state to prevent the use of its territory to inflict harm upon its neighbors. There are also new principles and obligations derived from formal agreements, usually treaties, regarding subjects hitherto untouched by the law of nations. Examples may be found in the conventions governing international spaces—notably outer space and the deep sea bed—where technology has at last enabled some nations to establish an operational presence beyond territorial jurisdictional claims.

An uncertain source of international law, potential but portentous, may now be found in resolutions and declarations of the General Assembly of the United Nations and of major international conferences such as the United Nations Conference on the Human Environment. Actions of majorities in the General Assembly, in the governing bodies of the specialized agencies, or in conferences within the UN system do not make law in the traditional sense. Yet, the official statements and pronouncements of United Nations bodies, especially when adopted by overwhelming majorities, have the potential of becoming recognized as obligatory rules of national conduct. The barrier to realizing this possibility is that the newer developing states, while most inclined to "legislate" through UN declarations and resolutions, are also jealous of their national sovereignties and, besides, they do not possess the economic, technical, or military capabilities of enforcing their policy preferences upon the greater powers. The legal influence of UN declarations and resolutions depends greatly upon the degree of actual consensus and the extent to which those pronouncements can be applied in practice. The World Charter for Nature, for example, elicited a high degree of at least formal consensus, but it provided no means whatever to make its precepts operational.

A process leading toward international lawmaking has now become established. The first step is a proposal to ECOSOC [United Nations Economic and Social Council] or the General Assembly to establish a committee or to convene an international conference to consider a major issue of international concern. Then preparatory committees, working groups, regional meetings, and symposia are organized and opinions of governments and of nongovernmental organizations invited. Usually, but not invariably, preconference preparation assures substantial agreement on official conference action. In the end it is not the action of an international conference or of the General Assembly that confers legal status; it is rather the effective consensus of nations that make it law. But the conferences and (potentially) the deliberations of the General Assembly provide the forums in which consensus building can occur.

A secondary aspect of the United Nations role in the development of international law derives from Article 13 of the UN Charter, which states that the General Assembly "shall initiate studies and make recommendations for the purpose of . . . encouraging the progressive development of international law and its codification." In 1947, the General Assembly established the International Law Commission, which undertakes the "progressive development of international law" through preparation of draft conventions on issues still in a formative stage and through "codification of international law" for topics upon which substantial agreement exists in practice among states. Drafts and recommendations prepared by the commission are reviewed by the Sixth (Legal) Committee of the General Assembly and may be laid before the assembly for action. An alternative route may be the convening of a United Nations conference to consider an international convention or treaty. The United Nations system thus provides the principal, although not exclusive, structure through which international law is developed and codified. The World Court has not played a significant role in the development of international environmental laws. The UN specialized agencies and the IAEA [International Atomic Energy Agency] have been far more important actors, and the principal forum for the initiation of new international legal concepts has been the UN General Assembly.

The very circumstances that necessitate the growth of an international legal order for the environment make its development difficult. The traditional structure of international law emerged out of a political configuration far simpler than that now existing. The issues were fewer and more manageable; the forms of political and economic action much less diverse. Even so, a body of positive environmental law has been adopted, notably since the 1960s, through a large number of treaties [for example, the Montreal Protocol—see Chapter 13].

. . . Some environment-related agreements among states have been in effect for more than a century. . . . UNEP [United Nations Environmental Program] has published a compendium of its own legislative authority and a register of international environmental treaties, supplemented periodically to indicate new agreements and new accessions to prior agreements. Collections of the texts of treaties have been published by the British Institute of International Law and Comparative Law, by UNEP, and by the IUCN [World Conservation Union] Environmental Law Center in Bonn. . . .

An important step in the development of international environmental law was taken by a UNEP-sponsored Ad Hoc Meeting of Senior Government Officials Expert in Environmental Law convened in Montevideo, Uruguay, from 28 October to 6 November 1981. The purpose of this meeting was "to establish and frame methods and programs, including global, regional, and national efforts, for the development and periodic review of environmental law." A secondary objective was to contribute to preparation and implementation of the environmental law component of UNEP's medium-term environment program.

The conclusions of the Montevideo meeting specified three subject areas in particular in which guidelines, principles, or agreements "should be developed. They were: (a) marine pollution from land-based sources, (b) protection of the stratosphere ozone layer, and (c) transport, handling, and disposal of toxic and dangerous wastes." In addition, eight subject areas were identified for action "in accordance with agreed objectives." They were (1) international cooperation in environmental emergencies, (2) coastal zone management, (3) soil conservation, (4) transboundary air pollution, (5) international trade in potentially harmful chemicals, (6) protection of rivers and other inland waters against pollution, (7) legal and administrative mechanisms for the prevention and redress of pollution damage, and (8) environmental impact assessment. . . .

INTERNATIONAL LIABILITY FOR THE ENVIRONMENT

The principle of liability for environmental harm has been recognized even by governments which least favor international action in the field of the protection of the environment. There seems to be no doubt about the liability of states for damages which they may cause (e.g., through negligence) to the environment of other states. Principle 21 of the Stockholm Declaration [1972 international accord that recognizes the right to a healthy environment] applies not only to this kind of damage, but also to the common spaces, to the environment of areas beyond the limits of national jurisdiction (e.g., the upper atmosphere, the oceans and deep sea bed, outer space, and Antarctica), and the same principle is embodied in Article 21(d) of the World Charter for Nature, and in many marine pollution conventions including the UN Law of the Sea (1982). However, under the present state of international law, this liability is not regarded as absolute except for nuclear damage or in relation to outer space. Yet even in these cases the acceptance or enforcement of liability is highly uncertain—the transboundary effects of the Chernobyl nuclear explosion is a case in point. [This was actually a *steam* explosion that spread radiation; see Chapter 3.]

There is an important difference between traditional views of national rights and obligations and the new perspective, which would extend liability for damage done not to a given state or its nationals but to what may be regarded as the common property of all nations. Hence, it is necessary to examine the two different aspects of international liability for environmental damages: the traditional one, which redresses harm suffered by states, and new interpretations of legal responsibilities. Each type provides occasions for international controversy.

International environmental law regarding a state's liability for damage caused beyond its borders, although not new in principle, is still in a developmental state, with many points of law unsettled. Advanced technologies allowing human penetration of outer space and the deep sea complicate traditional assumptions. Even so, there are specific treaties between nations from which the principle of liability may be invoked for new or changing technologies as they affect, for example, transfrontier pollution. One of the more familiar is the 1909 Treaty Relating to Boundary Waters and Questions Arising Between the U.S. and Canada, signed by the U.S. and Great Britain (which was still responsible for Canada's foreign affairs). This treaty established the International Joint Commission (IJC). The commission, in its fact-finding role, provided the factual basis for the Trail Smelter arbitration, which as yet is the only international "adjudication" on the merits of a pollution dispute. But provisions for arbitration have been written into the Nordic Convention on the Protection of the Environment (1974) and the Convention on the Protection of the Rhine against Chemical Pollution (1976).

In 1935 Trail Smelter, a metal refinery in British Columbia, was found to have discharged sulfurous gases that damaged farm crops across the border in the state of Washington. Mainly on evidence presented by the IJC, an international tribunal comprised of an American, a Canadian, and a Belgian found the Canadian government liable for damages of $350,000. The Trail Smelter decision set an important precedent in international environmental law on two counts. First, it established the precedent of an international tribunal to investigate a case concerning transboundary pollution damages. Second, the legal principles of the Trail Smelter decision were recognized at the Stockholm Conference on the Human Environment in 1972.

The growth in international trade and transport, and the transboundary effects of modern technology, have led to increased attention to liability for environmental damages. Oil spills in coastal waters have caused nations to demand some form of redress from offending polluters; such incidents led to the Convention on Civil Liability for Oil Pollution Damage. But establishing the origins and fixing responsibilities for transboundary pollution of air and water often have been difficult, as demonstrated by the controversies in Europe and North America over liability for acid [rain] precipitation. The Convention on Long-Range Transboundary Air Pollution, sponsored by the UN Economic Commission for Europe (ECE) and in force since 1983, attempts to clarify this problem. Scientific investigations of the flow of pollutants through the environment may in time yield findings that will open new subject areas to international law. Formal adjudication under the present world political order may not be the most effective way to resolve international environmental disputes.

The Nuclear Test Cases

The problem that the question of liability presents to the building of an international structure of environmental law is illustrated by the controversy over the French atmospheric nuclear tests. From 1966 to 1972 the French government conducted several series of these tests from a base on Mururoa in their South Pacific territory of New Caledonia. In May 1973 it seemed that France was about to engage on a new series, to last until 1975 or later. Australia and New Zealand asked France to put an end to the atmospheric tests, which they contended had resulted in an increase in radiation doses to their populations. They further argued that any radioactive material deposited in their territories constituted a potential danger. France refused to cancel or modify its scheduled test program, and the Australian and New Zealand governments applied to the International Court of Justice, asking it to declare that further atmospheric tests of nuclear weapons were inconsistent with applicable rules of international law, and to order France to abstain from carrying out any such tests. In addition, considering that irreparable damage might be caused by any test which would be carried out before the court's judgment, the Australian government requested the court to order interim measures of protection.

The World Court examined the record on the basis of Article 41 of the Statute of the International Court of Justice which states: "The court shall have the power to indicate, if it considers that circumstances so require, any provisional measures which ought to be taken to preserve the respective rights of either party." . . .

The nuclear test cases demonstrate that not all states regard seriously international responsibilities as reflected in Principle 21 of the Stockholm Declaration. It is nevertheless true that the declaration was an effective catalyst for the development of an environmental ethic for the global community. But there is obviously a difference between the affirmation of principles by national delegations in the euphoria of international assemblies and practical implementation where perceived national interests are involved.

C. M. Abraham and Sushila Abraham, from "The Bhopal Case and the Development of Environmental Law in India" (1991)

On 3 December 1984, methyl isocyanate gas leaked from a Union Carbide plant in Bhopal, India, killing 3,800 people and permanently injuring another 2,700. The legal ramifications of this disaster were profound and in many ways remain unresolved. One of the far-reaching outcomes of this case was an internal reassessment of the Indian legal system, documented by Abraham and Abraham. Despite the settlement outlined in this article, many feel that justice was not done. Birth defects continue to be reported.

Although the Bhopal case focused initially on the cases that were filed in the United States by the American lawyers, about 800 individual suits for damages were also initially filed by individual plaintiffs in the Bhopal District Court, claiming damages totalling 1,620 million rupees (about US$125 million). One representative [class action] suit has been filed in which damages of 12,500 million rupees (about US$1 billion) have been claimed. Pursuant to the Bhopal Act, the government of India assumed the exclusive right to represent the victims and impleaded itself as plaintiff in all the suits filed in India. All earlier suits were stayed pending further proceedings and the Union of India filed a fresh case before the District Court of Bhopal on 9 September 1986. This latest case before the Bhopal District Court was beset with many hurdles for more than two and a half years, with no relief in sight. The first judge, who began hearing the case in the Bhopal District Court, was promoted to the High Court, necessitating a fresh start by a second judge, who had to withdraw when it was revealed that he was himself one of the claimants. The third judge met with an accident and died after holding the first hearing. M. W. Deo was the fourth judge to begin hearing the case. But the progress of the case was much delayed by raising the hopes of an out-of-court settlement.

Negotiations for an out-of-court settlement were pursued by the Union of India even before any suit was filed in the United States. It was also reported that UCC [Union Carbide Corporation] had "active discussions" for a "concrete proposal" for settlement to the Indian Parliament. The offer made by UCC initially was for US$200 million, an amount reportedly equal to UCC's insurance cover, and it was reported that India's then Prime Minister, Rajiv Gandhi, rejected UCC's offer as insufficient. From the very beginning the proceedings were stopped on several occasions to allow the parties to come to a settlement. Once Judge Keenan reacted sharply in open court when he felt that his earnest efforts to bring about a settlement were subjected to criticism for dilatoriness. An out-of-court settlement resumed greater impetus when the case came before the Indian courts. Both the Union of India's Attorney-General and UCC's counsel were reported to be making "genuine" efforts and at one stage a settlement appeared imminent. However, these negotiations again broke down, and on 27 November 1987 both sides submitted to the court that they would proceed with the case. The fact that an agreement did not emerge at that time was probably due to the uproar the negotiations had caused in Parliament, and a "sell out" sentiment was voiced in the press and among the victims.

On 2 April 1987, M. W. Deo, judge of the Bhopal District Court, made an order proposing reconciliatory interim relief. Both the Union of India and UCC initially responded with proposals for an out-of-court settlement, but when their attempts failed the court proceeded to hear both parties on the matter for interim relief on its *suo moto* proposal. Then, on 17 December 1987, Judge Deo issued the order for substantial interim compensation in the sum of 3,500 million rupees (about US$270 million).

Judge Deo's order made unprecedented use of the courts' inherent power to render justice. The judge based his decision entirely on the exercise of the court's jurisdiction under section 151 of the Code of Civil Procedure, taken together with section 94(e) of the Code. The use of these provisions was justified by Judge

Figure 22.1 Young victim of the Bhopal tragedy. Courtesy of Greenpeace / Raghu Rai.

Deo by the paramount need for justice in the case. He questioned the very basis of the forensic praxis:

> Can the gas victims survive until the tangible data with meticulous exactitude is collected and proved and adjudicated in fine forensic style for working out the final amount of compensation with precision of quality and quantity? Will it not be prudent to order payment of a relative sum bearing in mind all the progress in the case so far, the facts and figures (though not undisputed), which have come on record and the material furnished during settlement efforts made by Judge Keenan?

As to the arguments of UCC's counsel, F. S. Nariman, against the exercise of the court's jurisdiction under section 151 of the Code for grant of interim relief, Judge Deo stated: "The inherent powers have not been conferred upon the courts. It is a power inherent in the court by virtue of its duty to do justice between the parties before it." He followed the law laid down by the Supreme Court in *Monohar Lal* on this aspect and went further in adding:

> Inherent powers are born with the creation of the court, like the pulsating life coming with a child, born into the world. Without inherent powers, the

> court will be like a stillborn child. The powers invested in the court after its creation are like many other acquisitions of faculties which the child acquires after birth during its life. Thus inherent powers are of primordial nature. They are almost plenary except for the restriction that they shall not be exercised in conflict with any express provision to the contrary.

He also found that the power under residuary section 94(e) of the Code is again somewhat of the same nature in relation to the interlocutory stages of the suit. Section 94 provides for making such interlocutory order as may appear to be just and convenient, and no provision is expressed or implied prohibiting such an order in a suit for pecuniary damages. Although he was aware that such an order had not so far been shown to have been made formally in a suit in a civil court he nevertheless held: "The principle under section 94 of the CPC is to recognize and grant powers to the court to make an interlocutory order of the same nature as that of the main claim in the suit." He also relied upon the law declared by the Supreme Court in *M. C. Mehta* for the limited principle that in an action for tortious liability a grant of interim compensation is permissible. . . .

Judge Deo felt that there was no need for him to go into the legal point of lifting the corporate veil as he was dealing with the matter only at an interlocutory level. He nevertheless referred to two cases, *DHN* and *Escorts*, and accepted the submission of the Union's Attorney-General that the admitted fact of UCC owning 50.9 percent of UCIL's [Union Carbide of India Limited] shares was enough to show that UCC always had the power and the capacity to control the working of UCIL. A sum of 3,500 million rupees was ordered by him as "substantial measures" for the gas victims. This, according to the order, was without prejudice to the rights and defense of the parties to the suit and counterclaim that might be finally adjudicated. . . .

On 14 and 15 February 1989 the Supreme Court of India made orders for payment of US$470 million (about 7,150 million rupees or £274 million) by UCC to the Union of India in full settlement of all claims. The Court took into consideration the different proposals that UCC and the Union had offered for a settlement and found the case pre-eminently fit for an overall settlement solely because of the "enormity of human suffering" occasioned by the disaster and the pressing urgency to provide immediate and substantial relief to the victims. On that basis the Court held it "just, equitable, and reasonable" to order UCC to pay that sum in full settlement of all claims including the quashing of all criminal proceedings.

The Supreme Court orders appear quite unusual in the sense that they seem to lack the expected judicial discourse and elaborate reasoning required in disposing of such an important case by the highest court. In other words, the Court justified them entirely on humanitarian considerations. Several petitions were filed before the Supreme Court to review the settlement orders. Thereupon the

Supreme Court ordered that both UCC and the Union of India would continue to be subject to the jurisdiction of the courts in India until further orders. On 4 May 1989, the Court gave the reasoning for its orders of 14 and 15 February 1989, reiterating the compelling duty, both judicial and humane, to secure immediate relief to the victims. The Court observed that "the compulsions of the need for immediate relief to tens of thousands of suffering victims could not, in our opinion, wait till these questions, vital though they may be, are resolved in the course of judicial proceedings."

On 22 December 1989, the Supreme Court of India upheld the constitutional validity of the Bhopal Act. . . .

ENVIRONMENTALISM IN INDIA

Law for environmental protection is not a new phenomenon in India, although the terms "environment" and "pollution" are of recent origin. *Dharma*, the fundamental Indian concept of law, is itself based on a recognition of the importance of harmonizing human activities with Nature in order to maintain a universal order. The duty to maintain a clean environment can be found in various provisions in the ancient laws of India. Kautilya in his *Arthasastra* laid down "environmental" precepts for city administration and prescribed penalties for anyone making the city dirty.[1] We also find relevant provisions in the prescription of punishment of public offenses in other ancient Hindu works on legal matters.[2]

The environmental consciousness which we find in India today is to a great extent an import of the Western culture concerning such issues. The present environmental consciousness of the West, which could be traced as an emergent culture of the mid-1960s in the United States, followed the publication of *Silent Spring*, the book by Rachel Carson, in 1962 [see Chapter 15]. The doctrinal roots of modern environmental law could be based on the law of nuisance: nuisance actions could challenge every major industrial and municipal activity which is today a subject of comprehensive environmental legislation.[3] The law of nuisance can be divided into public nuisance and private nuisance; public nuisance, which could encompass most environmental issues, falls mainly in the purview of the criminal law. . . .

THE ENVIRONMENT (PROTECTION) ACT 1986

After the accident at Bhopal, the Department of Environment came under considerable pressure from both the Prime Minister's office and the general public to decide on "comprehensive legislation" for controlling toxic and hazardous substances. A new umbrella statute, the Environment (Protection) Act, was enacted in May 1986. Only the broad outlines of the law can now be discussed here, since the work on implementing the rules has not been completed to date. The basic thrust of the Act is to empower the central government to correct deficiencies of policy-making and enforcement in the [various Indian] States through action not specifically permitted under earlier laws. New powers were conferred on the central government to set standards for pollution emitted or discharged into the environment and also to regulate the handling of hazardous substances. The Act established environmental laboratories responsible for analyzing air and water samples collected by the enforcement authorities, and substantially strengthened the government's capacity to penalize polluters. Even though the Act has closed some of the loopholes in the earlier laws, it is too early to say how effectively the environmental policies will be implemented through this legislation. . . .

THE BHOPAL ACT 1985

Apart from the early statutory modification brought about in the field of personal injury law, a major legislative breakthrough was achieved through the Bhopal Act 1985. The Bhopal Act was an immediate legislative reaction to the Bhopal disaster. The Act replaced an earlier Ordinance that had been promulgated as an urgent measure to meet an unprecedented situation created by the filing of individual suits by American lawyers. Under the Bhopal Act the government of India assumed the role of *parens patriae*, which gave to the Union government considerable powers to deal with the legal and administrative problems created by the disaster. The Bhopal Act was initially assessed by many as an executive maneuver that not only enabled the government to participate in the United States litigation but also avoided any litigation in India. It empowers the government to interpose an administrative compensation process as the exclusive primary resort of victims. The Bhopal Act gives exclusive rights of representation to the central government in all claims arising out of the disaster as well as powers of review and investigation and the framing of a scheme for the registration, processing, and enforcement of claims, including the creation and administration of a fund to meet the costs of the scheme.

Initially fears were expressed over the constitutionality of this statute, until it was upheld by the Indian Supreme Court. The Bhopal Act represents further evidence of the ability of the Indian legal system to respond effectively to the new challenges posed by mass disasters of the kind that occurred in Bhopal. The Act has been considered as the only realistic method of protecting the victims' rights of action. When a large number of claims cannot be handled effectively by what is generally acknowledged to be a slow legal system, a statutory claim and compensation scheme may provide a faster remedy. The Bhopal Act enables the government to take the required steps to administer and distribute the large sums of money involved either through litigation or an out-of-court settlement or through any other form of contribution.

Notes

1. R. P. Kangle, *The Kautilya Arthasastra* (2nd ed., 1970).
2. M. K. Sharma, *Court Procedure in Ancient India* (1978), 221–222.
3. William H. Rogers, Jr., *Handbook on Environmental Law* (1977), 100.

Warren Burger, *TVA v. Hill*, U.S. Supreme Court (1978)

The United States Endangered Species Act of 1973 is one of the strongest environmental laws in the world, thanks to the language of Section 7 and the interpretation of that language by the Supreme Court in the TVA v. Hill *case. The TVA is the Tennessee Valley Authority. Although the case enjoined the completion of the Tellico Dam, later action by Congress allowed the dam to be finished, which extinguished the population of snail darters, an endangered species, in the Little Tennessee River. However, other populations of the species have been discovered—accordingly, both the dam and the species still exist. Unfortunately, none of the benefits of the dam claimed by the TVA bore fruit, nor should they have been expected to.*

The Little Tennessee River originates in the mountains of northern Georgia and flows through the national forest lands of North Carolina into Tennessee, where it converges with the Big Tennessee River near Knoxville. The lower thirty-three miles of the Little Tennessee takes the river's clear, free-flowing waters through an area of great natural beauty. Among other environmental amenities, this stretch of river is said to contain abundant trout. Considerable historical importance attaches to the areas immediately adjacent to this portion of the Little Tennessee's banks. To the south of the river's edge lies Fort Loudon, established in 1756 as England's southwestern outpost in the French and Indian War. Nearby are also the ancient sites of several native American villages, the archeological stores of which are to a large extent unexplored. These include the Cherokee towns of Echota and Tennase, the former being the sacred capital of the Cherokee Nation as early as the sixteenth century and the latter providing the linguistic basis from which the State of Tennessee derives its name.

In this area of the Little Tennessee River the Tennessee Valley Authority, a wholly owned public corporation of the United States, began constructing the Tellico Dam and Reservoir Project in 1967, shortly after Congress appropriated initial funds for its development. Tellico is a multipurpose regional development project designed principally to stimulate shoreline development, generate sufficient electric current to heat 20,000 homes, and provide flatwater recreation and flood control, as well as improve economic conditions in "an area characterized by underutilization of human resources and outmigration of young people." Of particular relevance to this case is one aspect of the project, a dam which TVA determined to place on the Little Tennessee, a short distance from where the river's waters meet with the Big Tennessee. When fully operational, the dam would impound water covering some 16,500 acres—much of which represents valuable and productive farmland—thereby converting the river's shallow, fast-flowing waters into a deep reservoir over thirty miles in length.

The Tellico Dam has never opened, however, despite the fact that construction has been virtually completed and the dam is essentially ready for operation. Although Congress has appropriated monies for Tellico every year since 1967, progress was delayed, and ultimately stopped, by a tangle of lawsuits and administrative proceedings. After unsuccessfully urging TVA to consider alternatives to damming the Little Tennessee, local citizens and national conservation groups brought suit in the District Court, claiming that the project did not conform to the requirements of the National Environmental Policy Act of 1969 (NEPA). After finding TVA to be in violation of NEPA, the District Court enjoined the dam's completion pending the filing of an appropriate environmental impact statement. The injunction remained in effect until late 1973, when the District Court concluded that TVA's final environmental impact statement for Tellico was in compliance with the law.

A few months prior to the District Court's decision dissolving the NEPA injunction, a discovery was made in the waters of the Little Tennessee which would profoundly affect the Tellico Project. Exploring the area around Coytee Springs, which is about seven miles from the mouth of the river, a University of Tennessee ichthyologist, Dr. David A. Etnier, found a previously unknown species of perch, the snail darter, or *Percina (Imostoma) tanasi.* This three-inch, tannish-colored fish, whose numbers are estimated to be in the range of 10,000 to 15,000, would soon engage the attention of environmentalists, the TVA, the Department of the Interior, the Congress of the United States, and ultimately the federal courts, as a new and additional basis to halt construction of the dam.

Until recently the finding of a new species of animal life would hardly generate a cause celebre. This is particularly so in the case of darters, of which there are approximately 130 known species, eight to ten of these having been identified only in the last five years. The moving force behind the snail darter's sudden fame came some four months after its discovery, when the Congress passed the Endangered Species Act of 1973. This legislation, among other things, authorizes the Secretary of the Interior to declare species of animal life "endangered" and to identify the "critical habitat" of these creatures. When a species or its habitat is so listed, the following portion of the Act—relevant here—becomes effective:

> The Secretary [of the Interior] shall review other programs administered by him and utilize such programs in furtherance of the purposes of this chapter. All other Federal departments and agencies shall, in consultation with and with the assistance of the Secretary, utilize their authorities in furtherance of the purposes of this chapter by carrying out programs for the con-

> servation of endangered species and threatened species listed pursuant to section 1533 of this title and *by taking such action necessary to insure that actions authorized, funded, or carried out by them do not jeopardize the continued existence of such endangered species and threatened species or result in the destruction or modification of habitat of such species* which is determined by the Secretary, after consultation as appropriate with the affected States, to be critical. (emphasis added)

In January 1975, the respondents in this case and others petitioned the Secretary of the Interior to list the snail darter as an endangered species. After receiving comments from various interested parties, including TVA and the State of Tennessee, the Secretary formally listed the snail darter as an endangered species on 8 October 1975. In so acting, it was noted that "the snail darter is a living entity which is genetically distinct and reproductively isolated from other fishes." More important for the purposes of this case, the Secretary determined that the snail darter apparently lives only in that portion of the Little Tennessee River which would be completely inundated by the reservoir created as a consequence of the Tellico Dam's completion. The Secretary went on to explain the significance of the dam to the habitat of the snail darter:

> The snail darter occurs only in the swifter portions of shoals over clean gravel substrate in cool, low-turbidity water. Food of the snail darter is almost exclusively snails which require a clean gravel substrate for their survival. *The proposed impoundment of water behind the proposed Tellico Dam would result in total destruction of the snail darter's habitat.* (emphasis added)

Subsequent to this determination, the Secretary declared the area of the Little Tennessee which would be affected by the Tellico Dam to be the "critical habitat" of the snail darter. Using these determinations as a predicate, and notwithstanding the near completion of the dam, the Secretary declared that pursuant to Section 7 of the Act, "all Federal agencies must take such action as is necessary to insure that actions authorized, funded, or carried out by them do not result in the destruction or modification of this critical habitat area." This notice, of course, was pointedly directed at TVA and clearly aimed at halting completion or operation of the dam. . . .

One would be hard pressed to find a statutory provision whose terms were any plainer than those in Section 7 of the Endangered Species Act. Its very words affirmatively command all federal agencies "to *insure* that actions *authorized, funded*, or *carried out* by them do not *jeopardize* the continued existence" of an endangered species or "*result* in the destruction or modification of habitat of such species" (emphasis added). This language admits of no exception. . . .

Concededly, this view of the Act will produce results requiring the sacrifice of the anticipated benefits of the project and of many millions of dollars in public funds. But examination of the language, history, and structure of the legislation under review here indicates beyond doubt that Congress intended endangered species to be afforded the highest of priorities. . . .

The legislative proceedings in 1973 are, in fact, replete with expressions of concern over the risk that might lie in the loss of *any* endangered species. Typifying these sentiments is the Report of the House Committee on Merchant Marine and Fisheries on H. R. 37, a bill which contained the essential features of the subsequently enacted Act of 1973; in explaining the need for the legislation, the Report stated:

> As we homogenize the habitats in which these plants and animals evolved, and as we increase the pressure for products that they are in a position to supply (usually unwillingly) we threaten their—and our own—genetic heritage.
>
> *The value of this genetic heritage is, quite literally, incalculable.*
>
> From the most narrow possible point of view, *it is in the best interests of mankind to minimize the losses of genetic variations.* The reason is simple: they are potential resources. They are keys to puzzles which we cannot solve, and may provide answers to questions which we have not yet learned to ask. . . . (emphasis added)

As it was finally passed, the Endangered Species Act of 1973 represented the most comprehensive legislation for the preservation of endangered species ever enacted by any nation. Its stated purposes were "to provide a means whereby the ecosystems upon which endangered species and threatened species depend may be conserved," and "to provide a program for the conservation of such . . . species." In furtherance of these goals, Congress expressly stated in Section 2(c) that "all Federal departments and agencies *shall seek to conserve endangered species* and threatened species" (emphasis added). Lest there be any ambiguity as to the meaning of this statutory directive, the Act specifically defined "conserve" as meaning "to use and the use of *all methods and procedures which are necessary* to bring *any endangered species* or threatened species to the point at which the measures provided pursuant to this chapter are no longer necessary" (emphasis added). Aside from Section 7, other provisions indicated the seriousness with which Congress viewed this issue: Virtually all dealings with endangered species, including taking, possession, transportation, and sale, were prohibited, except in extremely narrow circumstances. . . .

One might dispute the applicability of these examples [foreseen effects of the Act on policies and practices of the Air Force and the Park Service as discussed during Congressional deliberations, omitted here] to the Tellico Dam by saying that in this case the burden on the public through the loss of millions of unrecoverable dollars would greatly outweigh the loss of the snail darter. But neither the Endangered Species Act nor Article III of the Constitution provides federal courts with authority to make such fine utilitarian calculations. On the contrary, the plain language of the Act, buttressed by its legislative history, shows clearly that Congress viewed the value of endangered species as "incalculable." Quite obviously, it would be difficult for a court to balance the loss of a sum certain—even $100 million—against a congressionally declared "incalculable" value, even assuming we had the power to engage in such a weighing process, which we emphatically do not. . . .

But these principles take a court only so far. Our system of government is, after all, a tripartite one, with each branch having certain defined functions delegated to it by the Constitution. While "it is emphatically the province and duty of the judicial department to say what the law is,"[1] it is equally—and emphatically—the exclusive province of the Congress not only to formulate legislative policies and mandate programs and projects, but also to establish their relative priority for the Nation. Once Congress, exercising its delegated powers, has decided the order of priorities in a given area, it is for the Executive to administer the laws and for the courts to enforce them when enforcement is sought.

Here we are urged to view the Endangered Species Act "reasonably," and hence shape a remedy "that accords with some modicum of common sense and the public weal." But is that our function? We have no expert knowledge on the subject of endangered species, much less do we have a mandate from the people to strike a balance of equities on the side of the Tellico Dam. Congress has spoken in the plainest of words, making it abundantly clear that the balance has been struck in favor of affording endangered species the highest of priorities, thereby adopting a policy which it described as "institutionalized caution."

Our individual appraisal of the wisdom or unwisdom of a particular course consciously selected by the Congress is to be put aside in the process of interpreting a statute. Once the meaning of an enactment is discerned and its constitutionality determined, the judicial process comes to an end. We do not sit as a committee of review, nor are we vested with the power of veto. The lines ascribed to Sir Thomas More by Robert Bolt are not without relevance here:

> The law, Roper, the law. I know what's legal, not what's right. And I'll stick to what's legal. . . . I'm *not* God. The currents and eddies of right and wrong, which you find such plain-sailing, I can't navigate, I'm no voyager. But in the thickets of the law, oh there I'm a forester. . . . What would you do? Cut a great road through the law to get after the Devil? . . . And when the last law was down, and the Devil turned round on you—where would you hide, Roper, the laws all being flat? . . . This country's planted thick with laws from coast to coast—Man's laws, not God's—and if you cut them down . . . d'you really think you could stand upright in the winds that would blow then? . . . Yes, I'd give the Devil benefit of law, for my own safety's sake.[2]

We agree with the Court of Appeals that in our constitutional system the commitment to the separation of powers is too fundamental for us to pre-empt congressional action by judicially decreeing what accords with "common sense and the public weal." Our Constitution vests such responsibilities in the political branches.

Affirmed.

Notes

1. *Marbury v Madison*, 1 Cranch (1803): 137, 177.
2. Robert Bolt, *A Man for All Seasons*, Act I, in *Three Plays* (London: Heinemann Educational, 1967), 147.

Antonin Scalia, *Lucas v. South Carolina Coastal Council*, U.S. Supreme Court (1992)

The Fifth Amendment to the United States Constitution, which prohibits the government's taking of property without due compensation, has come to the forefront of environmental law. The Lucas *case considers the effect an environmental regulation is required to have on private property before it becomes equivalent to a government taking of that property. This opinion represents an extremely conservative interpretation of the Fifth Amendment. The Sax essay recommended for further reading, above, offers a critique of this decision.*

In 1986, petitioner David H. Lucas paid $975,000 for two residential lots on the Isle of Palms in Charleston County, South Carolina, on which he intended to build single-family homes. In 1988, however, the South Carolina Legislature enacted the Beachfront Management Act, which had the direct effect of barring petitioner from erecting any permanent habitable structures on his two parcels. A state trial court found that this prohibition rendered Lucas's parcels "valueless." This case requires us to decide whether the Act's dramatic effect on the economic value of Lucas's lots accomplished a taking of private property under the Fifth and Fourteenth Amendments requiring the payment of "just compensation." . . .

In the late 1970s, Lucas and others began extensive residential development of the Isle of Palms, a barrier island situated eastward of the city of Charleston. Toward the close of the development cycle for one residential subdivision known as "Beachwood East," Lucas in 1986 purchased the two lots at issue in this litigation for his own account. . . . At the time Lucas acquired these parcels, he was not legally obliged to obtain a permit from the Council in advance of any development activity. His intention with respect to the lots was to do what the owners of the immediately adjacent parcels had already done: erect single-family residences. He commissioned architectural drawings for this purpose.

. . . As we have said on numerous occasions, the Fifth Amendment is violated when land-use regulation "does not substantially advance legitimate state interests or denies an owner economically viable use of his land."

. . . Regulations that leave the owner of land without economically beneficial or productive options for its use—typically, as here, by requiring land to be left substantially in its natural state—carry with them a heightened risk that private property is being pressed into some form of public service under the guise of mitigating serious public harm. As Justice Brennan explained: "From the government's point of view, the benefits flowing to the public from preservation of open space through regulation may be equally great as from creating a wildlife refuge through formal condemnation or increasing electricity production through a dam project that floods private property." The many statutes on the books, both state and federal, that provide for the use of eminent domain to impose servitudes on private scenic lands preventing developmental uses, or to acquire such lands altogether, suggest the practical equivalence in this setting of negative regulation and appropriation. . . .

We think, in short, that there are good reasons for our frequently expressed belief that when the owner of real property has been called upon to sacrifice all economically beneficial uses in the name of the common good, that is, to leave his property economically idle, he has suffered a taking. . . .

Where the State seeks to sustain regulation that deprives land of all economically beneficial use, we think it may resist compensation only if the logically antecedent inquiry into the nature of the owner's estate shows that the proscribed use interests were not part of his title to begin with. This accords, we think, with our "takings" jurisprudence, which has traditionally been guided by the understandings of our citizens regarding the content of, and the State's power over, the "bundle of rights" that they acquire when they obtain title to property.

. . . Any limitation so severe [as to deprive the owner of all economically beneficial use of the land] cannot be newly legislated or decreed (without compensation), but must inhere in the title itself, in the restrictions that background principles of the State's law of property and nuisance already place upon land ownership. A law or decree with such an effect must, in other words, do no more than duplicate the result that could have been achieved in the courts—by adjacent landowners (or other uniquely affected persons) under the State's law of private nuisance, or by the State under its complementary power to abate nuisances that affect the public generally, or otherwise. . . .

It seems unlikely that common-law principles would have prevented the erection of any habitable or productive improvements on petitioner's land; they rarely support prohibition of the "essential use" of land. . . . A "State . . . may not transform private property into public property without compensation." . . .

The judgment is reversed.

Robert D. Bullard, "Environmental Justice for All" in *Unequal Protection: Environmental Justice and Communities of Color* (1994)

Robert Bullard details the effects that living near toxic waste, garbage transfer stations, heavy industry, and other noxious facilities can have on communities. From Los Angeles to Louisiana to Chicago to New York, African Americans, Hispanics, and Native Americans live with air and water pollution and the diseases they bring at a significantly higher rate than the general population. The presence of these facilities undermines the cohesion of the communities that live with them, thus making it even more difficult for people lacking political power to begin to work together and fight back.

The hazardous waste problem continues to be one of the most "serious problems facing the industrial world."[1] Toxic time bombs are not randomly scattered across the urban landscape. In New Jersey (a state with one of the highest concentrations of uncontrolled toxic waste dumps), hazardous waste sites are often located in communities that have high percentages of poor, elderly, young, and minority residents.

Few national studies have been conducted on the sociodemographic characteristics of populations living around toxic waste sites. Although the federal EPA has been in business for more than two decades, it has yet to conduct a national study of the problems of toxic wastes in communities of color. In fact, the United Church of Christ Commission for Racial Justice, a church-based civil rights organization, conducted the first national study on this topic.

The Commission for Racial Justice's landmark study, *Toxic Wastes and Race in the United States*, found race to be the single most important factor (i.e., more important than income, home ownership rate, and property values) in the location of abandoned toxic waste sites.[2] The study also found that (1) three out of five African Americans live in communities with abandoned toxic waste sites; (2) sixty percent (fifteen million) of African Americans live in communities with one or more abandoned toxic waste sites; (3) three of the five largest commercial hazardous waste landfills are located in predominantly African American or Latino American communities and account for forty percent of the nation's total estimated landfill capacity; and (4) African Americans are heavily overrepresented in the populations of cities with the largest number of abandoned toxic waste sites.

In metropolitan Chicago, for example, more than 81.3 percent of Latino Americans and 76 percent of African Americans live in communities with abandoned toxic waste sites, compared with 59 percent of whites. Similarly, 81.3 percent of Latino Americans and 69.8 percent of African Americans in the Houston metropolitan area live in communities with abandoned toxic waste sites, compared with 57.1 percent of whites. Latino Americans in the Los Angeles metropolitan area are nearly twice as likely as their Anglo counterparts to live in a community with an abandoned toxic waste site.

The mounting waste problem is adding to the potential health threat to environmental high-impact areas. Incineration has become the leading technology for disposal of this waste. This technology is also becoming a major source of dioxin, as well as lead, mercury, and other heavy metals released into the environment. For example, millions of pounds of lead per year will be emitted from the nation's municipal solid waste incinerators in the next few years. All of this lead is being released despite what we know about its hazards to human health.

Hazardous waste incinerators are not randomly scattered across the landscape. A 1990 Greenpeace report, *Playing with Fire*, found that (1) the minority portion of the population in communities with existing incinerators is 89 percent higher than the national average; (2) communities where incinerators are proposed have minority populations 60 percent higher than the national average; (3) average income in communities with existing incinerators is 15 percent less than the national average; (4) property values in communities that are hosts to incinerators are 38 percent lower than the national average; and (5) average property values are 35 percent lower in communities where incinerators are proposed.[3]

Environmental scientists have not refined their research methodologies to assess the cumulative and synergistic effects of all of society's poisons on the human body. However, some health problems cannot wait for the tools to catch up with common sense. For example, the nation's lead contamination problem demands urgent attention. An environmental strategy is needed to address childhood lead poisoning. It is time for action.

THE POLITICS OF LEAD POISONING

Why has so little been done to prevent lead poisoning in the United States? Overwhelming scientific evidence exists on the ill effects of lead on the human body. However, very little has been done to rid the nation of lead poisoning—a preventable disease tagged the "number one environmental health threat to children" by the federal Agency for Toxic Substances and Disease Registry.[4]

Lead began to be phased out of gasoline in the 1970s. It is ironic that the "regulations were initially developed to protect the newly developed catalytic converter in automobiles, a pollution-control device that happens to be rendered inoperative by lead, rather than to safeguard human health."[5] In 1971, a child was not considered at risk for lead poisoning unless he or she had 400 micrograms of lead per liter of blood (or 40 micrograms per deciliter [μg/dl]). Since that time, the amount of lead that is considered safe has continually dropped. In 1991, the U.S. Public Health Service changed the official definition of an unsafe level to 10 μg/dl. Even at that level, a child's IQ can be slightly diminished and physical growth stunted. Lead poisoning is correlated with both income and race (Table 22.1).

TABLE 22.1

Estimated Percentages of City-Dwelling U.S. Children with Blood Lead Levels Greater than 15 µg/dl, by Race and Income (1988)

	Income		
Race	<$6,000	$6,000-$15,000	>$15,000
African American	68%	54%	38%
White	36%	23%	12%

Note: These statistics represent children between the ages of 6 months and 5 years who live in cities with population greater than 1 million.
Source: Agency for Toxic Substances and Disease Registry, *The Nature and Extent of Lead Poisoning in Children in the United States: A Report to Congress* (Atlanta: U.S. Department of Health and Human Services, 1998).

A coalition of environmental, social justice, and civil libertarian groups are now joining forces to address the lead problem. The Natural Resources Defense Council, the NAACP Legal Defense and Education Fund, the American Civil Liberties Union, and the Legal Aid Society of Alameda County, California, won an out-of-court settlement worth $15 million to $20 million for a blood lead-testing program. The lawsuit, *Matthews v Coye*, involved the failure of the state of California to conduct federally mandated testing for lead of some 557,000 poor children who receive Medicaid. This historic agreement will probably trigger similar actions in other states that have failed to live up to federally mandated screening requirements. [In the decade following 1994 significant progress in the abatement of lead poisoning was achieved in the United States.]

Despite the recent attempts by federal environmental and health agencies to reduce risks to all Americans, environmental inequities still persist. Some children, workers, and communities are disproportionately affected by unhealthy air, unsafe drinking water, dangerous chemicals, lead, pesticides, and toxic wastes.

If this nation is to achieve environmental justice, the environment in urban ghettos, barrios, reservations, and rural poverty pockets must be given the same protection as that provided to the suburbs. All communities—African American or white, rich or poor—deserve to be protected from the ravages of pollution.

The current emphasis on waste management and pollution control regulations encourages dependence on disposal technologies, which are themselves sources of toxic pollution. Pushing incinerators and risk technologies off on people under the guise of economic development is not a solution to this nation's waste problem. It is imperative that waste reduction programs mandated by federal, state, and local government be funded that set goals for recycling, composting, and using recycled materials.

An environmental justice framework needs to be incorporated into a national policy on facility siting. In addition to the standard technical requirements, environmental justice proposals will need to require implementation of some type of "fair share" plan that takes into account sociodemographic, economic, and cultural factors of affected communities. It is clear that current environmental regulations and "protectionist" devices (zoning, deed restrictions, and other land use controls) have not had the same impact on all segments of society.

The federal EPA needs to take the lead in ensuring that all Americans are protected. It is time for this nation to clean up the health-threatening lead contamination problem and prevent future generations from being poisoned. No segment of society should be allowed to become a dumping ground or be sacrificed because of economic vulnerability or racial discrimination.

In order for risk reduction strategies to be effective in environmental high-impact areas and for vulnerable populations, there need to be sweeping changes in key areas of the science model and environmental health research. At minimum, these changes must include a reevaluation of the attitudes, biases, and values of the scientists who conduct environmental health research and risk assessment and the officials who make policy decisions.

Acceptance of the public as an active and equal partner in research and environmental decision making is a first step toward building trust within affected communities. Government agencies and other responsible parties need to incorporate principles of environmental justice into their strategic planning of risk reduction.

We need a holistic methodology in documenting, remediating, and preventing environmental health problems. Prevention is the key. Environmental justice demands that lead poisoning—the number one environmental health problem affecting children—be given the attention and priority it deserves. It is the poorest among the nation's inhabitants who are being poisoned at an alarming rate. Many of these individuals and families have little or no access to regular health care.

The solution lies in leveling the playing field and protecting all Americans. Environmental decision makers have failed to address the "justice" questions of who gets help and who does not, who can afford help and who cannot, why some contaminated communities get studied while others are left off the research agenda, why some communities get cleaned up at a faster rate than others, why some cleanup methods are selected over others, and why industry poisons some communities and not others.

Finally, a national environmental justice action agenda is needed to begin addressing environmental inequities that result from procedural, geographic, and societal imbalances. Federal, state, and local legislation is needed to target resources for those areas where societal risk burdens are the greatest. States that are initiating fair share plans to address interstate waste conflicts need also to begin addressing intrastate environmental siting imbalances. It is time for environmental justice to become a national priority.

Notes

1. Samuel S. Epstein, Lester O. Brown, and Carl Pope, *Hazardous Waste in America* (San Francisco: Sierra Club Books, 1983), 33–39.
2. United Church of Christ Commission for Racial Justice, *Toxic Wastes and Race in the United States.*
3. Pat Costner and Joe Thornton, *Playing with Fire* (Washington, D.C.: Greenpeace, 1990).
4. Agency for Toxic Substances and Disease Registry, *The Nature and Extent of Lead Poisoning in Children in the United States: A Report to Congress* (Atlanta: U.S. Department of Health and Human Services, 1988).
5. Peter Reich, *The Hour of Lead* (Washington, D.C.: Environmental Defense Fund, 1992), 42.

Christopher H. Foreman, Jr., "Political and Policy Limitations of Environmental Justice" in *The Promise and Peril of Environmental Justice* (1998)

Christopher H. Foreman, Jr., acknowledges the problem set forth by Robert Bullard, above, that the environmental justice movement must overcome obstacles inherent to many grassroots movements. Moreover, in litigating environmental justice cases, there is the legal burden of proving that the industries and government agencies involved in creating the unequal distribution of environmental hazards actually had the intent to do so. Foreman advocates collective action and self-help over litigation for the future of the environmental justice movement.

Environmental justice suffers from serious *limitations.* Although it has emerged as a significant theme in environmental policy discourse, environmental justice does not dominate or define environmentalism. The movement is too weak, has too few resources, and has too strong a local orientation to be a significant separate presence on such national and international matters as global warming, acid rain, airborne particulates, or the future of the electric car. On issues like these, environmental justice at best only adds to the general clamor for emission and waste reductions, and citizen involvement, on all fronts. . . . Environmental justice cannot yet be described as a clear, durable, and primary goal for any national agency or significant interest group. It serves instead as an occasional, and maddeningly vague, political constraint, something to "take into account" (again, especially through mechanisms of community consultation) in the pursuit of other objectives. Even where environmental justice has found a few footholds in the federal bureaucracy, it has stimulated mainly institutional promises of access and a fair hearing. Such promises are important but it will be hard to redeem them to the satisfaction of local activists.

Unlike the major traditional environmental organizations, the movement has little institutionalized presence in the nation's capital outside the federal agencies that have been ordered to pay attention to it by President Clinton. This is partly a matter of activist preference; many regard a major Washington identity as incompatible with the antibureaucratic and fundamentally populist thrust of the movement. A large Washington organization would tend to draw funds and attention away from the grassroots, where the real battles are waged. Such an entity would likely come into tension with communities and local activists insistent on speaking for themselves and unwilling to see "their" issues submerged within, or dropped from, a national organizational agenda.

It is thus understandable that when former EPA attorney Deeohn Ferris created the Washington Office on Environmental Justice, it remained tiny (Ferris and a couple of assistants) and focused on helping local communities formulate and implement effective strategies and find their bearings in Washington. In this respect Ferris's organization resembles the more elaborate Citizens Clearinghouse for Hazardous Waste (CCHW) in nearby Virginia, created by Lois Gibbs to provide "information, assistance, and solidarity" for grassroots groups. Ecopopulism, whether black or white, generally does not really need regular Washington access to succeed at its crucial community-level advocacy function; local organizing has proved far more effective at blocking and curtailing targeted projects than recourse to Washington could ever be. As Gibbs and her CCHW assert: "the [federal] government won't stop the poisoning, but [local] organizing will."

But on the federal stage environmental justice finds itself a warily regarded guest. As noted earlier, neither congressional Democrats nor a reasonably hospitable president (much less Republicans) have been anxious to disturb the existing environmental statutory regime on behalf of the movement. Even sympathetic institutional actors generally see assuring management sensitivity and a voice for communities, not grand statutory change, as the real challenge.

Environmental justice also has limited recourse to the judiciary or to conventional regulatory criteria. As previously noted, current equal protection doctrine elevates the bar for proving discrimination too high for almost any community. The EPA's Office of Civil Rights, which began reviewing dozens of local complaints for possible Title VI violations, had a hard time identifying them. Four years after the executive order, the office had yet to find an environmental justice complaint that appeared viable under the Civil Rights Act.

"Racism" makes a superb rallying cry, but it is unlikely to move a court unless considerable evidence is offered, more than is likely to be available in most instances, even if the Supreme Court opens the way for environmental justice lawsuits under Title VI of the Civil Rights Act. This is not just because racists have been careful to cover their tracks but because the actual historical dynamics of siting, and of the proximity of particular communities to particular sites, are genuinely complex and ambiguous. Communities of color may spring up over time near worrisome sites, lured by the prospect of jobs or cheap land. Even where past racist intent might have been a factor in siting, current policy offers limited tools for rectifying present-day effects. Regulators may often have scant justification for shutting down, or racheting down, discharges at a facility that complies with its current permits, whatever deficiencies might have characterized the siting process at some earlier time. Wholesale bans on new siting are not a plausible answer; at least from a rationalizing point of view, each new proposed facility must be evaluated on its merits. And . . . those merits inevitably turn into a discussion of tradeoffs. Does the facility provide jobs and needed treatment or disposal capacity? Can the community be effectively compensated for hosting it? Can community concerns about possible negative impacts on health, property values, or overall quality of life be addressed early in the game through a relatively open dialogue that builds at least rudimentary trust? As stressed already, environmental justice really specifies no answer to the question of what is best for a given community beyond the conviction that rigorous democratic practice and accountability are essential to a just outcome.

Although environmental justice has succeeded politically by repeated reference to several studies, empirical analysis has also proved difficult terrain. The evidence for racial inequity in site distribution is mixed, at best. And there is little reliable empirical evidence that people of color generally suffer adverse health effects related to industrial pollution at greater rates than whites. Moreover, *anyone* living in a big city is likely breathing dirtier air than anyone living elsewhere. Reasonable public concern, grounded in multiple scientific studies, exists about lead intake sufficient to impair intellectual function and development, especially among black and low-income children. It is also clear that, as with occupational chemicals generally, farmworker exposure is a serious matter. Such workers are likely to encounter pesticides and other agricultural chemicals at levels that are orders of magnitude above those affecting the public at large. The difficulty of effectively according such issues the focus they deserve is one of the movement's most significant health-related weaknesses.

The social welfare aspirations that sustain much of the environmental justice agenda face perhaps the toughest sledding of all. Well-intended but limited and uncertain efforts are under way on a number of environment-related fronts (such as job creation, job training, post-secondary education, and workforce diversity). Some identifiable individuals will doubtless benefit from each of these efforts, but there is no established fund of experience to suggest that environmental programs can be bent very much or very effectively to the service of targeted social welfare goals. Society does not know how to do a number of these things reliably, especially among the kinds of populations of greatest concern to environmental justice advocates. Moreover, it is all too obvious that the political will for major new redistributive efforts, even those carried out under an environmental rubric, is limited. The great untouchable "third rail" of American politics is not social security reform, as is so often claimed, but rather residential segregation, which doubtless underlies at least some of what ends up being perceived or discussed as environmental injustice. In a useful development, some environmental justice activists have begun to talk about this. Unfortunately, they are also prone to a misleading conception of the status quo as a kind of "apartheid," an emotionally charged term that may only help mask the complexity of residential patterns and the forces underlying them.

The common vision of achieving integrated purposes simultaneously—enabling community residents to remediate local environments while gaining marketable skills for the longer term and paving the way for local jobs—is a fond one, but no one knows if it could be done with regular success. Very possibly it could not be, given the uncertainties, the uncontrollable variables, the multiplicity of actors, and the inadequate budgets that would be involved. The bipartisan appeal of "bricks and mortar" brownfields redevelopment may well have endowed it with political staying power. But the incentives and opportunities that initiate local redevelopment projects, and drive them to successful completion, will only rarely have much to do with helping environmental justice constituencies, especially the poor.

23 Economics

Economics (from the Greek for "management of the household"), like ecology ("science of the household"), is a study of exchanges and interdependencies, equilibria and disequilibria, growth and development. Over the years, exchanges between the two sister disciplines have informed their own growth and development. Malthus's forays into the realm of population ecology, for example, earned economics the nickname "the dismal science" in the nineteenth century, and early practitioners of ecology fancied that they were studying "the economy of nature." Even today, cultural critics such as David Korten urge that human economies should "mimic the behavior of healthy living organisms and ecosystems."

Modern economic thought is divided on the proper relationship between the two disciplines. Conventional market economists generally view the environment as an aggregation of resources whose preservation and distribution are governed by the same fundamental economic laws that apply to other resources such as labor and capital. In contrast, "ecological" economists view human economic activity as embedded within Earth's own vast and intricate system of cycles and exchanges, and thus subject to fundamental physical and biological laws and limitations. The application of market economic thought to environmental issues is usually called "environmental economics" to distinguish it from the more radical "ecological economics." Both schools of thought have valuable insights to offer, and both are represented in this chapter.

Environmental economists and others in the neoclassical tradition focus on the idealized marketplace, an aggregation of independent actors freely buying and selling, as the most efficient way of distributing society's scarce resources. In the ideal marketplace, prices reflect the real value and scarcity of resources, and thrift encourages conservation. When ideal conditions do not obtain in the real world (e.g., when prices are distorted by subsidies or externalities, or manipulated by monopolies), a "market failure" is said to have occurred. Economists may offer public policy remedies to correct the market failure, such as fees and taxes on environmentally destructive activities, the revocation of hidden subsidies that distort true costs, and government management of "public goods" like

Figure 23.1 *The Bulls and Bears in the Market* by William H. Beard, 1879. © Collection of The New-York Historical Society, 1971.104.

clean water and air that do not lend themselves to market distribution. Environmental economists may also prescribe market-based tools to help society reach environmental protection goals: for instance, regulators can efficiently cap and reduce pollution from a geographic region or an industry by issuing tradable emissions permits.

Assigning dollar values to environmental goods can be challenging. The value of a forest, for example, is not merely the market price of lumber. Recreational value, educational value, scientific value, the value of ecosystem services such as clean air and water purification, the value of species, ecosystems, and human cultural diversity sustained by the forest—all these contribute. When market prices fail to capture the whole value of a resource, decision-makers may need to employ questionnaires, scientific analyses, and economic models to estimate true costs and benefits.

While conventional economics presupposes that supply and demand will hold market prices in a dynamic equilibrium, it also embraces an important disequilibrium, namely the prospect of unbounded economic growth. Ecological economists challenge the possibility and desirability of unbounded economic growth, emphasizing the limits placed on human economic activity by the finitude of the world's natural resources and the fragility of its ecosystems—as well as the negative social impacts of overproduction and overconsumption.

The extent to which economic growth and the creation of financial wealth can be "decoupled" from increased throughput of resources, and thus overcome environmental limitations, is a matter of professional debate. But one area where environmental limitations are incontrovertible is energy supply. The laws of thermodynamics dictate that concentrated energy, once dissipated, can not be recovered in as useful a form. The limited supply of fossil fuels that power modern economies is thus a temporary bonus; in the long run, the scope of human economic activity will be constrained by the modest energy allowance afforded us by so-called "renewable" sources. (See Chapter 14, Energy.)

The debate between conventional and ecological economics has implications for the sustainable development community. Some mainstream economists hold that environmental protection is a luxury commodity, one that developing nations will be able to afford once poverty and overpopulation are relieved through economic growth. (Taking this viewpoint to an extreme, an infamous leaked 1992 World Bank memo suggested that "impeccable . . . economic logic" requires the migration of hazardous and polluting industries to poorer countries, where wages and health care costs are lower and the air is "inefficiently" underpolluted.) Ecological economists counter that no development scheme will succeed unless it ensures proper stewardship of the natural resources and ecosystems upon which the population and local economies depend. (See Chapter 5, Sustainable Development.)

Business leaders, meanwhile, are making practical forecasts and decisions, and they appear to be reaching a gradual consensus that "business as usual" will have to adapt to a new era of fossil fuel scarcity. Automakers have begun producing and marketing hybrid and hydrogen-powered cars in earnest, and oil companies like British Petroleum, which now markets itself as "Beyond Petroleum," are diversifying into renewable energy technologies. Niche markets for organic and environmentally friendly products are growing, and sustainability-minded entrepreneurs are finding creative ways to fill these needs, as well as to improve the material and energy efficiency of traditional manufacturing processes. Although in academia environmental and ecological economics have until now maintained their own distinct niches, we may find soon that the very market mechanisms touted by the one will help guide our economy onto the sustainable path envisioned by the other.

FURTHER READING

Brandt, Barbara. *Whole Life Economics: Revaluing Daily Life.* Philadelphia: New Society Publishers, 1995.

Daily, Gretchen, and Katherine Ellison. *The New Economy of Nature.* Washington, D.C.: Island Press, 2002.

Daly, Herman E., and Joshua Farley. *Ecological Economics: Principles and Applications.* Washington, D.C.: Island Press, 2004.

Jacobs, Jane. *The Nature of Economies.* New York: Modern Library, 2000.

Korten, David. *When Corporations Rule the World.* West Hartford, Conn.: Kumarian Press, 1995.

Myers, Norman, and Jennifer Kent. *Perverse Subsidies.* Washington, D.C.: Island Press, 2001.
Schumacher, E. F. *Small Is Beautiful: Economics as If People Mattered.* London: Blond & Briggs, 1973.
Stavins, Robert N. *Economics of the Environment: Selected Readings.* New York: W. W. Norton, 2000.

Herman Daly, from *Beyond Growth: The Economics of Sustainable Development* (1996)

Ecological economist Herman Daly is well-known for promoting the idea of a "steady-state" economy, one that evolves and adapts to changing human needs, but does not grow in size or scope beyond the capacity of the environment to sustain services and absorb wastes. In these two passages, he distinguishes between growth and development and describes the intellectual rift he observed between the worldviews of conventional "growth" economics and ecological economics during his time as senior economist of the environmental department of the World Bank.

Much confusion is generated by using the term "sustainable growth" as a synonym for sustainable *development.* Respect for the dictionary would lead us to reserve the word "growth" for quantitative increase in physical size by assimilation or accretion of materials. "Development" refers to qualitative change, realization of potentialities, transition to a fuller or better state. The two processes are distinct—sometimes linked, sometimes not. For example, a child grows and develops simultaneously; a snowball or a cancer grows without developing; the planet Earth develops without growing. Economies frequently grow and develop at the same time but can do either separately. But since the economy is a subsystem of a finite and non-growing ecosystem, then as growth leads it to incorporate an ever larger fraction of the total system into itself, its behavior must more and more approximate the behavior of the total system, which is development without growth. It is precisely the recognition that growth in scale ultimately becomes impossible—and already costs more than it is worth—that gives rise to the urgency of the concept of sustainable development. Sustainable development is development without growth in the scale of the economy beyond some point that is within biospheric carrying capacity. . . .

Sustainable development, development without growth, does not imply the end of economics—if anything, economics becomes even more important. But it is a subtle and complex economics of maintenance, qualitative improvement, sharing, frugality, and adaptation to natural limits. It is an economics of better, not bigger.

* * *

If development means anything concretely it means a process by which the South becomes like the North in terms of consumption levels and patterns. But current Northern levels and patterns are not generalizable to the whole world, assuming anything remotely resembling even our best existing technology, without exceeding ecological carrying capacity—that is, without consuming natural capital and thereby diminishing the capacity of the Earth to support life and wealth in the future. It is clear that we already consume natural capital and count it as current income in our national accounts. One need only try to imagine 1.2 billion Chinese with automobiles, refrigerators, washing machines, and so on, to get a picture of the ecological consequences of generalizing advanced Northern resource consumption levels across the globe. . . .

Certainly the World Bank would be the proper institution to recognize the ecological contradictions in the world's economic development plans, and to call attention to the need for the North to stop growth in resource throughput in order to both reserve for the people of the South the remaining ecological space needed for growth to satisfy their vital needs and set a generalizable and replicable example of sustainable development. The World Bank's best opportunity to date for doing this was through its 1992 World Development Report, entitled *Development and the Environment*. . . .

While the 1992 report made a number of contributions, especially in calling attention to the public health consequences of the environmental degradation of water and air, it nevertheless failed to address the biggest question. Environmental deterioration was held to be mainly a consequence of poverty, and the solution proposed was the same as the World Bank's solution to other economic problems, namely more growth. And this meant not only growth in the South, but also in the North, for how else could the South grow if it could not export to Northern markets and receive foreign investments from the North? And how could the North provide foreign investment and larger markets for the South if it in turn did not grow? While the World Bank's report acknowledged a few conflicts between growth and environment here and there, the world was seen to be full of "win-win" opportunities for both increasing growth as usual and improving the environment. The message was both a reaffirmation of the Bank's faith in economic growth and a denial of the existence of any fundamental ecological limits to that growth: problems reside mainly in the South, solutions are to be found mainly in the North. This formulation is politically convenient, at the very least, since the Bank is creditor to the South and debtor to the North. It is always easier to preach to your debtors than to your creditors.

The evolution of the manuscript of *Development and the Environment* is revealing. An early draft contained a diagram entitled "The Relationship Between the Economy and the Environment." It consisted of a square labeled "economy," with an arrow coming in labeled "inputs" and an arrow going out labeled "outputs"—nothing more. I suggested that the picture failed to show the environment, and that it would be good to have a large box containing the one depicted, to represent the environment. Then the relation between the environment and

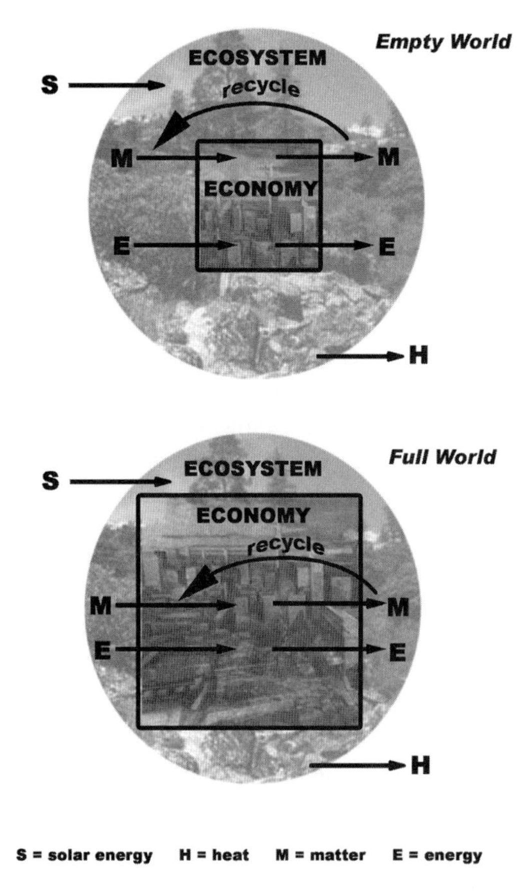

Figure 23.2 The economy as an ecological subsystem.

the economy would be clear—specifically, that the economy is a subsystem of the environment and depends upon the environment both as a source of raw material inputs and as a "sink" for waste outputs.

The next draft included the same diagram and text, but with an unlabeled box drawn around the economy like a picture frame. I commented that the larger box had to be labeled "environment" or else it was merely decorative, and that the text had to explain that the economy is related to the environment as a subsystem within the larger ecosystem and is dependent on it in the ways previously stated. The next draft omitted the diagram altogether.

By coincidence, a few months later the chief economist of the World Bank, Lawrence H. Summers, under whom the report was being written, happened to be on a conference panel at the Smithsonian Institution, discussing the book *Beyond the Limits* (Donella H. Meadows et al.), which Summers considered worthless. In that book there was a diagram showing the relation of the economy to the ecosystem, a diagram exactly like the one I had suggested (and like the one in Figure 23.2). During the question-and-answer time I asked the chief economist if, looking at that diagram, he felt that the question of the size of the economic subsystem relative to the total ecosystem was an important one, and whether he thought economists should be asking the question, What is the optimal scale of the macro economy relative to the environment? His reply was immediate and definite: "That's not the right way to look at it."

Reflecting on these two experiences has reinforced my belief that the main issue in the sustainable development controversy truly does revolve around what economist Joseph Schumpeter called "preanalytic vision." My preanalytic vision of the economy as subsystem leads immediately to the questions, How big is the subsystem relative to the total system? How big can it be without disrupting the functioning of the total system? How big should it be? What is its optimal scale beyond which further growth would be antieconomic, would cost more than it's worth? The World Bank's chief economist had no intention of being sucked into addressing these subversive questions, so he dismissed the viewpoint that gave rise to them.

Summers's dismissal was rather peremptory, but so, in a way, was my response to the diagram showing the economy receiving inputs from nowhere and exporting wastes to nowhere. That is not the right way to look at it, I felt, and any questions arising from that incomplete picture—say, how to make the economy grow as fast as possible by speeding up the flow of energy and materials through it—were not the right questions. Unless one has the preanalytic vision of the economy as subsystem, the whole idea of sustainable development—of a subsystem being sustained by a larger system whose limits and capacities it must respect—makes no sense whatsoever. On the other hand, a preanalytic vision of the economy as a box floating in infinite space allows people to speak of "sustainable *growth*"—a clear oxymoron to those who see the economy as a subsystem. The difference between these two visions could not be more fundamental, more elementary, or more irreconcilable.

It is interesting that such a huge issue should be at stake in a simple picture. Once you draw the boundary of the environment around the economy, you have said that the economy cannot expand forever. You have said that John Stuart Mill was right, that populations of human bodies and accumulations of capital goods cannot grow forever, that at some point quantitative growth must give way to qualitative development as the path of progress. . . .

When we draw that containing boundary of the environment around the economy we move from "empty-world" economics to "full-world" economics—from a world where inputs to and outputs from the economy are unconstrained, to a world in which they are increasingly constrained by the depletion and pollution of a finite environment. Economic logic stays the same—economize on the limiting factor. But the perceived pattern of scarcity changes radically—the identity of the limiting factor shifts from man-made capital to our remaining natural capital, from fishing boats to the populations of fish remaining in the sea—therefore policies must change radically. That is why there is such resistance to a simple picture. The fact that the picture is both so simple and so obviously realistic explains why it cannot be contemplated by the growth economists, why they must continue to insist, "That's not the right way to look at it!"

Cutler J. Cleveland, Robert Costanza, Charles A. S. Hall, and Robert Kaufmann, from "Energy and the U.S. Economy: A Biophysical Perspective" (1984)

Disruptions of the global oil supply in the 1970s sparked a debate over the economic effects of energy scarcity—a topic of greater concern than ever at the beginning of the twenty-first century, as unrest rocks the oil-extracting nations of the Middle East, and the inevitable peak and permanent decline of global oil production draw nearer. In the analysis presented here, the authors use historical macroeconomic data from the United States to show that energy is far more than "just another factor" of economic production.

ENERGY AND ECONOMIC PRODUCTION

The economic process is frequently depicted in basic economic texts as a closed system in which the flow of output is "circular, self-feeding, and self-renewing."[1] This model is seriously incomplete. In reality, the human economy is an open system embedded in a global environment that depends on a continuous throughput of solar energy. The global system produces the environmental services, foodstuffs, fossil and atomic fuels derived from solar and radiation energies, and various other resources that are essential inputs to the human economy. The human economy uses fossil and other fuels to support and empower labor and to produce capital. Fuel, capital, and labor are then combined to upgrade natural resources to useful goods and services. Economic production can therefore be viewed as the process of upgrading matter into highly ordered (thermodynamically improbable) structures, both physical structures and information. Where one speaks of "adding value" at successive stages of production, one may also speak of "adding order" to matter through the use of free energy.

Fuel quality as well as quantity limits economic production because fuels differ in the amount of economic work they can do per unit heat equivalent (kilocalorie). Petroleum, for example, can perform a more versatile array of tasks and do many of them more efficiently than coal. Per kilocalorie, petroleum is estimated to be 1.3 to 2.45 times as valuable as coal. Similarly, electricity can be converted to mechanical and heat energy at the point of application and can be controlled precisely, reducing the heat equivalents required to perform many tasks. . . .

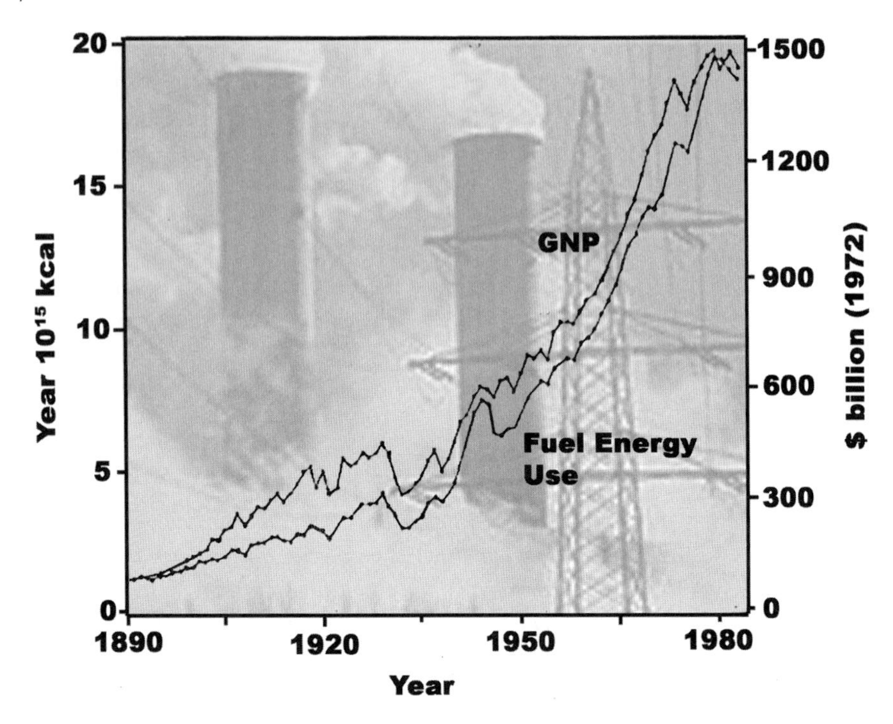

Figure 23.3 Fuel use and real GNP per year.

Another important quality of fuels is the amount of energy required to locate, extract, and refine them to a socially useful state. This aspect of fuel quality is measured by a fuel's energy return on investment (EROI), which is the ratio of gross fuel extracted to economic energy required directly and indirectly to deliver the fuel to society in a useful form. As the EROI for fuel declines, the energy opportunity costs of securing additional amounts increase, and increasing amounts of already extracted energy must be diverted from the production of nonenergy goods to extract a given quantity of new fuel. Net energy is a more relevant measure of fuel supply than gross energy because it represents the energy available to produce final-demand goods and services. At an absolute minimum, the aggregate EROI for fuels must be greater than 1 for an economic system to function, and probably much greater for it to grow. *Ceteris paribus* [other things being equal], economies with access to higher quality natural resources, particularly fuels with higher EROI, can do more economic work than those with lower EROI fuel resources. . . .

Fuel Use and Economic Output

Fuel use and economic output in the United States have been highly correlated for at least the past ninety years. This relation is shown in Figure 23.3. . . .

While a causal relation from fuel use to GNP [gross national product] or vice versa cannot be verified, a strong contemporaneous link between the two variables is supported. . . .

LABOR PRODUCTIVITY AND TECHNICAL CHANGE

In many economic models, technological advance is presented as an exogenous driving force powered by advances in human knowledge that increase labor and capital productivity. Denison states that advances in "human knowledge of how to produce things at low cost" are the most important causes of the increase in per capita national income observed between 1948 and 1973.[2] In this and other analyses, technological change is not measured directly, but rather is assigned the residual of increases in per capita income after all "tangible" factors have been accounted for. Because energy and natural resources are not considered tangible factors by most analysts, a large residual remains. Griliches and Jorgenson stated that relabeling changes in factor productivity as "technical progress" or "advance in knowledge leaves the problem of explaining growth in total output unresolved."[3]

From an energy perspective, productivity gains are facilitated by technical advances that enable laborers to empower their efforts with greater quantities of high-quality fuel embodied in and used by capital structures. As Cottrell observed, "productivity increases with the per capita increase in available energy."[4] Various empirical analyses support this view, and cross-sectional and temporal changes in labor productivity are correlated with the quantity of fuel used to empower a worker's efforts. Boretsky noted that higher labor productivity rates in the United States than in Western Europe nations were associated with the substantially greater quantities of fuel used per employee in the United States.[5] We found that in the U.S. manufacturing sector, output per worker-hour is closely related to the quantity of fuel used per worker-hour (Figure 23.4). A similar relation exists in the U.S. agricultural industry. . . .

NATURAL RESOURCE QUALITY FROM AN ENERGY PERSPECTIVE

The issue of natural resource scarcity has received considerable attention in recent years. Many suggest that the negative economic effects of depleting high-quality mineral deposits can be mitigated indefinitely through technical innovation and/or the use of more energy and capital structures to mine vast quantities of low-quality ore.[6] Evidence for this hypothesis is that capital and labor inputs per unit output in the extractive sectors have either declined or remained stable throughout most of this century, a trend attributed to technical advance in those industries.

When analyzed from a physical perspective, the trend in the scarcity of some

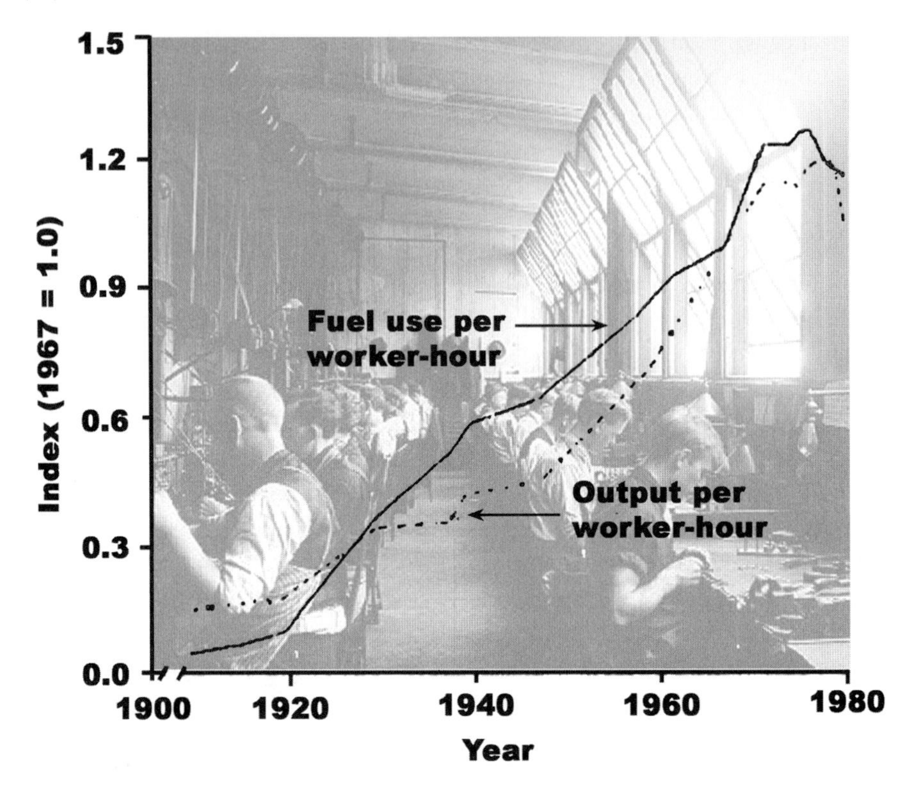

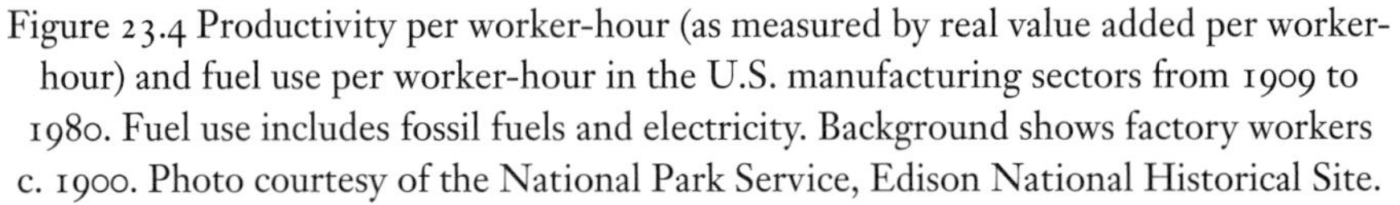

Figure 23.4 Productivity per worker-hour (as measured by real value added per worker-hour) and fuel use per worker-hour in the U.S. manufacturing sectors from 1909 to 1980. Fuel use includes fossil fuels and electricity. Background shows factory workers c. 1900. Photo courtesy of the National Park Service, Edison National Historical Site.

important natural resources is less reassuring. Technical improvements in the extractive sectors have made available previously uneconomic deposits only at the expense of more energy-intensive forms of capital and labor inputs. Physical output per kilocalorie of direct fuel input in the U.S. metal mining industries has declined sixty percent since 1939 (Figure 23.5, top), although a few exceptions to the trend are known.[7] The energy cost per ton of metal at the mine mouth for industrially important metals such as copper, aluminum, and iron has risen sharply as their average grade declined. For all U.S. mining industries (including fossil fuels), output per unit input of direct fuel has declined thirty percent since 1939. . . .

U.S. oil discoveries peaked in about 1930 and oil production in 1970. For natural gas these dates were 1950 and 1973, respectively. As we have increasingly exhausted the possibilities of finding new large petroleum deposits, the rate at which we find new oil per unit of drilling effort in the lower 48 states has de-

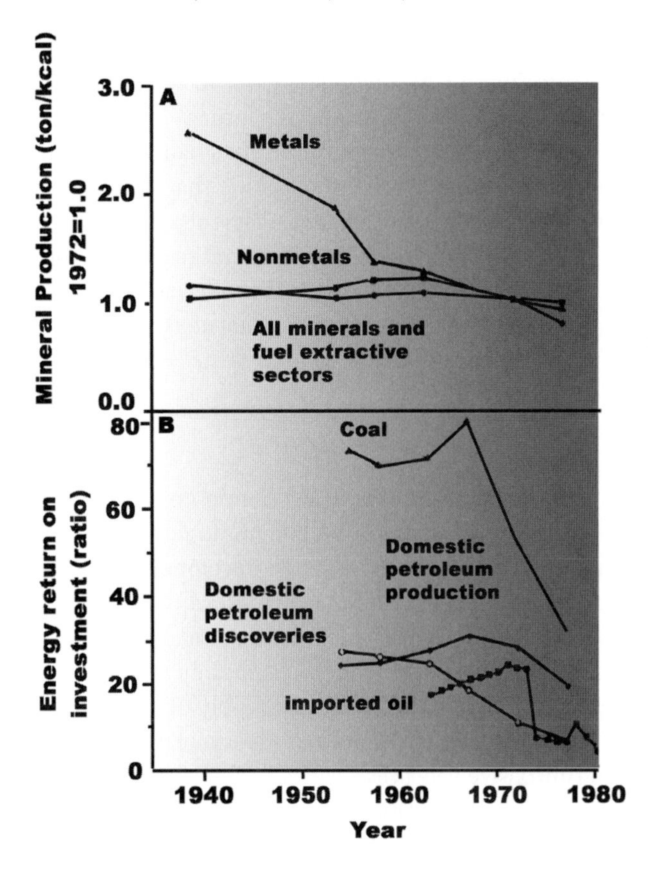

Figure 23.5 (top) Ratio of output per unit fuel (tons per kilocalorie) used by extractive sectors of the U.S. economy; (bottom) Net energy return (EROI) over time for U.S. fossil fuel sources.

clined precipitously. The large increase in drilling effort since 1973 has not reversed this decline. As a result, the running average EROI for oil and gas at the wellhead has declined precipitously (Figure 23.5, bottom). In Louisiana, a region that has accounted for seventeen percent of all domestic oil and gas discovered and produced to date, the EROI for natural gas extraction declined from 100:1 in 1970 to 12:1 in 1981. There has been a similar decline in the ratio of the energy in the petroleum we obtain from foreign sources compared to the energy required to make the goods and services we exchange for that petroleum (Figure 23.5, bottom). . . . The bituminous coal industry shows a similar but less dramatic trend over the past fifteen years. The EROI for coal at the mine mouth has decreased from about 80:1 in the mid-1960s to about 30:1 in 1977 (Figure 23.5, bottom).

Declining resource quality and higher fuel prices impede economic growth by diverting increasing amounts of capital and labor to the extractive and resource processing sectors. Throughout most of this century, the real dollar value of the mining sector share of real GNP was relatively small and constant, averaging three to four percent. This led some to conclude that natural resources were a small and unimportant factor of production. By 1982, however, more than ten percent of real GNP was needed to extract mineral resources from the environment. Most of this increase was for fossil fuel purchases, which in 1981 were 4.5 times greater in inflation-corrected dollars than in 1972, despite the fact that total fossil fuel use was about the same in both years. . . .

CONCLUSIONS

If we are to sustain current levels of economic growth and productivity as minimum long-run goals, alternative fuel technologies with EROI ratios comparable to that of petroleum today must be developed, or there must be unprecedented improvements in the efficiency with which we use fuel to produce economic output. Many discount the decreasing availability of high-quality fossil fuel deposits, stating that such depletion is merely a signal of our impending transition to a society based on a "boundless supply of energy at reasonably low cost"[8] such as breeder or fusion reactors or direct solar power. But past experience with capital-intensive ventures such as fission and synfuels suggests that it would be unwise to assume a priori that fusion or any other proposed fuel source will necessarily have a large EROI. Although we should research aggressively all potential fuel technologies, particularly in regard to their potential EROI, we should also plan for the contingency that new high-EROI sources might not be found.

. . . Our ability to cope with any economic contingencies will depend on the ability of our economic models to account for the biophysical constraints on human economic activity, and on the ability of our citizenry to accept and adapt to the realities of physical constraints imposed on our economic possibilities.

Notes

1. R. L. Heilbroner and L. C. Thurow, *The Economic Problem* (Englewood Cliffs, N.J.: Prentice-Hall, 1981).
2. E. F. Denison, *Survey of Current Business* 59 (1979): 1.
3. A. Griliches and D. W. Jorgenson, *The Review of Economic Studies* 34 (1967): 244.
4. F. Cottrell, *Energy and Society* (New York: McGraw-Hill, 1955).
5. M. Boretsky, *American Scientist* 63 (January 1975): 70.
6. J. L. Simon, *The Ultimate Resource* (Princeton: Princeton University Press, 1981).
7. P. J. Kakela, *Science* 202 (1978): 1151.
8. J. E. Stiglitz, in *Scarcity and Growth Reconsidered*, V. K. Smith, ed. (Baltimore: Johns Hopkins University Press, 1979).

Theodore Panayotou, from *Green Markets: The Economics of Sustainable Development* (1993)

Here, environmental economist Theodore Panayotou outlines various types of market failures that can affect the environment, and the principles that should guide policy makers when they respond.

The question is not how to prevent or eliminate environmental degradation altogether but how to minimize it or at least to keep it to a level consistent with society's objectives. When environmental degradation is seen in the context of the society's development objectives, not all deforestation, soil erosion, or water pollution is worth preventing. Some deforestation is necessary and beneficial when the forest land is put to a superior use, which may be agricultural, industrial, or residential. As long as all costs, including those arising from diminished quantity and quality and lost diversity of forests, have been accounted for; as long as both the productivity and the sustainability of the alternative uses have been considered with a due margin of error; and as long as any side effects of the forest conversion have been paid for by those who generated them, deforestation should be acceptable.

The problem is that decision makers usually consider the short-term benefits of forest conversion but not the long-term costs. As a result, too much conversion takes place in areas where the present value of costs outweighs any short-term benefits. Even worse, forests are converted to wastelands with little current benefit and enormous current and future costs.

. . . Physical manifestations of environmental degradation, such as rates of deforestation and soil erosion and levels of water pollution and urban density, tend to overstate the problem, because they seem to suggest that all degradation is preventable or worth abating. Because they are based on observed symptoms rather than underlying causes, they tend to be devoid of analytical insight about how to deal with the problem, other than banning the activities that appear to be responsible. For example, if logging leads to deforestation, it is common sense that banning logging will solve the problem. As Thailand is gradually discovering, however, a logging ban does not stop logging (let alone deforestation), any more than Prohibition in the United States several decades ago stopped drinking. . . .

Unlike physical manifestations and symptoms, which are devoid of analytical insight, the economic manifestations of environmental degradation raise analytical questions about cause and effect. Why are increasingly scarce resources being inefficiently used and wasted instead of economized and conserved? Why are valuable resources being put to inferior uses when superior uses exist? Why are renewable resources being mined rather than managed for a perpetual stream of benefits when the latter would generate a higher net present value? . . .

The answers to these questions are found in the disassociation of scarcity and price, benefits and costs, rights and responsibilities, actions and consequences. This disassociation exists because of a combination of market and policy failures. The prevailing configuration of markets and policies leaves many resources outside the domain of markets, unowned, unpriced, and unaccounted for. More often than not, it subsidizes their excessive use and destruction despite their growing scarcity and rising social cost. The result is an incentive structure that induces people to maximize their profits not by being efficient and innovative but by appropriating other people's resources and shifting their own costs onto others. . . .

MARKET FAILURES . . .

Insecurity of Resource Ownership

A fundamental condition for the efficient operation of markets is the existence of well-defined, exclusive, secure, transferable, and enforceable property rights over all resources, goods, and services. Property rights are a precondition for efficient use, trade, investment, conservation, and management of resources. No one would economize on, pay for, invest in, or conserve a resource without an assurance that he has secure and exclusive rights over it, that he can recover his costs through use, lease, or sale, and that such rights can and will be enforced. . . .

For historical and sociocultural reasons, property rights over many natural resources in developing countries are ill-defined, insecure, and unenforceable, and in a number of cases totally absent. Insecurely held resources can include private agricultural land, public forest land and forest resources, irrigation systems and water resources, coastal zone and fishery resources, and environmental resources. Resources over which property rights do not exist and to which everyone has free access are known as open-access or common property resources or, in layman's terms, "no-man's-land." *Common* property must be distinguished from *communal* property, which is well-defined and enforceable.

Unpriced Resources . . .

[An] obvious case of an unpriced resource is irrigation water. Here, the state has made a deliberate decision to provide farmers with irrigation water free of charge or at a nominal fee. In this case, not only is the water, a scarce natural resource with a positive opportunity cost, left unpriced (or priced at zero), but so is the scarce capital invested in the irrigation systems. The consequences are many and far reaching. Water is inefficiently and wastefully used without any attempt to conserve it, even when its scarcity is obvious to the user. The state is unable to recover capital, operation, and maintenance costs, with the result that watersheds remain unprotected and the irrigation system is poorly maintained. Serious environmental problems such as sedimentation, soil salinization, and waterlogging result from watershed degradation and from overirrigation, while other potentially irrigable areas receive insufficient quantities of water to grow dry-season crops. In the final analysis, better-off farmers near the irrigation canals are indirectly subsidized by worse-off farmers who pay taxes but have little or no access to irrigation water. . . .

Pricing is at the heart of natural resource policy and management. Almost all resource problems can be traced to discrepancies between private and social valuation of resource commodities and resource stocks. In the case of irrigation water, the private cost of both the *commodity* water and the *resource* water is constant at zero, while the social cost of both is positive and rising. Similarly, the cost to the private sector of using the environment (water, land, and air) for waste disposal is zero, while the cost to the society is positive and rising. Rapid deforestation and slow reforestation, even in securely owned forest land, is partly the consequence of the failure of the market to price forest products to capture the effects on watersheds, wildlife, and other nonmarketed services of the forest. . . .

Spillover Effects or Externalities

A major factor that drives a wedge between private and social valuation of resources and leads to inefficient pricing is the presence of external costs or spillover effects known as externalities. An externality is an effect of one firm's or individual's actions on other firms or individuals who are not parties in those actions. Externalities may be positive or negative. A positive externality, for example, is the benefit that upstream forest owners provide to downstream farmers in the form of a steady water supply made possible by a forested watershed. It is to the society's (and the farmer's) benefit that more of such positive externalities are provided, but since the forest owners receive no payment for their watershed service, they have no incentive to provide more of this service by logging less and planting more. The result is that more logging and less planting than is socially optimal takes place.

From another perspective, logging has negative externalities for downstream activities such as farming, irrigation, transport, and industry in the form of flooding, sedimentation, and irregular water supply. These are real costs to downstream activities and to the society as a whole, but not to upstream loggers or shifting cultivators, who have no cause or incentive to consider them because they do not affect the profitability of logging or shifting cultivation. In fact, taking such costs into account voluntarily amounts to a conscious decision to lower one's profit and price oneself out of the market. Unless every logger and every shifting cultivator takes such external costs into account, those who do are certain to lose to competitors who do not. This is precisely why government intervention is necessary to establish and enforce similar standards and incentives or disincentives for all competitors. . . .

Transaction Costs

Markets emerge to make possible beneficial exchanges or trade between parties with different resource endowments and different preferences. Establishing and operating markets, however, is not costless. Transaction costs—the costs of information, coordination, bargaining, and enforcement of contracts —are involved. Usually such costs are trivial compared with the benefits from trade that such markets make possible. Markets fail to emerge, however, if there are very high setup costs, if the costs per unit transacted exceed the difference between the supply and demand price, or if there are only a small number of buyers and sellers.

. . . If . . . transaction costs are high relative to the benefits of secure and exclusive ownership, property rights and the related markets will fail to emerge. For example, the costs of parceling out the sea to individual fishermen and enforcing property rights over a mobile resource are prohibitively high. Analogous is the case of externalities. There are costs related to identifying the afflicted and generating parties and to negotiating a mutually agreeable solution. The more parties involved, the less likely that a bargaining solution will be arrived at voluntarily because the transaction cost tends to exceed the benefits from internalizing the externality. The government, however, either through its collective or coercive power, may be able to internalize externalities at a lower transaction cost than the free market. . . .

Public Goods . . .

A public good is characterized by jointness in supply, in that to produce the good for one consumer it is necessary to produce it for all consumers. . . . Because nobody can or should be excluded from the benefits of a public good, consumers will not freely pay for it; hence, no firm would be able to cover its production cost through the market. The free market will therefore fail to supply a public good.

. . . This situation provides a rationale for many government activities aimed at providing public goods. . . .

The environment involves many public goods, ranging from environmental quality and watershed protection to ecological balance and biological diversity. . . .

Uncompetitive Markets

Even when markets do exist and are active, market failures may occur in the form of insufficient competition. For markets to be efficient there should be a large number of buyers and sellers of a more or less homogeneous commodity, or at least a lack of barriers to entry and a large number of potential entrants as insurance against monopolistic practices by existing firms. In reality, however, economies are ridden with monopolistic elements.

. . . A major source of monopolistic tendencies, affecting some resource-related sectors such as water and energy supply, is their decreasing industry cost feature. Because of the indivisibility of the necessary investment, the average cost of the service falls continuously as more and more customers are served until the whole market is dominated by a single firm (known as a natural monopoly). To prevent [exploitative] monopolistic practices, a government monopoly may be established, as is often the case with utilities and the postal service. . . .

A common monopolistic practice is to withhold supplies in order to raise prices. . . . [Thus] monopoly is not altogether bad for conservation. . . . This is not to imply that a monopoly is a solution to resource depletion: replacing one market failure with another does not usually improve welfare. . . .

Myopic Planning Horizons and High Discount Rates

Natural resource conservation and sustainable development ultimately involve sacrificing present consumption for the promise of future benefits. Because people tend to prefer immediate over future benefits, such an exchange appears unattractive unless one dollar of sacrifice today yields more than one dollar of benefits tomorrow. Future benefits are therefore discounted, and the more heavily they are discounted the less attractive they are. . . .

Environmental and market uncertainties coupled with a short and uncertain life span lead people to adopt myopic time horizons and discount rates, which result in shortsighted decisions in pursuit of survival or quick profits at the expense of long-term sustainable benefits. At subsistence levels of living, when people's very survival is at stake, a hand-to-mouth economy prevails in which the future is infinitely discounted. The results of such myopia are overexploitation of natural resources and underinvestment in their conservation and regeneration, which ultimately lead to their depletion. . . .

Uncertainty . . .

While uncertainty affects all sectors of the economy, natural resource sectors are more seriously affected for a variety of reasons. First, there are more uncertainties about ownership of and access to natural resources. Second, there are more potential spillovers from other activities. Third, natural resource investments such as tree planting tend to have much longer gestation periods than investments in agriculture or industry, and the longer the gestation period the more the uncertainties and risks involved. Fourth, natural resource commodity prices are subject to more violent fluctuations than other commodities and thus they are difficult to forecast. Last, most resource commodities are under the constant threat of being replaced by cheaper substitutes developed by continuous but unpredictable technological change.

Uncertainty about the future should make people more conservative in natural resource exploitation. Therefore it should work in favor of conservation of at least those resources that are least likely to be replaced by technology, such as biological diversity. After all, one reason people save is to provide themselves with a cushion against future uncertainty. Insecurity of tenure and pervasive externalities, however, create uncertainty about the benefits of conservation compared with the benefits of current exploitation. For the individual, it makes good economic sense to cut down the forest and mine the land to generate income, which he can then consume or invest in more secure assets. From society's point of view, it makes more sense to preserve the long-term productivity of the resource base both as a source of income in perpetuity and as insurance against uncertainty. . . .

Irreversibility

Market decisions about the future (such as the choice between consumption and investment) are made with the best available, yet incomplete, information about future developments, on the assumption that such decisions can be reversed if they prove unwise in the light of new information. This assumption of reversibility does not hold in many decisions involving natural resources. Consider the choice between preserving a tropical rain forest with some unique features and developing the site for logging and mining concessions. If the social benefits from development exceed the social benefits from conservation even marginally, the decision maker should choose logging and mining, except for the fact that conservation is reversible, while logging and mining are not. Choosing logging and mining forecloses his options; if he or future generations were to have a change of mind there would be no way to reproduce the uniqueness and authenticity of the original tropical forests and any species that became extinct. In contrast, choosing conservation preserves his option to reverse his decision. Clearly, there is a social value or shadow price for the preservation of options, though it is difficult to estimate. . . .

POLICY FAILURES . . .

The tendency of free markets to fail in the allocation and efficient use of natural resources and the environment opens an opportunity and provides a rationale for government intervention. Market failure by itself, however, is a necessary but not a sufficient condition for intervention. To be truly worthwhile, a government intervention must meet two other conditions. First, the intervention must outperform the market or improve its function. Second, the benefits from such intervention must exceed the costs of planning, implementation, and enforcement, as well as any indirect and unintended costs of distortions introduced to other sectors of the economy by such intervention.

Ideally, government intervention aims at correcting or at least mitigating market failures through taxation, regulation, private incentives, public projects, macroeconomic management, and institutional reform. For example, if the market fails to allocate land to its best possible use because of insecurity of land ownership, government intervention ought to be the issuance of secure land titles through cadastral [i.e., public] surveys and land registration, provided the ensuing benefits exceed the costs. If, on the other hand, the market fails to allocate land to its best possible use because of severe flooding due to upstream deforestation, the government ought to explore the costs and benefits of taxation on upstream logging or downstream agriculture and the use of the proceeds to subsidize upstream reforestation. If economic analysis that considers all costs and benefits involved leads to the conclusion that such an intervention can make both upstream loggers (or shifting cultivators) and downstream farmers better off, and no one else worse off (including the government treasury), it would be a policy failure not to act. Such an intervention is not a distortion, but a mitigation or correction of a distortion introduced by a failing market.

In practice, however, government policies tend to introduce additional distortions in the market for natural resources rather than correct existing ones. The reasons are many and varied. First, correction of market failure is rarely the sole or even the primary objective of government intervention. Other objectives such as national security, social equity, macroeconomic management, and political expediency may dominate. Second, government intervention often has unintended consequences and unforeseen or underestimated side effects. Third, policies such as subsidies and protection against imports or competition often outlive their usefulness because they become capitalized into people's expectations and property values, creating vested interests that make their removal politically difficult. Fourth, policy interventions tend to accumulate and interact with each other in subtle but profound ways to distort private incentives away from socially beneficial activities. Finally, policies that are seemingly unrelated to natural resources and the environment may have more pronounced effects on the environment than environmental and resource policies. For example, capital subsidies, tax and tariff exemptions for equipment, and minimum wage laws that displace labor lead to increased pressures on forest, marginal lands, coastal areas, and urban slums. . . .

Thus, environmental degradation results not only from overreliance on a free market that fails to function efficiently (market failure), but also from government policies that intentionally or unwittingly distort incentives in favor of overexploitation and against conservation of valuable and scarce resources (policy failure).

Policy failures may be classified into four basic types. One type involves distortions of otherwise well-functioning markets through taxes, subsidies, quotas, regulations, inefficient state enterprises, and public projects with a low economic return and a high environmental impact. This is a case of fixing what is not broken.

A second type is the failure to consider and internalize any significant environmental side effects of otherwise warranted policy interventions. . . . For example, in its drive for rice self-sufficiency, Indonesia provided generous subsidies for a variety of pesticides. These subsidies led to overuse, which, in turn, decimated the predators of the brown planthopper, an insect that threatened the country's self-sufficiency in rice. In a dramatic move, the government turned what was threatening to be a policy failure into a policy success by abolishing the subsidy and promoting the lower-cost strategy of IPM [integrated pest management]. . . .

A third type of policy failure is government intervention that aims to correct or mitigate a market failure but ends up generating a worse outcome than a free and failing market would have produced. . . . For example, if the free market fails to contain deforestation because the forests are open-access resources and the negative externalities of deforestation are not internalized (that is, paid for by the parties responsible), a logging ban is unlikely to be effective. First, higher prices are likely to stimulate illegal logging, and second, concessionaires may log illegally to recover sunk costs, or they may abandon their concessions to encroachers and slash-and-burn cultivators, as Thailand [discovered] following the January 1989 logging ban in response to the catastrophic landslides of November 1988.

The last type of policy failure is a lack of intervention in failing markets when such intervention is clearly needed to improve the functioning of the market and could be made at costs fully justified by the expected benefits. For example, it would have been a policy failure for the government of Thailand not to issue secure land titles to its farmers, when it was established that the cost of titling was only a small percentage (less than ten percent) of the potential benefits. By intervening in the land market to establish secure property rights, a precondition for well-functioning markets, the government has turned a market failure into a policy success. In contrast, the issuance of twenty-five-year usufruct or stewardship rights to squatters on public lands in Thailand and the Philippines is a half-measure that does not go to the root of the problem. It is likely to stimulate continued encroachment without significantly improving farmers' security of ownership, access to credit, and incentives to invest. Such half-measures risk turning a market failure into a policy failure of possibly greater dimensions.

. . . Policy failures are not the exclusive domain of governments. Development assistance agencies, through their project and program lending and policy dialogue, may introduce or exacerbate a policy failure. For example, liberalization as part of a structural or sectoral adjustment loan, in the absence of secure property rights and other legal foundations of markets, may simply transform a policy failure into a market failure, an outcome not uncommon with African liberalization programs.

Thomas H. Tietenberg, from "Using Economic Incentives to Maintain Our Environment" (1990)

In the 1970s, the United States Environmental Protection Agency began experimenting with market-based incentives as a means of bringing polluters into compliance with Clean Air Act standards. The use of such incentives was formalized and expanded in the Clean Air Act Amendments of 1990. In an article from the time of that crucial juncture, Thomas Tietenberg offers insight into the successes of the program, and ways it could be emulated on a global scale.

Environmental regulators and lobbying groups with a special interest in environmental protection in the United States have traditionally looked upon the market system as a powerful and potentially dangerous adversary. . . .

[However, in recent years] environmental groups and regulators have come to realize that the power of the market can be harnessed by economic incentive policies for the achievement of environmental goals.

. . . In the mid-1970s, several geographic U.S. regions were in violation of air quality standards designed to protect human health.

At that point, the law provided that new industries would not be allowed to move into these areas if they added any more of the pollutant responsible for the standard being violated. Since even those potential entrants adopting the most stringent control technologies would typically add some of the pollutant, this was a serious political blow to mayors eager to expand their employment and tax base. How could they allow economic growth while assuring that air quality would steadily improve to the level dictated by the ambient standard?

Regulators adopted the economic incentive approach known as the "offset policy." Under this policy, firms already established in these polluted areas that chose to voluntarily control their emissions more than required under the prevailing regulations would be allowed to have those excess emission reductions certified as "emission reduction credits." Once certified, their operating permits would be tightened to assure that the reductions were permanent. These emission reduction credits could then be sold to new firms seeking to move into the city providing that the acquiring firm bought 1.2 emission reduction credits for each 1.0 units of emissions added by the new plant. Air quality improved every time a new firm moved into the area.

Though definitive data are not available, several estimates suggest that some 2,000 to 2,500 offset transactions have taken place. With this policy, the confrontation between economic growth and environmental protection was diffused. New firms were not only allowed to move into polluted cities, but they became one of the main vehicles for improving the quality of the air. Economic growth facilitated, rather than blocked, air quality improvement.

Not long after this episode, it became clear that much tighter controls were needed on older, established sources of pollution in those areas still in violation of the standards. How would the responsibility for additional control be allocated among the sources? Because the number of emitters was extremely diverse, the menu of technologies was vast, and regulatory staffs too small to pick out the best technology in each setting, providing a satisfactory answer to this question was no easy task. How were regulators going to solve the problem cost-effectively when they had so little hard information about the choices?

THE BUBBLE POLICY

Their solution was to put the market to work for them under what is now known as the "bubble policy." They started by imposing emission standards on the established sources in the most reasonable manner they could, recognizing that in practice many of these standards would inevitably turn out to be unreasonable in the glare of hindsight.

Established sources were then encouraged to create "emission reduction credits" for emission reductions over and above those required by the standards. Once certified, these credits could be stored for subsequent use by the creating source when it wished to expand, or transferred to another source for its use in meeting its standard.

Sources that could control their emissions relatively cheaply generally created and sold these credits, and sources that found themselves confronting very expensive options bought the credits in lieu of installing unreasonably expensive equipment. The result was that the costs of meeting ambient standards were considerably lower than they would have been if the standards had been imposed and no transfers permitted.

This policy has also introduced some extremely beneficial flexibility into the regulatory process. One of the emission reduction credit trades, for example, involved a firm that planned to shut down operations at one plant and build a new one. Under the old regulations, either it would be required to install very expensive pollution control equipment at its old plant, equipment that would be absolutely useless once the new plant was built, or it would simply be let off the hook until the new plant was built, possibly encouraging some foot-dragging in the construction process.

With the bubble policy, the plant was allowed to lease emission reduction credits from another nearby facility. By acquiring these credits the source guaranteed that the region benefited from the cleaner air that it would have received if it had been forced to install the temporary equipment, but at much lower cost and [lower] waste of resources. Additionally, since financial outlays were required to acquire these credits for the entire time until the new plant was operating, no incentive for foot-dragging was created.

Another source of flexibility introduced by this approach involves the capability to control previously unregulated sources. This may be a particularly appealing attribute for those countries with currently underdeveloped environmental regulatory systems. For a number of reasons, ranging from the financial difficulties faced by the firm to a failure by regulators to recognize a particular pollution source, some sources are either not regulated or are regulated less than would be socially desirable. What makes this situation particularly appalling is the fact that many of these sources could be controlled at a lower cost than those sources that have to be controlled to a correspondingly higher degree to compensate for their immunity.

The Emissions Trading Program, the umbrella program that contains both the bubble and offset policies described above, provided a solution. Immune sources could voluntarily control their emissions more than required by their standards and sell the credits to other sources. These other sources would gladly purchase the credits because this was significantly cheaper than controlling their own emissions to a higher degree.

In this way those firms that were already subject to a high degree of control avoided having to ratchet their own controls up ever further to unreasonable levels to compensate for the lack of control in immune firms. In essence the responsibility for identifying additional sources of control had been transferred to the market.

The characteristic of emissions trading that makes the flexibility discussed in the previous paragraph possible is its ability to separate the question of what control is undertaken from the question of who ultimately pays for it. This characteristic is potentially extremely important, not only in terms of domestic environmental policy, but also in terms of international environmental policy.

GLOBAL ISSUES

. . . One proposal for [combating global warming] envisions creating a version of the offset policy for use in holding the emissions of greenhouse gases at their current levels.

Under this proposal, major new sources of greenhouse gases would be required to offset their emissions, assuring that the total emissions would not increase. Offsets could be generated by conservation, recycling, or by retiring older, heavily emitting plants. In the case of carbon dioxide, major new emitters could even invest in tree plantations that would tend to absorb the excess gas. By creating a new market for these offsets, an incentive to invest in offset-creating activities would be stimulated. Meanwhile, the higher prices associated with creating the greenhouse gases, due to the need to acquire offsets, would stimulate the creators to search for ways to reduce their emissions. Transferable offsets would therefore simultaneously encourage both source control and mitigating strategies.

Because the greenhouse gas pollutants are global in nature (the location of the emissions does not matter), global markets in offsets would be possible. This is a tremendously powerful characteristic because larger markets offer the opportunity for larger cost savings. Furthermore, by selling offsets, Third World countries could undertake environmentally sound investments (such as reforestation or protecting forests scheduled for harvest) that were financially underwritten by the developed world. [Precisely such offset credits are now in force for those nations that have ratified the Kyoto Protocol.]

Robert Costanza, "Valuation of Ecosystem Services and Natural Capital" (1996)

In order for regulators and market actors to make decisions about natural resources, monetary values must be assigned to environmental goods. Here, Robert Costanza surveys some valuation methods available to environmental and ecological economists.

VALUATION OF ECOSYSTEM SERVICES AND NATURAL CAPITAL

A prerequisite for achieving sustainability is an economic system that values natural capital and accounts for ecosystem goods and services. The first step is to determine values for them comparable to those of economic goods and services. In determining values we must also consider how much of our ecological life support systems we can afford to lose. To what extent can we substitute manufactured for natural capital, and how much of our natural capital is irreplaceable? For example, could we replace the radiation screening services of the ozone layer which are currently being destroyed?

Some argue that we cannot place economic value on such "intangibles" as human life, environmental aesthetics, or long-term ecological benefits. But, in fact, we do so every day. When we set construction standards for highways, bridges, and the like, we value human life—acknowledged or not—because spending more money on construction would save lives. To preserve our natural capital, we must confront these often difficult choices and valuations directly rather than denying their existence.

Because of the inherent difficulties and uncertainties in determining values, there is no consensus on the correct approach to valuation—perhaps because different approaches reflect particular value judgments and are associated with different social and environmental implications. Historically, resources have been managed individually, for particular values, without considering how they are linked in the ecosystem and how the costs and benefits are distributed among social groups. For example, forests have been valued and managed for timber resources, without considering the effects of forestry practices on fisheries and their role in regulating water flow, providing habitat for wildlife, and providing nontimber and nonmarket resources to people. Conflicts among these values and management priorities need to be resolved through informed social dialogue in which the interests of all stakeholders are represented, rather than on the basis of market behavior and partial values.

The conventional economic view defines value as the expression of individualistic human preferences, with the preferences taken as given and with no attempt to analyze their origins or patterns of long-term change. For goods and services produced with relatively few long-term impacts (like tomatoes or bread) that are traded in well-functioning markets with adequate information, market ("revealed preference") valuations may work relatively well.

But ecological goods and services (like wetland sewage treatment or global climate control) are long-term by nature, are generally not traded in markets (no one owns the air or water), and information about their contribution to individuals' well-being is poor. To determine their value, economists try to get people to reveal what they would be willing to pay for ecological goods and services in hypothetical markets. For example, we can ask people the maximum they would pay to use national parks, even if they don't have to actually pay it. The quality of results in this method depends on how well informed people are; and it does not adequately incorporate long-term goals since it excludes future generations from bidding in the markets. Also, it is difficult to induce individuals to reveal their true willingness to pay for natural resources when the question is put directly. Contingent referenda (willingness to be taxed as a citizen along with other citizens, as opposed to willingness to pay as an individual) are superior to ordinary willingness-to-pay studies in this regard. . . .

An alternative method for estimating ecological values assumes a biophysical basis for value. This theory suggests that in the long run humans come to value things according to how costly they are to produce, and that this cost is ultimately a function of how organized they are relative to their environment. To organize a complex structure takes energy, both directly in the form of fuel and indirectly in the form of other organized structures like factories. For example, a car is a much more organized structure than a lump of iron ore, and therefore it takes a lot of energy (directly and indirectly) to organize iron ore into a car. The amount of solar energy required to grow forests can therefore serve as a measure of their energy cost, their organization, and hence, according to this theory, their value.

Table 23.1 shows the results of applying these two radically different approaches, one based on human perceptions (willingness to pay or WTP) and one based on biophysical production (energy analysis or EA) to the valuation of wetlands in Louisiana. The striking feature is just how close the results are to each other. They can in fact be interpreted as setting the range within which the true value probably falls. The WTP method sets the low end of the range since it must enumerate all the individual nonmarketed services of the ecosystem and

TABLE 23.1
Summary of Wetland Value Estimates

	Per acre present value at specified discount rate	
Method	8%	3%
WTP based		
Commercial fishery	$317	$846
Trapping	$151	$401
Recreation	$46	$181
Storm protection	$1,915	$7,549
Option and Existence values	?	?
Total	$2,429+	$8,977+
EA based		
GPP conversion	$6,400–10,600	$17,000–28,200
"Best Estimate"	$2,429–6,400	$8,977–17,000

Values are estimated in 1983 dollars. GPP = Gross Primary Productivity.
Source: R. Costanza, S. C. Farber, and J. Maxwell, "The Valuation and Management of Wetland Ecosystems," *Ecological Economics* 1 (1989): 335–61.

develop pseudomarkets (via questionnaires or observations of behavior) to evaluate each one. This process will almost certainly miss some important services. The EA method, on the other hand, assumes that all the production of the ecosystem is valuable, directly or indirectly, and to the extent that some ecosystem services are not ultimately valuable to humans it overestimates.

The point that must be stressed, however, is that the economic value of ecosystems is connected to their physical, chemical, and biological role in both the short-term and the long-term global system—whether the present generation of humans fully recognizes that role or not. If it is accepted that each species, no matter how seemingly uninteresting or lacking in immediate utility, has a role in natural ecosystems (which *do* provide many direct benefits to humans), it is possible to shift the focus away from our imperfect short-term perceptions and derive more accurate values for long-term ecosystem services. Using this perspective we may be able to better estimate the values contributed by, say, maintenance of water and atmospheric quality to long-term human well-being.

ECOLOGICAL ECONOMIC SYSTEM ACCOUNTING

GNP, as well as other related measures of national economic performance, have come to be extremely important as policy objectives, political issues, and benchmarks of the general welfare. Yet GNP as presently defined ignores the contribution of nature to production, often leading to peculiar results.

For example, a standing forest provides real economic services for people: by conserving soil, cleaning air and water, providing habitat for wildlife, and supporting recreational activities. But as GNP is currently figured, only the value of harvested timber is calculated in the total. On the other hand, the billions of dollars that Exxon spent on the *Valdez* cleanup—and the billions spent by Exxon and others on the more than 100 other oil spills in the last sixteen months—all actually *improved* our apparent economic performance. Why? Because cleaning up oil spills creates jobs and consumes resources, all of which add to GNP. Of course, these expenses would not have been necessary if the oil had not been spilled, so they shouldn't be considered "benefits." But GNP adds up all production without differentiating between costs and benefits, and is therefore not a very good measure of economic health.

In fact, when resource depletion and degradation are factored into economic trends, what emerges is a radically different picture from that depicted by conventional methods. For example, Herman Daly and John Cobb [in *For the Common Good*] have attempted to adjust GNP to account mainly for depletions of natural capital, pollution effects, and income distribution effects by producing an "index of sustainable economic welfare" (ISEW). They conclude that while GNP in the United States rose over the 1956–86 interval, ISEW remained relatively unchanged since about 1970. When factors such as loss of farms and wetlands, costs of mitigating acid rain effects, and health costs caused by increased pollution are accounted for, the U.S. economy has not improved at all. If we continue to ignore natural ecosystems we may drive the economy down while we think we are building it up. By consuming our natural capital, we endanger our ability to sustain income.

Paul Hawken, Amory Lovins, and L. Hunter Lovins, from *Natural Capitalism: Creating the Next Industrial Revolution* (1999)

In Natural Capitalism, *Paul Hawken joins with Amory Lovins and L. Hunter Lovins, longtime energy efficiency advocates and founders of the Rocky Mountain Institute, to map out a "green revolution" in corporate culture, in which diminishing resources and smart competition will force businesses to be more savvy about the materials and methods they employ. In this selection, the authors discuss environmentally friendly innovations that are already being made by visionary entrepreneurs, including innovations borrowed from and inspired by nature.*

Process innovations in manufacturing help cut out steps, materials, and costs. They achieve better results using simpler and cheaper inputs. In practically every industry, visionaries are improving processes and products by developing highly resource-efficient materials, techniques, and equipment. Even in iron- and steelmaking, one of the oldest, biggest, and most resource-intensive of the

Figure 23.6 and Figure 23.7 Nature's engineering and human engineering. © Newscom / ZUMA Photos and © Newscom / Photographers Showcase.

industrial arts, researchers have discovered ways to reduce energy use by about four-fifths with better output quality, less manufacturing time, less space, often less investment, and probably less total cost.

A particularly exciting area of leapfrog improvements is the potential to replace high-temperature processes with gentler, cheaper ones based on biological models that often involve using actual microorganisms or enzymes. Such discoveries come from observing and imitating nature. Ernie Robertson of Winnipeg's Biomass Institute remarked that there are three ways to turn limestone into a structural material. You can cut it into blocks (handsome but uninteresting), grind it up and calcine it at about 2,700°F[1] into Portland cement (inelegant), or feed it to a chicken and get it back hours later as even stronger eggshell. If we were as smart as chickens, he suggested, we might master this elegant near-ambient-temperature technology and expand its scale and speed. If we were as smart as clams and oysters, we might even do it slowly at about 40°F, or make that cold seawater into microstructures as impressive as the abalone's inner shell, which is tougher than missile-nosecone ceramics.[2]

Or consider the previously noted sophisticated chemical factory within every humble spider. Janine Benyus contrasts arachnid with industrial processes:

> The only thing we have that comes close to [spider] silk . . . is polyaramid Kevlar, a fiber so tough it can stop bullets. But to make Kevlar, we pour petroleum-derived molecules into a pressurized vat of concentrated sulfuric acid and boil it at several hundred degrees Fahrenheit in order to force it into a liquid crystal form. We then subject it to high pressures to force the fibers into alignment as we draw them out. The energy input is extreme and the toxic byproducts are odious.
>
> The spider manages to make an equally strong and much tougher fiber at body temperature, without high pressures, heat, or corrosive acids. . . . If we could learn to do what the spider does, we could take a soluble raw material that is infinitely renewable and make a superstrong water-insoluble fiber with negligible energy inputs and no toxic outputs.[3]

Nature's design lessons can often be turned to an unexpected purpose. Watching a TV report on sea otters soaked by the 1989 *Exxon Valdez* oil spill, Alabama hairdresser Philip McCrory noticed that otter fur soaked up oil extremely well.

This was a good trait for keeping the otter dry in clean water, but for the same reason, fatal when the otter had to swim through oil. Could the characteristic be exploited to help pull oil *out* of the water? Could comparably oil-prone human hair do the same thing? McCrory took hair swept from his salon floor, stuffed it into a pair of tights to make a dummy otter, and threw it into a baby pool filled with water and a gallon of motor oil. In two minutes, he reported, "the water was crystal clear." Salon clients who worked for NASA put him in touch with an expert there who ran a larger-scale test. It found that "1.4 million pounds of hair contained in mesh pillows could have soaked up the entire *Exxon Valdez* oil spill in a week," saving much of the $2 billion Exxon spent to capture only twelve percent of the eleven million gallons spilled.[4]

In nature, nothing edible accumulates; all materials flow in loops that turn waste into food, and the loops are kept short enough that the waste can actually reach the mouth. Technologists should aim to do the same. One of most instructive of such loop-closings occurred in 1988 when the University of Zürich decided to revise the 1971-vintage elementary laboratory course accompanying the lectures in introductory inorganic, organic, and physical chemistry.[5] Each year, students' lab exercises turned $8,000 worth of pure, simple reagents into complex, nasty, toxic goop that cost $16,000 to dispose of. The course was also teaching the students once-through, linear thinking. So Professors Hanns Fischer and C. H. Eugster decided to *reverse the process*—redesigning some exercises to teach instead how to turn the toxic wastes back into pure, simple reagents. This would save costs at both ends and encourage "cycle thinking": "A few generations of science students trained in this domain," they suggested, "are the best investment for environmental protection by chemistry." Students volunteered vacation time for recovery, and by 1991, their demand for residues had outstripped the supply. Since then, the course has produced only a few kilograms of chemical waste annually—less than 100 grams per student per year, a 99 percent reduction—and cut net annual operating costs by around $20,000, or about $130 per student. . . .

Ultimately, there's every indication that large-scale, specialized factories and equipment designed for product-specific processes may even be displaced by "desktop manufacturing." Flexible, computer-instructed "assemblers" will put individual atoms together at a molecular scale to produce exactly the things we want with almost zero waste and almost no energy expended. The technology is a feasible one, not violating any physical laws, because it is exactly what happens whenever nature turns soil and sunlight into trees, bugs into birds, grass into cows, or mothers' milk into babies. We are already beginning to figure out how to do this molecular alchemy ourselves: such "nanotechnologies" are doing surprisingly well in the laboratory.[6] When they take over at a commercial scale, factories as we know them will become a thing of the past, and so will about 99 percent of the energy and materials they use. . . .

Materials efficiency is just as much a lesson of biological design as the making of spider-silk: biomimicry can inform not just the design of specific manufacturing processes but also the structure and function of the entire economy. As Benyus notes, an ecologically redesigned economy will work less like an aggressive, early-colonizer sort of ecosystem and more like a mature one. Instead of a high-throughput, relatively wasteful and undiversified ecosystem, it will resemble what ecologists call a Type Three ecosystem, like a stable oak-hickory forest. Its economy sustains a high stock of diverse forms of biological wealth while consuming relatively little input. Instead, its myriad niches are all filled with organisms busily sopping up and remaking every crumb of detritus into new life. Ecosystem succession tends in this direction. So does the evolution of sustainable economies. Benyus reminds us, "We don't need to invent a sustainable world—that's been done already."[7] It's all around us. We need only to learn from its success in sustaining the maximum of wealth with the minimum of materials flow.

Notes

1. Alternatively, at about 480°F by autoclaving olivine in steam (P. G. Kihlstedt, "Samhällets Råvaruförsörjning under Energibrist," IVA-Rapport 12 [Stockholm: Ingenjörsvetenskapsakademien, 1977]).
2. J. M. Benyus, *Biomimicry: Innovations Inspired by Nature* (New York: William Morrow, 1997), 98ff.
3. Ibid., 135.
4. "Oil Spills? Ask a Hairdresser," *New York Times*, 9 June 1998.
5. Prof. Dr. Hanns Fischer (personal communication, 4 December 1997), Physikalish-Chemisches Institut der Universität Zürich. We are indebted to Dow Europe's vice president, Dr. Claude Fussler, for this example.
6. Foresight Institute, http://www.foresight.org.
7. This paragraph is paraphrased from Dr. Benyus's keynote address to the E Source Members' Forum, 8 October 1998, Aspen, Colorado (www.esource.com).

Garrett Hardin, from "The Tragedy of the Commons" (1968)

This classic paper lays out a classic paradox. When multiple individuals pursue their economically "rational" self-interest in a commons (an open-access resource like the ocean or atmosphere), the result is not the greatest good for the greatest number, but disaster for all—far from rational. Privatization of the commons is one solution, at least in theory. Under what circumstances is privatization feasible? Hardin's own view is that the only sure way to prevent the tragedy of the commons is "mutual coercion, mutually agreed upon"—in other words, an active public sector. Are there other solutions? How have societies in other times and places governed their commons? Keep an eye out for instances of the tragedy in this chapter and throughout the book.

The tragedy of the commons develops in this way. Picture a pasture open to all. It is to be expected that each herdsman will try to keep as many cattle as possible on the commons. Such an arrangement may work reasonably satisfactorily for centuries because tribal wars, poaching, and disease keep the numbers of both man and beast well below the carrying capacity of the land. Finally, however, comes the day of reckoning, that is, the day when the long-desired goal of social stability becomes a reality. At this point, the inherent logic of the commons remorselessly generates tragedy.

As a rational being, each herdsman seeks to maximize his gain. Explicitly or implicitly, more or less consciously, he asks, "What is the utility *to me* of adding one more animal to my herd?" This utility has one negative and one positive component.

1) The positive component is a function of the increment of one animal. Since the herdsman receives all the proceeds from the sale of the additional animal, the positive utility is nearly +1.

2) The negative component is a function of the additional overgrazing created by one more animal. Since, however, the effects of overgrazing are shared by all the herdsmen, the negative utility for any particular decision-making herdsman is only a fraction of −1.

Adding together the component partial utilities, the rational herdsman concludes that the only sensible course for him to pursue is to add another animal to his herd. And another; and another. . . . But this is the conclusion reached by each and every rational herdsman sharing a commons. Therein is the tragedy. Each man is locked into a system that compels him to increase his herd without limit—in a world that is limited. Ruin is the destination toward which all men rush, each pursuing his own best interest in a society that believes in the freedom of the commons. Freedom in a commons brings ruin to all.

24 Human Population

Since Thomas Malthus published his 1798 *Essay on the Principle of Population*, observers have been haunted by the prospect that human population may grow to a point where the Earth's resources cannot sustain it. Some, like Paul Ehrlich in *The Population Bomb*, have envisioned various forms the crisis might take and how it might be avoided. The degrading effects of human overpopulation on the planet's environment are widely recognized. Yet the degree of probability—let alone inevitability—of a demographic catastrophe is debated, as is the speed with which it might come about and the geographic patterns in which it might occur. The specific environmental impacts of overpopulation are similarly discussed and contested; the following selections reflect this debate.

Malthus wrote in reaction to widespread optimism about the human future during the eighteenth-century Enlightenment. A mathematician, he argued that population increases geometrically while food resources increase arithmetically, until what he called the "positive checks" of war, famine, and disease reduce human numbers to a sustainable level. Later versions of his *Essay* did allow that individuals might avert catastrophe by deciding not to have children or postponing parenting. (This strategy was given the force of law in China and, for a brief time, India, during the late twentieth century.) Yet Malthus had a largely grim and deterministic vision of the human future. As Anup Shah demonstrates, he failed to foresee to what degree emigration and colonization, the rise of new technology and means of production, and, above all, changed attitudes toward the family and reproduction would alter the demography and resource base of what would later be called the industrialized world. In particular, the development of effective birth control devices, the education of women, and the change in their social status and ambitions contributed in the twentieth century to the demographic transition to prosperity with low fertility rates that occurred, first in Western Europe and the former British colonies in North America and Australasia, and later in Japan and other developed countries. For many today, a Malthusian future hardly seems inevitable.

At the same time, Shah and Barry Commoner both note that the spread of this demographic transition is also not inevitable. Overpopulation vis-à-vis the means of subsistence and high fertility combined with poverty persist throughout much of the world; the future is unclear. This is all the more true since the prosperity of the industrialized nations rests upon rates of resource consumption that are, in the long run, unsustainable. Add to this the environmental costs of the transition in demographic structure, questions of justice in terms of resource distribution, and the sheer pressure of numbers in the underdeveloped world, and one must admit that Malthus's vision should not be easily dismissed.

Similar debates and conflicting interpretations characterize the understanding of the past. Historians agree that one of the great human catastrophes was the devastation of the native populations of the Americas in the centuries after the arrival of Columbus in 1492. Jared Diamond's *Guns, Germs, and Steel* presents the numbers in stark detail. His account, which blames the diseases introduced by contact with Europeans, and then discusses why the American natives were vulnerable, whereas the Europeans were not, is an exercise in epidemiological history. By contrast, J. H. Parry's classic, *The Spanish Seaborne Empire*, roots the disaster in political, economic, and institutional history, with particular focus on the environmental factors and impacts at play.

The chapter in Joel E. Cohen's *How Many People Can the Earth Support?* from which the selection here has been taken, illustrates the importance of self-awareness about one's approach when discussing population and the environment. It surveys and evaluates four models of species "carrying capacity" taken from basic (i.e., descriptive, analytic, and predictive) ecology, and five models from applied ecology, before choosing one of the latter as the best lens for viewing the prospect of future human overpopulation and its impact on the environment. The very thoroughness with which Cohen goes about his task should remind us, however, that the "dissensus" we have been describing is in many ways a good thing, as it encourages complex thinking about this important problem.

FURTHER READING

Andelson, Robert V., ed. *Commons Without Tragedy: Protecting the Environment from Overpopulation—A New Approach.* London: Shepheard-Walwyn, 1991.

Ehrlich, Paul R. *The Population Bomb.* 1968. Rev. Ed. Rivercity, Mass.: Rivercity Press, 1971.

Evans, L. T. *Feeding the Ten Billion: Plants and Population Growth.* Cambridge: Cambridge University Press, 1998.

Thomas R. Malthus, from *An Essay on the Principle of Population* (1798, revised 1826)

Malthus originally wrote the Essay to urge reform of the English Poor Laws, which he thought encouraged the indigent to have children, thus worsening their economic condi-

Figure 24.1 *T. R. Malthus* by John Linnell, 1833.
Courtesy of the Master of Haileybury College, Haileybury, England.

tion. It asserts that population growth will always outpace resources until brought into balance either by "positive" checks (wars, famine, disease) or "preventive" checks (principally, refusing to have more children). Controversial from its first appearance, Malthus's work influenced Darwin, among others.

The principal object of the present essay is to examine the effects of one great cause intimately united with the very nature of man; which, though it has been constantly and powerfully operating since the commencement of society, has been little noticed by the writers who have treated this subject. The facts which establish the existence of this cause have, indeed, been repeatedly stated and acknowledged; but its natural and necessary effects have been almost totally overlooked; though probably among these effects may be reckoned a very considerable portion of that vice and misery, and of that unequal distribution of the bounties of nature, which it has been the unceasing object of the enlightened philanthropist in all ages to correct.

The cause to which I allude is the constant tendency in all animated life to increase beyond the nourishment prepared for it.

It is observed by Dr. [Benjamin] Franklin that there is no bound to the prolific nature of plants or animals but what is made by their crowding and interfering with each other's means of subsistence. Were the face of the earth, he says, vacant of other plants, it might be gradually sowed and overspread with one kind only;

as, for instance, with fennel; and were it empty of other inhabitants, it might in a few ages be replenished from one nation only; as, for instance, with Englishmen.[1]

This is incontrovertibly true. Through the animal and vegetable kingdoms Nature has scattered the seeds of life abroad with the most profuse and liberal hand; but has been comparatively sparing in the room and the nourishment necessary to rear them. The germs of existence contained in this spot of earth, with ample food, and ample room to expand in, would fill millions of worlds in the course of a few thousand years. Necessity, that imperious all-pervading law of nature, restrains them within the prescribed bounds. The race of plants and the race of animals shrink under this great restrictive law; and the race of man cannot by any efforts of reason escape from it.

In plants and animals the view of the subject is simple. They are all impelled by a powerful instinct to the increase of their species; and this instinct is interrupted by no reasoning or doubts about providing for their offspring. Wherever, therefore, there is liberty, the power of increase is exerted; and the superabundant effects are repressed afterwards by want of room and nourishment, which is common to plants and animals; and among animals, by their becoming the prey of each other.

The effects of this check on man are more complicated. Impelled to the increase of his species by an equally powerful instinct, reason interrupts his career, and asks him whether he may not bring beings into the world for whom he cannot provide the means of support. If he attend to this natural suggestion, the restriction too frequently produces vice. If he hear it not, the human race will be constantly endeavoring to increase beyond the means of subsistence. But as by that law of our nature which makes food necessary to the life of man, population can never actually increase beyond the lowest nourishment capable of supporting it; a strong check on population, from the difficulty of acquiring food, must be constantly in operation. This difficulty must fall somewhere, and must necessarily be severely felt in some or other of the various forms of misery, or the fear of misery, by a large portion of mankind. . . .

The checks to population, which are constantly operating with more or less force in every society, and keep down the number to the level of the means of subsistence, may be classed under two general heads; the preventive, and the positive checks. . . .

The positive checks to population are extremely various, and include every cause, whether arising from vice or misery, which in any degree contributes to shorten the natural duration of human life. Under this head therefore may be enumerated all unwholesome occupations, severe labor and exposure to the seasons, extreme poverty, bad nursing of children, great towns, excesses of all kinds, the whole train of common diseases and epidemics, wars, pestilence, plague, and famine. . . .

Of the preventive checks, the restraint from marriage which is not followed by irregular gratifications may properly be termed moral restraint. . . .

In every country some of these checks are, with more or less force, in constant operation; yet notwithstanding their general prevalence, there are few states in which there is not a constant effort in the population to increase beyond the means of subsistence. This constant effort as constantly tends to subject the lower classes of society to distress, and to prevent any great permanent amelioration of their condition.

These effects, in the present state of society, seem to be produced in the following manner. We will suppose the means of subsistence in any country just equal to the easy support of its inhabitants. The constant effort towards population, which is found to act even in the most vicious societies, increases the number of people before the means of subsistence are increased. The food, therefore, which before supported eleven millions, must now be divided among eleven millions and a half. The poor consequently must live much worse, and many of them be reduced to severe distress. The number of laborers also being above the proportion of work in the market, the price of labor must tend to fall, while the price of provisions would at the same time tend to rise. The laborer therefore must do more work to earn the same as he did before. During this season of distress, the discouragements to marriage and the difficulty of rearing a family are so great, that population is nearly at a stand. In the meantime, the cheapness of labor, the plenty of laborers, and the necessity of an increased industry among them, encourage cultivators to employ more labor upon their land, to turn up fresh soil, and to manure and improve more completely what is already in tillage, till ultimately the means of subsistence may become in the same proportion to the population as at the period from which we set out. The situation of the laborer being then again tolerably comfortable, the restraints to population are in some degree loosened; and, after a short period, the same retrograde and progressive movements, with respect to happiness, are repeated.

This sort of oscillation will not probably be obvious to common view; and it may be difficult even for the most attentive observer to calculate its periods. Yet that, in the generality of old states, some such vibration does exist, though in a much less marked, and in a much more irregular manner, than I have described it, no reflecting man who considers the subject deeply can well doubt.

Notes

1. Franklin, *Miscell.* 9. [For bibliographical details of this reference see Patricia James's edition published by Cambridge University Press for the Royal Economic Society in 1990.]

Anup Shah, from *Ecology and the Crisis of Overpopulation: Future Prospects for Global Sustainability* (1998)

Shah points out some of Malthus's faulty assumptions, particularly concerning the influence of material prosperity or dearth on the decision to have children, and the extent to which the means of subsistence can be increased. While allowing that the world may yet face a Malthusian catastrophe, he argues that both the potential for technology to expand resources and the precedent of the European "demographic transition" to prosperity with low fertility rates make such a future uncertain.

MALTHUS

Preferences: The Connection Between Malthus and Behavioral Ecology

Accidentally perhaps, Malthus [see selection above] forged a connection between economics and behavioral ecology and, at the micro level, that connection is in the description of human preferences. Thus when Malthus wrote about the human tendency to reproduce he meant that preferences are such that, under a vast range of circumstances, couples would prefer to have more children than fewer. When behavioral ecologists refer to Darwinian breeding habits they allude to these very same preferences. However, like conventional economists, Malthus emphasized the resource constraint on behavior.

In the comparative statics part of his analysis, an increase in the resources of the household translates into more children. Thus, if for some reason the husband's wage increases, the couple opts for a larger family. In the above reasoning, Malthus was assuming a virtual absence of a trade-off between children and other parental consumption in the preferences of the masses. Contrary to Malthus, modern economists would contend that people's preferences are such that there is, in fact, such a trade-off. Malthus's assumption makes sense if children are viewed as primarily a vehicle for hereditary immortality. A trade-off between children and other parental consumption makes sense if parents largely view children as consumer goods, that is, having, watching, interacting with their children makes parents happy and is the sole purpose for having them. Incidentally, a trade-off between children and public services (such as social security provision) makes sense if parents regard children as producer goods (for example, children serve as insurance).

Boulding's Dismal Theorem

Malthus went on to trace the aggregative consequences of household behavior. He first noted that human populations, like all natural populations, tend to increase in an exponential manner. He next assumed that food production grows much more slowly. As a consequence, he reasoned that population would outstrip food supply. With an "oversupply" of people competing directly or indirectly (via jobs) for food, wages would fall.

It is at this point that Malthus's analysis can be connected with that of [David] Ricardo. According to Malthus, an increase in population leads to greater competition for jobs which means lower wages. That would lead to an increase in the rent for land. Ricardo disagreed. He contended that the fall in wages resulting from a rise in population would lead to an increase in profits. However, both agreed that the working poor would grow poorer and the rich would become richer as a result of an increase in population (see below for elaboration).

A lower wage should reduce fertility rates. But Malthus thought that a time lag would be involved. During the transitional period, there would be starvation, disease, and war and these non-price mechanisms would check population growth and could even reduce population size. The influence of starvation on reducing population is best illustrated by the failure of the Irish potato crop in the late 1840s causing about half a million people (about 6 percent of the total population) to die of starvation. As for disease, the "Black Death" of the fourteenth century was responsible for killing at least 25 percent of the population. Finally, war, and in the Second World War about 25 percent of Poland's population was annihilated.[1] *If* starvation, disease, and war ultimately result as a consequence of overpopulation then clearly these are inhuman mechanisms for regulation. Furthermore, it is likely that some families could also practice infanticide in order to regulate family size. [Kenneth] Boulding[2] has encapsulated this outcome in his neat Dismal Theorem: if the only check on population growth is starvation, then population will grow until it starves. The Dismal Theorem and all that it entails should alert modern welfare economists to the possibility of welfare improvement. Human suffering (and possible waste of resources) due to starvation, illness, and wars need not occur and infanticide should not be necessary. A task of the welfare economist is to reduce such hardships and thereby increase human welfare.

Boulding's Utterly Dismal Theorem

It is worth noting that Malthus allowed for some technological improvements to take place. But his microeconomic analysis implied that such technical change would lead to increased reproduction and so, in the long run, its benefits would be soaked up. There would be short run benefits to humans but Malthus anticipated the ecologist's argument that small "aristocracies" would be able to capture a sizeable share of the benefits and enjoy a higher standard of living over the long run. However, the masses would eventually sink back to their narrow niche-spaces. Once again, Boulding has neatly captured this result in a corollary to the Dismal Theorem, the Utterly Dismal Theorem: if the only check on population growth is starvation, then any technological improvement will have the ultimate effect of increasing starvation as it allows a larger population to live in precisely the same narrow niche-spaces.

Malthus and Economic Inequality

Making use of Boulding's mesa metaphor . . . , we can explore the topic of Malthusian population analysis and inequality. In the Malthusian model, the minority of the rich occupy the affluent territory centered in the middle of the mesa, with a barbed wire fence around them. In some unspecified way, they are never short of resources with which to prop up their affluent territories. Meanwhile the mass of the poor all around them breed to the point where they push each other off the edge of the mesa.

There is, however, an alternative scenario that should help put Malthus's thinking into perspective. Suppose the poor manage to capture the affluent territories and eventually the whole mesa becomes one huge mass of narrow niche-spaces with lives of poverty and misery. Malthus had, in fact, witnessed the French Revolution and had been greatly disturbed by it. As a consequence, he thought that a society characterized by inequality was preferable. Better still would be a society in which the masses had smaller families. However, he did not think that government could help the poor accomplish the goal of smaller families. He placed the burden of responsibility on the individual household. If households behave responsibly then, in the visual image of the mesa, the rich would still hold on to their affluent territories in the middle, but the masses would not fall over the edge and, in fact, enjoy a higher standard of living.

THE INFLUENCE OF MALTHUS ON DARWIN

In October 1836, a twenty-eight-year-old Darwin, having returned from the round the world trip in the *Beagle*, happened upon Malthus's essay on population. (In 1854 Alfred Russel Wallace, a naturalist and collector fourteen years younger than Darwin, also read Malthus's essay as Darwin had done eighteen years before. Independently, he also hit upon natural selection of variable organisms as the driving force of evolution.)

Some observers say that Darwin was the ideological heir to Malthus. What struck Darwin about Malthus's population analysis was the reasoning that since human beings reproduce themselves many times more rapidly than they can increase their necessary food supply, the binding food constraint implies competition for existence among members of the human race and the consequent slaughter of the weak. Darwin was aware that there is competition for existence among animals and plants. Reading Malthus, the idea came to him that competition would preserve favorable variations in wildlife and destroy unfavorable ones. Combined with the accessional occurrence of chance mutation, Darwin reasoned that natural selection would give rise to new species.

MALTHUSIAN OUTCOME POSTPONED?

Malthus strongly believed that there was a tendency for human numbers to outstrip resources. This belief was based on two basic assumptions. One was our tendency to breed. Given the persuasive evidence, most observers would agree that, up to a point, geometric population growth is the innate result of all healthy living. However, Malthus was wrong in assuming that food production arithmetically increases. There is the evidence of the past two centuries which shows that increases in food production depend on land availability and technical progress and it is difficult to predict the path of technical progress. Malthus also underestimated the agricultural potential of the New World lands.

At the time that Malthus was writing, there existed vast tracks of fertile but uninhabited land in North and South America, as well as Australasia and Africa. As this land was put to use, food and raw materials poured into nineteenth-century Europe, encouraging late Victorian economists like Alfred Marshall to virtually ignore natural resources and to concentrate on labor and capital in economic analysis. As a consequence, the Malthusian outcome appears to have been postponed.

If Malthus was correct in his analysis but wrong in his timing, will the problem of mass poverty and misery occur after human population has colonized every tract of useful land? It all depends on technical progress, which has had an unimpeded run, and environmental resilience. Further technological advances appear to be on the cards. What we are not sure about today is the extent to which we can make technical progress without impoverishing the resilience, adaptability, and regenerative power of the environment.

In any case, something significant happened in European history; as Western Europe grew prosperous, fertility rates fell. Such a phenomenon, known as the demographic transition, is highly appealing, a win-win occurrence, and thus deserves closer examination.

Notes

1. R. Douglas, "The Commons and Property Rights: Towards a Synthesis of Demography and Ecology," in R. V. Andelson, ed., *Commons Without Tragedy* (London: Shepheard-Walwyn, 1991), 1–26.
2. K. E. Boulding, "Commons and Community: The Idea of a Public," in G. Hardin and J. Baden, ed., *Managing the Commons* (San Francisco: W. H. Freeman, 1977).

Barry Commoner, from *The Closing Circle: Nature, Man, and Technology* (1971)

Commoner summarizes the interacting economic, physiological, sociological, and political—though not the cultural—factors that affect the level and distribution of world population. He is more detailed than Shah about the "demographic transition" now almost complete in the industrialized world, and believes that it can be extended to underdeveloped countries. He therefore recommends a dual strategy of reducing population growth and raising living standards in order to avert environmental and human disaster.

The world population problem is the subject of a vast and complex literature. The field is one which cuts across a broad range of disciplines: the physiology of reproduction, the biology of sex and race, the sociology of families and larger groups, economics of agricultural and industrial production, world trade, and international politics.

Demographers have delineated an intricate network of interactions among these factors. These involve complex circular relationships, in which, as in an ecological cycle, every step is connected to several others. Thus, while the rate of population growth is, of course, the result of the birth rate and the death rate, so that population size would tend to increase with birth rate and decrease with death rate, there is also an opposite effect mediated by social factors, in particular the desired family size. If death rate, especially infant mortality, is high, there may be an increased desire for children in order to make up for the expected loss as the family tries to achieve the desired number of surviving children. Because of this relationship, a drop in infant mortality can *reduce* population growth by reducing the birth rate. On the other hand, if the birth rate is high, the number of children borne per woman is also high, a condition that tends to *increase* infant mortality.

Economic factors are also involved. If economic resources are limited, then a rising population reduces the resources available per capita, lowering the standard of living and tending thereby to raise the death rate. At the same time, a high standard of living may lower the age at marriage, which in turn raises the birth rate. On the other hand, if adequate resources are available, a rising population—by adding to the labor force—can lead to increased economic activity, which in turn elevates the standard of living and affects the influence of the latter on both birth rate and death rate. In addition, with a rising level of economic activity, educational level may improve, which in turn can increase the age at marriage, the employment of women, and the use of modern contraceptive techniques—and thereby reduce the birth rate. And added to all these relationships is the effect of deliberate government policies, which may increase the birth rate by propaganda or economic incentives or reduce it by similar means.

In view of these complexities, it is not surprising that demographers often differ in their views about the expected course of population growth or about the most effective ways of controlling it. Indeed, the network of effects may operate differently in each nation and within each cultural or economic group. One does, however, gain a general impression that most demographers agree that tendencies for self-regulation are characteristic of human population systems.

It is also generally recognized that the world population cannot grow indefinitely because there is some limit to crucial resources, such as food, available from the global ecosystem. However, the size of the world's ultimate food-producing capacity is also subject to disagreement and cannot be accurately estimated at this time. Hence, while it can be concluded that there is *some* limit to the possible size of the world population, where that limit might lie is only a more or less educated guess.

Demographers have learned a good deal about the connection between population growth and economic and social factors in industrialized countries. This is characterized by the "demographic transition". . . . Generally, in the industrialized nations, there is a tendency for population growth to level off, apparently as a natural social response to prosperity. This seems to be the way in which these societies have reacted to the acquisition of wealth, provided that this wealth is available in ways that improve their well-being and encourage confidence in the future.

The actual historical course of the demographic transition in the industrialized nations is particularly significant. With improved living conditions, over-all death rates and rates of infant mortality decline steadily over the years. At first, birth rates remain high, so that population growth is rapid. Later, birth rate and death rate decline more or less in concert, so that the population continues to grow, but less rapidly than before. In the final, most recent stages of the demographic transition, the birth rate drops particularly fast, narrowing the excess over death rate, and population growth becomes relatively slow. In nearly every advanced country, this rapid drop in birth rate occurs when death rates are from about 10 to 12 per thousand and infant mortality rates are about 20 per thousand. Thus, in industrialized countries, birth rates have been brought to the closest match with death rates only through a historical process—which may extend over a 50 to 100 year period—of achieving a minimal death rate largely as a result of improved living conditions. In other words, population balance has been approached through the material progress of the society.

When this course of events is compared to population trends in developing countries, certain important similarities and differences turn up. As a whole, the situation may be summed up as follows. Everywhere in the world where they have not yet reached the minimal rates characteristic of advanced countries, over-all death rates and rate of infant mortality have been dropping year by year. As it does in industrialized countries, the birth rate seems to decline particularly fast when the minimal death rate of from about 10 to 12 per thousand is approached. This situation can be seen in the demographic behavior of Taiwan, the Pacific Islands, Japan, Cyprus, Israel, and Singapore. In some Latin-American countries, such as Venezuela and Costa Rica, the minimal death rates have been approached, but birth rates have *not* declined. However, in every country in which the *rate of infant mortality*—which has important effects on the desired family size—has approached the minimal value that is characteristic of industrialized countries (about 20 per thousand or less), birth rates also begin to approach the low rates now found in advanced countries (from about 14 to 18 per thousand). Wherever infant mortality rates, despite their decline, remain high, birth rates are also high. Thus, in India, infant mortality in the 1951–61 period was about 139 per thousand, and the birth rate was about 42 per thousand; the Latin American countries with persistent, high birth rates (45 to 50 per thousand) have infant mortality rates in the range of 50 to 90 per thousand. In most developing countries, birth rates are still well in excess of death rates and population growth is rapid.

The scientific evidence regarding the *future* course of world population growth is by no means unambiguous or conclusive. Any conclusion relevant to the future represents an extrapolation from past trends. Depending on the past data that are chosen as a base, strikingly different extrapolations can be made.

On the one hand, it is possible to conclude from extrapolations of present trends in birth rates and death rates that in many of the less developed nations of the world, the present high rates of population growth will continue until there are catastrophic disparities between population size and needed resources. In this case, the only possible means to bring these nations close to demographic balance would be a campaign to reduce fertility, although death rate and infant mortality remain high. On the other hand, if some of the more subtle social and economic factors that influence fertility are taken into account, and if credence is given to the evidence that societies that do successfully bring birth rate nearly into balance with death rate do so only when death rate and infant mortality are reduced to minimal levels, then a more optimistic outlook emerges. In particular, according to this view, the strongest effort should be made to eliminate overt and hidden hunger in developing countries, so powerfully described by the Bolivian nutritionist Josué de Castro. Taking these views into account, there is reason to expect that nations that now exhibit a high rate of population growth will reduce their birth rate much more rapidly than they have in the past when they reach the critically low levels of death rate and of infant mortality. In this case, demographic balance can be approached through efforts to improve living standards and, in particular, to reduce infant mortality.

A conservative view of the available demographic evidence would be that *neither* of these two alternative interpretations is as yet exclusively supported by the data. The two approaches are not mutually exclusive, and obviously *both* direct efforts to reduce birth rate and improvement in living standards and in infant mortality can contribute to reduction in population growth. This dual approach is exemplified by a recent proposal to develop, world-wide, clinics that simultaneously provide prenatal and child care and contraceptive services.

J. H. Parry, from *The Spanish Seaborne Empire* (1966)

Parry focuses on Mexico as an example of the post-Columbian depopulation of the Americas. While crediting the devastating effect of European diseases on native population numbers, he concentrates on the agricultural and ecological causes and impacts of this human tragedy. In particular, he shows how shifting Spanish political and economic policies, which both continued and departed from previous Aztec practices, drove the pace of depopulation by changing the natural environment and resource base.

In every province of the Indies the European invasion was followed by a steep decline in the numbers of the native population. In the greater Antillean islands the decline began very shortly after the invasion and was clearly perceptible to Spanish observers within a decade or so of the first settlements. Missionary chroniclers attributed the high death-rate to ill-treatment and over-work. They hoped that legislation and stricter royal control would arrest the decline. . . .

On the mainland the *conquistadores* found, to all appearance, stronger peoples. Cortés and his companions were greatly struck by the warlike prowess of the Aztecs and their tributaries, and by the remarkable feats of Indian runners in carrying messages between the coast and the capital. In the area which became New Spain—roughly the area between the isthmus of Tehuántepec and the Chichimec frontier running in a sagging curve from Pánuco to Culiacán—the inhabitants at the time of the conquest formed sophisticated, settled, highly regimented societies, vastly different from the small and primitive groups in the islands. To the invaders the population appeared not only strong and well organized but also very numerous. All the early Spanish accounts stress the size of the towns and their proximity one to another, especially on the shores of the lake of Texcoco, the crowds thronging the markets, the streams of passers-by on the causeways, the great fleets of canoes on the lake. The lakeside towns, it is true, with their fertile *chinampas* [artificial islands used for agriculture] and their abundant tribute income, were exceptional; but the countryside also was populous. Indian peasants lived on a simple and monotonous diet, consisting chiefly of maize; they had no domestic animals of consequence, and ate very little meat. Maize is a productive crop, and on such a diet two or three acres of reasonably good land sufficed to maintain an average family. Moreover, the methods of intensive hoe cultivation made it possible to till marginal land, which with more developed tools would have been unusable. Most cultivable land was in fact cultivated. Throughout most of the area, except in parts too high, rough, and steep even for hoe cultivation, all the evidence indicates an extremely dense rural population; denser, indeed, in many places than it is today. Peasants were confined to a low level of subsistence, not only by pressure of population on land but also by the tribute exactions of their local nobilities and of the Aztec triple alliance [alliance of the three major Aztec city-states], which took from them most of their surplus of food and almost all their production of valuable goods such as cotton cloth and cacao. In the last few decades before the conquest population probably was tending to increase, despite low living standards and despite the checks imposed by war and human sacrifice. These checks were powerful; wars were regularly undertaken not only to acquire land, slaves, and tribute but also to procure captives for sacrifice. Major ceremonies often involved thousands of victims. Human life, then, was plentiful and cheap, and so was human labor. By means of closely organized communal and tributary labor, using hand methods, without the help of draught animals or mechanical appliances, Indian societies accomplished an astonishing amount of heavy construction, including irrigation systems, flood controls, aqueducts and causeways, and elaborate and massive public buildings.

Here was what the *conquistadores* had been looking for: not only land, food, and gold but an apparently inexhaustible supply of docile labor. In the early years of settlement the Spaniards, having seized the main centers of government, took over the native system of tributes and services. They began almost at once to modify the system, partly because their *encomienda* [system of tributary labor imposed on native peoples] grants did not always coincide with pre-conquest political or tribute groupings, partly because some forms of pre-conquest tributes—feather mantles, for example—were of no interest to them. In general, however, both from policy and through force of circumstances, they based their demands roughly upon the tributes and services formerly enjoyed by the Aztec triple alliance and by the Indian local nobility—still enjoyed, indeed, by many of the latter. Their declared policy was to demand somewhat less than precedent permitted, in order to reconcile the Indians to their rule; their practice, to drive the best bargain they could with the Indian headmen. Like their predecessors, they were great builders, and, like them also, they grew accustomed to an extremely lavish use of labor; but the reservoir of population from which the labor was drawn seemed boundless. Relieved of the drain of human sacrifice, indeed, and—once the conquest was accepted—of constant war, the Indian population, already vast, might have been expected to increase. Instead, just as it had done in the islands, it quickly began to decline.

. . . An astonishingly large population at the time of the conquest suffered a catastrophic decline in the course of the century. The Indian population in the early seventeenth century was probably less than one-tenth of what it had been 100 years before.

The pre-conquest population of New Spain has been roughly but credibly estimated at about 25,000,000. In 1532, when the second *audiencia* [royal court] was trying to systematize the assessment of tribute and to restrict *encomiendas*, it had probably declined to some 17,000,000. So large a decline could not have been caused simply by ill-usage and over-work; though these were common enough, especially in the feverish rebuilding of the city of Mexico. The number of Spaniards was then small and their penetration of the country very uneven. Most of them were settled in highland areas, especially in the central valley, in or near Mexico City. In the coastal lowlands Spanish settlers were many fewer. Many areas, even fertile areas with large Indian populations, were hardly touched by Spanish settlement. The extent of the decline in the Indian population, as might be expected, varied greatly from one region to another. It was far more drastic in the coastal lowlands than on the plateau; on both the Gulf and Pacific coasts the available evidence suggests a decline of about one-half between 1519 and 1532. Even in areas where there was, as yet, no permanent Spanish settlement, and no forced labor exactions, populations declined. The decline bore no obvious and direct relation, therefore, to the amount of Spanish activity; it was caused chiefly by pestilence, by contagious diseases introduced by the invaders, against which the natives had no immunity. Smallpox entered Mexico with Cortés's army, and caused many deaths in the city in the early 1520s. The extraordinarily high mortality on the coast may have been caused by malaria or yellow fever, introduced from Europe but spread by indigenous insects. The precise nature of the lethal factors can only be guessed; what is certain is that they operated with increasing severity throughout most of the century.

The progressive decline of the Indian population caused marginal land to be abandoned and much even of the relatively fertile areas to be left untilled. The vacuum so caused was filled, for the most part, by grazing animals, horned cattle chiefly in the lowlands and valleys, sheep in the highlands, horses, mules, and goats almost everywhere. These animals were all alien to the Americas and were introduced by Spaniards. Imports of livestock were particularly numerous in the 1530s; and, once introduced, the beasts multiplied prodigiously on land which had never before been grazed. They brought a new diversity into the economy of New Spain, and caused a major revolution in the use of land. Indians participated in this revolution to some extent. Apart from those who were employed as herdsmen or shepherds by Spanish masters, some rich individuals and a few communities, especially among the privileged and enterprising people of Tlaxcala, ran flocks of sheep of their own. Reference has already been made to Indian muleteers; and probably, though not recorded, backyard goats were not uncommon even in the 1530s. On the other hand, very few Indians went in for cattle ranching. Possibly dislike and even fear of large, unfamiliar animals deterred them. More probably, however, they were prevented from running cattle by difficulty in securing title to areas of grazing large enough to make ranching profitable.

Pastoral farming produced new commodities and new services which became available, to a limited extent, to Indians. The use of pack animals progressively replaced the wasteful demand for human carriers, as depopulation reduced the supply of porters. For those who acquired sheep and goats, mutton was a new source of protein in a hitherto almost meatless diet. (In this connection it is curious that although the pre-conquest Indians relied chiefly on fish for their supply of animal protein, their range of devices for fishing—nets, traps, weirs, hooks, spears, and so forth—was very limited and lacking in ingenuity. The Iberian peoples for many centuries have possessed a remarkable range of these devices. Indian efficiency in fishing and in snaring *chichicuilotes*—lakeside wading birds—improved greatly as a result of contact with Europeans.) Beef, the mainstay of the Spanish community, was little eaten by Indians except where they could afford to buy it or, more commonly, contrive to steal it. Spanish ranchers on the northern frontier sometimes complained of illicit slaughtering by "wild" Indians. Milk was unimportant. The Indians either did not like it or could not get it; scrawny backyard goats and half-wild scrub cattle cannot have yielded very much. More important than the change in diet was the change in clothing. Many woolen mills were set up from the 1530s onwards, a few owned by Indians, but most by Spaniards. In the highlands, woolen capes or blankets, forerunners of the modern *serape*, soon began to displace the traditional *manta* woven from cotton or other vegetable fiber. Cotton cloth had been a major item in many tribute assessments at the time of the conquest. In the second half of the century, as tributes were commuted to money payments, this form of assessment tended to disappear. The area of land planted to cotton in the warmer parts of the country grew progressively less through the century.

In settled New Spain, in the 100 years between 1520 and 1620, the viceregal government made formal grants totaling more than 17,000 square miles for cattle *estancias*, almost all to Spaniards, and more than 12,000 square miles for sheep farms, some to Indians, but the greater part to Spaniards. In addition, at least 2,000 square miles were granted to Spaniards for arable purposes; for the production, that is, either of crops such as wheat intended for Spanish consumption, or of fodder for animals. Well over 30,000 square miles, therefore, were officially converted to new uses. Most of this great area at the time of the conquest had been farmed by Indians and had subsequently either been taken from them or, more commonly, vacated by them because of the diminution in their numbers. In the course of the sixteenth century, then, a vast agrarian revolution took place; a wholesale substitution of an animal for a human population. The destructive effect of this revolution upon traditional Indian agriculture was in fact far greater than the figures suggest, for flocks and herds grazed over much bigger areas than the official grants. A cattle *estancia* was in theory a square tract, each side of which was one Castilian league, about 2.6 miles; its area was thus 6.76 square miles. On this, the grantee was entitled—indeed expected—to run 509 head of cattle, or seventy-four to the square mile. The corresponding figure for sheep was 666 to the square mile, which—even for scrub animals—meant heavy over-grazing. Each *estancia* was supposed to be one league distant from its nearest neighbor and one league from any Indian village. It was to be surrounded, therefore, by a belt of empty land. The large numbers of stock carried, and the rapid natural increase, made these territorial "cushions" inadequate. The area actually grazed by Spanish stock was probably two or three times as great as that comprised in the official grants. The animals strayed freely through the peripheral belts and into Indian cultivations beyond. In Spanish practice this unrestricted grazing at certain times of year was customary and necessary. In Spain the law required most arable land to be opened to grazing after the harvest was gathered. In New Spain the same rule was enacted; but in practice vast, half-wild, untended flocks and herds, on unfenced range, might invade cultivated land at any time. Indians constantly complained to the courts about the resulting damage to their crops. Both the *Juzgado general* [court for native peoples] and the *audiencias* made many orders forbidding trespass and awarding damages, but often the means of enforcement were lacking. The cactus hedges now characteristic of the Mexican rural scene were a tardy and inadequate defense against this destructive invasion. Many Indian communities, seeing their crops repeatedly destroyed by great herds of grazing, trampling beasts, abandoned their cultivations in despair.

Against this background of destruction the depopulation of New Spain is easier to understand. To the steady attrition caused by debilitating disease and near starvation, moreover, were added the heavy and irrecoverable losses caused by major epidemics. One such epidemic ravaged the whole of New Spain in 1545–6, while the colony was still in a state of angry uncertainty about the future of the *encomienda* system. The precise nature of the disease cannot be certainly identified, nor can the mortality caused by it be accurately assessed, but reports reaching Spain so alarmed the government that Prince Philip, in the absence of the Emperor, ordered a detailed inspection of the entire colony, and reports on the tributary population and resources of every Indian community, with the express intention of reassessing tributes and services. A large number of *visitas* were actually made in or about the year 1548. The surviving summary of the reports, the *Suma de Visitas*, covers about half the area of New Spain. From the information contained in it and from other sources a careful estimate has recently been made of the total Indian population of New Spain in 1548. The figure is 6,300,000.

Jared Diamond, from "Lethal Gift of Livestock" in *Guns, Germs, and Steel: The Fates of Human Societies* (1997)

In contrast to Parry, Diamond treats the post-1492 depopulation of the Western Hemisphere in general, and in the context of epidemiological rather than political, ecological, and economic history. He argues that infectious disease was the overwhelming cause of the catastrophe; and that such disease-spreading factors as contrasting patterns of settlement in the Americas versus Eurasia, and the "extreme paucity of domestic animals in the New World" explain why Europe was not similarly devastated.

The importance of lethal microbes in human history is well illustrated by Europeans' conquest and depopulation of the New World. Far more Native Americans died in bed from Eurasian germs than on the battlefield from European guns and swords. Those germs undermined Indian resistance by killing most Indians and their leaders and by sapping the survivors' morale. For instance, in 1519 Cortés landed on the coast of Mexico with 600 Spaniards, to conquer the fiercely militaristic Aztec Empire with a population of many millions. That Cortés reached the Aztec capital of Tenochtitlán, escaped with the loss of "only" two-thirds of his force, and managed to fight his way back to the coast demonstrates both Spanish military advantages and the initial naïveté of the Aztecs. But when Cortés's next onslaught came, the Aztecs were no longer naive and fought street by street with the utmost tenacity. What gave the Spaniards a decisive advantage was smallpox, which reached Mexico in 1520 with one infected slave arriving from Spanish Cuba. The resulting epidemic proceeded to kill nearly half of the Aztecs, including Emperor Cuitláhuac. Aztec survivors were demoralized by the mysterious illness that killed Indians and spared Spaniards, as if advertising the Spaniards' invincibility. By 1618, Mexico's initial population of about 20 million had plummeted to about 1.6 million.

Pizarro had similarly grim luck when he landed on the coast of Peru in 1531 with 168 men to conquer the Inca Empire of millions. Fortunately for Pizarro and unfortunately for the Incas, smallpox had arrived overland around 1526, killing much of the Inca population, including both the emperor Huayna Capac and his designated successor. . . . The result of the throne's being left vacant was that two other sons of Huayna Capac, Atahuallpa and Huascar, became embroiled in a civil war that Pizarro exploited to conquer the divided Incas.

When we in the United States think of the most populous New World societies existing in 1492, only those of the Aztecs and the Incas tend to come to our minds. We forget that North America also supported populous Indian societies in the most logical place, the Mississippi Valley, which contains some of our best farmland today. In that case, however, *conquistadores* contributed nothing directly to the societies' destruction; Eurasian germs, spreading in advance, did everything. When Hernando de Soto became the first European *conquistador* to march through the southeastern United States, in 1540, he came across Indian town sites abandoned two years earlier because the inhabitants had died in epidemics. These epidemics had been transmitted from coastal Indians infected by Spaniards visiting the coast. The Spaniards' microbes spread to the interior in advance of the Spaniards themselves.

De Soto was still able to see some of the densely populated Indian towns lining the lower Mississippi. After the end of his expedition, it was a long time before Europeans again reached the Mississippi Valley, but Eurasian microbes were now established in North America and kept spreading. By the time of the next appearance of Europeans on the lower Mississippi, that of French settlers in the late 1600s, almost all of those big Indian towns had vanished. Their relics are the great mound sites of the Mississippi Valley. Only recently have we come to realize that many of the mound-building societies were still largely intact when Columbus reached the New World, and that they collapsed (probably as a result of disease) between 1492 and the systematic European exploration of the Mississippi.

When I was young, American schoolchildren were taught that North America had originally been occupied by only about one million Indians. That low number was useful in justifying the white conquest of what could be viewed as an almost empty continent. However, archaeological excavations, and scrutiny of descriptions left by the very first European explorers on our coasts, now suggest an initial number of around 20 million Indians. For the New World as a whole, the Indian population decline in the century or two following Columbus's arrival is estimated to have been as large as 95 percent.

The main killers were Old World germs to which Indians had never been exposed, and against which they therefore had neither immune nor genetic resistance. Smallpox, measles, influenza, and typhus competed for top rank among the killers. As if these had not been enough, diphtheria, malaria, mumps, pertussis, plague, tuberculosis, and yellow fever came up close behind. In countless cases, whites were actually there to witness the destruction occurring when the germs arrived. For example, in 1837 the Mandan Indian tribe, with one of the most elaborate cultures in our Great Plains, contracted smallpox from a steamboat traveling up the Missouri River from St. Louis. The population of one Mandan village plummeted from 2,000 to fewer than 40 within a few weeks.

While over a dozen major infectious diseases of Old World origins became established in the New World, perhaps not a single major killer reached Europe from the Americas. The sole possible exception is syphilis, whose area of origin remains controversial. The one-sidedness of that exchange of germs becomes even more striking when we recall that large, dense human populations are a prerequisite for the evolution of our crowd infectious diseases. If recent reappraisals of the pre-Columbian New World population are correct, it was not far below the contemporary population of Eurasia. Some New World cities like Tenochtitlán were among the world's most populous cities at the time. Why didn't Tenochtitlán have awful germs waiting for the Spaniards?

One possible contributing factor is that the rise of dense human populations began somewhat later in the New World than in the Old World. Another is that the three most densely populated American centers—the Andes, Mesoamerica, and the Mississippi Valley—never became connected by regular fast trade into one huge breeding ground for microbes, in the way that Europe, North Africa, India, and China became linked in Roman times. Those factors still don't explain, though, why the New World apparently ended up with no lethal crowd epidemics at all. (Tuberculosis DNA has been reported from the mummy of a Peruvian Indian who died 1,000 years ago, but the identification procedure used did not distinguish human tuberculosis from a closely related pathogen [*Mycobacterium bovis*] that is widespread in wild animals.)

Instead, what must be the main reason for the failure of lethal crowd epidemics to arise in the Americas becomes clear when we pause to ask a simple question. From what microbes could they conceivably have evolved? We've seen that Eurasian crowd diseases evolved out of diseases of Eurasian herd animals that became domesticated. Whereas many such animals existed in Eurasia, only five animals of any sort became domesticated in the Americas: the turkey in Mexico and the U.S. Southwest, the llama/alpaca and the guinea pig in the Andes, the Muscovy duck in tropical South America, and the dog throughout the Americas.

In turn, we also saw that this extreme paucity of domestic animals in the New World reflects the paucity of wild starting material. About 80 percent of the big wild mammals of the Americas became extinct at the end of the last Ice Age, around 13,000 years ago. The few domesticates that remained to Native Americans were not likely sources of crowd diseases, compared with cows and pigs. Muscovy ducks and turkeys don't live in enormous flocks, and they're not cuddly species (like young lambs) with which we have much physical contact. Guinea pigs may have contributed a trypanosome infection like Chagas' disease or leishmaniasis to our catalog of woes, but that's uncertain. Initially, most surprising is the absence of any human disease derived from llamas (or alpacas), which it's tempting to consider the Andean equivalent of Eurasian livestock. However, llamas had four strikes against them as a source of human pathogens: they were kept in smaller herds than were sheep and goats and pigs; their total numbers were never remotely as large as those of the Eurasian populations of domestic livestock, since llamas never spread beyond the Andes; people don't drink (and get infected by) llama milk; and llamas aren't kept indoors, in close association with people. In contrast, human mothers in the New Guinea highlands often nurse piglets, and pigs as well as cows are frequently kept inside the huts of peasant farmers.

The historical importance of animal-derived diseases extends far beyond the collision of the Old and the New Worlds. Eurasian germs played a key role in decimating native peoples in many other parts of the world, including Pacific islanders, Aboriginal Australians, and the Khoisan peoples (Hottentots and Bushmen) of southern Africa. Cumulative mortalities of these previously unexposed peoples from Eurasian germs ranged from 50 percent to 100 percent. For instance, the Indian population of Hispaniola declined from around 8 million, when Columbus arrived in A.D. 1492, to zero by 1535. Measles reached Fiji with a Fijian chief returning from a visit to Australia in 1875, and proceeded to kill about one-quarter of all Fijians then alive (after most Fijians had already been killed by epidemics beginning with the first European visit, in 1791). Syphilis, gonorrhea, tuberculosis, and influenza arriving with Captain Cook in 1779, followed by a big typhoid epidemic in 1804 and numerous "minor" epidemics, reduced Hawaii's population from around half a million in 1779 to 84,000 in 1853, the year when smallpox finally reached Hawaii and killed around 10,000 of the survivors. These examples could be multiplied almost indefinitely.

However, germs did not act solely to Europeans' advantage. While the New World and Australia did not harbor native epidemic diseases awaiting Europeans, tropical Asia, Africa, Indonesia, and New Guinea certainly did. Malaria throughout the tropical Old World, cholera in tropical Southeast Asia, and yellow fever in tropical Africa were (and still are) the most notorious of the tropical killers. They posed the most serious obstacle to European colonization of the tropics, and they explain why the European colonial partitioning of New Guinea and most of Africa was not accomplished until nearly 400 years after European partitioning of the New World began. Furthermore, once malaria and yellow fever did become transmitted to the Americas by European ship traffic, they emerged as the major impediment to colonization of the New World tropics as well. A familiar example is the role of those two diseases in aborting the French effort, and nearly aborting the ultimately successful American effort, to construct the Panama Canal.

. . . There is no doubt that Europeans developed a big advantage in weaponry, technology, and political organization over most of the non-European peoples that they conquered. But that advantage alone doesn't fully explain how initially so few European immigrants came to supplant so much of the native population of the Americas and some other parts of the world. That might not have happened without Europe's sinister gift to other continents—the germs evolving from Eurasians' long intimacy with domestic animals.

Joel E. Cohen, from *How Many People Can the Earth Support?* (1995)

Cohen adapts the "open-access" model of the Earth's maximum human "carrying capacity" from applied ecology in order to argue that we must see the planet as a "temporal commons," in which any one generation must avoid depleting natural resources, lest future generations suffer a "second tragedy of the commons." He believes this the most adequate conceptual scheme by which to analyze future problems of overpopulation.

OPEN-ACCESS REPRODUCTION AND THE SECOND TRAGEDY OF THE COMMONS

The model of an open-access resource has some relevance to human carrying capacity. However, the size of the fishing fleet, rather than the size of the fished population, is the analogue of the human population. . . . In an open-access fishery, additional vessels enter the fleet until the total revenues of the fishery coincide with the total costs. This level of fishing effort is called the bionomic equilibrium. The bionomic equilibrium is economically inefficient because the potential profit that could be extracted from the fishery is lost. Each fishing boat imposes a cost on all the other boats by lowering the other boats' opportunity to make a profit from the fishery, but each boat does not have to bear that cost. Each boat has to absorb only its own fishing costs, and receives the full benefit of its catch, without any deduction for the amounts by which it reduces the catch of others.

The entry of additional fishing vessels into the fishery is analogous to the entry of additional people into the human population. Children don't support themselves: the new individual reaps the benefits of living, and the family and community may eventually reap benefits from the additional individual, but the parents and the child do not bear the full costs that his or her presence imposes on others already in the population. The Berkeley demographer Ronald D. Lee called the right of free access to common resources through reproduction "the second tragedy of the commons."

> In the real world . . . externalities to childbearing are pervasive. Among the most important are environmental externalities, since many enjoyable and

productive aspects of the environment are not privately owned: air shed, water shed, ozone layer, parks, climate, freedom from noise, and so on. . . .

The optimal level of use per person [of an open-access resource] depends on the number of people. Under optimal management, when the population is larger each person will be entitled to use the resource less—to visit Yosemite less often, to turn up the volume on his stereo less high, to burn less firewood, to discharge less waste. Furthermore, with a larger population, the optimal level of total use will generally be slightly higher (although less than proportionately so) and so each person will have to live with slightly more congestion, pollution, and degradation. Because each additional person is born with a birthright to public resource use, each birth inflicts costs on all others by reducing the value of their environmental birthright. Free access through reproduction is the second tragedy of the commons.[1]

Open access to reproduction affects employment. While each household makes hardly any difference to the supply of workers in a developing country, when all households hope to increase their income by having many children, the rapid increase in the number of available workers tends to depress pay, lower the quality and availability of education and health care, and dilute the existing capital stock among a larger number of workers. Poorly educated, unhealthy and undercapitalized workers have lower productivity, and it becomes more difficult to modernize the country's economy.[2]

In fact, human reproduction has always been socially regulated to some extent, and potential parents have probably never been able to reproduce freely. Even among nonhuman primates, the reproductive success of males and females is linked to social status. In the Babylonian epic of Atrahasis . . . certain classes of women were given responsibilities in temples and prohibited from reproducing, so that the gods would not be disturbed by the noise of an excessively large human population on the Earth.[3] Ecclesiastical constraints on reproduction in Western cultures underlie the notion of a "legitimate birth." Chinese social control of reproduction is new in degree but not in kind.[4] Because fertility is and has long been socially regulated, the analogy between the reproduction of the human population and an ideal free-access resource is less than perfect from the point of view of parents, but is widely applicable from the point of view of the child. Except where infanticide is still practiced, the birthright of the child, once born, is generally unchallenged.

The economist Julian L. Simon and others have argued that the so-called externalities of each additional birth are positive on balance.[5] From this point of view, a child confers benefits on his or her family, community, country, and the world that outweigh whatever costs he or she may impose. To illustrate this point of view, I offer the remark of a scientific visitor from Tokyo, who went first to a scientific meeting in sparsely populated northern Florida, then came to my laboratory in crowded Manhattan, New York. I asked him how he liked

Florida. He said, "It was lonely. There weren't enough people on the sidewalk. Here in New York, I feel at home, as in Tokyo." Even after allowing for his politeness (he was, after all, in New York), I cannot discount the empirical observation that, for various reasons, many people are attracted to the high population densities of cities.

Many people are also repelled by those same high densities. Opposing Simon, the World Bank economist Herman Daly and the theologian J. B. Cobb asserted that "the scale of human activity relative to the biosphere has grown too large. . . . Further growth beyond the present scale is overwhelmingly likely to increase costs more rapidly than it increases benefits, thus ushering in a new era of 'uneconomic growth' that impoverishes rather than enriches."[6]

It is possible for economists, demographers, and others to take passionately held and opposite positions about the net external benefits or net external costs of each additional birth because, to my knowledge, these externalities have never been quantified precisely. There are many difficulties in doing so.

Our present reproduction and our use of the Earth affect past and future generations as well as the present generation. The present generation affects the welfare of past generations by giving their projects meaning and bringing them to fruition, or alternatively by nullifying and frustrating them, according to the British philosopher John O'Neill at the University of Sussex.[7] For successive generations, O'Neill suggested, the Earth may be viewed as a temporal "commons" even if all resources are privately owned at any one time. If any generation exploits the Earth's resources to the limit within the lifetime of that generation, without regard to the future, that generation receives the benefit and all successive generations share the loss. By this logic, each self-interested generation would be expected to deplete the resources it bequeaths to the following generations. To avoid one generation's depleting the resources of the Earth, O'Neill argued, each generation must view its own identity as part of a chain of generations, and must express its identity through projects that require to be continued and given meaning by future generations. The never-completed enterprise of science is an example of such a project. Land viewed as the shared property of a family that endures across generations will be conserved better than land viewed as a parcel to be bought and sold like a sack of potatoes. "The mobilization of labor by the market, like the mobilization of land, has undermined a sense of community across generations. Both lie at the basis of the temporal myopia of modern society."[8]

This generation inherited the Earth and will surely leave it to future generations. The view that your generation and mine take of the role and importance of future generations will influence how we treat the Earth today.

Notes

1. Ronald D. Lee, "Comment: The Second Tragedy of the Commons," in *Resources, Environment, and Population: Present Knowledge, Future Options*, ed. Kingsley Davis and Mikhail S. Bernstam (New

York: Oxford University Press, 1991), 315–17. [Supplement to *Population and Development Review* 16 (1990): 315–22.]

2. Robert H. Cassen, "Economic Implications of Demographic Change," *Transactions of the Royal Society of Tropical Medicine and Hygiene* 87.suppl. 1 (1993): 15.
3. Erle Leichty, "Demons and Population Control," *Expedition* 13.2 (1971): 22–26 [University Museum of the University of Pennsylvania, Philadelphia]; Anne D. Kilmer, "The Mesopotamian Concept of Overpopulation and Its Solution as Reflected in the Mythology," *Orientalia* 41 n.s. (1972): 160–76.
4. Shulamith H. Potter and Jack M. Potter, *China's Peasants: The Anthropology of a Revolution* (Cambridge: Cambridge University Press, 1990); Susan Greenhalgh, "The Peasantization of the One-child Policy in Shaanxi," in *Chinese Families in the Post-Mao Era*, ed. Deborah David and Stevan Harrell (Berkeley: University of California Press, 1993), 219–50; Susan Greenhalgh et al., "Restraining Population Growth in Three Chinese Villages: 1988–93," *Population and Development Review* 20.2 (1994): 365–95.
5. Julian L. Simon, *Population Matters: People, Resources, Environment, and Immigration* (New Brunswick, N.J.: Transaction Publishers, 1990).
6. Herman E. Daly and J. B. Cobb, *For the Common Good: Redirecting the Economy Toward Community, the Environment, and a Sustainable Future* (Boston: Beacon Press, 1989), 2.
7. John O'Neill, "Future Generations: Present Harms," *Philosophy* 68.263 (1993): 46.
8. Ibid., 50.

25 Anthropology

The table of contents of this volume reflects the diverse ways in which the environment and its relationship to humans may be conceived. A strong connection exists between an interdisciplinary approach and the concept of diversity: an *interdisciplinary* book emphasizes the diversity of approaches to the subject of the environment. Yet this diversity, for the most part, reflects the diversity of ideas within just one subculture: a Northern, industrial, and academic one. The diversity of ways of thinking about nature and the environment increases exponentially when we contemplate the full spectrum of human cultures and the possibility that within *each* there also exists a diversity of ways of thinking about the environment.

Anthropology studies human cultural diversity and similarity. As such, it provides a deep understanding of the various ways that people can relate to their environment. The readings in this chapter offer insights into how diverse peoples—Africans, South Americans, Native North Americans—have evolved environmental attitudes, practices, and prohibitions. The discipline of anthropology asks, among other things, what we can learn about proper treatment of the environment from studying the beliefs and practices of the great diversity of human cultures. It also asks how to value this diversity of human cultures itself. From the standpoint of native species and ecosystem conservation, for instance, perhaps the most important lesson from anthropological studies is that, in many parts of the world, we cannot treat national parks and wildlife reserves as we have in the United States—as places where people are prohibited from establishing permanent settlements. Most native cultures enjoy a long-standing relationship with the land and have used it in a sustainable manner for generations. Imposing a North American view of conservation on them may not only be ethically wrong, it may be bad for the wildlife. Because their stake and knowledge is usually so much higher than that of a distant central government, native peoples often have turned out to be the best stewards of conservation areas.

In Chapter 11, "Biodiversity and Conservation Biology," we saw that an important part of any valuation of the environment concerns biodiversity: its diversity of ecosystems, species, and genes. A parallel normative question arises in anthropology: are human cultures units of biodiversity to be conserved, just as ecosystems, species, and genes are? If so, how do we value those units?

The readings in this chapter also ask whether a human culture is an entity possessing political and natural rights. Put another way, is it unethical to take an action that would tend to cause a culture to vanish? This question is deeper and more difficult than it first appears. Given the transportation and communication technologies of the twenty-first century, interactions between human cultures are unavoidable. The Maasai culture at the center of Raymond Bonner's selection "Whose Heritage Is It?" has had interaction forced upon it by German, English, and mainstream Tanzanian cultures (as it, in its time, has also forced interaction on—and conquered—other African cultures, such as the Datoga). Many of the interactions between the German, English, and mainstream Tanzanian cultures and the Maasai have disenfranchised the Maasai from their land. In the twenty-first century, is it a better strategy for the Maasai to establish connections with outside cultures or to withdraw and avoid contact? The problem with the first strategy is that the more interaction the Maasai have with outside cultures, the more their own will be diluted. The problem with the second is that unless the Maasai obtain some measure of political power, the disenfranchisement of their culture from the land will continue unabated.

A central question in determining the rights of human cultures is, the right to *what?* Is there a right to persist as an autonomous culture? Is there a right to cultural purity? Perhaps Helen Corbett puts it best in "The Rights of Indigenous Peoples": cultures should have a right to self-determination, especially relative to the authority of the sovereign states in which they find themselves embedded geographically and politically. Self-determination is difficult to codify and hard to enforce, but it allows the people of the cultures themselves to answer the problematic questions of cultural interaction and purity.

A corollary to the endangerment of human cultures is the endangerment and loss of human languages. Nettle and Romaine's "Where Have All the Languages Gone?" provides examples of languages that have died out or are on the verge of extinction. We lose not only scientific knowledge about the possible diversity of human language patterns when we lose one language, but also modes of perception of the environment—subtle distinctions among species, weather patterns, or other environmental phenomena expressed in only one or a few languages. Nettle and Romaine compare the loss of languages to the loss of species. How is such a comparison apt or not?

Certainly human cultures and biological species are different in one important respect closely related to language: human cultures can give voice to their own needs in a way that non-human species never can. Those who work to preserve species may not know exactly what is in the best interest of that species, but if they advocate the most likely survival strategy for that species based on the best

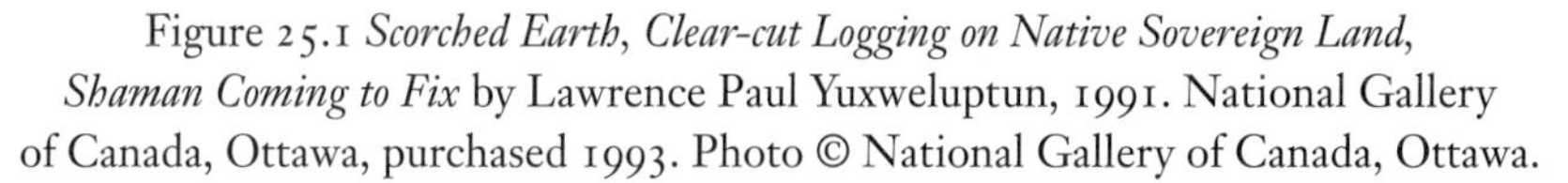

Figure 25.1 *Scorched Earth, Clear-cut Logging on Native Sovereign Land, Shaman Coming to Fix* by Lawrence Paul Yuxweluptun, 1991. National Gallery of Canada, Ottawa, purchased 1993. Photo © National Gallery of Canada, Ottawa.

science, then they are not guilty of usurping the self-determination of anyone or anything. This is not true for human cultures: those in first-world organizations dedicated to cultural survival always do so at the risk of misinterpreting or misrepresenting the best interests and desires of those they are seeking to help.

All the readings prompt these questions: If I were a member of a particular native culture, how would I want sympathetic people from the outside to treat me and my culture? Would I want well-meaning human-rights groups to speak for me in the corridors of power, knowing that there will necessarily be some mistranslation, as happened with the Huaorani depicted in Joe Kane's selection "With Spears from All Sides"? Or would I want no communication with outsiders? The problem with the second choice is that interaction with outsiders is going to occur eventually—if no one stands up for the rights of the native culture, it may occur in the context of an environmental depredation or disaster, such as deforestation or an oil spill.

When we realize that our own culture is not a monolithic entity, perhaps the most important corollary is that we not consider other cultures monolithic either. It is too easy to think that the "Maasai want this" or the "Huaorani don't want that," without recognizing that these cultures comprise individuals with varying desires and opinions. It is wise to remember this description in "With Spears from All Sides" of Enqueri, one of the Huaorani leaders:

> Enqueri is like Condorito because his head has too many ideas. . . . One day, he wants to be a schoolteacher, and the next day he wants to devote his life to the People, and the next day he is working for the Company, and the next day he wants to kill the Company, and the next day he is going to be a nurse, and the next day he wants to go to the sky and be with God.

Each of us, American, Huao, Maasai, or Mayan, has conflicting desires, and the diversity of these conflicting desires is multiplied a thousand- or a million-fold within each culture. More than anything else, anthropological studies teach us to be uneasy with many conceptions of human nature—other than the concept of diversity as a core characteristic of all human cultures.

FURTHER READING

Alvard, Michael. "Evolutionary Ecology and Resource Conservation." *Evolutionary Anthropology* 7 (1998): 62–74.

Cartmill, Matt. *A View to a Death in the Morning.* Cambridge: Harvard University Press, 1993.

Crumley, Carole L., ed. *New Directions in Anthropology and the Environment: Intersections.* Walnut Creek, Calif.: Altamira Press, 2001.

Grim, John A., ed. *Indigenous Traditions and Ecology.* Cambridge: Harvard University Press, 2001.

Grimes, Barbara F., ed. *Ethnologue: Languages of the World.* Dallas: Summer Institute of Linguistics, 2000.

George Catlin, from *Letters and Notes on the . . . North American Indians* (1841)

George Catlin was a self-taught artist who specialized in painting American Indians. His letters on the nature and the native inhabitants of the American West are classic descriptions of the region and its indigenous peoples. In this excerpt, Catlin provides a detailed account of how the Indians obtained their food — in this case the American bison. He combines this with a plea for the conservation of that great beast of the plains.

Mouth of the Teton River, Upper Missouri

The Indians in these parts are all mounted on small, but serviceable horses, which are caught by them on the prairies, where they are often running wild in numerous bands. The Indian, then, mounted on his little wild horse, which has been through some years of training, dashes off at full speed amongst the herds of buffalos, elks, or even antelopes, and deals his deadly arrows to their hearts from his horse's back. The horse is the fleetest animal of the prairie, and easily brings his rider alongside of his game, which falls a certain prey to his deadly shafts, at the distance of a few paces.

In the chase of the buffalo, or other animal, the Indian generally "strips" himself and his horse, by throwing off his shield and quiver, and every part of his dress, which might be an encumbrance to him in running; grasping his bow in his left hand, with five or six arrows drawn from his quiver, and ready for instant use. In his right hand (or attached to the wrist) is a heavy whip, which he uses without mercy, and forces his horse alongside of his game at the swiftest speed.

These horses are so trained, that the Indian has little use for the rein, which hangs on the neck, whilst the horse approaches the animal on the right side, giving his rider the chance to throw his arrow to the left; which he does at the instant when the horse is passing—bringing him opposite to the heart, which receives the deadly weapon "to the feather." When pursuing a large herd, the Indian generally rides close in the rear, until he selects the animal he wishes to kill, which he separates from the throng as soon as he can, by dashing his horse between it and the herd, and forcing it off by itself; where he can approach it without the danger of being trampled to death, to which he is often liable by too closely escorting the multitude. . . .

The Indians generally kill and dry meat enough in the fall, when it is fat and juicy, to last them through the winter; so that they have little other object for this unlimited slaughter, amid the drifts of snow, than that of procuring their robes for traffic with their Traders. . . .

The poor buffalos have their enemy *man*, besetting and besieging them at all times of the year, and in all the modes that man in his superior wisdom has been able to devise for their destruction. They struggle in vain to evade his deadly shafts, when he dashes amongst them over the plains on his wild horse—they plunge into the snowdrifts where they yield themselves an easy prey to their destroyers, and they also stand unwittingly and behold him, unsuspected under the skin of a white wolf, insinuating himself and his fatal weapons into close company, when they are peaceably grazing on the level prairies, and shot down before they are aware of their danger. . . .

The buffalo is a very timid animal, and shuns the vicinity of man with the keenest sagacity; yet, when overtaken, and harassed or wounded, turns upon its assailants with the utmost fury, who have only to seek safety in flight. In their desperate resistance the finest horses are often destroyed; but the Indian, with his superior sagacity and dexterity, generally finds some effective mode of escape. . . .

Reader! listen to the following calculations, and forget them not. The buffalos . . . have recently fled before the appalling appearance of civilized man, and taken up their abode and pasturage amid the almost boundless prairies of the West. An instinctive dread of their deadly foes, who made an easy prey of them whilst grazing in the forest, has led them to seek the midst of the vast and treeless plains of grass, as the spot where they would be least exposed to the assaults of their enemies; and it is exclusively in those desolate fields of silence (yet of beauty) that they are to be found—and over these vast steppes, or prairies, have they fled, like the Indian, towards the "setting sun" until their bands have been crowded together, and their limits confined to a narrow strip of country on this side of the Rocky Mountains.

This strip of country, which extends from the province of Mexico to Lake Winnepeg on the North, is almost one entire plain of grass, which is, and ever must be, useless to cultivating man. It is here, and here chiefly, that the buffalos dwell; and with, and hovering about them, live and flourish the tribes of Indians, whom God made for the enjoyment of that fair land and its luxuries.

It is a melancholy contemplation for one who has traveled as I have, through these realms, and seen this noble animal in all its pride and glory, to contemplate it so rapidly wasting from the world, drawing the irresistible conclusion too, which one must do, that its species is soon to be extinguished, and with it the peace and happiness (if not the actual existence) of the tribes of Indians who are joint tenants with them, in the occupancy of these vast and idle plains.

And what a splendid contemplation too, when one (who has traveled these realms, and can duly appreciate them) imagines them as they *might* in future be seen (by some great protecting policy of government) preserved in their pristine beauty and wildness, in a *magnificent park*, where the world could see for ages to come, the native Indian in his classic attire, galloping his wild horse, with sinewy bow, and shield and lance, amid the fleeting herds of elks and buffalos. What a beautiful and thrilling specimen for America to preserve and hold up to the view of her refined citizens and the world, in future ages! A *nation's Park*, containing man and beast, in all the wild and freshness of their nature's beauty. . . .

There are, by a fair calculation, more than 300,000 Indians, who are now subsisted on the flesh of the buffalos, and by those animals supplied with all the luxuries of life which they desire, as they know of none others. The great variety of uses to which they convert the body and other parts of that animal, are almost incredible to the person who has not actually dwelt among these people and closely studied their modes and customs. Every part of their flesh is converted into food, in one shape or another, and on it they entirely subsist. The robes of the animals are worn by the Indians instead of blankets—their skins when tanned are used as coverings for their lodges, and for their beds; undressed, they are used for constructing canoes—for saddles, for bridles—l'arrêts, lasos, and thongs. The horns are shaped into ladles and spoons—the brains are used for dressing the skins—their bones are used for saddle trees—for war clubs, and scrapers for graining the robes—and others are broken up for the marrow-fat which is contained in them. Their sinews are used for strings and backs to their bows—for thread to string their beads and sew their dresses. The feet of the animals are boiled, with their hoofs, for the glue they contain, for fastening their arrow points, and many other uses. The hair from the head and shoulders, which is long, is twisted and braided into halters, and the tail is used for a fly brush. In this wise way do these people convert and use the various parts of this useful animal, and with all these luxuries of life about them, and their numerous games, they are happy (God bless them) in the ignorance of the disastrous fate that awaits them.

Yet this interesting community, with its sports, its wildnesses, its languages, and all its manners and customs, could be perpetuated, and also the buffalos, whose numbers would increase and supply them with food for ages and centuries to come, if a system of non-intercourse could be established and preserved. But such is not to be the case—the buffalo's doom is sealed, and with their extinction must assuredly sink into real despair and starvation the inhabitants of these vast plains, plains which afford for the Indians no other possible means of subsistence; and they must at last fall a prey to wolves and buzzards, who will have no other bones to pick.

It seems hard and cruel (does it not?) that we civilized people with all the luxuries and comforts of the world about us, should be drawing from the backs of these useful animals the skins for our luxury, leaving their carcasses to be devoured by the wolves—that we should draw from that country some 150 or 200,000 of their robes annually, the greater part of which are taken from animals that are killed expressly for the robe, at a season when the meat is not cured and preserved, and for each of which skins the Indian has received but a pint of whiskey!

Such is the fact, and that number or near it, are annually destroyed, in addition to the number that is necessarily killed for the subsistence of 300,000 Indians, who live entirely upon them. It may be said, perhaps, that the Fur Trade of these great western realms, which is now limited chiefly to the purchase of buffalo robes, is of great and national importance, and should and must be encouraged. To such a suggestion I would reply by merely enquiring (independently of the poor Indians' disasters) how much more advantageously would such a capital be employed, both for the weal of the country and for the owners, if it were invested in machines for the manufacture of *woollen robes*, of equal and superior value and beauty; thereby encouraging the growers of wool, and the industrious manufacturer, rather than cultivating a taste for the use of buffalo skins; which is just to be acquired, and then, from necessity, to be dispensed with, when a few years shall have destroyed the last of the animals producing them.

Joe Kane, from "With Spears from All Sides" (1993)

Exploration and extraction of natural resources — especially oil — has had devastating effects on natural ecosystems and native cultures. Joe Kane describes his experiences with the Huaorani, an Amazonian rain forest tribe in eastern Ecuador, as they try to fight off the encroachment of "The Company," a term that designates anyone or anything that portends further oil exploration.

In 1967, Texaco made the first major discoveries of commercial oil in the Oriente [eastern Ecuador]. In 1972, it completed a 312-mile oil pipeline from the Oriente to Ecuador's Pacific coast. Over the next seventeen years—until 1989, when Petroecuador assumed operational control of the pipeline—it shipped 1.4 billion barrels of oil through the pipeline. The pipeline ruptured twenty-seven times, spilling what was estimated to be 16.8 million gallons of raw crude (more than the *Exxon Valdez*), most of it into the Oriente's delicate web of rivers, creeks, and lagoons.

Although the Oriente is a biological treasure chest—about the size of Alabama, it is home to what are estimated to be eight to twelve thousand species of plants, or up to five percent of all the plant species on earth—almost no attempt was made to assess the environmental impact of oil development until 1989, when a woman named Judith Kimerling came to Ecuador and began to stick her nose into things. A graduate of Yale Law School, and a former environmental litigator in the office of the New York State Attorney General (she helped to prosecute the Love Canal case), Kimerling has spent most of the last four years investigating the Company. Traveling by foot, canoe, and truck, sleeping in Indian homes, she has visited producing wells, exploratory sites, and seismic trails. She estimates that in addition to the pipeline spills—the pipeline is now all but worn out and ruptures with increasing frequency—the petroleum industry is spilling ten thousand gallons a week from secondary flow lines and is dumping some 4.3 million gallons of untreated toxic waste directly into the watershed every day. Where the Company is at work, she says, malnutrition rates in the local population are as high as ninety-eight percent, and health workers report high rates of cancer, birth defects, and other health problems linked to contaminants. Kimerling has little doubt where the blame resides: nearly all the oil produced in the Oriente was produced by Texaco and Petroecuador. "Texaco was the teacher," Kimerling says. "Petroecuador was the student."

What this means for the local people can be seen in almost any of the hamlets and well sites outside Coca. I was able to visit perhaps twenty such sites. The Quichua village of San Carlos is typical. The president of the community, Ignacio Sequihua, took me to see an oil well drilled by Texaco in 1977 and turned over to Petroecuador in 1990. It sits on one bank of a small creek that drains into the Manduro River, which, until Texaco arrived, was a principal source of fish and of drinking and bathing water for San Carlos. There is a waste pit next to the well. The pit, about the size of an Olympic swimming pool, is simply a hole dug in the forest floor. According to Kimerling, raw crude oil, toxic drilling wastes, formation water (water that is pumped up from the ground along with petroleum and carries such heavy metals as arsenic, cadmium, cyanide, lead, and mercury), maintenance wastes (including industrial solvents and acids), and the related effluvia of the oil-extraction process are regularly dumped into such pits, and the pits regularly wash out in the Oriente's heavy rains. Here black, tarlike pit waste coated both banks of the creek. It covered every rock and branch jutting out of the creek, and it spread in a glistening black carpet down the banks and as far back into the forest as I could see. When I reached into the creek, my hands came out covered with black goo, and I couldn't rinse it off.

"Now Texaco is gone, and the Company says the wells are cleaned up," a woman said. "How does it look to you?"

"When Texaco came, they didn't ask us for permission to enter," Ignacio Sequihua said. "They didn't talk to us at all. They just came." The villagers cut down trees and blocked Texaco's trucks for a month, until the military threatened them. "We were afraid they would attack us at night, so we stopped the blockade and signed an agreement with the Company. We had no choice." . . .

Beyond the camp, the Vía Auca died into a small, muddy track nearly overrun by vegetation. Enqueri [a Huaorani leader] and I walked a few hundred yards down the track, into a gully, and there we met a Huao family returning from what they considered a wildly successful hunting trip. They were a father, a mother, a young son, and a toddling daughter; they carried a shotgun, a spear, and a blowgun; and their prize, which they were rolling uphill with great effort and greater glee, was a badly dented fifty-five-gallon drum that had been discarded in the bush when Unocal [another oil company] abandoned its camp. The drum was clearly full, but of what no one knew. "It could be diesel," Enqueri said, "or extra," which is what the Huaorani call all gasoline, "or it could be forty-forty-one-hundred"—motor oil—"or it could be something else," which in all likelihood would be chemical waste. Whatever was in the drum, the family hoped to burn it in the kerosene lamps that the Company had left behind. "The Company hid many of these barrels around here," Enqueri said. "At least, they thought they hid them. But you cannot hide anything from the Huaorani." . . .

Quemperi [a local chief] usually sat beside the fire when we spoke, but when he discussed the Conoco road he stood up and declaimed loudly and forcefully, sometimes stamping his feet for emphasis. It was the only time I saw him angry—indeed, it was the only time I saw any Huao angry—and the display was a frightening one. His rhetoric was direct: a road means bad hunting; game won't cross it; colonists will come and cut down the forest and kill the animals. A road, in other words, means hunger; it means the end of abundance, and the end of the self-reliance and independence the Huaorani value above all else.

The ninety-mile road that Conoco proposed to build, and along which it would develop all its hundred and twenty wells, would terminate in the heart of Quemperi's traditional hunting territory. By then, it would have traversed the territory of the five or six clans that inhabit the area north of Quemperi, and it would have opened up to petroleum exploitation—by Conoco and by the growing list of companies counting on the infrastructure Conoco would build—the homelands of the Tagaeri, the Taromenane, and the Oñamane, if they exist. The dangers the road poses for the Huaorani are, of course, immense—exposure to a

Figure 25.2 A Huaorani native walks along an oil pipeline. From *Savages* © 1995 by Joe Kane. Used by permission of Joe Kane and Alfred A. Knopf, a division of Random House, Inc.

simple flu, for example, could wipe out the groups never before reached. To its credit, Conoco, while finishing its exploration phase, sent an anthropologist, James A. Yost, to speak with the Huaorani living in and around its concession. (Yost had been affiliated with the Summer Institute of Linguistics, and had worked among the Huaorani on and off for a decade.) In a report Yost told Conoco that although these Huaorani had already had some contact with *cowode* [non-natives], mainly from oil exploration, their social, economic, and political structures had changed very little, because they had been able to absorb the inevitable changes on their own terms. He warned, however, that "to lose any more land or resources will result in deculturation and ethnocide."

Yost's chief fear was colonization along the road, and Conoco had incorporated into its management plan such preventive measures as gates, guards, patrols, and satellite surveillance. But none of these measures had ever been tested. Obviously, there was no proof they would work. Conoco executives couldn't provide a single example of a road anywhere in the Oriente—indeed, anywhere in the Amazon—that had not been colonized.

Inevitably, in such discussions someone brings up the fact that colonization of the Conoco road would be illegal, because it would be inside the Huao reserve, but that point is never pursued with conviction. Even the Company recognizes that in Ecuador such laws, like many other good ones, are irrelevant, because the government has little inclination to enforce them. Colonization is illegal inside the Cuyabeno Wildlife Reserve, which is about twenty miles north of the Huaorani, yet more than four thousand colonist families have settled along its oil roads since 1979 and the government has not made even a pretense of trying to remove them. . . .

Huao culture has no precedent for speaking in the united voice required for political power. Traditionally, the clans are autonomous, and, even within the clans, loyalty beyond one's immediate family is tenuous. To put the welfare of the nation above that of the family is a huge leap—and an area of vulnerability readily comprehended by the Company. Seiscom Delta, the company that performs most of the seismic testing, reportedly enters the Indian territories without informing the inhabitants, and without doing even the minimal paperwork required by the government. Last year, one of the Quichua federations staged a strike against Seiscom Delta, blocking all access roads; at about the same time, the Cofán kidnapped an entire seismic crew. Last October, Seiscom Delta started testing inside the Huao protectorate, within a mile of Toñampare, but two days later the Huaorani raided the camp and scared off the workers. The military escorted the workers back; the Huaorani raided the camp again. Petroecuador officials flew in to try to smooth things over. They spoke with Nanto [another Huaorani leader]. "What do you want?" one of the Petroecuador men asked, and his co-workers reached into the Company helicopter and hauled out bags of pots and pans and rice and candies. Torn between the political position of ONHAE [the local organization for the rights of indigenous people] (absolutely opposed to the Company) and the old loyalties of clan and family (what do you have to eat?), Nanto opted for the latter, accepting the gifts and distributing them in Toñampare.

Nanto's decision meant that the Company would be back, with more things; and they would look for their oil; and, if they found it, one day the rivers in [missionary] Rachel Saint's protectorate—where eighty percent of the Huaorani live, where the resources they depend on are already stretched precariously thin, and where the Huaorani have no choice but to stay, since there is nowhere for them to go—one day, surely, those rivers will run black with Amazon crude.

Maxus [an American oil company] began the construction of its road last November [1992]. In protest, Moi and Enqueri led two hundred Huao men, women, and children on a long trek, by foot, canoe, truck, and bus, from the deep Amazon to the high Andes—to Quito, where they laid siege to the Maxus offices. . . . The ONHAE mantle has come to rest on the thin shoulders of Condorito, Enqueri—with whom, last month, Maxus announced that it had signed an agreement. . . .

On September 7th, eleven days after the agreement was signed, Maxus spilled thirty-eight thousand gallons of raw crude in the Oriente. What was perhaps most remarkable about this spill—only the latest in a string of accidents, violations, and confrontations that have occurred since Maxus began building its road—is that it happened before Maxus had opened its first well. Somehow, in the course of building its own pipeline, Maxus managed to rupture a line operated by Occidental Petroleum.

Of this I am sure: what the Company believes Enqueri has agreed to and what Enqueri believes he has agreed to are two entirely different things. This belief was reinforced just a few weeks ago, when, through the mysterious and difficult channels such communications take, I (and others) received a public letter from Moi. It had been written two weeks after the Maxus contract was signed. Maxus, he said, had bribed Enqueri "with an outboard motor, lots of gas, a new set of teeth, and glory." The contract had been kept secret until a few days before the signing, and even after the signing most Huaorani still did not know its contents, or what it meant for them. The tribal elders were "vehemently opposed" to the contract, and bitterly angry that it had been signed without benefit of legal counsel. "Several Maxus personnel have threatened to control the Huaorani with use of violence," Moi wrote. "The last thing Maxus told me was that if we bother them there will always be a military presence in Huaorani territory."

The Ecuadorian government has announced that it is accelerating oil development in the Oriente; it is opening new concessions throughout the Oriente, some in lands that the Huaorani have long considered theirs. (Petroecuador, in developing a field that will take advantage of the Maxus infrastructure, has reportedly dumped seven thousand gallons of raw crude in the Yasuní River.) By most estimates, the Oriente will be pumped dry in ten or fifteen years.

Raymond Bonner, from "Whose Heritage Is It?" in *At the Hand of Man* (1993)

The American ideal of a national park is not necessarily transportable to the developing world, as Raymond Bonner's study of the Maasai people and the Ngorongoro Crater in Tanzania shows. Many developing countries hope that ecotourism can be one of their sustaining industries for the twenty-first century; however, if not managed fairly and carefully, ecotourism can have negative effects on both wildlife and native cultures.

Ngorongoro is often acclaimed—justifiably—as one of the natural wonders of the world. It arose from volcanic eruptions five to ten million years ago, and at one time, its cone may have reached nearly three miles in height. An eruption blew off the top, then a million or so years ago, the cone collapsed; today, it is the largest unbroken, unflooded caldera in the world. Its thirty-five-mile rim is almost perfectly symmetrical. The walls, forested with dark green lichen-draped acacia trees and tangled shrubs, slope 2,000 feet down to the crater floor, a 100-

square-mile bowl, or nearly five times the size of Manhattan. In the midst of the floor's natural zoo is a profusion of pink—thousands of flamingos standing on their slender legs in the shallow water of Lake Magadi, a soda lake that at one time was more than fifty feet deep. It is almost impossible to be in the crater for more than ten minutes without spotting lions stretched in the sun. There are rhinoceros and elephants, and bat-eared foxes, a nocturnal animal the size of a large house cat. Spotted hyenas camouflage themselves in shallow niches dug into the sides of mounds. Buffalo herds graze with wildebeest and zebra.

When you drive out of the crater and head west, you are soon on the grassy Serengeti Plain. They [the plains] go on forever and ever—until finally, following the curvature of the planet, they meet up with the lower regions of the sky. Here, on these plains, where the sun's rays seem to arrive unfiltered, the numbers of wildlife are staggering; not thousands of wild animals, but hundreds of thousands—260,000 zebra and 400,000 Thomson's gazelle, with their brownish-red coats, dark stripes, and horns that curve back and then up. And 1,600,000 wildebeest—ungainly-looking antelope with buffalo-like horns, large heads, scraggly whiskers, and the hindquarters of a racehorse.

. . . No definitive history of the Maasai has been written and much of their past is still a mystery, but the name means someone who speaks Maa, and they are believed to have arrived in what is now Kenya in the fifteenth century, after migrating south up the Nile. Along the way, they conquered other tribes and acquired a reputation as fierce warriors; the Arab slave traders were careful to go around Maasailand. At the peak of their power, at the end of the nineteenth century, the Maasai ranged over an area that stretched between Lake Victoria on the west and the Indian Ocean on the east and from north of Mount Kenya to south of Mount Kilimanjaro, some 80,000 square miles altogether. In Kenya, they occupied some of the best land, including the Great Rift Valley, the mile-high region of today's Nairobi—the word means "cold" in Maasai—and the territory over which the British, in 1895, began building a rail line to connect the Indian Ocean coast with their territory of Uganda. The British slowly forced the Maasai off these lands, and onto reserves in the southern part of Kenya. . . . It is thought that the Maasai came to the Serengeti Plain and Ngorongoro Crater about the middle of the last century. At the time, the crater was occupied by another tribe, the Datoga, whom the Maasai defeated and forced out.

Today, there are about 300,000 Maasai in Tanzania and 200,000 in Kenya. And many times that number of cattle. The Maasai, like other pastoralist tribes, believe that all the world's cattle belongs to them, and some Maasai still do not take seriously the strictures of the modern state against theft; Maasai warriors stole more than 300 head of cattle and killed four people during raids around the village of Sakasaka, just south of the Serengeti, in early 1990. Maasai worship their cattle. When they meet, they ask, "How are the children? How are the cattle?" They sing to their cattle, young calves have a special place inside their small huts, and a man's status is determined by the number of cattle he has.

Westerners have long been enamored of the Maasai, who seem to fulfill the image of the "noble savage." Traditionally warriors coated their bodies with ochre, plaited their hair, stretched their earlobes, carried spears and, as part of the rite of passage, they hunted lions. "Physically they are among the handsomest of mankind, with slender bones, narrow hips and shoulders and most beautifully rounded muscles and limbs," Norman Leys, a doctor who spent twenty years in Africa at the beginning of [the twentieth] century, wrote about the Maasai. Isak Dinesen described the Maasai warrior as "chic," "daring," and "wildly fantastical." Many of the Maasai customs have been abandoned or prohibited by law, but it is still possible to observe Maasai who have not adapted to Western ways, particularly in Tanzania.

"Don't be romantic about the Maasai," a Tanzanian businessman and former government minister admonished when I asked about the forced removal of the Maasai from the Serengeti and their plight in Ngorongoro today. In part, his rebuke was a reflection of the tribal loyalties still prevalent in Africa. It was also good advice. There is nothing charming about much of the Maasai way of life, particularly for women. Maasai society is male-dominated, organized by age-sets: boyhood, warriors, elders. The manner in which women are treated falls somewhere between slavery and criminal abuse. Women gather the firewood, which they balance on their heads, and they fetch the water in ten- or twenty-liter containers, which they carry on their backs held by a strap around their foreheads. Women build the boma, or house, putting up the latticework of sticks, filling it with wattle and plastering it with dung, then putting on a roof of dung or thatch. What do the men do? "Supervise," says Kasiaro. And beat their wives if they do not perform their responsibilities—if the boma leaks, if there is not enough firewood. Women accept their status. "If I do wrong, my husband has the right to beat me," said a twenty-six-year-old Maasai woman as she waited in a long line with other women to fetch water that was trickling from a pipe.

But whether the Maasai are viewed romantically or harshly should not affect a judgment on how they have been treated by conservationists and governments who have coveted their land. And their story is one of how Africa's spaces have been set aside for preservation and tourism. Africa may be a vast continent, but the land that has been turned into parks wasn't idle land. Scores of tribes and millions of people have been dispossessed. And it is still happening. In 1991, 170 square miles of dense forest in southern Madagascar was set aside as Ranamafano National Park, created under pressure from American conservationists. Their desire was understandable—the forests are filled with exotic plants and animals found nowhere else on the planet, including several species of lemur, a cute, furry primate with a long tail and beady eyes. But 72,000 people lived in villages in or on the edges of the park. Many of them went into the park to cut wood, one of the few wage-paying jobs available; even more people took their cattle into the park. Now they are barred from it. . . .

The law creating Serengeti National Park had a special provision that allowed the Maasai who were living there at the time to remain. Even Major Hingston [who arrived in 1930 as a representative of Britain's Society for the Preservation of the Fauna of the Empire] would have accepted this. "The existence of Maasai in a National Park offers little difficulty since they neither hunt game nor cultivate the soil, and are thus better adapted than other native tribes to live amicably amongst the game," he wrote in the report of his mission. "Nor is it an important objection that the Maasai kill lions, since it has been found in the Kruger National Park that a certain amount of lion destruction has to be periodically carried out." (This is because lions do not stay inside a park's boundaries but wander onto ranches, where they kill cattle.)

Tanganyika government officials frequently, and publicly, reiterated the rights of the Maasai to remain in the park. "The rights of the Maasai are protected by law and cannot be abrogated," the district commissioner said during a meeting of Maasai elders and provincial officials at Ngorongoro Crater in 1952. Two years later, in a speech of the Maasai Federal Council, the governor of Tanganyika said, "I wish at once to reassure you that all Maasai and other pastoralists who have been normally resident within the area of the Park will not be turned out." The promises didn't survive the decade.

This is what we gave up! Kasiaro thought as he was riding through Serengeti National Park. His deep-set eyes sparkling, he exclaimed with wonderment when we passed herds of topi and hartebeest; gazelle leaped across the road in front of us, and we saw a klipspringer, a tiny antelope, perched like a ballet dancer on a few centimeters of rock. "My father used to tell me about the beauty of Serengeti, but I didn't know it myself," Kasiaro said at the end of the day, his first in the Serengeti. He grew up on the periphery of the park and had watched tourist vehicles stir up dust as they hurried from Ngorongoro to Serengeti, but until November 1990, he had never entered it. Spotting a lion with a radio collar, he asked, wonder-struck, "Who put that on him?" He was not aware that Serengeti has been one of the premier laboratories for the study of lions; today when you see a lion pride, you are just as likely to see a research vehicle with an antenna on it, and a student inside making sketches and notes. All of this was strange to Kasiaro, who as a young warrior had killed lions with a spear. After several hours' driving, as we crossed the Seronera River, where hippos were submerged up to their backs and their beady little eyes, Kasiaro had an idea. "Why didn't they give us part of this?" He laughed lightly. "Why couldn't they divide it, give us the right side of the road, and they take the left?"

From the outset, the park's trustees had hoped to entice the Maasai into leaving the park voluntarily by providing them with better grazing areas, along with wells and boreholes outside. It didn't work. Areas of the Serengeti were dear to the Maasai, rich with grass and water. One of these was the Moru Kopjes, a collection of granite boulders on the western edge of the plains; it was here that young warriors set up their camps, where they sang and danced and feasted, and set out on their lion-hunting forays. Another choice spot was Seronera, where the park headquarters is today. And the Maasai did not want to give up Ngorongoro, because that was one of their best grazing areas, one of the few places where there was permanent water. When the Maasai refused to leave voluntarily, the local authorities ordered them out, in the early 1950s. The Maasai are a tough people, and they remained intransigent. "We told them you better shoot us together with the cows; we are not going to leave the Serengeti," one of the Maasai who had been living in the Serengeti at the time, Tendemo ole Kisaka, recalled nearly forty years later. . . .

Underlying much of the campaign to get the Maasai out of the Serengeti was, of course, the colonial prejudice against Africans, which was particularly strong when it came to the Maasai. While many Westerners were awed by the Maasai, just as many, maybe more, found them arrogant and dirty. "They never wash," Alan Moorehead wrote in *No Room in the Ark*, which is deservedly considered a classic about African wildlife, but which reflects a colonial contempt for African people. "At certain times they streak their faces with colored grease and dirt, and braid their hair so as to give it the appearance of a mop," he wrote. He described some of their tribal ceremonies as "barbaric," and said that the Maasai "disdains all forms of trade and ordinary labor." So pervasive was the antipathy toward the Maasai, particularly among the white wardens in the Serengeti and in Kenya, that [Louis] Leakey felt compelled to stress in one of his many papers advocating the removal of the Maasai from the Serengeti, "I am not one of those who consider that the Maasai should be treated with contempt and disdain or that they are unfit to survive." . . .

The Maasai remain poor while living amidst a very valuable resource, the wildlife—which is the case for most rural Africans. Tourism has steadily increased in the Ngorongoro Conservation Area, and in 1989 the number of visitors reached a record level, as did the Conservation Authority's revenues—nearly $1 million. (This is in addition to what the hotels and safari companies earned and the income generated by Serengeti National Park, which is separately administered.) Obviously almost none of it has been spent to help the people. Nor have there been many, if any, indirect benefits for the Maasai. . . .

Though the Maasai receive no benefits, they pay the price of living with the wildlife, as a story Clemence [ole Rokango, a junior elder] told reveals.

"I was sitting here near the boma," Clemence began; to ward off the chill, he was wearing a red-checkered shuka, a robe-like garment, over a red toga. "My brother's child and my sister's child were walking to meet other children in that boma," he said, pointing across a shallow draw, where two of his sisters and their husbands and children live. "The leopard came—he was just laying in the long grass there—and caught one child near that boma. The other child cried. We heard this, and we saw the leopard running in the bushes there with the child.

We took our spears and we ran and chased it. We tried to find it, but it became very dark. We turned back home. It was raining and misty."

The next morning they took up the pursuit again, following bloodstains on the grass. They found the girl's skull. They continued looking for the leopard. "We saw him and he jumped, and he made a sound"—here Clemence made a low, rumbling sound from deep in his throat, which sounds different from the one he makes when imitating lions he has cornered. "We tried to follow it in the bushes, but we didn't see it again." During the next month, the leopard continued raiding people's bomas, killing calves and goats, and eluding Conservation Authority rangers who had rifles. Then one day the leopard returned and killed three goats near Clemence's boma. "We saw him, and we ran after him. He ran away. We said, 'We must kill it today.'" Clemence and his friends cut some branches and made a small hut, and tied a goat to it. A man hid inside. That evening, the leopard came and tried to seize the goat. The man threw his spear, hitting the leopard in its side, killing it.

When the Conservation Authority officials came to take the dead leopard—it would sell the skin—Clemence and other elders talked to them. "We said, 'The leopard killed many goats, and we said nothing. But now it takes a ten-year-old girl. Now the government must pay.' But they said no." . . .

The appeal of tourism is understandable: it could finance the country's development and could do so more quickly and with fewer capital costs than probably any other industry. But tourism is necessary not only to finance Tanzania's general development. As the director of Tanzania's National Parks, David Babu, put it in 1992, "What is the future of wildlife in Tanzania without tourism?" In other words, tourism can generate the revenues for park maintenance.

Many Third World countries see tourism as their economic savior; most are dreaming. Not Tanzania. The country has more elephants, more rhinoceros, more lions, and more of just about every other species, than Kenya. Four Tanzanian areas are World Heritage sites, a designation under a 1972 United Nations convention calling for the recognition and protection of natural and cultural areas of "outstanding universal value." Throughout the world, only seventy-seven natural sites have been granted this status, and none are in Kenya. Ngorongoro Crater was the first place in Tanzania to go on the World Heritage list (in 1979), and it was included for its cultural significance as well—there are Stone Age ruins buried deep under the crater's floor. The Serengeti was proclaimed a World Heritage site two years later. The next year the Selous Game Reserve, in the southern part of Tanzania, made the list. It is one of the largest protected areas in Africa—22,000 square miles—and one of the wildest. In 1986, Mount Kilimanjaro achieved the honor. Rising more than three and a half miles, it is the highest mountain in Africa, and a mile higher than any peak in the continental United States. It is a freestanding inactive volcano, not far from the equator, yet perpetually crowned with snow. It is an awesome and majestic sight.

Underlying the World Heritage Convention is the premise that some places are the heritage not just of the political entity in which they happen to be located, but of all mankind. It is a noble ideal. But if places are part of the "world heritage," then the world has an obligation to come up with the millions and millions of dollars needed to preserve them.

The white settlers in Africa and the men who set up the conservation organizations like AWF [African Wildlife Federation] and WWF were certain that once the Africans became independent and were free to run their own governments, they would destroy their natural heritage. How wrong they were: 8 percent of Zambia, 7 percent of Zimbabwe and 6 percent of Kenya are set aside as National Parks and Reserves. In contrast, in the United States, 7 percent of the land is National Parks and Reserves; in the United Kingdom, 6 percent; in Australia, 5 percent, and in Canada, 4.5 percent. New Zealand has the best record in the developed world, and still it has set aside only 9 percent. Twelve percent of Tanzania is National Parks and Reserves, according to IUCN [World Conservation Union], which maintains a worldwide list for the United Nations. (Tanzania says that twenty-five percent of its land is protected to some degree, but like nearly all countries Tanzania uses its own definition of what is a protected area. Under the IUCN definition, a National Park is a relatively large area that has not been materially altered by human exploitation and that contains plants or animals of special scientific, educational, or recreational interest or contains a landscape of great beauty. A Reserve is usually a smaller area, in which a specific species or habitat is protected.) . . .

More than thirty years ago, in *Serengeti Shall Not Die*, [Bernhard] Grzimek cried, "Must *everything* be turned into deserts, farmland, big cities, native settlements, and dry brush? One small part of the continent at least should retain its original splendor so that the black and white men who follow us will be able to see it in its awe-filled past glory." Then followed his prayerful declaration: "Serengeti, at least, shall not die."

On the book's final pages, he wrote:

> Men are easily inspired by human ideas, but they forget them again just as quickly. Only Nature is eternal, unless we senselessly destroy it. In fifty years' time nobody will be interested in the results of the conferences which fill today's headlines.
>
> But when, fifty years from now, a lion walks into the red dawn and roars resoundingly, it will mean something to people and quicken their hearts whether they are bolsheviks or democrats, or whether they speak English, German, Russian, or Swahili. They will stand in quiet awe as, for the first time in their lives, they watch twenty thousand zebras wander across the endless plains.

His cry has more resonance today, on a planet more crowded and more polluted, with fewer and fewer spaces to escape to. We need silence, places where we can stand in awe of that which God or Nature created. But who is going to pay for these places? Is it fair to put the burden on the Maasai, on the people of Africa, which has been the effect of the policies in the past? These glorious spaces *are* part of our heritage, and so we—New Yorkers and Berliners, Japanese and French, Canadians and Australians—must pay. Yes, Africans, too, must pay, but they are paying already. If we in the West don't meet our responsibility, if we don't match our rhetoric about the urgency of preserving Africa's wild animals and wilderness with money, then the future for the grandeur of the continent is grim. "The advance of civilization . . . is fatally disturbing to the primitive forms of animal life," the *Times* of London noted in calling upon the European nations to take action to preserve Africa's wildlife. That was at the beginning of this century, in 1900; now we are at its end. The dire prediction may become a sad reality, as modern civilization, which has marched slowly in Africa throughout most of this century, begins to gallop.

Helen Corbett, "The Rights of Indigenous Peoples" (1996)

Helen Corbett argues for self-determination as the most important political goal for indigenous people. This concept had its origin in the Fourteen Points that Woodrow Wilson brought to the Paris Peace Conference at the end of World War I. It was a difficult concept to apply then and it is difficult to apply now. Long-standing unrest in the Balkans and the Middle East, two areas whose boundary lines were drawn, in part, to protect the self-determination of the people within them, illustrates this. Most countries in the world today contain a diversity of indigenous groups, very few of which have political power in the national government.

The principle of the right to self determination is a major concern in the draft declaration [U.N. Draft Declaration on the Rights of Indigenous Peoples]. The problem lies in the fact that there is no clear agreement upon the definition, its content, applicability nor implementation of this right. The age old argument that the right applied only to European colonies in other continents fails to take into account the successful claims of nations and peoples such as those in the former Soviet Union, the former Yugoslavia, Eritrea, and Slovakia.

States argue that the granting of this right will impair territorial integrity and thus State's sovereignty. Yet increasingly today, we see a global pattern of the exclusive sovereign authority of states evaporating. Peoples are fast becoming increasingly linked into broader communities where there is shared sovereignty. For example the European Union has sovereignty distributed through fifteen European and Nordic States. In Hong Kong, there exist two distinct economies while it waits to revert to China's control in 1997 and [we witness] the increase

Figure 25.3 A Saami man leads a reindeer. © Newscom/Digital Press.

of local regional resolution of internal conflicts in Papua New Guinea, Haiti, Cambodia, and Australia's White Australia policy which has ended with half of Australian immigrants coming from Asia. The Nordic governments' joint action on some matters affecting Saami peoples is another case in mind.

To support their alarmist views, some States point to global regions where a protracted bloody and violent war has been used as a strategy to seek a people's right to self determination. Yet the UN community itself must bear some of the responsibilities for such human tragedies that arise from those battles.

Unfortunately it has given out a clear message that it will respond much quicker to those peoples using violence as a method to seek their right to self determination. The claims of those who do not choose violence are either not dealt with or very little progress is made over decades.

Indigenous peoples attending the WGIP [Working Group of Indigenous Populations] forums have demonstrated that they are willing to enter into peaceful negotiations with UN Member States on their right to self determination. As the WGIP Chairperson has said on a number of occasions, if they (indigenous peoples) wanted to dismember State's sovereignty, then they would not be sitting down in Geneva each year having dialogue with respective States. Instead, they would be on home battlefields.

Rigoberto Queme Chay, from "The Corn Men Have Not Forgotten Their Ancient Gods" (1993)

The Mayan culture of Mesoamerica predated the Aztecs and still survives today in Guatemala. Rigoberto Queme Chay, a communal leader of the Quiche group of Mayans, explains in this essay that the ceremonies and beliefs of his people instill a continuing respect for their land and the bounty it provides.

The Mayan culture, still alive today, has managed to survive five hundred years of colonization, a process that started with the conquistadores and continued at the hands of their descendants. Over all these years, we have been on the receiving end of a constant offensive against our forms of production and social organization, our culture and our religion, in reality a cult of nature.

We Mayas have invented forms of resistance to conserve our values. In some cases, this has enabled us to reject the values of the invaders, in others to adapt them, and in still others to accept them.

The Guatemala of today is a country torn by serious social conflicts rooted in two very different forms of life, two cultures tied together by a bond of domination-subordination. The aim of the conquistadores was to introduce Christianity as a form of ideological control and justification for new forms of oppression and exploitation of both man and nature.

As far as religion is concerned, the "cosmovision" of the Mayas differs fundamentally from the Jewish-Christian ideology forced on us from 1492 onward. The essence of the globalizing character of the Mayan religion lies in the way man sees himself and his relation to the world that surrounds him.

In the first place, time and space for the Mayas are primogenital gods, and nature is the superior force from which emanates the authority that gives direction to life and to the reproduction of all beings. Mayas believe that all nature is life: each animal, stone, and river has its own *nahual* or "divine personification."

The earth and water are superior to all other elements of nature because they are the origins of life. "Thus, for example, Tlaloc (in Nahuatl) or Chaks (in Yucateca Mayan) are the guardian gods of the rain, who pay particular attention to the milpas (the areas of land dedicated to corn growing) where tribute is paid to them, their intervention sought to ensure good harvests, and thanks given when that happens."[1]

Man bends to the design of nature, which he does not consider alien to himself and which he cannot exploit without mercy. The irrational use of the natural resources made available to man is held to be a sin.

A Mayan year is divided into 260 days, further divided into thirteen months, representing the cycle of corn in the bowels of the earth and the gestation cycle of the human being in the womb. Tradition has it that man was made from the dough of yellow corn after many attempts by his creators. Corn is omnipresent in all human activity, as food, as decoration, and as religious symbol. Most meals are based on corn, which, with beans and chili, makes up the basic diet. The dominant colors in women's clothes are yellow and green, the colors of grain and the leaves of corn, and the designs used for decoration recall the leaves of the plants.

The Mayan farmer observes certain propitiatory rituals and asks forgiveness for the wound he is about to inflict on Mother Earth. . . .

Many argue that the Mayas are happy living in their current condition of poverty, ignorance, lack of health care, and shortage of land because we carry out our social and productive activities accompanied by ceremonies replete with entertainment, music, color, and movement. We are considered folkloristic and picturesque.

The Mayas always have occasion for ceremony, whether for a burial or a wedding. That gives some people the idea that we accept the way things are. The reality belies such judgments. Our people have struggled without respite for their survival. Our demands for land and our protests against abuses and injustice have been violently suppressed for the last five hundred years.

Our resistance continues at various levels and in various geographic regions. Our claims are part of an alternative social model that envisages a type of economic development that is compatible with the preservation of nature.

The Mayas have been excluded from the running of the state and from the formulation of national policies that can express their point of view. The state does not recognize the customary rights of the Mayas nor our forms of social organization, which still remain in place in the rural communities.

We believe that any proposal aimed at finding a solution to the problems of the environment must recognize changes in land ownership patterns. The most important is to preserve and revitalize what remains of the indigenous communal ownership, which the state wants to do away with so that it can throw what little land still remains onto the market.

At the same time, the problem of poverty affecting eighty percent of the population has to be overcome, and there has to be a decentralization to combat the constant centralization of power, services, and investments that runs counter to the interests of rural communities.

Strategies and programs for conserving the environment should be less concerned with technology and education and more with effective community participation in their design. Only in this way can we overcome the technocratic point of view that touches only the superficial level and fails to deal with the deep-rooted and fundamental problems. . . .

What is most important is that everything be rooted in a political context. As long as the Mayas are a demographic majority but a political minority, the longed-for development tied to environmental preservation will not be possible, for the following reasons:

1. Development has to be based on peace, and this can occur only where there is participatory and representative democracy. This is not the case in Guatemala.

2. Socioeconomic problems (among which those of the environment are of importance) are the result of intolerance, exclusion, the excessive desire for profit, and an irrational approach to progress that is not shared by everyone—despite the "frustrated attempts to impose a code and a unique (colonial) political and cultural design on a pluralistic social grouping that, from the moment of its breaking away from the invasion, declared itself in opposition and took refuge inside a culture of resistance."[2]

3. The logic of capitalist productive and economic organization, as industrial and productive versions of historic socialism, is accumulation. The logic of ethnic-Indian economies, whether in the form of home or small market production, is anti-accumulation.[3]

4. To these, add the fact that for the Mayan culture Mother Earth is sacred and man is part of nature, and one can understand why we claim that the complete integration of the Mayas—with all their values and forms of organization—at the level of the state will lay the foundation for a new structure of society that is the precondition for harmonious and environmentally sound development. The Mayas' territorial claims and demands for self-determination should create the national objective of delimiting the space necessary for developing alternatives to the productive processes that hold sway today.

In conclusion, we have to stress that the final word has not yet been said, and we must listen to the indigenous people before it is possible to construct a new society that can resolve the apparent contradiction between economic development and conservation of the environment.

Notes

1. Carlos Guzman Bockler, *Donde Enmudecen las Conciencias* (Where the Consciences Are Silent), (Mexico City: Secretaria de Educación Pública, 1986).
2. Stefano Varese, *El Rey Despedazado* (The Torn King), quoted in *Utopia y Revolución* (Mexico D.F.: Editorial Nueva Imagen, 1981).
3. Ibid.

Daniel Nettle and Suzanne Romaine, from "Where Have All the Languages Gone?" in *Vanishing Voices* (2000)

Nettle and Romaine compare the loss of languages to the loss of species. This comparison is apt in many ways. Both languages and species evolve in direct response to their environments and, as a result, provide insights for understanding those environments. However, there are significant differences between species and languages that the authors do not mention: species need other species to survive, languages do not need other languages to survive; one person can be bilingual, but no organism can be a member of two different species; and there can be a value to loss of languages, that is, increased communication between cultures, whereas there is generally no value to loss of species.

The United States alone is a graveyard for hundreds of languages. Of an estimated 300 languages spoken in the area of the present-day U.S. when Columbus arrived in 1492, only 175 are spoken today. Most, however, are barely hanging on, possibly only a generation away from extinction.

A survey of the North American continent done some time ago in 1962 revealed that there were seventy-nine American Indian languages, most of whose speakers were over fifty (for example, the Pomo and Yuki languages of California). There were fifty-one languages with fewer than ten speakers, such as the Penobscot language of Maine; thirty-five languages had between ten and 100 speakers. Only six languages—among them Navajo, Cherokee, and Mohawk—had more than 10,000 speakers. It is almost certain that at least fifty-one of these languages have all but disappeared. Languages with under 100 speakers are so close to extinction that revival for everyday use seems unlikely. The remaining native American languages in California are not being taught to children. Among the many native American languages already lost are some which gave the Pilgrims their first words for the new things they found in America, such as *moose* and *raccoon*. Our only reminders of them now are these words and state names such as *Massachusetts*.

We have used terms such as "death" and "extinction" in relation to languages just as a biologist would in talking about species. This may sound strange or inappropriate. What justification is there for this? After all, languages are not living things which can be born and die, like butterflies and dinosaurs. They are not victims of old age and disease. They have no tangible existence like trees or people. In so far as language can be said to exist at all, its locus must be in the minds of the people who use it. In another sense, however, language might be regarded as an activity, a system of communication between human beings. A language is not a self-sustaining entity. It can only exist where there is a community to speak and transmit it. A community of people can exist only where there is a viable environment for them to live in, and a means of making a living. Where communities cannot thrive, their languages are in danger. When languages lose their speakers, they die.

. . . Should we be any less concerned about Taiap [a native New Guinean language] than we are about the passing of the California condor? Although the greatest threat is posed to the languages spoken by peoples whose cultures and traditional lifestyles are also at risk, language death is a problem found within modern nations as well. . . . In the Hawaiian islands, for instance, the majority of native plants and animals are, like the Hawaiian language, found nowhere else on earth and face impending extinction. Although the island state represents less than one percent of the U.S. total land mass, it has 363 (over thirty percent) of 1,104 species federally listed as threatened or endangered, including the yellow hibiscus, the state flower, and the Hawaiian goose (nēnē), the state bird. It is not coincidental that language endangerment has gone hand in hand with species endangerment. Languages are like the miner's canary: where languages are in danger, it is a sign of environmental distress.

We think there are many reasons why all of us—not just linguists, or those whose languages are under threat—should be alarmed at what is happening and try to do something to stop it. As a uniquely human invention, language is what has made everything possible for us as a species: our cultures, our technology, our art, music, and much more. In our languages lies a rich source of the accumulated wisdom of all humans. While one technology may be substituted for another, this is not true of languages. Each language has its own window on the world. Every language is a living museum, a monument to every culture it has been vehicle to. It is a loss to every one of us if a fraction of that diversity disappears when there is something that can have been done to prevent it. Moreover, every people has a right to their own language, to preserve it as a cultural resource and to transmit it to their children.

It is hard for most English speakers to imagine what it might mean if the English language were to die and they would no longer be able to speak it as they went about their daily activities. How would it feel to be the last speaker of English on Earth? Marie Smith, the last Eyak Indian of Cordova, Alaska, explained how she felt at being the only full-blooded Eyak and the only speaker of her language: "I don't know why it's me, why I'm the one. I tell you, it hurts. It really hurts. . . . My father was the last Eyak chief, and I've taken his place. I'm the chief now, and I have to go down to Cordova to try to stop the clear cutting on our land."

English has always seemed such a secure possession, despite the fact that after the Norman Conquest, its future was actually in some doubt. Yet it was likewise difficult for English speakers at that time to imagine that their language would one day spread all over the globe. Most English speakers take the present position and status of English for granted, and do not realize that English was very much once a minority language initially in all of the places where it has since become the mother tongue of millions. It has gained its present position by replacing the languages of indigenous groups such as Native Americans, the Celts, and the Australian Aborigines.

. . . Some of the detailed knowledge of the natural environment encoded in human languages spoken by small groups who have lived for centuries in close contact with their surroundings may provide useful insights into management of resources on which we all depend. At the moment, as many as one-quarter of the prescription medicines used in the United States are derived from plants which grow in the world's rain forests. We know that many more plants and trees growing in tropical rain forests may contain remedies and even cures for human diseases, but we may never learn about some of them because the rain forests are being destroyed.

Moreover, traditional knowledge tends not to be valued as a human resource unless it makes an economic contribution to the West. Even though the United States government recognized the Pacific yew as the most valuable tree in American forests because its bark can be processed to yield taxol, a drug useful in the treatment of ovarian cancer, the bark is still being burned as scrap or left to rot on the forest floor in the aftermath of wasteful logging operations. The next great steps in scientific development may lie locked up in some obscure language in a distant rain forest.

The Inuit people who inhabit northern Arctic regions developed ways for surviving in an extremely cold and adverse climate. Knowledge of which kinds of ice and snow could support the weight of a man, a dog, or a kayak was critical for the continued survival of the Inuit, so they were named individually. In the Native American language Micmac, trees are named for the sound the wind makes when it blows through them during the autumn, about an hour after sunset when the wind always comes from a certain direction. Moreover, these names are not fixed but change as the sound changes. If an elder remembers, for example, that a certain stand of trees used to be called by a particular name seventy-five years ago but is now called by another, these terms can be seen as scientific markers for the effects of acid rain over that time period. One Palauan traditional fisherman born in 1894 and interviewed by marine biologist R. E. Johannes had names for more than 300 different species of fish, and knew the lunar spawning cycles of several times as many species of fish as have been described in the scientific literature for the entire world.

Today scientists have much to learn from the Inuit people about the Arctic climate, and from Pacific Islanders about the management of marine resources. Much of this indigenous knowledge has been passed down orally for thousands of years in their languages. Now it is being forgotten as their languages disappear. Unfortunately, much of what is culturally distinctive in language—for example vocabulary for flora, fauna—is lost when language shift takes place. The typical youngster today in Koror, Palau's capital, cannot identify most of Palau's native fish; nor can his father. The forgetting of this knowledge has gone hand in hand with over-fishing and degradation of the marine environment.

. . . The areas with the greatest biological diversity also have the greatest linguistic/cultural diversity. These correlations require close examination and must be accounted for. Extinctions in general, whether of languages or species, are part of a more general pattern of human activities contributing to radical alterations in our ecosystem. In the past, these extinctions took place largely without human intervention. Now they are taking place on an unprecedented scale through our intervention—in particular, through our alteration of the environment. The extinction of languages can be seen as part of the larger picture of worldwide near total ecosystem collapse. Our failure to recognize our intimate connection with the global ecosystem lies behind what we will call the biolinguistic diversity crisis facing us today. What has brought us to this brink?

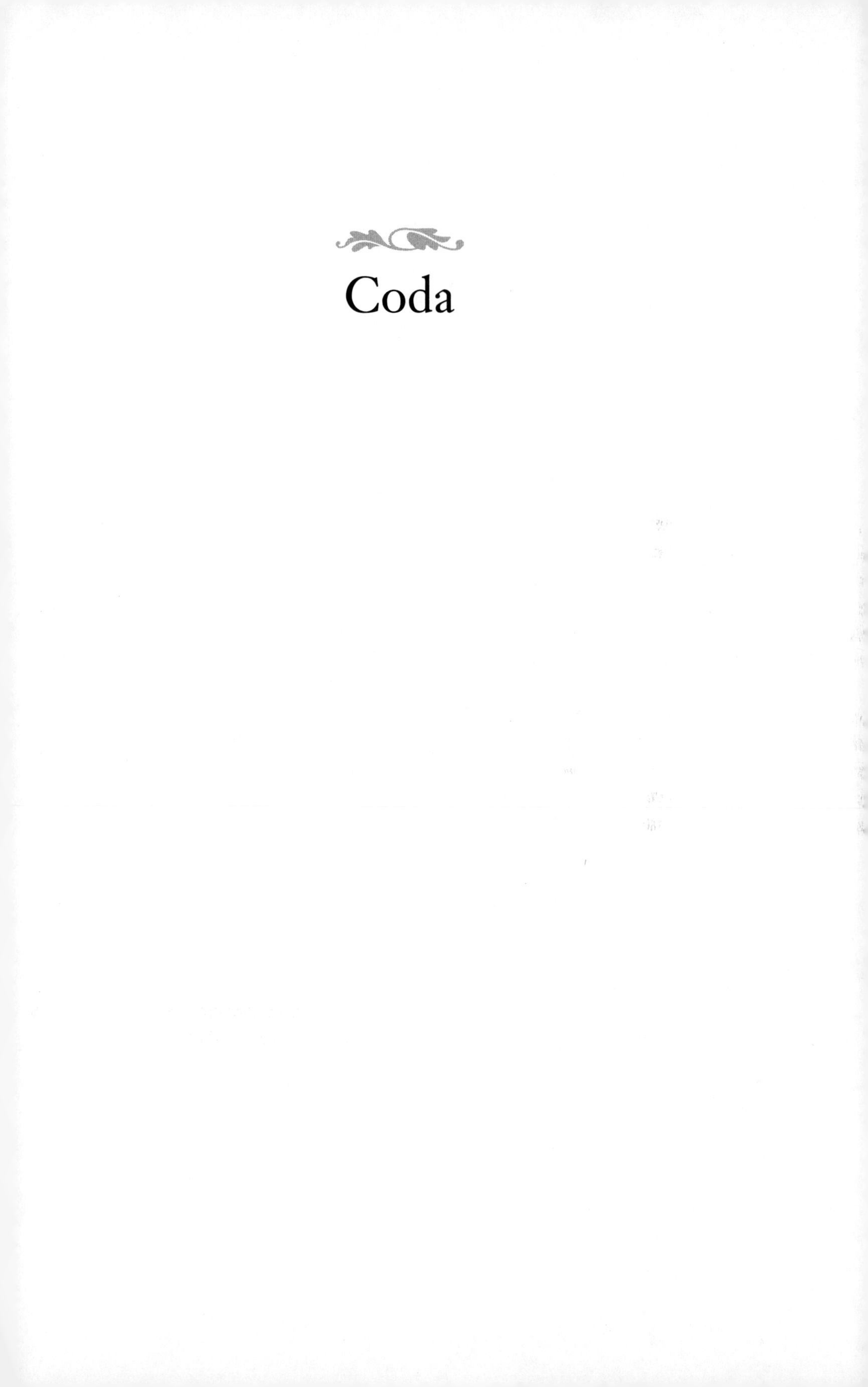

Coda

26 Conviction and Action

For the first time in two billion years, a single species has become so "successful" and technologically advanced that it is changing Earth's environment markedly. These changes are happening rapidly, in one ten-millionth of the planet's time of existence. Antarctic ice cores, which run back tens of thousands of years, reveal that the unprecedented changes of the past two centuries have come more swiftly than all but the most catastrophic of previous alterations, such as those following major meteor impacts. If current human habits continue—and, as William James notes, "habit is twice nature"—then the future will be, as Archie Carr reflects in his essay below on the Florida Everglades, "dubious." Yet, many environmental changes caused by humans are under conscious control: we can do something, even a great deal, about those changes. We have become chief stewards of the entire globe and its biosphere. This is now the single most important fact about both the human and natural conditions.

In contrast with environmental impacts produced by other species, the ones we generate now are by far the most profound and widespread. These consequences stretch back and forward in time. In burning fossil fuels and hydrocarbons we have, in 150 years, cremated hundreds of millions of years of animal and plant life. None of it is renewable, nor is much of the biodiversity we now are reducing, a diversity that required eons to develop. We are eradicating many species, even when it is unnecessary for, and may even diminish, our own preservation and health.

Norbert Wiener, godfather of modern computer technology and a humanistic as well as scientific visionary, argues in *The Human Use of Human Beings* (1954) that we have, in a few centuries, entered a radically new phase. Mastery over nature now is common and moves at "a still increasing pace." Yet Wiener cautions that "the more we get out of the world the less we leave, and in the long run we shall have to pay our debts at a time that may be very inconvenient for our own survival. We are the slaves of our technical improvement." We cannot, for example, "return a New Hampshire farm to the self-contained state in which it was maintained in 1800." In other words, "Progress imposes not only new possibilities for the future but new restrictions." However, Wiener is not pessimistic. He believes that "the new means of meeting these [new] needs" may be within our grasp—if we do not act blindly and ignore ultimate ends. It is this larger awareness of ends, of consequences beyond the present generation, that environmental education cultivates. We *can* achieve proper management and reverence. Because ignoring or downplaying ecological consequences can make doing certain things easier and seemingly cheaper, environmental conviction can seem alarmist and environmental action can appear to delay gratification needlessly. But, as Wiener notes, "The simple faith in progress is not a conviction belonging to strength, but one belonging to acquiescence and hence to weakness." To trust *only* further advances in technology is a form of ignorance. Real gratification follows a deeper sensibility.

Just as "progress" and technical applications do not happen simply because all the necessary elements are in place, the same is true of environmental consciousness and action, both individual and collective. They flourish through a willed, educated series of actions that revolutionize habitual thinking and habitual business. It's easy to see that, with the sobering exception of possible nuclear self-extermination, we are in general much better off today than two hundred years ago. We have shrunk time and space, fashioned new material goods, extended life spans, and sustained population growth to nearly seven billion people. But in the face of mounting evidence of the unsustainable character of many current environmental practices, blind faith in such progress masks a lack of imagination and a lack of concern for future generations.

Environmental issues can be issues of life or death; they *always* determine what kind of life or death. We can imagine the planet ravaged by toxins and radioactivity, its seas polluted to their depths, the air made permanently visible with suspended waste, a violent atmosphere, impoverished soils, and zoos struggling to perpetuate a miniscule number of the species eradicated elsewhere. There would still be life. There might be large, crowded human communities successful in adapting to conditions that we would consider alien or abysmal. Will it be a life our descendants find worth living?

We won't suffer the worst consequences of our environmental misjudgments. Our children will. Yet, caring and commitment, the desire to make a positive difference yoked with sound education, can advance a spiritual relationship with Earth, its life, and all its elements. In "Buckeye," below, Scott Sanders explores this kind of relationship. It all boils down to this: How much do we care?

We have created many "needs" that now we are re-examining. Do we "need" massive vehicles that consume gallons of gasoline to transport one person and one grocery bag a few miles? Do we "need" hundreds of pounds of packaging for one person's food each year? Do we "need" to keep buildings and cars at 20°C, about 70°F, winters and summers? Do we "need" a meat-rich diet predicated on rainforest destruction, excessive use of antibiotics and hormones, and water usage three times greater than the usage required by a diet that is healthier? Preferences rapidly become "needs." So, we can test and change preferences and choices. In the end, political action is often required, locally, nationally, and globally. Such action occurs, and exerts lasting results, when enough individuals commit to changing their own lives first. The information, knowledge, and judgments conveyed in this book (and in thousands of book and web pages) in essence convey one message. As interpreted by the poet Rainer Maria Rilke, it is the message voiced by the archaic yet entirely present statue of the ancient god Apollo—and a message also voiced by every other great spiritual teacher: "*You must change your life.*" And this is something we can do.

A correlation exists between healthy human styles of living and a sustainable environment. Put another way, environmental consciousness includes awareness of the condition of our own bodies. A connection between human and environmental health deserves keen attention. If someone breathes toxic air or drinks polluted water, these hazards will, sooner or later, damage that person's health. Yet, what logicians call the contrapositive turns out to be true, too: what promotes human health is good for the environment. In industrial nations many people have been abusing their bodies and the environment through excess. They often rescue themselves by drugs and medical advances, but the best and cheapest way to preserve health is preventive. Walking or cycling instead of driving is good for the body and reduces fossil fuel use; eating more grains and vegetables and less meat promotes longevity and reduces the environmental footprint of agribusiness; quitting smoking enhances health and improves indoor air quality. Environmentally conscious actions are, in part, health-conscious ones, and health-conscious decisions turn out to aid the environment, or at least they inflict less damage than unhealthy choices. This extends to farming, food processing and packaging, even the way healthcare systems set priorities for treatment and research. For example, should we continue to pour more money into treating diseases that are avoidable, or should we not only treat them but also spend money to prevent them? Per capita, the countries with the longest life expectancies do not spend the most on healthcare. They spend the most on prevention.

As consumers and citizens gain an informed understanding of environmental issues, they will participate more effectively in political action and public policy. The readings in this book also reveal that the humanities, sciences, and social professions together form their own ecosystem. They interact; they evolve with and through each other. Each has something important to contribute to an understanding of the environment and our responsibilities to it. As with all previous ideas of comprehensive general education, this reconfiguration of all the sciences and the arts seeks to shape conduct and to sharpen advocacy, both for personal ethics and for larger communities. This learning promotes personal welfare; and there is every reason that such learning can also promote a sustainable, globalized welfare embracing the human and the natural. Those embarking on a career, whether in anthropology or art, business management or medicine, law or zoology, will have opportunities to act as an informed friend and advocate of the environment. The opportunity for a better world, a better Earth, is ours.

FURTHER READING

Easterbrook, Gregg. *A Moment on the Earth: The Coming Age of Environmental Optimism.* New York: Viking, 1995.

Elgin, D. *Voluntary Simplicity.* New York: William Morrow, 1993.

McDonough, William, and Michael Braungart. *Cradle to Cradle: Remaking the Way We Make Things.* New York: North Point Press, 2002.

McKibben, Bill. *The End of Nature.* New York: Anchor, 1999 [1989].

———. *Enough: Staying Human in an Engineered Age.* New York: Times Books, 2003.

Moeller, Dade W. *Environmental Health.* Cambridge: Harvard University Press, 2004.

Steffen, Alex. *Worldchanging: A User's Guide for the Twenty-first Century.* New York: Harry N. Abrams, 2006.

Scott Russell Sanders, "Buckeye" in *Writing from the Center* (1995)

Among the foremost nature writers in English today, Scott Russell Sanders often describes in his work landscapes touched, sometimes destroyed, by humans. In one essay, "At Play in the Paradise of Bombs," he describes learning to love nature while growing up on a missile base. In "Buckeye," he finds deep meaning in the degraded forest where his father first taught him to love trees. Many nature writers speak for the pristine wilderness, but some need to speak for the altered yet still lovely land, or else there will be little chance of recovering the natural world that is close to home.

Years after my father's heart quit, I keep in a wooden box on my desk the two buckeyes that were in his pocket when he died. Once the size of plums, the brown seeds are shriveled now, hollow, hard as pebbles, yet they still gleam from the polish of his hands. He used to reach for them in his overalls or suit pants and click them together, or he would draw them out, cupped in his palm, and twirl them with his blunt carpenter's fingers, all the while humming snatches of old tunes.

"Do you really believe buckeyes keep off arthritis?" I asked him more than once.

He would flex his hands and say, "I do so far."

My father never paid much heed to pain. Near the end, when his worn knee often slipped out of joint, he would pound it back in place with a rubber mallet. If a splinter worked into his flesh beyond the reach of tweezers, he would heat the blade of his knife over a cigarette lighter and slice through the skin. He sought to ward off arthritis not because he feared pain but because he lived through his hands, and he dreaded the swelling of knuckles, the stiffening of fingers. What use would he be if he could no longer hold a hammer or guide a plow? When he was a boy he had known farmers not yet forty years old whose hands had curled into claws, men so crippled up they could not tie their own shoes, could not sign their names.

"I mean to tickle my grandchildren when they come along," he told me, "and I mean to build doll houses and turn spindles for tiny chairs on my lathe."

So he fondled those buckeyes as if they were charms, carrying them with him when our family moved from Ohio at the end of my childhood, bearing them to new homes in Louisiana, then Oklahoma, Ontario, and Mississippi, carrying them still on his final day when pain a thousand times fiercer than arthritis gripped his heart.

The box where I keep the buckeyes also comes from Ohio, made by my father from a walnut plank he bought at a farm auction. I remember the auction, remember the sagging face of the widow whose home was being sold, remember my father telling her he would prize that walnut as if he had watched the tree grow from a sapling on his own land. He did not care for pewter or silver or gold, but he cherished wood. On the rare occasions when my mother coaxed him into a museum, he ignored the paintings or porcelain and studied the exhibit cases, the banisters, the moldings, the parquet floors.

I remember him planing that walnut board, sawing it, sanding it, joining piece to piece to make foot stools, picture frames, jewelry boxes. My own box, a bit larger than a soap dish, lined with red corduroy, was meant to hold earrings and pins, not buckeyes. The top is inlaid with pieces fitted so as to bring out the grain, four diagonal joints converging from the corners toward the center. If I stare long enough at those converging lines, they float free of the box and point to a center deeper than wood.

I learned to recognize buckeyes and beeches, sugar maples and shagbark hickories, wild cherries, walnuts, and dozens of other trees while tramping through the Ohio woods with my father. To his eyes, their shapes, their leaves, their bark, their winter buds were as distinctive as the set of a friend's shoulders. As with friends, he was partial to some, craving their company, so he would go out of his way to visit particular trees, walking in a circle around the splayed roots of a sycamore, laying his hand against the trunk of a white oak, ruffling the feathery green boughs of a cedar.

"Trees breathe," he told me. "Listen."

I listened, and heard the stir of breath.

He was no botanist; the names and uses he taught me were those he had learned from country folks, not from books. Latin never crossed his lips. Only much later would I discover that the tree he called ironwood, its branches like muscular arms, good for axe handles, is known in the books as hophornbeam; what he called tuliptree or canoewood, ideal for log cabins, is officially the yellow poplar; what he called hoop ash, good for barrels and fence posts, appears in books as hackberry.

When he introduced me to the buckeye, he broke off a chunk of the gray bark and held it to my nose. I gagged.

"That's why the old-timers called it stinking buckeye," he told me. "They used it for cradles and feed troughs and peg legs."

"Why for peg legs?" I asked.

"Because it's light and hard to split, so it won't shatter when you're clumping around."

He showed me this tree in late summer, when the fruits had fallen and the ground was littered with prickly brown pods. He picked up one, as fat as a lemon, and peeled away the husk to reveal the shiny seed. He laid it in my palm and closed my fist around it so the seed peeped out from the circle formed by my index finger and thumb. "You see where it got the name?" he asked.

I saw: what gleamed in my hand was the eye of a deer, bright with life. "It's beautiful," I said.

"It's beautiful," my father agreed, "but also poisonous. Nobody eats buckeyes, except maybe a fool squirrel."

I knew the gaze of deer from living in the Ravenna Arsenal, in Portage County, up in the northeastern corner of Ohio. After supper we often drove the Arsenal's gravel roads, past the munitions bunkers, past acres of rusting tanks and wrecked bombers, into the far fields where we counted deer. One June evening, while mist rose from the ponds, we counted three hundred and eleven, our family record. We found the deer in herds, in bunches, in amorous pairs. We came upon lone bucks, their antlers lifted against the sky like the bare branches of dogwood. If you were quiet, if your hands were empty, if you moved slowly, you could leave the car and steal to within a few paces of a grazing deer, close enough to see the delicate lips, the twitching nostrils, the glossy, fathomless eyes.

The wooden box on my desk holds these grazing deer, as it holds the buckeyes and the walnut plank and the farm auction and the munitions bunkers and the breathing forests and my father's hands. I could lose the box, I could lose the polished seeds, but if I were to lose the memories I would become a bush without roots, and every new breeze would toss me about. All those memories lead back to the northeastern corner of Ohio, the place where I came to consciousness, where I learned to connect feelings with words, where I fell in love with the earth.

It was a troubled love, for much of the land I knew as a child had been ravaged. The ponds in the Arsenal teemed with bluegill and beaver, but they were also laced with TNT from the making of bombs. Because the wolves and coyotes had long since been killed, some of the deer, so plump in the June grass, collapsed on the January snow, whittled by hunger to racks of bones. Outside the Arsenal's high barbed fences, many of the farms had failed, their barns caving in, their topsoil gone. Ravines were choked with swollen couches and junked washing machines and cars. Crossing fields, you had to be careful not to slice your feet on tin cans or shards of glass. Most of the rivers had been dammed, turning fertile valleys into scummy playgrounds for boats.

One free-flowing river, the Mahoning, ran past the small farm near the Arsenal where our family lived during my later years in Ohio. We owned just enough land to pasture three ponies and to grow vegetables for our table, but those few acres opened onto miles of woods and creeks and secret meadows. I walked that land in every season, every weather, following animal trails. But then the Mahoning, too, was doomed by a government decision; we were forced to sell our land, and a dam began to rise across the river.

If enough people had spoken for the river, we might have saved it. If enough people had believed that our scarred country was worth defending, we might have dug in our heels and fought. Our attachments to the land were all private. We had no shared lore, no literature, no art to root us there, to give us courage, to help us stand our ground. The only maps we had were those issued by the state, showing a maze of numbered lines stretched over emptiness. The Ohio landscape never showed up on postcards or posters, never unfurled like tapestry in films, rarely filled even a paragraph in books. There were no mountains in that place, no waterfalls, no rocky gorges, no vistas. It was a country of low hills, cut over woods, scoured fields, villages that had lost their purpose, roads that had lost their way.

"Let us love the country of here below," Simone Weil urged. "It is real; it offers resistance to love. It is this country that God has given us to love. He has willed that it should be difficult yet possible to love it." Which is the deeper truth about buckeyes, their poison or their beauty? I hold with the beauty; or rather, I am held by the beauty, without forgetting the poison. In my corner of Ohio the gullies were choked with trash, yet cedars flickered up like green flames from cracks in stone; in the evening bombs exploded at the ammunition dump, yet from the darkness came the mating cries of owls. I was saved from despair by knowing a few men and women who cared enough about the land to clean up trash, who planted walnuts and oaks that would long outlive them, who imagined a world that would have no call for bombs.

How could our hearts be large enough for heaven if they are not large enough for earth? The only country I am certain of is the one here below. The only paradise I know is the one lit by our everyday sun, this land of difficult love, shot through with shadow. The place where we learn this love, if we learn it at all, shimmers behind every new place we inhabit.

A family move carried me away from Ohio thirty years ago; my schooling and marriage and job have kept me away ever since, except for visits in memory and in flesh. I returned to the site of our farm one cold November day, when the trees were skeletons and the ground shone with the yellow of fallen leaves. From a previous trip I knew that our house had been bulldozed, our yard and pasture had grown up in thickets, and the reservoir had flooded the woods. On my earlier visit I had merely gazed from the car, too numb with loss to climb out. But on this November day, I parked the car, drew on my hat and gloves, opened the door, and walked.

I was looking for some sign that we had lived there, some token of our affection for the place. All that I recognized, aside from the contours of the land, were two weeping willows that my father and I had planted near the road. They had been slips the length of my forearm when we set them out, and now their crowns rose higher than the telephone poles. When I touched them last, their trunks had been smooth and supple, as thin as my wrist, and now they were furrowed and stout. I took off my gloves and laid my hands against the rough bark. Immediately I felt the wince of tears. Without knowing why, I said hello to my father, quietly at first, then louder and louder, as if only shouts could reach him through the bark and miles and years.

Surprised by sobs, I turned from the willows and stumbled away toward the drowned woods, calling to my father. I sensed that he was nearby. Even as I called, I was wary of grief's deceptions. I had never seen his body after he died. By the time I reached the place of his death, a furnace had reduced him to ashes. The need to see him, to let go of him, to let go of this land and time, was powerful enough to summon mirages; I knew that. But I also knew, stumbling toward the woods, that my father was here.

At the bottom of a slope where the creek used to run, I came to an expanse of gray stumps and withered grass. It was a bay of the reservoir from which the water had retreated, the level drawn down by engineers or drought. I stood at the edge of this desolate ground, willing it back to life, trying to recall the woods where my father had taught me the names of trees. No green shoots rose. I walked out among the stumps. The grass crackled under my boots, breath rasped in my throat, but otherwise the world was silent.

Then a cry broke overhead and I looked up to see a red-tailed hawk launching out from the top of an oak. I recognized the bird from its band of dark feathers across the creamy breast and the tail splayed like rosy fingers against the sun. It was a red-tailed hawk for sure; and it was also my father. Not a symbol of my father, not a reminder, not a ghost, but the man himself, right there, circling in the air above me. I knew this as clearly as I knew the sun burned in the sky. A calm poured through me. My chest quit heaving. My eyes dried.

Hawk and father wheeled above me, circle upon circle, wings barely moving, head still. My own head was still, looking up, knowing and being known. Time scattered like fog. At length, father and hawk stroked the air with those powerful wings, three beats, then vanished over a ridge.

The voice of my education told me then and tells me now that I did not meet my father, that I merely projected my longing onto a bird. My education may well be right; yet nothing I heard in school, nothing I've read, no lesson reached by logic has ever convinced me as utterly or stirred me as deeply as did that red-tailed hawk. Nothing in my education prepared me to love a piece of the earth, least of all a humble, battered country like northeastern Ohio; I learned from the land itself.

Before leaving the drowned woods, I looked around at the ashen stumps, the wilted grass, and for the first time since moving from this place I was able to let it go. This ground was lost; the flood would reclaim it. But other ground could be saved, must be saved, in every watershed, every neighborhood. For each home ground we need new maps, living maps, stories and poems, photographs and paintings, essays and songs. We need to know where we are, so that we may dwell in our place with a full heart.

Archie Carr, "A Dubious Future" (1963) in *A Naturalist in Florida: A Celebration of Eden* (1994)

In the end, one's own life depends on nature, and it's good to experience what Thoreau did near the summit of Mt. Katahdin in Maine: "Talk of mysteries!—Think of our life in nature—daily to be shown matter, to come in contact with it—rocks, trees, wind on our cheeks! the solid *earth! the* actual *world! the* common sense! Contact! Contact!" *Archie Carr, a twentieth-century naturalist, lover and defender of the Florida Everglades, worried that the natural world was under attack in what Rachel Carson called "man's war against nature." Since the 1960s, Carr's Everglades have suffered more degradation and encroachment. Damage was done, but recent legislation and environmental engineering are restoring some beneficial water flows and wildlife. Carr, peering into the future, stands like the older, wise deer at the edge of the herd who warns the less wary of what they cannot, or choose not, to see.*

If the world goes on the way it is going, it will one day be a world without reptiles. Some people will accept this calmly, but I mistrust the prospect. Reptiles are a part of the old wilderness of Earth, the environment in which man got the nerves and hormones that make him human. If we let the reptile go it is a sign we are ready to let all wilderness go. When that happens we shall no longer be exactly human.

One of the awesome enigmas of today is how to slow the ruin of the natural earth while our breeding continues. There is no more need to multiply with the old fever. Breeding is good business, but it is herding our race toward a tragic impasse. When this is clearly seen and the reproduction is slowed down it will be because thoughtful people have taken charge; and these people will look about for what has been left of old values. One of the values is what the human spirit gets from wilderness—from all kinds of wild original landscapes and beings. The way we are going, what we keep of the old Earth will not be enough to save our honor with our descendants.

Writing this, I felt one of the qualms you cannot keep down when in your mind you weigh new industries against rough country empty of all but unused beasts and vegetables. I have no real doubts myself, mind you, but to many others in the world, especially the Florida world, to question the complete goodness of population growth is a perverse and sinister sort of iconoclasm that probably should be investigated by a committee. Thinking that way, I scared myself a little, and to get over it I called off the writing for a spell and went over to Lake Alice. Lake Alice is one of the solid assets of the University of Florida. It is a sinkhole lake with tree-swamp at one end and open water at the other, and all through it a grand confusion of marsh creatures and floating and emergent plants. The place is a little relic of a vanishing past, and, incredibly, it lies on the campus of a university with thirteen thousand students and less than half a mile from where I am writing now. It is there to go to when euphoria spreads through the press over some new gain the state has made in people.

I went this time to where an alligator called Crooked-Jaw has her nest beside a wire fence at one edge of the swampy end of the lake. I stopped the car and walked over to the nest and looked at it closely. I had taken a picture of it the day before, and I could see that Crooked-Jaw had made some changes during the night. They were not drastic—only small, fastidious adjustments to show she knew the heap was warming a new generation of her kind. A root-mass of buttonbush had been added, along with a few live switches of *Decadon* and some scooped-up slush of coontail from the bottom. On the top of the pile was a single balled-up pink paper towel; and though it seems unlikely, I am sure I had seen this lying six feet to one side of the nest the day before. I can say that because I was roused at the sight of it, at the idea of anybody defiling with pink the premises of an alligator nest. Crooked-Jaw clearly failed to share my resentment. There is no accounting for tastes. The nest did not look as good to me with the paper towel on it, but the matter was not in my hands.

The alligator was not in her usual station, her lying-in pool, as it were—the little dredged-out hole of water a mother gator waits in for eggs to hatch. She was off somewhere among the floating islands, and I started croaking—*eer-rump, eer-rump*—like a little gator. A long way out through the flooded willows a floating island began to quake; and then all at once water surged out from the frogbit raft beside the waiting pool, and Crooked-Jaw came up looking at me. A gallinule whined from a bonnet patch, and in the high haze to the west, the sandhill cranes were bugling. I croaked some more, but the alligator had lost interest. She sank into the water till her chin rested on the mud, and only the bumps of her eyes and nose and the big scales of her back stuck out.

Looking at her there in her fragment of a doomed landscape, I was sure again that the saving of parts of the primitive earth has got to be done, and that it has got to be done without trying to justify it on practical grounds. Species and landscapes must be kept because it pleases people to contemplate them and because freer men of future times will be appalled if we irresponsibly let them go. Not facing that fact seems to me the great weakness in the outlook for wilderness preservation today.

It will take resolute people to put abstract values in place of material progress. In testing the mettle and conscience of recruits for the work, the reptile—particularly the unloved, legless snake—may serve as a sort of shibboleth. A man who feels in his bones that snakes must be kept in the woods will be proper stuff for the struggle coming.

Snakes are probably disappearing at a more rapidly rising rate than any other group of vertebrates. Besides the widespread antipathy they get from man, marshes are drained, country is reforested in pure stands of unsuitable cover, poisons spread abroad kill off the food supplies of the creatures snakes eat and even kill the snakes themselves. But the most spectacular thing happening to snakes is the onslaught of cars on the roads. In his book *That Vanishing Eden*, Thomas Barbour spoke of the passing of snakes before cars on the roads of Florida, but he never saw the big change. It came with the many-laned highways of the fifties and sixties.

The worst snake traps are the causeways across marshes and the streams of cars that cross them. Snakes are lured to them to enjoy the warm pavement or to escape flooded habitat, or they encounter them merely in the course of their foraging. I remember a vast dying of snakes on the road across Paynes Prairie decades ago, when man and weather chanced to move together against the creatures of the marsh. On October 18, 1941, a hurricane moved in from the Gulf and spun in the vicinity for thirty-six hours, bringing fourteen inches of rain during five days. The prairie changed from a marsh to a lake, and the water rose so high that only the tips of the tallest grasses showed. On the twenty-fifth some students brought in two hundred snakes they had caught along the road-fill and told of a great hegira of snakes and of congregations of buzzards squabbling over the ones mashed by passing cars. There was clearly something extraordinary going on, and four of us from the biology department went out to investigate. We started at the northern edge of the prairie and walked abreast down the road with flashlights, one of us at each guardrail and two along the middle of the pavement. The road over the marsh was two miles long. We counted every snake dead or alive between the guardrails, which in those single-lane days were twenty feet apart. We picked up 723 snakes in the two miles, about two-thirds of them dead or injured.

As an accumulation of several days, this number of casualties would not have been unprecedented. But these were the accumulation of no more than the four hours or so since sundown. During the daylight hours buzzards—black vultures and turkey vultures—had been attracted to the killing by the hundreds and had carried the dead snakes away almost as fast as they were run over. So the snakes we counted had been killed after dark. The tally was: 184 red-bellied snakes; 200

ribbon snakes; 85 green water snakes; 64 banded water snakes; 55 garter snakes; 19 Allen's mud snakes; 6 brown snakes; 4 cottonmouth moccasins; 3 horn snakes; 3 king snakes.

The slaughter had no noticeable effect on the levels of snake populations in the prairie. For a decade afterward the road remained a mecca for snake collectors, and they kept coming from distant places to walk along it with bag and stick. But in recent years the prairie snakes have declined. Although the roadside was made a wildlife sanctuary, and the snakes in it are now immune to people who used to take them away in sacks, the cars keep going by, and snakes have no immunity to them.

No significant preserving of nature can be done with slight sacrifice. The true test will come when great sacrifices are needed, when it becomes necessary to fight the indifference of most of the world and the active opposition of much of it, to surmount man's ingrained determination to put the far future out of his mind in matters of current profit.

Besides the inherent technical difficulties of wilderness conservation, the effort to save original nature faces a whole constellation of other kinds of problems. The easiest obstacle to recognize is the opposition by people who for material reasons oppose the keeping of wilderness. There is another block of humanity that simply does not care and an unsorted lot made up of those who think of themselves as conservationists—and who in one way or another are, but who are not facing the real tough obligation at all. I refer to all people who think of saving nature for meat, water, timber, or picnic grounds for the future; and to the hunters who hope their grandsons will get red blood by shooting things; and to the reverence-for-life cultists who are foredoomed to inconsistency; and to the biologists who resist the loss of material for study; and to keepers of zoological gardens who preserve nature in cages. Putting this mixture of motives and aspirations together under the label conservation has made, in some cases, a temporarily stronger front. But it has muddied the real issue, hidden the dimensions of the long job and kept everybody from articulating the awful certainty that the hard saving has got to be done for the sake of abstract values.

For several years I have been involved in a preservation program that has been atypically feasible. This is a campaign to rehabilitate the green sea turtle, *Chelonia mydas*, in the Caribbean Sea, where its once extensive nesting range has been reduced to only two rookery beaches.

In the Caribbean, the way things were going a short while ago, the green turtle was facing complete extirpation. Now I believe there is no such danger. The change in outlook was made possible by a combination of circumstances such as cannot be counted on in most preservation projects. In the first place, the suspected migratory feats of green turtles focused scientific interest on them and brought support from research foundations—the National Science Foundation and the Office of Naval Research—for studies of their basic natural history. A major factor that has greatly eased the way for preservation is the lucky circumstance that the single nesting beach remaining in the western Caribbean is located at Tortuguero, on the coast of that gem of a small nation, the Republic of Costa Rica. In former times exploitation of the Tortuguero colony brought Costa Rica a steady small revenue in the form of a fee paid by the concessionaire, who parceled out the beach to the turtlers and sold their catch to the Cayman schooners or sent it away as deck-loads on freight boats going back to Florida. But in 1957 the government closed the beach to exploitation. The move saved the green turtle for the western Caribbean, but it also deprived Costa Rica of all profit from its green turtles because there is no good turtle pasture along the Costa Rican shore and no turtles go there except during the breeding season. The refuge will repopulate the pastures from Colombia to Mexico and will increase the yield of the turtle grounds of the Nicaraguan Miskito Coast to schooners turtling for the markets of New York and Europe. For Costa Rica itself there is only the satisfaction of having faced the choice between quick gain and a better future—and having chosen with characteristic wisdom.

In 1955, when the first of a series of grants from the National Science Foundation was made, a tagging camp was established at Tortuguero. The information accumulated helped stimulate the founding of the Caribbean Conservation Corporation, a nonprofit undertaking dedicated to restoring the Atlantic green turtle in American waters.

The world is responsible for reptiles. The inadvertent saving of scraps will never keep off the ruin of the earth. The only way is to name the real obligation clearly, to say without hedging that no price can be set for the things that have to be preserved. Basically, what must be done are the harder jobs, like justifying a future for snakes, which have no legs, hear no music, and badly clutter subdivisions. Bore through to the core of what is required and you see that it is an aggressive stewardship of relics, of samples of original order, of objects and organizations of cosmic craft. This work will take staunch people, and the reptile can be the shibboleth by which they pass.

To get the real feel of the problem, I conjure up a man of some far future time walking in a last woods lying unruined among launching pads of a planetary missile terminal and coming astounded upon the last of all living individuals of *Crotalus adamanteus*, the great unruly diamondback rattlesnake. It is a full-grown female snake that I see, two yards long, stern of face, and all marked off in geometric velvet. It is the sort of being that always, inadvertently and without malice, has been a thorn in the flesh of Americans, one of the novel terrors the land held for humans whether they came in caravels or wandered down into the New World out of the snake-free Siberian cold. Seeing the man, this last diamondback begins readying the steel of its coils, and they ebb and flow behind the thin neck holding the broad head steady and still, except for the long tongue waving. By the girth of her I judge that this is a pregnant snake, heavy with some dozens of prehatched perfect little snakes the same as herself, all venomous and indignant from the start, all intractable and, like their mother, unable to live except as free snakes.

The snake that confronts the imagined man is a moving thing to see. It is not easy to understand all the feelings aroused by such a sight, and the snake I think forward to is the last in all the pabulum agar culture of the purified world. The coils of her body rise and fall in slow spirals, the keen singing of her rattle sounds, and she waits there, testing with the forks of her tongue the whole future of her kind. In my thought the man then stoops with an old urge and picks up a stick. It is almost the only stick left lying in the eastern half of North America, and the man takes it up and moves in closer to the wondering snake. He raises the stick, then somehow lowers it as if in thought, then halfway brings it up again. And then the conjuring fails for me, and the snake song falls away, like the song of cicadas losing heart, one by one. The woods grow dark and fade off into distant times.

David S. Wilcove and Thomas Eisner, "The Impending Extinction of Natural History" (2000)

Modern science moves so rapidly that one area can seem obsolete or quaint almost overnight. Only later does its value re-emerge in a changed, urgent form. The world has many museums of natural history; as a discipline of modern science natural history is venerable. But few now train as natural historians. Even some biologists call the field outmoded. Yet to others its unique understanding of the living world has become essential not merely to taxonomy and history but also to deeper environmental consciousness and a vision of the future. Faced with competition, valued species can die out. Confronting more sophisticated and recent specialties, the active understanding of natural history can become extinct too. It's already endangered.

Imagine you are a naturalist with a liking for insects. You are interested in how insects make a living, in how they are fit for survival. You marvel at how protected they are as adults, when they are able to fly. And you think of how helpless they are as eggs and pupae, when they are stuck in place, unable to take evasive action. True, pupae are sometimes enclosed in protective cocoons, or hidden in dugouts in the soil, but some live out in the open, where they are exposed to a world of predators. How, for instance, do the pupae of ladybird beetles (family Coccinellidae) manage to survive? They are typically affixed to stems or leaves, where one would imagine they don't stand a chance against ants. Might they have special weaponry? You look closely and find that they do. They have what are essentially biting devices, in the form of clefts along the backs of their abdomens that they can open and close and use to snap at ants that come too close.

As a naturalist with a Darwinian bent, you wonder whether such snapping devices are present in every ladybird-beetle pupa or whether, in the best evolu-

Figure 26.1 *Course of Empire: Desolation* by Thomas Cole, 1836. © Collection of The New-York Historical Society, 1858.5.

tionary tradition, different ladybird species have come to possess variants of this defense. You look at different species and find that, yes indeed, the beetles of one genus, *Epilachna*, which includes among others the Mexican bean beetle and the squash beetle, have evolved a remarkable alternative defense. Instead of the pinching devices, *Epilachna* pupae have a dense covering of tiny glandular hairs, the secretion of which forms a potent deterrent to ants.

You get in touch with chemists, whom you provide with a sample of the secretion, and in due course you find out that you have stumbled upon a unique group of chemicals. The substances include some fascinating new ring structures of enormous size—so novel, in fact, that the paper you eventually write on the secretion with your colleague chemists attracts wide attention.

The discovery may look serendipitous, but it was not. It was driven by rational inference from pure, old-fashioned natural history, the close observation of organisms—their origins, their evolution, their behavior, and their relationships with other species. That kind of close, scrupulous observation of nature has a long and illustrious history, but it is now sliding into oblivion.

The scenario we describe actually happened to one of us (Thomas Eisner). The impending extinction of natural history is very real as well. In schools and universities, in government agencies and research foundations, natural history has fallen out of favor. What was once considered a noble field of inquiry—no less a figure than Charles Darwin proudly called himself a natural historian—is now viewed as a relict discipline, a holdover from the era of Victorian cabinets and private butterfly collections. A knowledge of, or even an avowed interest in, natural history is no longer a prerequisite for admission to a graduate program in ecology or any other branch of biology. Financial support for basic natural-history research has all but evaporated. Even the field trip, long a staple of science education from the primary grades through graduate school, has become increasingly uncommon.

This deinstitutionalization of natural history looms as one of the biggest scientific mistakes of our time, perpetrated by the very scientists and institutions that depend upon natural history for their well-being. What's at stake is the continued vibrancy of ecology, of animal behavior and botany, of much of molecular biology, and even of medicine and biotechnology. A knowledge of natural history enables the professional ecologist to see functional relationships in nature, to uncover the broader patterns that lead to critical scientific advances. Natural history also provides the "nuts and bolts" information necessary for managing wildlife and other natural resources. As the president of the Society for Conservation Biology recently lamented, "How can we possibly construct . . . a successful recovery plan for an endangered bird when we lack basic information on such things as what it eats, where it nests, and so on?" For the molecular biologist, natural history is often the path to finding something truly strange and wonderful, like the elaborate chemicals that protect the pupae of certain ladybird beetles. Even the search for new medicines can benefit from natural history. Was it not in his capacity as a natural historian that Alexander Fleming saw significance in the observation of a zone of bacterial inhibition around a *Penicillium* mold growing in a petri dish, a discovery that launched the era of antibiotics?

Perhaps the strongest argument in support of natural history is simply the magnitude of our current ignorance about nature. To date, scientists have discovered and described approximately 1.5 million species. That tally represents only a small fraction of the total number, perhaps less than a tenth. Even in the United States, where approximately 200,000 species (terrestrial, freshwater, and marine) have been described to date, an additional 100,000 to 400,000 may await discovery. And only a tiny fraction of the described species have been studied in any detail. Given how little we know about nature, it hardly makes sense to discourage its further exploration.

Several factors have contributed to the demise of natural history. As any field of scientific inquiry matures, it has a tendency to become more theoretical. Previously unconnected observations are brought together under the mantle of a set of unifying principles. Scientists who contribute to that body of theory emerge as the leaders in the field: they are the ones who are hired by research universities, who receive tenure, and who then encourage their graduate students to follow in their footsteps. (This is not to say that one cannot be both a first-rate natural historian and a first-rate theoretician, but such individuals are the exception rather than the rule. Most scientists tend to be strong in one or the other.) No one can blame the universities for wanting to hire the rising stars in each discipline, but with respect to the natural sciences, the practice has led to an unanticipated but regrettable result: The traditional natural historian has been pushed to the margins of academe. Moreover, the institutions that finance scientific research, be they governmental or private, are drawn to the leaders in any given field and may wrongly assume that the natural historian has comparatively little to contribute. Unable to obtain support for their research, the natural historians drop even lower in the academic pecking order.

At universities, the key to reversing the situation lies in hiring (and eventually granting tenure to) scientists with an abiding affection for natural history. Unfortunately, a Catch-22 applies here. Administrators and senior professors who are uninterested in or even hostile to natural history are not likely to value it when judging candidates for junior faculty positions. And without access to entry-level positions, a new generation of natural historians will never emerge to become tomorrow's administrators and senior faculty members. The institutions that pay for research, however, could assume a leadership role in rescuing natural history. Were more money available for basic natural-history studies, we are convinced that more graduate students and faculty members would incorporate natural history into their researching and teaching.

An even more fundamental step would be to reinstate natural-history studies in elementary and secondary school. Most children are fascinated by plants and animals—from dandelions to dinosaurs. That seemingly innate interest, if nurtured by adults, can become a lifelong joy or even the path to a career. Untended, it usually atrophies as a child grows older. For the price of a stereo microscope, now less than $250, a science teacher can turn a pinch of soil into a bustling world of springtails, oribatid mites, and nematodes, creatures as bizarre and engaging as anything to appear in a *Star Wars* movie.

The current push to connect every classroom in America to the Internet demonstrates how quickly elected leaders and the public can be galvanized to address what is rightly perceived to be a critical educational need. Meanwhile, the demise of natural history goes unnoticed, increasing the likelihood that future generations of schoolchildren will spend even more time indoors, clicking away on their plastic mice, happily viewing images of the very plants and animals they could be finding in the woods, streams, and meadows they no longer visit.

David Orr, "Is Conservation Education an Oxymoron?" (1990)

A positive correlation exists between how well-educated a society is and how much environmental damage it has done. Does that correlation imply causation? David Orr's seemingly simple essay undermines some of our most cherished ideas about education, the nature of knowledge, and the knowledge of nature.

For those calling themselves conservation educators it is sobering to note that the only people who have lived sustainably in the Amazon rain forests, the desert Southwest, or anywhere else on earth, couldn't read (which is not to say that they were uneducated). And those in the United States living closest to the ideal of sustainability, the Amish for example, don't make a fetish of education, seeing it as another source of deadly pride. On the other hand, those whose decisions are wreaking havoc on the planet are not infrequently well educated, armed with B.A.s, B.S.s, LL.B.s, M.B.A.s, and Ph.D.s. Elie Wiesel has made the same point in a different context, noting that the designers and perpetrators of Auschwitz, Buchenwald, and Dachau, the heirs and kin of Kant and Goethe, also possessed quite substantial academic credentials.[1] It would seem, then, that the relationship between education and decent behavior of any sort is not exactly straightforward. Three possibilities about this relationship are worth considering.

First, perhaps education is part of the problem. Cultures capable of generating an alphabet and written language have tended to become environmentally destructive. Written language is implicated in the rise of cities, agricultural surpluses and soil erosion, fanatical belief systems, irate and well-armed pilgrims, armies, usury, institutionalized greed, notions of collective aggrandizement, and eventually pro-growth hucksters. All of which take a toll on soils, forests, wildlife, and landscape—hence Chateaubriand's observation that forests precede civilization and deserts follow it. In the larger scheme of things, education may only have made us more clever, not ecologically wiser.

As circumstantial evidence I offer the observation that the time and expense of higher education is most often excused on the grounds that it increases lifetime earnings, a crude but useful measure of the total amount of carbon the scholar is able to redistribute from the earth's crust to the atmosphere. It is somewhat rarer for education to be extolled on the grounds that it reduces the graduate's impact on the biosphere, or because it hones their skills in the art of living simply. Such claims are indeed sufficiently rare that we may reasonably surmise that, on average, those whose lifetime earnings are enhanced by degrees do more damage to the planet than those who are less encumbered.

Second, it may be that beyond some fairly minimal level, education is simply not an important determinant of behavior, ecological or other. There is a shelf of dust-laden studies about the difference education makes. And what difference does four years of higher learning make? The conclusions, given present tuition rates, are remarkably ambivalent. For the majority, peer influences seem to be a more important source of ideas and behavior than professors or courses. Most students seem to regard education as a ticket to a high-paying job, not as a path to a richer interior life, let alone to saving the planet. We also have reason to believe that television, the automobile, and cheap energy have had a greater influence on ecological behavior than formal schooling.

A third possibility is that under certain conditions, education might exert a positive influence on ecological behavior, but that these conditions by and large do not now prevail. Higher education, particularly as it takes place in prestigious universities, is often animated by other forces including those of pecuniary advantage and prestige. "Academic professionalism, specialism, and careerism," in Bruce Wilshire's words, "have taken precedence over teaching, and the education and development of both professors and students has been undermined." The "moral collapse" that he describes results from the separation of the professionalized intellect from the personhood of the scholar. Moreover, the university "exists in strange detachment from crucial human realities, and perpetuates the implicit dogma that there is no truth about the human condition as a whole."[2] . . .

Defenders of the generic university tend to justify it not on grounds of the quality of teaching or the moral refinement and ecological rectitude of its faculty and graduates, but rather on its contributions to what, with suitable gravity, is called the "fund of human knowledge," i.e., research. And what can be said of this form of human activity? Historian Page Smith, for one, writes that

> the vast majority of research turned out in the modern university is essentially worthless. It does not result in any measurable benefit to anything or anybody. It does not push back those omnipresent "frontiers of knowledge" so confidently evoked; it does not *in the main* result in greater health or happiness among the general populace or any particular segment of it. It is busywork on a vast, almost incomprehensible scale. It is dispiriting; it depresses the whole scholarly enterprise; and most important of all, it deprives the student of what he or she deserves—the thoughtful and considerate attention of a teacher deeply and unequivocally committed to teaching.[3]

There is more to be said. Most research is aimed to further human domination of the planet. Considerably less is directed at understanding the effects of domination. Less still is aimed to develop ecologically sound alternatives that enable us to live within natural limits. . . . Ultimately our survival may depend as much on *re*discovery of what Erwin Chargaff once called "old and solid knowledge" as on *re*search.[4] In this category I would include knowledge of justice, appropriate scale, the synchronization of morally solvent ends and means, sufficiency, and the vernacular knowledge of how to live well in a place.

The university's preoccupation with research rests on the belief . . . that ignorance is a solvable problem. Ignorance is not solvable because we simply cannot know all of the effects of our actions. As these actions become more extensive and varied through "research and development," knowledge grows. But like the circumference of an expanding circle, ignorance multiplies as well. (This is not true, I think, for what is called wisdom, which has to do with knowledge about the limits and proper uses of knowledge.) The relationship between ignorance and knowledge is not zero-sum. For every research victory there is a corresponding increase in ignorance. The discovery of CFCs, for example, "created" the ignorance of their effects on climate and stratospheric ozone. In other words, what was until 1930 a trivial, hypothetical area of ignorance became, with the "advance of knowledge," a critical and possibly life-threatening gap in human understanding of the biosphere. Likewise our ignorance of how to safely and permanently store nuclear waste did not exist as an important category until we discovered how to make a nuclear reactor. This is not an argument against knowledge or for ignorance. It is, rather, a statement about the physics of knowledge and the peril of thinking ourselves smarter than we are, and smarter than we can ever become. In Wendell Berry's words:

> If we want to know and cannot help knowing, then let us learn as fully and accurately as we decently can. But let us at the same time abandon our superstitious beliefs about knowledge: that it is ever sufficient; that it can of itself solve problems; that it is intrinsically good; that it can be used objectively or disinterestedly.[5]

The belief that we are currently undergoing an explosion of knowledge is a piece of highly misleading and self-serving hype. The fact is that some kinds of knowledge are growing while others are in decline. Among the losses are vast amounts of genetic information from the wanton destruction of biological diversity, due in no small part to knowledge put to destructive purposes. As David Ehrenfeld has observed, we are losing whole sections of the university curriculum in areas such as taxonomy, systematics, and natural history.[6] [See also Wilcove and Eisner, above.] We are also losing the intimate and productive knowledge of our landscape. In Barry Lopez's words: "year by year, the number of people with firsthand experience in the land dwindles . . . herald[ing] a society in which it is no longer necessary for human beings to know where they live except as those places are described and fixed by numbers."[7] On balance, I think, we are becoming more ignorant, because we are losing knowledge about how to inhabit our places on the planet sustainably, while impoverishing the genetic knowledge accumulated through millions of years of evolution. And some of the presumed knowledge we are gaining, given our present state of social, political, and cultural evolution, is dangerous; much of it is monumentally trivial.

Conservation education need not be an oxymoron. But if it is to become a significant force for a sustainable and humane world, it must be woven throughout the entire curriculum and through all of the operations of the institution, not confined to a few scattered courses. This will require a serious effort to rethink the substance and process of education, the purposes and use of research, the definition of knowledge, and the relationship of institutions of higher education to human survival. All of which will require courageous and visionary leadership. In the mounting battle for a habitable planet it is time for teachers, college and university presidents, faculty, and trustees to stand up and be counted.

Notes

1. Elie Wiesel, *On Global Education: Address Before the Global Forum,* Moscow, January 18, 1990.
2. Bruce Wilshire, *The Moral Collapse of the University* (Albany: State University of New York Press, 1990), xiii, 40.
3. Page Smith, *Killing the Spirit: Higher Education in America* (New York: Viking, 1990), 7.
4. Erwin Chagroff, "Knowledge Without Wisdom," *Harper's,* May 1980, 47.
5. Wendell Berry, *Standing By Words* (San Francisco: North Point Press, 1983), 66.

6. David Ehrenfeld, "Forgetting," *Orion Nature Quarterly* 8:4 (Autumn 1989): 5–7.
7. Barry Lopez, "The American Geographies," *Orion Nature Quarterly* 8:4 (Autumn 1989): 57.

Richard White, from "Are You an Environmentalist or Do You Work for a Living?" (1995)

An environmental historian of the American West, Richard White here uses a phrase from an anti-environmentalist bumper sticker to ask vital questions about the relationships between work, play, and nature. In White's view, an environmental ethic that bans human work in nature and allows only leisured humans to play in nature is bound to fail. He addresses the difficult task of determining what the proper mix of work and nature should be in an age of environmental fragility and degradation.

In Forks, Washington, a logging town badly crippled by both over-cutting and the spotted owl controversy, you can buy a bumper sticker that reads "Are You an Environmentalist or Do You Work for a Living?" It is an interesting insult, and one that poses some equally interesting questions. How is it that environmentalism seems opposed to work? And how is it that work has come to play such a small role in American environmentalism?

Modern environmentalists often take one of two equally problematic positions toward work. Most equate productive work in nature with destruction. They ignore the ways that work itself is a means of knowing nature while celebrating the virtues of play and recreation in nature. A smaller group takes a second position: certain kinds of archaic work, most typically the farming of peasants, provides a way of knowing nature. Whereas mainstream environmentalism creates a popular imagery that often harshly condemns all work in nature, this second group is apt to sentimentalize certain kinds of farming and argue that work on the land creates a connection to place that will protect nature itself. Arguments that physical labor on the land establishes an attachment that protects the earth from harm have, however, a great deal of history against them.

There are, of course, numerous thoughtful environmentalists who recognize fruitful connections between modern work and nature, but they operate within a larger culture that encourages a divorce between the two. Too often the environmental movement mobilizes words and images that widen the gulf. . . .

There is no avoiding questions of work and nature. Most people spend their lives in work, and long centuries of human labor have left indelible marks on the natural world. From pole to pole, herders, farmers, hunters, and industrial workers have deeply influenced the natural world, so virtually no place is without evidence of its alteration by human labor. Work that has changed nature has simultaneously produced much of our knowledge of nature. Humans have known nature by digging in the earth, planting seeds, and harvesting plants. They have known nature by feeling heat and cold, sweating as they went up hills, sinking into mud. They have known nature by shaping wood and stone, by living with animals, nurturing them, and killing them. Humans have matched their energy against the energy of flowing water and wind. They have known distance as more than an abstraction because of the physical energy they expended moving through space. They have tugged, pulled, carried, and walked, or they have harnessed the energy of animals, water, and wind to do these things for them. They have achieved a bodily knowledge of the natural world.

Modern environmentalism lacks an adequate consideration of this work. Most environmentalists disdain and distrust those who most obviously work in nature. Environmentalists have come to associate work—particularly heavy bodily labor, blue-collar work—with environmental degradation. This is true whether the work is in the woods, on the sea, in a refinery, in a chemical plant, in a pulp mill, or in a farmer's field or a rancher's pasture. Environmentalists usually imagine that when people who make things finish their day's work, nature is the poorer for it. Nature seems safest when shielded from human labor.

This distrust of work, particularly of hard physical labor, contributes to a larger tendency to define humans as being outside of nature and to frame environmental issues so that the choice seems to be between humans and nature. "World War III," Andy Kerr of the Oregon Natural Resources Council likes to say, "is the war against the environment. The bad news is, the humans are winning."[1] The human weapon in Kerr's war is work. It is logging, ranching, and fishing; it is mining and industry. Environmentalists, of course, also work, but they usually do not do hard physical labor, and they often fail to think very deeply about their own work and its relation to nature.

Like Kerr, most Americans celebrate nature as the world of original things. And nature may indeed be the world we have not made—the world of plants, animals, trees, and mountains—but the boundaries between this world of nature and the world of artifice, the world of things we have made, are no longer very clear. Are the cows and crops we breed, the fields we cultivate, the genes we splice natural or unnatural? Are they nature or artifice? We seek the purity of our absence, but everywhere we find our own fingerprints. It is ultimately our own bodies and our labor that blur the boundaries between the artificial and the natural. Even now we tamper with the genetic stuff of our own and other creatures' bodies, altering the design of species. We cannot come to terms with nature without coming to terms with our own work, our own bodies, our own bodily labor.

But in current formulations of human relations with nature there is little room for such a reconciliation. Nature has become an arena for human play and leisure. Saving an old-growth forest or creating a wilderness area is certainly a victory for some of the creatures that live in these places, but it is just as certainly a victory for backpackers and a defeat for loggers. It is a victory for leisure and a defeat for work.

Work and play are linked, but the differences matter. Both our work and our play, as Elaine Scarry has written, involve an extension of our sentient bodies out into the external world. Our tools, the products of our work, become extensions of ourselves. Our clothes extend our skins; our hammers extend our hands. Extending our bodies into the world in this manner changes the world, but the changes are far more obvious in our work than in our play. A logger's tools extend his body into trees so that he knows how the texture of their wood and bark differs and varies, how they smell and fall. The price of his knowledge is the death of a tree.

Environmentalists so often seem self-righteous, privileged, and arrogant because they so readily consent to identifying nature with play and making it by definition a place where leisured humans come only to visit and not to work, stay, or live. Thus environmentalists have much to say about nature and play and little to say about humans and work. And if the world were actually so cleanly divided between the domains of work and play, humans and nature, there would be no problem. Then environmentalists could patrol the borders and keep the categories clear. But the dualisms fail to hold; the boundaries are not so clear. And so environmentalists can seem an ecological Immigration and Naturalization Service, border agents in a socially dubious, morally ambiguous, and ultimately hopeless cause.

I have phrased this issue so harshly not because I oppose environmentalism (indeed, I consider myself an environmentalist) but precisely because I think environmentalism must be a basic element in any coherent attempt to address the social, economic, and political problems that confront Americans at the end of the [twentieth] century. Environmentalists must come to terms with work because its effects are so widespread and because work itself offers both a fundamental way of knowing nature and perhaps our deepest connection with the natural world. If the issue of work is left to the enemies of environmentalism, to movements such as wise use, with its single-minded devotion to propertied interests, then work will simply be reified into property and property rights. If environmentalists segregate work from nature, if they create a set of dualisms where work can only mean the absence of nature and nature can only mean human leisure, then both humans and nonhumans will ultimately be the poorer. For without an ability to recognize the connections between work and nature, environmentalists will eventually reach a point where they seem trivial and extraneous and their issues politically expendable. . . .

Coming to terms with modern work and machines involves both more complicated histories and an examination of how all work, and not just the work of loggers, farmers, fishers, and ranchers, intersects with nature. Technology, an artifact of our work, serves to mask these connections. There are clearly better and worse technologies, but there are no technologies that remove us from nature. We cannot reject the demonization of technology as an independent source of harm only to accept a subset of technologies as rescuing us from the necessity of laboring in, and thus harming, nature. We have already been down this road in the twentieth century.

In the twentieth century technology has often become a container for our hopes or our demons. Much of the technology we now condemn once carried human hopes for a closer and more intimate tie to nature. Over time the very same technology has moved from one category to another. Technology that we, with good reason, currently distrust as environmentally harmful—hydroelectric dams, for example—once carried utopian environmental hopes. To Lewis Mumford, for instance, dams and electricity promised an integration of humans and nature. Mumford saw technology as blurring the boundaries between humans and nature. Humans were "formed by nature and [were] inescapably . . . part of the system of nature." He envisioned a Neotechnic world of organic machines and "ecological balance."[2]

In an ironic and revealing shift, Mumford's solution—his liberating technology, his union of humans and nature—has become redefined as a problem. It is not just that dams, for example, kill salmon; they symbolize the presence of our labor in the middle of nature. In much current environmental writing such blurred boundaries are the mark of our fall. Nature, many environmentalists think, should ideally be beyond the reach of our labor. But in taking such a position, environmentalists ignore the way some technologies mask the connections between our work and the natural world. . . .

I do modern work. I sort, compile, analyze, and organize. My bodily movement becomes electrical signals where my fingers intersect with a machine. Lights flicker on a screen. I expend little energy; I don't sweat, or ache, or grow physically tired. I produce at the end of this day no tangible product; there are only stored memories encoded when my fingers touched keys. There is no dirt or death or even consciousness of bodily labor when I am done. Trees still grow, animals still graze, fish still swim.

But I cannot see my labor as separate from the mountains, and I know that my labor is not truly disembodied. If I sat and typed here day after day, as clerical workers type, without frequent breaks to wander and to look at the mountains, I would become achingly aware of my body. I might develop carpal tunnel syndrome. My body, the nature in me, would rebel. The lights on this screen need electricity, and this particular electricity comes from dams on the Skagit or Columbia. These dams kill fish; they alter the rivers that come from the Rockies, Cascades, and Olympics. The electricity they produce depends on the great seasonal cycles of the planet: on falling snow, melting waters, flowing rivers. In the end, these electrical impulses will take tangible form on paper from trees. Nature, altered and changed, is in this room. But this is masked. I type. I kill nothing. I touch no living thing. I seem to alter nothing but the screen. If I don't think about it, I can seem benign, the mountains separate and safe from me. . . . But, of course, the natural world has changed and continues to change to allow me to sit here. My separation is an illusion. What is disguised is that I—unlike loggers, farmers, fishers, or herders—do not have to face what I alter, and so I learn nothing from it. The connection my labor makes flows in only one direction.

My work, I suspect, is similar to that of most environmentalists. Because it seems so distant from nature, it escapes the condemnation that the work that takes place out there, in "nature," attracts. I regularly read the *High Country News*, and its articles just as regularly denounce mining, ranching, and logging for the very real harm they do. And since the paper's editors have some sympathy for rural people trying to live on the land, letters from readers denounce the paper for not condemning these activities enough. The intention of those who defend old growth or denounce overgrazing is not to denounce hard physical work, but that is, in effect, what the articles do. There are few articles or letters denouncing university professors or computer programmers or accountants or lawyers for sullying the environment, although it is my guess that a single lawyer or accountant could, on a good day, put the efforts of Paul Bunyan to shame.

Most humans must work, and our work—all our work—inevitably embeds us in nature, including what we consider wild and pristine places. Environmentalists have invited the kind of attack contained in the Forks bumper sticker by identifying nature with leisure, by masking the environmental consequences of their own work. To escape it, and perhaps even to find allies among people unnecessarily made into enemies, there has to be some attempt to come to terms with work. Work does not prevent harm to the natural world—Forks itself is evidence of that—but if work is not perverted into a means of turning place into property, it can teach us how deeply our work and nature's work are intertwined.

And if we do not come to terms with work, if we fail to pursue the implications of our labor and our bodies in the natural world, then we will return to patrolling the borders. We will turn public lands into a public playground; we will equate wild lands with rugged play; we will imagine nature as an escape, a place where we are born again. It will be a paradise where we leave work behind. Nature may turn out to look a lot like an organic Disneyland, except it will be harder to park.

There is, too, an inescapable corollary to this particular piece of self-deception. We will condemn ourselves to spending most of our lives outside of nature, for there can be no permanent place for us inside. Having demonized those whose very lives recognize the tangled complexity of a planet in which we kill, destroy, and alter as a condition of living and working, we can claim an innocence that in the end is merely irresponsibility.

If, on the other hand, environmentalism could focus on our work rather than on our leisure, then a whole series of fruitful new angles on the world might be possible. It links us to each other, and it links us to nature. It unites issues as diverse as workplace safety and grazing on public lands; it unites toxic sites and wilderness areas. In taking responsibility for our own lives and work, in unmasking the connections of our labor and nature's labor, in giving up our hopeless fixation on purity, we may ultimately find a way to break the borders that imprison nature as much as ourselves. Work, then, is where we should begin.

Notes

1. William Dietrich, *The Final Forest: The Battle for the Last Great Trees of the Pacific Northwest* (New York: Simon and Schuster, 1992), 209.
2. Lewis Mumford, *Technics and Civilization* (New York: Harcourt, Brace, 1934), 256–57.

U.N. Convention on Environment and Development, *Earth Charter* Preamble (1991)

After World War II, the U.N. Charter established an international body of nations and a governing structure intended to protect each member country. Members soon supported a Universal Declaration of Human Rights. In 1991, the U.N. took the unprecedented step of according Earth a chartered status worthy of similar protection. The Earth Charter *enjoins human restraint and humility. It articulates the views of representatives from across the globe. It links environmental sustainability to economic justice and human rights. While it affords little in the way of genuine enforcement powers, more than 725 organizations with active memberships approaching fifty million people have endorsed it.*

We stand at a critical moment in Earth's history, a time when humanity must choose its future. As the world becomes increasingly interdependent and fragile, the future at once holds great peril and great promise. To move forward we must recognize that in the midst of a magnificent diversity of cultures and life forms we are one human family and one Earth community with a global destiny. We must join together to bring forth a sustainable global society founded on respect for nature, universal human rights, economic justice, and a culture of peace. Towards this end, it is imperative that we, the peoples of Earth, declare our responsibility to one another, to the greater community of life, and to future generations.

EARTH, OUR HOME

Humanity is part of a vast evolving universe. Earth, our home, is alive with a unique community of life. The forces of nature make existence a demanding and uncertain adventure, but Earth has provided the conditions essential to life's evolution. The resilience of the community of life and the well-being of humanity depend upon preserving a healthy biosphere with all its ecological systems, a rich variety of plants and animals, fertile soils, pure waters, and clean air. The global environment with its finite resources is a common concern of all peoples. The protection of Earth's vitality, diversity, and beauty is a sacred trust.

THE GLOBAL SITUATION

The dominant patterns of production and consumption are causing environmental devastation, the depletion of resources, and a massive extinction of species. Communities are being undermined. The benefits of development are not shared equitably and the gap between rich and poor is widening. Injustice, poverty, ignorance, and violent conflict are widespread and the cause of great suffering. An unprecedented rise in human population has overburdened ecological and social systems. The foundations of global security are threatened. These trends are perilous—but not inevitable.

THE CHALLENGES AHEAD

The choice is ours: form a global partnership to care for Earth and one another or risk the destruction of ourselves and the diversity of life. Fundamental changes are needed in our values, institutions, and ways of living. We must realize that when basic needs have been met, human development is primarily about being more, not having more. We have the knowledge and technology to provide for all and to reduce our impacts on the environment. The emergence of a global civil society is creating new opportunities to build a democratic and humane world. Our environmental, economic, political, social, and spiritual challenges are interconnected, and together we can forge inclusive solutions.

UNIVERSAL RESPONSIBILITY

To realize these aspirations, we must decide to live with a sense of universal responsibility, identifying ourselves with the whole Earth community as well as our local communities. We are at once citizens of different nations and of one world in which the local and global are linked. Everyone shares responsibility for the present and future well-being of the human family and the larger living world. The spirit of human solidarity and kinship with all life is strengthened when we live with reverence for the mystery of being, gratitude for the gift of life, and humility regarding the human place in nature.

James Gustave Speth, from *Red Sky at Morning: America and the Crisis of the Global Environment; A Citizen's Agenda for Action* (2004)

James Speth heads the Yale School of Forestry and Environmental Studies. He helped to found the Natural Resources Defense Council. His book title comes from the old mariner's rhyme: "Red sky at night, sailor's delight. Red sky at morning, sailor's warning." (In the Northern Hemisphere, prevailing westerly winds mean that a red sunset promises that dry air, more refractive than moist, will come; a red sunrise warns that dry air has passed and wet or stormy weather is on its way.) Speth says that we now stand warned. But we can change our habits and actions, and we can encourage others, too. The word "crisis," too often loosely invoked, Speth employs advisedly. Even as he provides a superb overview of the global environment, he laments America's role in its degradation. His volume is a "wake-up call." Yet, no matter how bad the situation seems, he believes "there are solutions."

In the end, what can reliably be said about the prospect for humans and nature? A pessimist might conclude that the drivers of deterioration are too powerful to counter, that our economy is too dependent on unguided growth and laissez-faire, that our politics cannot accommodate long-term thinking, and that our society responds only to major crises and in this case the crises will come too late.

Weighed against this, there are hopeful signs and encouraging developments in each of the eight areas of transition. Scientific understanding is greatly improved. Population growth is slowing, and the proportion of the world's people in poverty is being reduced. Technologies that can bring a vast environmental improvement in manufacturing, energy, transportation, and agriculture are either available or close at hand. We are learning how to harness market forces for sustainability, and major schemes for capping and trading the right to emit climate-changing gases are emerging. International environmental law has expanded and is ready for a new phase. Environmental and other civil society organizations have developed remarkable new capacities for leadership and effectiveness. Private businesses, environmental organizations, and local governments the world over are taking impressive initiatives often far ahead of international agreements or other government requirements. Environment is emerging as a force in business strategic planning. . . . Europe, at least, is providing real leadership on the policy front. So, despite the gravity of our predicament, the situation is far from hopeless, and some areas such as a green consumer movement and emissions trading to control greenhouse gases may be poised for takeoff. Solutions—including the policy prescriptions and other actions needed to move forward—abound. We need but use them.

Helpful trends outside the environmental arena are becoming discernable. Globalization of many descriptions is eroding sovereignty. We are seeing the slow but steady emergence of a global civil society, as like-minded organizations in many countries come together. The nation-state, it has been said, is too little for the big things and too big for the little things. "Glocalization" is emerging, with action shifting to local and global levels. In many places, especially in Europe, one can see psychological disinvestment in the nation-state and the strengthening of local and global citizenship. These trends should enhance prospects for international cooperation, both official and unofficial.

Figure 26.2 *Wanderer Above the Sea of Mist* by Caspar David Friedrich, 1817. Hamburger Kunsthalle, Dauer Leihgabe der Stiftung zur Förderung der Hamburgischen Kunstsammlungen. Photo by Elke Walford © Hamburger Kunsthalle.

One thing is clear: the needed changes will not simply happen. No hidden hand is guiding technology or the economy toward sustainability. The issues on the global environmental agenda are precisely the type of issues—long-term, chronic, complex—where genuine, farsighted leadership from elected officials is at a premium. But we have not seen this leadership emerge, and we have waited long enough. What we need now is an international movement of citizens and scientists, one capable of dramatically advancing the political and personal actions needed for the transition to sustainability. We have had movements against slavery and many have participated in movements for civil rights and against apartheid and the Vietnam War. Environmentalists are often said to be part of "the environmental movement." We need a real one. It is time for we the people, as citizens and as consumers, to take charge.

Eventually, leaders in the political and business worlds will see that it is powerfully in their self-interest to promote the eight transitions. But the clear evidence to date is that, absent some new force in the picture, they will be much too late in coming to this realization. The best hope we have for this new force is a coalescing of a wide array of civic, scientific, environmental, religious, student, and other organizations with enlightened business leaders, concerned families, and engaged communities, networked together, protesting, demanding action and accountability from governments and corporations, and taking steps as consumers and communities to realize sustainability in everyday life.

A new movement of consumers and households committed to sustainable living could drive a world of change. Young people will almost certainly be centrally involved in any movement for real change. They always have been. New dreams are born most easily when the world is seen with fresh eyes and confronted with impertinent questions. The Internet is empowering young people in an unprecedented way—not just by access to information but by access to each other, and to a wider world.

One goal should be to find the spark that can set off a period of rapid change, like the flowering of the domestic environmental agenda in the early 1970s. Part of the challenge is changing the perception of global-scale concerns so that they come alive with the immediacy and reality of our domestic challenges of the 1970s. In the end, we need to trigger a response that in historical terms will come to be seen as revolutionary—the Environmental Revolution of the twenty-first century. Only such a response is likely to avert huge and even catastrophic environmental losses.

Web Connections

PART ONE. CONCEPTS AND CASE STUDIES

1. *Climate Shock*

http://www.ipcc.ch Intergovernmental Panel on Climate Change.

http://yosemite.epa.gov/oar/globalwarming.nsf/content/ResourceCenterPublicationsUSClimateActionReport.html U.S. Climate Action Report.

http://www.pewclimate.org Pew Center on Global Climate Change, industry-sympathetic non-profit, helps companies work together; level-headed reports on global warming; statements from CEOs of Fortune 500 companies.

http://web.mit.edu/globalchange/www/reports.html Contains complete contents of the more than one hundred reports published by the Massachusetts Institute of Technology Joint Program on the Science and Policy of Global Change.

http://www.hm-treasury.gov.uk/independent_reviews/stern_review_economics_climate_change/stern_review_report.cfm Human impact on climate change, its economic repercussions, challenges, and opportunities.

2. *Species in Danger: Three Case Studies*

http://www.cites.org The Convention on International Trade of Endangered Species fights against illegal ivory sales. Information on Monitoring the Illegal Killing of Elephants (MIKE) and Elephant Trade Information System (ETIS).

http://www.panda.org/about_wwf/what_we_do/species/our_solutions/endangered_species/elephants/african_elephants/index.cfm Concise, informative WWF page on elephants and the ivory ban.

http://www.natureserve.org/explorer NatureServe Explorer, an Online Encyclopedia of Life. Conservation status, distribution information, life histories for over 50,000 species and subspecies of animals and plants in U.S. and Canada, 4,500 ecological communities.

http://eelink.net/EndSpp/endangeredspecies-mainpage.html Public resources on endangered species. An extremely well-constructed database of links to other sites, FAQs, basic data.

http://www.redlist.org The IUCN Red List of Threatened Species is only on CD-ROM or online, here. Convenient database of threatened species and biological and conservation information about them.

3. Nuclear Power: Three Mile Island, Chernobyl, and the Future

http://www.world-nuclear.org The World Nuclear Association, an industry page on nuclear power. Explanations of the Chernobyl and Three Mile Island nuclear incidents are informative.

http://www.pbs.org/wgbh/amex/three/index.html A site associated with a PBS program on Three Mile Island; features a timeline of American nuclear power, diagrams of the Three Mile Island reactor, biographies of key personalities, and map of all the nuclear reactors in the country.

http://www-bcf.usc.edu/meshkati/chernobyl.html Personal page on Chernobyl, complete directory to Web resources. Original introduction and interpretation of Chernobyl.

4. Biotechnology and Genetically Manipulated Organisms: Bt *Corn and the Monarch Butterfly*

http://www.safe-food.org The site of "Mothers for Natural Law." Views the genetic engineering of food as a vast experiment in which we are all uninformed volunteers.

http://www.biotech-info.net Ag BioTech InfoNet, sponsored by a consortium of scientific, environmental, and consumer organizations.

http://www.gene-watch.org The Council for Responsible Genetics. Dedicated to public participation in the debates over genetically based technology.

5. The Paradox of Sustainable Development

http://sdgateway.net Integrates online information developed by members of the Sustainable Development Communications Network.

http://www.ulb.ac.be/ceese/meta/sustvl.html While visually unexciting, this is perhaps the best site for links to hundreds of other sites grouped by topics and categories; this site could be a starting place for Web-based research on this topic.

http://www.maweb.org The home of the Millennium Ecosystem Assessment, which monitors the health of ecosystems worldwide, and evaluates their ability to sustain services to regional populations.

6. Deforestation

http://www.forests.org Portal to online forest conservation info. Aims to end deforestation, preserve primary and old-growth forests, conserve and sustainably manage other forests, and commence ecological restoration; news items, listing of links for forest destruction and conservation.

http://www.rainforestweb.org/Rainforest_Regions/South_America This is the South American page of the World Rainforest Information Portal. Contains up-to-date news, action alerts, and links to other sites involved in protection of South American rain forests.

7. War and Peace: Security at Stake

http://www.ppu.org.uk/learn/infodocs/st_environment.html The Peace Pledge Union's site on "War and the Environment" provides examples of military engagements that have destroyed environments and the cultures that depended on them.

http://www.library.utoronto.ca/pcs/database/libintro.htm The Environmental Security Database provides information on the links between environmental stress and violent conflict in developing countries.

http://www.envirosecurity.net/index.php The Institute for Environmental Security promotes research on tackling global environmental security risks.

8. Globalization Is Environmental

http://www.unep.org This U.N. site for its environmental program provides substance on globalization issues of environment, trade, and sustainability.

http://www.ifg.org The International Forum on Globalization, a progressive think tank with a fine site: go to "programs" to access information on connections between globalization and environmental issues.

http://www.iied.org The International Institute for Environment and Development. Aside from globalization issues, this site is strong on sustainable development.

http://www.globalexchange.org Fosters increased global awareness among the U.S. public and builds ties for education and action on issues of social and economic justice around the world. Information on "fair trade coffee" and other initiatives.

9. What Is Wilderness and Do We Need It?

http://www.wilderness.net National Wilderness Preservation System; list of all National Wilderness Areas and links to their home pages; searchable map of all the Wilderness Areas.

http://www.wildwilderness.org Wild Wilderness: non-profit group for "undeveloped recreation" like hiking, birdwatching, and stream-fishing. Wilderness philosophy, and postings of news items offending the group's pro-wilderness sensibilities.

10. The Urban Environment: Calcutta and Los Angeles

http://www.plannersweb.com/sprawl/home.html *Planning Commissioners Journal*; sprawl-related articles.

http://www.urbanecology.org San Francisco Bay Area non-profit organization "works to build cities that are ecologically thriving and socially just." Sparse site dominated by extensive list of links to urban ecology organizations.

http://www.treelink.org A Web site devoted to trees in the urban environment.

PART TWO. FOUNDATIONAL DISCIPLINES AND TOPICS

I. Biological Interactions

11. Biodiversity and Conservation Biology

http://tolweb.org/tree/phylogeny.html Massive, multi-authored phylogenetic tree of all life; photographs and information on phylogeny and diversity; useful Web links, standard bibliography.

http://www.pollinator.com Eclectic portal site for pollination ecology, one of the most involved mutualisms known; terrestrial ecosystems depend on pollinators.

http://www.trfic.msu.edu The Tropical Rain Forest Information Center. A variety of resources connected with rain forests, the world's most diverse ecosystems; NASA data and satellite images are also available for a fee.

12. Soil and Agriculture

http://plants.usda.gov PLANTS Database is comprehensive: classification for all plants, practical applications such as an erosion predictor, crop requirement information.

http://www.primalseeds.org NGO dedicated to fighting monoculturing. Excellent links; intriguing pages like "guerrilla gardening."

http://www.agnic.org/agnic The Agricultural Network Information Center. Directory of agricultural links, many with local or state focus; offers access to question experts on agricultural topics.

13. Air and Water

http://www.ace.mmu.ac.uk/eae/english.html Encyclopedia of the Atmospheric Environment: comprehensive coverage of the atmosphere, excellent links.

http://jwocky.gsfc.nasa.gov NASA's Total Ozone Mapping Spectrometer. Maps of ozone and aerosol levels in almost real-time, and satellite images from every day for the last five years.

http://www.noaa.gov The National Oceanic and Atmospheric Administration site includes information on fisheries, air quality, weather, and other air and water topics.

http://www.reefbase.org Data and information on coral reefs and associated shallow tropical habitats; reports on the world's reefs; nearly two thousand maps.

http://www.indigenouswater.org Site promotes an indigenous perspective on water and development.

http://www.unicef.org/wes UNICEF site devoted to water, environment, and sanitation.

http://www.worldwater.org Data from the annual volume *The World's Water*, including access to safe drinking water and sanitation.

14. Energy

http://www.eia.doe.gov The U.S. Government's Energy Information Administration. National and international data, analyses and forecasts for conventional and renewable fuels; links to government and industry.

http://www.eere.energy.gov The Department of Energy's green wing, the Office of Energy Efficiency and Renewable Energy; links to hundreds of international energy sites; all major types of renewable energy explained in detail; energy-efficiency tips.

http://www.crest.org The Internet face of the Renewable Energy Policy Project; an excellent, annotated directory of alternative energy resources on the Internet.

http://facultystaff.vwc.edu/~gnoe/avd.htm The Alternative-Fuel Vehicle Directory links to alternative fuel sites on electric vehicle conversions, the electrochemistry of batteries and fuel cells, U.S. Government research sites, university hybrid projects, regional electric vehicle associations, and international automobile manufacturers.

15. Toxicology

http://www.scorecard.org Index of polluters and emissions throughout the United States, searchable by zip code. Hypertext maps of pollution sources; information on individual pollutants.

http://www.chem.unep.ch Home of the United Nations Environmental Program's Chemicals branch. Inventory of Internet information on chemicals and Pollutant Release and Transfer Registers (PRTRs).

http://members.aol.com/rccouncil/ourpage/index.htm Rachel Carson Council; offers pesticide-related information, both technical and more philosophical.

http://www.greenpeace.org/international/campaigns/toxics Greenpeace's Toxics Campaign page has all necessary resources for armchair activism.

II. Human Dimensions

16. The Inner Life

http://environment.harvard.edu/religion Harvard University Forum on Religion and Ecology. Essays on major religions' relationships to ecology; research links, educational resources.

http://www.religionandnature.com Gateway to learning about the role of religion and spirituality in human ecology. Includes references to a comprehensive *Encyclopedia of Religion and Nature*.

17. Ethics, Philosophy, Gender

http://www.earthcharter.org Declaration espousing "respect and care for the community of life, ecological integrity, social and economic justice, and democracy, nonviolence, and peace"; links to ethical ecology sites.

http://trumpeter.athabascau.ca An online journal devoted to Deep Ecology. Archives and current issues available.

http://www.ecofem.org A hub of ecofeminist theory and activism.

18. Poetry

http://www.robertfrostoutloud.com Site contains selected poems, audio files of Frost reading his poems, and links.

See also http://www.asle.umn.edu/index.html in Nature Writing, below.

19. History and the Environment

http://lcweb2.loc.gov/ammem/amrvhtml/conshome.html Evolution of the Conservation Movement, 1850–1920: an online collection of Library of Congress documents detailing the ideological roots of conservationism. Texts, photographs, and two motion pictures.

http://ecotopia.org/ehof/index.html The Ecology Hall of Fame, an excellent source of information on figures like Henry David Thoreau, Alan Chadwick, and Aldo Leopold, among others. Some contemporary figures included.

20. Nature Writing

http://www.sej.org Society of Environmental Journalists, with perhaps the best directory of links to environmental information on the Internet.

http://www.asle.umn.edu/index.html Association for the Study of Literature and Environment; links to ecocriticism resources, online nature writing, and sites about nature writers. Popular discussion groups.

http://www.vcu.edu/engweb/eng385/natweb.htm Web page for an English class at Virginia Commonwealth University on Nature Writing; an all-encompassing online resource list for lovers of nature writing.

III. Social Connections

21. Politics and Public Policy

http://www.lcv.org League of Conservation Voters database of congressional voting records on environmental bills; links to congressional e-mail addresses.

http://www.cnie.org/cms.cfm?id=518 State of the Environment Reports from all over the world, maintained by National Council for Science and the Environment.

http://www.nrdc.org Natural Resources Defense Council, a New York–based non-profit organization. Site is similar to www.greenpeace.org in comprehensiveness of coverage, but more policy oriented and less radical.

http://www.sej.org The site for the Society of Environmental Journalists. Contains late-breaking news on environmental issues.

22. *Law and Environmental Justice*

http://www.ecolex.org ECOLEX, a "gateway to environmental law," global in scope; treaties, national legislation, court decisions, and legal writing.

http://www.ccej.org The Community Coalition for Environmental Justice, a Seattle community activist group. Web site contains environmental justice movement links.

23. *Economics*

http://www.american.edu/TED/ted.htm Trade and Environment Database examines interactions between trade, environment, and culture. 700 cross-referenced and cross-linked case studies.

http://www.ecosystemvaluation.org/default.htm Non-technical examination of ecosystem valuation. Algorithms used by economists and links to online resources.

24. *Human Population*

http://www.census.gov/main/www/popclock.html Population clocks: U.S. Census Bureau estimates of the current populations of United States and the world.

http://www.unfpa.org United Nations Population Fund links to raw population data; emphasizes importance of social issues like gender equality in improving the lives of the earth's billions.

http://www.prb.org Population Reference Bureau; provides population information to policymakers and "concerned citizens around the world"; Web page features original articles by demographers on trends.

25. *Anthropology*

http://www.survival-international.org Survival: international organization which advocates for and aids tribal peoples worldwide. Site details the plight of tribal peoples around the world and includes petitions, campaigns, and other ways to help.

http://cnie.org/NAE Native Americans and the Environment: insight on environmental problems in Native American communities, relevant history, and values of Native Americans. Enormous database of online documents is sortable by region and subject.

CODA

26. Conviction and Action

http://www.solutions-site.org/artman/publish HORIZON Solutions Site: hundreds of case studies of environmental problems with a forum for replies and suggestions on how to address them.

http://www.nwf.org/campusecology National Wildlife Foundation's Campus Ecology program promotes environmentalism on college campuses.

http://www.loe.org Archives of Public Radio International program *Living on Earth;* various and wide-ranging environmental topics.

http://www.envirolink.org A great site for general resources about the environment.

About the Editors

Glenn Adelson is the chair of the Environmental Studies Department at Lake Forest College. He is co-director of the Mamoní Valley Preserve in Panama, a biocultural preservation and restoration project. He taught conservation biology at Harvard University for thirteen years, a course that used field trips—both in New England and in the tropics—as well as poetry, art, history, economics, philosophy, and mathematics to explain the concepts of "Nature" and "Biodiversity." He is recipient of the Phi Beta Kappa teaching prize and is twice the recipient of the Levenson teaching award at Harvard. Adelson has a Ph.D. in organismic and evolutionary biology from Harvard and a J.D. from the University of Michigan Law School. He is co-author of *Biodiversity: Exploring Values and Priorities in Conservation* (1996).

James Engell is Gurney Professor of English at Harvard University, where he currently serves as department chair of English & American Literature & Language. Previously, he chaired the Department of Comparative Literature as well as the degree program in History & Literature, and served on the Committee on the Study of Religion. In addition to publishing nine books and numerous journal articles, Engell is well-known for essays such as "Imagining into Nature: 'This Lime-Tree Bower My Prison'" and "Only This: Connect," on the need to teach the liberal arts *and* sciences together. He has long advocated the integration of the humanities and social sciences into environmental education and policy decisions. Engell has won four faculty-wide teaching prizes and offers courses in romantic poetry, eighteenth-century studies, general education, rhetoric, and (for the Gilder Lehrman Institute) environmental issues. He is co-author of *Saving Higher Education in the Age of Money*, which won the Association of American Colleges and Universities 2007 Frederic W. Ness Book Award for best book on liberal education. He is secretary of the Cambridge Scientific Club, a faculty member of the Harvard University Center for the Environment, and a member of the American Academy of Arts and Sciences.

Brent Ranalli is a senior analyst at The Cadmus Group, an employee-owned environmental consulting firm, where he specializes in drinking water and ground water policy and innovative regulatory strategies for clients including the U.S. EPA and states. He earned a B.A. in history and science from Harvard University and an M.Sc. in environmental science and policy from the Central European University, with additional coursework at the Harvard School of Public Health. His writings have been featured in *News from Below* and at *TheGlobalist.com.*

K. P. Van Anglen has degrees in English from Princeton, Cambridge, and Harvard Universities. He has taught English and American literature at Harvard, the University of Pennsylvania, Boston College, and Boston University, where he is now a faculty member. He has also been a postdoctoral fellow of the Institute for Research in the Humanities at the University of Wisconsin–Madison. He currently serves as an editor of the Princeton University Press Edition of *The Writings of Henry D. Thoreau*, a consultant for the Gilder Lehrman Institute of American History, the coeditor of *Religion and the Arts*, and a member of the board of directors of the Thoreau Society. Van Anglen's publications include *The New England Milton: Literary Reception and Cultural Authority in the Early Republic* (1993), the *Translations* volume (1986) of the Princeton Thoreau Edition, and *"Simplify, Simplify" and Other Quotations from Henry David Thoreau* (1996).

Acknowledgments

We express profound indebtedness to many individuals, foundations, and institutions.

The following scholars and professionals generously gave their time, advice, and suggestions. To them we owe deep and personal debts: Daniel Aaron, Mary Sarah Bilder, Patricia Birnie, Lawrence Buell, Luis Campos, David Cannadine, Cheng Man-lung, Jan E. Dizard, Brian Donahue, John Elder, Sarah Frederick, Erica Brown Gaddis, Madhav Gadgil, David Galbraith, Paul Giles, Barbara Goldoftas, David Grewal, Charles A. S. Hall, George Hallberg, F. Gene Hampton, Scott Hess, John Hildebidle, Alan Hodder, William L. Howarth, Susan Hurley, James Jones, Gwynne Kennedy, Richard Lewontin, Leo Marx, James McCarthy, Michael B. McElroy, Arnico Panday, Dan Perlman, Jay Phelan, Naomi Pierce, Zygmunt Plater, Nina Rabin, Douglas Rand, Matthew Reeve, Eliza Richards, Paul Rodhouse, Ilene Rosin, Charles Rzepka, James Siemon, Marco Simons, James W. Skehan, S.J., James Smith, Robert Stavins, Peter Taylor, Tu Wei-Ming, Elizabeth Hall Witherell, Mark Wormald, and Yang Xiaoshan.

We thank the directors and staff members who opened to the fullest extent possible collections housed in: The Harvard College Library, comprising or associated with the Harry E. Elkins Memorial Library, the Cabot Science Library, the Gordon McKay Library of Engineering and Applied Science, the Kummel Library of Geological Sciences, the Littauer Library of Economics, the Tozzer Library of Anthropology, and the Gray Herbarium Library; also at Harvard, the Harvard Law School Library (Langdell), the Countway Library of Medicine, the Schlesinger Library at the Radcliffe Institute, the Baker Library of Business, the Andover-Harvard Library of Divinity, and the Kennedy School of Government Library; the O'Neill Library of Boston College; at Boston University the Mugar Memorial, School of Theology, and Science and Engineering Libraries; and the Geisel Library of Saint Anselm College.

This book could not have been completed without the support of the following editors and staff of Yale University Press. Our gratitude to Jean E. Thomson Black is immeasurable and cannot adequately be rendered. She supported this project as few if any other editors would or could, and we cannot sufficiently express our thanks. Hers is the invisible hand here. Laura Davulis, Molly Egland, Jessie Hunnicutt, and Noreen O'Connor assisted in editorial and production matters with dispatch and professionalism. Bob Land created the index. Lara Heimert first encouraged this project and was our initial, welcoming contact at the press. Anonymous readers for the press provided astute suggestions for improvement.

Our personal indebtedness extends to: Maude Emerson for scanning, proofing, research, and permissions assistance; Joshua Dunn and Rebecca Schneider for keeping financial records; Sol Kim-Bentley for research and permissions assistance; Ruth Murray and Jake Fleming for research; and Paul Beaulieu, Harriet Lane, and Margaret Johnson for office support at Boston University.

We thank The Cadmus Group, Inc., for making Brent Ranalli available and supporting his participation in this project.

For costs associated with research, compilation, editing, permissions fees, and assistance for producing this anthology, we gratefully acknowledge support of these foundations, institutions, and individuals: the Baker Foundation of Cambridge, Massachusetts; the Hyder E. Rollins and Fred Norris Robinson Funds of Harvard University; the Clark/Cooke Funds of Harvard University; Yale University Press, particularly Jean E. Thomson Black; former Deans Dennis D. Berkey and Jeffrey Henderson, Senior Associate Dean Susan K. Jackson, and Professors James Winn and Bonnie Costello for the subvention of two research funds from the College of Arts and Sciences Dean's Research Fund, Boston University.

Selection Credits

Excerpt from *I and Thou* by Martin Buber, translated by Walter Kaufmann (Charles Scribner's Sons, 1970), copyright © 1970 by Charles Scribner's Sons. Reprinted with permission of Scribner, an imprint of Simon & Schuster Adult Publishing Group, and permission of T & T Clark International.

"The Fusion of Lines of Thought," a.k.a "Wilderness," by Aldo Leopold, undated, University of Wisconsin Archives, Aldo Leopold Papers 10–6, 16, as it appears in *Aldo Leopold: His Life and Work* by Curt Meine (University of Wisconsin Press, 1988). Reprinted courtesy of the Aldo Leopold Foundation.

Excerpt from *Breaking New Ground* by Gifford Pinchot (Harcourt, Brace and Company, 1947), © 1947 by Harcourt, Brace and Company, Inc. Reprinted by kind permission of Peter C. Pinchot.

Excerpts from *Global Warming: The Complete Briefing* by John Houghton (Cambridge University Press, 1997), first edition © John Houghton 1994, second edition © Cambridge University Press 1997. Reprinted with the permission of Cambridge University Press.

"Modern Global Climate Change" by Thomas R. Karl and Kevin E. Trenberth, reprinted from *Science*, 302, no. 5651 (5 December 2003), pp. 1719–1723.

Excerpts from *Hothouse Earth: The Greenhouse Effect and Gaia* by John Gribbin (Grove Weidenfeld, 1990), © 1990 John and Mary Gribbin. Reprinted by permission.

"Höfn," in *District and Circle* by Seamus Heaney (Farrar, Straus and Giroux, 2006), © 2006 Seamus Heaney. Used by permission of the poet.

Excerpts from *International Trade in Ivory from the African Elephant: Issues Surrounding the CITES Ban and SACWM's Chances of Overturning It* by Mafaniso Hara (Centre for Southern African Studies, 1997), © 1997 Centre for Southern African Studies. Reprinted by permission of the CSAS at the School of Government, UWC.

Excerpts from "Conservation of Large Mammals in Africa: What Lessons and Challenges for the Future?" by Philip Muruthi, Mark Stanley Price, Pritpal Soorae, Cynthia Moss, and Annette Lanjouw, in *Priorities for the Conservation of Mammalian Diversity*, ed. Abigail Entwistle and Nigel Dunstone (Cambridge

University Press, 2000), © 2000 Cambridge University Press. Reprinted with the permission of Cambridge University Press and the principal author.

"The Author of *American Ornithology* Sketches a Bird, Now Extinct" from *Travelling Light: Collected and New Poems* (University of Illinois Press, 1999). Copyright 1999 by David Wagoner. Used with permission of the poet and the University of Illinois Press.

Excerpts from "The Ghost Bird" by Jonathan Rosen. This material originally appeared in the May 14, 2001 issue of the *New Yorker.* Reprinted by permission.

Excerpts from "Recovery Outline for the Ivory-billed Woodpecker" by the United States Fish and Wildlife Service (Atlanta, Ga.), September 2005.

Excerpts from "It's Not the Only Alien Invader" by Alan Burdick, *New York Times* (November 13, 1994), pp. 49–87. © 1994, Alan Burdick. Reprinted by permission.

Excerpts from "Extinction of an Island Forest Avifauna by an Introduced Snake" by Julie A. Savidge, in *Ecology*, 68, no. 3 (1987), pp. 660–668. Reproduced by permission of The Ecological Society of America, Inc.

Excerpts from "Nuclear Reactor Accidents" in *The Nuclear Lion: What Every Citizen Should Know About Nuclear Power and Nuclear War* by John Jagger (Plenum Press, 1991), pp. 113–118 and 128–129. © 1991 John Jagger. Reprinted with kind permission of Springer Science and Business Media.

Excerpts from *Fallout from Chernobyl* by L. Ray Silver (Deneau, 1987), © 1987. Reprinted by permission.

Excerpts from David R. Marples' Introduction to *No Breathing Room* by Grigori Medvedev (Basic Books, 1993). Copyright © 1993 by Basic Books, a Member of the Perseus Books Group. Reprinted by permission of Basic Books, a member of Perseus Books, L.L.C.

Excerpts from *Normal Accidents* by Charles Perrow, © 1999 Princeton University Press. Reprinted by permission of Princeton University Press.

Excerpts from "Nuclear Power and the Environment" by Hans Blix, from the Conference on Nuclear Power and the Changing Environment, British Nuclear Forum, 4 July 1989 (International Atomic Energy Agency). Reprinted by permission.

Excerpts from *Corn: Its Origin, Evolution, and Improvement* by Paul C. Mangelsdorf, pp. 207–210, 211–214, Cambridge, Mass.: Belknap Press of Harvard University Press, Copyright © 1974 by the President and Fellows of Harvard College. Reprinted by permission of the publisher.

Excerpt from "Transgenic Pollen Harms Monarch Larvae" by John E. Losey, Linda S. Raynor, and Maureen E. Carter, in *Nature*, 399, no. 6733 (20 May 1999), p. 214. © 1999 Macmillan Magazines, Ltd.

Excerpt from "Canary in the Cornfield: The Monarch and the *Bt* Corn Controversy" by Lincoln Brower. This article originally appeared in the Spring 2001 issue of *Orion* magazine, 187 Main Street, Great Barrington, Mass. 01230, www.oriononline.org. For a free copy, please visit the Web site.

"Seeds of Change" by Molly Lesher, as revised by the author. A version of the paper originally appeared in the *Regional Review*, published by the Federal Reserve Bank of Boston, 14, no. 2 (2004 2nd–3rd Quarter), pp. 12–20. Reprinted by permission.

Excerpts from "Towards Sustainable Development" in *Our Common Future* by the World Commission on Environment and Development (Oxford University Press, 1987), © World Commission on Environment and Development 1987. Used by permission of Oxford University Press.

Excerpts from "Sustainable Development: A Critical Review" by Sharachchandra M. Lélé, in *World Development*, 19, no. 6 (1991), pp. 607–621. Copyright 1991, with permission from Elsevier.

Excerpts from *Our Ecological Footprint: Reducing Human Impact on the Earth* by Mathis Wackernagel and William E. Rees (New Society Publishers, 1996), © 1996 Mathis Wackernagel and William Rees. Reprinted by permission.

Excerpts from *The Economy of Permanence* by J. C. Kumarappa (Sarva Seva Sangh Prakashan, 1945). Reprinted by permission.

Excerpts from *The Collapse of Complex Societies* by Joseph A. Tainter (Cambridge University Press, 1988), © Cambridge University Press 1988. Reprinted with the permission of Cambridge University Press.

Excerpts from *Americans and Their Forests: A Historical Geography* by Michael Williams (Cambridge University Press, 1989), © 1989 Cambridge University Press. Reprinted with the permission of Cambridge University Press.

Excerpts from "Preface: A Turning Point" and "Chapter 1: Forests at the End of the Second Millennium" by George M. Woodwell in *Forests in a Full World* by George M. Woodwell, with contributions from Ola Ullsten, Richard A. Houghton, Sten Nilsson, Peter Kanowski, Eric D. Larson and Thomas B. Johansson, and Brian Kerr (Yale University Press, 2001), © 2001 Yale University. Reprinted by permission.

Excerpt from *Forests: The Shadow of Civilization* by Robert Pogue Harrison (The University of Chicago Press, 1992), © 1992 by The University of Chicago. Reprinted by permission.

Excerpts from *Democracy in America* by Alexis de Toqueville, translated by Henry Reeve, copyright 1945 and renewed 1973 by Alfred A. Knopf, a division of Random House, Inc. Used by permission of Alfred A. Knopf, a division of Random House, Inc.

Excerpts from *The Final Forest* by William Dietrich (Simon & Schuster, 1992). Copyright © 1992 by William Dietrich. Reprinted with the permission of Simon & Schuster Adult Publishing Group.

Excerpts from "Wall Street Sleaze: How the Hostile Takeover of Pacific Lumber Led to the Clear-Cutting of Coastal Redwoods," by Robert K. Anderberg. Copyright 1988 by Robert K. Anderberg. First published in *The Amicus Journal*, 10, no. 2 (spring 1988), pp. 8–10 (www.nrdc.org/onearth). Reprinted by permission.

"Brazil, January 1, 1502" from *The Complete Poems, 1927–1979* by Elizabeth Bishop. Copyright © 1979, 1983 by Alice Helen Methfessel. Reprinted by permission of Farrar, Straus and Giroux, LLC.

Excerpt from *The Earth Summit at Rio: Politics, Economics, and the Environment* by Ranee K. L. Panjabi (Northeastern University Press, 1997). Copyright 1997 by Ranee K. L. Panjabi. Reprinted with the permission of Northeastern University Press.

Excerpt from *Tropical Deforestation: Small Farmers and Land Clearing in the Ecuadorian Amazon* by Thomas K. Rudel with Bruce Horowitz (Columbia University Press, 1993), © 1993 Columbia University Press. Reprinted with the permission of the publisher.

Excerpts from "Valuation of an Amazonian Rainforest" by Charles M. Peters, Alwyn H. Gentry, and Robert O. Mendelsohn, in *Nature*, 339, no. 6227 (29 June 1989), pp. 655–656. © 1989 Nature Publishing Group.

Excerpts from "The Coming Anarchy" by Robert D. Kaplan, in *The Atlantic Monthly* (February 1994), pp. 44–76. © Robert D. Kaplan. Reproduced by permission.

"Water and Security in the Middle East," excerpted from OECD, "State-of-the-Art Review on Environment, Security and Development Co-operation," 2000. © OECD, 2000. Reproduced by permission of the OECD, http://www.oecd.org.

Excerpts from "Biodiversity, War, and Tropical Forests" by Jeffrey A. McNeely, pp. 1–20 in *War and Tropical Forests: Conservation in Areas of Armed Conflict*, ed. Steven V. Price (Food Products Press, 2003), © 2003 The Haworth Press, Inc., Binghamton, New York. Reprinted by permission.

Excerpts from "A New Vigilance: Identifying and Reducing the Risks of Environmental Terrorism," by Elizabeth L. Chalecki, in *Global Environmental Politics*, 2, no. 1 (February 2002), pp. 46–64. © 2002 by the Massachusetts Institute of Technology. Reprinted by permission.

"What Is Real Security?" by Amory B. Lovins and L. Hunter Lovins, in YES! magazine, 21 (Spring 2002), pp. 50–51. Reprinted by permission. YES! magazine, P.O. Box 10818, Bainbridge Island, Wash. 98110. Web: www.yesmagazine.org.

Excerpts from *The Hydrogen Economy* by Jeremy Rifkin (Jeremy P. Tarcher/Putnam, 2002), copyright © 2002 by Jeremy Rifkin. Used by permission of Jeremy P. Tarcher, an imprint of Penguin Group (USA) Inc.

"Politics for the Age of Globalization," excerpted from "Demolition Man" and "There Is a Way Forward" from *The Lexus and the Olive Tree: Understanding Globalization* by Thomas L. Friedman (Farrar, Straus and Giroux, 1999). Copyright © 1999, 2000 by Thomas L. Friedman. Reprinted by permission of Farrar, Straus and Giroux, LLC and International Creative Management, Inc.

Excerpts from "The WTO: Inside, Outside, All Around the World" by Paul Hawken, © 2000 Paul Hawken. Reprinted by permission.

Excerpts from "The World Trade Organization: The Debate in the United

States" by Arlene Wilson, Congressional Research Service. CRS Report RL30521, April 12, 2000.

"Economic Globalization Has Become a War Against Nature and the Poor," excerpted from "Globalization and Poverty" by Vandana Shiva, in *Resurgence* no. 202 (September–October 2000), www.resurgence.org. Reprinted by permission.

Excerpts from "The Problem of the Wilderness" by Robert Marshall, reprinted with permission from *Scientific Monthly*, 30, no. 2 (February 1930), pp. 141–148. Copyright 1930 American Association for the Advancement of Science.

Excerpts from "The Value of Wilderness" by Roderick Nash republished with permission of the American Society for Environmental History, from *The Environmental Review*, 3 (1977), pp. 14–25, © 1977 The American Society for Environmental History; permission conveyed through Copyright Clearance Center.

Excerpts from "The Trouble with Wilderness; or, Getting Back to the Wrong Nature" by William Cronon, from *Uncommon Ground*, edited by William Cronon (W. W. Norton & Company, 1995). Copyright © 1995 by William Cronon. Used by permission of W. W. Norton & Company, Inc.

Excerpts from "Getting Back to the Right Nature: A Reply to Cronon's 'The Trouble with Wilderness'" by Donald M. Waller, in *The Great New Wilderness Debate*, ed. J. Baird Callicott and Michael P. Nelson (The University of Georgia Press, 1998), © 1998 The University of Georgia Press. Reprinted by permission.

"Trail Crew Camp at Bear Valley, 9000 Feet" from *The Back Country* by Gary Snyder, copyright © 1968 by Gary Snyder. Reprinted by permission of New Directions Publishing Corp.

Excerpts from "The Urban Challenge" in *Our Common Future* by the World Commission on Environment and Development (Oxford University Press, 1987), © World Commission on Environment and Development 1987. Used by permission of Oxford University Press.

Excerpt from *Ecology and the Crisis of Overpopulation: Future Prospects for Global Sustainability* by Anup Shah (Edward Elgar, 1998), © 1998 Anup Shah. Reprinted by permission.

Excerpts from *Sharing the World: Sustainable Living and Global Equity in the Twenty-first Century* by Michael Carley and Philippe Spapens (St. Martin's Press, 1988), © 1998 Friends of the Earth Netherlands (Vereniging Milieudefensie). Reproduced with permission of Palgrave Macmillan.

Excerpts from *Land Mosaics: The Ecology of Landscapes and Regions* by Richard T. T. Forman (Cambridge University Press, 1995), © Cambridge University Press 1995. Reprinted with the permission of Cambridge University Press.

Excerpt from *The City of Joy* by Dominique Lapierre, translated from the French by Kathryn Spink (Doubleday & Company, Inc., 1985). Originally published in French as *La Cité de la Joie* by Éditions Robert Laffont. © 1985 by Dominique Lapierre. English translation © 1985 by Pressinter, S.A.

Excerpts from *Calcutta in the Twentieth Century: An Urban Disaster* by Manimanjari Mitra (Asiatic Book Agency, 1990), © 1990 Manimanjari Mitra.

Excerpts from "How Eden Lost Its Garden" in *Ecology of Fear: Los Angeles and the Imagination of Disaster* by Mike Davis (Metropolitan Books, 1998), © 1998 by Mike Davis. Reprinted by permission of Henry Holt and Company, LLC, the author, and the Sandra Dijkstra Literary Agency.

Excerpts from *Cadillac Desert, Revised and Updated* by Marc P. Reisner (Penguin Books, 1993), copyright © 1986, 1993 by Marc P. Reisner. Used by permission of Viking Penguin, a division of Penguin Group (USA) Inc.

"Night Song of the Los Angeles Basin" by Gary Snyder, from *Mountains and Rivers Without End* by Gary Snyder (Counterpoint, 1996). The poem first appeared in *The Ten Directions*, 7, no. 1 (spring–summer 1986) , p. 15, published by the Zen Center of Los Angeles.

Excerpts from "The Ecozoic Era," lecture delivered by Thomas Berry in 1991, anthologized in *People, Land, and Community: Collected E. F. Schumacher Society Lectures*, ed. Hildegarde Hannum (Yale University Press, 1997), © 1997 E. F. Schumacher Society. Reproduced by permission.

Genesis 1:20–31. Revised Standard Version of the Bible, copyright 1952 [2nd edition, 1971] by the Division of Christian Education of the National Council of the Churches of Christ in the United States of America. Used by permission. All rights reserved.

Excerpts from "Human Domination of the Earth's Ecosystems" by Peter M. Vitousek, Harold A. Mooney, Jane Lubchenco, and Jerry M. Melillo in *Science*, 277, no. 5325 (25 July 1997), pp. 494–499. Copyright 1997 AAAS. Reprinted with permission.

Excerpt reprinted by permission of the publisher from *The Diversity of Life* by Edward O. Wilson, pp. 188–192, Cambridge, Mass.: The Belknap Press of Harvard University Press, Copyright © 1992 by Edward O. Wilson.

Excerpt from "The View from Walden" from *Changes in the Land* by William Cronon (Hill and Wang, 1983). Copyright © 1983 by William Cronon. Reprinted by permission of Hill and Wang, a division of Farrar, Straus and Giroux, LLC.

Excerpts from *The Wealth of Nature: Environmental History and the Ecological Imagination* by Donald Worster (Oxford University Press, 1993), copyright © 1993 by Donald Worster. Used by permission of Oxford University Press, Inc.

Excerpts from *Cod: A Biography of the Fish That Changed the World* by Mark Kurlansky (Walker and Company, 1997), © 1997 Mark Kurlansky. Published by arrangement with Walker & Co.

Excerpts from "What Is Conservation Biology?" by Michael E. Soulé, in *Bioscience*, 35, no. 11 (December 1985), pp. 727–734. Copyright 1985 by American Institute of Biological Sciences (AIBS). Reproduced with permission of American Institute of Biological Sciences (AIBS) in the format Textbook via Copyright Clearance Center.

Excerpts from "The Invaders" in *The Ecology of Invasions by Animals and Plants* by Charles S. Elton, pp. 15–26, 29–32 (University of Chicago Press, 2000).

© 1958 by Charles S. Elton, Copyright 1969, 1977, Methuen & Co. / Chapman & Hall, Kluwer Academic Publishers B.V. Reproduced with kind permission from Springer Science and Business Media.

Excerpts from *The Sunflower Forest: Ecological Restoration and the New Communion with Nature* by William R. Jordan III (University of California Press, 2003), © 2003 by the Regents of the University of California. Reprinted with the permission of the University of California Press and the author. William R. Jordan III is Director, New Academy for Nature and Culture; and Co-director, Institute for Nature and Culture, DePaul University.

Excerpt from *An Introduction to Soils for Environmental Professionals* by Duane L. Winegardner (Lewis Publishers, 1996), © 1996 CRC Press, Inc. Reprinted with permission from CRC Press.

Excerpts from *Land and Soil Management: Technology, Economics, and Institutions* by Alfredo Sfeir-Younis and Andrew K. Dragun (Westview Press, 1993), © 1993 Westview Press, Inc. Reprinted by permission.

Excerpts from "The Oil We Eat: Following the Food Chain back to Iraq" by Richard Manning, in *Harper's Magazine* (February 2004). Copyright © 2004 by *Harper's Magazine*. All rights reserved. Reproduced from the February issue by special permission.

Excerpts from "Science and Progress: Seven Developmentalist Myths in Agriculture" by Richard Levins, in *Monthly Review* 38, no. 3 (July–August 1986), pp. 13–20. Copyright © 1986 by MR Press. Reprinted by permission of Monthly Review Foundation.

Excerpt from *Africa South of the Sahara: A Geographical Interpretation* by Robert Stock, Second Edition (The Guilford Press, 2004), © 2004 The Guilford Press. Reprinted by permission.

"Oda al vino" from *Odas Elementales* by Pablo Neruda (Editorial Losada, 1954), © 1954 Editorial Losada, S. A., Buenos Aires. New English translation © 2007 James Engell. Translated by permission of the heirs of Pablo Neruda.

Excerpts from *Atmospheric Chemistry and Physics: From Air Pollution to Climate Change* by John H. Seinfeld and Spyros N. Pandis (John Wiley & Sons, Inc., 1998). Copyright © 1998 John Wiley & Sons, Inc. This material is used by permission of John Wiley & Sons, Inc.

Excerpts from S. George Philander; "The Ozone Hole, A Cautionary Tale" in *Is the Temperature Rising?* © 1998 Princeton University Press. Reprinted by permission of Princeton University Press.

Excerpts from "Stratospheric Sink for Chlorofluoromethanes: Chlorine Atom-Catalysed Destruction of Ozone" by Mario J. Molina and F. S. Rowland. Reprinted with permission from *Nature*, 249, no. 5460 (June 28, 1974), pp. 810–812. Copyright 1974 Macmillan Magazines Limited.

"The Ozone Layer: Burnt by the Sun Down Under" by Kathryn S. Brown. Reprinted with permission from *Science*, 285, no. 5434 (10 September 1999), pp. 1647–1649. Copyright 1999 AAAS.

Excerpts from *The Water We Drink: Water Quality and Its Effects on Health* by Joshua I. Barzilay, Winkler G. Weinberg, and J. William Eley (Rutgers University Press, 1999), copyright © 1999 by Joshua I. Barzilay, Winkler G. Weinberg, and J. William Eley. Reprinted by permission of Rutgers University Press.

Excerpts from *Global Water Supply and Sanitation Assessment 2000 Report* by World Health Organization and UNICEF, © 2000 World Health Organization and United Nations Children's Fund. Reprinted by permission.

Excerpts from "Water Supply" in *North, South, and the Environmental Crisis* by Rodney R. White (University of Toronto Press, 1993), © University of Toronto Press Incorporated 1993. Reprinted with permission of the publisher.

Excerpt from "Ending the Energy Stalemate: A Bipartisan Strategy to Meet America's Energy Challenges" by the National Commission on Energy Policy © 2004 National Commission on Energy Policy. Reprinted by permission.

"Stabilization Wedges: Solving the Climate Problem for the Next Fifty Years with Current Technologies" by S. Pacala and R. Socolow, reprinted with permission from *Science*, 305, no. 5686 (13 August 2004), pp. 968–972. Copyright 2004 AAAS.

Excerpts from "Industrial Growth, Air Pollution, and Environmental Damage: Complex Challenges for China" by Michael B. McElroy, in *Energizing China: Reconciling Environmental Protection and Economic Growth*, ed. Michael B. McElroy, Chris P. Nielsen, and Peter Lydon (Harvard University Committee on Environment, 1998), © 1998 The President and Fellows of Harvard College. Reprinted by permission.

Excerpts from *Exploring the Dangerous Trades* by Alice Hamilton (Atlantic Monthly Press and Little, Brown and Company, 1943), © 1942, 1943 by Alice Hamilton.

Excerpt from "And No Birds Sing," from *Silent Spring* by Rachel L. Carson. Copyright © 1962 by Rachel L. Carson, renewed 1990 by Roger Christie. Reprinted by permission of Frances Collin, Trustee, and by permission of Houghton Mifflin Company. All rights reserved.

Excerpts from "What the World Needs Now Is DDT" by Tina Rosenberg, in *The New York Times Magazine* (April 11, 2004), pp. 38–43, © 2004 The New York Times. Reprinted by permission.

"Arctic Life Threatened by Toxic Chemicals, Groups Say" by Sharon Guynup, *National Geographic Today* (October 8, 2002). Reprinted by permission.

Excerpts from "The Real World Around Us" in *Lost Woods: The Discovered Writing of Rachel Carson*, ed. Linda Lear (Beacon Press, 1998). Copyright © 1998 by Roger Allen Christie. Reprinted by permission of Frances Collin, Trustee.

Excerpt from *Mundaka Upanishad*, translated by Sanderson Beck, © 1996 Sanderson Beck. Reprinted by permission.

Matthew 6:24–34. Revised Standard Version of the Bible, copyright 1952 [2nd edition, 1971] by the Division of Christian Education of the National

Council of the Churches of Christ in the United States of America. Used by permission. All rights reserved.

"Naga Sutta" ("The Bull Elephant"), Udana IV.5 from *The Udana: Inspired Utterances of the Buddha*, trans. John D. Ireland (Buddhist Publication Society, 1997), © 1997 Buddhist Publication Society. Reprinted courtesy of Buddhist Publication Society, Sri Lanka.

"The Canticle of Brother Sun" by St. Francis of Assisi, translated by Benen Fahy, O.F.M., in *St. Francis of Assisi: Writings and Early Biographies*, ed. Marion A. Habig, © 1983 Franciscan Herald Press.

Excerpts from "The Historical Roots of Our Ecologic Crisis" in *Machina Ex Deo: Essays in the Dynamism of Western Culture* by Lynn White, Jr. (The MIT Press, 1968), © 1968 The Massachusetts Institute of Technology. Reprinted by permission.

Excerpts from "The Gift of Good Land" from *The Gift of Good Land* by Wendell Berry (North Point Press, 1981). Copyright © 1981 by Wendell Berry. Reprinted by permission of North Point Press, a division of Farrar, Straus and Giroux, LLC.

Excerpts from *The Need for a Sacred Science* by Seyyed Hossein Nasr, the State University of New York Press, © 1993, State University of New York. All rights reserved. Reprinted by permission.

Excerpts from "The Continuity of Being" by Tu Wei-Ming from *On Nature*, ed. Leroy Rouner. Copyright 1984 University of Notre Dame Press. Notre Dame, Indiana 46556. Used by permission of the publisher.

Excerpts from "The Ecology of Grief" by Phyllis Windle, as it appears in *Ecopsychology: Restoring the Earth, Healing the Mind*, ed. Theodore Roszak, Mary E. Gomes, and Allen D. Kanner (Sierra Club Books, 1995). The original article appeared in *BioScience*, 42, no. 5 (May 1992), pp. 363–366, © 1992 BioScience. Republished with permission of BioScience; permission conveyed through Copyright Clearance Center, Inc.

Excerpts from *What Is Nature? Culture, Politics, and the Non-Human* by Kate Soper (Blackwell, 1995), © 1995 Kate Soper.

Excerpt from "Formation and Transformation" in *Goethe's Botanical Writings* by Johann Wolfgang von Goethe, translated by Bertha Mueller (University of Hawaii Press, 1952), © 1952 University Press of Hawaii. Reprinted by permission.

Excerpts from "The Land Ethic," pp. 201–226 from *A Sand County Almanac and Sketches from Here and There* by Aldo Leopold, copyright 1949, 1953, 1966, renewed 1977, 1981 by Oxford University Press, Inc. Used by permission of Oxford University Press, Inc.

Excerpts from "Organism and Environment" by R. C. Lewontin in *Learning, Development, and Culture: Essays in Evolutionary Epistemology*, ed. H. C. Plotkin (John Wiley & Sons, 1982), © 1982 John Wiley & Sons Ltd. Reproduced by permission of John Wiley & Sons Limited.

Excerpts from *Earthcare: Women and the Environment* by Carolyn Merchant (Routledge, 1996), © 1996 Routledge, Inc. Reproduced by permission of Routledge / Taylor & Francis Books, Inc. and the author.

"Alma Venus" from *De rerum natura* by Lucretius, new English translation © 2007 James Engell.

"A Traveler at Night Writes His Thoughts" by Tu Fu, from *The Columbia Book of Chinese Poetry*, translated and edited by Burton Watson (Columbia University Press, 1984) © 1984 Columbia University Press. Reprinted with the permission of the publisher.

"Stopping by the Temple of Incense Massed" by Wang Wei, translated by Stephen Owen, from *An Anthology of Chinese Literature: Beginning to 1911* by Stephen Owen, editor and translator (W. W. Norton & Company, 1996). Copyright © 1996 by Stephen Owen and The Council for Cultural Planning and Development of the Executive Yuan of the Republic of China.

"A Visit to the South Mountains" by Meng Jiao, translated by Stephen Owen, from *An Anthology of Chinese Literature: Beginning to 1911* by Stephen Owen, editor and translator (W. W. Norton & Company, 1996). Copyright © 1996 by Stephen Owen and The Council for Cultural Planning and Development of the Executive Yuan of the Republic of China. Used by permission of W. W. Norton & Company, Inc.

"You Summer Grasses!" and "Into the Old Pond" by Matsuo Bashō in *One Hundred Famous Haiku*, selected and translated into English by Daniel C. Buchanan, Ph.D. (Japan Publications, 1973). Copyright © 1973 by Japan Publications, Inc. Reprinted with the permission of Japan Publications, Inc.

Excerpts from *Beautyway* by Maud Oakes. © 1957 Princeton University Press, 1985 renewed. Reprinted by permission of Princeton University Press.

"Spring Pools" from *The Poetry of Robert Frost* edited by Edward Connery Lathem. Copyright 1928, 1959 by Henry Holt and Company. © 1956 by Robert Frost. Reprinted by permission of Henry Holt and Company, LLC.

"The Planet on the Table" from *The Collected Poems of Wallace Stevens* by Wallace Stevens, copyright 1954 by Wallace Stevens and renewed 1982 by Holly Stevens. Used by permission of Alfred A. Knopf, a division of Random House, Inc., and by permission of Faber and Faber Ltd.

"Corsons Inlet," Copyright © 1963 by A. R. Ammons, from *The Selected Poems, Expanded Edition* by A. R. Ammons. Used by permission of W. W. Norton & Company, Inc.

"The Kingfisher" from *House of Light* by Mary Oliver (Beacon Press, 1990). Copyright © 1990 by Mary Oliver. Reprinted by permission of Beacon Press, Boston.

Excerpts from "Transformations of the Earth: Toward an Agroecological Perspective in History" by Donald Worster, and "Modes of Prophecy and Production: Placing Nature in History" by William Cronon, in *Journal of American His-*

tory, 76, no. 4 (March 1990), Copyright © Organization of American Historians. Reprinted with permission.

Excerpts from "Modes of Prophecy and Production: Placing Nature in History" by William Cronon. Reprinted with permission from *Journal of American History*, 76, no. 4, (March 1990).

Excerpts from *The Machine in the Garden: Technology and the Pastoral Ideal in America* by Leo Marx (Oxford University Press, 1964), copyright © 1964 by Oxford University Press, Inc. Used by permission of Oxford University Press, Inc.

Excerpt from *The Environmental Imagination: Thoreau, Nature Writing, and the Formation of American Culture* by Lawrence Buell, pp. 53–55, Cambridge, Mass.: The Belknap Press of Harvard University Press, Copyright © 1995 by the President and Fellows of Harvard College. Reprinted by permission of the publisher.

Excerpts from *The Columbian Exchange: Biological and Cultural Consequences of 1492*, by Alfred W. Crosby, Jr. (Greenwood Press, 1972). Copyright © 1972 by Alfred W. Crosby, Jr. Reproduced with permission of Greenwood Publishing Group, Inc., Westport, CT.

Excerpts from "Disorientation and Reorientation: The American Landscape Discovered from the West," in *Something in the Soil: Legacies and Reckonings in the New West* by Patricia Nelson Limerick (W. W. Norton & Company, 2000). The essay originally appeared as "Disorientation and Reorientation: The American Landscape Discovered from the West" by Patricia Nelson Limerick, in *Journal of American History*, 79, no. 3 (December 1992), pp. 1021–1049. Reprinted with permission of the copyright holder, Organization of American Historians. All rights reserved.

Excerpts from *Beauty, Health, and Permanence: Environmental Politics in the United States, 1955–1985* by Samuel P. Hays in collaboration with Barbara D. Hays (Cambridge University Press, 1987), © Cambridge University Press 1987. Reprinted with the permission of Cambridge University Press.

Excerpts from "Prose Poem on the Unreal Dwelling" by Matsuo Bashō, translated by Donald Keene, in *Anthology of Japanese Literature from the Earliest Era to the Mid-Nineteenth Century*, compiled and edited by Donald Keene (Grove Press, 1955), © 1955 Grove Press. Reprinted by permission.

"Thinking Like a Mountain" from *A Sand County Almanac and Sketches from Here and There* by Aldo Leopold, copyright 1949, 1953, 1966, renewed 1977, 1981 by Oxford University Press, Inc. Used by permission of Oxford University Press, Inc.

Excerpts from *The Log from the Sea of Cortez* by John Steinbeck, copyright 1941 by John Steinbeck and Edward F. Ricketts. Copyright renewed © 1969 by John Steinbeck and Edward F. Ricketts, Jr. Used by permission of Viking Penguin, a division of Penguin Group (USA) Inc.

"Teaching a Stone to Talk," pp. 67–76 from *Teaching a Stone to Talk: Expeditions and Encounters* by Annie Dillard (Harper & Row, Publishers, 1982), Copyright © 1982 by Annie Dillard. Reprinted by permission of HarperCollins Publishers and Blanche C. Gregory Inc.

"On Water," from *The Solace of Open Spaces* by Gretel Ehrlich, copyright © 1985 by Gretel Ehrlich. Used by permission of Viking Penguin, a division of Penguin Putnam Inc., and courtesy of Darhansoff, Verrill, Feldman Literary Agents.

Excerpts from *Reading the Mountains of Home* by John Elder, pp. 94–95, 97–98, 98–99, Cambridge, Mass.: Harvard University Press, Copyright © 1998 by the President and Fellows of Harvard College. Reprinted by permission of the publisher.

Excerpt from *Genocide in Nigeria: The Ogoni Tragedy* by Ken Saro-Wiwa (Saros International Publishers, 1992), © Ken Saro-Wiwa.

Excerpts from *Managing the Environment, Managing Ourselves: A History of American Environmental Policy* by Richard N. L. Andrews (Yale University Press, 1999), © 1999 Yale University. Reproduced by permission.

Excerpts from *Environmental Law and Policy: Nature, Law, and Society* by Zygmunt J. B. Plater, Robert H. Abrams, William Goldfarb, and Robert L. Graham, Esq., Second Edition (West Group, 1998), © 1992 West Publishing Co., © 1998 by West Group. Reprinted with permission of the West Group.

Excerpts from *Modern Environmentalism: An Introduction* by David Pepper (Routledge, 1996), © 1996 David Pepper. Reprinted by permission.

Excerpt from *The Earth Summit at Rio: Politics, Economics, and the Environment* by Ranee K. L. Panjabi (Northeastern University Press, 1997). Copyright 1997 by Ranee K. L. Panjabi. Reprinted with the permission of Northeastern University Press.

Excerpts from *Rethinking Green Politics: Nature, Virtue and Progress* by John Barry (Sage Publications, 1999), © 1999 John Barry. Reprinted by permission of Sage Publishing Ltd. and the author.

Excerpts from "Ecofeminism and the Politics of Resistance," by Lee Quinby, and "Women, Home and Community: The Struggle in an Urban Environment" by Cynthia Hamilton, in *Reweaving the World: The Emergence of Ecofeminism*, edited by Irene Diamond and Gloria Orenstein (Sierra Club Books, 1990). Copyright © 1990 by Irene Diamond and Gloria Orenstein. Reprinted by permission of Sierra Club Books.

Excerpts from pp. 118–124 of *International Environmental Policy: Emergence and Dimensions, Second Edition* by Lynton Keith Caldwell (Duke University Press, 1990), © 1990 Duke University Press. All rights reserved. Used by permission of the publisher.

Excerpts from "The Bhopal Case and the Development of Environmental Law in India" by C. M. Abraham and Sushila Abraham, in *International and Comparative Law Quarterly*, 40, part 2 (April 1991), pp. 334–365. © 1991 British Institute of International and Comparative Law. Used by permission of the British Institute of International and Comparative Law.

Excerpts from "Environmental Justice for All" by Robert D. Bullard, in *Unequal*

Protection: Environmental Justice and Communities of Color, ed. Robert D. Bullard (Sierra Club Books, 1994), © 1994 Robert D. Bullard. Reprinted by permission.

Excerpts from *The Promise and Peril of Environmental Justice* by Christopher H. Foreman, Jr. (Brookings Institution Press, 1998), © 1998 The Brookings Institution. Reprinted by permission.

Excerpts from *Beyond Growth* by Herman E. Daly (Beacon Press, 1996). Copyright © 1996 by Herman E. Daly. Reprinted by permission of Beacon Press, Boston.

Excerpts from "Energy and the U.S. Economy: A Biophysical Perspective" by Cutler J. Cleveland, Robert Costanza, Charles A. S. Hall, and Robert Kaufmann. Reprinted with permission from *Science*, 225, no. 4665 (August 31, 1984), pp. 892–896. Copyright 1984 AAAS.

Excerpts from *Green Markets: The Economics of Sustainable Development* by Theodore Panayotou (ICS Press, 1993), © 1993 Theodore Panayotou. Reprinted by permission of the Institute for Contemporary Studies.

Excerpts from "Using Economic Incentives to Maintain Our Environment" by T. H. Tietenberg, from *Challenge*, 33, no. 2 (March–April 1990), pp. 42–45. Copyright © 1990 by M. E. Sharpe, Inc. Reprinted with permission.

"Valuation of Ecosystem Services and Natural Capital," excerpted from "Ecological Economics: Creating a Transdisciplinary Science" by Robert Costanza, in *Pricing the Planet: Economic Analysis for Sustainable Development*, ed. Peter H. May and Ronaldo Serôa da Motta (Columbia University Press, 1996), © 1996 Columbia University Press. Reprinted with the permission of the publisher.

Excerpts from *Natural Capitalism* by Paul Hawken, Amory Lovins, and L. Hunter Lovins, first published in 1999 by Little, Brown & Co. Copyright © 1999 by Paul Hawken, Amory Lovins, and L. Hunter Lovins. Reprinted by permission of Little, Brown and Company, (Inc.) and the authors.

Excerpt from "The Tragedy of the Commons" by Garrett Hardin, reprinted with permission from *Science*, 162, no. 3859 (13 December 1968), pp. 1243–1248. Copyright 1968 AAAS.

Excerpts from *Ecology and the Crisis of Overpopulation: Future Prospects for Global Sustainability* by Anup Shah (Edward Elgar, 1998), © 1998 Anup Shah. Reprinted by permission.

Excerpts from *The Closing Circle* by Barry Commoner (Alfred A. Knopf, 1971), copyright © 1971 by Barry Commoner. Used by permission of Alfred A. Knopf, a division of Random House, Inc.

Excerpts from *The Spanish Seaborne Empire* by J. H. Parry (Hutchinson, 1966), © 1966 J. H. Parry.

Excerpts from "Lethal Gift of Livestock" in *Guns, Germs, and Steel: The Fates of Human Societies* by Jared Diamond (W. W. Norton, 1997 and Jonathan Cape, 1997). Copyright © 1997 by Jared Diamond. Used by permission of W. W. Norton & Company, Inc. and The Random House Group Limited.

Excerpts from *How Many People Can the Earth Support?* by Joel E. Cohen

(W. W. Norton & Company, 1995). Copyright © 1995 by Joel E. Cohen. Used by permission of W. W. Norton & Company, Inc.

Excerpts from "With Spears from All Sides" by Joe Kane (*New Yorker*, September 27, 1993) and *Savages* by Joe Kane (Knopf, 1995). Reprinted by permission.

Excerpts from *At the Hand of Man* by Raymond Bonner (Alfred A. Knopf, 1993), copyright © 1993 by Raymond Bonner. Used by permission of Alfred A. Knopf, a division of Random House, Inc.

Excerpts from "The Rights of Indigenous Peoples" by Helen Corbett, in *Essays on Indigenous Identity and Rights*, edited by Irja Seurujärvi-Kari and Ulla-Maija Kulonen (Helsinki University Press, 1996), © 1996 Yliopistopaino ja kirjoittajat (Helsinki University Press). Reprinted by permission.

Excerpts from "The Corn Men Have Not Forgotten Their Ancient Gods" by Rigoberto Queme Chay in *Story Earth: Native Voices on the Environment*, © 1993 by Inter Press Service. Published by Mercury House, San Francisco, and reprinted by permission.

Excerpts from *Vanishing Voices: The Extinction of the World's Languages* by Daniel Nettle and Suzanne Romaine (Oxford University Press, 2000), copyright © 2000 by Daniel Nettle and Suzanne Romaine. Used by permission of Oxford University Press, Inc.

"Buckeye" from *Writing from the Center* by Scott Russell Sanders (Indiana University Press, 1995). The essay was originally published as "The Buckeyes: An Introduction," the introduction to *In Buckeye Country*, ed. John Moor and Larry Smith (Bottom Dog Press, 1994), © 1994 Bottom Dog Press. Used by permission.

Excerpt from "A Dubious Future" by Archie Carr in *A Naturalist in Florida: A Celebration of Eden*, ed. Marjorie Carr (Yale University Press, 1994). The essay first appeared in *The Reptiles* by Archie Carr and The Editors of LIFE (Time Incorporated, 1963), © 1963 Time, Inc. Reprinted by permission of the Carr Family Corporation.

"The Impending Extinction of Natural History" by David S. Wilcove and Thomas Eisner, from *The Chronicle of Higher Education* (September 15, 2000). Reprinted by permission.

"Is Conservation Education an Oxymoron?" by David W. Orr, from *Conservation Biology*, 4, no. 2 (June 1990), pp. 119–121. Reprinted by permission of Blackwell Publishing.

Excerpts from "'Are You an Environmentalist or Do You Work for a Living?': Work and Nature" by Richard White, from *Uncommon Ground*, edited by William Cronon (W. W. Norton & Company, 1995). Copyright © 1995 by William Cronon. Used by permission of W. W. Norton & Company, Inc.

Earth Charter Preamble reprinted by permission of Earth Charter International Secretariat, www.earthcharter.org.

Excerpts from *Red Sky at Morning: America and the Crisis of the Global Environment, A Citizen's Agenda for Action*, by James Gustave Speth (Yale University Press, 2004), © 2004 James Gustave Speth. Reprinted by permission.

Index

Numbers in italics indicate figures and artwork.